Cerebral Blood Flow and Metabolism Measurement

Edited by A. Hartmann and S. Hoyer

With 320 Figures

Springer-Verlag   Berlin Heidelberg New York Tokyo

Professor Dr. ALEXANDER HARTMANN
Universitäts-Nervenklinik und Poliklinik, Abteilung Neurologie,
Sigmund-Freud-Straße 25, D-5300 Bonn 1

Professor Dr. SIEGFRIED HOYER
Institut für Pathochemie und Allgemeine Neurochemie
im Zentrum Pathologie der Universität, Im Neuenheimer Feld 220–221,
D-6900 Heidelberg 1

ISBN-13:978-3-642-70056-9   e-ISBN-13:978-3-642-70054-5
DOI: 10.1007/978-3-642-70054-5

Reproduction of the figures: Gustav Dreher GmbH, Stuttgart

2122/3130-543210

Preface

At the present time several techniques are available for studying quantitatively global and regional blood flow and metabolism of the human brain. However, many scientists working in the clinical and research field who would like to use these tools for their investigations may be less familiar with the indications and limitations of the individual methods. The rapid development of both modern imaging techniques and new tracers may have led to some confusion in answering the question as to which method is appropriate to solve the diagnostic problem of an individuum with brain disease.

Scepticism and ignorance as to the methods to be used as tools in differential diagnosis of brain disorders may have prevented their widespread introduction into clinical practice. Thus, the significance of circulatory and metabolic parameters involved in the majority of diseases of the central nervous system may have been overlooked.

The contributions compiled in this book describe in detail the individual techniques, outline their indications and limitations and deal in particular with newer methods such as the atraumatic ^{133}Xe technique, stable xenon tomography, three-dimensional techniques such as ^{133}Xe single photon emission tomography and N-isopropyl-I^{123}-iodoamphetamine. Positron emission tomography studies provide information on function and metabolism, particularly that of oxygen and glucose, in regional brain areas of interest. Nuclear magnetic resonance may be a promising method for studying metabolic parameters; however, accurate circulation measurements can not be performed at present. Sonographic procedures are highly important to diagnose vessel disorders under clinical and practice conditions.

Therefore, the aim of the international symposium on Cerebral Blood Flow and Metabolism Measurement in Man, held in Heidelberg, FRG, from September 29 to October 1, 1983, was to give more insight into the basic principles of the procedures used. Furthermore, it aimed to present the methods which are available today for measuring global and focal brain blood flow and metabolism in man.

The organizers would like to thank the chairmen for helping to make the symposium successful. The editors wish to express their gratitude to Springer-Verlag for its technical assistance in the preparation and publication of this book.

Spring 1985

A. HARTMANN
S. HOYER

Contents

Contents   IX

Three-Dimensional Xenon 133 Technique (SPECT)

Contents

Studies Related to CBF and Metabolism

List of Contributors

You will find the addresses at the beginning of the respective contribution

Styles, P. 519
Szyszkowitz, R. 92
Takagi, S. 442
Takahashi, K. 481
Takakura, K. 350
Takamatsu, M. 130
Takamatsu, S. 130
Thal, H.-U. 430, 608
Timms, W. E. 419
Tremoulet, M. 153
Trockel, U. 593
Tsuda, Y. 23
Turski, P. 356

Ulrich, P. 136, 145, 172
Uzzell, B. P. 88
Valetitsch, H. 123, 205
Verhas, M. 109
Vidošić, S. 81
Vorstrup, S. 234, 253, 264
Wagner, R. 391
Wassmann, H. 487, 603
Weiss, H. 598
Wenzel, E. 598
Wienhard, K. 67, 391, 403
Wilkinson, W. E. 30
Winkler, C. 295, 305

Winkler, S. 356
Wise, R. J. S. 452
Wolf, R. 202
Wolfson, L. 459
Wolfson, S. K., Jr. 361, 459
Wüllenweber, R. 487
Yokouchi, H. 130
Yonas, H. 361
Yoneda, S. 23
Yoshimasu, N. 350
Zeumer, H. 566
Zimmermann, R. A. 523

The Intraarterial Xenon 133 Method: Principles and Clinical Application

K. KOHLMEYER

The intraarterial (intracarotid) inert gas clearance method was introduced experimentally and elaborated for clinical application during the 1960s and the beginning of the 1970s (Agnoli et al. 1968; Brock et al. 1969a, b, 1970; Cronquist 1968, 1969; Cronquist and Greitz 1969; Cronquist and Laroche 1967; Cronquist and Lundberg 1968; Cronquist et al. 1965, 1966; Fieschi 1968; Fieschi et al. 1964, 1966; Glass and Harper 1963; Greitz and Cronquist 1968; Harper and Glass 1965; Harper et al. 1961; Heiss 1972; Heiss et al. 1968, 1969; Herrschaft 1972, 1973; Herrschaft and Gleim 1972; Höedt-Rasmussen 1964, 1965, 1967, 1969; Höedt-Rasmussen et al. 1966a, b, 1967, 1968; Ingvar 1964, 1967; Ingvar and Lassen 1961, 1962, 1965; Ingvar et al. 1965, 1968; Kohlmeyer 1972, 1973; Lassen 1964, 1965, 1966, 1968; Lassen and Höedt-Rasmussen 1966a, b; Lassen and Ingvar 1961, 1972; Lassen and Munck 1955; Lassen et al. 1963; McHenry et al. 1968, 1969; Olesen et al. 1971; Paulson 1968, 1969, 1970; Rees et al. 1971b; Skinhöj et al. 1964, 1970; Sveinsdottir et al. 1969, 1971a, 1971b). At first krypton 88 was used in animal experiments and in clinical studies performed on the open skull during brain surgery. The introduction of xenon 133 emitting high energies of gamma-radiation enabled studies of cerebral blood flow (CBF) in man under clinical conditions. Xenon is a freely diffusible tracer and as an inert gas does not take part in cerebral metabolism. When injected into an artery it is eliminated very rapidly and almost completely through the lung. Therefore recirculation of the isotope and a second passage through the brain, which would influence the accuracy of the measurement, can be neglected. If xenon 133 dissolved in saline is administered into the internal carotid artery (ICA) it will enter the brain vessels and the brain tissue exclusively. The wash-out of the isotope from the hemisphere to be studied is followed by several individual scintillation detectors placed lateral to the skull or by a gamma- camera. The clearance curves are recorded either linearly or logarithmically. Some possibilities for calculating regional cerebral blood flow (rCBF) values from the curves are available. The height-over-area formula, well known from the studies of Kety and Schmidt (1945), and the two-compartmental analysis considering a fast and a slow part of flow require linear registration of the clearance over 10 – 15 min. Under normal conditions the first part of the curve (which indicates the fast part of flow) corresponds to the perfusion of the cortex, and the second (the slow part) represents the flow of the

Zentralinstitut für Seelische Gesundheit, Abteilung Neuroradiologie, Postfach 5970 D-6800 Mannheim 1

Cerebral Blood Flow and Metabolism Measurement. Eds. Hartmann/Hoyer

white matter. But this can change fundamentally under pathologic conditions, especially in infarcts and intracranial bleedings. Based on the experience that the most important, especially clinical, information is obtained by mathematical analysis of the first part of the clearance curves, the so-called initial slope index (ISI) was evaluated and proven in clinical studies of rCBF by Sveinsdottir (1965), Sveinsdottir et al. (1971 a, 1971 b), Olesen et al. (1971), and Rees et al. (1971 b). Using this formula a study of 2 min of the clearance only is necessary, recording the curves logarithmically. The great advantage is that a patient studied can maintain a steady state of pCO_2 and blood pressure for 2 min without difficulties, whereas this would not be ensured during a time of 10 – 15 min. This ISI gives approximate values of rCBF of the cortical perfusion but is influenced by the immediate subcortical layers of the white matter. Therefore the flow values obtained by using the ISI for calculation are slightly lower than the values for the fast flow using the two-compartmental analysis and somewhat higher than those obtained by calculating the height-over-area value. To obtain clearance curves suitable for reliable calculations, the maximal, i.e., the initial counting rate per region must be in the order of 25 000 counts per minute or more. Such an amount of activity can be achieved in a short time by an injection of the isotope very rapidly and intracarotideally.

Since the half-thickness of tissue absorptions of xenon is about 4 cm, only a small part of the radioactivity coming from deep layers of the brain can reach the corresponding scintillation detector placed beside the lateral surface of the skull. This means that when counting from the side of the head, as is usual in clinical rCBF studies, the convexity of the brain is recorded more than twice than the deeper structures. Therefore the depth resolution of this method is very poor, and pathologic flow changes in the basal ganglia, the inner capsule, or the centrum semiovale are not seen clearly by the detectors. In clinical practice, only two-dimensional resolution is available with the systems we have used. But the resolution in the other two dimensions of the hemisphere at a level 3 cm from the surface of the collimators is 2 cm. This distance corresponds approximately to the convexity of the hemisphere to be studied. Since the scintillation crystals' diameter is 11.5 mm, one can conclude that information on regional flow abnormalities should be based on data obtained by more than one detector.

For clinical purposes the rCBF data must be available very quickly. This was a problem using the early equipment with 16 detectors. The impulses measured by the 16 channels were stored on a tape. After the study the information was replayed as curves channel by channel logarithmically, and evaluated by manual analysis according to the ISI (Fig. 1). This very time-consuming procedure was unsatisfactory for the clinician. The problem was solved by an instrument which did display the initial part of the logarithmically recorded rate meter curves for each region on an oscilloscope screen during the study (Fig. 2) (Lassen 1968). Documentation of the curves was made by a Polaroid photo taken from the screen. The rapidity and simplicity of the information supplied are the advantages of such a system. Possible loss of information may be a disadvantage since curves for calculating values other than the ISI are not available.

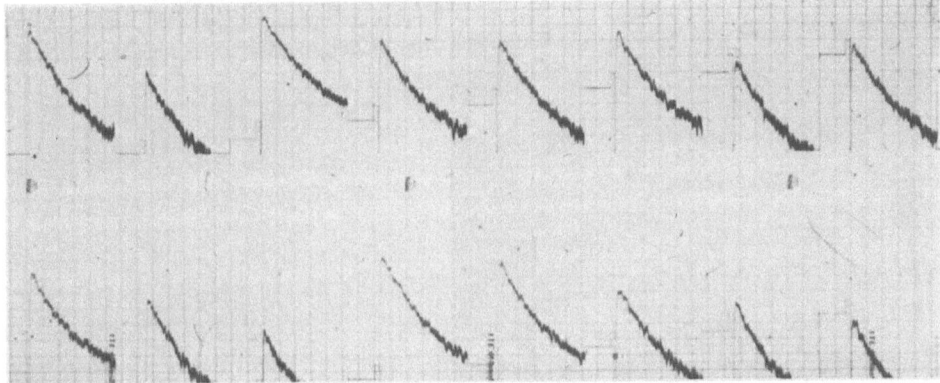

Fig. 1. Wash-out curves of the 16-channel instrument recorded semilogarithmically after being stored on a tape

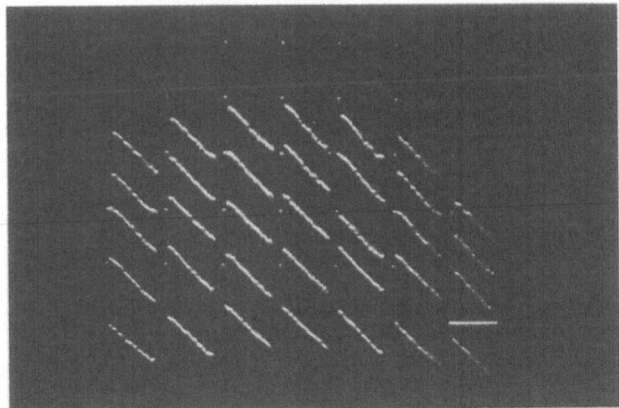

Fig. 2. Polaroid photo of the curves of the 35-channel instrument taken from an oscillocope screen

At first the 16-detector instrument was used by our group, but later we did employ the 35-channel device (Sveinsdottir 1971 b). To inject 3−4 mCi xenon-133 dissolved in saline, a puncture of the common carotid artery was performed. Then a small polyethylene catheter was inserted in the ICA according to Seldinger's method. The tip of the catheter must be placed as near as possible to the base of the skull, as shown in Fig. 3. If the catheter is not moved far enough, reflux of radioactivity into the external carotid artery cannot be excluded, as demonstrated by Fig. 4 with injected contrast medium. The correct placement of the catheter is easy to determine. Injection of about 5 ml of saline into the ICA effects a discoloration of a small zone above the eye, corresponding to a skin branch of the ophthalmic artery, which originates from the siphon of the ICA. On the other hand when the saline enters the external carotid artery pallor of that half of the face will follow. But to exclude even a small reflux into the external carotid artery, which would falsify the results and the exact localization of regional flow disturbances, a radiograph with contrast medium injected into the catheter was performed routinely following the rCBF study. It is

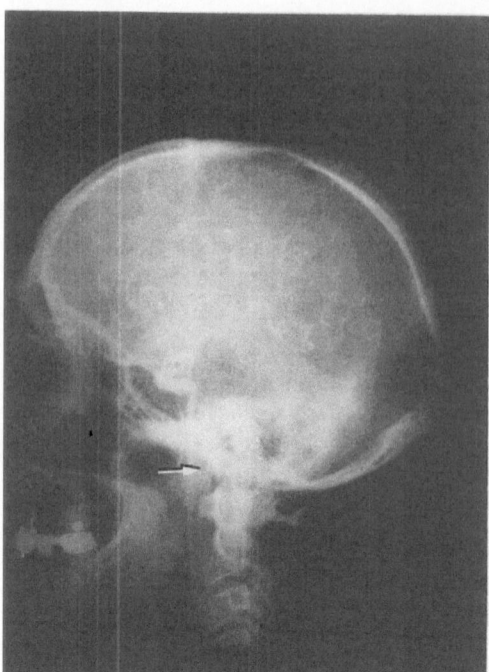

Fig. 3. Correct placement of the catheter near the basis of the skull (*arrow*)

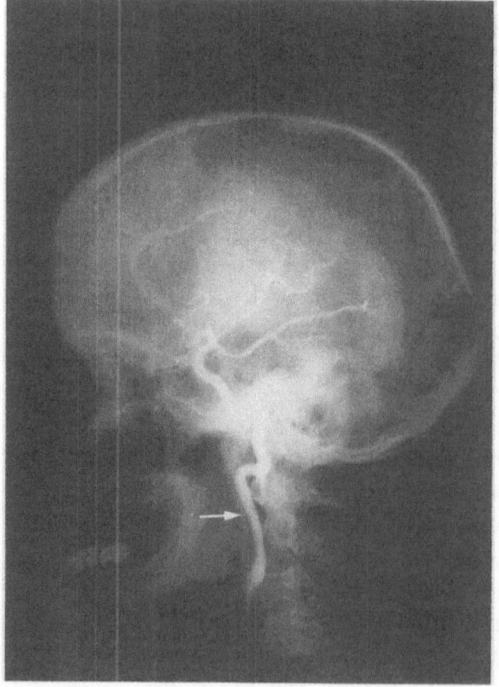

Fig. 4. Incorrect placement of the catheter. Tip at the level of the body of C2. Reflux into the external carotid artery (*arrow*)

very important that this test is not done before the flow study since we have found that 5–10 ml of the usual contrast media injected into the ICA effects an increase of mean and rCBF by 40% immediately; these values decrease slowly, reaching the baseline after 30 min (Kohlmeyer 1976).

To obtain normal values we investigated 30 volunteers without a history of signs and symptoms of vascular or cerebral diseases and aged between 18 and 76 years. These studies revealed a normal mean value of CBF of 60 ml/100 g/min, with a standard deviation 7.8 ml. There was no remarkable difference in flow values between persons of different ages, probably due to the fact that the elderly people investigated were completely healthy. For calculating focal abnormalities of rCBF, the interregional coefficient of variation is important; in our experiments this value amounts to 6.8%. Using double the normal SD, we considered an ischemic or hyperemic focal abnormality of flow (measuring local flow values by two detectors) as being at least 20% below or above the hemispheric mean value. It should be mentioned especially that our data regarding normals are in agreement with those of Ingvar (1976) and Lassen et al. (1978), showing a pattern of rCBF during rest whose values are 15%–20% higher frontally than the mean value of the hemisphere studied (Fig. 5).

Our interest was chiefly directed toward changes in mean and rCBF of patients suffering from acute cerebrovascular diseases: cerebral infarcts, transient ischemic attacks (TIA), spontaneous intracerebral hematomas, and subarachnoid hemorrhages (SAH). 988 such patients were studied and the details are listed in Table 1. Each patient underwent carotid angiography (CAG);

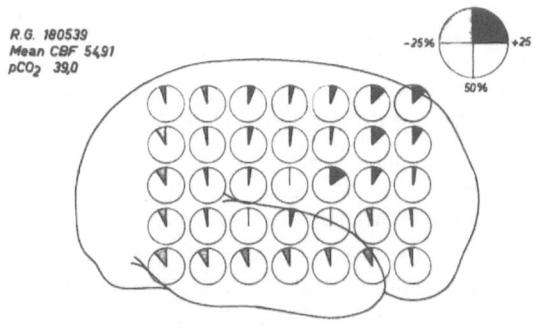

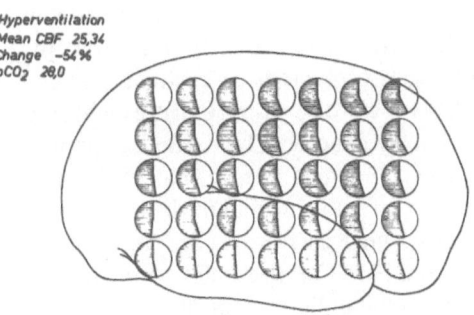

Fig. 5. *Top:* Normal figures of an rCBF study of the right hemisphere. Note the higher regional values than the mean flow within the frontal and precentral areas. *Bottom:* Normal metabolic response of rCBF during voluntary hyperventilation causing a decreased mean flow distributed in the whole hemisphere equally

Table 1. 988 patients with acute cerebrovascular diseases studied by rCBF using the intracarotideal Xenon[133]-clearance method

Diagnosis	No.	%	CAG	%	
Infarction	701	71	Normal 266	38	Occlusion and stenosis of the ex- tracranial ICA are excluded
			Occlusion or stenosis 435	62	
TIA	119	12	Normal	100	
Intracerebral hematoma	82	8	Signs of Space- occupying lesion	100	
Subarachnoid hemorrhage	86	9	No spasm 34	38	
			Spasm 52	62	
Total	988	100			

again, the results are mentioned in Table 1. Patients with extracranial occlusions or stenoses of the ICA must be excluded from rCBF studies since such vascular lesions do not allow insertion of a catheter into the ICA. According to our experience it is not reliable to measure the rCBF of the affected hemisphere due to such lesions by injecting the isotope into the contralateral ICA, even though the collateral cross-flow through the anterior communicating artery seems to be sufficiently shown by bilateral CAG. Cerebral angiograms and studies of rCBF were performed during the first day and the third week after the onset of the strokes with the exception of patients with TIAs, in whom the time of the rCBF studies was put back, in some cases up to 4 weeks after the attack. With regard to the results of the rCBF studies, it seems interesting that 38% of the patients diagnosed as having suffered cerebral infarction clinically presented with a normal CAG.

Table 2 shows the data of the rCBF studies on 701 patients with cerebral infarction clinically. The most frequent vascular lesions in cerebral angiography concerned the middle cerebral artery (MCA), either totally or one or two cortical branches of it (57%). The 266 patients (38%) with a normal cerebral angiogram presented with clinical signs indicating vascular-dependent damage within the territory of the MCA. While in cases with trunk occlusions of the MCA the whole hemisphere or a large part of it was affected by severe flow disturbances, occlusions of cortical branches exclusively predominantly caused only slight focal flow abnormalities which corresponded to the supply of the occluded artery exactly. The mean hemispheric flow value was not decreased severely in such cases. Comparing the figures of the lines "MCA total" and "Normal", there are no significant differences in the numbers of small focal, extended focal, or general flow disorders, whether or not a cerebral artery was found to be occluded. The presence of focal ischemia or focal hyperemia is not dependent on the enlargement of flow abnormalities and it plays no role, whether

Table 2. Localisation of stenoses and occlusions in CAG and findings in studies on rCBF of 701 patients with cerebral infarcts

CAG	No.	rCBF				Overall mean CBF
		Small focal abnormality	Extended focal abnormality	Whole hemisphere affected	Normal	
MCA trunk	157 22%	5 3%	18 11%	134 86% severe		20.01
MCA branches	244 35%	219 90%	12 5%	7 3% mild to moderate	6 2%	42.83
MCA total	401 57%	224 56%	30 8%	141 35%	6 1%	
PCA	30 4%	26	2	1		38.52
ACA	4 ~1%	2		1	1	34.28
Normal	266 38%	157 59%	29 11%	80 30%		39.64

angiography be normal or pathologic. The time during the course of an infarction is the only parameter governing whether or not focal hyperemia will be found by rCBF studies — for in agreement with the results of Paulson (1969), focal hyperemia, the so-called luxury perfusion syndrome (Lassen 1966), was disclosed only during the first 4 days after onset of the stroke.

Table 3 shows the results of functional tests performed during voluntary hyperventilation (HV) or CO_2 inhalation following an initial study during the resting state to ascertain the metabolic response and/or during induced high

Table 3. CO_2 response and autoregulation of patients with cerebral infarct tested by hyperventilation, CO_2 inhalation, and induced high blood pressure

CAG	No. CO_2	pCO₂ response				Autoregulation			
		Normal	Global dimin.	Focal dimin.	Steal or count. st.	No.	Preserved	Global loss	Focal loss
MCA	82	Ø	67 82%	9 11%	6 7%	21	Ø	17 81%	4 19%
MCA branches	156	Global 100%	Ø	75 48%	81 52%	39	Global 100%	Ø	34 87%
PCA ACA	18	Global 16 89%	2 11%	14 78%	2 11%	4	Global 4	Ø	2
MCA-syndr. clinically	44	Ø	34 77%	6 14%	4 9%	12	Ø	9 76%	3 24%
MCA's branches syndr. clinically	152	Global 100%	Ø	64 42%	88 48%	55	Glob. 100%	Ø	44 80%

blood pressure to test the autoregulation of cerebral circulation. 583 such studies were carried out, 452 during hypo- or hypercapnia and 131 during artificial hypertension. In patients with total occlusion of the MCA, global deterioration of the regulatory functions predominated, and just the same occurred if there was a clinical syndrome suspected of total occlusion of MCA but presenting with normal angiography. On the other hand in each case with occlusion of MCA branches or with corresponding clinical signs but normal angiography, CO_2 response and autoregulation were preserved globally while focal disorders of these functions were common. Global or focal loss of autoregulation as well as steal and counter steal syndromes were observed only in studies performed soon after onset of the disease. But the occurrence of these disorders is not limited to such a brief period as is focal hyperemia. Thus we have seen such flow abnormalities 10–14 days after onset of the disease, a finding also mentioned by Paulson (1969). The CO_2 response can, however, be diminished focally for some weeks.

Looking at the results demonstrated by Tables 2 and 3 it becomes obvious that global or focal disorders of cerebral blood flow correspond to the degree of loss of cerebral functions and do not depend on the findings of cerebral angiography. Therefore studies of rCBF are more sensitive for detecting the severity and increase of damaged cerebral blood flow than is angiography (Meyer et al. 1970).

Not only completed strokes due to cerebral ischemia are marked by focal abnormalities of rCBF and its regulation; rather pathologic foci of the cerebral circulation were also found in patients who had suffered from the mildest course of cerebral ischemia, although the flow studies were performed between the first day and the fourth week after complete restitution of the focal neurologic deficits. Table 4 shows the results obtained by studies of rCBF on 119 patients with TIAs. A normal or slightly decreased cerebral blood flow without focal abnormalities resulted in 40% only. In nearly half the material studied, regional disorders occurred during the functional tests while ischemic foci during rest were observed only rarely. The regional flow abnormalities concerned those areas of the hemispheres which were suspected of having been damaged by ischemia during the attacks on the basis of the past clinical signs. These findings are controversial: Using the same method Skinhöj (1969) and Paulson (1976) have never found focal disorders of rCBF after TIAs. On the other hand Rees et al. (1971) and Heiss (1976) have mentioned persistent focal

Table 4. Findings of studies on rCBF in 119 patients with TIA

	No.	%	
Focal abnormality during resting state	17	14	} 60%
Focal abnormality during changes of pCO₂ and/or blood pressure	55	46	
No focal abnormality decreased mean CBF	14	12	
Normal	33	28	
Total	119	100	

Table 5. Details of studies on rCBF in 82 patients with intracerebral haematoma. Overall mean CBF: 26.68 ml/ 100 g (12.31–38.96)

Left	Right	Voluntary HV disturbed	CO_2 inhalation disturbed	Induced HT	No functional test	Focal abnormality during rest	
						Ischemic	Hyperemic.
47	35	32	5	∅	45	6	1
57%	43%	39%	6%		55%		6% bis 9%

disturbances of rCBF in some cases when measuring rCBF by the intracarotid method after TIAs. Recently three-dimensional methods like single photon emission tomography (Henriksen et al. 1981) and positron emission tomography (Heiss 1981) were able to show focal disorders of CBF or metabolism following TIAs. Finally Hartmann (1981), using the xenon inhalation technique and two-dimensional imaging of rCBF, has published some studies on patients after TIAs showing persistent decreased blood flow of the damaged hemisphere, though this concerned the mean flow only. This difference could be an example of the high accuracy of the intraarterial method for localizing flow abnormalities.

Another cause of a severe stroke can be a spontaneous intracerebral haematoma (ICH). 82 such patients were studied, mostly during the acute phase (Table 5). In contrast to cerebral infarction and TIA, focal abnormalities of rCBF were very rare. The result generally was a severe decreased mean flow (50% or more) of the hemisphere affected by the bleeding, with values comparable to those obtained in studies of very large infarcts caused by occlusion of the trunk of the MCA. The metabolic response of cerebral circulation during HV or CO_2 inhalation was impaired severely and globally in all cases examined by these tests. Since the hematomas are situated within the deep layers of the hemisphere which mostly cannot be measured by the method discussed here, it is obvious that focal flow abnormalities did not appear in the flow patterns. The global impairments of the hemispheric blood flow are to be explained by the raised intracranial pressure and the extensive cerebral edema well known in most cases of ICH. It should be mentioned that the majority of these patients were in a condition of unconsciousness, and that each case showed signs of space-occupying lesions angiographically.

A last group of cerebrovascular diseases studied were patients suffering from SAH. The occurrence of vasospasm during the early period of SAH is more dangerous for the clinical condition and the outcome than is the bleeding itself. To obtain results on the influence of vasospasm upon CBF we have studied patients suffering from SAH with and without spasm angiographically (Kohlmeyer 1976, 1979). In Table 6 are listed the results of 86 cases: 33 without, 23 with moderate, and 30 with severe vasospasm. The flow studies were performed immediately before or at most 1 day after angiography to reveal whether or not spasms were present. Spasms of MCA were taken into consideration only because measurements of the supplies of the ACA and the PCA are uncertain.

Table 6. Details and findings of studies of rCBF in 86 patients with subarachnoid hemorrhage

CAG	No.	Mean CBF	Focal abnormality		CO₂ Inhal. or HV	Response to changes in pCO₂			
			Ischemic	Hyper-emic		Nor-mal	Dimin-ished	No re-sponse	In-verse
No spasm	33 38%	48.63	2		33	31	2	∅	∅
Moderate spasm	23 27%	32.96	2 4	2	23	16	5	2	∅
Severe spasm	30 35%	21.75	9 14	5	12	∅	∅	4	8

The mean flow values of the hemisphere investigated by angiography were clearly dependent on the presence of spasm and its degree. Thus they were normal or insignificantly diminished in cases without spasms, decreased by 37% of the normal value in cases with moderate spasms, and decreased by 58% in cases with severe spasms of the MCA. The numbers of either ischemic or hyperemic focal flow abnormalities increased dramatically in cases with severe spasms, and the response to changes in pCO₂ was increasingly impaired according to the degree of vasospasm. An inverse reaction means that there is a decreased mean flow during hypercapnia and vice versa. Two further facts related to the figures of Table 6 are not mentioned in the table:

1. In a few cases the rCBf studies were carried out bilaterally. The results of mean and rCBF were dependent on the different condition of the intracranial arteries of the two hemispheres; for instance normal mean flow, no focus, and preserved CO₂ response were found in the hemisphere without spasm, whereas there were considerable disturbances of mean and focal flow as well as of the metabolic response in the hemisphere affected by severe spasm.
2. The clinical condition and the outcome were in agreement with the results obtained by angiography and flow studies. Cerebral focal signs were observed only in cases with spasm and regional flow disturbances. Persistence of such severe defects and outcome in an appallic syndrome or in death were significantly high in the group with severe spasm and remarkable flow disorders globally and focally (Koester 1976; Kohlmeyer 1979). Our results have been confirmed by many authors using atraumatic methods (Meyer 1979; Brawanski et al. 1981; Ferguson et al. 1981; Kawase and Mizukami 1981; Koike et al. 1981), even though these studies concerned the mean flow more than the demonstration of focal disorders.

An important example proving the excellent spatial resolution in two dimensions of the intraarterial method is illustrated by our results obtained by rCBF studies on patients with different types of aphasia due to cerebral infarction (Kohlmeyer 1971, 1972, 1973, 1975, 1976a, 1976b, 1976c, 1977, 1979, 1984). Table 7 shows the localization of focal abnormalities of rCBF in 223 aphasics. Broca's, Wernicke's, and conduction aphasia are marked by small ischemic or

Table 7. Focal findings in studies on rCBF of 223 stroke patients with aphasia

Aphasia	No.	Precentral central	Temporal	Temporo-parietal	Parieto-occipital	Occipito-temporal	Ø focus mean CBF	CBF normal
Broca	55	51 93%					1 2%	3 5%
Wernicke	32		29 91%	3 9%				
Global	48				← 7 → 15%		41 85%	
Conduction	55			50 91%	← 4 → 7%		1 2%	
Amnestic	33			← 15 → 45%		13 39%		5 15%

hyperemic foci situated typically in nearly 100% of cases. On the other hand there are no focal flow disturbances in global aphasia. A decreased mean flow in the left hemisphere is a common finding. A typical localization of a focal flow disturbance is missed in amnestic aphasia even though the retrorolandic zone is affected by a flow abnormality. Our principles of localization of aphasia by the intracarotid xenon wash-out method were confirmed by Soh et al. (1977) using a 254-detector instrument with much higher spatial resolution (Lassen 1980). It should be mentioned that we were able to localize several other neuropsychologic disorders by rCBF studies of both the major and the minor hemisphere (Kohlmeyer 1975a, 1976, 1977, 1979). The xenon inhalation method cannot reveal such accurate focussing findings as yet.

A kind of reflected image follows from Table 8, which contains the results of rCBF studies on nonaphasics and aphasics during activation of the cerebral cortex by talking or an effort to talk respectively (Kohlmeyer 1975b). When rCBF is measured in normals during talking a marked increase of flow is to be observed within the whole speech area (Déjérine 1906) and the frontal cortex, indicating a higher neuronal activity of those regions than during the resting

Table 8. Regional cortical activation during talking or effort to talk in aphasics and non aphasics respectivity

Aphasia	No.	Frontal	Precentral	Centro-parietal	High parietal	Parieto-occipital	Temporal	Occipito-temporal
Ø aphasia	5	↑↑	↑↑	↑↑	↑	↑↑	↑↑	↑
Broca	5	↑↑↑	Ø	Ø	↑↑	↑↑	↑↑	↑
Wernicke	3	↑↑↑	↑↑	↑↑	↑	↑	Ø	Ø
Global	4	↑	Ø	Ø	↑	Ø	Ø	Ø
Conduction	4	↑↑	↑↑	↑↑	↑	Ø	↑	Ø
Amnestic	3	↑↑	↑↑	↑	↑	Ø	Ø	Ø

Table 9. Comparative study on CT and rCBF in 49 patients with stroke

CT	No.	Focal abnormality	No. focal abnormality
Infarct	33	Corresponding 24	∅
		More extended 9	
No infarct	16	12	4

state. The frontal cortical activation is obviously a nonspecific phenomenon since it occurs in all patients with aphasia, regardless of the type. However, an increase of rCBF is absent in certain areas, indicating the localization of aphasia; for instance there is no increase of focal flow within the precentral region when a patient with Broca's aphasia is trying to talk whereas this region is activated in patients with Wernicke's aphasia. However, no focal increase of flow is seen in the posterior temporal region during a conversation with such a patient. Similar results are described during talking, reading, calculating, etc. by Ingvar and Risberg (1965, 1967), Risberg and Ingvar (1971/72, 1973), Ingvar and Schwartz (1974), and Larsen et al. (1977), and during movements of the hand by Olesen (1971) and Ingvar and Philipson (1977).

After the introduction of CT into routine clinical practice we wondered whether the sensitivity of CT for detecting ischemic lesions in the brain was superior to studies of rCBF (Kohlmeyer and Graser 1978, 1979; Kohlmeyer 1980). 49 stroke patients were analyzed in a comparative study (Table 9). 33 showed infarcts in CT, and in 24 of these rCBF studies revealed of focal abnormality of flow corresponding in site and enlargement to the focal hypodensity in CT. But in nine cases the focal ischemias detected by rCBF studies were more enlarged than the infarcted zone in CT. A normal CT was found in 16 cases in spite of the presence of clear cerebral focal signs clinically. But the rCBF studies on these cases were able to show ischemic or hyperemic foci in 12 patients. In both the group with the more extended focal flow disorder than the lesion in CT and the group with regional flow disturbances despite a normal CT, the findings of the rCBF studies were localized exactly as one would expect on the basis of the clinical cerebral focal signs. Only patients with clinical signs indicating a cortical vascular lesion and with cortical infarcts in CT were included in this comparative study to avoid mistakes in the rCBF investigation due to the poor depth resolution of the method. Similar results were published in papers dealing with single photon emission tomography (Hendriksen et al. 1981) and with positron emission tomography (Heiss 1981; Ilsen and Heiss 1983). But in comparative studies by Halsay (1982) using the xenon inhalation method, CT was superior to the flow studies in imaging ischemic lesions of stroke patients.

Conclusions

There is no doubt that the intraarterial xenon-133 clearance method is far superior to the inhalation or intravenous application in localizing small focal changes and abnormalities of rCBF, and just these small foci and changes of rCBF are the most important in understanding local cerebral functions, local cerebral activities, and the pathophysiology of disorders of such functions and activities within the brain. This is demonstrated especially by our studies on aphasic patients and by studies on normals during talking and other mental activities by the Scandinavian groups of Ingvar and Lassen. Further advantages are the very short time of one investigation and the small dosage of radiation. The main disadvantage − that it is a traumatic method − should not be overestimated. We have performed more than 1000 studies and observed only two complications; and one of these could have been prevented, for we failed to inform ourselves of the condition of the ICA prior to the rCBF study. The problem in this case was an unexpected occlusion of the ICA, which was perforated by the guide wire, causing an embolization into the MCA. Other disadvantages concerning the scientific and diagnostic information gained: results are obtained on one hemisphere only, we cannot study the conditions of flow in patients with stenoses or occlusions of the ICA (and precisely these are the most frequent causes of strokes and TIAs), we can hardly perform repeat studies to control drug effects during long-term treatment, and we can study the occipital and the temporal-occipital regions in only a minority of cases since the supplying posterior cerebral artery originates from the ICA in just 20% − 25% of patients. Despite several advantages compensating for the aforementioned disadvantages of the intraarterial xenon-133 clearance method, the atraumatic two-dimensional methods could not achieve approximate results on the localization of cerebral functions and disorders thereof. However, it seems that the use of the sophisticated three-dimensional techniques will be able to yield results on focal cerebral blood flow comparable to those of the intracarotid method in the future (Kanno and Lassen 1979).

References

Agnoli, A., C. Fieschi, L. Bozzao, N. Battistini, M. Prencipe (1968) Autoregulation of cerebral blood flow studies during drug-induced hypertension in normal subjects and in patients with cerebral vascular diseases. Circulation 38, 800

Brawanski, A., J. Bockhorn, M. R. Gaab, V. A. Maximilian (1981) Case studies on non-invasive regional blood investigations in patients with vasospasm after subarachnoid haemorrhage. rCBF Bulletin 1, 9

Brock, M., A. A. Hadjidimos, M. Ellger, K. Kohlmeyer, K. Schürmann (1969a) Die örtliche Hirndurchblutung und Gefäßreaktivität bei intrakraniellen Gefäßerkrankungen. Radiologe 9, 451

Brock, M. A. A. Hadjidimos, K. Schürmann, M. Ellger, F. Fischer (1969b) Zur klinischen Messung der örtlichen Hirndurchblutung nach der intraarteriellen Isotopen-Clearance-Methode. Dtsch. med. Wschr. 94, 1377

Brock, M., A. A. Hadjidimos, J. P. Deruaz, F. Fischer, H. Dietz, K. Kohlmeyer, W. Pöll, K. Schürmann (1970) The effect of hyperventilation on regional cerebral blood flow. In R. G. Siekert (ed.): Cerebral Vascular Diseases, Grune and Stratton, New York

Cronquist, S. (1968) Regional cerebral blood flow and angiography in apoplexy. Acta Radiol. (Diagn.) 7, 521

Cronquist, S. (1969) Regional cerebral blood flow and angiographic findings in 61 cases with cerebrovascular disorders. Acta Radiol. (Diagn.) 8, 878

Cronquist, S., F. Laroche (1967) Transitory hyperemia in focal cerebral vascular lesions, studied by angiography and regional cerebral blood flow measurements. Brit. J. Radiol. 40, 270

Cronquist, S., N. Lundberg (1968) Regional cerebral blood flow in intracranial tumors with special regard to cases with intracranial hypertension. Scand. J. clin. Lab. Invest. Suppl. 102, XV:A

Cronquist, S., T. Greitz (1969) Cerebral circulation time and cerebral blood flow − a comparison of angiography and the 133 xenon method. Acta Radiol. (Diagn.) 8, 296

Cronquist, S., R. Ekberg, D. H. Ingvar (1965) Regional cerebral blood flow related to neuroradiological findings. Acta neurol. scand. Suppl. 14, 176

Cronquist, S., D. H. Ingvar, N. A. Lassen (1966) Quantitative measurements of regional cerebral blood flow related to neuroradiological findings. Acta Radiol. (Diagn.) 5, 760

Déjérine, J. (1906) L'aphasie sensorielle. Sa localisation et son anatomie pathologique. Presse méd. 55, 437

Déjérine, J (1906) L'aphasie motrice. Sa localisation et son anatomie pathologique. Presse méd. 55, 453

Ferguson, G. G., J. K. Farrar, K. Meguro, S. J. Peerless, C. G. Drake, H. J. M. Barnett (1981) Serial measurements of CBF as a guide to surgery in patients with ruptured intracranial aneurysms. Journal of Cerebral Blood Flow and Metabolism 1, Suppl. 1, 518

Fieschi, C (1968) Regional cerebral blood flow in acute apoplexy, including pharmacodynamic studies. Scand. J. clin. Lab. Invest. Suppl. 102, XVI:E

Fieschi, C., A. Agnoli, L. Bozzao (1964) Blood flow measurements in the brain of cats by analysis of the clearance-curves of hydrogen gas with implanted electrodes, and krypton 85 with external counting of gamma activity. In: O. Eichhorn, H. Lechner (eds.). Der Hirnkreislauf in Forschung und Klinik. Brüder Hollinek, Wien

Fieschi, C., A. Agnoli, N. Battistini, L. Bozzao (1966) Regional cerebral blood flow in patients with brain infarcts: a study with the 85 Kr clearance technique. Arch. Neurol. (Chic.) 15, 653

Glass, H. I., A. M. Harper (1963) Measurement of regional blood flow in cerebral cortex of man through intact skull. Brit. Med. J. I, 593

Greitz, T., S. Cronquist (1968) Angiographic evaluation of cerebral circulation time and regional cerebral blood flow. A comparative study. Scand. J. clin. Lab. Invest. Suppl. 102, XI:A

Halsey, J. (1981) Is there a clinical value in measurements of rCBF. rCBF Bulletin 1, 5

Harper, A. M., H. I. Glass (1965) Effect of alterations in the arterial carbon dioxide tension on the blood flow through the cerebral cortex at normal and low arterial blood pressures. J. Neurol. Neurosurg. Psychiat. 28, 449

Harper, A. M., H. I. Glass, M. M. Glover (1961) Measurement of blood flow in the cerebral cortex of dogs by the clearance of krypton 85. Scot med. J. 6, 12

Hartmann, A. (1981) Preservation of rCBF abnormalities in patients with TIA's. Journal of Cerebral Blood Flow and Metabolism 1, Suppl. 1, 540

Heiss, W.-D. (1972) Regional cerebral blood flow measurement with scintillation camera. Internat. J. Neurol. 11, 129

Heiss, W.-D (1976) Discussion. In S. Hoyer (ed.): Hirnstoffwechsel und Hirndurchblutung, Excerpta Medica, Amsterdam, Oxford

Heiss, W.-D., V. Kvicala, P. Prosenz, H. Tschabitscher (1969) Gamma-camera and multichannel analyzer for multilocular rCBF measurements. In: M. Brock, C. Fieschi, D. H. Ingvar, N. A. Lassen, K. Schürmann (eds.): Cerebral blood flow. Springer Verlag, Berlin-Heidelberg-New York

Heiss, W.-D., P. Prosenz, A. Roszouczky, H. Tschabitscher (1981) A quantitative gamma-camera technique. Scand. J. clin. Lab. Invest. Suppl. 102, XI:L

Heiss, W.-D., G. Kloster, K. Vyska, H. Traupe, C. Freundlieb, V. Becker, L. E. Feinendegen, G. Stoecklin (1981) Regional cerebral distribution of 11-C-methyl-D-Glucose compared with CT perfusion pattern in stroke. Journal of Cerebral Blood Flow and Metabolism. Vol. 1, Suppl. 1, 506

Hendriksen, L., O. B. Paulson, N. A. Lassen (1981) Regional cerebral blood flow in ischemic cerebrovascular disease recorded by emission computerized tomography of inhaled xenon-133. Journal of Cerebral Blood Flow and Metabolism 1, Suppl. 1, 469

Herrschaft, H. (1972) Die quantitative Messung der regionalen Hirndurchblutung beim Menschen. Electromedica 1, 6

Herrschaft, H (1973) Regional cerebral blood flow changes effected by vasoactive substances. In: J. S. Meyer, M. Reivich, H. Lechner, O. Eichhorn (eds.): Cerebral Vascular Disease. Thieme, Stuttgart

Herrschaft, H., F. Gleim (1972) Relationship between circulation time and regional cerebral blood flow in cerebral vascular disease. Neuroradiology 3, 199

Höedt-Rasmussen, K. (1964) Regional cerebral blood flow in man: In: O. Eichhorn, H. Lechner (eds.). Der Hirnkreislauf in Forschung und Klinik. Brüder Hollinek, Wien

Höedt-Rasmussen, K. (1965) Regional cerebral blood flow in man measured externally following intra-arterial administration of 85 krypton or 133 xenon dissolved in saline. Acta neurol. scand. Suppl. 14, 65

Höedt-Rasmussen, K. (1967) Regional cerebral blood flow: The intra-arterial injection-method. Acta. neurol. scand. 43, Suppl. 27, 1

Höedt-Rasmussen, K. (1969) Regional variations in cerebral blood flow in cerebrovascular disease studied under variations of blood pressure and arterial pCO_2. In: J. S. Meyer, H. Lechner, O. Eichhorn (eds.): Research on the Cerebral Circulation. Ch. C. Thomas, Springfield

Höedt-Rasmussen, K., N. A. Lassen, E. Sveinsdottir (1966a) The inert gas intra-arterial injection method for determining regional blood flow in man through the intact skull. J. clin. Invest. 45, 488

Höedt-Rasmussen, K., E. Sveinsdottir, N. A. Lassen (1966b) Regional cerebral blood flow in man determined by intra-arterial injection of radioactive inert gas. Circulat. Res. 18, 237

Höedt-Rasmussen, K., E. Skinhöj, O. B. Paulson, J. Ewald, J. K. Bjerrum, A. Fahrenkrug, N. A. Lassen (1967) Regional cerebral blood flow in acute apoplexy. The luxury perfusion syndrome of brain tissue. Arch. Neurol. (Chic.) 17, 271

Höedt-Rasmussen, K., O. B. Paulson, E. Skinhöj, N. A. Lassen (1968) Regional cerebral blood flow in apoplexy (acute hemiparesis) without arterial occlusion. Scand. J. clin. Lab. Invest. Suppl. 102, XVI:F

Ilsen, H. W., W.-D. Heiss (1983) Nuclear diagnostic methods for measuring regional cerebral blood flow and metabolism in man. In W. H. Gispen, J. Traber (eds.): Aging of the Brain. Developments in Neurology Vol. 7. Elsevier Science Publishers. Amsterdam, New York, Oxford

Ingvar, D. H. (1964) Regional cerebral blood flow in focal disorders. In: O. Eichhorn, H. Lechner (eds.). Der Hirnkreislauf in Forschung und Klinik. Brüder Hollinek, Wien

Ingvar, D. H. (1967) The pathophysiology of the stroke related to findings in EEG and to measurements of regional cerebral blood flow. In: A. Engel, T. Larsson (eds.): Stroke. Nord. Bokhandelns Förlag. Stockholm

Ingvar, D. H. (1976) Die Funktionsverteilung in der dominanten Hemisphäre, untersucht mit Messungen der regionalen Hirndurchblutung. In S. Hoyer (ed.): Hirnstoffwechsel und Hirndurchblutung. Excerpta Medica. Amsterdam-Oxford

Ingvar, D. H. (1976) Functional landscapes of the dominant hemisphere. Brain 107, 181

Ingvar, D. H. N. A. Lassen (1961) Quantitative determination of regional cerebral blood flow in man. Lancet II, 806

Ingvar, H. D., N. A. Lassen (1962) The blood flow of the cerebral cortex determined by krypton-85. Acta physiol. scand. 54, 325

Ingvar, D. H., N. A. Lassen (1965) Methods for cerebral blood flow measurements in man. Brit. J. Anaesth. 37, 216

Ingvar, D. H., J. Risberg (1965) Influence of mental activity upon regional cerebral blood flow in man. Acta neurol. scand. Suppl. 14, 183

Ingvar, D. H., J. Risberg (1967) Increase of regional cerebral blood flow during mental effort in normals and in patients with focal brain disorders. Exp. Brain. Res. 3, 195

Ingvar, D. H., M. S. Schwartz (1974) Blood flow patterns induced in the dominant hemisphere by speech and reading. Brain 97, 273

Ingvar, D. H., L. Philipson (1977) Distribution of cerebral blood flow in the dominant hemisphere during motor ideation and motor performance. Ann. neurol. 2, 230

Ingvar, D. H., S. Cronquist, R. Ekberg, J. Risberg, K. Höedt-Rasmussen (1965) Normal values of regional cerebral blood flow in man, including flow and weight estimates of gray and white matter. Acta neurol. scand. 14, 72

Ingvar, D. H., T. Lundmark, J. Risberg, E. Sabsay, U. Burklint, U. Sundelin (1968) Recording of multiple clearance curves by means of a magnetic core memory. Scand. J. clin. Lab. Invest. Suppl. 102, XI:H

Kanno, I., N. A. Lassen (1979) Two methods for calculating regional cerebral blood flow from emission computed tomography of inert gas concentration. Journ. Comput. Assist. Tomogr. 3 (1), 71

Kawase, T., M. Mizukami (1981) Time course and critical cerebral blood flow in vasospasm. rCBF Bulletin 1, 7

Kety, S. S., C. F. Schmidt (1945) The determination of cerebral blood flow in man by the use of nitrous oxide in low concentrations. Amerc. J. Physiol. 143, 53

Koester, V. (1974) Die Bedeutung der Angiokontrakturen für Symptomatologie und Prognose der primären Subarachnoidalblutung unter besonderer Berücksichtigung der Herdbefunde. Thesis, University of Giessen

Kohlmeyer, K. (1971) Rekanalisation intracranieller Gefäßverschlüsse und vergleichende Messungen der regionalen Hirndurchblutung. Radiologe 11, 479

Kohlmeyer, K. (1972) Relevance of regional perfusion measurements (regional cerebral blood flow=rCBF). J. Nucl. Biol. Med. 16, 283

Kohlmeyer, K. (1973) Lokalisationsprobleme der Aphasie in moderner Sicht. In: A. Leischner (ed.): Beiträge zur klinischen Hirnpathologie. Thieme, Stuttgart

Kohlmeyer, K. (1975) Studies of rCBF on neuropsychological disorders caused by acute cerebro-vascular accidents in the major hemisphere. In T. W. Langfitt, L. C. McHenry, Jr., M. Reivich, H. Wollmann (eds.): Cerebral circulation and metabolism. Springer, Berlin-Heidelberg-New York

Kohlmeyer, K. (1975) Dynamic speech studies by measurements of regional cerebral blood flow in aphasic and non-aphasic cases. In: A. M. Harper, W. B. Jennet, J. D. Miller, J. O. Rowan (eds.): Blood flow and metabolism in the brain. Churchill Livingstone, Edinburgh-London-New York

Kohlmeyer, K. (1976) Untersuchungen der Hirndurchblutung mit Xenon-133. Therapiewoche 25, 4516

Kohlmeyer, K. (1976) Aphasia due to focal disorders of cerebral circulations: Some aspects of localisation and of spontaneous recovery. Neurolingguistics 4, 79

Kohlmeyer, K. (1976) Untersuchungen der regionalen Hirndurchblutung bei Patienten mit neuropsychologischen Störungen infolge Schlaganfall. In S. Hoyer (ed.): Hirnstoffwechsel und Hirndurchblutung. Excerpta Medica, Amsterdam-Oxford

Kohlmeyer, K. (1977) Studies of regional cerebral blood flow in stroke patients with focal cerebral disorders of the minor hemisphere. Acta Neurol. Scand. 56, Suppl. 64, 170

Kohlmeyer, K. (1977) Regional cerebral blood flow in stroke patients. Acta Clinica Belgica 32, 2

Kohlmeyer, K. (1979) Disorders of brain function due to stroke. Correlates in regional cerebral blood flow and in computed tomography. In: F. Hoffmeister, C. Müller (eds.): Brain function in old age, Springer, Berlin-Heidelberg-New York

Kohlmeyer, K. (1979) Regional cerebral blood flow in subarachnoid hemorrhage. In: H. W. Pia, C. Landmaid, J. Zierski (eds.): Cerebral aneurysms. Springer, Berlin-Heidelberg-New York

Kohlmeyer, K. (1980) The sensitivity of CT and rCBF-studies for the pathology of stroke. In: J. M. Caillé and G. Salamon (eds.): Computerized tomography. Springer, Berlin-Heidelberg-New York

Kohlmeyer, K. (1981) Comparison between results of rCBF- and CT studies. rCBF Bulletin 2, 33

Kohlmeyer, K. (1984) Die Lokalisation vaskulär entstandener Aphasien mit radiologischen Methoden. In: H. J. Bochnik, Richtberg (eds.): Sprache, Sprechen, Verstehen. Perimed-Verlag, Erlangen

Kohlmeyer, K., C. Graser (1978) Comparative studies of computed tomography and measurements of regional cerebral blood flow in stroke patients. Neuroradiology 16, 233

Kohlmeyer, K., C. Graser (1979) Correlates of disorders of brain functions due to stroke in regional cerebral blood flow (rCBF) and in computed tomography (CT). Acta Neurol. Scand. Suppl. 72 Vol. 60, 426

Koike, T., R. Ishii, I. Ihara, S. Kameyama, S. Takeuchi, K. Kobayashi (1981) Cerebral circulation and metabolism in patients with ruptured aneurysms – with special reference to vasospasm and infarction. Journal of Cerebral Blood Flow and Metabolism 1, Suppl. 1, 522

Larsen, B., E. Skinhöj, K. Soh, H. Endo, N. A. Lassen (1977) The pattern of cortical activity provoked by listening and speech revealed by rCBF measurement. Acta neurol. scand. 56, Suppl. 64, 268

Lassen, N. A. (1964) Regional cerebral perfusion studies by intra-arterial injection of Kr^{85} and Xe^{133}: Theoretical considerations. In: O. Eichhorn, H. Lechner (eds.): Der Hirnkreislauf in Forschung und Klinik. Brüder Hollinek, Wien

Lassen, N. A. (1965) Blood flow of the cerebral cortex calculated from [85]krypton-beta-clearance recorded over the exposed surface; evidence of inhomogeneity of flow. Acta neurol. scand. Suppl. 14, 24

Lassen, N. A. (1966) The luxury-perfusion syndrome and its possible relation to acute metabolic acidosis localised within the brain. Lancet II, 1113

Lassen, N. A. (1968) Preliminary experience with oscilloscope and polaroid camera as recorded unit in a multichannel scintillation detector instrument. Scand. J. clin. Lab. Invest. Suppl. 102, XI:I

Lassen, N. A. (1980) Regional cerebral blood flow studied by gamma camera and gamma tomograph. In: J. M. Caillé, G. Salamon (eds.): Computerized tomography. Springer, Berlin-Heidelberg-New York

Lassen, N. A., O. Munck (1955) The cerebral blood flow in man determined by the use of radioactive krypton. Acta physiol. scand. 33, 30

Lassen, N. A., D. H. Ingvar (1961) The blood flow of the cerebral cortex determined by radioactive krypton-85. Experientia (Basel) 17, 42

Lassen, N. A., K. Höedt-Rasmussen (1966a) Human cerebral blood flow measured by two inert gas techniques. Neurology (Minneap.) 19, 681, 46

Lassen, N. A., K. Höedt-Rasmussen (1966b) Human cerebral blood flow measured by two inert gas techniques: Comparison of the Kety-Schmidt method and the intra-arterial injection method. Circulat. Res. 19, 681

Lassen, N. A., D. H. Ingvar (1972) Radioisotopic assessment of regional cerebral blood flow in man. In: E. J. Potchen, V. R. McCready: Progress in Nuclear Medicine I, Karger, Basel

Lassen, N. A., K. Höedt-Rasmussen, S. C. Sörensen, E. Skinhöj, S. Cronquist, B. Bodforss, R. Eng, D. H. Ingvar (1963) Regional cerebral blood flow in man determined by krypton[85]. Neurology (Minneap.) 13, 719

Lassen, N. A., D. H. Ingvar, E. Skinhöj (1978) Brain function and blood flow. Sci. Americ. 30, October

Lassen, N. A., E. Sveinsdottir, J. Kanno, E. M. Stokely, P. Rommer (1978) A fast single photon-emission tomograph for regional cerebral blood flow studies in man (Abstr.). Journ. Comput. Assist. Tomogr. 2, 260

McHenry, jr. L. C., M. E. Jaffe, H. I. Goldberg (1969) Regional cerebral blood flow measurement with small probes. I. Evaluation of the method. Neurology (Minneap.) 19, 1198

McHenry, jr. L. C., M. E. Jaffe, H. I. Goldberg (1969) Evaluation of the rCBF method of Ingvar and Lassen. In: M. Brock, C. Fieschi, D. H. Ingvar, N. A. Lassen, K. Schürmann (eds.): Cerebral blood flow. Springer, Berlin-Heidelberg-New York

Meyer, J. S. (1979) Noninvasive regional cerebral blood flow measurement in subarachnoid haemorrhage. In: H. W. Pia, C. Langmaid, J. Zierski (eds.): Cerebral aneurysms. Springer, Berlin-Heidelberg-New York

Meyer, J. S., Y. Shinohara, F. Kanda, Y. Fukuucki, A. D. Ericsson, N. K. Kok (1970) Abnormal hemispheric blood flow and metabolism despite normal angiograms in patients with stroke. Stroke 1, 219

Olesen, J. (1971) Contralateral focal increase of cerebral blood flow during arm work. Brain 94, 635

Olesen, J., O. B. Paulson, N. A. Lassen (1971) Regional cerebral blood flow in man determined by the initial slope of the clearance of intra-arterially injected 133 Xe. Stroke 2, 519

Paulson, O. B. (1968) Regional cerebral blood flow in middle cerebral artery occlusion. Scand. J. clin. Lab. Invest. Suppl. 102, XVI:C

Paulson, O. B. (1969) Regional cerebral blood flow at rest and during functional tests in occlusive and non-occlusive cerebrovascular disease: In: M. Brock, C. Fieschi, D. H. Ingvar, N. A. Lassen, K. Schürmann (eds.): Cerebral blood flow. Springer, Berlin-Heidelberg-New York

Paulson, O. B. (1976) Regionale Hirndurchblutung bei Patienten mit ischämischen Attacken. In: S. Hoyer (ed.): Hirnstoffwechsel und Hirndurchblutung. Excerpta Medica. Amsterdam-Oxford

Paulson, O. B., S. Cronquist, J. Risberg (1969) Regional cerebral blood flow: A comparison of 8-detector and 16-detector instrumentation. J. nucl. Med. 10, 164

Rees, J. E., G. du Boulay, J. W. D. Bull, J. Marshall, R. W. Russel, L. Symon (1971a) Persistence of disturbance of regional cerebral blood flow after transient ischaemic attacks. In: R. W. Ross Russel (ed.): Brain and blood flow. Pitman Medical and Scientific Publishing Company Ltd. London

Rees, J. E., J. W. D. Bull, G. H. du Boulay, J. Marshall, R. W. Russel, L. Symon (1971b) A comparison of stochastic, compartmental and flow initial analysis of rCBF data in a group of normal and ischaemic patients. Europ. Neurol. 6, 213

Skinhöj, E. (1965) Bilateral depression of CBF in unilateral cerebral diseases. Acta neurol. scand. Suppl. 14, 161

Skinhöj, E. (1968) Cerebral blood flow adaptation in man to chronic hypo- and hypercapnia and its relation to cerebrospinal fluid pH. Scand. J. clin. Lab. Invest. Suppl. 102, VIII:A

Skinhöj, E. (1969) rCBF at rest and during functional tests in transient ischemic attacks. In: M. Brock, C. Fieschi, D. H. Ingvar, N. A. Lassen, K. Schürmann (eds.): Cerebral blood flow. Springer, Berlin-Heidelberg-New York

Skinhöj, E., N. A. Lassen, K. Höedt-Rasmussen (1964) Cerebral blood flow in man. Arch. Neurol. (Chic.) 10, 464

Skinhöj, E., K. Höedt-Rasmussen, O. B. Paulson, N. A. Lassen (1970) Regional cerebral blood flow and its autoregulation in patients with transient focal cerebral ischemic attacks. Neurology (Minneap.) 20, 485

Soh, K., B. Larsen, E. Skinhöj, N. A. Lassen (1977) rCBF in aphasia. Acta neurol. scand, 56, Suppl. 64, 270

Sveinsdottir, E. (1965) Clearance curves of Kr^{85} or Xe^{133} considered as a sum of monoexponential outwash functions. Acta neurol. scand. Suppl. 14, 69

Sveinsdottir, E., N. A. Lassen, J. Risberg, D. H. Ingvar (1969) Regional cerebral blood flow measured by multiple probes: An oscilloscope and a digital computer system for rapid data processing. In: M. Brock, C. Fieschi, D. H. Ingvar, N. A. Lassen, K. Schürmann (eds.): Cerebral blood flow. Springer, Berlin-Heidelberg-New York

Sveinsdottir, E., P. Thorlöf, J. Risberg, D. H. Ingvar, N. A. Lassen (1971a) Calculation of regional cerebral blood flow (rCBF): Initial-slope-index compared to height-over-total-area values. In: R. W. Ross Russel (ed.): Brain and blood flow. Pitman Medical and Scientific Publishing Company Ltd. London

Sveinsdottir, E., P. Thorlöf, J. Risberg, D. H. Ingvar, N. A. Lassen (1971b) Monitoring regional cerebral blood flow in normal man with a computer controlled 32-detector system. Europ. Neurol. 6, 228

Theoretical Limitations in the Use of Two-Dimensional rCBF for the Diagnosis of Cerebrovascular Disease

E. RYDING and B. NILSSON

Introduction

Since there is a clinical demand for laboratory techniques for the assessment of ischemic cerebral lesions, regional cerebral blood flow (rCBF) measurements with ^{133}Xe has frequently been used for this purpose. For a correct interpretation of the results from such rCBF measurements, knowledge of the limitations in the ability of two-dimensional rCBF to visualize ischemic cerebral areas is essential.

Method

A theoretical model of the rCBF measurement (Fig. 1) was used. The cerebral blood flow distribution seen by a ^{133}Xe detector was derived from the perfusion rates of the cerebral tissues in the detector field (Ryding 1984). The intraarterial (IA) clearance curve for ^{133}Xe was then calculated from the blood flow distribution. The CBF values were computed from the IA clearance curve both by the use of monoexponential analysis of the initial segment of the curve (Olesen et al. 1971) and by the use of biexponential analysis of the curve (Ingvar and Lassen 1962; Hoedt-Rasmussen et al. 1966).

The theoretical model was used to determine the theoretical rCBF value of the detector directed at the lesion in the following pathologic conditions:

1. A general flow decrease in the fast flow compartments
2. A small cortical gray matter infarct (Fig. 2)
3. A large infarct of subcortical white matter with preserved normal cortical gray matter blood flow (Fig. 3)

The results of clinical intravenous rCBF measurements in patients with cerebral lesions similar to those of the theoreticl models further illustrated the relation between rCBF values and cerebral pathology, both for monoexponential analysis (Wyper et al. 1976) and for biexponential analysis (Obrist et al. 1975) of the ^{133}Xe clearance curves.

Department of Clinical Neurophysiology, University Hospital, S-221 85 Lund

Cerebral Blood Flow and
Metabolism Measurement.
Eds. Hartmann/Hoyer
© Springer-Verlag Berlin Heidelberg 1985

Normal

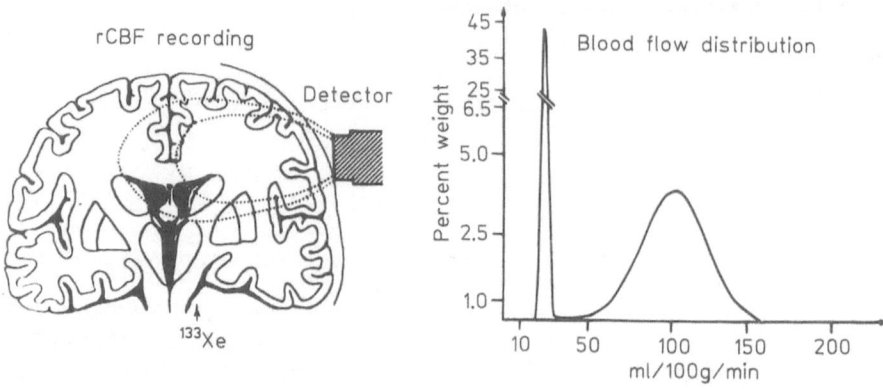

rCBF recording

Detector

^{133}Xe

Blood flow distribution

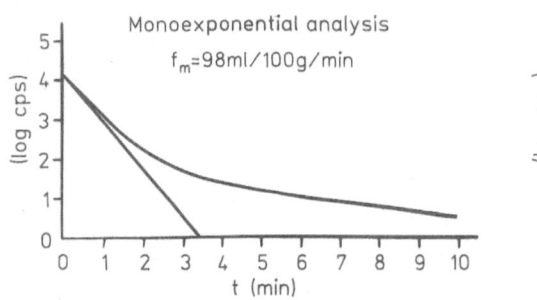

Monoexponential analysis

$f_m=98ml/100g/min$

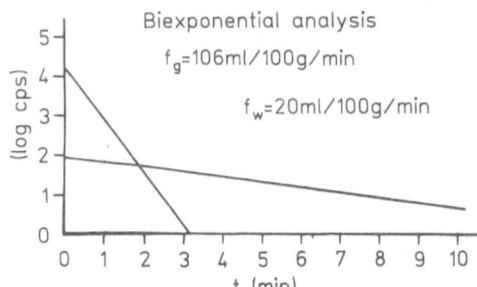

Biexponential analysis

$f_g=106ml/100g/min$

$f_w=20ml/100g/min$

Fig. 1. The theoretical model used to calculate the effects of regions of low cerebral blood flow on the rCBF values. The rCBF values are determined by mono- and biexponential analysis of the intraarterial clearance curve for ^{133}Xe, which is calculated from the normal bimodal blood flow distribution

Results

1. When there is a flow decrease in all the fast flow compartments, the decrease is correctly reproduced by rCBF, both at monoexponential and at biexponential analysis, provided that the flow distribution is still bimodal. If the flow distribution is no longer bimodal, the biexponential analysis by definition is no longer applicable.
2. A small infarct of the cortical gray matter immediately beneath the ^{133}Xe detector gives an 8% decrease of IA rCBF at monoexponential analysis, but no change in the flow values at biexponential analysis (Fig. 2).
3. A large subcortical infarct of cerebral white matter with preserved blood flow in the cortical gray matter gives a 4% IA rCBF increase at monoexponential analysis and no change in the flow values at biexponential analysis (Fig. 3).

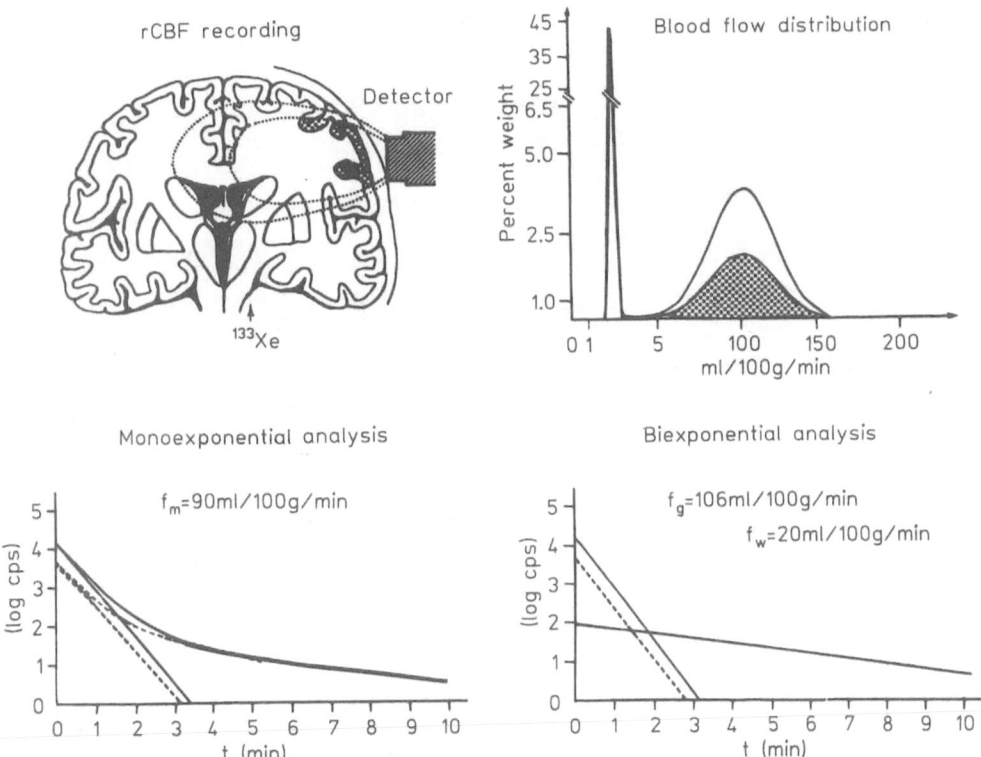

Fig. 2. The effect of cortical gray matter infarct on the cerebral blood flow distribution and on the rCBF values from mono- and biexponential analysis of the ^{133}Xe clearance curve

Discussion

The two-dimensional rCBF shows serious limitations in the ability to detect small regions of low or no flow. Especially a cortical – subcortical mixed gray and white matter infarct may at monoexponential analysis be entirely undetectable by two-dimensional rCBF since the two components in the infarct will have opposite and cancelling effects on the recorded blood flow values. The use of a biexponential analysis of the ^{133}Xe clearance curves in the diagnosis of cerebrovascular disease is of doubtful value since the recorded flow values are entirely insensitive to the existence of small cortical or subcortical infarcts ("look through," Donley et al. 1975; Bolmsjö 1984). When a cerebral infarct gives a regional decrease of gray matter flow, these low blood flow values are derived from regions with a preserved blood flow which is decreased as a secondary effect of the infarct. Furthermore, the cerebral blood flow distribution may not be bimodal, in which case the use of biexponential analysis gives unstable and unreliable flow values due to "slippage" (Risberg et al. 1975). The regions of flow decrease in cortical infarcts are detectable if a monoexponential analysis of the clearance curves is used, but the amount of flow reduction is grossly underestimated.

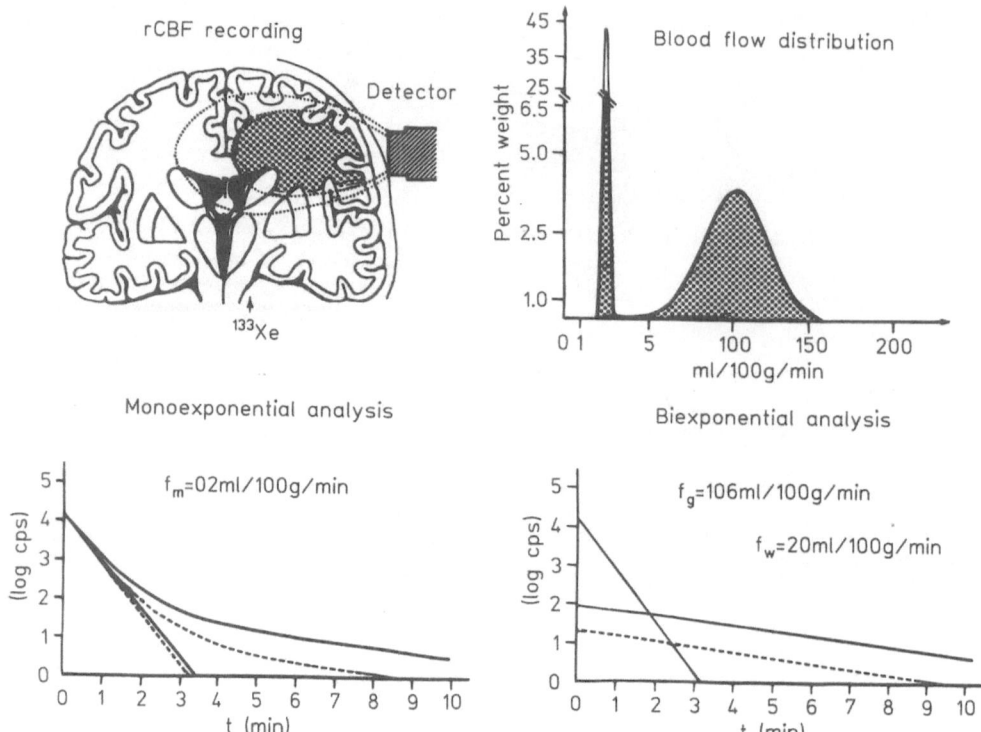

Fig. 3. The effect of a white matter infarct on the cerebral blood flow distribution and on the rCBF values from mono- and biexponential analysis of the ^{133}Xe clearance curves provided that the gray matter blood flow is unchanged

Acknowledgments. The authors were aided by the Swedish Medical Research Council (project no.: B84-04X-00084-20C) and by the Wallenberg Foundation, Stockholm.

References

Bolmsjö M (1984) Hemisphere cross-talk and signal overlapping in bilateral rCBF measurements using Xe-133. Eur J Nucl Med 9:1–5

Donley FR, Sundt TM, Anderson RE (1975) Blood flow measurements and the "look through artefact" in focal cerebral ischemia. Stroke 6:121–131

Hoedt-Rasmussen K, Sveinsdottir E, Lassen NA (1966) Regional cerebral blood flow in man determined by intra-arterial injection of radioactive inert gas. Circ Res 43:237–247

Ingvar DH, Lassen NA (1962) Regional blood flow of the cerebral cortex determined by krypton[85]. Acta Physiol Scand 54:325–338

Obrist WD, Thompson HK, Wang HS, Wilkinson WE (1975) Regional cerebral blood flow estimated by 133-xenon inhalation. Stroke 6:245–256

Olesen J, Paulson OB, Lassen NA (1971) Regional cerebral blood flow in man determined by the initial slope of the clearance of intra-arterially injected 133-Xe. Stroke 2:519–540

Risberg J, Ali Z, Wilson EM, Wills EL, Halsey JH (1975) Regional cerebral blood flow by 133-xenon inhalation. Stroke 6:142–148

Ryding E (1984) Monoexponential analysis of ^{133}Xe clearance curves for regional cerebral blood flow measurements. J Cer Blood Flow Metabol 4:250–258

Wyper DJ, Lennox GA, Rowan JO (1976) Two minute slope inhalation technique for cerebral blood flow measurement in man. J Neurol Neurosurg & Psychiat 39:141–146

Effect of STA-MCA Bypass on Hemispheric CBF and CO$_2$ Reactivity in Patients with Hemodynamic TIAs and Watershed-Zone Infarctions

Y. Tsuda[1], K. Kimura[1], Y. Iwata[2], T. Hayakawa[1], H. Etani[1], T. Asai[1], M. Nakamura[1], S. Yoneda[1], and H. Abe[1]

Introduction

Information concerning the effectiveness of superficial temporal artery to middle cerebral artery (STA-MCA) anastomosis on cerebral blood flow (CBF) and its CO$_2$ reactivity is still limited and incomplete, and an objective assessment of the postoperative effects on cerebral hemodynamics of the procedure in inaccessible occlusive carotid artery disease is needed. In this study, our initial experience concerning the pre- and postoperative CBF measurements, together with CO$_2$ reactivity in patients undergoing STA-MCA anastomosis, is reported.

Clinical Materials and Methods

Ten patients, 14 days to 3 months after onset, with TIAs and watershed infarctions (mean = 55.7 years) and with evidence of occlusive carotid lesions angiographically, were selected for STA-MCA anastomosis, and a total of 20 CBF measurements, together with CO$_2$ reactivity at hypocapnia, were investigated pre- and postoperatively. Three patients with TIAs and three with watershed minor complete stroke had proximal occlusion of the internal carotid artery (ICA) as the lesion responsible for symptoms (Occl-TIA and WS-MCS, respectively). The remaining four patients with TIAs had > 50% stenosis in diameter of the ICA as the lesion responsible for symptoms (> 50% S-TIA). All ten patients had a hemodynamically patent contralateral ICA angiographically. Clinical disability was mild in the three patients with watershed infarctions, and no significant hypodense lesions were found on CT scan as evidence of cerebral infarction in the seven patients with TIAs. Three-vessel angiography was performed in all patients by Seldinger's technique, in association with CBF measurements (with the informed consent of the patient) by the intraarterial ^{133}Xe injection method. Preoperatively, CBF was measured in the subacute-chronic stage of the clinical course, from the ipsilateral ICA in > 50% S-TIA with a lateral aspect and from the contralateral ICA in WS-MCS and Occl-TIA

1 Division of Neurology, First Department of Internal Medicine, and Department of Neurosurgery, Osaka University Medical School, Osaka, Japan
2 Department of Neurosurgery, Minoh Municipal Hospital, Osaka, Japan

Cerebral Blood Flow and
Metabolism Measurement.
Eds. Hartmann/Hoyer
© Springer-Verlag Berlin Heidelberg 1985

with a vertex aspect. Postoperatively, CBF was measured within an average of 3 months after surgery, from the ipsilateral common carotid artery in all patients with a lateral aspect. Hemispheric mean CBF (mCBF) was measured with dynamic washout studies of ^{133}Xe, i.e., serial images of 5 s duration were acquired for 130 s after injection of ^{133}Xe and computed by initial slope analysis. The region of interest (ROI) was set at the total perfused area of the brain parenchyma in the affected hemisphere. In our patients with ICA occlusions, the cross-filling via the anterior communicating artery was well developed and the activity of ^{133}Xe in the occluded hemisphere was similar to that in the non-occluded hemisphere, and the mCBF in the affected hemisphere could be calculated with sufficient activity (Nillson et al. 1979; Prosenz et al. 1974; Tsuda et al. 1983). As the normal mCBF value with our method, a value of 50 ml/100 g/min ± 10% was considered to represent the normal range. After a rest period of 15 min or more, the procedures were repeated during hyperventilation as a physiologic test for decreasing $PaCO_2$. The CO_2 reactivity (%) was calculated as the ratio of the change in mCBF before and during hyperventilation to the measured change in $PaCO_2$, and expressed as $\Delta CBF(\%)/\Delta PaCO_2$. The diagnosis of TIA was based upon the criteria for TIA of the Joint Committee for Stroke Facilities (Heyman et al. 1974) and the diagnosis of water-

Table 1. Effect of STA-MCA bypass on hemispheric mCBF, CO_2 reactivity (%), and clinical symptoms

Group	Case	mCBF (ml/100 g/min) preop. → postop.	CO_2 reactivity (%) (%/mmHg) preop. → postop.	Clinical symptoms (postop.)
WS-MCS ($n=3$)	1.	34 → 43 ↑	2.03 → 2.80 ↑	→ improved
	2.	34 → 48 ↑	3.27 → 4.65 ↑	
	3.	25 → 40 ↑	3.92 → 4.65 ↑	
		31.0 ±2.4 → 43.7 ±1.9 ↑	3.07 ±0.45 → 4.17 ±0.57 ↑	
Occl-TIA ($n=3$)	4.	38 → 51 ↑	4.39 → 3.57	→ asymptomatic
	5.	37 → 40 ↑	2.57 → 2.50	
	6.	53 → 59 ↑	2.83 → 2.92	
		42.7 ±4.2 → 50.0 ±4.5 ↑	3.26 ±0.46 → 2.99 ±0.25	
>50% S-TIA ($n=4$)	7.	42 → 44	2.51 → 3.41 ↑	→ asymptomatic
	8.	65 → 60	2.44 → 2.78 ↑	
	9.	50 → 49	2.92 → 3.51 ↑	
	10.	46 → 47	2.14 → 2.96 ↑	
		50.8 ±2.2 → 50.1 ±3.1	2.50 ±0.14 → 3.17 ±0.15 ↑	

WS-MCS and Occl-TIA: watershed minor complete stroke and TIA, due to ICA occlusion as the lesion responsible for symptoms; *>50% S-TIA:* TIA due to >50% stenosis in diameter of ICA as the lesion responsible for symptoms; *preop.:* preoperatively, *postop.:* postoperatively. Values are expressed as means ± SEM

shed infarctions was judged by CT scan evidence according to the study by Wo-dartz (1980).

Results

With respect to the postoperative change in mCBF and/or CO_2 reactivity (%) in each group of patients (Table 1), in patients with WS-MCS both mCBF and CO_2 reactivity (%) increased and tended to return to near-normal values [mCBF: $31.0 \rightarrow 43.7$; ml/100 g/min; CO_2 reactivity (%): $3.07 \rightarrow 4.17$; %/mmHg]; in each of the three patients, and this was accompanied by improved clinical disability. In patients with Occl-TIA, increased mCBF within the normal range ($42.7 \rightarrow 50.0$; ml/100 g/min) was observed postoperatively in each of the three patients despite relatively constant values of CO_2 reactivity (%). On the other hand, in patients with > 50% S-TIA, a definitely increased CO_2 reactivity (%) ($2.50 \rightarrow 3.17$; %/mmHg) was observed postoperatively in each of the four patients despite relatively constant normal mCBF values. All the seven patients with TIAs have remained asymptomatic for an average of 2 years after surgery up to the present time. From the results, it can be said that postoperative changes in mCBF and/or CO_2 reactivity (%) showed definite trends with different attitudes in each group of patients, although experience here is in a limited number of patients.

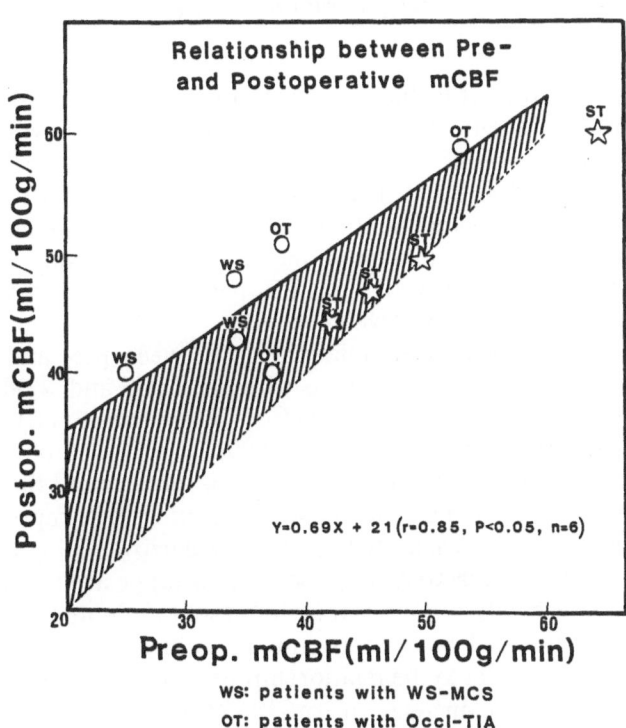

Fig. 1. The *solid line* is the regression line of preoperative mCBF against postoperative mCBF values in WS-MCS and Occl-TIA. The *dashed line* represents an equal level between pre- and postoperative mCBF values. The *shaded area* represents the postoperative increase in mCBF values

WS: patients with WS-MCS
OT: patients with Occl-TIA
ST: patients with >50%S-TIA

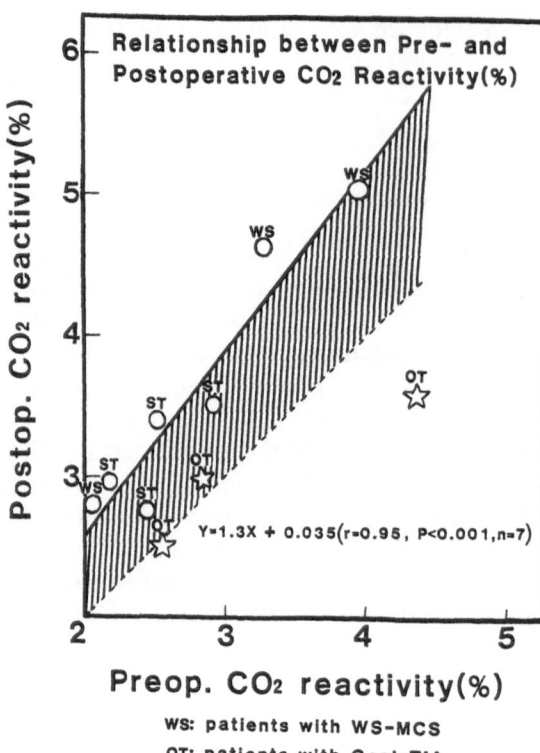

Fig. 2. The *solid line* is the regression line of preoperative CO_2 reactivity (%) against postoperative CO_2 reactivity (%) in WS-MCS and > 50% S-TIA. The *dashed line* represents an equal level between pre- and postoperative values of CO_2 reactivity (%). The *shaded area* represents the postoperative increase in CO_2 reactivity (%)

WS: patients with WS-MCS
OT: patients with Occl-TIA
ST: patients with >50%S-TIA

With respect to the relationship between the pre- and postoperative mCBF (Fig. 1) in patients with WS-MCS and Occl-TIA, who postoperatively showed increased mCBF values, a significant positive correlation ($r = 0.85$, $P < 0.05$, $n = 6$) was observed. The improvement of the mCBF, moreover, was greater in those with poorer mCBF preoperatively, i.e., the shaded area in Fig. 1, representing the postoperative increase in mCBF values, shows a wider range at the lower preoperative mCBF level and a narrower range at the higher preoperative mCBF level.

With respect to the relationship between pre- and postoperative CO_2 reactivity (%) (Fig. 2), in patients with WS-MCS and > 50% S-TIA, who postoperatively showed increased values of CO_2 reactivity (%), a highly significant positive correlation ($r = 0.95$, $P < 0.001$, $n = 7$) was observed. Moreover, the improvement of CO_2 reactivity (%) is greater in those with better reactivity preoperatively, i.e., the shaded area in Fig. 2, representing postoperative increase in CO_2 reactivity (%), shows a narrower range at the lower preoperative level of CO_2 reactivity (%) and a wider range at the higher preoperative level of CO_2 reactivity (%), i.e., an inverse trend as compared with the relationship between pre- and postoperative mCBF values in WS-MCS and Occl-TIA.

With respect to the relationship between preoperative CO_2 reactivity (%) and postoperative increase in mCBF (Fig. 3), a significant positive correlation ($r = 0.72$, $P < 0.02$, $n = 10$) was observed in patients who underwent STA-

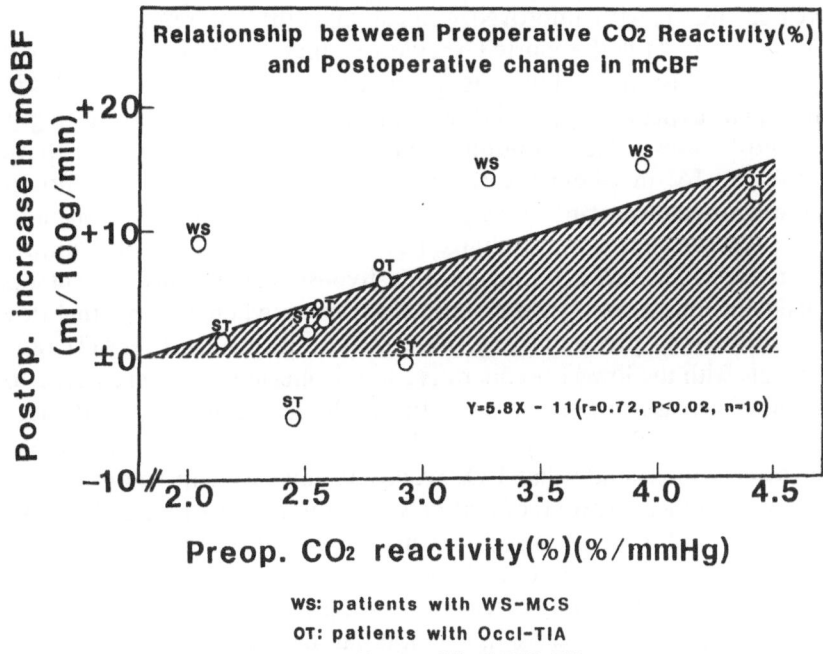

WS: patients with WS-MCS
OT: patients with Occl-TIA
ST: patients with >50%S-TIA

Fig. 3. The *solid line* is the regression line of preoperative CO_2 reactivity (%) against postoperative increase in mCBF values in WS-MCS, Occl-TIA, and > 50% S-TIA. The *dashed line* represents a zero level of postoperative change in mCBF. The *shaded area* represents the postoperative increase in mCBF values against preoperative CO_2 reactivity (%)

MCA anastomosis, i.e., the shaded area in Fig. 3 represents the postoperative increase in mCBF values against the preoperative CO_2 reactivity (%). The postoperative increase in mCBF against the preoperative CO_2 reactivity (%) is the greatest in patients with WS-MCS and greater in patients with Occl-TIA than in patients with > 50% S-TIA, who preoperatively showed rather constrained CO_2 reactivity (%), i.e., preoperatively better CO_2 reactivity (%) brings about a greater increase in mCBF postoperatively, particularly in patients with WS-MCS and Occl-TIA.

Discussion

From our results, potentially viable tissue might be expected preoperatively, surrounding watershed necrotizing tissue or existing in patients with TIAs, e.g., "ischemic penumbra," a concept first proposed by Symon and Astrup (1979). On the other hand, Baron et al. (1981) concluded in their study that bypass anastomosis resulted in a dramatic improvement in the clinical state and disappearance of the "misery perfusion syndrome," and was one way of improving the cerebral perfusion pressure and, in turn, the CBF. Similar results were seen in our patients with WS-MCS or Occl-TIA, who postoperatively showed increased mCBF with improved clinical disability or remained asymptomatic. Halsey et al. (1982) reported previously that the resting flow preoperatively was not constrained but was adequate for the brain's metabolic demand and that bypass surgery did not substantially effect the resting flow level in most cases. These findings may be consistent with our patients with > 50% S-TIA, in whom the resting flow was not constrained preoperatively, showing normal mCBF values preoperatively and remaining constant after surgery; however, the CO_2 reactivity (%) in such patients clearly improved postoperatively. Therefore, we believe that bypass surgery may correct a part of the metabolic impairment which may exist even when the resting flow level is not constrained preoperatively. Moreover, Laurent et al. (1982) reported that patients with the lowest preoperative hemispheric flows were noted to have the greatest increase in postoperative flow, which is consistent with our results in patients with WS-MCS and Occl-TIA. On the other hand, Halsey et al. (1982) stressed in their study that CO_2 reactivity increased in some of their patients (more so in those who preoperatively had the poorest reactivity), and from the observation of a highly significant negative relationship between preoperative reactivity and postoperative change in reactivity, they concluded that the surgery did not significantly affect the resting flow level but augmented the collateral reserve in those cases in which it had been most severely impaired by the arterial lesion. However, we observed a somewhat different, and even contradictory result, i.e., a more obvious trend toward enhancement of the hemispheric mCBF and/or CO_2 reactivity (%) postoperatively, both in watershed infarctions and in TIAs due to occlusive carotid artery disease, i.e., improvement of CO_2 reactivity (%) is apparent in > 50% S-TIA and even in WS-MCS, showing a greater increase of CO_2 reactivity (%) in those patients with better reactivity preoperatively, while the improvement of mCBF is clearly seen in WS-MCS and also in Occl-TIA, showing a greater increase of mCBF in those patients with poorer mCBF preoperatively, which is an inverse trend as compared with the improvement of the CO_2 reactivity (%) in WS-MCS and > 50% S-TIA. From our study, a postoperative greater increase in mCBF values might also be expected from the preoperative better CO_2 reactivity (%), particularly in patients with WS-MCS and Occl-TIA. In fact, surgery affected the resting flow level in some patients, and the effect of bypass procedure augmented not only the CBF but also the collateral reserve capacity in the potentially viable tissue, so that the CO_2 reactivity (%) or the hemispheric mCBF was corrected to adequate levels before the occurrence of permanent tissue damage. Therefore, preoperatively poorer mCBF and/or better CO_2 reactivity (%) might be considered to be a favorable indication for bypass anastomosis. The preoperative evaluation of CO_2 reactivity moreover could also be a reliable indicator for forecasting the postoperative increase in hemispheric mCBF, except in patients with > 50% S-TIA, in whom a postoperative increase in CO_2 reactivity (%) might be more expected than in CBF.

In conclusion, potentially viable tissue, surrounding watershed necrotizing tissue or existing in patients with TIAs, might be expected preoperatively, and bypass anastomosis might augment the CBF and/or CO_2 reactivity (%) in various forms of the occlusive carotid artery disease, with different attitudes being appropriate in each group of patients. Bypass surgery might be beneficial both prophylactically and for improvement of inadequate hypoperfusion of potentially viable tissue, i.e., preoperatively poorer mCBF and/or better CO_2 reactivity might be considered a favorable indication for bypass anastomosis. Pre- and postoperative evaluation of the hemispheric mCBF together with its CO_2 reactivity is believed to be a useful measure for evaluating the improvement of cerebrovascular reserve capacity after STA-MCA bypass. Moreover, preoperative evaluation of this might be an informative tool in the selection of candidates for bypass anastomosis by predicting the postoperative improvement of cerebral hemodynamics in occlusive carotid artery disease. However, for more definitive hemodynamic evaluation, further study in larger patient groups must be conducted.

References

Baron JC, Bousser MG, Rey A, et al. (1981) Reversal of focal "misery perfusion syndrome" by extra-intracranial arterial bypass in hemodynamic cerebral ischemia. A case study with ^{15}O positron emission tomography. Stroke 12:454−459

Halsey JH Jr, Morawetz RB, Blauenstein UW (1982) The hemodynamic effect of STA-MCA bypass. Stroke 13:163−167

Heyman A, Leviton A, Millikan LH, et al. (1974) Report of the joint committee for stroke facilities. XI. Transient focal cerebral ischemia: epidemiological and clinical aspect. Stroke 5:277−284

Laurent JP, Lawner PM, O'Connor M (1982) Reversal of intracerebral steal by STA-MCA anastomosis. J Neurosurg 57:719−724

Nillson B, Cronqvist S, Ingvar DH (1979) Regional cerebral blood flow (rCBF) studies in patients to be considered for extracranial-intracranial bypass operations. In: Meyer JS, Lechner H, Reivich M (eds) Cerebral vascular disease 2, 9th Salzburg conference. Excerpta Medica, Amsterdam, p 295−300

Prosenz P, Heiss WD, Tschabitscher H, et al. (1974) The value of regional cerebral blood flow measurements compared to angiography in the assessment of obstructive neck vessel disease. Stroke 5:19−31

Symon L, Astrup J (1979) I. Physiological aspects. Phenomena associated with focal ischaemia in the cental nervous system. J Acta Neurochirurgica suppl. 28:215−217

Tsuda Y, Kimura K, Yoneda S, et al. (1983) Bi-hemispheric CBF and its CO_2 reactivity of TIAs and completed strokes in ICA occlusions. Neurol Res 5 (3):1−15

Tsuda Y, Kimura K, Yoneda S, et al. (1983) Cerebral blood flow and CO_2 reactivity in transient ischemic attacks: Comparison between TIAs due to the ICA occlusion and ICA mild stenosis. Neurol Res 5 (3):17−37

Wodarz R (1980) Watershed infarctions and computed tomography. A topographical study in cases with stenosis or occlusion of the carotid artery. Neuroradiology 19:245−248

Stability and Sensitivity of CBF Indices in the Noninvasive ^{133}Xe Method

W. D. Obrist[1] and W. E. Wilkinson[2]

Although the two-compartment model for ^{133}Xe inhalation/IV injection studies (Obrist et al. 1975) has proven useful in a variety of applications, under certain pathologic conditions the computed CBF parameters become quite unstable due to shifts in compartment size (Risberg et al. 1975). This phenomenon, referred to as "slippage," applies not only to the noninvasive method, but also to compartmental variables of the intracarotid ^{133}Xe technique (Waltz et al. 1972; Iliff et al. 1974; Heiss et al. 1975; Enevoldsen and Jensen 1977).

Adequate compartmentalization requires distinctly separate blood flow components that are within the resolving power of multiexponential analysis, a situation that prevails in healthy subjects where gray and white matter have two different distributions of flows (Reivich et al. 1969). Unfortunately, pathologic tissue (or extreme physiologic conditions) may not provide such a clear separation of tissue components, so that some gray matter may "slip" into the slow CBF compartment, or vice versa. When this happens, compartmental flows vary inversely with the relative weight of the fast compartment. This not only presents difficulties in the anatomic designation of flow as gray or white and in the assignment of proper partition coefficients, but the computed blood flow values become unstable, showing large interregional and intertest variability. Although such instability can be reduced by obtaining higher count rates and/ or extending the length of curve analyzed (Obrist et al. 1975), even these procedures may not solve the basic problem when tissue components are overlapping.

One solution to this problem is the use of a noncompartmental index similar to the height-over-area analysis of the intracarotid technique (Hoedt-Rasmussen et al. 1966). Indeed, Bruce and co-workers (1975) found that CBF_{15} (height-over-area to 15 min) was relatively immune to shifts in compartment size.

Another approach is the use of slope measurements such as the "initial slope index" (ISI), introduced by Risberg and co-workers (1975) for the ^{133}Xe inhalation method. Although considerably more stable than compartmental variables, this index lacks the sensitivity to high flows obtained by the initial slope of the intracarotid technique (Olesen et al. 1971).

The present study was designed to assess the effect of slippage on both compartmental and noncompartmental CBF indices, and to compare the sensitivity

---

1 University of Pennsylvania, Philadelphia, PA, USA
2 Duke University, Durham, NC, USA

Cerebral Blood Flow and
Metabolism Measurement.
Eds. Hartmann/Hoyer
© Springer-Verlag Berlin Heidelberg 1985

of the several indices with respect to changes in fast compartment flow. Relevant information was also obtained on the length of clearance curve analysis.

Methods

Three sets of computer simulated head curves were constructed, each based on 1 min of ^{133}Xe inhalation and containing 100 random patterns of Poisson-distributed noise. The three curves represented: (1) a normal resting CBF, (2) an elevated flow typical of cortical activation, and (3) a pathologically reduced flow with overlapping tissue components. Except for noise, count rates for the three curves were equivalent (peak = 1000/s). A three-compartment model was used to generate the curves, consisting of gray matter, white matter, and extracerebral (scalp) contamination, which had relative tissue weights of 0.4, 0.4, and 0.2, respectively. The individual compartments were constructed from a Gaussian distribution of rate constants, as described previously (Obrist et al. 1975). Air passage artifact (6% at 1 min) was added to each curve. True cerebral values for the several CBF variables (independent of noise and extracerebral contamination) are given in Tables 1 and 2.

A two-compartment analysis was performed on the three different head curves using the 20% point of the end-tidal air curve as a start-fit time (Obrist et al. 1975). Curve fitting ended at 11.0 min, except in the case of pathologically reduced flow, where the analysis time was extended to 14.0 min for comparison. The traditional compartmental variables, f_1 (fast flow), f_2 (slow flow), w_1 (relative tissue weight), and FF_1 (fast compartment fractional flow) were compared with two noncompartmental estimates proposed by the authors (Obrist and Wilkinson 1980). The latter, designated CBF_∞ and CBF_{15}, are mathematically equivalent to the height-over-area index employed in the intracarotid technique, where the upper limit of integration is infinity and 15 min, respectively. Each of these variables was also compared with the initial slope (IS), defined as the tangent at time zero of an equivalent bolus injection (Obrist and Wilkinson 1980), and with the initial slope index (ISI) of Risberg and co-workers (1975).

In terms of the four unknown parameters (k_1, k_2, P_1, P_2) determinend by least squares fit:

$$FF_1 = P_1/(P_1 + P_2) \quad \text{and} \quad FF_2 = 1.0 - FF_1,$$

$$CBF_\infty = \bar{\lambda}/(FF_1/k_1 + FF_2/k_2),$$

$$IS = FF_1 \cdot k_1 + FF_2 \cdot k_2,$$

where $\bar{\lambda}$ is the assumed mean tissue−blood partition coefficient. CBF_{15} is the same as CBF_∞, except for correction factors that adjust the integration time.

Results

Table 1 presents the findings for normal resting and activated flows, while Table 2 shows the findings for reduced flow with overlapping tissue com-

Table 1. Comparison of CBF indices: Normal resting and activated flows[a]

	Compartmental indices				Noncompartmental indices			
	f_1	f_2	w_1	FF_1	CBF_r	CBF_{15}	IS	ISI
Normal resting flow								
True cerebral value	80.0	21.0	50.0	79.2	50.5	–	82.1	55.4
Mean of 100 computed values	79.5	18.3	42.8	76.5	45.2	49.2	78.9	50.6
Coef. var. (%)	3.5	3.4	1.2	0.6	2.7	2.2	3.3	1.2
Activated flow								
True cerebral value	104.0	21.0	50.0	83.2	62.5	–	110.5	64.2
Mean of 100 computed values	102.7	18.8	42.5	80.1	55.2	60.1	105.4	57.3
Coef. var. (%)	3.2	2.4	0.8	0.5	2.8	2.5	3.3	1.1

[a] f_1, f_2, CBF_r, and CBF_{15} are expressed in ml/100 g/min; w_1 and FF_1 are in percent; IS and ISI are × 100

Table 2. Comparison of CBF indices at two end-fit times: Overlapping compartments[a]

	Compartmental indices				Noncompartmental indices			
	f_1	f_2	w_1	FF_1	CBF_r	CBF_{15}	IS	ISI
Overlapping compartments								
True cerebral value	30.0	21.0	50.0	58.8	25.5	–	27.8	25.5
Mean of 100 computed values								
End fit 11 min	32.0	17.1	41.8	57.2	23.4	25.0	27.5	24.2
End fit 14 min	30.5	16.5	44.9	59.9	23.4	24.9	27.1	24.1
Coef. variation (%)								
End fit 11 min	14.8	20.3	19.0	17.3	8.0	2.3	5.0	1.9
End fit 14 min	8.2	9.2	11.2	8.6	2.4	1.3	3.4	1.6

[a] f_1, f_2, CBF_r, and CBF_{15} are expressed in ml/100 g/min; w_1 and FF_1 are in percent; IS and ISI are × 100

partments. In each table the "true cerebral value" is the flow level used to generate the synthetic curves. The computed values, based on a two-compartment analysis, are presented as the mean and coefficient of variation (SD/mean) of 100 simulations. It should be emphasized that the simulations for a given curve differ from each other only in random noise.

As seen in Tables 1 and 2, most of the computed CBF indices underestimate the true flow, but to variable degrees. This is explained by the presence of a slow extracerebral component in the simulated curves, which depresses the flow value of the second compartment (f_2) and decreases the relative size of the first compartment (w_1). In agreement with previous findings (Obrist et al. 1975), f_1 is only minimally affected.

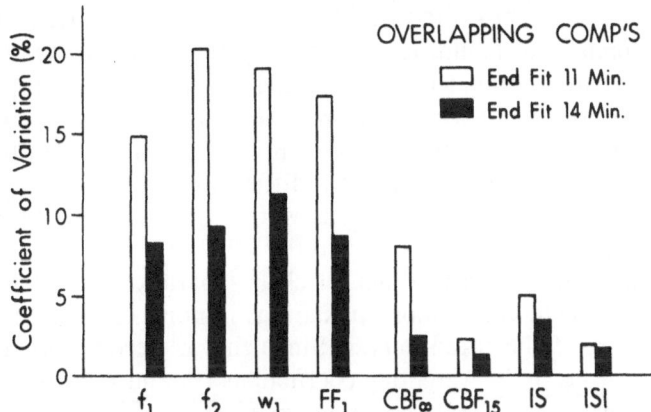

Fig. 1. Coefficients of variation (SD/mean) for different CBF indices computed with two end-fit times. The analysis is based on 100 simulations of a clearance curve representing pathologically reduced flow with overlapping tissue compartments. The 100 simulations differ only in random noise. See Table 2 for details

Of particular interest in Tables 1 and 2 is the close approximation of mean CBF_{15} to the true cerebral value of CBF_∞. This confirms the intended effect of CBF_{15} in compensating for extracerebral contamination, achieved by subtracting area under the tail of the head curve.

The influence of cortical activation on the several CBF indices is revealed in Table 1. The most sensitive variables are IS and f_1, where the mean values undergo increases of 34% and 29%, respectively, relative to the normal resting level. In contrast, CBF_{15} increases by 22%, and ISI by only 13%.

As shown in Table 1, coefficients of variation for all CBF indices are less than 4% in both the normal and activated conditions. This is in sharp contrast to the higher values obtained for overlapping compartments (Table 2). Two findings in Table 2 deserve special attention: (1) the difference in coefficient of variation between compartmental and noncompartmental indices, the latter having much lower values; and (2) the favorable influence of extending curve analysis from 11 to 14 min, which in all instances reduces the coefficient of variation. Figure 1 summarizes these findings in graphic form.

Discussion

The results have clear implications for the choice and interpretation of CBF indices in patients with pathologically reduced flow, where clearance rates of the tissue compartments approach each other or their distributions are overlapping. Whereas both compartmental and noncompartmental variables were highly stable in the normal and activated conditions (Table 1), only the non-compartmental indices had acceptable coefficients of variation for curves with overlapping components (Table 2). Although extending the analysis time for these curves improved the stability of all CBS indices, compartmental estimates continued to be quite variable (Fig. 1). Thus, in patients with severe disease, interpretations should be based on the more stable noncompartmental indices derived from extended end-fit times. Reliance on compartmental variables in such cases could actually be misleading (Obrist and Wilkinson 1980).

The increased variation of computed parameters in the reduced flow condition cannot be explained by differences in signal-to-noise ratio, since both count rates and noise patterns were equivalent for the several types of curve. Rather, it can be attributed to the inherent instability of exponential rate constants when there is a lack of clear separation between compartments, a situation where even minor fluctuations in noise can produce large variations in k_1 and k_2 of the fitted curves. Since this is accompanied by compensatory adjustments in the weighting coefficients (P_1 and P_2), the relative size of the compartments may undergo pronounced shifts. Such "slippage" is manifested by increased variability of w_1 and FF_1, both of which correlate negatively with f_1. Because the noncompartmental indices are derived from the combined flow of the two compartments, being based on weighted averages, they are relatively insensitive to shifts in compartment size, and hence possess greater statistical stability.

Aside from the distinction between compartmental and noncompartmental variables, there is a difference between indices that directly estimate flow (f_1, f_2, CBF_∞, CBF_{15}) and those that are derived from clearance curve slopes (IS, ISI). Both types of index have proven useful in specific applications. As shown in Tables 1 and 2, IS is comparable to f_1 in sensitivity to high flow, but has the advantage of greater stability in low flow situations. On the other hand, ISI is less sensitive than CBF_{15}, although slightly more stable. The primary advantage of CBF_{15} over ISI is that it describes the level of flow in standard units of ml/ 100 g/min and, when averaged over multiple brain regions, can be used for computation of cerebral metabolic rate (Obrist et al. 1979).

One of the characteristics of ^{133}Xe is the large difference between tissue – blood partition coefficients for gray and white matter and their variation with hemoglobin level (Veall and Mallett 1965). Whereas measurements of flow (f_1, f_2, CBF_∞, CBF_{15}) require the explicit assumption of a partition coefficient (λ), the slope indices (IS, ISI) do not require such an assumption. It should be emphasized, however, that the rate constants employed in the computation of slopes are nevertheless influenced by λ, which is included in their definition: $k_i = f_i/\lambda_i$, $i = 1, 2$. The finding that partition coefficients may vary with certain types of disease (O'Brien and Veall 1974) indicates that caution should be exercised in interpreting flow values when marked tissue alterations occur. The recently developed stable xenon-CT scan method (Drayer et al. 1980) offers the possibility of evaluating the extent of partition coefficient changes in pathologic conditions.

The relative sensitivity of the CBF indices to higher flows is of particular interest in studies of cortical activation. As might be expected, IS showed the largest increase in the activated condition, due to its dependence on the early portion of the reconstructed impulse-equivalent curve. In this respect, it is similar to the initial slope of the intracarotid ^{133}Xe technique (Olesen et al. 1971). Although considerably less sensitive, ISI (the slope between 2 and 3 min) had a lower coefficient of variation. Thus, in terms of their ability to discriminate small statistically significant changes in flow, the two indices appear to be comparable (Gur et al. 1983). A possible advantage of IS over ISI is that it more accurately reflects the magnitude of fast compartment CBF changes.

Recognizing the insensitivity of ISI to cortical activation, Risberg and Prohovnik (1981) proposed fitting the early portions of the head curve in order to give greater representation to fast clearing components. They argued that starting the fit at the beginning of the curve would facilitate detection of fast compartment blood flow changes, relative to the standard procedure of delaying the start-fit time. Such "total curve analysis" required correction for scattered radiation from the air passages, which was achieved by introducing a fifth unknown into the analytic model. Recently, Hazelrig and co-workers (1981) attempted a similar analysis that included a sixth unknown to account for the time displacement of the head and air curves.

Based on the simulated activation data in Table 1, the present authors made a comparison between delayed (standard) and early start-fit times, using the previous four and new six parameter models, respectively. Almost identical values ($< 1\%$ difference between means) were obtained by the two types of analysis for each of the CBF indices. This result can be explained by the continued presence of fast clearing components at the delayed start-fit time, due to recent isotope input and to concomitant recirculation. As might be expected, inclusion of additional data points in the total curve analysis resulted in some improvement in stability of the computed parameters. When applied to real data, however, this gain was partly offset by increased sensitivity to air curve artifact at the earlier times. Assessment of the relative merits of these two types of analysis awaits further investigation.

References

Bruce DA, Schutz H, Vapalahti M, Langfitt TW (1975) Pitfalls in the interpretation of xenon CBF studies in head-injured patients. In: Langfitt TW, McHenry LC, Reivich M, Wollman H (eds) Cerebral circulation and metabolism. Springer-Verlag, New York, p 406−408

Drayer BP, Gur D, Yonas H, Wolfson SK, Cook EE (1980) Abnormality of the xenon brain: blood partition coefficient and blood flow in cerebral infarction: An in vivo assessment using transmission computed tomography. Radiology 135:349−354

Enevoldsen EM, Jensen FT (1977) Compartmental analysis of regional cerebral blood flow in patients with acute severe head injuries. J Neurosurg 47:699−712

Gur RE, Skolnick BE, Gur RC, Caroff S, Rieger W, Obrist WD, Younkin D, Reivich M (1983) Brain function in psychiatric disorders: I. Regional cerebral blood flow in medicated schizophrenics. Arch Gen Psychiatry 40:1250−1254

Hazelrig JB, Katholi CR, Blauenstein UW, Halsey JH, Wilson EM, Wills EL (1981) Total curve analysis of regional cerebral blood flow with ^{133}Xe inhalation: Description of method and values obtained with normal volunteers. IEEE Trans Biomed Eng 28:609−615

Heiss WD, Reisner T, Hoyer J (1975) Changes of compartmental weight in the course of stroke. In: Harper M, Jennett B, Miller D, Rowan J (eds) Blood flow and metabolism in the brain. Churchill Livingstone, Edinburgh, p 13.10−13.11

Høedt-Rasmussen K, Sveinsdottir E, Lassen NA (1966) Regional cerebral blood flow in man determined by intra-arterial injection of radioactive inert gas. Circulat Res 18:237−247

Iliff L, Zilkha E, Bull JWD, Du Boulay GH, McAllister VL, Marshall J, Ross Russell RW, Symon L (1974) Effect of changes in cerebral blood flow on proportion of high and low flow tissue in the brain. J Neurol Neurosurg Psychiat 37:631−635

O'Brien MD, Veall N (1974) Partition coefficients between various brain tumors and blood for ^{133}Xe. Phys Med Biol 19:472−475

Obrist WD, Wilkinson WE (1980) The non-invasive Xe-133 method: Evaluation of CBF indices. In: Bes A, Geraud G (eds) Cerebral circulation. Excerpta Medica, Amsterdam, p 119−124

Obrist WD, Thompson HK, Wang HS, Wilkinson WE (1975) Regional cerebral blood flow estimated by ^{133}xenon inhalation. Stroke 6:245−256

Obrist WD, Gennarelli TA, Segawa H, Dolinskas CA, Langfitt TW (1979) Relation of cerebral blood flow to neurological status and outcome in head-injured patients. J Neurosurg 51:292−300

Olesen J, Paulson OB, Lassen NA (1971) Regional cerebral blood flow in man determined by the initial slope of the clearance of intra-arterially injected ^{133}Xe. Stroke 2:519−540

Reivich M, Slater R, Sano N (1969) Further studies on exponential models of cerebral clearance curves. In: Brock M, Fieschi C, Ingvar DH, Lassen NA, Schürmann K (eds) Cerebral blood flow. Springer-Verlag, Berlin, p 8−10

Risberg J, Prohovnik I (1981) rCBF measurements by ^{133}Xe inhalation: Recent methodological advances. Prog nucl Med 7:70−81

Risberg J, Ali Z, Wilson EM, Wills EL, Halsey JH (1975) Regional cerebral blood flow by ^{133}xenon inhalation: Preliminary evaluation of an initial slope index in patients with unstable flow compartments. Stroke 6:142−148

Veall N, Mallett BL (1965) The partition of trace amounts of xenon between human blood and brain tissues at 37 °C. Phys Med Biol 10:375−380

Waltz AG, Wanek AR, Anderson RE (1972) Comparison of analytic methods for calculation of cerebral blood flow after intracarotid injection of ^{133}Xe. J Nucl Med 13:66−72

Two-Dimensional Measurements of rCBF by the Intravenous Xenon 133 Method

J. Seylaz and P. Meric

Since the original method was developed in 1965 by Mallet and Veall, different noninvasive techniques for measuring cerebral blood flow have been widely used by various laboratories. The aims of this paper are: first, to describe the method; second, to underline the improvements made over the years; and third, to situate our own work in this field. The reproducibility, the sensitivity, and the limits of these methods will also be discussed, as will their usefulness at the present time when sophisticated, three-dimensional, but more expensive methods are available.

The principle of CBF measurement in man is based on the determination of the rate of elimination of a radioactive tracer gas, previously introduced into the brain. Before noninvasive techniques were proposed, radioactive tracers were injected via the internal carotid artery. Several studies (Lassen et al. 1963; Hoedt-Rasmussen et al. 1966; Olesen et al. 1971) have demonstrated the validity of such approaches. Nevertheless, these methods have the disadvantages of (1) requiring carotid puncture (which cannot be considered to be without danger), and (2) allowing only unilateral determination of flow. These limitations restricted their general application in clinical work and explain why other means of tracer administration were sought.

Two methods have been proposed: inhalation of an air–xenon 133 mixture (Mallet and Veall 1965; Obrist et al. 1967) and intravenous injection of ^{133}Xe dissolved in physiologic solution (Agnoli et al. 1969; Austin et al. 1972; Meric et al. 1975; Meric et al. 1979). These methods, which basically are similar, differ from the intracarotid injection method in that the head clearance curves are influenced by (1) the noninstantaneous arrival of the tracer in the brain (mixed with recirculating tracer) and (2) the contamination of extracerebral tissues, which constitutes a source of parasitic radiation.

In the 13 years that we have been developing an intravenous method of measuring local CBF, most of the technical improvements we have made have been aimed at minimizing the influence of these two factors.

Theoretical Models

Modelization of cerebral blood flow in man has been established from animal experiments which enable the distribution of flow values within the whole

Laboratoire de Physiologie et Physiopathologie Cérébrovasculaire U. 182 I.N.S.E.R.M., E.R.A. 361 C.N.R.S., Université Paris VII, 10, avenue de Verdun, F-75010, Paris

Cerebral Blood Flow and
Metabolism Measurement.
Eds. Hartmann/Hoyer
© Springer-Verlag Berlin Heidelberg 1985

brain to be determined. These experiments, based on the quantitative simultaneous multiregional determination of flow (Reivich et al. 1969; Sakurada et al.
1978; Lacombe et al. 1979) have led to the development of a model involving
two simple compartments in parallel (one fast, one slow) representing respectively the blood flow in the gray matter and the blood flow in the white matter.
The difference between these flows (one being four to five times higher than
the other) is sufficient for them to be easily differentiated. The validity of such
a simple model has been widely demonstrated in the analysis of clearance
curves obtained by intracarotid injection of the tracer (Lassen et al. 1963;
Hoedt-Rasmussen et al. 1966; Olesen et al. 1971).

In the case of intravenous administration or inhalation of the tracer this
simple model is no longer valid. Various attempts have been made to allow for
the extracerebral circulation. They will be discussed in the section "Influence of
the Different Error Factors in the Flow Determination." Our experience led us
to apply a model similar to that used for the intracarotid method, i.e., a model
with two compartments in which the fast compartment again represents gray
matter flow, and the slow compartment the white matter and extracerebral
flows combined. In this model, only the gray matter flow has a physiologic
meaning. Nevertheless, as discussed later, the low flow rate, as well as the respective weights of the two compartments, is also of physiologic or pathologic
significance.

The use of such a model, although developed independently in our laboratory, is in agreement with the work done in this field by others (Obrist et al.
1971; Austin et al. 1972). As already indicated, because of the recirculation of
the tracer when introduced intravenously (or by inhalation) the count rate $N(t)$
registered by the scalp detector cannot be represented by the simple equation:

$$N(t) = A_i \exp(-K_i t) \quad i = 1, 2$$

where A_i is the weight factor of the exponential corresponding to compartment
i, and $K_i = F_i / \lambda_i$, F_i being flow per unit mass in compartment i and λ_i the
blood/tissue i partition coefficient (Kety 1951).

The corrected equation must take into account the fact that the arrival of the
tracer cannot be considered instantaneous. The new equation is

$$N(t) = \int_0^t A_i \exp - K_i (t - \tau) \, Ca(\tau) \, d\tau \quad i = 1, 2.$$

Computer solutions for A_1, A_2, K_1, and K_2 were obtained by means of a
weighted least-square method. The sum of the squared deviations was developed in Taylor series in order to overcome its nonlinearity in K_1 and K_2. Details of the computation have been published previously (Meric et al. 1979). We
should emphasize the fact that, using this type of computation, the expression
of the estimated recirculation curve $(Ca(t))$ appears only inside integrals. This
minimizes the inevitable perturbations in the determination of $Ca(t)$.

Methods

As shown above, the determination of blood flow necessitates simultaneous measurements of head clearance curves and the recirculation curve (Ca(t)) of the tracer.

Measurement of the Head Clearance Curves

Clearance curves were obtained in 13 areas of each hemisphere following an intravenous injection of $10-15$ mCi of ^{133}Xe. The 26 probes containing NaI (Tl) crystals ($\frac{3}{4}''$ diameter $\times \frac{3}{4}''$ long) and lead collimators $\frac{3}{4}''$ ID $\times 1''$ long were arranged perpendicularly to the scalp. Each detector was connected to an amplifier and an adjustable energy discriminator in series; the whole system was driven by either a multi 4 Intertechnique computer or an Apple II microcomputer. The detector position and the window discrimination selected ($75-100$ keV) allowed count rates of about 60×10^3 counts/minute (at the peak of the curve). Such count rates limit statistical fluctuations to $1\%-2\%$ for the part of the curve used in calculating rCBF.

Apart from the detector position and the energy window used in this study (see section "Influence of the Different Error Factors in the Flow Determination") the measurement protocol is quite similar to those developed in other laboratories.

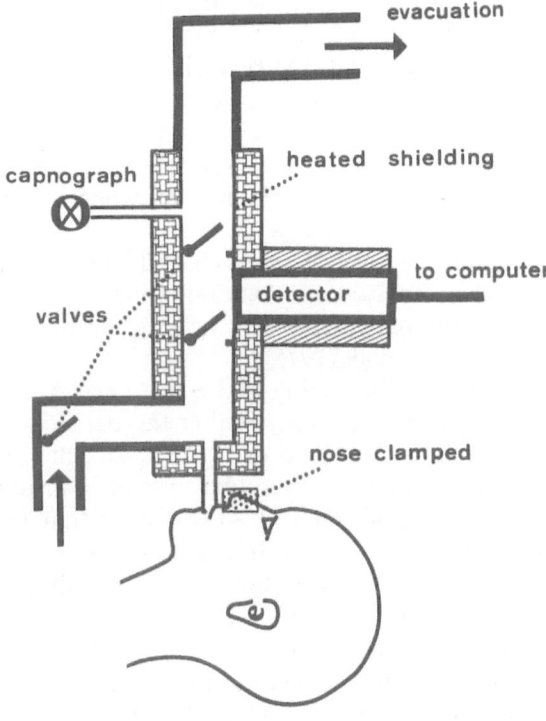

Fig. 1. Diagram of the device used to measure the xenon concentration in the end-tidal air. Outside air enters via the lower gravity valve during the inspiration (the patient breathes through the mouth only, the nose being clamped). At the same time, the two upper valves, being closed, trap a portion of the previously expired air, the activity of which is measured by a detector. During the expiration the lower valve remains closed while the two upper valves open for the evacuation of the contaminated air; the portion corresponding to the end of the expiration remains captured when the two valves close again at the beginning of a new respiratory cycle. The shielding of the measuring chamber, between the two upper valves, is heated to avoid condensation of contaminated water vapour, which otherwise could disturb the measurements. (Meric et al. 1979)

Measurement of the Tracer Recirculation, Ca(t)

The determination of Ca(t) is based on the hypothesis that the end tidal concentration of xenon is identical to the concentration in pulmonary venous blood. This assumption has been demonstrated to be valid except in cases of serious respiratory disorders (Isbister et al. 1965; Obrist et al. 1967; Agnoli et al. 1969). The expired air was collected by a face mask connected to a chamber enclosed within nonreturn valves (Fig. 1). The activity was measured by means of a detector similar to the head probes.

Validation of the Method: Comparison with the Results of the Intracarotid Method

Previous studies have shown that in the case of young, healthy volunteers, the mean values of rCBF obtained by atraumatic methods did not significantly differ from those obtained by the intracarotid method (Obrist et al. 1967; Kuikka et al. 1977; Austin et al. 1978). Although these results validate these atraumatic measurements in normal conditions, they are insufficient to validate the method for pathologic cases. This must be done by comparison of the two methods in the same patient. Only Reivich et al. (1975) have carried out such a comparison for the inhalation method.

In our laboratory this comparison has been made in atheromatous subjects: two measurements were performed successively using the intravenous injection method and the intracarotid method. As noted by others (Obrist et al. 1967, 1975), we found that the real difficulties in clearance curve analysis were the perturbations caused by the extracerebral contamination at the end of the clearance and the various factors enumerated below which occur during the first $10-30$ s of the clearance. A systematic investigation of the part of the curve to be used showed that the results vary when t_0 (start fit time) and t_1 (end fit time) are modified.

The changes in rCBF values found when t_1 is varied are due to the slow clearance from extracerebral contamination. The influence of this factor was found to be minimized when t_1 was set at $9-13$ min. This is in accordance with findings by Obrist et al. (1975).

As far as the beginning of the clearance curve is concerned, particular efforts were made to determine whether the perturbations were related to the type of function introduced into the system, i.e., the arterial input function, $C_a(t)$. To this purpose we performed two sorts of injection in two different groups of patients: a bolus injection $(2-3$ s$)$ similar that used by Austin et al. (1972) or a slow regular injection lasting 1 min, which is comparable to the 1-min air–xenon inhalation used in the atraumatic inhalation methods. The results are illustrated by Figs. 2 and 3. Values of K_1 (clearance rate of the gray compartment) obtained by intravenous injection (K_1iv) were compared with values of K_1 obtained by intracarotid injection (K_1ic).

The ratio K_1iv/K_1ic is close to 1.0 when t_0 attains, respectively, 1.5 min for the bolus injection and 2.5 min for the slow injection. However, the variance of

Fig. 2. Plots of K_1iv/K_1ic against t_0, where t_0 is the start fit time used in the calculations of K_1iv (the value of K_1 obtained by intravenous injections), K_1ic being the value of K_1 obtained by intracarotid injection in the same patient. Each point represents the mean K_1iv/K_1ic for the group of seven patients given a bolus injection (2–3 s) of the tracer. *Bars* indicate $\pm$ 1 SD

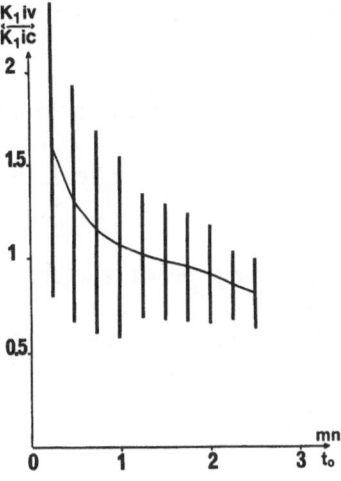

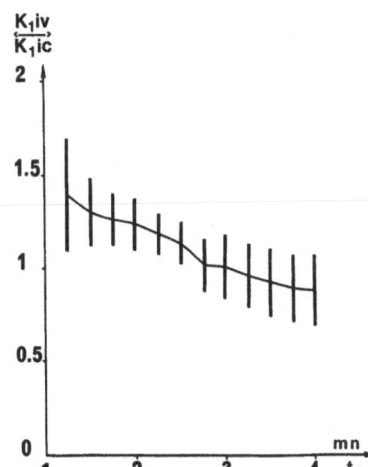

Fig. 3. Same analysis as in Fig. 2 for a group of seven patients given a slow (1 min) injection of tracer

this ratio is smaller in the case of a long-lasting injection, resulting in more reliable CBF determinations. The linear regression performed in these conditions between K_1iv and K_1ic is

$$K_1\text{iv} = 1.126\ K_1\text{ic} - 0.086$$

This result is not significantly different from the result of the inhalation method of Reivich et al. (1975):

$$K_1\text{iv} = 1.149\ K_1\text{ic} - 0.014$$

In summary, it appears that even when applied to patients with cerebrovascular disorders, if t_0 and t_1 are correctly fixed, we obtain results sufficiently precise for clinical use. These choices of t_0 and t_1 may seem quite arbitrary, but examination of the factors disturbing the clearance curves, as discussed later, corroborates these experimentally obtained results.

Influence of the Different Error Factors in the Flow Determination

Influence of the Extracerebral Contamination

This chapter essentially deals with the justification of the method used in this study to minimize the extracerebral contamination. Several approaches have been proposed to this purpose. First, Veall and Mallet (1965) suggested minimizing the extracerebral contamination by decreasing blood circulation through this vascular bed. However, scalp compression left the contribution from cranial bones intact.

Another attempt was made by Crawley et al. (1968). It was based on the characteristics of the xenon 133 spectrum, which has two main peaks at 30 keV and 81 keV. The attenuation of the 30 keV radiation being greater than that at 81 keV, a clearance curve registered at this energy level would correspond to superficial layers only of the detected area. Thus such a clearance curve subtracted from the clearance curve registered at 81 keV in the same area would provide a curve practically free from extracerebral contamination.

This solution was tested by Risberg et al. (1977) and Meric et al. (1975). It was shown that this procedure effectively gave accurate results for the atraumatic measurement but the narrow windows of energy discrimination needed by this method necessitate a very high injected dose of ^{133}Xe in order to maintain count rates at a usable level. Under these conditions the main value of the atraumatic methods, their ready repeatability, is strongly limited. For this reason this attractive solution was discarded.

A third solution was proposed by Obrist et al. (1967). It consists of modeling the extracerebral circulation by addition of a third, slowly clearing compartment parallel to the first two, these representing the gray matter and the white matter. We tested this model by comparing clearance curves obtained in patients who had received successively an injection of the tracer into the common carotid artery (supplying brain and extracerebral tissues) and the internal carotid artery (supplying the brain alone). As shown in Table 1, analysis of the clearance curves obtained from common carotid injections by two- and three-compartment models leads to the following conclusions.

It can be seen that in the case of the bicompartmental model, the results obtained for K_1, the clearance rate for the fast compartment, are not significantly different for the two types of injection. This is also true for K_2 using the two-compartmental model, despite a larger standard deviation. In contrast, however, when a tricompartmental model is used, K_2 is greatly overestimated, and K_1, although not significantly different from the internal carotid value, displays greater variance. It would therefore seem of dubious value to use the tricompartmental model to eliminate the extracerebral contamination. Moreover a series of injections in the external carotid artery revealed that the extracerebral tissue clearances are poorly represented by a single low-flow compartment (Meric et al. 1979).

Furthermore, an extra low-flow compartment in the model implies first, a longer recording time (35−40 min), and second, an increased uncertainty in the

Table 1. Comparison of the results obtained in the same patients by: (1) injections into the internal carotid artery and use of a two-compartment model. Results are noted: $K_{1(IC)}/K_{2(IC)}$; (2) injections into the common carotid artery (CC 2 comp) when a two-compartment model is used in analysing the first 10 min of the curves, a three-compartment model (CC 3 comp) is used with analysis over 35 min, and clearance rates of the first two compartments are $K_{1(CC)}$ and $K_{2(CC)}$. The ratio $K_{1(CC)}/K_{1(IC)}$ is significantly different from 1 when a three-compartmental model is used, showing the inadequacy of this model in the case of cerebrovascular disease with low flow values. It is impossible in this case to distinguish the low flow of the white matter from the low flow of the extracerebral tissues

	$\dfrac{K_{1(CC)}}{K_{1(IC)}} \pm SD$	$\dfrac{K_{2(CC)}}{K_{2(IC)}} \pm SD$
CC 2 comp	0.97 ± 0.08[a]	0.92 ± 0.14[a]
CC 3 comp	1.11 ± 0.15[a]	1.62 ± 0.59[b]

[a] Not significantly different from 1
[b] Significantly different from 1 ($P < 0.05$)

curve analysis due to the greater number of parameters (six instead of four) to be determined (Glass and Garreta 1967).

The difficulties of modelization of cerebral circulation by multicompartmentalization were also described by Risberg et al. (1975). These authors even considered that there are limits to such a model as the bicompartmental modelization used in our study. They introduced an "initial slope index" (ISI) based on a one-compartmental analysis of the early part of the clearance curve. The ISI apparently provides no particular information in addition to that given by F_1, F_2 and the relative weights of the compartments. In fact, this parameter is useful mainly in highly pathologic states where computations fail to distinguish between lowered gray matter flows and white matter flows.

Influence of the Scattered Radiation

Obrist et al. (1975) suggested that the respiratory passage is highly contaminated at the beginning of the clearance, producing scattered radiation which contributes to the registered clearance curve. The presence of this contamination would explain the systematic overestimation of K_1 for low start fit times. On the other hand, Wilkinson et al. (1969) and Potchen et al. (1969) found that, without energy discrimination, nearly 50% of the radiation arriving at a detector comes from outside the collimated zone. As demonstrated by Meric et al. (1977), when the lower level of the discriminating gate was set at 75 keV, the diffused radiation represented only 10% of the total count rate. This energy discrimination resulted in a somewhat diminished count rate, but statistical fluctuations remained within suitable limits. This was possible because, in our study, all the head detectors were set perpendicularly to the skull, and most groups using the atraumatic method have now adopted this procedure. Thus, using this energy discrimination, the measurement of CBF became more truly regional and the influence of extracerebral regions such as the respiratory tract was reduced.

Influence of Error in Measurement of the Recirculation of the Tracer

Several factors may introduce error in the determination of the tracer recirculation when estimated by end-tidal air contamination. The first one, a possible impairment of pulmonary function, may be discarded because the patients studied were selected on the basis of normal respiratory function. The second factor which may affect the clearance curve analysis consists of transient perturbations occurring at the beginning of the recirculation curve. Their influences are, however, minimized by an appropriate type of algorithm used in clearance curve analysis, i.e., Ca(t) does not appear in the calculations except within the integrations. It has been demonstrated by others (Obrist et al. 1975; Correia et al. 1979), and verified by us, that a third possible factor, i.e., the statistical fluctuations of the count rate representing Ca(t), cannot be responsible for errors in the CBF determinations.

The last factor of error introduced by this technique of determination of Ca(t) is the time lag existing between the end-tidal air contamination curve and the true arterial contamination of the blood arriving in the brain. Although no direct measurement has been made to determine this time lag, it can be estimated to be around 3 s in healthy volunteers (Obrist 1975) for the inhalation method. In pathologic cases, this lag may be more variable and can reach 10 s

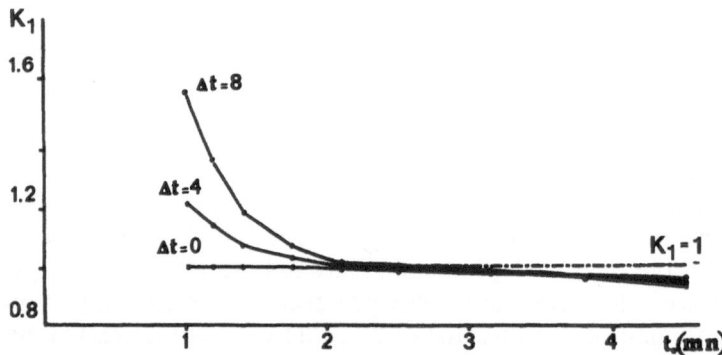

Fig. 4. Variations of K_1, clearance rate of the fast flow compartment, against t_0, start fit time, where K_1 is calculated from synthetic curves prepared with a typical authentic recirculation curve ca(t), using as values of the parameters $K_1 = 1.0$, $K_2 = 0.2$, $A_1/A_2 = 3$ and time lag $t = 0$ s, 4 s, 8 s. The ascending part of the curves for large t_0 is a perturbation due to the poor contribution of the fast clearing compartment to the clearance, so that when a high value of K_1 is used for the preparation of the curves, a large distortion of the calculated value of K_1 occurs. (Seylaz et al. 1980)

(Seylaz et al. 1980). In order to determine the possible effects of this time lag on calculation of rCBF we performed a computer simulation. Artificial clearance curves were generated with authentic Ca(t) curves with various time lags and then analyzed by the usual program.

As shown in Fig. 4, the introduction of such a time lag results in an overestimation of K_1 (gray matter flow), whatever the flow levels, and the greater the time lag the larger the overestimation. However, when t_0 (start fit time) is increased, the influence of the time lag on CBF determination decreases. It even disappears when t_0 attains $2-3$ min, depending on the flow level used in preparing the arterial clearance curves. These data are perfectly consistent with the experimental findings described above concerning the sensitivity of the part of the curve to be used in CBF calculation.

Therefore, as far as pathologic flows are concerned, it seems that this factor cannot be neglected. Since it cannot in practice be determined, t_0 has to be adapted to the flow level first estimated by a rough analysis performed systematically with t_0 set at 2.5 min, followed by a more accurate analysis performed with t_0 equal to 2.0, 2.5, and 3.0 min for high, medium, and low flow levels respectively.

Reproducibility and Sensitivity of the Method

Many of the clinical applications of this technique require repeated rCBF measurements to be made in the same patient, which implies precise knowledge of the reproducibility and the sensitivity of the method. However, the available information in the literature on these points is essentially based on results on healthy volunteers. We therefore performed a study designed to measure these parameters of the technique in the case of patients with cerebrovascular disorders, and thus to determine to what extent the results obtained in healthy subjects can be extrapolated to those obtained in patients with focal or global perturbations of the cerebral circulation.

The three aspects of the technique studied were:

1. The reproducibility of two consecutive measurements and the influence of the time interval between them.
2. The value of corrections for the arterial partial pressure of CO_2 ($PaCO_2$), given that $PaCO_2$ is a powerful cerebral vasodilator.
3. The sensitivity of the method, estimated by comparing in the same subject a measurement at rest and a measurement during moderate unilateral hand work, which should induce a rise in rCBF in the corresponding sensory and motor areas of the contralateral hemisphere (Olesen 1971).

Reproducibility

The influence of the time interval between two measurements was investigated by dividing the patients studied into two groups: in group I, comprising 14 pa-

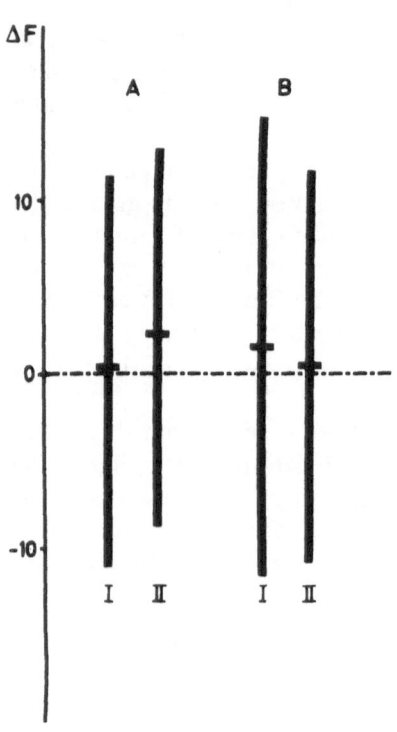

Fig. 5. Mean and standard deviation of the differences in rCBF between two measurements performed in 22 patients. The patients are divided into two groups according to two sets of criteria:

A. The two groups correspond to the time interval between the two measurements, as follows: group I: $n = 14$, $t = 30$ min; group II: $n = 8$, $t =$ several days.

B. The two groups correspond to the variation in $PaCO_2$ observed between the first and the second rCBF measurement, as follows: group I: $n = 9$, variation < 1 mmHg; group II: $n = 13$, variation ≥ 1 mmHg. There is no significant difference between either the means or the standard deviations of groups I and II in either A or B. (Meric et al. 1983)

tients, the measurements were performed with as small an interval as possible (30 min, eliminating any risk of residual gas contamination); in group II, comprising eight patients, the measurements were performed with an interval of several days. Figure 5 shows the mean ± SD of the differences between the measurements of rCBF for each group. There is no significant difference between the groups, either for the means (not significantly different from zero) or for the variances. Figure 6 shows that the reproducibility of the two groups taken together is inferior to that found by Blauenstein et al. (1977) in healthy subjects.

Corrections for $PaCO_2$ Variation

In these same two groups of patients, certain subjects showed appreciable variation in the $PaCO_2$ (estimated from end-tidal PCO_2), which could be thought to account for the relatively high variance of the rCBF differences (Fig. 6). Attempts were therefore made to improve the reproducibility by correcting for the $PaCO_2$ variations, but it was found that even when using the lowest correction coefficients proposed in the literature (Maximilian et al. 1980) the corrections led to a significant increase in the variance of the results in both groups. Furthermore, as Fig. 6 shows, the comparison of patients showing appreciable $PaCO_2$ changes with those showing little variation in $PaCO_2$ revealed no difference in either the mean or the variance of the change in rCBF.

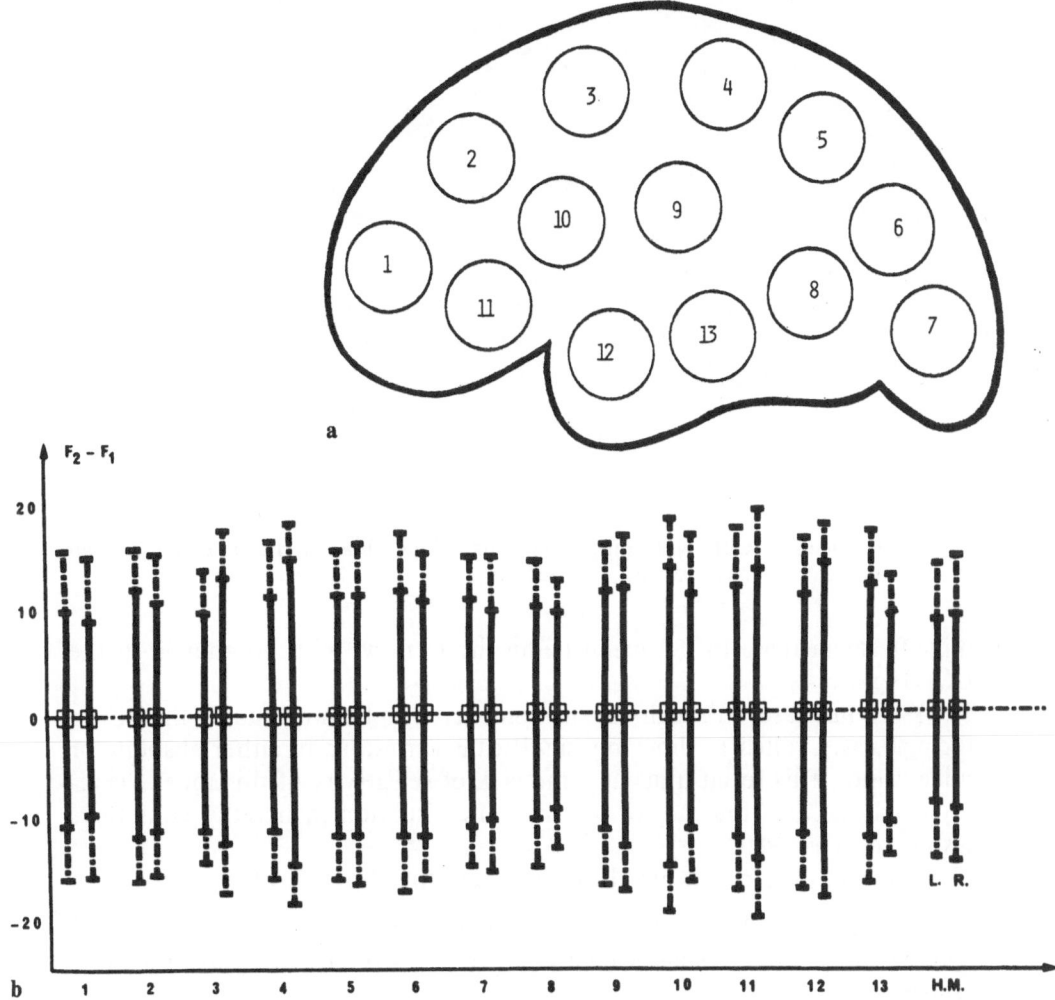

Fig. 6. a. Position of the head detectors with respect to the cerebral hemisphere. **b.** Mean ± SD expressed as ml/100 g/min of the differences in rCBF between two successive measurements (F_1 and F_2) in 22 patients. Differences for the 13 regions measured and the hemispheric mean (*H.M.*) are shown for both the left and the right hemisphere. The *dashed vertical lines* represent the increase in SD obtained when the rCBF values were corrected for the $PaCO_2$ variations in each patient. The mean differences between the two measurements, both corrected and uncorrected, are given by the same point since there was no significant difference between these means, which, moreover, were not significantly different from zero. (Meric et al. 1983)

Sensitivity to Focal Activation

An activation test was carried out in eight patients with only unilateral focal disorders of the cerebral circulation. A control measurement on the healthy hemisphere was followed by a measurement effected during unilateral hand work: a significant focal increase in rCBF was found, accompanied by a slight

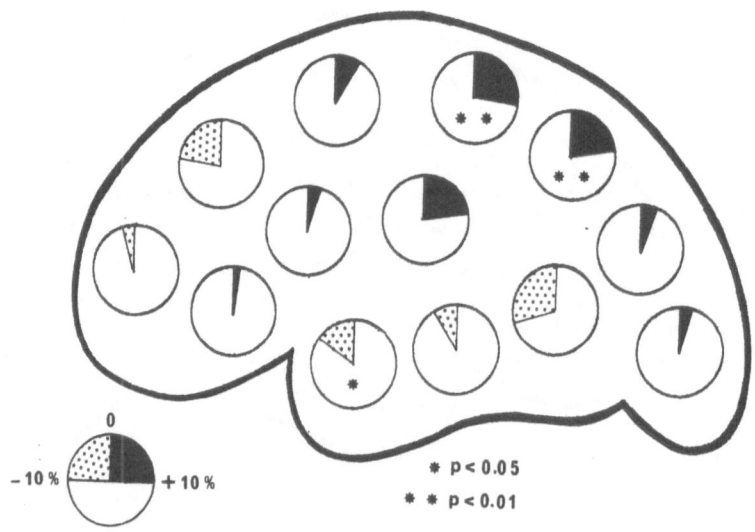

Fig. 7. Mean changes in rCBF, with respect to resting values, obtained during moderate contralateral hand exercise in eight patients. (Meric et al. 1983)

nonsignificant increase in the mean hemispheric flow, with no change in the $PaCO_2$ (Fig. 7).

The ISI in this case, although usually considered to be more stable than F_1 in pathologic cases, failed to show any significant variations in either absolute or relative values. This result hints at some lack of sensitivity of this index, as already emphasized in the case of a hyperemic situation in healthy volunteers (Maximilian et al. 1980).

The following conclusions may be drawn from the above results:

a) The time interval between two measurements has no influence on the reproducibility of the results, which is nonetheless inferior to that in healthy subjects.

b) Corrections for $PaCO_2$ differences are not appropriate for use in cases of cerebrovascular disorders. It seems preferable in routine work to use uncorrected values. Presumably the inappropriateness is due to perturbed regulation of CBF by the $PaCO_2$ in certain pathologic cases.

c) Although the resolution is less fine than that of the intracarotid method, the sensitivity is sufficient to allow this technique to be used in neurophysiologic tests in routine hospital examinations.

As far as the application of the method is concerned, most of the 1500 measurements we have performed were in cases of vascular pathology. The main fields of pathophysiologic studies concerned were subarachnoid hemorrhage, polycythemia, and diagnosis of the need for an IC-EC bypass. Neurologic studies were also performed on the normal hydrocephalus syndrome and the relations between ageing and rCBF patterns.

By way of a general conclusion, we would like to speculate on the future of the two-dimensional atraumatic rCBF techniques in the face of the development of techniques employing positron emitting isotopes which by tomographic analysis enable both the local CBF and oxygen consumption to be determined. It is even possible with such techniques to explore the deep structures which the ^{133}Xe clearance techniques cannot measure because of the greater attenuation of the radiation from these structures. However, the use of such methods is restricted by the enormous cost of the equipment required. Although these materialistic considerations may seem improper when appraising the scientific value of a technique, it is also clear that the value of a technique designed for investigating function in man is related to the limits of its applications and to the number of investigators who have access to the equipment. Thus, simpler methods, being less onerous, may prove their value in widespread routine utilization, even though they may be more limited in their aims.

Atraumatic rCBF measurement by ^{133}Xe clearance seems to correspond to this type of method, since it should enable widespread exploration of CBF in several clinical fields such as neurology, neurophysiology, and neuropsychiatry.

References

Agnoli A, Prencipe M, Priori AM, Bozzao L, Fieschi C (1969) Measurements of rCBF by intravenous injection of Xenon 133. A comparative study with the intraarterial injection method. In Brock M, Fieschi C, Ingvar DH, Lassen NA, Schümann K (Eds) Cereb blood flow, Springer-Verlag, Berlin 31 – 34

Austin G, Horn N, Rouhe S, Hayward W (1972) Description and early results of an intravenous radioisotope technique of measuring regional cerebral blood flow in man. In Cerebral blood flow and intracranial pressure, Proceedings of the 5th International Symposium, Roma, Siena 1971, part II, Europ Neurol 8:43 – 51

Blauenstein UW, Halsey JH, Wilson EM (1977) ^{133}Xenon inhalation method, analysis of reproducibility: some of its physiological implications. Stroke 8:92 – 102

Correia J, Chang J, Alpert N (1979) Stimulation studies of the statistical errors in inhalation and intracarotid 133 Xe studies. Acta Neurol Scand 60 (suppl 72):236 – 237

Crawley JCW, O'Brien MD, Veall N (1968) The gamma spectrum substraction technique applied to cerebral blood flow measurement by inhalation of 133-xenon. In: Brain and blood flow. Ross Russel RW (ed): Proceedings of the Fourth International Symposium on regulation of cerebral blood flow, London: Pittman:6 – 54

Glass HI, De Garreta AC (1967) Quantitative analysis of exponential curve fitting for biological applications. Phys Med Biol 2:379 – 388

Hoedt-Rasmussen K, Sveinsdottir E, Lassen NA (1966) Regional cerebral blood flow in man determined by intra-arterial injection of radioactive inert gas. Circ Res 18:237 – 247

Isbister WH, Schofield PF, Torrance HB (1965) A study of arterial clearance of xenon 133 in man. Brit J Anaesth 37:153 – 157

Kety SS (1951) Theory and applications of the exchange of inert gas at the lungs and tissues. Pharmacol Rev 3:1 – 41

Kuikka J, Ahonen A, Koivula A, Kallanranta T, Laitinen J (1977) An intravenous isotope method for measuring regional cerebral blood flow (rCBF) and volume (rCBV). Phys Med Biol 22:958 – 970

Lacombe P, Meric P, Reynier-Rebuffel AM, Seylaz J (1979) Critical evaluation of cerebral blood flow measurements made with ^{14}C-ethanol. Med and Biol Eng and Comput, 17:126 – 142

Lassen NA, Hoedt-Rasmussen K, Sorensen SC, Skinhoj E, Cronquist S, Bodforss B, Ingvar DH (1963) Regional cerebral blood flow in man determined by a radioactive inert gas (krypton 85). Neurology 13:719 – 727

Mallet BL, Veall N (1965) Measurement of regional clearance rates in man using xenon 133 inhalation and extracranial recording. Clin Sci 29:179–191

Maximilian VA, Prohovnik I, Risberg J (1980) Cerebral hemodynamic response to mental activation in normo- and hypercapnia. Stroke 11:342–347

Meric P, Seylaz J (1977) Radiation scattering and determination of regional cerebral blood flow. Med Progr Technol 5:41–46

Meric P, Seylaz J, Correze JL (1979) Measurement of regional cerebral blood flow by intravenous injection of Xe 133. Med Progr Technol 6:53–63

Meric P, Seylaz J, Rossi-Mori A, De Cosnac B, Luft A, Correze JL, Mamo H, Houdart R (1975) Non-traumatic measurement of cerebral blood flow by intravenous injection of xenon 133. In Harper AM, Jennett WB, Miller JD, Rowan JO (Eds) Blood flow and metabolism in the brain, Proceedings of 7th International Symposium on Cerebral Blood Flow and Metabolism, London, Churchill Livingstone: 8-3-6

Meric P, Luft A, Seylaz J, Mamo H (1983) Analysis of reproducibility and sensitivity of atraumatic measurements of regional cerebral blood flow in cerebrovascular diseases. Stroke 14:82–87

Obrist WD, Thompson HK, King CH (1967) Determination of regional cerebral blood flow by inhalation of ^{133}Xenon. Circ Res 20:124–135

Obrist WD, Thompson HK Jr, Wang HS (1971) A simplified procedure for determining fast compartment rCBFs by ^{133}xenon inhalation. In Russel RWR (ed): Brain and Blood Flow. Proceedings of the Fourth International Symposium. London, Pitman:11–15

Obrist WD, Thompson HK, Wang HS (1975) Regional cerebral blood flow estimated by ^{133}xenon inhalation. Stroke 6:245–256

Olesen J (1971) Contralateral focal increase of cerebral blood flow in man during arm work. Brain 94:635–646

Potchen EJ, Davis DO, Wharton T (1969) Regional cerebral blood flow in man. Arch Neurol 20:378–383

Reivich M, Jehle J, Sokoloff L, Kety SS (1969) Measurements of regional cerebral blood flow with antipyrine ^{14}C in awake cats. J Appl Physiol 27:296–300

Reivich M, Obrist WD, Slater R (1975) A comparison of the Xe 133 intracarotid injection and inhalation techniques for measuring regional cerebral blood flow. In Harper M et al. (Eds): Blood flow and metabolism in the brain. Proceedings of the Seventh International Symposium on Cerebral Blood Flow and Metabolism. Churchill Livingstone (Edinburgh):803–810

Risberg J, Ali S, Wilson EM (1975) Regional cerebral blood flow by ^{133}xenon inhalation. Stroke 6:142–148

Risberg J, Uzzell B, Obrist W (1977) Spectrum subtraction technique for minimizing extracranial influence on cerebral blood flow measurement by ^{133}xenon inhalation. Stroke 8:380–382

Sakurada O, Kennedy C, Jehle J, Brown JD, Carbin GL, Sokoloff L (1978) Measurements of local cerebral blood flow with iodo ^{14}C antipyrine. Am J Physiol 234:H59–H66

Seylaz J, Meric P, Correze JL (1980) Analytical problems associated with the non-invasive measurements of cerebral blood flow in cerebrovascular diseases. Med Biol Eng Comput 18:39–47

Veall N, Mallet BL (1965) The partition of tracer amounts of xenon between human blood and brain tissue at 37 °C. Phys Med Biol 10:375–380

Wilkinson IMS, Bull JWD, Du Boulay GH, Matshall J, Ross Russel WR, Symon L (1969) Regional cerebral blood flow in the normal cerebral hemisphere. J Neurol Neurosurg and Psych 32:367

Correlation Between CBF and pCO_2, pO_2, pH, Hemoglobin, Blood Pressure, Age, and Sex

P. Gündling, J. Haneder, and M. R. Gaab

Introduction

While many investigations concerning the influence of certain parameters on the CBF, like carbon dioxide pressure (pCO_2) and hemoglobin, have been carried out, such parameters as age and sex have not normally been taken into account. Therefore the existing correcting formulas seem to be too restrictive and, besides this, too inflexible. This means that they are valid for every cerebral disorder and depend on the absolute CBF values (higher values are corrected more than lower ones; a better formula should do the opposite. Thus the questions leading us to our study have been:

1. Which parameters do influence the cerebral blood flow in various groups?
2. Are these influences similar in various kinds of cerebral disorder?
3. Are the existing correcting formulas adequate?
4. If not, is it possible to find a formula to eliminate all these influences in order to obtain standardized results to enable, for example, a comparison of different methods of medical treatment?

Method

Our measurements were made with a Novo-Cerebrograph (Novo Diagnostic Systems, Hadsund/Denmark). This ^{133}Xe inhalation system consists of 32 scintillation detectors placed in parallel and is connected to a computer system with terminal and printout, which gives an immediate evaluation of the flow of the gray matter (F1) and of the initial slope index (ISI, as defined by Risberg 1975).

The measurements were carried out in a quiet and semidarkened room, while the patients were relaxed and with their eyes closed. Blood gases and pH from capillary blood were determined according to Astrup, blood pressure according to Riva-Rocci, and hemoglobin according to Sahli.

From our patients, who numbered about 500, we selected 153 with a total of 217 measurements according to certain criteria like clinically proven symptoms,

Neurochirurgische Universitätsklinik, Kopfklinik, Josef-Schneider-Strasse 11, D-8700 Würzburg

Cerebral Blood Flow and
Metabolism Measurement.
Eds. Hartmann/Hoyer
© Springer-Verlag Berlin Heidelberg 1985

interference-free measurements, and few attendant symptoms. These 92 males and 61 females aged 6–80 years – the men with a mean age of 49 and the women with a mean age of 48 years – were divided into groups of the same cerebral disorders (SAH, stroke, TIA, cerebral atrophy) and first evaluated all together and later separately according to their diagnosis.

For the evaluations we used the following statistical methods:

1. Product–moment correlation
2. Kendal's Tau
3. Multiple regression
4. t-Test

Results and Discussion

We found that the *total patient group* showed an *in*crease of the cerebral blood flow with rising pCO_2 and rising pO_2 and a *de*crease of the CBF with rising hemoglobin, pH, and age. We were also able to observe that women on average had a higher CBF than men (Fig. 1) and that there was no correlation between the blood pressure and the CBF, although patients with hypertension on average had a higher cerebral blood flow than patients with normal blood pressure.

These results mostly confirm those of other authors (Obrist et al. 1967; Olesen 1974; Deshmukh and Meyer 1978) while, however, the rising of the pO_2 is more likely a result than a cause of the increasing CBF (i.e., we did not prove the influence of various pO_2 values on the CBF in the same patient, but just noticed a *statistical correlation* between lower pO_2 and lower CBF values in an

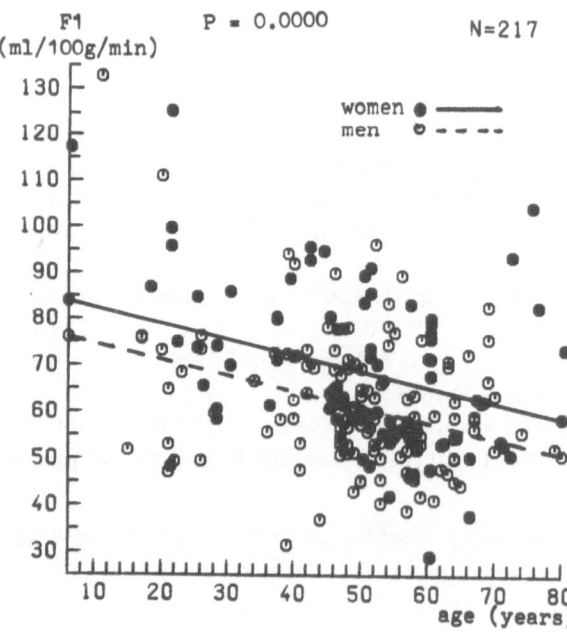

Fig. 1. Correlation between age and F1 arranged in groups of men and women. The F1 values show a significant negative correlation with advancing age and women have higher values than men

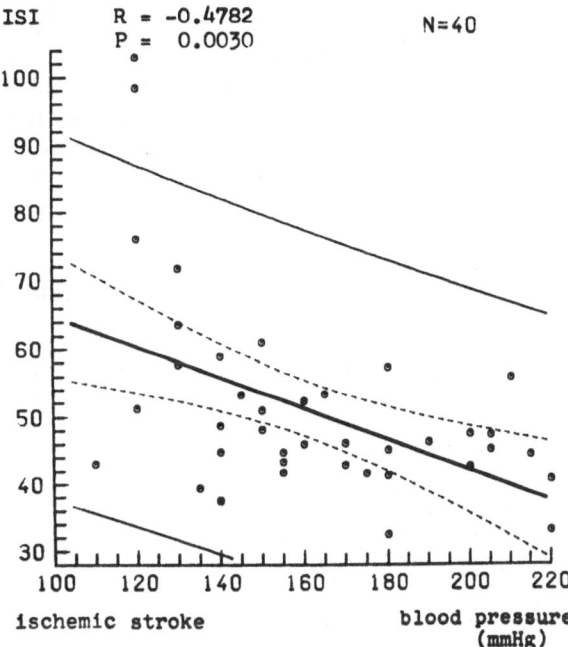

Fig. 2. Correlation between blood pressure and ISI in patients with ischemic stroke. The CBF shows a significant negative correlation with rising blood pressure

accidentally large group, so that patients with a worse CBF, e.g., as a result of a degenerative disease or arteriosclerosis, generally have a worse respiratory function and hence a lower pO$_2$ than those with a higher CBF).

The accepted theory explaining the higher CBF of women with their lower hemoglobin turned out to be unsatisfactory, because even after introducing a correcting factor for the hemoglobin the correlation was still significant.

Moreover, it is striking that the blood gases chiefly correlated with the initial slope index, while the hemoglobin chiefly correlated with the flow of the gray matter. Age and sex, in contrast, showed no difference in their correlation with ISI and F1.

In contrast to the results of the total sample, the single diagnosis groups revealed the following:

1. Patients with *subarachnoid hemorrhage* (SAH) — both with and without clinically proven vasospasm — showed an *increase* of the CBF with rising blood pressure. To our mind this phenomenon is most probably explained by the loss of autoregulation of the brain vessels in these patients. The reason why there is no difference in the behavior of the CBF between patients with and without clinically proven vasospasm might be the difficulty in proving vasospasm clinically in every case and state.
2. Patients with *ischemic stroke* showed a *decreasing* CBF with rising blood pressure (Fig. 2) on the one hand, and on the other no correlation between CBF and the other parameters. This phenomenon we suspect to be a result of a disturbed cerebral vasoregulation. In this case there may be an excess of autoregulation concerning the blood pressure and a weak or a lost reaction on other stimulants.

3. Patients with *transient ischemic attack* (TIA), as well as those with *cerebral atrophy*, did not show any correlation between the CBF and any of the examined parameters. This could be the result of the quite variable picture of the disease in the case of cerebral atrophy and the different time of measurement in relation to the events in the patients with TIA.

Finally we tested some commonly accepted correcting formulas, e.g.,

$$CBF_{corr.} = \frac{CBF}{1 + 0.025 \, (pCO_2 - 4\,O)} \quad \text{(Herrschaft 1975),}$$

and found that the preceding correlation became more significant instead of being levelled out (Fig. 3), which demonstrates the inadequacy of such formulas.

From these findings we tried to identify those parameters which are necessary for a statistically standardizing formula. With the help of multiple regression we were able to show that the influence of pO_2 and pH on the CBF is already represented sufficiently by the pCO_2 and that the hemoglobin (Hb) and Hk correlate absolutely. Thus we arrived at the following equation:

$$ISI_{corr.} = ISI - (53 - 0.4 \, age + 6.8 \, sex + 0.5 \, pCO_2 + 3.5 - 0.2 \, Hb)$$

(53 is about the ISI mean; sex: 1 = male, 2 = female; 3.5 is put in in the case of hypertension)

This equation allows a conclusion from the stated parameters on the CBF and can be used for statistical purposes to compare large groups of patients, but is of no practical use for the individual diagnosis.

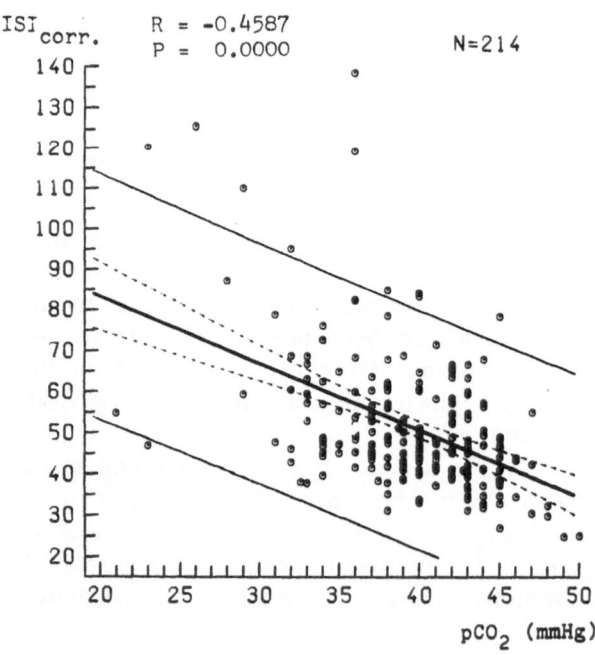

Fig. 3. Correlation between pCO_2 and $ISI_{corr.}$. This demonstrates the inadequacy of the usual correcting formulas, because the preceding correlation became more significant

Summary

Investigating the influences of the mentioned parameters on the CBF, which was measured with the ^{133}Xe inhalation method, in a group of 153 patients we found:

1. The total group showed an increase of the CBF with rising pCO$_2$ and rising pO$_2$ and a decrease of the CBF with rising pH, hemoglobin, and age. On average women had a higher CBF than men.
2. The differentiation according to diagnosis revealed different results:
   a) An increase of the CBF with rising blood pressure in patients with SAH
   b) An increase of the CBF by decreasing blood pressure in patients with ischemic stroke
   c) No correlation in patients with TIA or cerebral atrophy
3. The existing correcting formulas were found to be inadequate: the already existing correlations even became more significant.
4. A formula for interindividual standardization should consider pCO$_2$, hemoglobin, blood pressure, age, and sex, but is of no value for clinical practice.

References

Deshmukh VD, Meyer JS (1978) Noninvasive measurement of regional cerebral blood flow in man. SP Medical and Scientific Books, New York, London, p 129−151

Herrschaft HF (1975) Die regionale Gehirndurchblutung. Springer, Berlin, Heidelberg, New York

Obrist WD et al. (1967) Determination of regional cerebral blood flow by inhalation of 133-xenon. Circulation Research, Vol. XX, p 124−135

Olesen J (1974) Cerebral blood flow methods for measurement regulation, effects of drugs and changes in disease. Fadl, Kobenhaven, Århus, Odense, p 11−18

Risberg J et al. (1975) Stroke 6, p 142−148

Theoretical Evaluation and Simulation Test of the Initial Slope Index for Noninvasive rCBF

I. Prohovnik, E. Knudsen, and J. Risberg

The original analysis for the inhalation method of measuring regional Cerebral Blood Flow (rCBF) was described solely in terms of compartmental analysis, and did not mention other possible flow indices (Obrist et al. 1975). At the same time, however, a problem was discovered in the use of the k_1 (or f_1) parameter, and an Initial Slope Index (ISI) was adapted from the intraarterial injection method (Olesen et al. 1971) and suggested as more reliable than f_1, under certain conditions, for inhalation studies (Risberg et al. 1975).

Although several variants of noncompartmental indices have since been proposed, the ISI as described by Risberg et al. (1975) is most widely used in the published literature (cf., Prohovnik 1984, for a review). Despite its frequent use, this index is often incompletely understood; one purpose of the present article is to explain its calculation and characteristics. In addition, we have recently recognized some theoretical ambiguities in this index, which we shall present here; an alternative index will be proposed, which is theoretically less ambiguous and consists of a clearer analogy to the arterial injection method. Several versions of both indices have been extensively tested by means of computer simulations; the results will be presented here.

Since the ISI is usually described as dominated by fast compartment flow, our basic criterion is the validity of using it as an index of f_1. It must be realized, however, that the original derivation (Risberg et al. 1975) did not consider this f_1 sensitivity a major factor. Indeed, the ISI was conceived mostly as a "slippage" resistant index, and only with later use did it become known as a gray matter dominated flow index. The present communication addresses only this sensitivity issue; as will be discussed later, further work is necessary to test immunity to slippage.

Methods

Data Simulation and Analysis Programs

The model, parameters, and method of synthetic data generation have recently been published (Prohovnik et al. 1983). Briefly, we conducted two large simu-

Department of Biological Psychiatry, New York State Psychiatric Institute, 722 West 168th Street, New York, NY 10032, USA, Novo Diagnostic Systems, Copenhagen, and Lund University, Sweden

Cerebral Blood Flow and
Metabolism Measurement.
Eds. Hartmann/Hoyer
© Springer-Verlag Berlin Heidelberg 1985

lation studies. The first compared three compartmental models and three least-squares algorithms in both time and frequency domains. We have computed six different versions of the ISI, three for each variant discussed here. Based on the results of this experiment, we chose one algorithm in the time domain; we have also chosen one version of the ISI. The second experiment tested the single algorithm and single ISI version again, using all three physiologic models.

The ISI, as will be shown later, is not truly noncompartmental: its value is heavily dependent on, and its calculation directly utilizes, the compartmental parameters, p_1, k_1, p_2 and k_2. Therefore, to some extent, any error in the estimation of these primary parameters will propagate to the ISI. Using different models with different errors in calculation of the primary parameters, the ISI will reflect those errors as a function of the validity of the model, not the validity of the ISI itself. For this reason, in the present article we present the ISI results from only one model, the classical one derived by Obrist et al. (1975).

Definitions

ISIa

This index was proposed by Risberg et al. (1975). Given an inhalation curve defined and generated by the model above, the first step is the usual least-squares solution for the primary parameters p_1, k_1, p_2, k_2. The model also includes the start-fit time (SFT), usually defined as the point at which the end-tidal xenon concentration has dropped to 20% of its maximal value. The next step is to find the size (counts) of each compartment at SFT; these will be denoted $N_1(SFT)$ and $N_2(SFT)$, for the fast and slow compartment.

Next, assume a purely exponential clearance from both compartments, beginning at SFT. Construct the value of total tissue tracer concentration (both compartments) at $t = 2$ and $t = 3$ minutes (considering time zero to correspond to the beginning of inhalation); these two points will be denoted $N_T(2)$ and $N_T(3)$, respectively.

Then compute the monoexponential slope between $N_T(2)$ and $N_T(3)$; this slope, multiplied by 100 (equivalent to ml/100 g/min if the partition coefficient is assumed to be 1) is the ISIa:

$$ISIa = 100 * \{\ln[N_T(2)] - \ln[N_T(3)]\} \tag{1}$$

ISIb

This variant eliminates both the dependence on SFT and the consideration of recirculation-loaded compartment weights by directly utilizing the primary parameters. Using these parameters, construct a clearance curve that would result from an intraarterial bolus injection into the identical tissue volume:

$$N(t) = p_1 * e^{-k_1 t} + p_2 * e^{-k_2 t} \tag{2}$$

On this curve, one can again choose any two points for slope calculation.

Procedures

In the first experiment, we tested three versions of each ISI, with slope-calculation times (relative to SFT or the moment of carotid artery injection) of -0.5 min to $+0.5$ min, 0 to 1 min, and 1 to 2 min (the conventional ISI, which is fixed at $2-3$ minutes with a variable SFT, is approximately equivalent to an ISIa: $0.5-1.5$). In the second experiment, we only tested the ISIb at 0.5 to 1.5 min, which we chose as an optimal compromise based on the results of the first experiment. The first test was run on a Hewlett-Packard 9845, the second on a DEC PDP 11/23; total material consisted of 161280 curves, analyzed by three models and four algorithms.

Results

1. Under our standard condition, the absolute value of ISI varies between 10% and 90% of 100 k_1 (Fig. 1). ISI underestimates high flows as a function of its

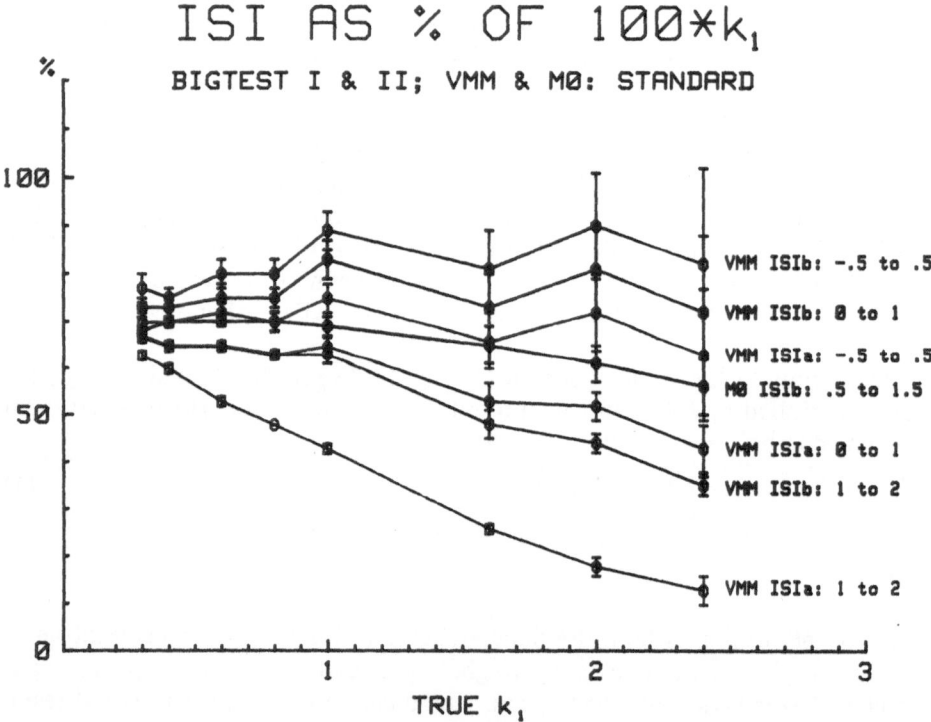

Fig. 1. Mean (*circles*) and standard deviations (*vertical lines*) of ISI values, each determined independently on 32 curves at eight values of simulated k_1 values (all other parameters held constant). An ideal ISI would consist of a straight horizontal line with minimal SDs. Note the unacceptable decline of sensitivity of ISIa: $1-2$, and the higher variance of all versions at high flow rates

Table 1. ISI values as % of true 100*k_1 values, averaged across eight k_1 levels, using the simple model under different simulated conditions

Condition:	Standard	$K_2 = 0.2$	$p_2 = 0.8$	$Td = -5$	$p_3 = 1$	$p_4 = 1$
ISI type:						
ISI a:						
-0.5 to 0.5	70 (4)	71 (7)	60 (6)	66 (4)	74 (8)	86 (16)
0 to 1	59 (9)	63 (11)	50 (9)	58 (8)	63 (12)	71 (17)
1 to 2	41 (19)	47 (21)	33 (17)	41 (18)	42 (20)	44 (23)
ISI b:						
-0.5 to 0.5	82 (5)	81 (6)	75 (12)	76 (6)	89 (11)	106 (24)
0 to 1	76 (4)	76 (6)	67 (8)	71 (5)	81 (9)	95 (18)
1 to 2	56 (12)	61 (13)	46 (13)	55 (11)	60 (15)	66 (20)
0.5 to 1.5	66 (5)	63 (9)	57 (5)	64 (4)	67 (6)	78 (11)

time of computation; in this repect, ISIa: $0-1$, ISIb: $1-2$, and ISIa: $1-2$ are most affected, showing a decline in sensitivity to k_1 values as low as $0.8-1$. The other variants are fairly constant at about $60\% - 90\%$. The ISIb: $-0.5-0.5$, ISIb: $0-1$, and ISIa: $-0.5-0.5$ show no decline of sensitivity at high flows, but considerable instability (indicated by their SDs). On these grounds, ISIb: $0.5-1.5$ seems the best compromise between sensitivity and stability.

2. To test the effects of physiologic parameters other than k_1, we averaged ISI values across eight values of k_1 (from 0.3 to 2.4) with the following results (Table 1). ISIb values are about $20\% - 30\%$ higher than the corresponding ISIa values. All variants are somewhat sensitive to p_2: in our experiments, a doubling of p_2 caused a decrease of $10\% - 20\%$ of ISI. This decrease, as expected, is larger with later slope computation times. A doubling of k_2, however, did not affect ISIa: $-0.5-0.5$, ISIb: $-0.5-0.5$, and ISIb: $0-1$. ISIb: $0.5-1.5$ was slightly decreased (5%) and the other versions were somewhat elevated (about 10%). Lung$-$brain delay is only a problem for the variants most sensitive to k_1: ISIa: $-0.5-0.5$, ISIb: $-0.5-0.5$, and ISIb: $0-1$. This is also the case with the two physiologic artifacts, but the overestimation effects caused by them are larger.

Comment

The ISI is intended to be an index of f_1, with somewhat lower sensitivity but much higher reliability, especially under slippage conditions. In the present article, we are only concerned with its sensitivity and theoretical properties, because we feel that it is inadequately understood despite its frequent use.

Theoretically, the ISIa is complicated by its dependence on SFT and by its derivation from values that include the influence of the input function. The ISIb avoids these difficulties and is an exact analogue of an intraarterial injection curve. Empirically, the ISIa has been found to be less sensitive to k_1 than

ISIb, and more sensitive to the second compartment. These features are easily predictable from their mathematical derivation. Based on those considerations, we find the ISIb, computed between 30 and 90 s, to allow the best combination of sensitivity to k_1 and stability under the conditions tested here. The common ISI, as originally defined (Risberg et al. 1975), corresponds approximately to an ISIa: 0.5–1.5. In Fig. 1, it would show results intermediate between ISIa: 0–1 and ISIa: 1–2. It would thus demonstrate a declining sensitivity to k_1 at values of approximately 0.8 and higher, and it is not an appropriate index of high, or even normal, flows. In addition, all ISIa versions include variance contributed by the shape of the input function and the inevitable variance of SFT. As shown in the figure, a 30-s shift of the SFT can exert a significant influence on the absolute value and k_1 sensitivity of ISIa. This reasoning indicates another advantage for the ISIb.

However, the slippage immunity of the various ISI variants has not been tested here. In the original description (Risberg et al. 1975), this immunity was the major feature of the proposal. Indeed, Risberg (1974, unpublished data) conducted preliminary testing for slippage sensitivity of four ISI variants, including the one we here call ISIb: 0–1 and what became known as IS (Obrist and Wilkinson 1980). Those data indicated that, as curves are shortened from the usual 11 min to about 5 min, k_1 was overestimated by 43% and k_2 by 27%. Under the same conditions, the original ISI (approximately equivalent to the present ISIa: 0.5–1.5) was elevated by only 3%, and the version equivalent to the present ISIb: 0–1 was elevated by 12% (the IS-equivalent version was elevated by 21%). It is theoretically predictable that the ISIb, due to its higher sensitivity to k_1, will be more affected by slippage than the ISIa. The challenge therefore remains to find an ISI version that achieves maximum sensitivity to k_1 while maintaining minimal sensitivity to other parameters and maximal slippage immunity.

Acknowledgment. This work was supported in part by NIMH grant MH-35636.

References

Obrist WD, Wilkinson WE (1980) The noninvasive Xe-133 method: Evaluation of CBF indices. In: Bes A, Geraud G (eds.) Cerebral circulation. Excerpta Medica, Amsterdam, pp. 119–124
Obrist WD, Thompson HK, Wang HS, Wilkinson WE (1975) Regional cerebral blood flow estimated by ^{133}xenon inhalation. Stroke 6:245–256
Olesen J, Paulson OB, Lassen NA (1971) Regional cerebral blood flow in man determined by the initial slope of the clearance of intra-arterially injected ^{133}Xe. Stroke 2:519–540
Prohovnik I (1984) Clinical studies of rCBF by two-dimensional noninvasive ^{133}Xe clearance. In Spencer RP (ed) Interventional nuclear medicine. Grune and Stratton, New York, pp. 63–108
Prohovnik I, Knudsen E, Risberg J (1983) Accuracy of models and algorithms for determination of fast-compartment flow by noninvasive ^{133}Xe clearance. In: Magistretti PL (ed) Functional radionuclide imaging of the brain. Raven Press, New York, pp. 87–115
Risberg J, Ali Z, Wilson EM, Wills EL, Halsey JH (1975) Regional cerebral blood flow by ^{133}xenon inhalation: Preliminary evaluation of an initial slope index in patients with unstable flow compartments. Stroke 6:142–148

Inaccuracies in the Calculation of CBF from Inert Gas Clearance

R. von Kummer, S. Herold, and F. von Kries

Introduction

The measurement of cerebral blood flow (CBF) is complicated by the brain's intracranial localization as well as by heterogeneity of the brain tissue with respect to local CBF rates. For CBF measurement in man, tracers must be used which can be detected extracranially. The calculation of regional CBF (rCBF) is then dependent on the interpretation of time – tracer concentration curves according to the theories of Kety (1951), Ingvar and Lassen (1962), and Obrist et al. (1967), which require the validity of several assumptions.

These assumptions are (1) that the tracer is inert and does not influence the cerebral circulation, (2) that the diffusion equilibrium of the tracer between blood and tissue is instantaneous, (3) that the tissue is homogeneous and contains no concentration gradients of the tracer, (4) that the tracer is removed only by blood from the site sensed, (5) that the arterial tracer concentration is known in the area observed, and (6) that the blood – tissue partition coefficient of the region of interest has been determined.

To investigate the validity of assumptions 3, 4, and 5, a technique is required which can detect a diffusible tracer directly and locally. Therefore polarographic tissue microelectrodes were used to measure the uptake and desaturation of hydrogen (H_2) in very discrete regions of the brain.

Methods

By means of stereotactic instruments, 5, 6, or 10 epoxy- and glass-insulated platinum wire electrodes were inserted into the brains of N_2O-anesthetized cats. The electrodes were 0.2 mm in diameter. Their conical tips were 0.2 mm long so that the sensitive surface measured approximately 0.07 mm². Only electrodes which showed an immediate response to H_2 were used. During the experiments the following physiologic parameters were controlled: body temperature, pH, PaO_2, $PaCO_2$, arterial blood pressure, and central venous pressure. In three cats repeated local CBF (lCBF) measurements were performed at the cortico – white matter junction before and after introducing the electrodes by increments

Neurologische Universitätsklinik, Voss-Strasse 2, D-6900 Heidelberg

Cerebral Blood Flow and
Metabolism Measurement.
Eds. Hartmann/Hoyer
© Springer-Verlag Berlin Heidelberg 1985

of 1 mm or 0.1 mm. In six cats the H_2 uptake and desaturation were registered simultaneously in cortex, white matter, and caudate nucleus, as well as in the external carotid artery after different periods of H_2 inhalation.

Results

Spatial Resolution of the H_2 Clearance Method

At the cortico – white matter junction significant differences of lCBF could be measured even within a distance of 0.1 mm. The sudden change of the exponential function of the clearance curves when the electrode was protruded from gray into white matter is demonstrated in Fig. 1. Despite the high spatial resolution of the method, a nonexponential decrement of the curves was registered in gray as well as in white matter. In the cortex a progressive *decrease* of the clearance rate was observed during desaturation, whereas in white matter a progressive *increase* of the clearance rate was registered.

H_2 Uptake

The rate of H_2 uptake varied considerably between different brain regions, depending on lCBF. Figure 2 shows the exponential relationship between lCBF

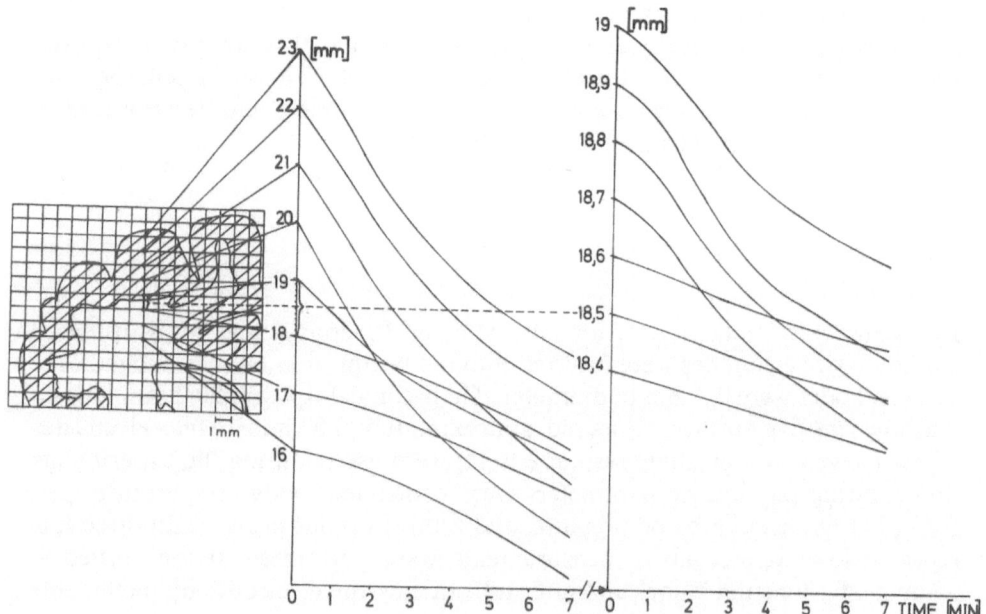

Fig. 1. Exponential function of H_2 clearance curves at different distances from the cortico – white matter junction. The curves are copied from the semilog plotting of the original registration. The drawing shows the different measuring points and the stereotactic coordinates, which are given in mm from the ear-to-ear line

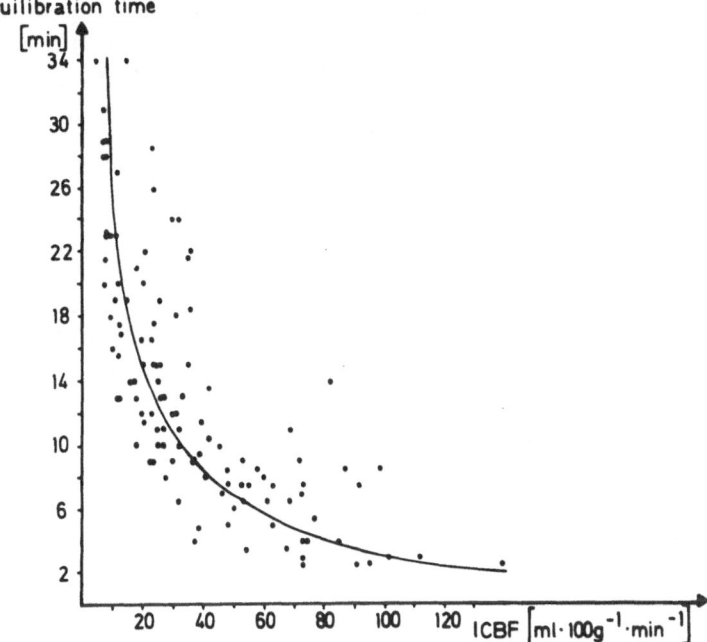

Fig. 2. Relationship between lCBF and the time which was needed to achieve concentration equilibrium in the region sensed. Repeated measurements were performed in six cats under different conditions with varying PaCO$_2$. lCBF was calculated from the initial slope index

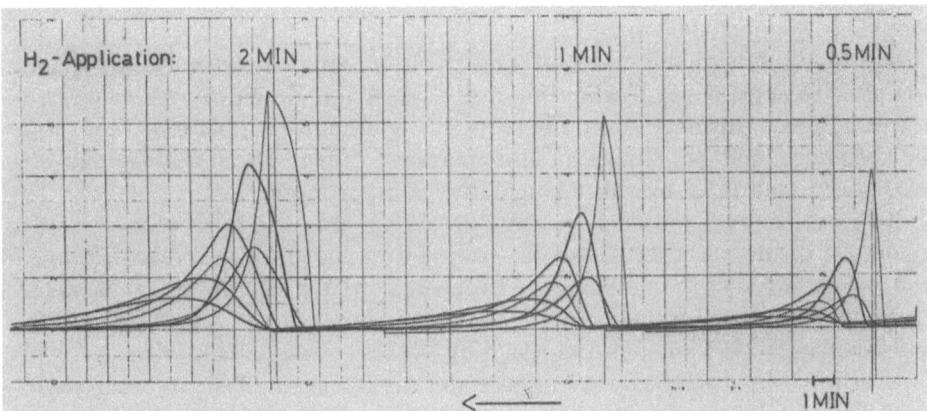

Fig. 3. Original registration of H$_2$ uptake and desaturation by seven electrodes in the external carotid artery (1), cortex (2), caudate nucleus (2), and subcortical white matter (2) after different periods of H$_2$ inhalation. The highest concentration and uptake rate is shown by the arterial electrode. The different uptake rates of the tissue are obvious, with lowest values in white matter. Note that the desaturation in white matter does not start until gray matter concentration has become lower than that of white matter, and that the clearance in different brain regions starts at different arterial concentrations

and the time which is required to reach concentration equilibrium. It varies between 3 min and 34 min. Thus, a short application time of a few seconds or minutes will inevitably lead to different degrees of tissue saturation with H_2.

H_2 Desaturation

Figure 3 shows the original registration of H_2 concentration in the external carotid artery, gray, and white matter after 0.5, 1, and 2 min of H_2 administration. It demonstrates that during initial clearance of H_2 after a short period of application, the degrees of H_2 saturation in the external carotid artery and in the individual brain regions are considerably different. Even if the displacement of the recorder's pens is taken into account, there is a clear delay in the onset of the desaturation phase which is dependent on lCBF. Analysis of the different clearance curves showed that desaturation in low flow regions does not start until H_2 concentration of higher perfused areas has become lower than that of the low flow region.

Discussion

The above data show that the spatial resolution of the H_2 clearance method is higher than previously suggested by Young (1980). Although he concluded from the nonexponential decrement of H_2 clearance curves that they could not represent a single flow compartment, he did not discuss the possibility that the distortion of the curves might be due to influences other than flow. If the spatial resolution is less than 0.1 mm, the volume of tissue sensed by the electrode would be less than 0.001 mm^3.

It is hardly imaginable that this small volume could contain both high and low flow compartments. Furthermore, no significant flow differences were detected at greater distances from the cortico − white matter junction. Thus, it can be concluded that the shapes of the clearance curves registered by the electrodes were controlled by only a single flow compartment.

If this conclusion is correct, it must be asked what the nonexponential decrement of the desaturation is due to. Taking into account the sudden change of the curves' shape at the boundary between gray and white matter, it seems unlikely that the tissue trauma produced by the electrodes is responsible for the nonexponential decrement as suggested by Senter et al. in 1978.

Following simultaneously the uptake and desaturation of H_2 in brain regions of different lCBF the considerable concentration differences due to the heterogeneity of flow are obvious. With respect to the immense surface area of the cortico − white matter junction, it seems likely that these concentration differences may act as diffusion gradients as suggested by Gillespie in 1967. During the first phase of desaturation after a short period of application, H_2 would diffuse from higher to lower perfused regions. Later the diffusion process would reverse when the H_2 concentration of gray matter had become lower than that of white matter. The situation is similar even after concentration equilibrium has been achieved after prolonged tracer application. The heterogeneity of flow will set up new diffusion gradients which approach a maximum when H_2 desaturation is nearly complete in gray matter and H_2 concentration in white matter is still high. Following both brief and prolonged H_2 administration, the intercompartmental diffusion will lead to nonexponential clearance curves, as were registered in this study.

With regard to heterogeneity of the tracer concentration at the onset of clearance, Ingvar and Lassen (1962) have suggested the compartmental analysis for rCBF calculation on the assumption that the tracer is cleared from the tissue simultaneously in high and low flow regions, even after brief indicator application. This study of H_2 clearance, however, shows that a simultaneous desaturation after a short period of H_2 application does not occur, and that the nonexponential decrement of the curves can be better explained by intercompartmental diffusion than by the influence of two or more different flow compartments.

From this the important question arises, whether other tracers used for rCBF measurement will also diffuse from or into the region sensed in the case of tissue — tissue disequilibrium. Using less diffusible substances like xenon or krypton it is imaginable that secondary equilibration does not occur since their clearance by blood flow is much faster than their diffusion from tissue to tissue (Halsey et al. 1977). This argument, however, does not take into consideration that the amount of a substance diffusing depends not only on its diffusion coefficient, but also on the concentration gradient and the extent of the diffusion area. Using xenon, heterogeneity of lCBF will also create considerable concentration differences, as previously demonstrated by Segawa et al. (1983), who used the stable xenon CT technique. It must also be considered that concentration differences might be more marked in the presence of pathologic impairment of rCBF. The diffusion area is substantial if the extent of the boundary zone between gray and white matter is considered. Furthermore it must be remembered that the high diffusibility of the tracer is one basic requirement of the Kety theory.

The influence of intercompartmental diffusion on xenon clearance has not yet been excluded by experimental study. Thus, the calculation of rCBF from xenon clearance curves could be erroneous. Comparing the H_2 concentration curves of the external carotid artery and of white matter in Fig. 3 it is obvious that the tissue is still taking up the tracer when the arterial concentration approaches zero. The experiment demonstrates that due to different distribution of flow the peak saturation of tracer occurs at different times within the carotid artery and individual brain regions. This observation indicates that the tracer concentration in the carotid artery is probably not representative of the arterial concentration in regions with differing lCBF values, and is certainly not a reliable reflection of tracer input in regions of low flow.

In conclusion, the measurement of H_2 uptake and desaturation in individual small flow compartments of the cat's brain shows that basic assumptions of the Kety theory are not valid for calculation of rCBF. The hypothesis of simultaneous desaturation in differently perfused brain regions, on which the compartmental analysis of Ingvar and Lassen (1962) is based, could not be confirmed by experimental data.

References

Gillespie FC (1967) Some factors influencing the interpretation of regional blood flow measurements using inert gas clearance techniques. In: Bain WH, Harper AM (eds) Blood flow through organs and tissues. Williams and Wilkins, Baltimore, p 195 – 209

Halsey JH, Capra NF, McFarland RS (1977) Use of hydrogen for measurement of regional cerebral blood flow – problem of intercompartmental diffusion. Stroke 8:351 – 357

Ingvar DH, Lassen NA (1962) Regional blood flow of. the cerebral cortex determined by krypton[85]. Acta Physiol Scand 54:325 – 338

Kety SS (1951) The theory and applications of the exchange of inert gas at the lungs and tissue. Pharmacol Rev 3:1 – 41

Obrist WD, Thompson HK, King CH (1967) Determination of regional cerebral blood flow by inhalation of 133-xenon. Circ Res 20:124 – 135

Segawa H, Wakai S, Tamura A, Yoshimasu N, Nakamura O, Ohta M (1983) Computed tomographic measurement of local cerebral blood flow by xenon enhancement. Stroke 14:356 – 362

Senter HJ, Burger DH, Metzler J (1978) An improved technique for measurement of spinal cord blood flow. Brain Res 149:197 – 200

Young W (1980) H_2 clearance measurement of blood flow: A review of technique and polarographic principles. Stroke 11:552 – 564

A Graphic Approach to Compartmental Slippage

K. Herholz, G. Pawlik, W.-D. Heiss, H. W. Ilsen, and K. Wienhard

The variance of rCBF data estimated from simulation of random influences or from retest procedures provides some insight into the general stability of the two-compartment model for noninvasive xenon 133 studies. In an actual patient study, however, it is not very informative with regard to the reliability of individual results. Determination of the accuracy of individual measurements is of special interest under conditions which may cause compartmental slippage (Reivich 1969; Waltz et al. 1972). As shown by Illif et al. (1975), Bruce et al. (1975), and Heiss et al. (1976), this may occur in most cerebrovascular disorders. Therefore, a graphic method for assessing this individual accuracy with particular reference to compartmental slippage was developed. Its properties are demonstrated in an exemplary rCBF measurement before and after extra-intracranial bypass surgery.

Method

According to the bicompartmental convolution equation

$$Q_{fit}(T) = \sum_{m=1}^{2} P_m \int_0^T C_a(t) \, e^{-k_m(T-t)} \, dt$$

a least-squares fit to the measured head activity data is obtained (Obrist et al. 1975). The parameters P_1, P_2, k_1, and k_2 thus determined form the basis of any computation of flow values.

The goodness of fit is then described by the coefficient of determination

$$r^2 = \frac{\displaystyle\sum_{i=i_0}^{i_n} (Q_{fit}(i) - \bar{Q}_{meas})^2}{\displaystyle\sum_{i=i_0}^{i_n} (Q_{meas}(i) - \bar{Q}_{meas})^2}$$

representing the ratio of estimated to measured variances. For a perfect fit it would equal unity.

Max-Planck-Institut für neurologische Forschung, Ostmerheimer Strasse 200, D-5000 Köln 91 (Merheim)

Cerebral Blood Flow and
Metabolism Measurement.
Eds. Hartmann / Hoyer
© Springer-Verlag Berlin Heidelberg 1985

Using the coefficient of determination as the dependent variable, a three-dimensional graph can be constructed comparing the goodness of fit of the actual least-squares solution with the goodness of fit obtained with other sets of k-values. In this graph k_1-values ranging from 0 to 1.52 min⁻¹ and k_2-values ranging from 0 to 0.36 min⁻¹ are plotted along the x- and y-axis, respectively. For each possible pair of k-values a least-squares fit with respect to P_1 and P_2 is obtained and the corresponding coefficient of determination (r^2) is computed. The reciprocal of the coefficient of nondetermination $(1 - r^2)^{-1}$, standardized to the individual maximum, is then plotted along the z-axis.

As shown by Herholz et al. (1983), a well-defined sharp peak in this graph indicates a stable measurement, while a flat, broad-based peak indicates poor reliability of results.

Results

Consequences of compartmental slippage and its detection by graphing the goodness of fit are demonstrated in repeated rCBF measurements obtained with a gamma camera in a right lateral view, using the intravenous injection method (Podreka et al. 1981) and the algorithm described by Obrist et al. (1975). Regional values of mean flow, weight of the first compartment, and flow values of the first and second compartment are shown in Fig. 1.

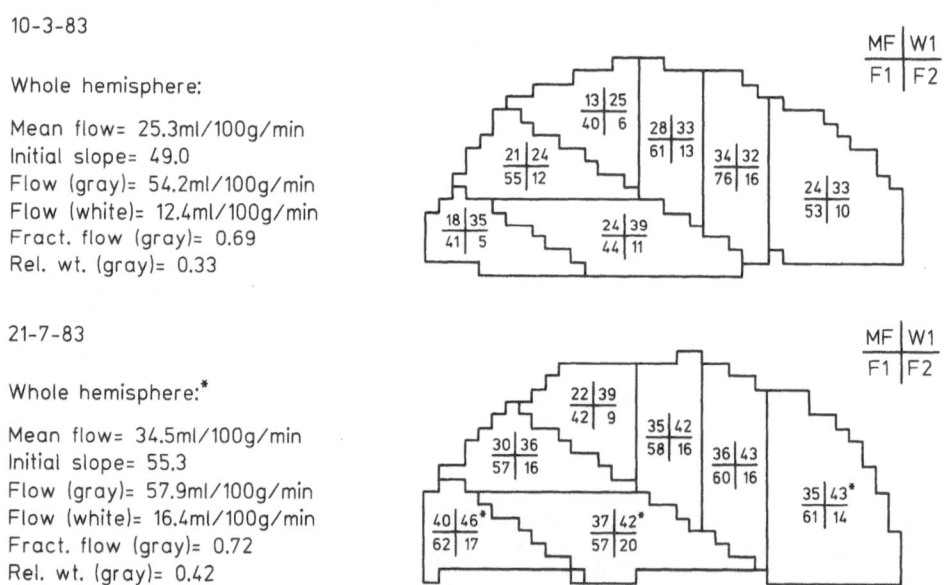

Fig. 1. rCBF measurements before and after extra-intracranial arterial bypass surgery, in a right lateral view. Parietal infarct due to severe stenosis of the middle cerebral artery. Regional values: *MF* = mean flow, *F1* = flow of first, *F2* = flow of second compartment (all in ml/100 g/min); *W1* = percentage weight of first compartment. Significant flow changes are marked by an *asterisk*

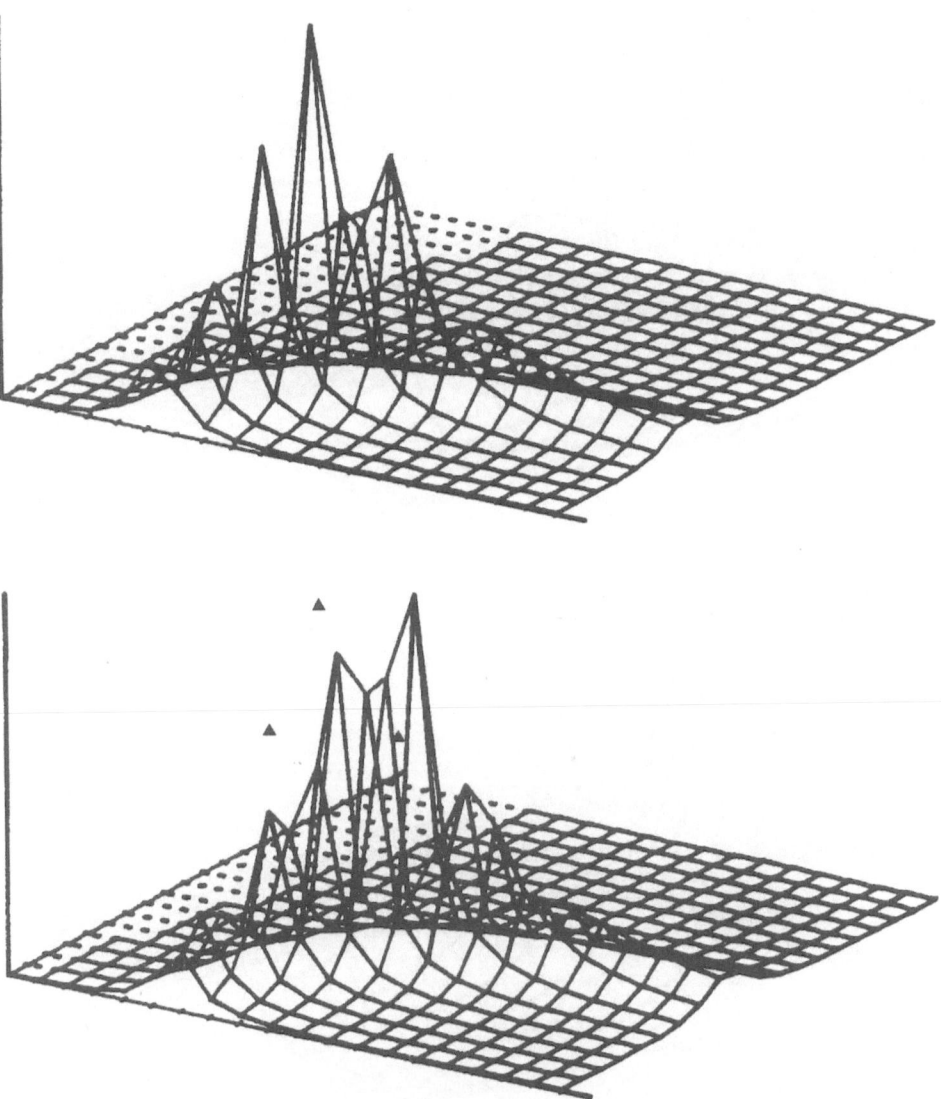

Fig. 2. Graphs of the goodness of fit for the whole hemisphere before (*top*) and after (*bottom*) surgery. k_1 (range from 0 to 1.52 min^{-1}) is plotted along the x-axis, k_2 (range from 0 to 0.36 min^{-1}) along the y-axis. For comparison the peak of the first graph is indicated also in the second graph by *small triangles*

Hemispheric peak count rates were 127 330 cpm and 180 620 cpm, respectively, and regional peak count rates ranged between 9950 and 30 040 cpm.

The 38-year-old female patient had suffered an ischemic stroke in the right parietal region due to a severe stenosis of the corresponding middle cerebral artery. Extra-intracranial bypass surgery was carried out successfully and, 3 months later, rCBF was measured again. This time higher mean flow values

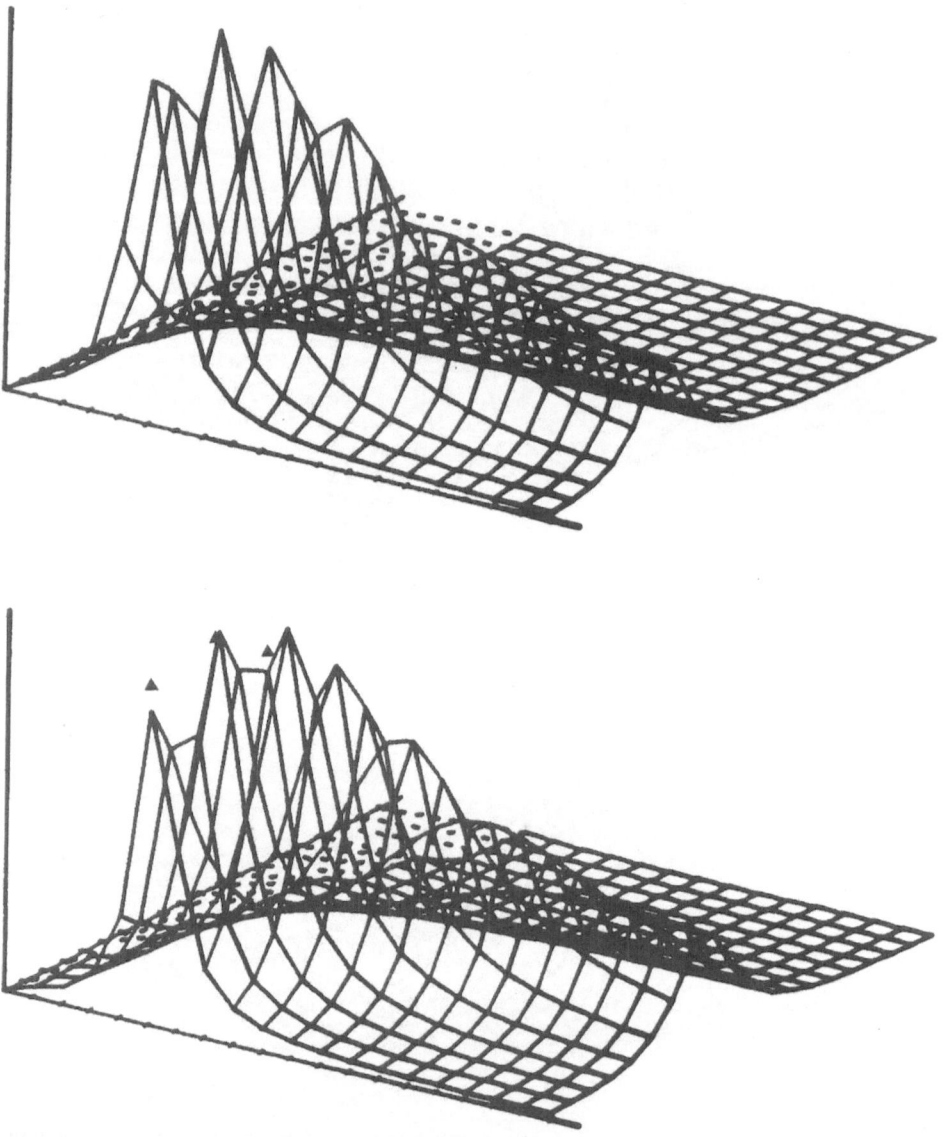

Fig. 3. Graphs of the goodness of fit for the infarcted parietal region before (*top*) and after surgery (*bottom*). Ruling and symbols are as described for Fig. 2

were obtained for the whole hemisphere and in all regions, even in the infarcted parietal regions. In frontoparietal regions, however, flow to the first compartment was decreased, and the weighting factor of the first compartment was changed in all regions, indicating compartmental slippage.

Graphs of the goodness of fit for the whole hemisphere before and after surgery are shown in Fig. 2. The peaks are well defined and located at distinctly different k-values.

By contrast, graphs for the infarcted parietal region (Fig. 3) exhibit broad and poorly defined peaks, indicating instability of rCBF results. Moreover, as to their localization the two peaks are almost indistinguishable, and therefore the apparent flow increase in this region very likely represents an artifact due to compartmental slippage rather than a real improvement in rCBF.

In Fig. 1, regions yielding clearly defined, separate peaks are marked by an asterisk (frontal, occipital, temporal regions, and whole hemisphere). rCBF increase appears to be significant there, although some compartmental slippage probably occurred. In other regions (frontoparietal, parietal, and temporal) a significant change in rCBF must not be assumed, since broad and overlapping peaks indicate a lack of accuracy of regional flow values.

Conclusion

Reliability of rCBF values obtained by the noninvasive bicompartmental xenon 133 clearance method may vary considerably from region to region. By constructing individual graphs of the goodness of fit, accurate results can be distinguished from unreliable ones, thus preventing overinterpretation of spurious flow changes, e.g., in therapeutic trials.

References

Bruce DA, Schutz H, Vapalahti M et al. (1975) Pitfalls in the interpretation of xenon CBF studies in head-injured patients. In: Cereb Circulat and Metab. Springer, Berlin New York, p 406–408

Heiss WD, Zeiler K, Turnheim M et al. (1976) Flow and compartmental weight in relation to the course of stroke. Stroke 7:399–403

Herholz K, Heiss WD, Pawlik G et al. (1983) Detection of compartmental slippage in noninvasive rCBF measurements. J Nucl Med 24:1188–1191

Iliff L, Zilka E, Bull JWD et al. (1975) The effect of changes in cerebral blood flow on compartmental weight. In: Cereb Circulat and Metab. Springer, Berlin New York, p 145–147

Obrist WD, Thompson HK, Wang HS et al. (1975) Regional blood flow estimated by ^{133}xenon inhalation. Stroke 6:245–256

Podreka I, Heiss WD, Brücke T (1981) Atraumatic CBF measurement with the scintillation camera. Stroke 12:47–53

Reivich M (1969) Observation on exponential models of cerebral clearance curves. In: Research on the cerebral circulation. CC Thomas, Springfield, p 135–141

Waltz AG, Waner AR, Anderson RE (1972) Comparison of analytic methods for calculations of cerebral blood flow after intracarotid injection of 133-Xe. J Nucl Med 13:66–72

Application of the Nontraumatic Xenon 133 Method in Neuropsychiatry

J. Risberg

Several conditions need to be met before a technique is suitable for general clinical use:

1. It has to provide valid (which is to say reliable) information.
2. This information must be relevant for the care of the patient.
3. The clinical gain to be realized must be sufficient to justify the hazards and discomfort for the patient and the cost of obtaining the data.
4. Similar information should not be obtainable with a less costly or hazardous technique.
5. The measurement procedure and interpretation of results must be standardized and reproducible − which is to say that competent workers anywhere can obtain comparable results by carefully following the instructions for obtaining the data.

With these conditions in mind, we will discuss the potential clinical usefulness of the ^{133}Xe inhalation technique for measurement of regional cerebral blood flow (rCBF) in the field of neuropsychiatry, with emphasis on its use for differential diagnosis of dementia. The need for better diagnostic methods in this area is evident. For example, distinction between dementia due to organic brain disease (e.g., Alzheimer's disease) and pseudodementia due to affective disease, psychosis, intoxication, etc. is sometimes very difficult using present routine clinical methods.

Measurement of rCBF by the ^{133}Xe Inhalation Method

Since the ^{133}Xe inhalation technique has been described in previous papers in this volume (Obrist and Wilkinson and others), only a short description will be needed here to indicate specific features of our system. For a more detailed description, see Risberg (1980) and Risberg and Prohovnik (1981).

The indicator, mixed with air (75 MBq/l), was inhaled during 1 min followed by 10 min of normal air breathing (Obrist et al. 1975). Radiation was recorded by 32 scintillation detectors placed in parallel in two detector holders positioned at right angles to the lateral surfaces of the head (Novo Diagnostic Sys-

Department of Psychiatry, University Hospital, S-221 85 Lund

Cerebral Blood Flow and
Metabolism Measurement.
Eds. Hartmann/Hoyer
© Springer-Verlag Berlin Heidelberg 1985

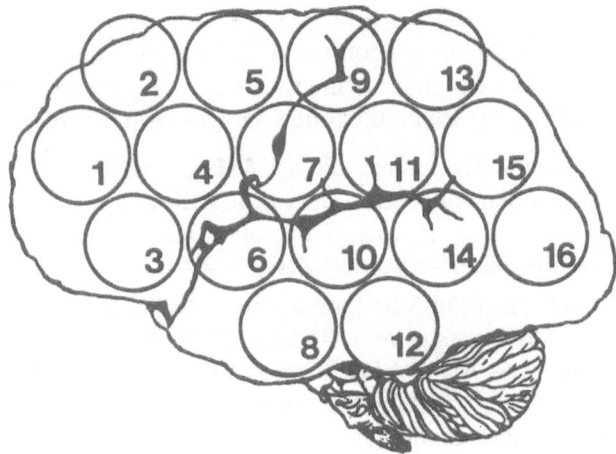

Fig. 1. Approximate localization of the 16 detectors covering each hemisphere

tems, Hadsund, Denmark). Sixteen detectors, grossly localized according to Fig. 1, covered each hemisphere. Changes in arterial concentration of ^{133}Xe were estimated by determining the end-tidal values of continuously recorded respiratory tracer concentrations. The recorded curves were analyzed by an algorithm for two-compartmental analysis developed by Obrist et al. (1975), with the addition of a correction routine for the air-passage artifact (Risberg 1980). The data presented here will be based on the initial slope index (ISI) (Risberg et al. 1975).

Measurement of rCBF in Dementia

Patients with symptoms of dementia due to suspected or verified organic brain disease have been subjected to extensive investigations with a variety of global and regional CBF and metabolic methods during the last four decades. The general finding has been a subnormal level, and that the lower the level, the more severe are the dementia symptoms (Freyhan et al. 1951; Lassen et al. 1957 and others). It has also been established that dementia is generally not a condition of deficient cerebral circulatory capacity. The patient with organic dementia is as a rule capable of increasing his CBF if provoked (e.g., by hypercapnia), but his baseline blood flow is low due to its coupling to a subnormal cerebral metabolic level (Simard et al. 1971; Frackowiak et al. 1981).

A number of studies have explored whether regional accentuations of the CBF decreases exist in different subgroups of patients with organic dementia. Some findings point to typical and specific patterns of rCBF disturbances in different subgroups of patients with organic dementia (such as Alzheimer's disease, multi-infarct dementia, and Pick's disease; Gustafson et al. 1977; Gustafson and Risberg 1979). Such findings indicate the possibility of using the rCBF method for differential diagnosis. Our continuing research, some of which will be presented here, provides further proof of the clinical usefulness of the nontraumatic ^{133}Xe method in neuropsychiatry.

Material

The present material consists of 121 patients pertaining to our ongoing prospective study of organic dementia. The patients were classified into four clinical subgroups for which the following were the tentative diagnoses:

1. Presenile Alzheimer's disease (AD), $n = 28$
2. Senile dementia, Alzheimer type (SDAT), $n = 27$
3. Pick's disease or diffuse frontotemporal degenerative disease (PD), $n = 22$
4. Multi-infarct dementia (MID), $n = 44$

Each diagnosis was based primarily on clinical ratings of symptoms of dementia (Gustafson and Risberg 1974; Gustafson and Nilsson 1982). To be classified as AD or SDAT (onset of symptoms before or after 65 years respectively) the patient had to display the characteristic symptoms of amnesia, agnosia, apraxia, etc. The PD patient was recognized by changes in personality, lack of planning and control of behaviour, and other symptoms which have been accepted as indicators of frontal lobe dysfunction. The MID patient was identified mainly on the basis of a high ischemic score on the scale of Hachinski and co-workers (1975). The clinical diagnoses were made without knowledge of the rCBF results. Patients who could not clearly be classified as belonging to any of the groups were not included in the present analysis.

Forty-four of the 121 patients are now deceased. Autopsy has been performed in 38 of the cases, providing definite diagnoses of AD (8), SDAT (11), PD (6), and MID (13). The clinical diagnosis and the neuropathologic diagnosis agreed in all cases.

The results from the demented groups will be compared with data from normal subjects (Risberg and Hagstadius 1983) and from depressed patients of similar age (Silfverskiöld et al. 1979).

Results

Clinical Diagnosis Compared with rCBF

The results from comparisons of hemispheric mean CBF between the four diagnostic groups of the total material and the two reference groups are shown in

Table 1. Mean hemispheric flow values (ISI) in patients with Alzheimer's disease, Pick's disease, MID, SDAT, and depression and in normal subjects

Diagnosis	n	Age	ISI		pCO_2
			Right	Left	(mmHg)
AD	28	64 ± 6	34.4 ± 6.1	33.6 ± 5.8	37.0 ± 5.0
PD	22	62 ± 8	37.3 ± 6.7	36.8 ± 6.9	36.1 ± 3.1
MID	44	68 ± 9	35.8 ± 7.1	35.4 ± 6.9	35.9 ± 4.5
SDAT	27	76 ± 7	33.5 ± 6.2	33.5 ± 5.9	36.5 ± 3.5
Depression	19 ·	60 ± 14	47.0 ± 6.8	47.1 ± 7.0	38.8 ± 3.4
Normals	16	53 ± 5	44.6 ± 7.6	43.8 ± 7.3	37.8 ± 3.4

Table 1. Mean flow levels were, as expected, significantly ($P < 0.01$ and less) lower for all the demented groups compared with the reference groups. The values for the depressed group were not significantly different from those of the normal group.

In comparisons of the flows in different regions of the brain we found marked and significant (by t-test) differences between the groups. First, in comparing the AD and PD groups (Fig. 2), the former group was found to have significant flow reductions in parietal, parieto-occipital, and parietotemporal regions, while for the PD group the lowest flow values were seen in premotor, supplementary motor, and prefrontal regions. In both groups the rCBF pathology generally involved both hemispheres to an equal extent. The regional flow patterns in both demented groups were also different from the patterns found in normal subjects: normals are characterized by slightly higher flows frontally and lower flows in the postcentral and temporal regions (5% − 10% above and below hemispheric averages respectively; Prohovnik et al. 1980).

For the AD and SDAT groups the regional flow pathologies were similar in many respects (Fig. 3). However, the AD patients generally had a more focalized postcentral flow disturbance, while the SDAT pathology was more wide-

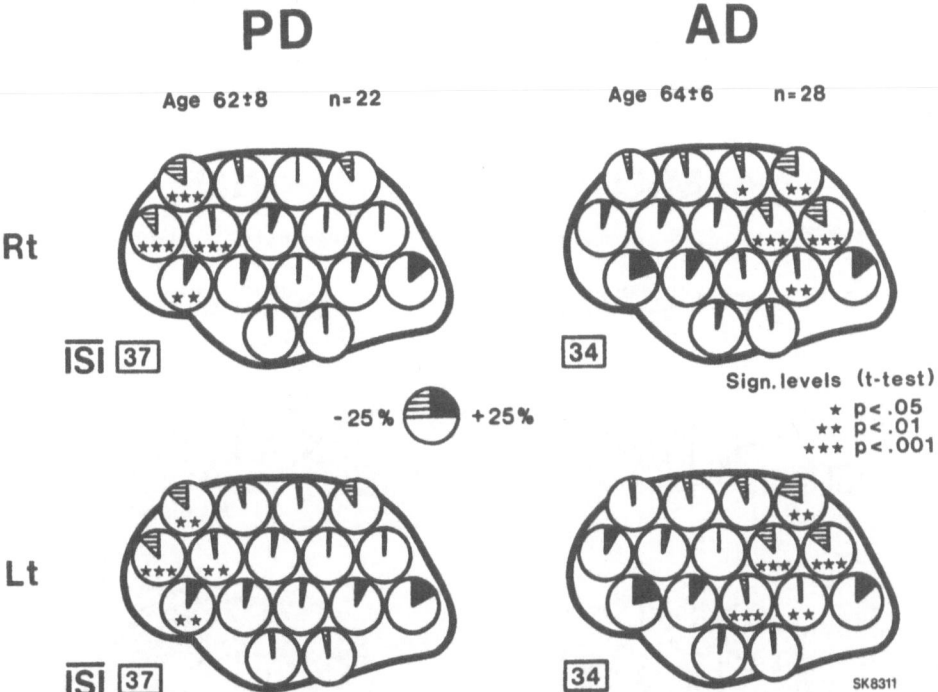

Fig. 2. Regional cerebral blood flow in 22 patients with a tentative clinical diagnosis of Pick's disease (*PD*) compared with 28 patients with a tentative clinical diagnosis of Alzheimer's disease (*AD*). Mean hemispheric flow is indicated in the *boxes*. *Clock symbols* indicate regional values in per cent of hemispheric mean. *Black shadowing* indicates a regional value above and a *striped field* a regional value below the hemispheric average (90 ° = 25%). Note the significant postcentral decreases in AD and the low frontal values in PD

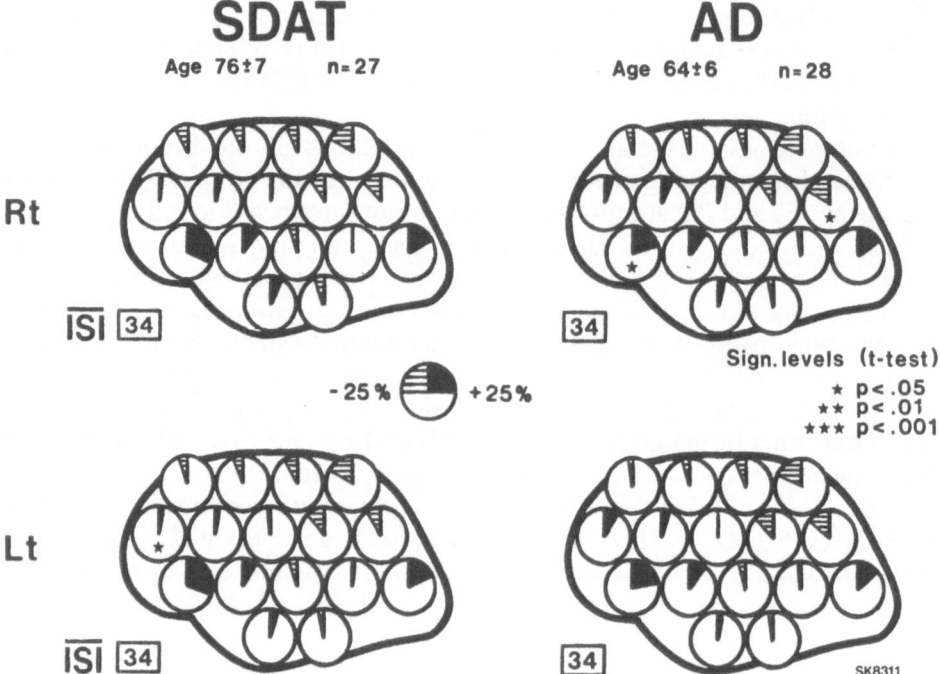

Fig. 3. Comparison of rCBF in 27 cases with a tentative clinical diagnosis of senile dementia of the Alzheimer type (*SDAT*) with 28 cases of presumed Alzheimer's disease (*AD*). Symbols as in Fig. 2

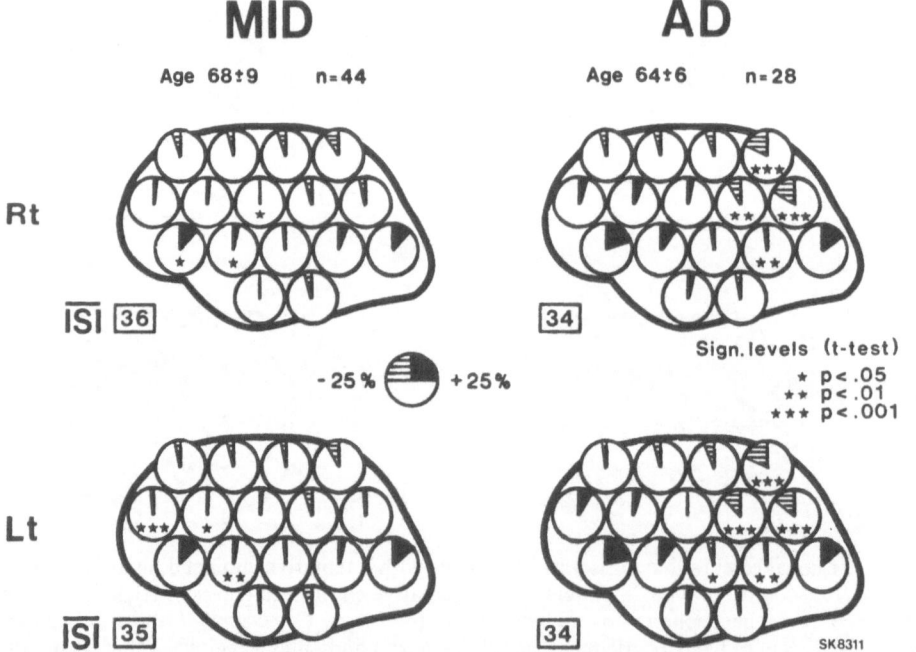

Fig. 4. Comparison of rCBF in 44 cases of presumed multi-infarct dementia (*MID*) with 28 cases of presumed Alzheimer's disease (*AD*). Symbols as in Fig. 2

spread and appeared to involve the frontal lobes to a greater extent. A typical finding in both groups was better preserved flow in Rolandic, occipital, and frontotemporal structures relative to other areas.

Averaging the flow values for the MID group resulted in a very flat mean rCBF pattern. There was large variance associated with the means, however. The flat pattern can thus be due to the heterogeneous and unsystematic flow abnormalities in this group (Fig. 4). The typical finding when there were only early clinical signs with small and/or deep infarcts was a normal flow level and a normal regional distribution. In the more advanced stages of MID right − left asymmetries of hemispheric means and regional patterns were often seen. Some of the MID cases can thus be clearly distinguished from the AD, PD, and SDAT cases by virtue of asymmetric and "spotty" flow patterns. Further statistical analyses of these discriminatory possibilities in the rCBF data are in progress.

Diagnosis by Autopsy Compared with rCBF

Since yearly reexaminations have been routinely made in the present group, the time period between the last rCBF measurement of a patient and autopsy was less than 1 year in most cases. Also, most of the patients had had several rCBF measurements throughout the progression of their disease. The rCBF diagnosis was made mainly on the basis of the last results in the series but some consideration was also given to earlier findings. The diagnosis was made by clinical evaluation of the rCBF data. The criteria for distinguishing diagnostic categories were derived from findings in the larger clinically diagnosed groups and from previous research (Gustafson et al. 1977; Gustafson and Risberg 1979).

Figure 5 shows the correspondence between diagnoses based on rCBF data alone and the final diagnoses made on the basis of autopsy. It can be seen that there were only four disagreements. Correct placement by rCBF was thus achieved in about 90% of the patients. Two of the four misplaced patients belonged to the AD and SDAT groups by autopsy but the rCBF data gave a di-

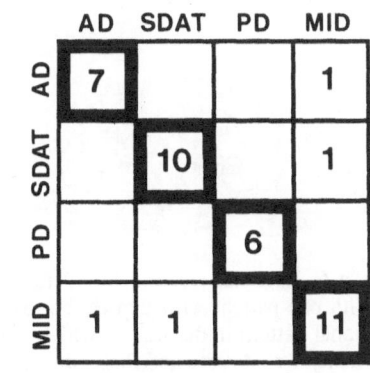

Fig. 5. Comparison of diagnosis by rCBF with diagnosis by autopsy in 38 deceased patients. *AD*, Alzheimer's disease; *SDAT*, senile dementia, Alzheimer type; *PD*, Pick's disease; *MID*, multi-infarct dementia

agnosis of MID. Autopsy did, however, show evidence of cerebrovascular disease in addition to the Alzheimer pathology. Two cases of MID were erroneously classified as AD and SDAT.

Other Applications in Psychiatry

Space does not permit a detailed description of our other applications of the method in psychiatry. The reader is referred to Risberg (1980) and Risberg et al. (1981) for a more detailed review.

One of our main series of studies has involved patients with affective disorders (Silfverskiöld et al. 1979). This group is of particular interest for comparison with the organic dementia group because "pseudodementia" is common – i.e., classifying a patient as demented when a depressive state exists. As is illustrated in Fig. 6, depression and dementia can be reliably distinguished using rCBF, because rCBF abnormality is not linked to affective disease. Changes in rCBF are, however, seen in conjunction with therapy, i.e., ECT (Silfverskiöld et al. 1979).

Other applications of the rCBF technique are found in our studies of toxic influences on brain function. Several investigations have described rCBF changes related to alcohol abuse. Subnormal flow values have been found during withdrawal from dependent use of alcohol (Berglund and Risberg 1981). Normalization of rCBF parallels periods of abstinence (Berglund et al. 1980). A persistent global CBF diminution is, however, seen in patients with sustained

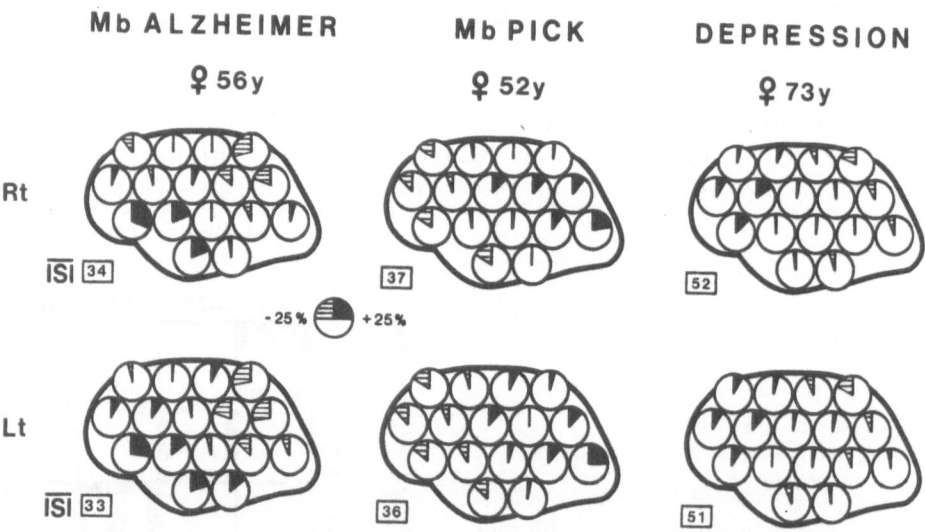

Fig. 6. Regional cerebral blood flow in two cases of presenile organic dementia compared with one patient with depression. Symbols as in Fig. 2. Note the normal flow level and regional pattern in depression which is in contrast to the global and regional flow pathology in Alzheimer's disease and Pick's disease

drinking and alcohol dementia. A slight but significant decrease of rCBF has also been found in subjects with long-term occupational exposure to organic solvents (Risberg and Hagstadius 1983). The largest decreases were seen in frontotemporal regions.

The repeatability of the rCBF method makes it well suited for clinical drug studies. Using a double-blind, cross-over design with placebo control, we have recently found significant improvements in rCBF during medication with vincamine and bromvincamine in ten patients with MID (Hagstadius et al. 1984).

Discussion

How does the present rCBF method meet the requirements for clinical usefulness that were listed in the introduction of this paper? Evidence that the first two conditions are met — provision of information that is valid and of relevance for the care of patients — is clearly provided by the results we have reviewed. Early and correct diagnosis of suspected organic dementia is important for proper care.

The hazards of obtaining rCBF data are minimal and the costs are reasonable. The radiation exposure is small (0.4 mGy to the critical organ, the lungs). The cost per measurement is competitive with other diagnostic techniques. In our laboratory, where about 1000 measurements are done each year, we have calculated the total cost per study of one patient to be about US $ 80.

This kind of information *cannot* be obtained by other safer or cheaper methods (but certainly with more advanced and expensive ones like PET and SPECT). The information offered by rCBF usually cannot be substituted by CT, EEG, or other methods.

Concerning our last condition — that the method be standardized and repeatable — there are still improvements to be made. Variation between laboratories exists in, for example, detector placement and flow parameters used. There is no standard manual for interpretation of results. Further development and more widespread use of the technique will probably bring about improvements.

In conclusion, then, we can say that there are sound and useful clinical applications for ^{133}Xe inhalation rCBF measurements in the differential diagnosis of dementia and other conditions within the field of psychiatry as well as for the evaluation of therapy.

Acknowledgments. The author is indebted to Lars Gustafson, MD, for the psychiatric examinations, to Arne Brun, MD, for the neuropathologic investigations, and to John Horn, PhD, for constructive criticism and improvements in the language of the manuscript. Siv Karlson, Aleksandra Walfisz, and Helena Fernö aided in the rCBF measurements and in the preparation of the illustrations. Maria Regnér aided in the preparation of the manuscript. The investigations were supported by the Swedish Medical Research Council (projects Nr. 04969 and 3950), The King Gustaf V and Queen Victoria's Foundation, the Swedish Council for Social Science Research, and the Swedish Work Environment Fund.

References

Berglund M, Risberg J (1981) Regional cerebral blood flow during alcohol withdrawal. Arch Gen Psychiatry 38:351−355

Berglund M, Bliding G, Bliding Å, Risberg J (1980) Reversibility of cerebral dysfunction in alcoholism during the first seven weeks of abstinence − a regional cerebral blood flow study. Acta Psychiat Scand 62: suppl 286, 119−128

Frackowiak RSJ, Pozzilli C, Legg NG, du Boulay GH, Marshall J, Lenzi GL, Jones T (1981) Regional oxygen supply and utilization in dementia. A clinical and physiological study with oxygen-15 and positron tomography. Brain 104:753−778

Freyhan FA, Woodford RB, Kety SS (1951) Cerebral blood flow and metabolism in psychoses of senility. J Nerv Ment Dis 113:449−459

Gustafson L, Risberg J (1974) Regional cerebral blood flow related to psychiatric symptoms in dementia with onset in the presenile period. Acta Psychiat Scand 50:516−538

Gustafson L, Risberg J (1979) Regional cerebral blood flow measurements by the ^{133}Xe inhalation technique in differential diagnosis of dementia. Acta Neurol Scand 60, suppl 72:546−547

Gustafson L, Nilsson L (1982) Differential diagnosis of presenile dementia on clinical grounds. Acta Psychiat Scand 65:194−209

Gustafson L, Brun A, Ingvar DH (1977) Presenile dementia: clinical symptoms, pathoanatomical findings and cerebral blood flow. In: Meyer JS, Lechner H, Reivich M (eds) Cerebral vascular disease. Excerpta Medica, Amsterdam, p 5−9

Hachinski VC, Iliff LD, Zilhka E, du Boulay GH, McAllister VL, Marshall J, Ross Russell RW, Symon L (1975) Cerebral blood flow in dementia. Arch Neurol 32:632−637

Hagstadius S, Gustafson L, Risberg J (1984) The effects of bromvincamine and vincamine on regional cerebral blood flow and mental functions in patients with multi-infarct dementia. Psychopharmacology 83:321−326

Lassen NA, Munck O, Tottey ER (1957) Mental function and cerebral oxygen consumption in organic dementia. Arch Neurol Psychiat 77:126−136

Obrist WD, Thompson HK, Wang HS, Wilkinson WE (1975) Regional cerebral blood flow estimated by 133-xenon inhalation. Stroke 6:245−256

Prohovnik I, Håkansson K, Risberg J (1980) Observations on the functional significance of regional cerebral blood flow in "resting" normal subjects. Neuropsychologia 18:203−217

Risberg J (1980) Regional cerebral blood flow measurements by ^{133}Xe-inhalation: Methodology and applications in neuropsychology and psychiatry. Brain Lang 9:9−34

Risberg J, Prohovnik I (1981) rCBF measurements by ^{133}Xe inhalation: Recent methodological advances. Prog Nucl Med 7:70−81, Karger, Basel

Risberg J, Hagstadius S (1983) Effects on the regional cerebral blood flow of long-term exposure to organic solvents. Acta Psychiat Scand 67: suppl 303, 92−99

Risberg J, Ali Z, Wilson EM, Wills EL, Halsey J (1975) Regional cerebral blood flow by [133]xenon inhalation. Stroke 6:142−148

Risberg J, Gustafson L, Prohovnik I (1981) rCBF measurements by ^{133}Xe inhalation: Applications in neuropsychology and psychiatry. Prog Nucl Med 7:82−94, Karger, Basel

Silfverskiöld P, Gustafson L, Johanson M, Risberg J (1979) Regional cerebral blood flow related to the effect of electro-convulsive therapy in depression. In: Obiols J, Ballús C, González E, Pujol J (eds) Biological psychiatry today, Elsevier/North Holland, Amsterdam, p 1178−1183

Simard D, Olesen J, Paulson OB, Lassen NA, Skinhøj E (1971) Regional cerebral blood flow and its regulation in dementia. Brain 94:302−311

Distinct rCBF Pattern During Different Types of Short-Term Memory Activation

Z. Mubrin, S. Knežević, B. Barac, N. Gubarev, M. Lazić, R. Liščić, and S. Vidošić

Introduction

Memory impairment is frequently seen in a great number of psychiatric and neurologic conditions. Even normal individuals often complain of forgetfulness, which becomes more prominent with advancing age. Although it is generally accepted that the limbic system plays a major role in memory processes, it has been demonstrated that cortical structures are also important; thus cortical changes were observed during memorization in subjects in whom regional cerebral blood flow measurements were performed (Maximilian et al. 1978).

As memory is defined as a mental process that allows the individual to store experience and perceptions for recall at a later time, it is also evident that it consists of several steps which include registration, retention, and recall.

Clinically, memory is subdivided into three basic types based upon the time span between stimulus presentation and memory retrieval. Short-term memory is of particular clinical interest because it is more prone to damage and can be easily tested, differentiating at least roughly between its two subtypes: immediate memory (recall) and recent memory (Strub and Black 1977). It is also known that cortical structures play an important role in immediate memory and recall, because defects can be seen clinically if lesions in the region of the sylvian fissure occur. However, as far as we know, differentiation between these two subtypes of short-term memory has not been made in the same individuals when employing the ^{133}Xe inhalation method for regional cerebral blood flow measurements.

Subjects and Methods

For reliable study in a variety of pathologic conditions, good, normal control results are necessary. We used the ^{133}Xe inhalation method to obtain regional cerebral blood flow measurements in 16 healthy volunteers of both sexes whose mean age was 27.7 ± 8.1 years ($\bar{x} \pm 1$ SD), ranging from 20 to 47 years. The first measurement, using the Novo Cerebrograph, was made with the patient in

Pliva Research Institute, Department of Neurology, University Hospital Center and Military Hospital, 41000 Zagreb, Yugoslavia

Cerebral Blood Flow and
Metabolism Measurement.
Eds. Hartmann/Hoyer
© Springer-Verlag Berlin Heidelberg 1985

the resting condition, and was followed immediately by two memory activating measurements randomly applied to each subject. For activation of immediate memory, subjects were instructed to remember a series of seven digits presented with a 1-s interval between each number. For recent memory activation, series of word pairs were used. Each word pair was presented for 10 s by means of a slide projector on a screen suspended above the subject's head, as described elsewhere (Risberg et al. 1977). During the measurements pCO_2 was monitored, and as it remained quite stable during the activation measurements no

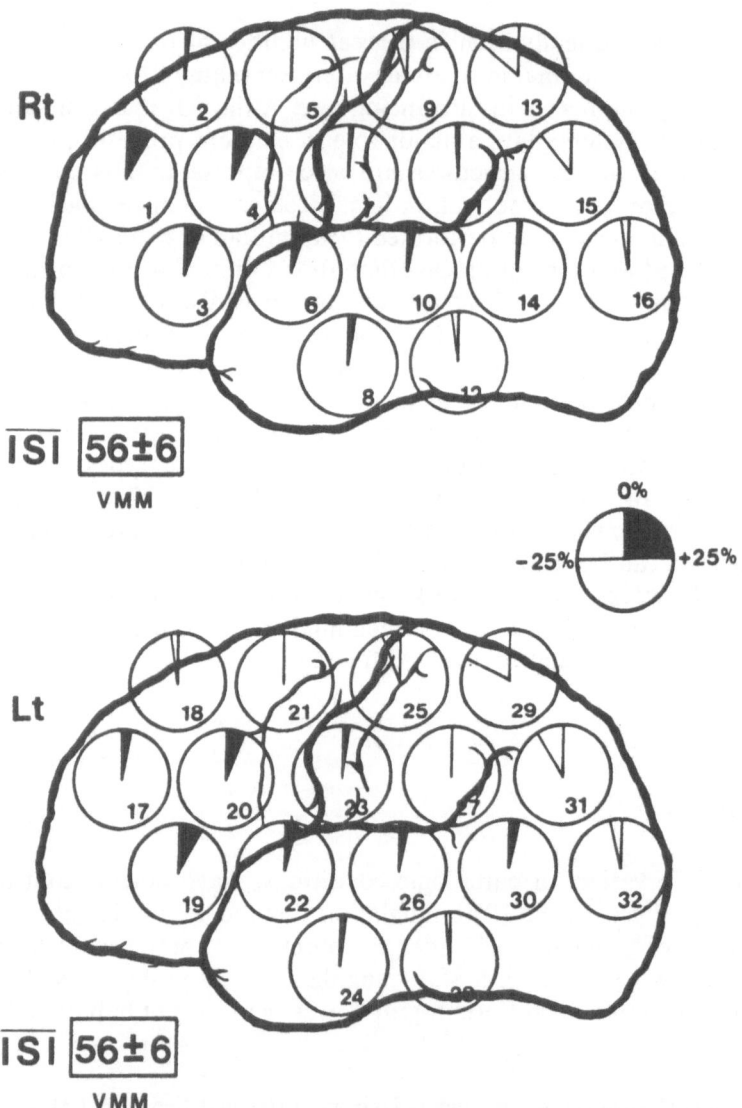

Fig. 1. Resting regional cerebral blood flow pattern expressed as percentage changes in mean hemispheric flow. The hemispheric average flow values are indicated in *boxes*. Regional deviations are shown as *"clock" symbols* in the *circles*. Normal hyperfrontal pattern

correction was made on the final results. For calculation, only reliable results were included, after careful analysis of each record. Statistical analysis was performed by the Wilcoxon matched-pairs signed-ranks test using ISI as a parameter of blood flow. All measurements were done during morning hours without any special preparation of the subjects.

Results

Resting measurements gave the normal pattern of hyperfrontal regional distribution of blood flow as described by many authors (Ingvar 1977; Ingvar 1979; Prohovnik et al. 1980). Mean hemispheric flow revealed symmetrical values for both hemispheres and was $56 \pm 6 (\bar{x} \pm 1 \text{ SD})$, as shown in Fig. 1.

Activation using the digit repetition test revealed a significant increase in mean hemispheric flows which was, when expressed as relative values in relation to the resting condition, somewhat greater on the left side, though interhemispheric differences were not significant. Regionally, a significant increase in blood flow was found mainly in the regions surrounding the sylvian fissure and the fissure of Rolando on both sides. A relatively slight decrease in blood flow was seen in the parieto-occipital region, and, in part, in some frontal regions, as presented in Fig. 2.

Activation measurement during word pair learning and recall (Fig. 3) indicated a slight, insignificant increase in mean hemispheric flow which was more pronounced on the left side. Regional changes were found mainly postcentrally, in both posterior temporal regions and parieto-occipitally on the left side only. While blood flow decreased in the majority of frontal regions, it remained stable in the regions surrounding the sylvian fissure and the fissure of Rolando.

Comparison of the two tests employed as activation procedures (Fig. 4) revealed similar changes in mean hemispheric flow although they were slightly greater during immediate memory activation. Significant differences were found in regional distribution; an increase took place mainly in the central regions on the left side, while in the right hemisphere changes were more scattered and localized in the posterior frontal and temporal regions. Parieto-temporo-occipitally, a relative decrease in blood flow was seen.

Discussion

Although neuropsychologists have shown particular interest in activation methods and have obtained interesting results in clinical medicine (Risberg and Ingvar 1973; Maximilian et al. 1980), widespread application of the ^{133}Xe inhalation method has not yet been established. The results of these activation measurements have highlighted several points. Cortical activation patterns are different during immediate and recent memory activation, as expected and as

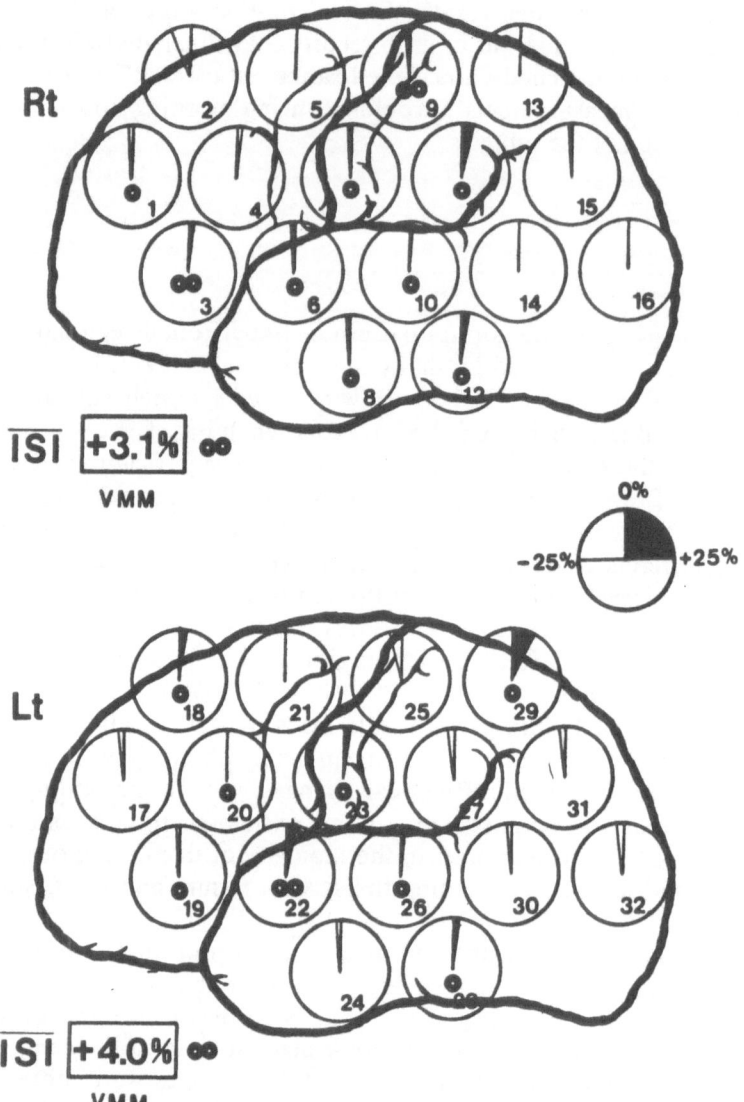

Fig. 2. Differences between mean hemispheric flow in the resting state and during immediate memory activation by means of the digit repetition test. Statistically significant changes are designated by *small circles*. Relative changes in the mean hemispheric flow from rest are shown in *boxes*. Other descriptions as in Fig. 1. The highest increase is evident in regions surrounding the sylvian fissure and the fissure of Rolando

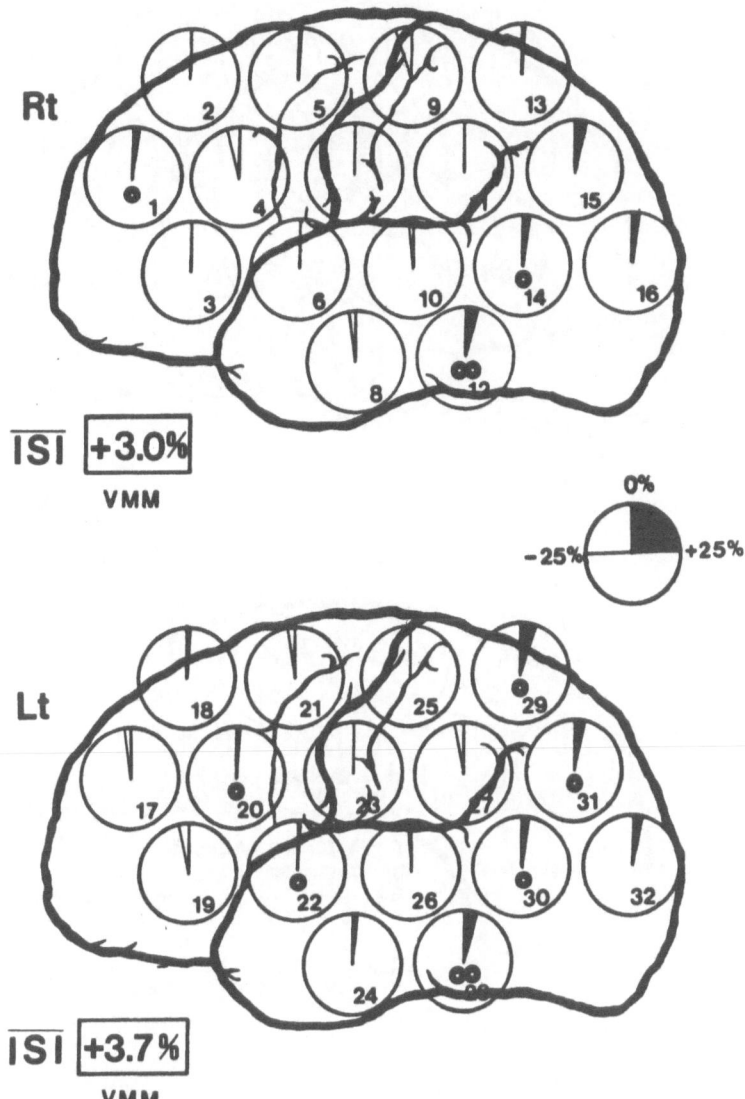

Fig. 3. Differences between mean hemispheric flow in the resting state and during recent memory activation by means of the word pair learning test. Descriptions as in Fig. 2. Regional changes are seen in postcentral regions, especially in the left hemisphere

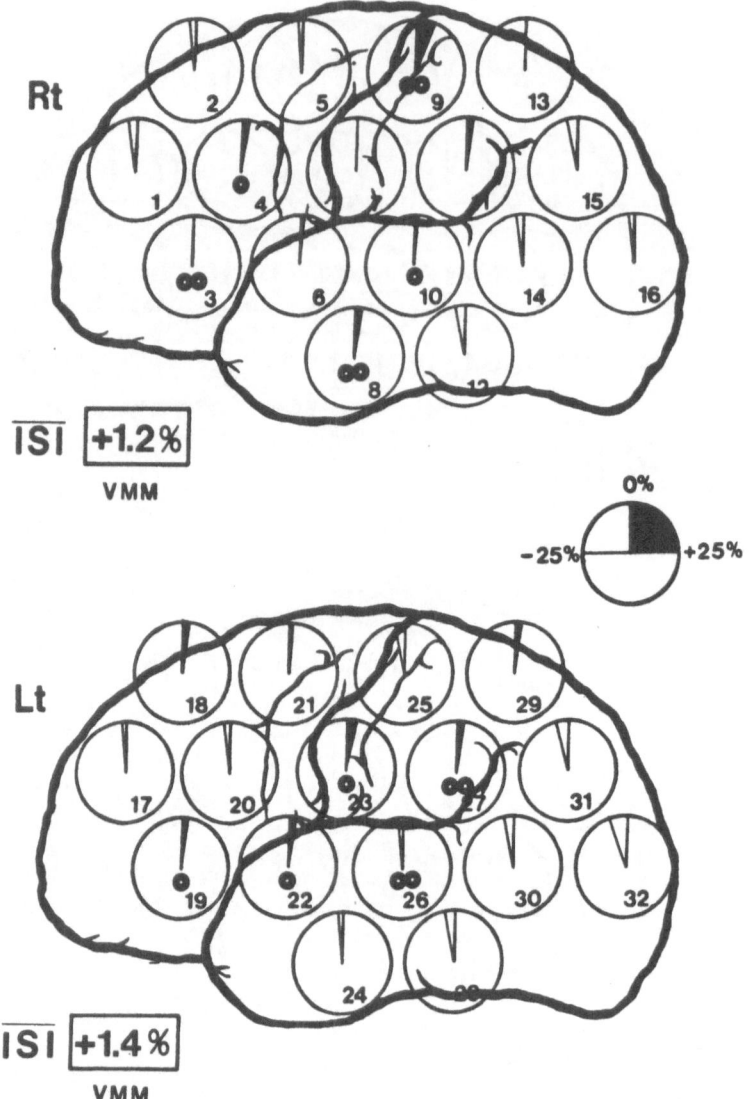

Fig. 4. Differences between the two memory activation tests. Descriptions as in Figs. 1 and 2. The highest increase is seen in central regions, while a decrease is most prominent post-centrally and in some frontal regions, although later changes are not statistically significant

shown previously by other researchers (Risberg and Ingvar 1973; Maximilian et al. 1978) who did not, however, compare results in the same individuals. We are aware that the activated cortical regions do not represent a part of the brain that is exclusively activated or solely responsible for memorization but rather the activity of cortical projections of deeper structures, e.g., the limbic system, including the hippocampus. Such different activation patterns, which reflect increased metabolic activity in particular regions (Raichle et al. 1976), can help in clinical work in testing patients' ability to activate these regions of the brain, which could be damaged without a visible cause. Furthermore, the effect of drugs on memorization could be examined in normal individuals as well as in patients suffering from various types and degrees of memory disturbance. Dissociation between immediate and recent memory has been known clinically for a long time and is the most frequently encountered memory disorder in psychogeriatric patients (Ban 1976).

As memory activation is fairly easily performed and the results are highly reproducible (Meyer et al. 1978), this method could be used as an ancillary tool in clinical neurophysiology for studying different characteristics of memory and memory disturbances.

References

Ban TA (1976) Psychopathology, psychopharmacology and organic brain syndrome. I. Psychosomatics 17:77–82

Ingvar DH (1977) Functional responses of the human brain studied by regional cerebral blood flow techniques. Acta clinica Belgica 32:68–83

Ingvar DH (1979) Hyperfrontal distribution of the cerebral grey matter in resting wakefulness: on the functional anatomy of conscious state. Acta neurologica Scandinavica 60:12–25

Maximilian VA, Prohovnik I, Risberg J, Håkansson K (1978) Regional blood flow changes in the left cerebral hemisphere during word pair learning and recall. Brain and Language 6:22–31

Maximilian VA, Prohovnik I, Risberg J (1980) Cerebral hemodynamic response to mental activation in normo- and hypercapnia. Stroke 11:342–347

Meyer JS, Ishihara N, Desmukh VD, Naritomi H, Sakai F, Hsu MC, Pollack P (1978) Improved method for noninvasive measurement of regional cerebral blood flow by [133]xenon inhalation. Part I. Stroke 9:195–205

Prohovnik I, Håkansson K, Risberg J (1980) Observations on the functional significance of regional cerebral blood flow in "resting" normal subjects. Neuropsychologia 19:203–217

Risberg J, Ingvar DH (1973) Patterns of activation in the grey matter of the dominant hemisphere during memorization and reasoning. A study of regional cerebral blood flow changes during psychological testing in a group of neurologically normal subjects. Brain 96:737–756

Risberg J, Maximilian VA, Prohovnik I (1977) Changes of cortical activity patterns during habituation to a reasoning test. Neuropsychologia 15:793–798

Strub RL, Black FW (1977) The mental status examination in neurology, Davis, Philadelphia

rCBF and Neuropsychological Findings in Head Injury

B. P. Uzzell, C. A. Dolinskas, J. L. Jaggi, and W. D. Obrist

A high incidence of CBF abnormality has been reported in patients following acute head injury (Obrist et al. 1983). Previous observations in such patients suggested that both lateralized CT scan lesions and hemispheral CBF differences are associated with distinctive psychological impairment (Uzzell et al. 1979, 1980). The present study examines the relationship between acute CBF findings and subsequent neuropsychological function in survivors of severe head injury.

Serial CBF measurements were obtained in 47 adult patients by the ^{133}Xe intravenous injection method from eight detectors over each hemisphere. The initial examination was performed within 96 h of injury; follow-up studies were performed between 6 and 12 months. CT scans were obtained in all patients on hospital admission. A battery of neuropsychological tests was administered between 3 and 6 months postinjury.

Using CBF_{15} as an index of flow (Obrist and Wilkinson 1980), the patients were classified in two ways according to: (1) the level of their acute CBF (hyperemia vs. reduced flow), and (2) the occurrence of significant hemispheral asymmetries (left vs. right). Lateralization of the CBF abnormality depended on blood flow level, i.e., the abnormal hemisphere was defined as having the lower CBF in patients with reduced flow, and the higher CBF in patients with hyperemia.

Results

As shown in Table 1, 23 patients revealed hyperemia and 24 had reduced flow during the acute phase of their illness. CBF abnormalities were lateralized to the left hemisphere in 11 cases and to the right hemisphere in 19. The CBF groups were comparable with respect to sex, handedness, age, education, and depth of coma (initial Glasgow Coma Score).

Table 2 compares the CBF groups with respect to Full Scale IQ, as determined by the Wechsler Adult Intelligence Scale (WAIS). A median IQ score of 71 was obtained in patients with previous hyperemia, which was significantly

Division of Neurosurgery, Hospital of the University of Pennsylvania, 3400 Spruce Street, Philadelphia, PA. 19104, USA

Cerebral Blood Flow and
Metabolism Measurement.
Eds. Hartmann/Hoyer
© Springer-Verlag Berlin Heidelberg 1985

Table 1. Description of blood flow groups classified in two ways

	CBF level		CBF laterality		
	Hyperemia	Reduced flow	Left	Right	Symmetry
No. cases (Total = 47)	23	24	11	19	17
No. males	17	18	9	13	13
No. rt-handed	19	21	9	16	15
Age (mdn yrs)	20	22	21	22	20
Education (mdn yrs)	10	12	12	10	12
Glasgow Coma Score (mdn)	6	7	6	7	6

Table 2. Median full scale IQ scores

CBF level	CBF laterality			Combined groups
	Left	Right	Symmetry	
Hyperemia	69[a]	67	79	71[a]
Reduced flow	84	72	90	85

[a] Significant difference between hyperemia and reduced flow ($P < 0.02$, Mann-Whitney U test)

Table 3. Median IQ scores

WAIS Scale	CBF laterality		
	Left	Right	Symmetry
Verbal	74	86[a]	94[a]
Performance	73	66	80
Full Scale	71	70	89

[a] Significant difference between verbal and performance IQ ($P < 0.01$, Mann-Whitney U test)

lower than the value of 85 obtained in patients who had acute reductions in flow ($P < 0.02$).

Median IQ scores for the CBF laterality groups are shown in Table 3. Performance IQ was significantly lower than Verbal IQ in both the right hemisphere and symmetrical CBF groups ($P < 0.01$). In contrast, Verbal and Performance scores were equally low among patients with left hemisphere CBF abnormalities.

Table 4. Aphasia screening test errors

CBF laterality	% of cases with errors				
	Naming	Reading	Calculation	Compre-hension	Visuographic
Left	100[a]	83	83	83	22[a]
Right	20	50	40	30	82
Symmetry	23	8	31	38	63

[a] Significant difference between left and right CBF groups ($P < 0.01$, chi-square test)

Table 4 compares findings on the Reitan Aphasia Screening Test for the left and right CBF groups. Whereas patients with left CBF abnormalities had significantly more naming errors (dysnomia), those with right CBF abnormalities showed greater impairment of visuographic skills ($P < 0.01$).

CT scans of the brain were rated according to a modified system developed for acute pathology (Ward 1978). Lesion types included hematomas (intracerebral, subdural, or epidural), hemorrhagic contusions, and cerebral swelling. In 35 of the patients (74%), the primary CT lesion was lateralized either to the left hemisphere (13 cases) or to the right (22 cases). This compares with lateralized CBF findings in 30 of the patients (64%) (see Table 1). CBF laterality agreed with the CT scan in all but one case, where assignment of the abnormality was to the opposite hemisphere.

At 6–12 months follow-up, CBF level had returned to normal in 64% of the patients; the remaining cases had a reduced flow. At this time, CBF was bilaterally symmetrical in 85%, but continued to show asymmetries in 15%.

Conclusions

1. Intellectual ability following head trauma is related to the level of blood flow during the acute phase of illness. Lower IQ scores were found in patients who experienced hyperemia compared with those who had reduced flow.
2. Patients with left CBF abnormalities showed different intellectual dysfunction than those with right or symmetrical hemispheric abnormalities. Verbal (as opposed to nonverbal) abilities were more intact in patients in the latter two groups.
3. The primary verbal deficit in patients with left CBF abnormalities was dysnomia. In contrast, the right hemisphere group showed greater visuographic impairment.

References

Obrist WD, Wilkinson WE (1980) The non-invasive Xe-133 method: Evaluation of CBF indices. In: Bes A, Geraud G (eds) Cerebral circulation. Excerpta Medica, Amsterdam, p 119–124

Obrist WD, Dolinskas CA, Jaggi JL, Cruz L, Steiman DL (1983) Serial cerebral blood flow studies in acute head injury: Application of the intravenous ^{133}Xe method. In: Magistretti PL (ed) Functional radionuclide imaging of the brain. Raven Press, New York, p 145–150

Uzzell BP, Zimmerman RA, Dolinskas CA, Obrist WD (1979) Lateralized psychological impairment associated with CT lesions in head injured patients. Cortex 15:391–401

Uzzell BP, Obrist WD, Gennarelli TA (1980) Lateralized CBF and neuropsychological findings in acute head injury. In Bes A, Geraud G (eds) Circulation cerebrale. Fournie, Toulouse, p 223–225

Ward C (1978) Terminology and extent measures for digital head and neck injury records. Civil Engineering Lab, Naval Construction Battalion Center, Port Hueneme, California

Alterations of Cerebral Blood Flow in Slight Head Injury

E. Körner, K. Marguc, E. Ott, H. Lechner, E. Flooh, R. Szyszkowitz, G. Bertha, and F. Fazekas

Treatment of craniocerebral injuries is based on careful clinical, neurophysiologic, and neuroradiologic investigations. The availability of a noninvasive method for cerebral blood flow (CBF) measurements using the intravenous 133 xenon clearance technique offers the possibility of performing serial observations to achieve a better understanding of the hemodynamic consequences during the course of illness.

Wide variations in CBF have been reported in comatose patients following acute head injuries, indicating that both low and elevated CBFs may occur in patients with similar neurologic findings (Fieschi et al. 1974; Langfitt et al. 1977; Obrist et al. 1979; Overgaard and Tweed 1974). Regarding the available literature, there are no reports about CBF measurements in slight disturbances of cerebral functions after head trauma, e.g., mild concussion of the brain.

The present study demonstrates changes in the CBF after slight closed head injuries in a group of relatively young, healthy patients to provide information about the relationship between clinical state and cerebral hemodynamics.

Methods

Measurements of CBF using the intravenous 133 xenon clearance technique with ten detectors over each hemisphere were carried out in 20 patients with closed head injuries ranging in age from 16 to 54 years (14 males, 6 females) with a mean age of 29 ± 15 years. By convention, CBF is expressed as milliliters of blood flow per 100 g of brain per minute (ml/100 g/min). We obtained a number of flow parameters from each clearance curve, but will primarily relate our observations in respect of the gray matter flow (F_1), because in every patient the other values paralleled changes in F_1. In normal age-matched control subjects the present method yields a mean value for F_1 of approximately 70 ml/100 g/min.

As to the severity of the injury, neurologic deficits could not be observed in any patient, but in all cases both retrograde (2 ± 2 min) and anterograde (36 ± 54 min) amnestic episodes were found. Consciousness was not impaired at the time of the first investigation, which was made within 48 h after the trau-

Universitäts-Nervenklinik, Abteilung Neurologie, Psychiatrie und Traumatologie, Auenbruggerplatz 22, A-8036 Graz

matic event. Control measurements were carried out 3 weeks afterwards in eight patients.

Results

In 7 of the 20 patients (35%) the results of CBF measurements were within the normal range (Fig. 1). Only in one case did we consider an elevated CBF, F_1 being increased to approximately 25% above the level in normals. Although $PECO_2$ showed a low level in this patient (32.9 mmHg) we desisted from correction, because in patients who have suffered a head injury disturbed autoregulation or CO_2 reactivity can be observed, as will be mentioned below. On the other hand in 60% of patients a reduction in total CBF could be demonstrated; this reduction amounted to about 26% when compared with an age-matched control group and showed no significant interhemispheric differences.

Regarding focal changes, in 65% there were no variations from normal. A reduced focal CBF concerning only one hemisphere was seen in one patient, while focal changes manifested over both hemispheres in six. These localized reductions occurred especially in the parieto-occipital regions, whereas hyperemic areas were particularly observed in frontotemporal fields. In three cases such areas were localized over both hemispheres, while in 25% (five patients) they concerned only one side, thus showing a distinct anterior-posterior CBF gradient.

Control investigations carried out 22 days after the first measurements revealed a tendency toward improvement of reduced CBF in eight patients

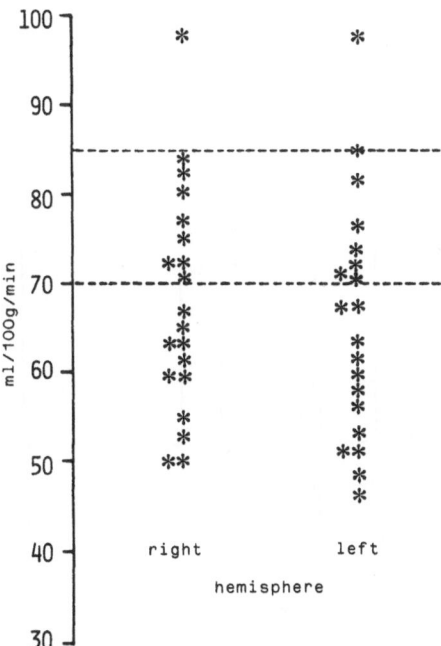

Fig. 1. F_1 in ml/100 g/min in 20 patients after head injury. Measurements were obtained within 48 h after trauma. The space between the two lines represents the normal range

E. Körner et al.

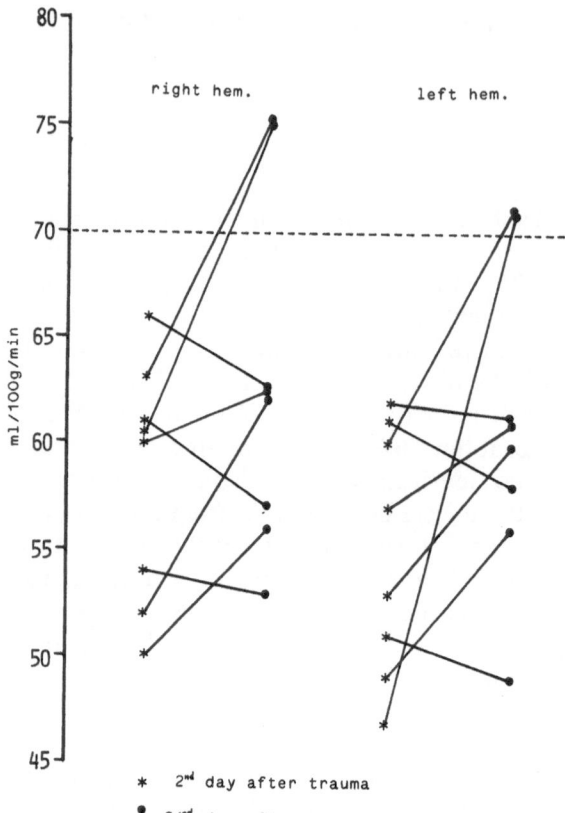

Fig. 2. Changes of F_1 in ml/100 g/min in eight patients shown by control measurements after 3 weeks. The space above the line represents the normal range

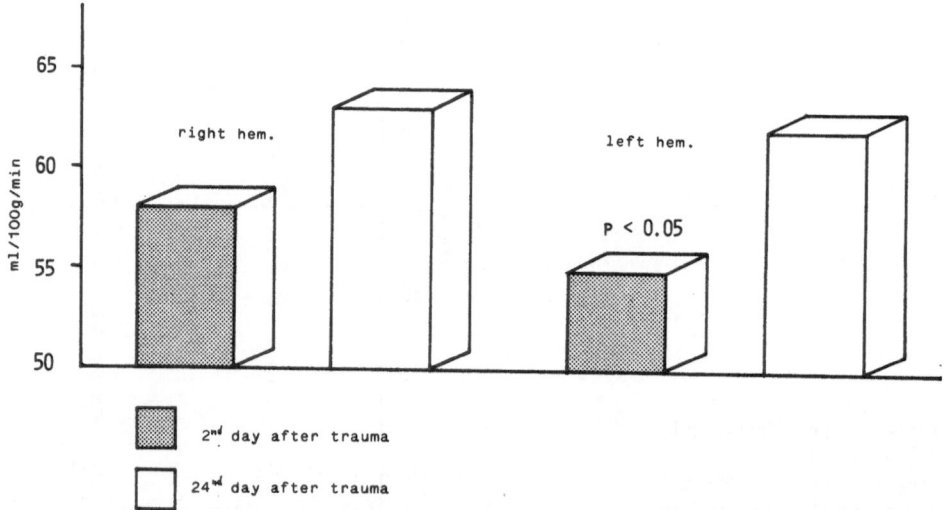

Fig. 3. Changes of the average value of F_1 in ml/100 g/min in eight patients shown by control measurements after 3 weeks. Statistical evaluation using the Student's paired t-test

(Fig. 2). In all eight the first results had ranged in lower fields (mean value 57 ± 5 ml/100 g/min). In the last investigation the measurements of the total CBF showed a distinct increase (Fig. 3), although in only two cases was the normal range reached. Altogether in five patients an improvement could be found, whereas in three the CBF had not changed. With regard to regional alterations, evaluation of the control measurements showed normalization in nearly all cases: both focal reductions and hyperemic fields could ultimately be found in only one patient.

Discussion

The present paper illustrates the value of the atraumatic intravenous 133 xenon clearance technique after slight head injuries. There are many reports relating to alterations in the CBF that show neurologic deficits following severe head trauma in comatose patients. Most of these authors obtained similar results, with decreases in CBF in the first measurements after the traumatic event (Brodersen and Jorgensen 1974; Obrist et al. 1982; Overgaard 1982). The outcome of the subsequent investigations depended on the clinical course of the illness, with slow increases in CBF when patients had a good recovery but further declines in patients showing clinical deteriorations with a poor prognosis (Langfitt et al. 1977; Obrist et al. 1979). In contrast to these findings, in some of the investigated patients hyperemic phases were also described, either with a latency of some hours or days after initial low CBF values (Fieschi et al. 1974; Obrist et al. 1982; Overgaard and Tweed 1974; Overgaard 1982) and distinct hyperemic asymmetries (Obrist et al. 1979) — higher blood flow concerning the more injured side — or connected directly with the head trauma (Langfitt et al. 1977; Obrist et al. 1979; 1982). Such hyperemic reactions can be attributed first to the distribution of age, insofar as they were especially seen in adolescents and children (Langfitt et al. 1977; Obrist et al. 1979; Overgaard and Tweed 1974; Overgaard 1982), partly independent of the prognosis. On the other hand, hyperemia occurred in the presence of diffuse cerebral swelling or systemic shock (Obrist et al. 1979) in additional complications like generalized seizures and development of brain edema connected with an uncoupling of blood flow and metabolism, as illustrated by the "luxury perfusion"; in general this signified a bad course.

In our investigation of patients with slight head injuries, hyperemia was also observed in one patient, but because of the low $PECO_2$ level we did not interpret this as a warning sign. This consideration was borne out by the stable course of the illness. Despite this case we disregarded theoretical correction of the $PECO_2$ values, because disturbances of CO_2 reactivity and autoregulation in the first 2 weeks after head injuries are known (Fieschi et al. 1974; Overgaard and Tweed 1974) and in all the other patients measurements of $PECO_2$ were anyway within a small range.

In regard to focal changes, both reductions in blood flow and hyperperfused compartments have been described in the literature (Obrist et al. 1979, 1982;

Overgaard and Tweed 1974; Overgaard 1982). In this connection the occurrence of low values concerning the frontal and parietal lobes bilaterally and suggesting a poor prognosis must be stressed (Obrist et al. 1982; Overgaard 1982). In contrast to these results, in our patients focal reductions in CBF occurred especially in parieto-occipital regions, whereas hyperemia was generally observed in frontotemporal fields, thus showing an anterior-posterior CBF gradient in agreement with the mild disturbances of the functional systems.

Despite these marked differences in results when comparing the CBF of patients with severe and slight head injuries, we should not overlook the clear reduction of CBF in our patients and the slow recovery to normal values, which demonstrates the cerebral participation and the distinct irritation of the brain. Such hints, generally not seized upon by other methods, should not make the diagnosis of brain concussion more difficult or complicate the classification of head trauma, but should account for aches lasting over a long period after the traumatic event.

In addition it may be concluded from the present results that even in slight head injuries global as well as regional CBF can be found to be impaired to a high degree for a period of some weeks, thus showing a dissociation of clinical symptoms and cerebral hemodynamics.

Summary

In this study, CBF in 20 patients who suffered a slight head injury without any neurologic deficits is described. With one exception all results out of the normal range (n = 13) showed a clear reduction in F_1; furthermore focal changes in the form of regional reductions generally in parieto-occipital areas and localized hyperemic fields especially in frontotemporal regions were observed.

In eight patients whose first CBF values were reduced, control investigations after 3 weeks showed an increase in CBF, but in only two cases was the normal range reached; regional changes, by contrast, disappeared in almost all cases.

In conclusion it may be stated that even in slight head injuries both global and regional CBF is impaired to a high degree over a period of some weeks, thus showing a dissociation between the clinical symptoms and cerebral hemodynamics.

References

Brodersen P, Jorgensen EO (1974) Cerebral blood flow and oxygen uptake, and cerebrospinal fluid biochemistry in severe coma. J Neurol Neurosurg Psychiatr 37:384–391

Fieschi C, Battistini N, Beduschi A, Boselli L, Rossanda M (1974) Regional cerebral blood flow and intraventricular pressure in acute head injuries. J Neurol Neurosurg Psychiatr 37:1378–1388

Langfitt TW, Obrist WD, Gennarelli TA, O'Connor MJ, Ter Weeme CA (1977) Correlation of cerebral blood flow with outcome in head injured patients. Annals of Surgery 186:411–414

Obrist WD, Gennarelli TA, Segawa H, Dolinskas CA, Langfitt TW (1979) Relation of cerebral blood flow to neurological status and outcome in head-injured patients. J Neurosurg 51:292–300

Obrist WD, Dolinskas CA, Gennarelli TA, Zimmerman RA (1979) Relation of cerebral blood flow to CT scan in acute head injury. In: Popp AJ et al. (eds) Neural Trauma. Raven Press, New York, p 41–50

Obrist W, Gennarelli T, Segawa H, Laurent J, Czernicki Z (1979) Uncoupling of cerebral blood flow and metabolism in acute head injury. In: Gotoh F, Nagai H, Tazaki Y (eds) Cerebral blood flow and metabolism. Munksgaard, Copenhagen, p 374–375

Obrist WD, Dolinskas CA, Jaggi JL, Cruz J, Steiman DL (1982) Serial CBF Studies in acute head injury: Application of the intravenous ^{133}Xe method. In Press: Proceedings of the Serono Symposium on Radionuclides and Brain Disease. Raven Press, New York

Obrist WD, Jaggi JL, Steiman DL, Cruz J (1982) Relationship of cerebral blood flow to metabolism in acute head injury. In Press: Proceedings of the 11th International Salzburg Conference on Cerebral Vascular Disease. Excerpta Medica, Amsterdam

Overgaard J (1982) The distribution of regional cerebral blood flow values in traumatic coma. In: Grossman RG, Gildenberg PL (eds) Head injury: basic and clinical aspects. Raven Press, New York, p 239–249

Overgaard J, Tweed WA (1974) Cerebral circulation after head injury. Part 1: Cerebral blood flow and its regulation after closed head injury with emphasis on clinical correlations. J Neurosurg 41:531–541

Is It Possible to Measure Regional Cerebral Blood Flow in Patients with Acute Cerebral Ischemia?

A. HARTMANN

Since the introduction of the atraumatic method for measurement of regional cerebral blood flow (rCBF) by Veall and Mallett (1966) and improvement of the technique and calculation procedures by Obrist et al. (1975) and Risberg et al. (1975), this method has been used widely in patients with different diseases of the brain. This is possible since in nonobstructive vascular diseases the inhaled xenon 133, which is trapped in the brain tissue, reaches the tissue within the time of the inhalation period (60 s). This prerequisite is not fulfilled if the entrance of the inert gas into the regional areas is inhibited by vascular narrowing or even obstruction. This might partially be eluded by the fact that the inhalation of xenon over 1 min enables the gas to reach most of the tissue by the collateral system. However, phenomena like "look-through" due to complete nonfilling of parts of the brain, no reflow (and therefore absence of filling) in tissue with open vessels, and very low flow (which does not allow separation of the activity of the brain tissue from activity trapped in the extracranial tissue) do exist and limit the practical usefulness of this method in patients with cerebrovascular diseases.

Nowadays it is possible to obtain information about the state of tissue perfusion and the vascular condition by computer tomography, brain scintigraphy, and cerebral angiography. It therefore should be possible to validate blood flow measurements gained with the inhalation technique by considering the results of the other diagnostic procedures. In cases where, for instance, angiography reveals complete occlusion of the middle cerebral artery and where computer tomography indicates an area with complete necrosis, all techniques which use saturation and desaturation procedures (xenon 133, positron emission tomography, single photon emission tomography) should be regarded with scepticism.

Our interest in recent years has focussed on measurement of rCBF in patients with acute onset of neurologic symptoms due to ischemic events in the brain. To evaluate the usefulness of the inhalation technique we have followed two groups of patients with measurements at different times during acute cerebral ischemia.

Particular interest was devoted to the following questions:

1. Is the atraumatic method able to differentiate between the ischemic and the contralateral side or is the perfusion difference masked by technical artifacts?
2. Are flow changes over time recognized by this technique?

Neurologische Universitätsklinik, D-5300 Bonn

Cerebral Blood Flow and
Metabolism Measurement.
Eds. Hartmann/Hoyer
© Springer-Verlag Berlin Heidelberg 1985

Fig. 1. Detector holder for two-dimensional rCBF measurement. The detectors which are positioned over the tissue close to the superior sagittal sinus and the base of the skull are slightly angled to point away from these structures

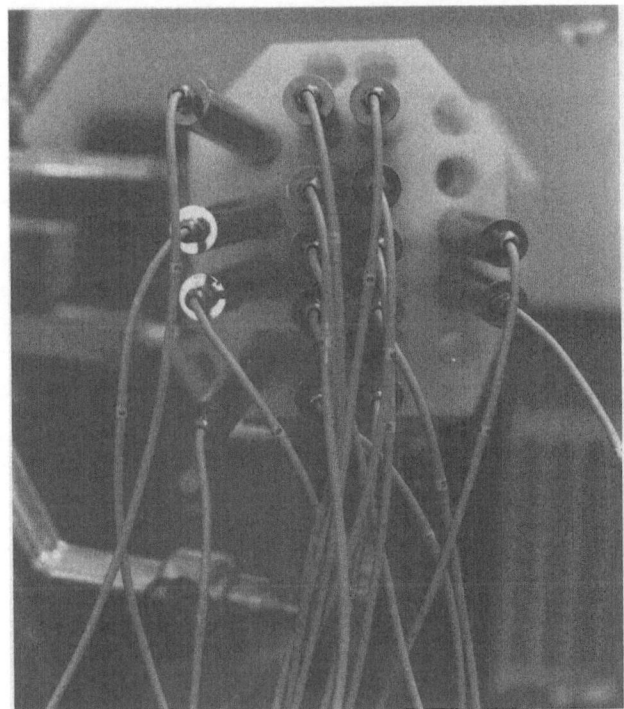

Method

rCBF was measured by the atraumatic method (Hartmann and v. Kummer 1983). A setup from NOVO was used with 32 detectors, 16 positioned over each side of the skull (NaI crystals of 0.75-in. diameter, collimation 20 mm, window 81 keV $\pm$ 20%). To avoid contamination of the uppermost and lowest detectors, we constructed a detector holder with slightly angled detectors close to the vertex and the skull (Fig. 1). Using this setup a "crosstalk" was measured of less than 15%.

The inhalation time was 60 s (from an 8-liter air bag with $3-4$ mCi xenon 133 per liter), and the desaturation was recorded for 10 min. Endexpiratory xenon 133 activity was continuously recorded for correction of recirculation. The printout gave data on fast flow (F1), mixed flow (F15), and the initial slope index (ISI). Since the latter does not require the partition coefficient lambda, this parameter was used for measurement of rCBF in patients with acute ischemia of the brain. Endexpiratory CO_2 (vol%) and blood pressure were recorded during the study.

Results

To evaluate the spontaneous course of rCBF in humans without acute cerebrovascular disease, so-called test-retest procedures were performed. In

Table 1. Test–retest measurement in patients without any acute cerebrovascular disease. The number indicates the percentage deviation of the second CBF measurement from the first for ISI for all right and all left hemispheres

	Second measurement after			
	20 min	180 min	24 h	7 days
Right	− 1.4±4.2	+ 1.2±1.8	− 2.9±1.7	− 2.2±2.5
Left	− 2.8±3.4	+ 2.4±2.4	− 3.0±1.7	− 2.4±1.4

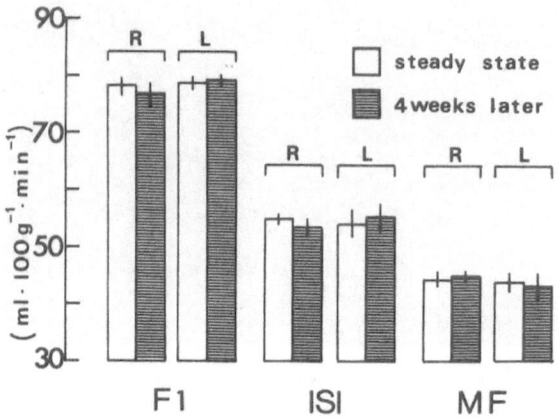

Fig. 2. Test-retest procedure. In normals rCBF was measured a second time after 4 weeks. All parameters (F1, ISI, and mean flow according to the corrected stochastic method) showed a stable flow over both (right and left) hemispheres

these patients a second CBF procedure was done either 20 min, 3 h, 1 day, 7 days, or 4 weeks after the first measurement.

Table 1 indicates the percentage deviation of the second mean rCBF from the first measurement in both hemispheres. In none of the test-retest procedures did the deviation exceed 3.0% ± 1.7%. Even after 4 weeks (Fig. 2) the deviation was rather small (if procedures had been carried out under the same conditions in respect of time of the day, hours since last meal, intake of medication, and laboratory environment vis-à-vis darkness and noise level).

From Table 1 it can be concluded that deviation of the second CBF value of more than 10% can be considered statistically significant in cases with nonobstructing disease of the cerebral vessels.

Another criterion for judging individual flow courses in patients is the stability of the normal flow map in healthy humans. Figure 3 indicates the percentage difference between corresponding regions in the right and left hemispheres in young normals. In none of the regions did the difference exceed 15% (mean − 0.6% ± 5.6%). From these data it can be concluded that any percentage difference between corresponding right and left regions of more than

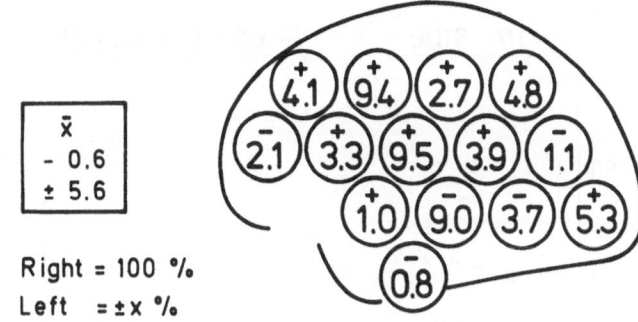

Fig. 3. Regional right-left differences in corresponding areas in young normals (28 – 36 years old). Each number indicates the difference of the left rCBF compared with the right rCBF (100%). The figure indicates the regional differences for the initial slope index (ISI)

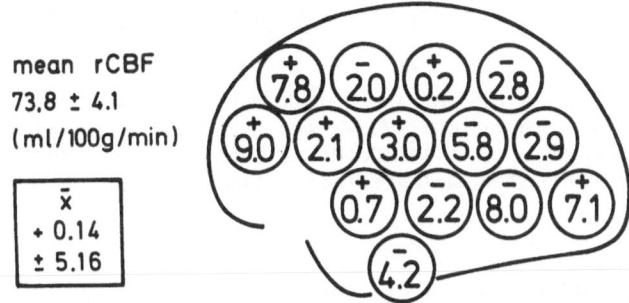

Fig. 4. Percentage deviation of rCBF from mean rCBF in ten young normals (aged 28 – 36 years). Expressed is the regional difference for the flow of the fast perfused tissue (F1). The figure indicates that rCBF deviates more from the mean flow over frontal areas than over other areas (hyperfrontality)

15% is statistically significant. Figure 4 indicates the percentage deviation of the mixed corresponding regional flow data (ISI) from the homolateral mean rCBF. The data indicate that a deviation of any regional flow value of more than 15% from the hemispheric value represents a pathologic flow (either *relative* ischemia or *relative* hyperemia).

Transient Ischemic Attacks (TIA)

There exist only a few studies on CBF measurement in patients suffering from TIA, none of which report early and repeated measurements of rCBF. Skinhøj et al. (1970) mentioned one patient in whom CBF was measured 30 min after an attack. All other authors using the two-dimensional xenon 133 technique (either intraarterial or after inhalation) performed the CBF measurement days or weeks after the last attack. There is only one report in the literature which mentions repeated measurements. This study (Vorstrup et al. 1983) used the newly developed three-dimensional single photon emission tomography with

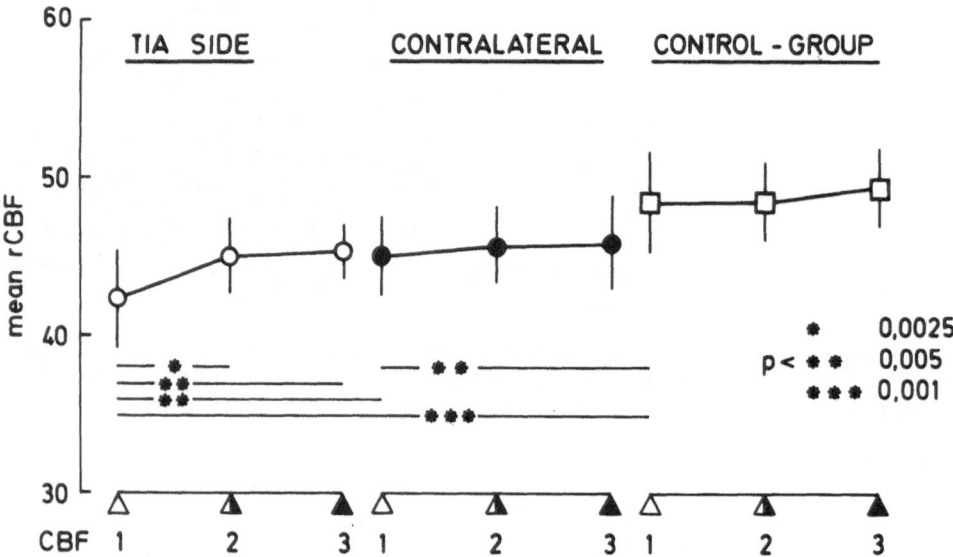

Fig. 5. Mean rCBF over the TIA side and over the contralateral hemisphere in patients with transient ischemic attacks (*TIA*). The mean rCBF of the "control group" (patients with two risk factors for cerebrovascular disease but no history of ischemic strokes of any duration) was stable throughout the time of the repeat measurements. rCBF as ISI in sec^{-1}

inhalation of xenon 133. Our goal was to perform follow-up studies in patients with TIA who presented during or shortly after an attack. 19 patients with unilateral TIA in the territory of the internal carotid artery completed the protocol. In 13 cases the clinical symptoms were still present during the first measurement, while in six the TIA had ceased not longer than 6 h earlier. CBF was measured on the day of the TIA (CBF 1), 1 day later (CBF 2) − when the symptoms had ceased − and on day 7 (CBF 3). Figure 5 indicates the mean rCBF for the TIA side and the contralateral hemisphere during CBF 1−3. The mean rCBF 1 on the TIA side was significantly reduced compared with the contralateral side and with mean rCBF of a control group with at least two risk factors but no history of acute ischemic attacks to the brain. Whereas CBF did not change between CBF 1 and CBF 3 on the contralateral side or in the control group, there was a significant increase in mean rCBF on the TIA side from CBF 1 to CBF 2 but not from CBF 2 to CBF 3. These data indicate that cerebral blood flow is reduced during and shortly after an ischemic attack. One day after the attack, mean blood flow is normalized.

However, this does not mean that *regional* flow is normal on day 2. Based on the normal flow map, all areas which deviate from the mean rCBF by more than 15% can be considered relatively abnormal (either ischemic or hyperemic). Using this classification in all areas of the TIA side, on day CBF 1 61 regions of interest (ROI) were ischemic, 17 were hyperemic, and 199 were within the normal distribution (Fig. 6). At CBF 2 29 of the former 61 ischemic ROI were still ischemic but 22 now had normal values. There was some further change by CBF 3: At this time only 19 ROI were ischemic. It can be concluded that be-

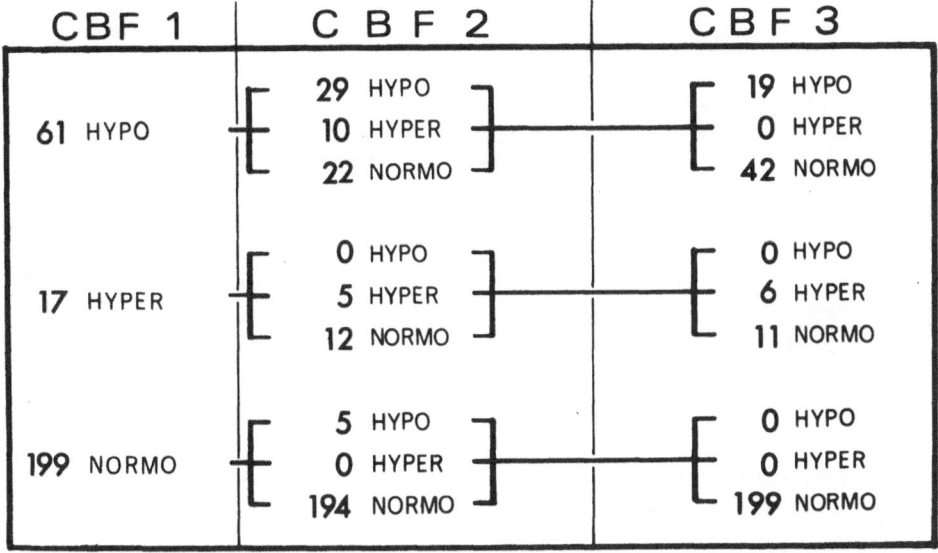

Fig. 6. Number of areas with so-called relative ischemia, hyperemia, or normoemia on the TIA side of TIA patients at CBF 1 − 3

Table 2. The percentage deviation of rCBF (ISI) of all relatively hypoperfused and all relatively hyperperfused tissue on the TIA side from the contralateral corresponding tissue

	Ischemic	Hyperemic
CBF 1	− 25.1 ± 11.7	+ 22.9 ± 11.2
CBF 2	− 14.9 ± 10.2	+ 20.0 ± 12.4
CBF 3	− 12.4 ± 10.8	+ 16.0 ± 15.5

tween CBF 2 and CBF 3 some change in the distribution of the flow map does occur. This indicates that on the day after the TIA (CBF 2) cerebral blood flow is not normal, despite the fact that symptoms have vanished.

The slowly changing flow pattern is reflected by the deviation of rCBF in all so-called relatively ischemic and hyperemic areas from the contralateral corresponding areas (Table 2). In terms of rCBF all ischemic areas deviate from contralateral areas maximally on day 1 (25.1% ± 11.7%). The hyperemic tissue shows a bigger difference from the contralateral areas on day CBF 2 than on CBF 3, again indicating that flow restoration is not completed 1 day after TIA.

For presentation of regional flow pattern, two patients have been selected here since on day 1 they presented with hyperemia above the normal flow and displayed only slow recovery by day 7 (Fig. 7). This was recorded only by those detectors which were localized over the area responsible for the clinical symptoms. The neighboring detectors showed ischemia with only slow recovery by day 7. This flow distribution (early luxury perfusion with perifocal steal

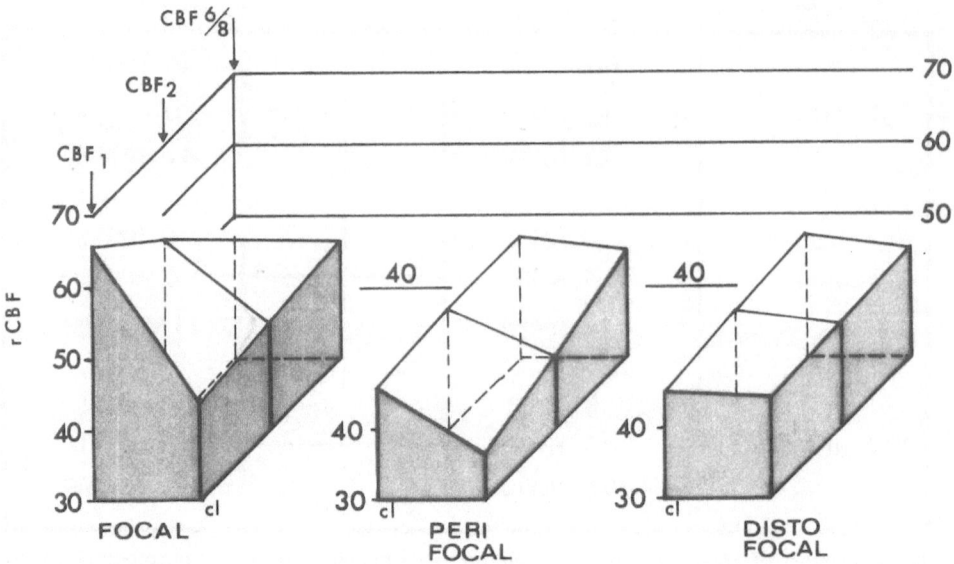

Fig. 7. rCBF (ISI: sec^{-1}) in two patients with TIA at CBF 1, CBF 2 (day 2), and day 6 or 8. The detectors over both the TIA side and the contralateral hemisphere were classified as "focal" (over the tissue responsible for the neurologic deficit), "perifocal" (the detectors neighboring the focal detectors), and "distofocal." Whereas the contralateral flow was normal throughout, there was luxury perfusion over the focal tissue with reduced flow over the perifocal tissue at CBF 1 and CBF 2

phenomenon and normal contralateral flow) is due to slow rCBF normalization. It might be assumed that the rCBF pattern at day 7 (CBF 3) indicates a rather stable flow.

Completed Stroke

In 30 patients with acute ischemic infarction and clinical symptoms of at least 2 weeks' duration, rCBF was measured on the day of onset of symptoms (CBF 1), 6−8 days later (CBF 2), 13−16 days later (CBF 3), and in some cases in the 4th week after onset of the ischemic attack (CBF 4).

Since the cause of the ischemic stroke varied, we were interested in two problems:

1. Whether it is possible to differentiate between both hemispheres using CBF values and CBF course. This might be of particular interest in patients with early reduction of CBF in the contralateral hemisphere (diaschisis), if regional patterns are masked by so-called crosstalk (Hartmann and v. Kummer, 1983).
2. Whether the nature of the acute cerebral ischemia has some influence on the CBF course. This seems to be of importance, since slow occlusion of a major artery gives rise to adaptation to the altered perfusion pressure whereas sudden occlusion as in cerebral emboli hits an "unprepared" vascular system.

Course of Mean rCBF in Both Hemispheres

In acute infarction of the brain the course of rCBF (and of the clinical symptomatology) depends on the degree of tissue damage, the metabolic demand, preservation of the vascular system, and its blood gas and perfusion pressure regulation. Furthermore the preexisting disease (generalized cerebral vascular disease versus a single vascular focus, as in cardiac emboli) influences the fate of the tissue. The hemisphere contralateral to the side with acute infarction might suffer from preexisting vascular disease which itself leads to a reduction in rCBF. Under these circumstances focal flow over corresponding areas of both hemispheres might be similar, which gives the impression that the homolateral ischemic focus has a mirror focus on the contralateral side. In other cases it cannot be excluded that the normal contralateral flow is being picked by the homolaterally positioned detectors (crosstalk), which then masks the ischemic focus.

Figure 8 illustrates the mean rCBF and its flow changes in both hemispheres of all 30 patients. At all three measurements the mean rCBF on the infarcted side was significantly lower than that on the contralateral side. There was no significant change in mean rCBF from measurement to measurement on the involved side, but a transient significant drop in blood flow over the contralateral side occurred at CBF 2.

The different course of mean rCBF over the two sides and the different mean values between them indicate that crosstalk contributes to the actual flow values only to a limited extent. This conclusion might also be derived from the measurements of CBF in those patients who presented with an initially reduced flow over both the infarcted and the contralateral side. This "diaschisis" was observed in 8 of 30 patients. CBF was measured in all patients four times. Table 3 indicates the actual final mean rCBF at CBF 4 and the percentage de-

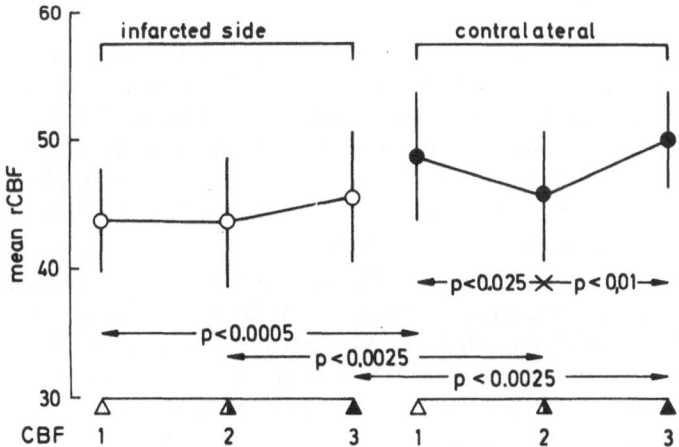

Fig. 8. Mean rCBF as ISI (sec^{-1}) in the involved and the contralateral hemisphere in 30 patients with acute ischemic brain infarction. In all 30 cases CBF was measured three times

Table 3. Mean rCBF in ischemic infarction. The table indicates the final mean rCBF as ISI (sec^{-1}) over the infarcted and the contralateral hemisphere at CBF4. Under CBF1–3 is the percentage deviation of the mean rCBF from the value of mean rCBF at CBF 4. The difference between the contralateral and the infarcted side was statistically significant for CBF 2 and CBF 3 (*)

	CBF 1 (%)	CBF 2 (%)	CBF 3 (%)	CBF 4
Contralateral side	– 14.8	– 6.4	+ 1.1	50.7
	± 3.9	±0.7	±1.5	±3.0
		*	*	
Infarcted side	– 16.8	– 12.4	– 3.7	48.7
	± 6.3	± 6.7	±2.7	±6.3

viation of mean rCBF at the previous measurements (CBF 1–3) from CBF 4. The difference in the courses of mean rCBF between the involved and the contralateral side was significant at CBF 2 and CBF 3. At CBF 1 and CBF 4 there was no difference between the sides in these eight cases. The different course of mean rCBF over the two hemispheres supports the hypothesis that the measured diaschisis is not a technical phenomenon due to "look-through" but a real flow pattern. Another finding that supports this interpretation is the fact that the diaschisis was not always limited to the corresponding detectors of the homolateral ischemic focus but rather was widespread throughout the contralateral hemisphere.

rCBF Changes in Respect to the Cause of Infarction

In patients with thrombosis of the internal carotid artery (ICA) the collateral circulation is quite different from that in patients with sudden occlusion of the middle cerebral artery (MCA) due to emboli. In the former the blood might be shunted via the circle of Willis from the contralateral side. However, in patients with emboli to the MCA, resolution of the emboli might normalize the vascular supply within a short time. To decide whether the differing blood flow in these two situations might be picked up by the two-dimensional CBF setup we separated all 12 patients with occlusion of one ICA and sudden onset of clinical symptoms from those five in whom the ischemic stroke was believed to be due to embolic infarction in the territory of the MCA. Cerebral angiography showed a permanent occlusion of the MCA in only one of these five cases. The diagnosis of embolic stroke to the MCA was based on internal cardiac and other diagnostic findings.

Figure 9 shows the mean rCBF in both groups. In the MCA group there was a significant increase of mean rCBF from CBF 1 to CBF 4. Such an alteration

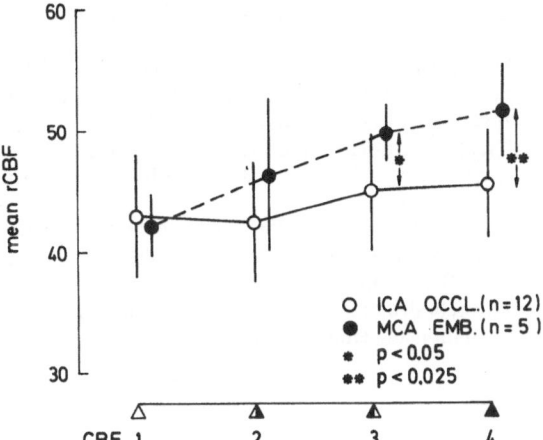

Fig. 9. Mean rCBF in 12 patients with acute occlusion of one internal carotid artery and 5 patients with cardiac emboli to the middle carotid artery after onset of symptoms (day of infarction; 1, 2, and 3–4 weeks later). Mean rCBF as ISI (sec^{-1})

was not observed in the patients with ICA occlusions. At the time of CBF 3 and CBF 4 there was a significant difference between the groups. This might be explained by the observation that at the time of the angiography the vascular system did not show any pathologic findings in the majority of patients with embolic stroke to the MCA.

Discussion

Test-retest procedures in normal volunteers and in patients with no history or signs of cerebrovascular disorders have indicated quite high stability of the cerebral blood flow. This may be shown using a conventional two-dimensional technique to measure rCBF.

When using this technique, certain conditions such as small windows (81 keV $\pm$ 20%), sufficient count rate (1000 counts per interval), and positioning of the detectors away from the venous sinus and the base of the skull have to be fulfilled. The results of angiography and computer tomography as an indication of vascular narrowing or occlusion and possible necrosis should be considered. Under these circumstances rCBF can be measured with the xenon 133 inhalation technique in the majority of patients in whom no major occlusion of the intracranial arteries exists. Different actual flow values and their changes over time may be recognized over both hemispheres.

New problems do evolve when intracerebral arteries are occluded; this condition requires filling of the tissue by the tracer via collaterals. At present these problems are under investigation.

References

Hartmann A, v Kummer R (1983) Die atraumatische Messung der regionalen Hirndurchblutung: Methodik und Zuverlässigkeit. Fortschr. Neurol. Psychiat. 51:57–68

Obrist WD, Thompson HK, Wang HS, Wilkinson WF (1975) Regional cerebral blood flow estimated by xenon 133 inhalation. Stroke 6:245−256

Risberg J, Ali Z, Wilson EM, Wills EL, Halsey JH (1975) Regional cerebral blood flow by 133 xenon inhalation. Preliminary evaluation of an initial slope index in patients with unstable flow compartments. Stroke 6:142−148

Skinhøj E, Høedt-Rasmussen K, Paulson OB, Lassen NA (1970) Regional cerebral blood flow and its autoregulation in patients with transient focal cerebral ischemic attacks. Neurology, Minneap. 20:485−493

Veall N, Mallett B (1966) Regional cerebral blood flow determination by 133 Xe inhalation and external recording: The effect of arterial recirculation. Clin Sci 30:353−369

Vorstrup S, Hemmingsen R, Henriksen L, Lindewald H, Engel HC, Lassen NA (1983) Regional cerebral blood flow in patients with transient ischemic attacks studied by xenon 133 inhalation and emission tomography. Stroke 14:903−910

The Prognostic Value of rCBF Measurements in Aphasic Stroke Patients

G. Demeurisse, M. Verhas, and A. Capon

In aphasic stroke patients, the prognosis can be deduced to a certain extent from the initial clinical assessment (severity and type of aphasia) and from CT scan data. Searching for further information concerning the prognosis, we performed rCBF studies in such patients. Resting rCBF measurements proved to have but a limited prognostic value. Only the most severely affected patients (those with total aphasia for which the prognosis was already known to be bad) could be distinguished from the others by their resting rCBF (Demeurisse et al. 1984). In the patients with a cortico-subcortical infarct, we also observed a relation between the mean CBF in the right hemisphere at rest and the final aphasiologic assessment (3 months after the stroke). This might suggest that recovery from aphasia in these cases depended on appropriate perfusion in the unaffected hemisphere (Demeurisse et al. 1982).

In the present work we analyze the prognostic value of rCBF responses to a linguistic task performed during the maximum recovery period (the first 3 months after stroke) (Sarno and Levita 1971; Kertesz and McCabe 1977; Demeurisse et al. 1980).

Material and Method

Thirty right-handed patients (mean age 68; range 35 – 84) with a left hemisphere infarction (demonstrated by CT scan) were studied at 3 and 10 weeks after the stroke. The type of aphasia was noted and the severity of the language disorders was assessed by means of quantitative scales exploring verbal expression and verbal comprehension separately (Demeurisse et al. 1979). The rCBF measurement method (xenon 133 inhalation) has been described previously (Capon et al. 1981; Demeurisse et al. 1983). The flow values were expressed as initial slope index. The rCBF measurement during the functional task (naming of familiar objects presented in the visual fields) took place 30 min after a first measurement made at rest.

In normal right-handed subjects, this test induced rCBF increases on the right side in the parieto-occipital region and on the left side at the parieto-temporo-occipital junction, in the central region, and in the vicinity of the supplementary motor area and of Broca's area.

Service de Revalidation Neurologique – Hôpital Universitaire Brugmann,
Place A. Van Gehuchten 4, B-1020 Brussels

Cerebral Blood Flow and
Metabolism Measurement.
Eds. Hartmann/Hoyer
© Springer-Verlag Berlin Heidelberg 1985

Results

In all groups of patients (Broca's, Wernicke's, and nominal aphasias) the activation patterns changed in the course of time (Figs. 1 – 3).

In the group of patients with the best prognosis (those with nominal aphasia), the rCBF increases were obviously more important and widespread than in the control group on the left side at 3 weeks, and in both hemispheres at 10 weeks (Fig. 1).

In Wernicke's aphasia, our group of patients who had a poor evolution also had weak activation patterns at 3 weeks. At 10 weeks their patterns resembled those observed in normal subjects on the left side but were slightly more extended on the right side (Fig. 2).

The patients with Broca's aphasia were divided into two groups according to the severity of their verbal expression disorders at the end of the observation period. The activation patterns in the bad prognosis group resembled those observed in our sample of patients with Wernicke's aphasia, whereas in the good prognosis group they resembled those obtained in nominal aphasia.

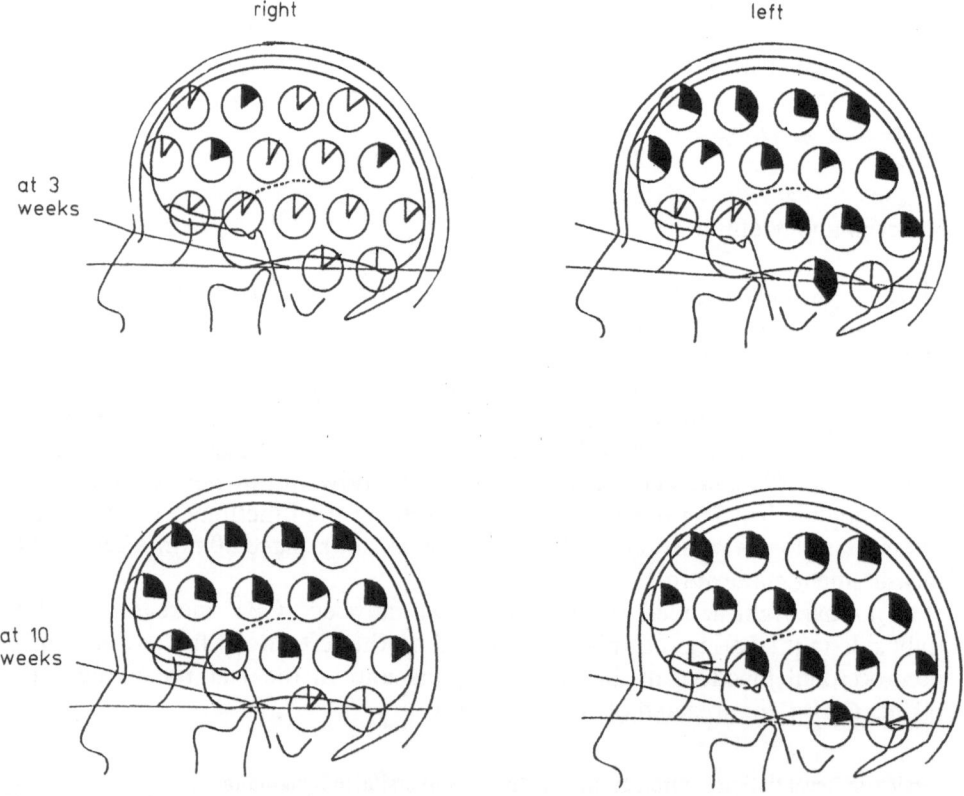

Fig. 1. Nominal aphasias ($n = 7$): rCBF changes during an object naming test at 3 and 10 weeks after stroke. A clock symbol of 90° represents a 15% variation (darkened when statistically significant at $P \le 0.05$)

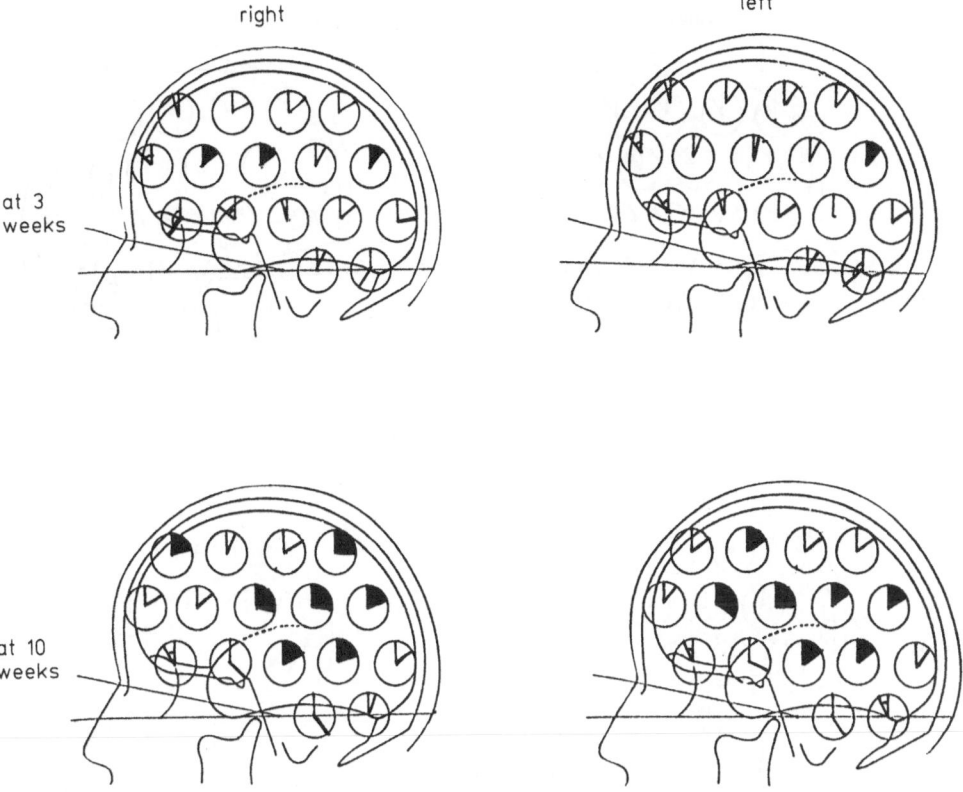

Fig. 2. Wernicke's aphasias ($n = 7$): rCBF changes during an object naming test at 3 and 10 weeks after stroke. A clock symbol of 90° represents a 15% variation (darkened when statistically significant at $P \leq 0.05$)

Figure 3 represents the differences between the activation patterns of the two groups of patients with Broca's aphasia. At 3 weeks, the flow increases were more important in the patients with a good prognosis but the observed differences were not statistically significant. At 10 weeks, the rCBF increases in the good prognosis group were greater, particularly in the left upper frontal and parietal areas (Fig. 3).

Conclusion

In aphasic stroke patients, rCBF measurements performed during a linguistic task allowed us to distinguish those patients with a good prognosis: they had important and widespread increases in flow in their left hemisphere. Moreover, in Broca's aphasia, marked regional increases in flow were observed (particularly in the upper frontal and parietal regions of the left hemisphere) in the patients with a good prognosis, but not in the others.

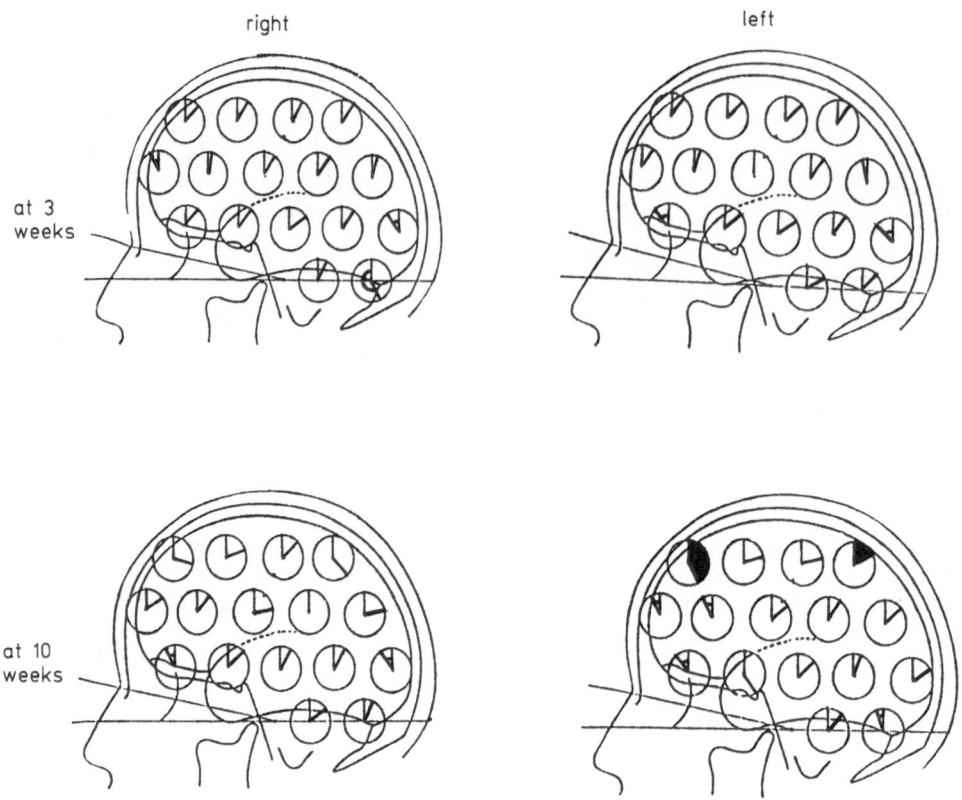

Fig. 3. Broca's aphasias ($n = 16$): Differences between the rCBF changes during an object naming test in more and less disabled patients at 3 and 10 weeks after stroke. A clock symbol of 90° represents a 15% difference (darkened when statistically significant at $P \leq 0.05$)

References

Capon A, Verhas M, Demeurisse G (1981) Evolution of blood flows in either hemispheres during the first three months after a stroke. In: Drugs and methods in CVD. Pergamon Press, New York Oxford Toronto Sydney Frankfurt Paris, p 455–459

Demeurisse G, Demol O, Robaye E, Coekaerts MJ, De Beuckelaer R, Derouck M (1979) Quantitative evaluation of aphasia resulting from a cerebral vascular accident. Neuropsychologica 17:55–65

Demeurisse G, Demol O, Derouck M, De Beuckelaer R, Coekaerts MJ, Capon A (1980) Quantitative study of the rate of recovery from aphasia due to ischemic stroke. Stroke 11:455–458

Demeurisse G, Verhas M, Capon A, Demol O, Strul S (1982) Apports du CTscan et des mesures des débits sanguins cérébraux au pronostic revalidatif de l'infarctus cérébral. Congrès Med. Phys. 47–50

Demeurisse G, Verhas M, Capon A, Paternot J (1983) Lack of evolution of the cerebral blood flow during clinical recovery of a stroke. Stroke 14:77–81

Demeurisse G, Verhas M, Capon A (1984) Resting CBF sequential study during recovery from aphasia due to ischemic stroke. Neuropsychologia 22:241–246

Kertesz A, McCabe P (1977) Recovery patterns and prognosis in aphasia. Brain 100:1–18

Sarno MT, Levita E (1971) Natural course of recovery in severe aphasia. Arch. phys. Med. 52:175–178, 186

The Detection of Cerebral Ischemia Using Xenon 133

J. K. FARRAR

Introduction

A major problem with the xenon 133 clearance methods has been their inability to reliably measure cerebral blood flow in ischemic regions — the "look-through" phenomenon. Relatively little xenon enters the ischemic area and the higher concentrations and normal clearance rate in the surrounding tissues tend to mask the region of reduced flow. Under these conditions, both stochastic and compartmental curve analysis overestimate the true flow rate. These methods also require that the tissue partition coefficient is known, which adds further uncertainty to the calculated flow value.

Both Risberg and Ingvar (1972) and Blauenstein et al. (1978) have suggested that the initial distribution of the isotope can be used as a relative measure of mean cerebral blood flow although the reliability of this technique in areas of focal ischemia has not been examined in detail. The following analysis suggests that a similar approach applied to the individual flow components may provide additional information about the flow rate and volume of ischemic regions and that the area of the clearance curves may be used to detect changes in the partition coefficient.

Theory and Methods

The peak count rate (P) resulting from an arterial injection of short duration (t) may be approximated by

$$P = (S_d/W_d) \, C_a \, t \sum_{i=1}^{n} W_i \, f_i \tag{1}$$

where S_d = detector sensitivity (relating count rate to concentration), W_d = weight or volume of tissue seen by detector d, C_a = arterial concentration, W_i and f_i are the weight and flow rate of tissue compartment i, and n is the number of tissue compartments. If a multidetector system is used, the dependence on arterial concentration and time of saturation (which change from one study to

Department of Clinical Neurological Sciences, University Hospital, P.O. Box 5339,
Terminal "A" London, Ontario N6A 5A5, Canada

Cerebral Blood Flow and
Metabolism Measurement.
Eds. Hartmann/Hoyer
© Springer-Verlag Berlin Heidelberg 1985

the next) can be removed by dividing the value obtained from an individual detector by the average peak count rate of all detectors. If the detectors are matched in terms of sensitivity (S_d = constant) and all view an equal volume of tissue (W_d = constant), then

$$P_d/P_{av} = (\sum W_i f_i)_d / (\sum W_i f_i)_{av} \tag{2}$$

which is the ratio of the mean flow in tissue volume d/average mean flow. Variations of the above formulas were used by Risberg and Ingvar (1972) to derive a multibolus injection technique and by Blauenstein et al. (1978) as an index of "total flow" for use with inhalation studies. This flow index will not be affected by the "look-through" artifact since the peak count rate will decrease in direct proportion to the flow rate and volume of the ischemic region. It has the additional benefit of being independent of changes in partition coefficient. A similar method of analysis may be applied to the area under the clearance curves. This area is given by

$$A = (P_i \lambda_i / f_i) \tag{3}$$

where λ_i is the partition coefficient of tissue compartment i. Substituting for P_i according to equation (1) and dividing by the average curve area gives

$$A_d/A_{av} = (\sum W_i \lambda_i)_d / (\sum W_i \lambda_i)_{av} \tag{4}$$

which is the ratio of the mean partition coefficient of tissue volume d/average mean partition coefficient and is independent of the flow rate.

Further information concerning the type and volume of tissue which has become ischemic may be obtained by analyzing the peaks and areas of the individual flow components obtained by compartmental analysis of the clearance curves since

$$(P_i)_d/(P_i)_{av} = (W_i f_i)_d / (W_i f_i)_{av} \tag{5}$$

and

$$(A_i)_d/(A_d)_{av} = (W_i \lambda_i)_d / (W_i \lambda_i)_{av} \tag{6}$$

It is important to note that W_i represents the actual weight (or volume) of tissue component i and is not the mathematical weighting factor in common usage. Thus, it is possible to obtain an estimate of the "absolute" weight of a particular tissue component. This technique may also be used to analyze curves obtained by inhalation or intravenous injection methods although the parameters P_i and K_i calculated during curve-fitting (Obrist et al. 1975) must be used in lieu of the measured curve peak and area (the measured values deviate substantially from those which would be obtained by an equivalent arterial injection due to the prolonged input function). The parameter P_i is substituted for the peak count rate of compartment i and the area is calculated as P_i/K_i.

Since most centers now use the inhalation technique, the theoretical accuracy of the peak and area analysis was tested on computer-simulated clearance curves generated using the arterial input function of a typical inhalation CBF measurement. The curves were constructed as described by Obrist et al. (1975); however, a fourth compartment was added to represent the ischemic region where appropriate, as outlined below. The normal values used for the gray matter, white matter, and extracerebral compartments were as follows: $K1 = 1.0$, $W1 = 0.4$, $\lambda1 = 0.8$; $K2 = 0.14$, $W2 = 0.4$, $\lambda2 = 1.5$; and $K3 = 0.35$, $W3 = 0.2$, $\lambda3 = 1.5$. "Detector" sensitivity was adjusted to give a peak count rate of approximately 1000 counts/s under normal flow conditions and was held constant. Random noise was added to the resulting curves, which were then analyzed using Obrist's method to obtain P1, P2, K1, and K2. Fifty curves were generated for each test condition and the mean, standard deviation, and percentage change from control values were calculated.

Results

Increasing Severity of Ischemia

The effects of increasing severity of ischemia were tested by assigning 50% of the gray matter compartment to the ischemic region (i.e., $W4 = W1 = 0.2$), and flow in the ischemic compartment (F4) was reduced from 80 to 0 ml/100 g/min in increments of 10%. The true weighted mean flow (gray matter + white matter + ischemic region) decreased by 4% at each increment. The total curve peak (P1 + P2) decreased from $100 \pm 1.4\%$ (mean $\pm$ SD) to 96.4 ± 1.5 with the initial decrease in F4 and progressively thereafter. The difference between the true mean flow and the total curve peak never exceeded 1.2%. In contrast, the mean flow as calculated by conventional analysis $(W1 \times F1 + W2 \times F2)$ consistently underestimated the true flow reduction by 8%–10% during severe ischemia $(F4 \leq 50\%\ F1)$. The total curve area (A1 + A2) remained constant to within 3%, indicating no change in mean partition coefficient. The gray matter flow rate $(F1 = 0.8\ ^{*}K1)$ decreased initially from $100 \pm 2.2\%$ to $82.2 \pm 2.9\%$ as F4 was reduced by 50%. This decrease was due to averaging of the F1 and F4 flow compartments by the curve-fitting algorithm. As flow in the ischemic region was decreased below 50%, F1 gradually returned to control levels. The ischemic compartment merged with the slow flow components and was overlooked by this type of analysis. The fast component peak (P1) and area (A1) decreased progressively to approximately $50 \pm 2\%$ as the weight of tissue with flow rate F1 decreased. The decline in P1 and A1 was nonlinear due to mixing of the F1, F2, and F4 flow components (i.e., forcing a two-component fit). As the ischemic compartment began to merge with the slow flow, P1 tended to underestimate the residual flow rate. This situation represents a "worst case" in terms of curve-fitting and yet the decline in both P1 and A1 clearly indicated the development of ischemia and the volume of gray matter involved.

Increasing Volume of Ischemia

The effects of an increasing volume of ischemic tissue were tested by setting the clearance constant of the ischemic region equal to that of white matter (K4 = K2) and its weight (W4) was increased from 0 to 0.36 while the weight of the gray matter compartment was reduced accordingly (i.e., W1 decreased from 0.4 to 0.04). The initial volume of ischemia (10% of W1) caused a significant reduction in total curve peak (to 93.5 ± 1.4%), P1 (to 90.4 ± 1.3%), and A1 (to 90.2 ± 1.4%), and all three parameters declined linearly thereafter. For all simulations, the total curve peak gave an accurate estimate of true mean flow to within ± 2% and the changes in P1 and A1 showed the reduction in gray matter weight to within ± 3%. The estimate of mean partition coefficient (A1 + A2) remained constant, indicating that the total tissue volume was unchanged. As expected, the conventional flow parameters demonstrated the "look-through" phenomenon. The weighted mean flow consistently underestimated the true flow reduction, and the gray matter flow rate (F1) was unchanged. The standard deviation of F1 increased markedly as the compartment volume decreased.

Alterations in Gray Matter Partition Coefficient

To test the stability of the method with respect to changes in λ, the gray matter partition coefficient was varied from 0.4 to 1.3 while all other parameters were held constant. The conventional flow indices (calculated assuming a normal value for λ) overestimated the true flow rate at reduced values of $\lambda 1$ and underestimated flow when $\lambda 1$ was high (at $\lambda 1 = 0.4$, F1 = 196 ± 8.5%, and mean flow = 127 ± 5.5%; at $\lambda 1 = 1.3$, F1 = 62 ± 1.3%, and mean = 78 ± 1.5%). The curve peak indices (P1, P2, and P1 + P2) all remained within ± 3% of the control value, correctly indicating that flow was unchanged. The area of the slow flow component was constant throughout while A1 and A1 + A2 increased linearly with increasing $\lambda 1$. The difference between the true partition coefficient and that predicted by area analysis never exceeded 2%.

Summary and Conclusions

These data indicate that the total curve peak provides an accurate estimate of mean flow under all conditions tested and is not affected by "look-through," "slippage," or changes in λ. In addition, more specific information about the volume and type of tissue which has become ischemic and/or undergone a change in partition coefficient may be obtained by the expanded peak and area analysis described above. Limitations and uses of the method will be described in subsequent communications.

References

Blauenstein UW, Halsey JH, Wilson EM, Wills EL (1978) 133-Xenon inhalation method: significance of indicator maldistribution for distinguishing brain areas with impaired perfusion. Stroke 9:57−66

Obrist WD, Thompson HK, Wang HS, Wilkinson WE (1975) Regional cerebral blood flow estimated by 133-xenon inhalation. Stroke 6:245−256

Risberg J, Ingvar DH (1972) Multibolus technique for measuring the distribution of cerebral blood flow over short intervals in man. Circ Res 31:889−898

Acknowledgment. This work was supported by the Ontario Heart Foundation.

Intravenous Xenon 133 rCBF Measurement with a New Mobile System

T. Schroeder, P. Holstein, and H. C. Engell

Introduction

Measurement of CBF with 133 xenon is currently being introduced into daily clinical practice as a diagnostic tool, as guidance for therapy, and as a predictor of prognosis (Lassen 1982). This is partly due to improvements in the noninvasive techniques, and also to the compactness of the equipment manufactured today, which makes the apparatus fully mobile and thus available in every clinical situation.

The present paper describes the preliminary results with such a mobile system – the Novo Cerebrograph – for measurement of cerebral blood flow using the intravenous xenon 133 method.

Materials and Method

The Cerebrograph is manufactured in two versions, equipped with either two or ten collimated brain scintillation detectors, in addition to an air sample detector. The detectors, being $3/4'' \times 3/4''$ NaI crystals, are integrated in a complete system with channel analyzers, a data collection unit, a microprocessor, and a matrix printer. The system is placed on a trolley which also carries an air sample pump and a trap for exhaled xenon.

The two-detector system, *the Cerebrograph 2a,* uses the classic bicompartmental analysis with a delayed start-fit time, as first suggested by Obrist. The flow values, expressed as gray matter flow (F1) and initial slope index (ISI), are calculated automatically by the incorporated computer, the ISI being determined between 0.5 and 1.5 min, as proposed by Risberg.

In addition to the extra detectors, *the Cerebrograph 10a* is provided with a new analytic model implemented in the time domain as developed by Prohovnik et al. Moreover the 10a system has an output for connection to an external computer, permitting use of alternative algorithms.

The reproducibility of each of the two systems was examined using double measurements within 2 h. Each group consisted of 12 patients admitted with carotid artery disease. The patients were examined awake in the supine position.

Section of Vascular Surgery and Laboratory of Surgical Circulation Research, Department D, Rigshospitalet, University of Copenhagen, DK-Copenhagen

Cerebral Blood Flow and
Metabolism Measurement.
Eds. Hartmann/Hoyer
© Springer-Verlag Berlin Heidelberg 1985

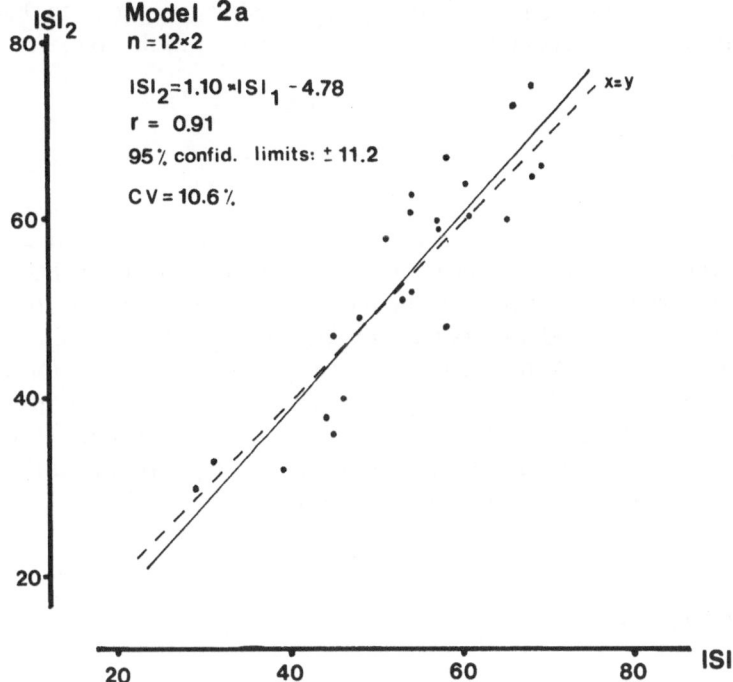

Fig. 1

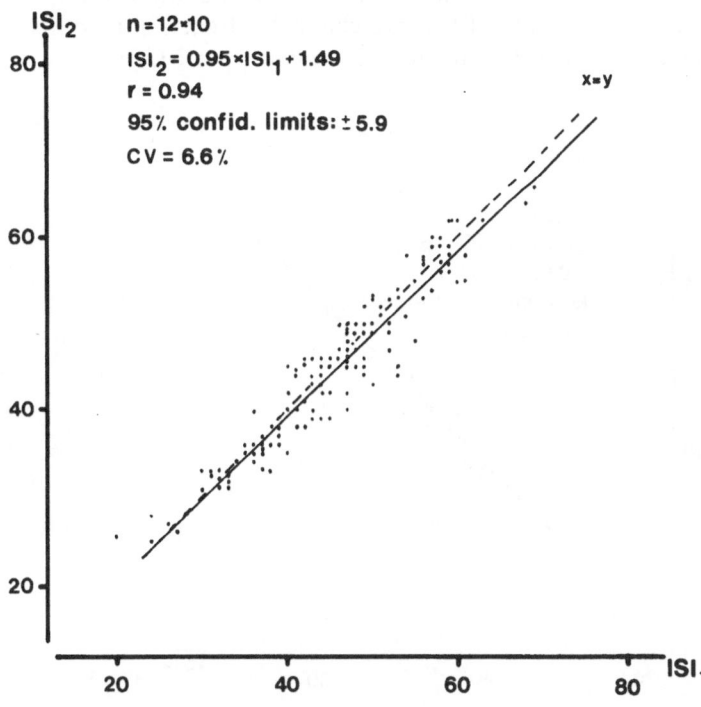

Fig. 2

In both groups the reproducibility of the ISI was superior to that of the F1; thus only the results concerning the ISI are presented in this paper.

Results

With *the Cerebrograph 2a* using the Obrist model we found a coefficient of variation (CV) for test-retest measurements of 10.6% (Fig. 1).

Corresponding values for the individual detectors in *the 10a system* using the time domain model averaged 6.6% (range 6% − 8%) (Fig. 2). The coefficient of variation of the hemispheric mean values was as low as 5.7% (Fig. 3). Thus the test-retest reproducibility lay within ± 5.9 for the individual detectors and within ± 5.2 for the hemispheric mean (95% confidence limits) (Figs. 2, 3). Moreover we analyzed the difference in hemispheric mean flow and found the 95% reproducibility interval to be within ± 2.8.

Discussion

Different indices have been suggested for calculating the CBF using the noninvasive xenon 133 technique. Our results with the Novo Cerebrograph demonstrated inadequate reproducibility with the two-detector model using Obrist analysis. The better results obtained with the ten-detector system could be due to the new algorithm used, i.e., the time domain analysis. The possibility of averaging the data of the five channels of each hemisphere will further reduce the error in determining the mean hemispheric CBF.

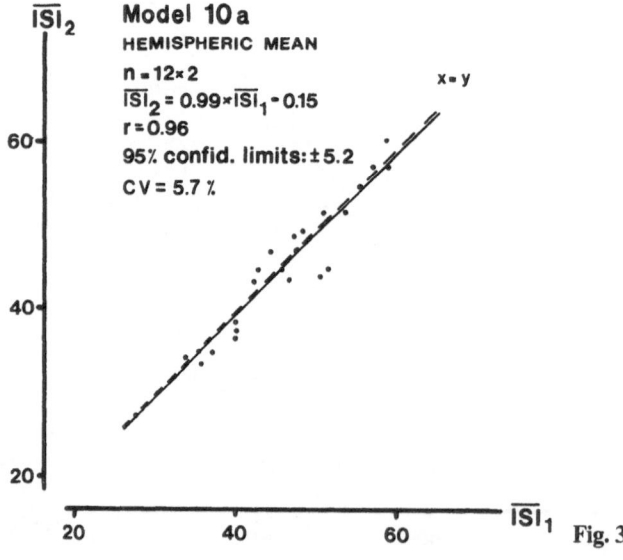

Fig. 3

Using the 10a system we have performed sequential perioperative CBF measurements in a patient undergoing carotid endarterectomy, as detailed below.

Case History

The patient, a 55-year-old male, presented with paresthesia of the left arm, and the neurologic examination revealed a dysdiadochokinesia of the arm and hand only. The CT scan was without any sign of infarction. As the angiography showed occlusion of the right and severe stenosis of the left internal carotid artery, a left-sided endarterectomy was decided upon. At operation a gradient of 65 mmHg over the stenosis was measured. The stump pressure during clamping of the common carotid artery was 18 mmHg, and as flattening of the continuously monitored EEG was registered a Javid shunt was inserted. Marked hyperperfusion was demonstrated immediately following endarterectomy, reaching 200% of the preoperative resting values 2–4 h postoperatively (see Fig. 4). During the following days the flow subsided to near baseline values. The neurologic condition of the patient remained stable, but on the second day he developed headache and confusion concomitantly with a marked rise in systolic blood pressure. Immediate treatment with antihypertensive drugs normalized the blood pressure as well as the clinical condition. The patient was discharged after 10 days without further events.

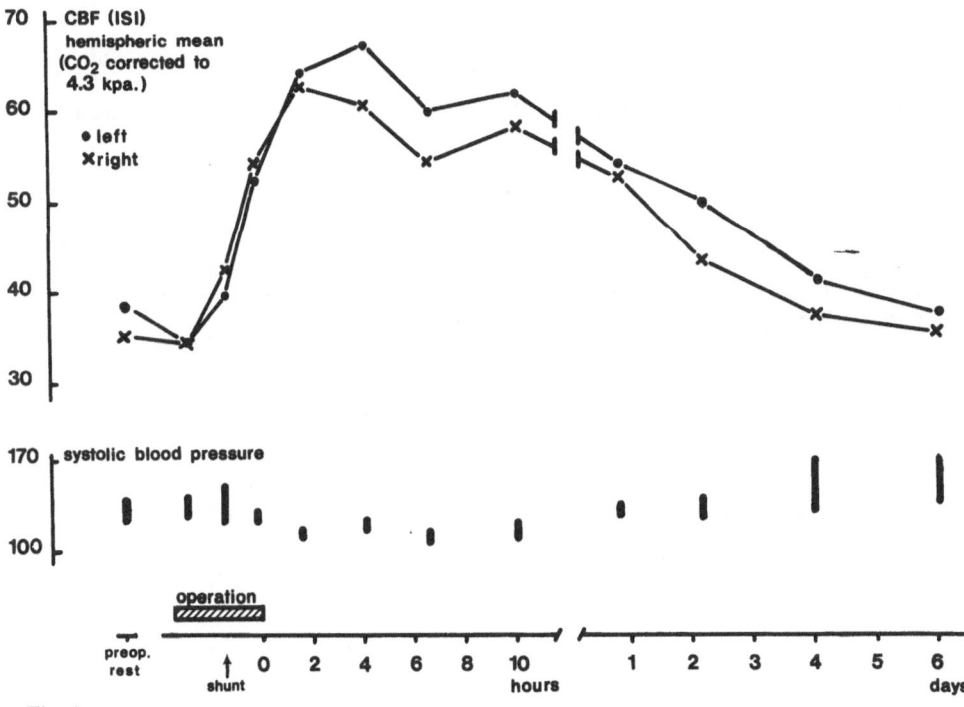

Fig. 4

Cerebral hyperperfusion following restoration of normal perfusion pressure to a vascular bed with impaired autoregulation has been described previously (Sundt 1983). Assuming hyperperfusion may lead to cerebral edema and hemorrhage, it could be a major factor in the development of neurologic deficits postoperatively. However, as former CBF measurements have been performed either immediately following surgery or several days after the operation only, the course and significance of postendarterectomy hyperperfusion have not previously been outlined. Neurologic complications related to hyperperfusion may be prevented by meticulous control of systemic blood pressure as guided by CBF measurements in selected patients.

Conclusion

The described mobile ten-detector system for intravenous CBF measurement has been proven reliable and easy to handle in bedside examinations as well as in perioperative measurements. The apparatus might be a valuable tool in carotid surgery for monitoring CBF in selected patients postoperatively.

References

Lassen NA (1982) Measurement of cerebral blood flow and metabolism in man (Editorial). Clinical Science 62:567–572

Obrist WD, Thompson KH, Wang HS, Wilkinson WE (1975) Regional cerebral blood flow estimated by 133-xenon inhalation. Stroke 6:245–256

Prohovnik I, Knudsen E, Risberg J (1983) Accuracy of models and algorithms for determination of fast compartment flow by non-invasive 133 xenon clearance. In: Functional radionuclide imaging of the brain, edited by Magistretti PL. Raven Press. New York, 87–115

Risberg J (1980) Regional cerebral blood flow measurement by 133 xenon-inhalation: Methodology and applications in neuropsychology and psychiatry. Brain and language 9:9–34

Sundt TM (1983) The ischemic tolerance of neural tissue and the need for monitoring and selective shunting during carotid endarterectomy. Stroke 14:93–98

Rheological Determinants of Cerebral Blood Flow [1]

E. Ott, F. Fazekas, G. Bertha, H. Valetitsch, and H. Lechner [2]

Several factors have been identified as reducing cerebral blood flow (CBF) following cerebral infarction, among which the role of blood viscosity has attracted increasing interest during the past few years. An increase of whole blood viscosity may be produced by several factors, such as high hematocrit (hct), enhanced aggregation of platelets (PAG) and of red cells (RCA), loss of red cell elasticity (RCE), and an increase of the fibrinogen content. Likewise, a close correlation between whole blood viscosity and CBF has been established in normals as well as in cerebrovascular disease (CVD).

The present paper reports the importance of the whole blood viscosity parameters for alteration in CBF in patients with CVD. The results which will be presented are based on investigations in different groups of patients with CVD and in part have been reported previously.

Patients and Methods

Determinations of whole blood viscosity parameters were carried out in a total of 199 patients (100 males, 99 females) with a mean age of 64 years who suffered from CVD with persistent and/or reversible neurologic deficit. The clinical diagnosis was confirmed by axial computerized tomography in most of the patients. Rheological investigations were performed upon venous blood which was drawn from an antecubital vein without stasis and consisted of determinations of whole blood viscosity (Contraves Low Shear 30), of hematocrit (Wintrobe technique), of red cell filterability (Reid and Dormandy), of platelet aggregation (Breddin et al. 1975), and of fibrinogen (standard method).

Cerebral blood flow (CBF) was measured by the noninvasive xenon clearance technique (Obrist et al. 1975) and the clearance of the isotope was followed by 20 collimators placed over both hemispheres following intravenous injection of $15-20$ mCi xenon 133.

End-tidal pCO_2 ($pECO_2$) was continuously measured and peripheral blood pressure was controlled repeatedly throughout the CBF measurements.

---

1 This paper is dedicated to Prof. Dr. H. Schurz, Chairman of the Institute for Physical Chemistry, University of Graz, on the occasion of his 60th birthday

2 Universitäts-Nervenklinik, Abt. Neurologie und Psychiatrie, Auenbruggerplatz 22, A-8036 Graz

Cerebral Blood Flow and
Metabolism Measurement.
Eds. Hartmann/Hoyer
© Springer-Verlag Berlin Heidelberg 1985

The results thus obtained have been compared with those obtained in 83 healthy volunteers (controls) with a similar age and sex distribution.

Results

Blood Viscosity

As reported previously, the average whole blood viscosity was significantly elevated at all shear rates investigated (230, 46, 23, 15 s^{-1}) in 104 patients with cerebral infarction. Individually, whole blood viscosity was increased in 48% of the patients. Moreover, calculations in 38 patients showed a significant correlation between mean hemispheric CBF and values of blood viscosity ($r = -0.468$).

Hematocrit (Hct)

In 104 patients with cerebral infarction Hct was determined and correlated with blood viscosity (Ott et al. 1979); Table 1 shows that in these patients Hct was found to be significantly higher (relative polycythemia) when blood viscosity was elevated. In 25 of these patients Hct was greater than 0.5. The correlation with CBF reached significance ($P < 0.05$) at a correlation coefficient of $r = -0.43$.

Table 1. Hematocrit in the presence of normal and increased blood viscosity in 104 patients with CVD (after Ott et al. 1979)

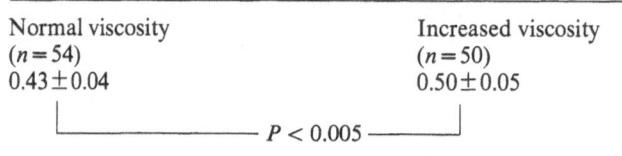

Normal viscosity ($n = 54$)	Increased viscosity ($n = 50$)
0.43 ± 0.04	0.50 ± 0.05

$P < 0.005$

Dependence of CBF on Hct

The dependence of CBF on Hct can be demonstrated in patients with CVD and with evidence of relative polycythemia undergoing phlebotomy and/or isovolemic hemodilution (IHDL). Table 2 displays the effect of lowering Hct on CBF and on viscosity-related parameters in 15 patients with CVD.

Fibrinogen

Fibrinogen levels were determined in a total of 148 patients with CVD (Ott et al. 1979) and compared with those obtained in 83 healty volunteers. Table 3

Table 2. Effect of IHDL on CBF and on viscosity-related parameters

	IHDL	
	Before	After
Blood viscosity (mPas) (11 sec^{-1})	11.9 ± 2.7	9.5 ± 2.4
	└── $P < 0.01$ ──┘	
Haematocrit (%)	0.51± 0.06	0.46± 0.07
	└── $P < 0.01$ ──┘	
Fibrinogen (mg/dl)	465 ±117	447 ±105
CBF$_{15}$ (ml/100 g/min)	29.9 ± 4.9	35.4 ± 8.0
	└── $P < 0.01$ ──┘	

Δ pECO$_2$ = – 2.0 mmHg Δ MABP = – 2 mmHg

Table 3. Mean fibrinogen levels in 148 patients with CVD (after Ott et al. 1979)

	Controls ($n = 83$)	TIA ($n = 44$)	Definite stroke ($n = 104$)
Fibrinogen (mg/dl)	307±41	368±59	416±93
	└── $P < 0.05$ ──┘	└── $P < 0.001$ ──┘	
	└──────── $P < 0.005$ ────────┘		

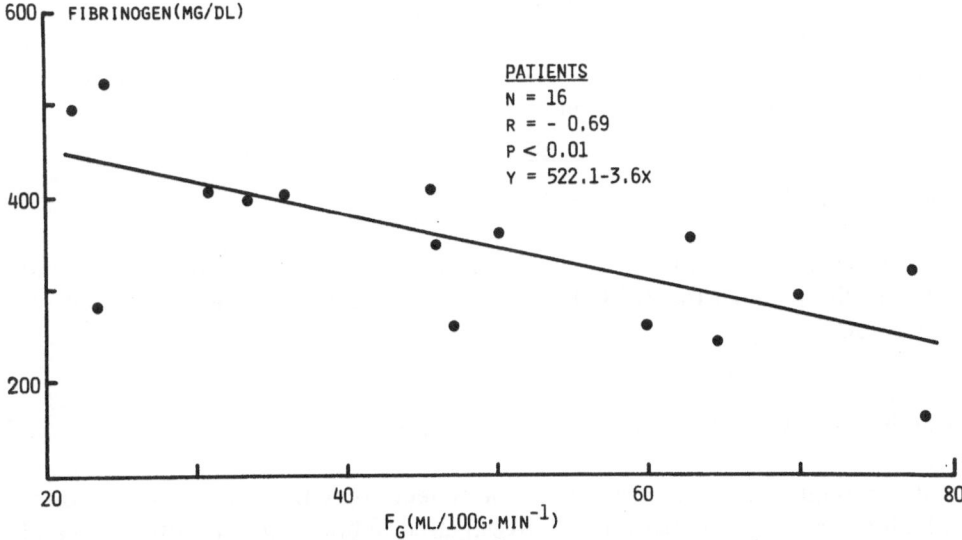

Fig. 1. Correlation of CBF with fibrinogen in 16 patients with CVD

displays that in relation to the control subjects mean fibrinogen concentrations were significantly higher in both patients with TIA and those with definite stroke. Individually fibrinogen levels greater than 400 mg/dl were noted in 51% of patients with definite stroke and in 30% of patients with TIA.

In the patients with CVD investigated, fibrinogen levels greater than 400 mg/dl were more frequent when whole blood viscosity was increased and the correlation coefficient ($r = 0.38$) reached a level of significance.

Dependence of CBF on Fibrinogen

The correlation of the fibrinogen data with the corresponding CBF in 11 controls revealed a correlation coefficient $r = -0.33$, while the correlation of fibrinogen with CBF in 16 patients revealed a correlation coefficient $r = -0.69$ and reached a level of significance (Fig. 1).

Platelet Aggregation (PAG)

The frequency of spontaneous PAG was greater in patients with completed stroke than in patients with TIA, and the frequency of spontaneous PAG was also different when compared with a control population (Table 4) (Ott et al. 1979).

The correlation of PAG with blood viscosity as well as with CBF was rather small ($r = 0.21$ and -0.31, respectively).

Red Cell Filterability (RCF)

Red cell filterability (RCF) was determined in 36 patients with completed stroke. As can be seen from Table 5, RCF was significantly lower in patients than in healthy volunteers; however, dependence of RCF on the fibrinogen content of the plasma was also evident.

Dependence of CBF on RCF

As can be seen from Fig. 2, there was a close correlation between CBF and RCF in the patients ($r = 0.58$) but the correlation was less good in controls ($r = 0.43$).

Correlation of Blood Viscosity Factors with CBF

Table 6 summarizes the correlation coefficients of CBF and blood viscosity with the blood viscosity parameters investigated in both patients with CVD and in controls.

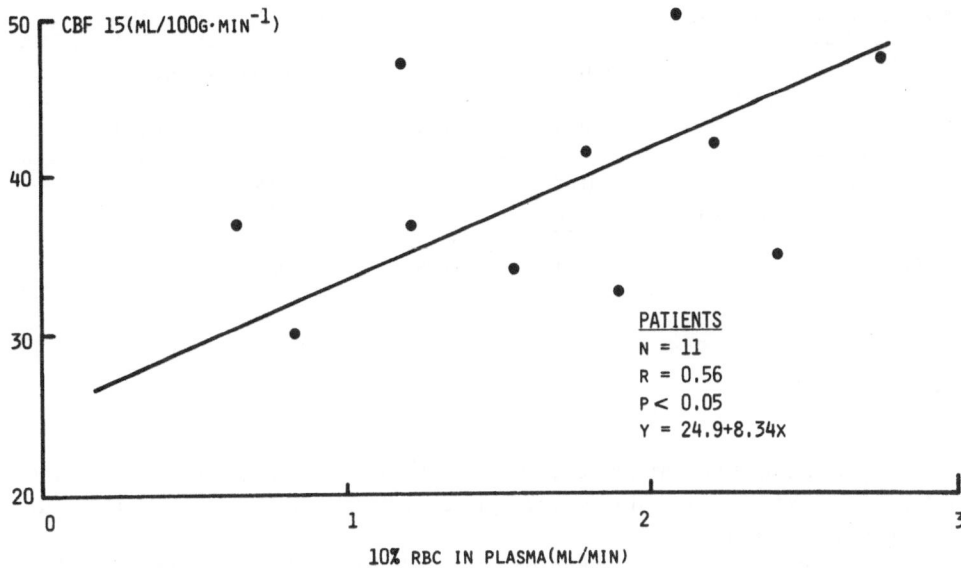

Fig. 2. Correlation of CBF with the filtration rate of a 10% red cell suspension (10% RBC) in 16 patients with CVD. Mean fibrinogen level: 353 mg/dl

Table 4. Frequency of spontaneous PAG in 148 patients with CVD (after Ott et al. 1979)

	Controls ($n = 83$)	TIA ($n = 44$)	Completed stroke ($n = 104$)
No evidence of spontaneous PAG	62/83	28/44	57/104
Evidence of spontaneous PAG	21/83	16/44	47/104

Table 5. Red cell filterability (RCF) in 36 patients with CVD versus 10 controls

	Controls ($n = 10$)	Patients ($n = 36$)
RCF (ml/min)	2.6 ± 0.5	1.2 ± 0.6
	$P < 0.01$	
Fibrinogen (mg/dl)	243 ± 10	349 ± 98
	$P < 0.01$	

Nucleopore filter $5\ \mu m$, $\Delta P = 50\ mmH_2O$, $37\,^{\circ}C$, 10% red cell suspension (RBC) in plasma

Table 6. Matrix of correlation coefficients of pairs of parameter in patients and controls

	CBF		Blood viscosity	
	Patients	Controls	Patients	Controls
Hematocrit	− 0.43	− 0.49	0.89	0.99
RCF (10% RBC in plasma)	0.58	0.43	− 0.69	− 0.59
Fibrinogen	− 0.69	− 0.33	0.38	0.47
PAG	− 0.31	− 0.08	0.21	0.11

Discussion

This paper again stresses the evidence of impaired rheological properties in patients with CVD and the importance of several related parameters for both blood viscosity and CBF.

Impairment of rheological behavior of the blood can be found in more than 40% of patients with CVD, being measurable as increased blood viscosity, and the factors responsible for this elevation of blood viscosity have been investigated recently.

In a previous paper we have established dependence of CBF on blood viscosity, and Grotta et al. demonstrated that both Hct and fibrinigen have to be considered important viscosity parameters influencing CBF. Their findings have been confirmed by the present investigation but in addition it has now been established that the filterability of the erythrocytes (RCF) must also be taken into consideration in both viscosity and CBF determinations.

If red blood cells are suspended in autologous plasma and filtered through a micropore filter with a pore diameter of 5 μm within 1 min the passing volume of red cells is dependent on the Hct, the elasticity of the red cells (RCE), and the plasma proteins, i.e., fibrinogen. Since the experimental conditions used in our investigation (Hct 0.1, $\Delta p = 50 \, \text{mmH}_2\text{O}$) resemble very closely a condition likely to exist in the microcirculatory bed, conclusions about flow conditions in the microvasculature may very well be drawn from the present results. Hence, it is not surprising that the correlation coefficient between RCF and CBF in our population sample amounted to nearly 0.6.

In conclusion it can be seen from the results presented in this paper that Hct, RCF, and fibrinogen have to be considered the most important parameters of both blood viscosity and CBF in patients as well as in healthy control subjects.

Summary

Measurements of whole blood viscosity and its related parameters were performed in a total of 199 patients with CVD and in 83 healthy control subjects. The values thus obtained have been correlated with CBF measurements.

The results clearly indicate that Hct, fibrinogen, and RCF have to be considered the most important parameters of blood viscosity and CBF in both patients with CVD and controls.

References

Breddin K, Grun H, Krzywanek H et al. (1975) Zur Messung der spontanen Thrombocytenaggregation. Klin. Wschr. 53:81–89

Grotta J, Ackermann R, Correia J et al. (1982) Whole blood viscosity parameters and cerebral blood flow. Stroke 13:296–301

Obrist WD, Thompson HD, Wang HS et al. (1975) Regional cerebral blood flow estimated by 133 xenon inhalation. Stroke 6:245–256

Ott E, Lechner H (1982) Haemorheologic and haemodynamic aspects of cerebrovascular disease. Path Biol 30:611–614

Ott E, Lechner H, Aranibar A (1974) High blood viscosity syndrome in cerebral infarction. Stroke 5:330–333

Ott E, Ladurner G, Lechner H (1977) Relationship between disturbed rheological properties and cerebral haemodynamics in recent cerebral infarction. Progr Biochem 14:349–352

Ott E, Lechner H, Aranibar A et al. (1979) Impairment of rheologic conditions and cerebral blood flow in patients with cerebral vascular disease. In: Cerebral vascular disease 2. Meyer JS, Lechner H, Reivich M (eds) Excerpta Medica, Amsterdam

Reid HC, Barnes AJ, Dormandy TC (1976) A simple method for measuring erythrocyte deformability. Clin Path 29:855–858

Fibrinogen/Albumin Ratio: A Useful Indicator of Hemorheological Abnormalities in Observation of Cerebrovascular Disorders

S. Takamatsu[1], K. Satoh[1], I. Osanai[1], Y. Kawamura[1], S. Mizuno[1], M. Takamatsu[2], B. Shōji[3], and H. Yokouchi[3]

Although the importance of local microcirculatory disturbances in thrombotic disorders has been generally recognized [1, 3, 5], clinical observations from the viewpoint of hemorheology have not been made intensively because of difficulties in the detection of changes occurring in the microcirculation. A high ratio of fibrinogen to albumin accelerates the erythrocyte aggregation in capillaries which is responsible for the occurrence of thrombotic diseases [1]. This study aims to establish the usefulness of the ratio of fibrinogen to albumin for observations of cerebrovascular disorders.

Materials and Methods

The subjects were cerebrovascular patients who had had strokes 3 or more months previously and healthy adults with normal physical findings, blood pressures, urinalyses, blood counts, chest films, ECGs, and liver function tests. As hematocrit affects hemorheological behavior greatly, subjects with similar levels of hematocrit were selected for investigation.

In order to clarify the significance of the fibrinogen/albumin ratio for the pathophysiologic status of cerebrovascular disorders, its relationship to patients' capacity to perform daily activities was observed. The perfect score of this test, abbreviated as ADL-T, is 48; patients with a score of $0-24$ were mostly bedridden, and those with a score of $37-48$ could walk of their own volition. The deformability index proposed by Reid et al. [4], whole blood viscosity, and ATP content were measured and related to the fibrinogen/albumin ratio in order to clarify its hemorheological significance.

Results

Hemorheological examinations in healthy subjects by age are shown in Table 1. Whole blood viscosity and plasma viscosity were higher in the older age group

1 Department of Pathologic Physiology, Institute of Cerebrovascular Diseases, Hirosaki University School of Medicine, Hirosaki 036, Japan
2 Training Course of School Nursing, Department of Education, Hirosaki University, Hirosaki 036, Japan
3 National Sanatorium, Aomori Hospital, Aomori 039-35, Japan

Cerebral Blood Flow and
Metabolism Measurement.
Eds. Hartmann/Hoyer
© Springer-Verlag Berlin Heidelberg 1985

Table 1. Hemorheological examinations in healthy subjects

Age group	Young	Middle aged	Old
Cases	13	15	19
Mean age	21.8 ± 3.1	53.0 ± 9.6	73.7 ± 4.6
Hematocrit (%)	42.2 ± 4.7	42.0 ± 5.1	42.4 ± 3.0
Whole blood viscosity (Cp)[a]	4.4 ± 0.5	4.3 ± 0.5	6.1 ± 0.9*
Plasma viscosity (Cp)	1.34 ± 0.11	1.36 ± 0.25	1.52 ± 0.17*
Deformability index (ml/min)	1.01 ± 0.4	0.91 ± 0.1	0.73 ± 0.1*
ATP (mg/dl)	31.5 ± 4.0	24.3 ± 4.1	21.0 ± 3.6*
Fibrinogen (mg/dl)	266 ± 61	322 ± 59	378 ± 86*
Albumin (g/dl)	4.1 ± 0.3	4.1 ± 0.2	3.8 ± 0.3*
Fibrinogen/albumin ($\times 10^{-3}$)	65 ± 14.3	79 ± 13.8	101 ± 24.6*

* $P < 0.01$
[a] 37°C, D = 225.00 s^{-1}

Table 2. Hemorheological examinations in cerebrovascular patients

Subjects	Healthy	Hemorrhagic	Infarct
Cases	24	12	16
Mean age	61.3 ± 13.7	57.1 ± 7.1	61.7 ± 14.8
Hematocrit (%)	40.8 ± 4.6	38.1 ± 3.8	39.5 ± 4.0
Whole blood viscosity (Cp)[a]	4.8 ± 1.6	3.9 ± 0.4*	4.1 ± 0.5*
Plasma viscosity (Cp)	1.41 ± 0.21	1.41 ± 0.20	1.45 ± 0.18
DI (ml/min)	0.84 ± 0.14	0.57 ± 0.08*	0.54 ± 0.13*
ATP (mg/dl)	23.1 ± 4.1	20.8 ± 4.0	20.4 ± 3.6*
Fibrinogen (mg/dl)	336 ± 63	441 ± 84**	382 ± 69**
Albumin (g/dl)	3.9 ± 0.3	4.0 ± 0.8	3.6 ± 0.6*
Fibrinogen/albumin ($\times 10^{-3}$)	86 ± 24.5	113 ± 28.3*	106 ± 24.6**

* $P < 0.01$, ** $P < 0.02$
[a] 37°C, 225.00 s^{-1}

than in the younger group. The fibrinogen level was also high in the older age group. By contrast the deformability index, ATP content, and albumin level were lower in the older age group than in the younger group. The ratio of fibrinogen to albumin in the older age group was greater than in the younger group. Hemorheological findings in cerebrovascular patients were compared with those in healthy subjects of almost the same age and hematocrit levels. As shown in Table 2, whole blood viscosity in infarct patients was higher than in healthy subjects. The deformability index and albumin concentration in infarct patients were lower than in healthy subjects. In hemorrhagic patients, whole blood viscosity and fibrinogen concentration were high. There were no differences in these hemorheological parameters between infarct and hemorrhagic patients. The ratio of fibrinogen to albumin in cerebrovascular patients, both infarct and hemorrhagic, was significantly greater than in healthy subjects. The

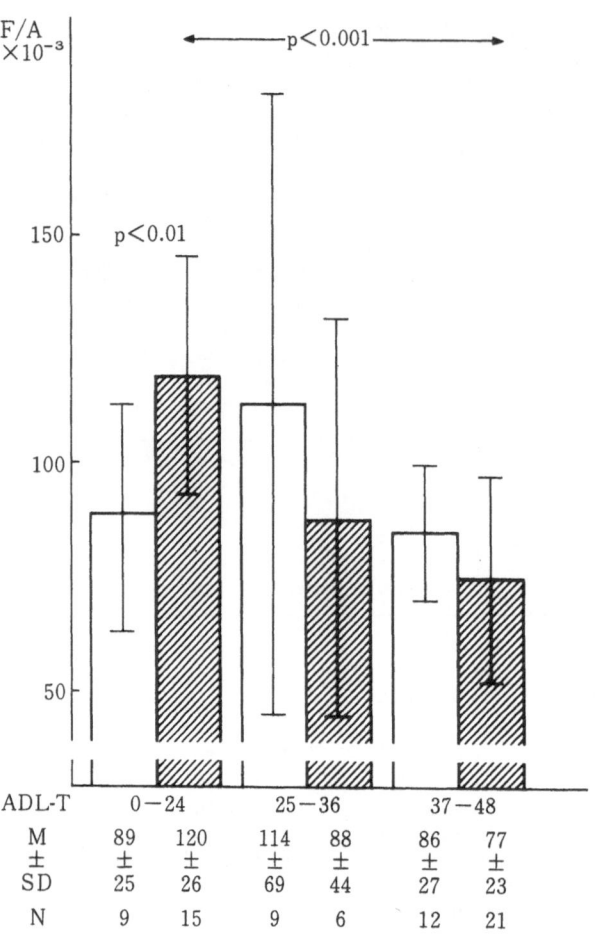

Fig. 1. The ratio of fibrinogen to albumin (F/A) and activity of daily living test (ADL-T) in cerebrovascular patients. *Open bars*: cerebral hemorrhage; *hatched bars*: cerebral infarction

ADL-T	0—24		25—36		37—48	
M ± SD	89 ± 25	120 ± 26	114 ± 69	88 ± 44	86 ± 27	77 ± 23
N	9	15	9	6	12	21

ratio of fibrinogen to albumin in infarct patients with an ADL-T score of 37—48 was significantly smaller than in those patients with a score of 0—24 (Fig. 1); in hemorrhagic patients such a relationship was not found. The significance of the ratio for hemorheological behavior was examined in healthy subjects. The ratio was directly proportional to whole blood viscosity, and inversely proportional to the deformability index and whole blood ATP (Fig. 2). In cerebrovascular patients, whole blood viscosity was not influenced by this index. The incidence of high ratios was greater than that of simple hypoalbuminemia or hyperfibrinogenemia in both healthy subjects and cerebrovascular patients (Fig. 3).

Comments

The close association between the ratio of fibrinogen to albumin and ADL-T in cerebrovascular patients indicates the former's usefulness in observations of the

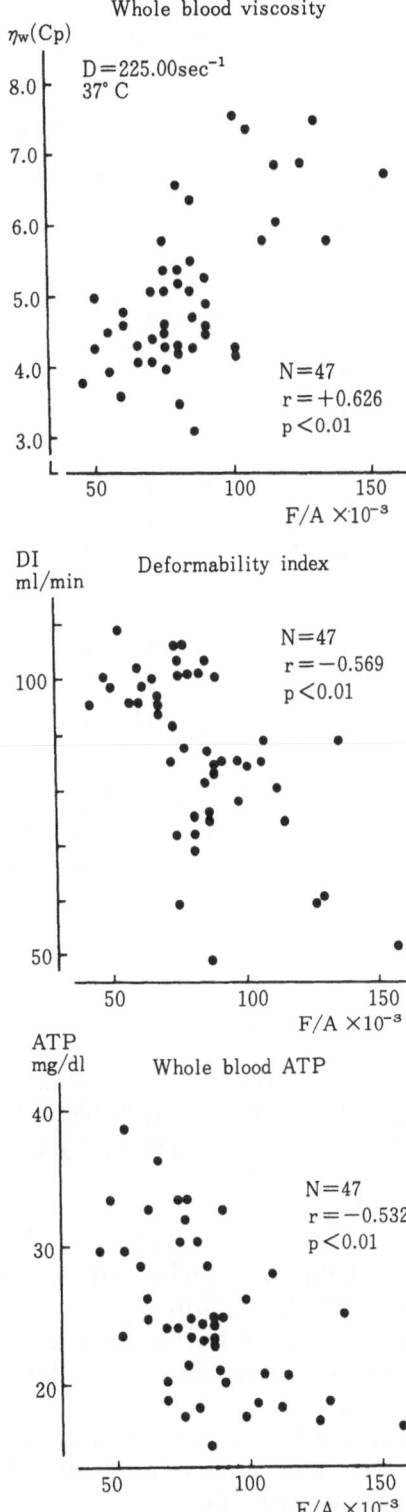

Fig. 2. Association between the fibrinogen/albumin ratio (F/A) and hemorheological examinations in healthy subjects. Results obtained from observations of whole blood viscosity by other shear rates demonstrated similar findings

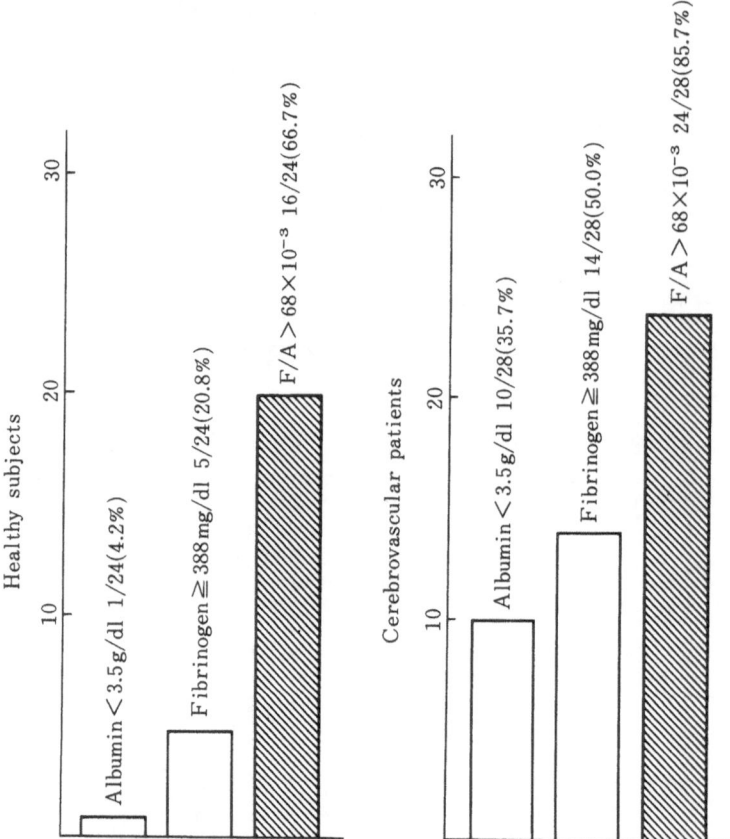

Fig. 3. Incidence of hypoalbuminemia, hyperfibrinogenemia, and the ratio of fibrinogen to albumin

pathophysiologic status of cerebrovascular disorders. Furthermore, significant relationships between the ratio and hemorheological parameters demonstrates its value in clinical observations of hemorheological behavior. The deformability index, which reflects blood filterability in the capillaries with smaller diameters, is affected greatly by erythrocyte ATP, and the majority of ATP in whole blood is contained in erythrocytes. Our previous observations [5] revealed a directly proportional relationship between whole blood ATP and the deformability index. In this study, close relationships between the ratio and (a) the deformability index and (b) whole blood ATP were found in both healthy subjects and cerebrovascular patients, and the ratio should be regarded as a useful indicator for the detection of hemorheological change in the microvasculature. Fibrinogen and albumin affect hemorheological behavior independently, and the measurement of each protein is appropriate clinically for hemorheological observations. However, in this study it was demonstrated that a high value of the ratio calculated from levels of fibrinogen and albumin was more frequent than the incidence of simple hyperfibrinogenemia or hypoalbuminemia in both healthy subjects and cerebrovascular patients. Thus it is evident that the ratio is more significant than either simple hyperfibrinogenemia or hypoalbuminemia in clinical observations of hemorheological changes in the microvasculature. The data also demonstrate the existence of microcirculatory disturbances even in healthy subjects, suggesting that the ratio might be used as a basis for preventive measures against cerebrovascular disorders by detecting slight disturbances in the microvasculature.

Conclusion

The ratio of fibrinogen to albumin, a simple index, relates closely to hemorheological behavior. This index is a valid indicator of the pathophysiologic condition in preventive and therapeutic measures against cerebrovascular disorders.

References

1. Dintenfass L, Lake B (1977) Blood viscosity factors in evaluation of submaximal work output and cardiac activity in men. Angiology 28:788−797
2. Nakao M et al. (1961) Adenosine triphosphate and shape of erythrocyte. J Biochem 42:487−492
3. Ott EO et al. (1974) High blood viscosity syndrome in cerebral infarction. Stroke 5:330−333
4. Reid HL et al. (1976) A simple method for measuring erythrocyte deformability. Clin Pathol 29:855−858
5. Sakuta S (1981) Blood filtrability in cerebrovascular disorders, with special reference to erythrocyte deformability and ATP content. Stroke 12:824−828

nrCBF for the Assessment of Vasoactive Drugs in Stroke Patients

G. Meinig and P. Ulrich

Introduction

Vasoactive drugs have appeared to be obsolete in the therapy of stroke patients since in the ischemic area an intracerebral steal effect may be induced. However, since the xenon inhalation technique now allows follow-up measurements of CBF on both hemispheres during acute and long-term treatment of stroke patients and since drugs or combinations of different drugs with a bivalent effect on cerebral vessels promise to improve CBF, especially in ischemic areas, without causing a steal effect [1], it seems appropriate that we again take up the question of whether therapy with vasoactive substances in stroke patients might be indicated under certain circumstances.

Methods

We used the NOVO cerebrograph with 16 detectors on each hemisphere. The flow calculations were performed according to Obrist et al. [2] by the bicompartmental analysis (F_1) and according to Risberg et al. by the initial slope index (ISI) [3]. For details of the methods and analysis with our NOVO cerebrograph, see Ulrich et al. 1983 [4].

In 16 stroke patients (1 × TIA, 1 × PRIND, 14 × CS) in whom the stroke was caused by ICA occlusion ($n = 14$) or MCA obstruction ($n = 2$) we measured CBF during rest. A few days later nrCBF measurement was repeated during rest, immediately after infusion of a test preparation "Defluina" (Nattermann Co.) containing 12.5 mg raubasine, 0.5 mg dihydroergocristine, and 0.125 mg dihydroergotamine. Control CBF was performed 30 min later.

Results

We observed an increase in mean CBF (F_1) of 15% in 11 of our 16 patients immediately after the end of Defluina infusion. In seven patients the F_1 levels were still 12% above the initial level 30 min after the end of the infusion.

Neurochirurgische Universitätsklinik, Langenbeckstraße 1, D-6500 Mainz

Cerebral Blood Flow and Metabolism Measurement.
Eds. Hartmann/Hoyer
© Springer-Verlag Berlin Heidelberg 1985

		AFFECTED HEMISPHERE				NON-AFFECTED HEMISPHERE				
		REST.FLOW	5 MIN. POST INFUSION	30 MIN. POST INFUSION	%	REST.FLOW	5 MIN. POST INFUSION	30 MIN. POST INFUSION	%	
DOS.GR. I (n=7)	MEAN F_1	61.3 ± 6.8	65.4 ± 7.1	63.2 ± 9.4	+6.7 +3.1	65.7 ± 6.4	66.2 ± 6.2	65.9 ± 8.0	+0.8	+0.3
DOS.GR. II (n=9)	MEAN F_1	58.0 ± 8.0	61.8 ± 7.8	60.1 ± 8.3	+6.6 +3.6	61.2 ± 9.0	65.5 ±10.0	63.3 ±11.1	+7.1	+3.4

nrCBF (F_1) FOLLOWING INFUSION OF A TEST PREPARATION (DEFLUINA[R]) I: 5 ML; II: 10 ML/20'

Fig. 1. Mean nrCBF (F_1) at rest, 5 min, and 30 min after infusion of Defluina at the affected and nonaffected hemisphere in two dose groups. Group I received 5 ml, group II, 10 ml Defluina during 20 min intravenously

Pat. D. A. ♂ (51); F_1 ⊕; $R_{1,2,3}$

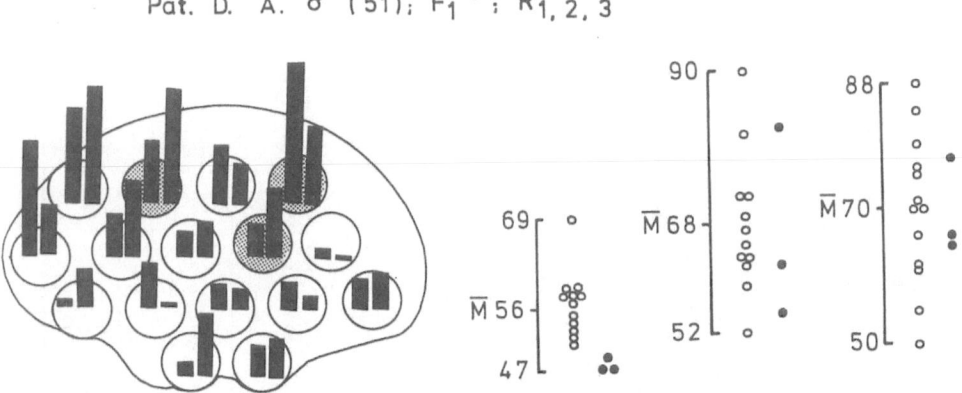

Fig. 2. nrCBF at rest, 5 min, and 30 min after infusion of 5 ml Defluina in a 51-year-old man suffering from a completed stroke due to a left-sided 2-month-old ICA occlusion

Figure 1 shows the average values of mean CBF (F_1) after a low dose (5 ml) of Defluina in seven patients and a high dose (10 ml) in nine patients. After the low dose we found an increase of CBF 5 min after infusion and a drop 30 min after infusion, whereby the initial value was not yet reached. In the nonaffected hemisphere we also found a minimal increase of CBF following the low dose. After the high dose we found an increase of CBF in both the affected and the nonaffected hemispheres 5 min after the infusion and only a slight decrease after 30 min.

More important are the regional changes of CBF in areas which showed low perfusion, as can be seen from the following examples:

Figure 2 shows the blood flow pattern of a 51-year-old male patient with ICA occlusion. Following infusion of Defluina, CBF changed significantly. Thus five minutes after the infusion an increase of CBF had occurred in all areas,

Pat. R. A. ♀ (66); $F_1^{\oplus}$; $R_{1,2,3}$

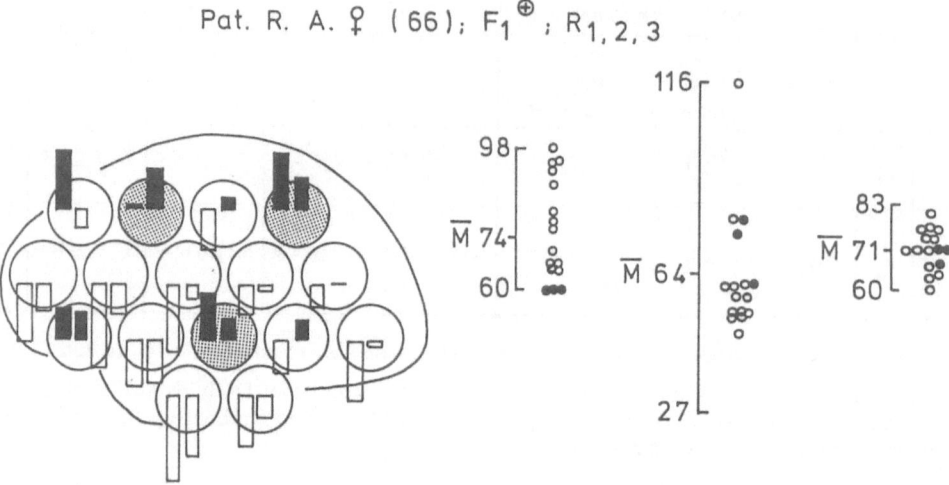

Fig. 3. nrCBF at rest, 5 min, and 30 min after infusion of 5 ml Defluina in a 66-year-old woman suffering from a completed stroke due to a 5-month-old ICA occlusion on the left side

Pat. K. D. ♂ (43); $F_1^{\oplus}$; $R_{1,2,3}$

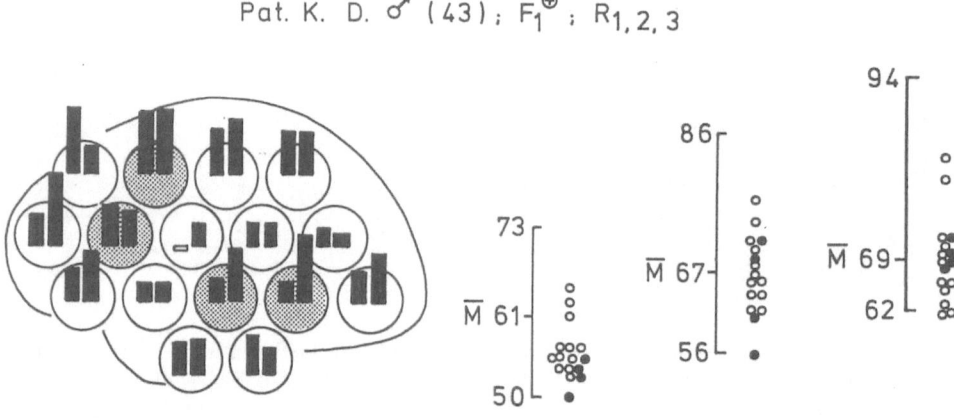

Fig. 4. nrCBF at rest, 5 min, and 30 min after infusion of 10 ml Defluina in a 43-year-old man suffering from a completed stroke due to a 2-month-old ICA occlusion on the right side

and especially in those with low perfusion. Even after 30 min a distinct increase was observed in the poorly perfused areas.

Figure 3 shows the flow pattern of a 66-year-old woman with occlusion of the left ICA. After infusion of 5 ml Defluina there was a small increase in mean CBF and 30 min postinfusion there was a further small increase; regionally this involved a decrease in CBF in areas with high perfusion and an increase in areas with low perfusion.

Figure 4 shows a 43-year-old patient with a 2-month-old right-sided ICA occlusion. After the infusion of 10 ml Defluina there was an increase in mean CBF as well as an increase in CBF in areas with low perfusion.

Pat. B. B. ♂ (31); $F_1^{\oplus}$; $R_{1,2}$

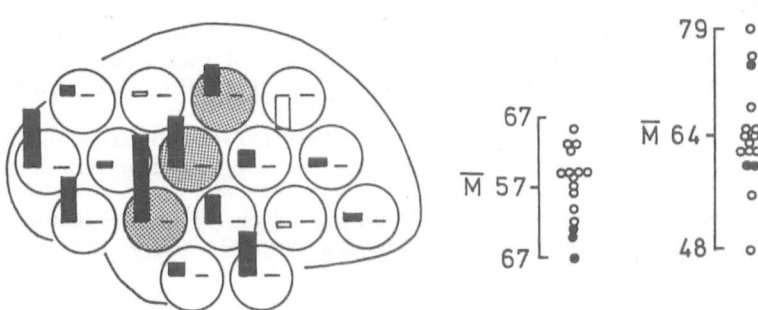

Fig. 5. nrCBF at rest, 5 min, and 30 min after infusion of 10 ml Defluina in a 31-year-old man suffering from a 6-week-old completed stroke due to an MCA occlusion on the left side

Figure 5 shows a 31-year-old patient with an MCA occlusion on the left of 6 weeks' duration. After the Defluina infusion we observed a significant increase of CBF in all areas (third measurement was refused).

Comment

We observed a global increase in CBF in most of our stroke patients following infusion of the test preparation Defluina. Moreover, we observed an increase of flow in poorly perfused areas, which seems a remarkable result. No patient displayed a steal phenomenon. As yet we do not know whether our preliminary results in 16 patients allow us to infer therapeutic consequences, because these results refer only to acute effects. In any case, we now have the possibility of testing the effect of so-called vasoactive drugs after acute and chronic administration and we feel encouraged to go on looking for drugs which may increase CBF, especially in the poorly perfused cerebral areas, without disadvantage for our stroke patients.

Summary

In 16 stroke patients (1 TIA, 1 PRIND, 14 completed stroke) in whom the stroke was caused by ICA ($n = 14$) or MCA ($n = 2$) obstruction, we measured nrCBF at rest, 5 min after infusion of a vasoactive test preparation Defluina, and 30 min postinfusion. In most of our stroke patients we observed a global increase in CBF during infusion of Defluina. Moreover, we observed an increase of flow in poorly perfused areas. No patient showed a steal phenomenon. We are not encouraged to draw therapeutic consequences from our findings, since the results refer only to acute effects.

References

1. Kohlmeyer, K., Blessing, J. (1978) Zur Wirkung von Dihydroergocristin − Methansulfonat auf den Hirnkreislauf des Menschen im Akutversuch. Untersuchungen mit der intracarotidialen Xenon-133-Clearance-Methode. Arzneim.-Forsch. Drug. Res. 28(II):1788−1797
2. Obrist, W. D., Thompson, H. K. , Wang, H. S. et al. (1975) rCBF estimated by xenon-133 inhalation. Stroke 6:245−256
3. Risberg, J., Ali, Z., Wilson, E. M. et al. (1975) rCBF by 133-xenon inhalation. Preliminary evaluation of an initial slope index in patients with unstable flow compartments. Stroke 6:142−148
4. Ulrich, P., Meinig, G. (1983) Nicht-invasive regionale Hirndurchblutungsmessung (nrCBF) bei ischämisch zerebro-vaskulären Erkrankungen. Akt. Neurol. 10:184−187

CBF and Clinical Findings in Patients with Cerebral Ischemia Treated with Extra-Intracranial Bypass Surgery

E. Højer-Pedersen, G. Gulliksen, E. Enevoldsen, and J. Haase

Extra-intracranial bypass surgery (EIAB) was performed in 39 patients, who were selected for the operation on the basis of clinical and angiographic criteria. CBF measurements were carried out in all patients before surgery and again 3 months after surgery. The results of the flow studies did not influence the decision whether to operate. However, the flow studies have been analyzed to evaluate whether CBF measurements might be used as a future criterion in helping to select candidates for EIAB.

Methods

The 39 patients, 11 women and 28 men, were 34−67 years old, the mean age being 53 years. They all had minor ischemic symptoms from the carotid territory, caused by occlusion of the internal carotid artery in 12 patients, and in the other patients by intracranial stenosing or occluding arterial lesions. Postoperative angiography was performed after 3 months in 35 patients and showed a patent anastomosis in 33 patients. Only the 33 patients with a proven patent anastomosis are considered in this paper. The clinical diagnoses were transient ischemic attacks (TIA) in 9 patients, reversible ischemic neurologic deficits (RIND) in 7 patients, and completed stroke (CS) in 17 patients. CBF was measured by the ^{133}Xe inhalation method (Obrist et al. 1975) and the flow values were calculated as initial slope index (Risberg et al. 1975). The flow measurements were carried out at rest and during mental activation with a test including memorizing, silent counting, finger movements, and listening to music. The activation studies were intended to measure the cerebrovascular functional reserve.

Results

Before operation the mean hemispheric values of resting CBF (mCBF) varied from 29 to 70 ml/100 g/min (Fig. 1) with no correlation to the clinical sub-

Neurological and Neurosurgical Departments, Odense University Hospital, DK-5000 Odense C

Cerebral Blood Flow and
Metabolism Measurement.
Eds. Hartmann/Hoyer
© Springer-Verlag Berlin Heidelberg 1985

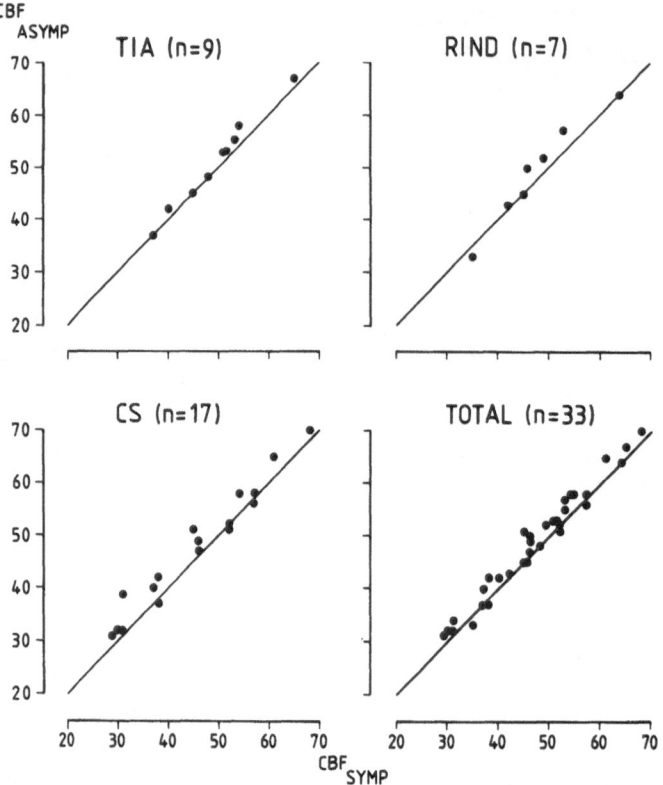

Fig. 1. Resting mean CBF in the symptomatic and asymptomatic hemisphere before operation

groups – TIA, RIND, and CS – nor to the angiographic lesions in the patients. A slight hemispheric asymmetry with lowest mCBF in the symptomatic hemisphere was a consistent finding.

After operation this slight hemispheric asymmetry persisted. The postoperative mCBF (Fig. 2) did increase in individual patients, especially in the TIA group; but mCBF also decreased in other patients, especially in the CS group, though by no more than random statistical variation. For the whole group of patients there was no significant change in mCBF and no correlation between the clinical course, the angiographic lesions, and the postoperative flow values. However, during the period from the preoperative flow studies to the postoperative studies 9 of the 17 CS patients improved clinically and among them 6 of 7 patients with occlusion of the internal carotid artery improved. In the TIA and RIND groups, 2 patients developed new symptoms, while the other 14 stayed free of further symptoms.

The ability to increase flow during mental activation varied considerably (Fig. 3). No correlation was found between flow response and level of resting mCBF, clinical symptoms, angiographic lesions, or infarctions at CT scan. 60% of the patients were able to increase flow by more than 10% in both hemispheres. However, in 9 patients an interhemispheric difference of more

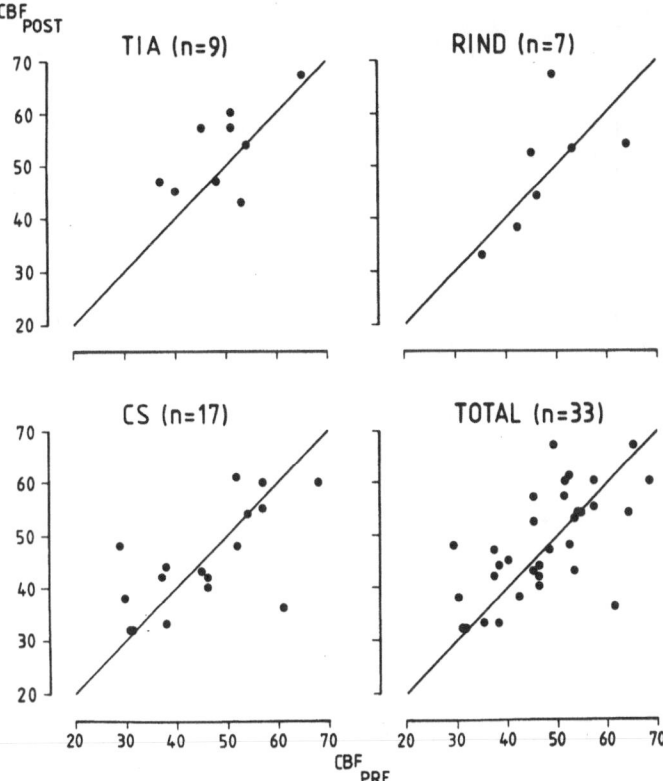

Fig. 2. Resting mean CBF in the symptomatic hemisphere before and after operation

than 5% showed up, with the poorest response being in the symptomatic hemisphere in 6 patients.

The bypass operation did not influence flow response during activation. Postoperative studies revealed that patients with a poor response during activation before operation continued to have a poor response after operation.

Conclusion

The bypass operation did not significantly affect the resting flow level. This is in accordance with the findings of Halsey et al. (1982) and de Weerd et al. (1982). No correlation was found between the preoperative resting flow level and the clinical course after operation. Thus, the findings of Schmiedek et al. (1976) – that general reduction of CBF contraindiates EIAB – could not be confirmed. Several of our patients improved clinically after the operation, but any beneficial clinical effect could not be predicted by either the level of resting flow or the flow response to mental activation.

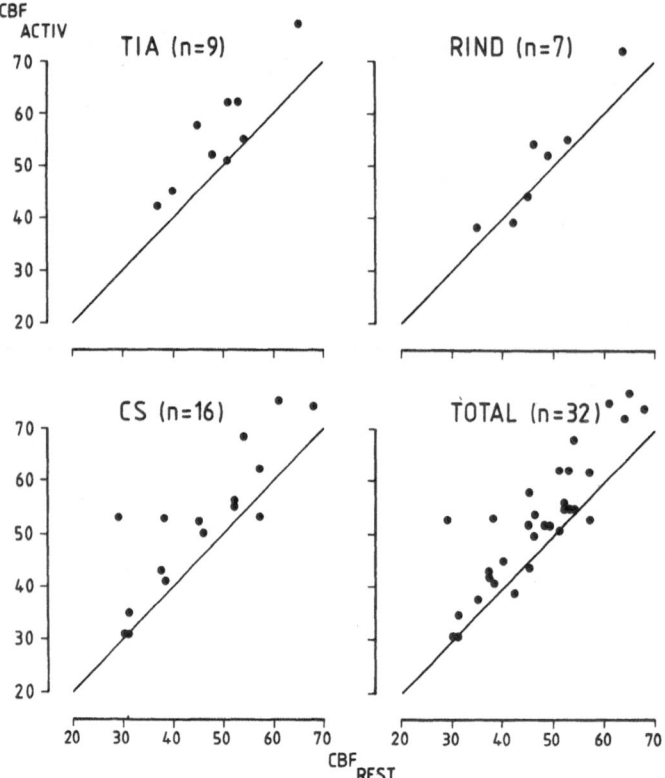

Fig. 3. Mean CBF in the symptomatic hemisphere before operation at rest and during mental activation

References

Halsey JH et al. (1982) The hemodynamic effect of STA-MCA bypass. Stroke 13:163–167

Obrist WD et al. (1975) Regional cerebral blood flow estimated by [133]xenon inhalation. Stroke 6:245–256

Risberg J et al. (1975) Regional cerebral blood flow by [133]xenon inhalation. Stroke 6:142–148

Schmiedek P et al. (1976) Selection of patients for extra-intracranial arterial bypass surgery based on rCBF measurements. J Neurosurg 44:303–312

Weerd AW de et al. (1982) Effect of the extra-intracranial (STA-MCA) arterial anastomosis on EEG and cerebral blood flow. A controlled study of patients with unilateral cerebral ischemia. Stroke 13:674–679

Long-Term Follow-Up nrCBF Studies After EC/IC Anastomoses, Including Different Autoregulation Tests

P. Ulrich, G. Meinig, E. Köster, and K. Schürmann

Introduction

It has been shown by Halsey et al. (1982) and ourselves (Meinig et al. 1982) that about 1 year after EC/IC bypass CBF rises in the operated as well as in the contralateral hemisphere and later declines slightly. The aim of the present study was to contribute to the following questions:

1. Which changes in global and regional CBF occur after STA and MCA bypass operations during a follow-up period of several years?
2. Is it useful to test the autoregulation in order to evaluate the bypass effect and is perhaps its improvement the crucial benefit derived from the operation?

Material and Methods

nrCBF measurements by a ^{133}Xe inhalation device with 32 detectors per hemisphere were done at least once preoperatively. The data of 86 (mean age 50 ± 13 yrs, range $19-78$, female to male ratio $1:2.3$) of the 145 patients operated on by one of us (G. M.) to date are represented in this work. 62 patients had a completed stroke (CS), while 19 had transient ischemic attacks (TIA) or reversible ischemic neurologic deficit (RIND) in their history due to an occlusion of the carotid artery (ICA) in the neck, stenosis, or partial or complete occlusion of the middle cerebral artery (MCA). In five patients the ICA had been occluded as therapy for giant aneurysms. CBF measurements were done postoperatively at least once every 6 months for the first 2 years, and later once a year. Pre- and postoperatively 7% CO_2 inhalation studies after a resting flow measurement were performed in ten patients to calculate the CO_2 reactivity factor. In a similar way we used a mixed preparation of raubasine 12.5 mg, dihydroergocristine 0.5 mg, and dihydroergotamine 0.125 mg (Defluina), which was intravenously applied within 20 min as a vasoactive stimulus to compare the response of CBF pre- and postoperatively.

Neurochirurgische Universitätsklinik, Langenbeckstrasse 1, D-6500 Mainz

Cerebral Blood Flow and
Metabolism Measurement.
Eds. Hartmann/Hoyer
© Springer-Verlag Berlin Heidelberg 1985

Results

In the group of patients with CS (Fig. 1) the preoperative differences in mean flow of the affected and the nonaffected hemisphere were about 14%. They shrank towards the end of the follow-up period to about 3%, while mean flow in both hemispheres declined again after reaching a maximum after 1 year. In patients with TIA or RIND the peak of mean CBF was reached after 18 months. Afterwards the course of CBF mean values was very similar to that in the CS group. If preoperative CBF (ISI = initial slope index according to Risberg) in the more affected hemisphere was 34 ml/100 g/min or lower, the improvement of flow seemed to be more striking and longer lasting. Younger patients (< 50 yrs.) showed a much steeper rise in resting flow than older ones; this was also the case in some cases of MCA occlusion as compared with ICA occlusion.

The regional interhemispheric flow differences (RIFD) declined in all groups gradually during the entire follow-up period (Fig. 2). After 1 year there was an impressive reduction of RIFD especially in the central and parietal areas. For the assessment of autoregulative capacity:

1. The CO_2 reactivity factor expressed as percent change of flow F_1 (fast compartment flow) per 1 mmHg CO_2 difference was calculated preoperatively and in the first 6 months postoperatively in eight patients. It rose in the operated hemisphere in six patients. In the less affected hemisphere the reactivity factor declined postoperatively in all analyzed cases.
2. The global and regional CBF changes were registered once pre- and once postoperatively after infusion of the vasoactive drug Defluina. The direction and degree of mean flow changes seemed nonuniform. Mostly the size of the

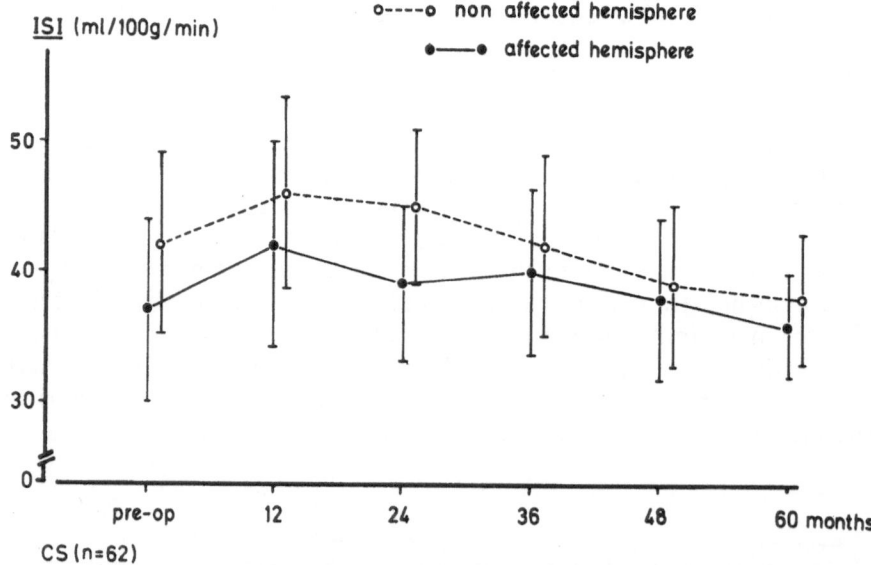

Fig. 1. Changes in mean hemispheric flow (*ISI*, initial slope index) up to 60 months after EC/IC anastomoses in patients with completed stroke (*CS*)

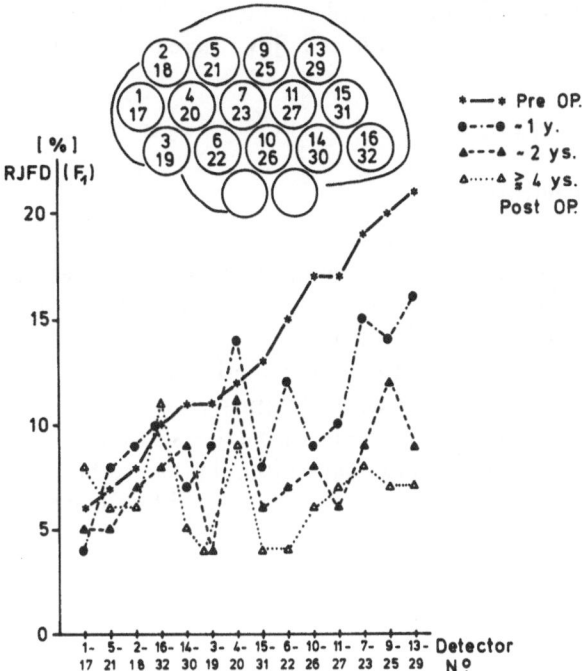

Fig. 2. Changes in regional interhemispheric flow differences (*RIFD*) expressed as percent of mean flow (F_1, flow in fast compartment) of the affected hemisphere after EC/IC anastomoses

changes was greater postoperatively. In four cases the regional flow increases postoperatively were more pronounced than preoperatively.

Figure 3 illustrates the case of a 45-year-old female patient with a CS in the MCA territory of the right hemisphere due to ICA occlusion. Preoperatively the changes in flow were very small, probably as a sign of vasoparalysis, and in areas close to the infarct a flow decrease was observed, which might be in-

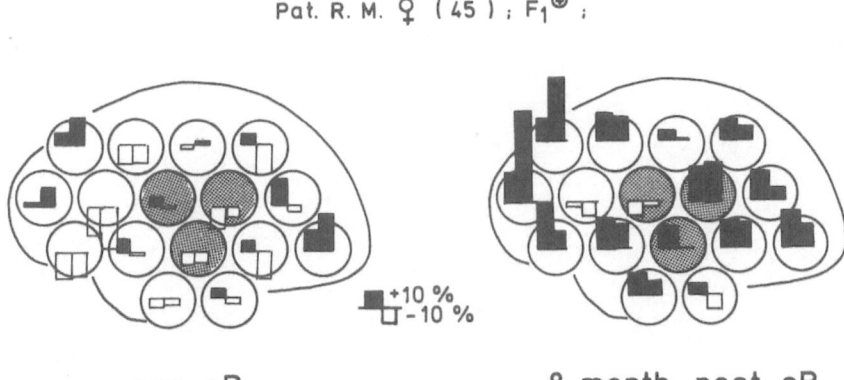

Fig. 3. Changes in rCBF immediately after Defluina infusion (*left bars*) and 30 min later (*right bars*), expressed as percent of resting flow, in a 45-year-old woman after CS (*hatched areas*, ischemic regions)

terpreted as a steal phenomenon. Eight months after the operation we saw a much more pronounced flow increase in most regions of the affected hemisphere.

Conclusions

1. In many cases the hemispheric mean values do not correlate with the clinical course in long-term follow up; their value for the assessment of EC/IC bypass effect seems limited.
2. The regional interhemispheric flow differences tend to normalize in accordance with the clinical improvement.
3. The CBF response to vasoactive substances seems to be improved even in the first month after operation. This may be an important effect of EC/IC bypass, though this has to be proved by further investigation.

References

Halsey, J. H., Morawetz, R. B., Blauenstein, U. W. (1982) The hemodynamic effect of STA-MCA bypass. Stroke 13(2):163–167
Meinig, G., Ulrich, P., Köster, E., Schürmann, K. (1982) Follow up nrCBF after EC/IC anastomoses. 6th International Symposium "On Microsurgical Anastomoses for Cerebral Ischemia". Kyoto 12.–15. 09. 1982

Use of Cerebral Blood Flow Measurements in the Prediction of Delayed Cerebral Ischemia Following Subarachnoid Hemorrhage

J. D. Pickard, D. H. Read, and A. H. J. Lovick

Introduction

Many different factors may contribute to the development of delayed cerebral ischemia following subarachnoid hemorrhage. In part, such delayed ischemic episodes may reflect the inability of part of the cerebral circulation in some individuals to compensate for periods of hypotension or hypoxia. For example, the deficit sometimes develops in association with a fall in the patient's blood pressure, and may be reversed by raising the patient's blood pressure and expanding his blood volume (Kosnik and Hunt 1976; Symon 1978). Patients with impaired autoregulation to moderate intraoperative hypotension induced with halothane have been shown to have an increased risk of developing postoperative neurologic deficits of late onset (Pickard et al. 1980). Farrar et al. (1981) have confirmed these findings. A preoperative test that gave the same clues as does the intraoperative measurement of autoregulatory capacity would be very helpful, and in this paper we report both the modifications to the intravenous ^{133}Xe technique required to make measurements in confused patients as well as the preliminary results from our cerebrovascular reactivity test.

Methods

The original intraoperative intravenous ^{133}Xe injection method (Wyper et al. 1979) was designed specifically so that neither surgeon nor anesthetist would in any way be impeded by the measurements. In the operative situation it is not realistic to expect physiologic parameters to remain constant for long periods and hence the data acquisition period was cut from 10 min to 3 min. The initial slope method of analysis was used, thereby avoiding any assumptions about the compartmental nature of blood flow in the diseased brain. Clearance of ^{133}Xe from the head following its intravenous injection was measured by a collimated thallium-activated sodium iodide crystal mounted unobtrusively under the operating table. No attempt at estimation of regional cerebral blood flow was made – our previous primate studies had provided the justification for this

Wessex Neurological Centre, Southampton General Hospital, Shirley, Southampton S09 4XY, Great Britain

Cerebral Blood Flow and
Metabolism Measurement.
Eds. Hartmann/Hoyer
© Springer-Verlag Berlin Heidelberg 1985

simplification (Pickard et al. 1979). The changes in cerebrovascular reactivity after subarachnoid hemorrhage are present globally in the brain although, of course, such changes may be more marked focally. The arterial ^{133}Xe concentration was estimated by using the end-tidal ^{133}Xe concentration. By using pulse–height analysis and restricting analysis of the head curve to the period $60 s < t < 180 s$, the most transient part of the expired air curve was avoided and the problem of scattered radiation from the airways minimized. We have subsequently modified and extended this technique in order to measure the cerebrovascular reactivity in patients preoperatively. The full details of our new equipment have been described elsewhere (Lovick et al. 1982). Unfortunately, our early attempts to measure cerebral blood flow in confused patients following subarachnoid hemorrhage were not entirely successful, mainly because such patients would not tolerate the use of a face mask in order to sample end-tidal ^{133}Xe and CO_2 concentrations. We have therefore examined an alternative method for the estimation of arterial ^{133}Xe concentration, namely, the clearance from the right lung. During the course of our work, Jaggi and Obrist (1981, 1983) reported on their use of lung clearance as an estimate of arterial recirculation of ^{133}Xe. In order to justify this technique, we have compared arterial ^{133}Xe concentrations following its intravenous injection with lung clearance and end-tidal clearance recorded from multiple sites within the respiratory tree in anesthetized rabbits.

Experimental Studies

Adult rabbits were anesthetized with nitrous oxide, oxygen, and halothane. The animals were artificially ventilated. Arterial blood pressure and blood gases were routinely monitored. Arterial ^{133}Xe concentration was measured by placing a well shielded scintillation detector over an abdominal aortic arterial shunt.

Following intravenous injection of ^{133}Xe in saline, the arterial clearance of ^{133}Xe follows a biexponential pattern ($T\frac{1}{2} = 352 s \pm 31$ SEM; $T\frac{1}{2} = 12 \pm 1 s$) whereas the end-tidal ^{133}Xe clearance shows a triexponential form ($T\frac{1}{2}$'s of $297 \pm 27 s$, $39 \pm 8 s$, and $8 \pm 1 s$). However, it was found that by moving the sampling point down the trachea the intermediate component disappeared and the curve assumed a biexponential form. In ten patients, a comparison of the end-tidal ^{133}Xe clearance with the activity recorded over the right upper lung field shows a similar pattern. The end-tidal activity shows a triexponential pattern ($T\frac{1}{2}$'s of $367 \pm 10 s$, $31 \pm 1 s$, and $10 \pm 0.3 s$), whereas the lung activity shows a biexponential pattern ($T\frac{1}{2}$'s of $269 \pm 8 s$ and $19 \pm 0.5 s$).

Cerebrovascular Reactivity Test

The code of practice of the local ethical committee was followed throughout during the performance of these tests. A neuroanesthetist and a senior

Fig. 1. Distribution of CBF values in 26 patients during baseline conditions, following sedation and during hypotension. The *hatched area* refers to the patients in whom CBF fell by more than 30% of control during hypotension

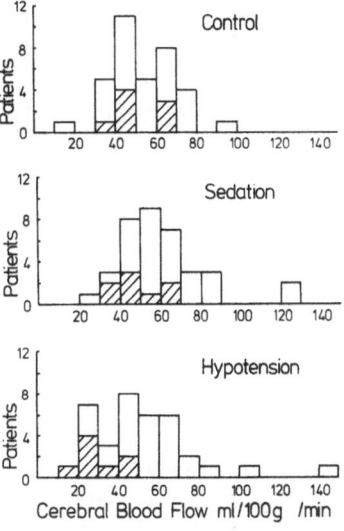

neurosurgical nurse were present at all times during the test. Head clearance of ^{133}Xe was recorded with a scintillation detector placed under the head and a small cadmium telluride detector placed over the right upper lung field to estimate arterial ^{133}Xe concentration. Arterial blood pressure and ECG were measured and end-tidal CO_2 obtained from a fine catheter placed in the postnasal space following a nasal spray with local anesthetic. Following at least two estimations of cerebral blood flow to establish a stable baseline, the patients were sedated with intravenous Diazemuls – the level of sedation was adjusted so that the patient was still able to respond readily to verbal commands. Following a further estimation of cerebral blood flow, the patient was tilted to approximately 30° head up and the mean arterial blood pressure reduced to 70% of the control value with an intravenous infusion of Trimetaphan. Cerebral blood flow was again measured (details of this technique have been reported elsewhere, see Read et al. 1983). Twenty-seven tests have been performed on 26 patients within 24 h of proposed surgery. Cerebral blood flow changed little overall with sedation although there was considerable individual variability (mean control CBF 50 ± 3 ml/100 g/min; sedation CBF 54 ± 3 ml/100 g/min; mean arterial blood pressures of 105 ± 3 mmHg and 100 ± 3 mmHg respectively). On reducing arterial blood pressure to a mean of 66 ± 3 mmHg, CBF overall in these 27 tests fell to 46 ± 5 ml/100 g/min. However, of greater significance was the finding that in eight of these tests, the fall in cerebral blood flow was greater than 30% (hatched areas in Fig. 1). The effects of sedation with Diazemuls do not appear to correlate with the response to hypotension.

Conclusions

Our new technique has enabled us to measure CBF reliably in confused patients. In particular, the use of the measurement of lung clearance of ^{133}Xe as an estimate of arterial recirculation has permitted our further clinical studies and has been justified by our animal work. Our test of cerebrovascular reactivity is feasible within the clinical context of patients with subarachnoid hemorrhage and can identify patients with marked impairment of cerebrovascular autoregulation.

Acknowledgment. We are very grateful to the Winham Foundation and the Chest, Heart and Stroke Association for their generous support of this work.

References

Farrar JK, Gamache FW, Ferguson GG, Barker J, Varkey GP, Drake CG (1981) Effects of profound hypotension on cerebral blood flow during surgery for intracranial aneurysms. J Neurosurg 55:857−864

Jaggi JL, Obrist WD (1981) External monitoring of the lung as a substitute for end-tidal 133Xenon sampling in non-invasive CBF studies: preliminary findings. rCBF Bull 2:25−28

Jaggi JL, Obrist WD (1983) External monitoring of the lung as a substitute for end-tidal 133xenon sampling in non-invasive CBF studies. J CBF Metab 3 Suppl 1:S123−S124

Kosnik EJ, Hunt WE (1976) Post-operative hypertension in the management of patients with intracranial arterial aneurysms. J Neurosurg 45:148−154

Lovick AHJ, Pickard JD, Goddard BA (1982) Prediction of late ischaemic complications after cerebral aneurysm surgery − use of a mobile microcomputer system for the measurement of pre-, intra- and post-operative cerebral blood flow. Acta Neurochirurgica 63:37−42

Pickard JD, Boisvert DPJ, Graham DI, Fitch W (1979) Late effects of subarachnoid haemorrhage on the response of the primate cerebral circulation to drug-induced changes in arterial blood pressure. J Neurol Neurosurg Psychiat 42:899−903

Pickard JD, Matheson M, Patterson J, Wyper D (1980) Prediction of late ischaemic complications after cerebral aneurysm surgery by the intra-operative measurement of cerebral blood flow. J Neurosurg 53:305−308

Read DH, Lovick AHJ, Pickard JD (1983) A pre-operative test of cerebrovascular autoregulation following subarachnoid haemorrhage. Brit J Anaesth 55:918P

Symon L (1978) Disordered cerebrovascular physiology in aneurysmal subarachnoid haemorrhage. Acta Neurochir 41:7−22

Wyper DJ, Pickard JD, Acar U (1979) Monitoring cerebral blood flow during intracranial operations: an intravenous injection method. Neurol Res 1:31−37

The Prognostic Value of Atraumatic CBF Measurement in Subarachnoid Hemorrhage

G. Geraud[1], A. Guell[1], P. Andrieu[1], M. Tremoulet[2], and A. Bes[1]

The development of atraumatic measurement of CBF by inhalation or intravenous injection of xenon 133 and the recent perfection of dependable and easy to handle apparatus have stimulated renewed interest in this problem and have resulted in new publications in the last few years [1, 2, 5, 6, 10].

The main question remaining is whether atraumatic measurement of CBF has prognostic value, and if so, what role should it play in regard to clinical evaluation and angiographic results? The present study, then, centers on this practical problem.

Materials and Method

Sixty-two patients (25 men, 37 women) with subarachnoid hemorrhage were studied. The mean age of the subjects was 47.5 ± 13 years (ranging from 21 to 69 years). Clinical grading was established according to the Hunt and Hess criteria (3): grades I and II, asymptomatic or headache patients without alteration of consciousness or localized neurologic symptoms (37 cases); grade III, drowsiness or confusion without focal signs (17 cases); grade IV, stupor with moderate-to-severe hemiparesis (8 cases). Grade V comatose patients were excluded from this study.

Angiography revealed an aneurysm in 46 patients and vasospasm in 26 patients (42%): 11 localized and moderate cases, 11 multifocal cases, and 4 diffuse cases. Tomodensitometry was performed on 43 patients and showed blood to be present in the basal cisterns in 24 cases, intracerebral hematoma in 6 cases, and active hydrocephaly in 6 cases.

Ninety-one measurements of CBF were taken in these 62 patients by the xenon 133 inhalation method perfected using a 32-detector multiprobe system (16 per hemisphere). Two parameters were calculated from the clearance curves:

- F_1 in ml/100 g/min, which is assimilated to the cortical flow (bicompartmental model). This value ranges from $70-80$ ml/100 g/min in the normal subject at rest and with the eyes closed.

1 Department of Neurology, Centre Hospitalier Universitaire de Rangueil, Chemin du Vallon, F-31054 Toulouse
2 Department of Neurosurgery, Centre Hospitalier Universitaire de Purpan, F-31094 Toulouse

Cerebral Blood Flow and
Metabolism Measurement.
Eds. Hartmann/Hoyer
© Springer-Verlag Berlin Heidelberg 1985

– ISI (initial slope index) calculated between the 2nd and 3rd minute of the clearance curve where compartmental instability is present (mono-compartmental model) (9). Normal values range from 40 to 50.

Focal oligemia was considered present when localized CBF decreased by more than 15% (compared with mean hemispheric flow) in at least three adjacent detectors. At the time of each CBF measurement, clinical grade, end-tidal pCO2, and mean arterial pressure (MAP) were recorded. In 54 out of 62 patients, the initial CBF measurement was performed within the first 12 days following the bleeding.

Surgical clipping of the aneurysm was performed on 31 patients. The remaining subjects did not undergo surgery, either because they presented no arterial malformation or because their condition was felt to be too serious.

The usual statistical methods were used, notably variance analysis, the Student t-test, and the χ^2 test. Results with error risk of less than 5% were considered significant.

Results

Correlations with Clinical Grade (Table 1)

The 91 CBF values were correlated with the clinical grade of the patients at the time of measurement: CBF values were significantly lower in grade IV patients, whereas there was no difference between the grade I–II and grade III groups. This decrease in CBF is not explained by modification in pCO2 or MAP or by age difference. Focal oligemia occurred more often in grade IV patients (85%); nevertheless, in 34% and 57% respectively of those grade I–II and III patients with no focal neurologic symptoms, localized hypoperfusion was present ("silent focal oligemia").

Table 1. Correlations between clinical grade and CBF values

Clinical grade	n	Age (years)	Cortical flow (F_1) ml/100 g/min	ISI	pCO$_2$ (mmHg)	MABP (mmHg)
I–II	43	47±13	68±18	39± 8	32±3	97±10
III	28	48±14	67±16]*	38±11	32±3	97±12
IV	20	50±14	53±10]*]*	31± 7	31±3	97± 9

* $P < 0.05$, ** $P < 0.01$

Correlations with Vasospasm

Considering the CBF measurement performed closest to the time of angiography for each patient, the 26 patients with vasospasm showed a decrease

Table 2. Correlations between vasospasm and CBF values

Vasospasm	n	Cortical flow (F$_1$) ml/100 g/min	ISI	pCO$_2$ (mmHg)	MABP (mmHg)	Oligemic focus 0	+
Absent	36	72±21	41±11	32±3	98±10	67%	33%
Focal	11	62±13	37± 7	31±3	97±10	55%	45%
Multifocal	15	60± 8	36± 5	32±3	102±13	30%	60%

* $P < 0.05$

of CBF compared with the 36 patients free of vasospasm, but the difference is significant only for cortical flow (Table 2). Neither the intensity nor the extent of vasospasm influenced CBF values. This poor correlation was perhaps due to too long an interval of time between angiography and CBF recording (3.5 ± 9 days).

Neither the presence or absence of blood in the basal cisterns nor the presence and location of the aneurysm correlated with hemodynamic parameters in this study.

We tried to compare the influence of vasospasm and clinical grade on CBF (Fig. 1): mean cortical values were around 70 ml/100 g/min in patients without

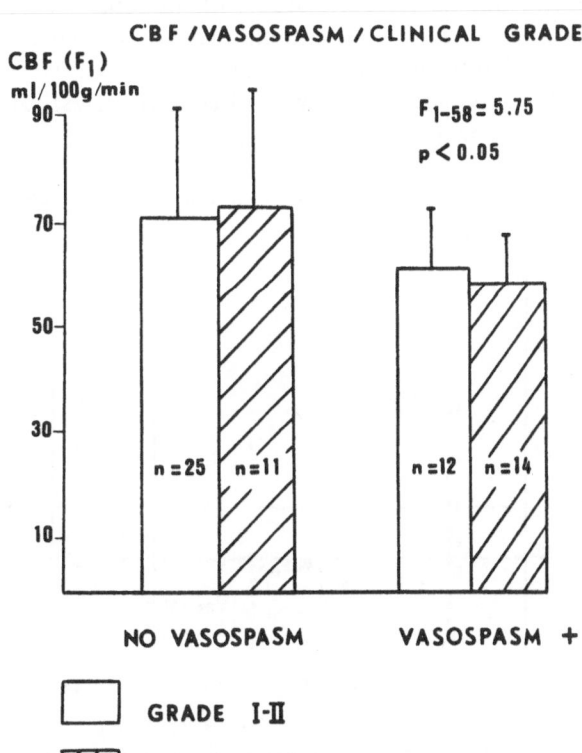

Fig. 1. Comparative correlation of clinical grade and vasospasm with the cortical CBF (F$_1$)

vasospasm, whatever the clinical grade at the time of CBF measurement. Conversely, cortical CBF measured around 60 ml/100 g/min in subjects with arterial spasm in all clinical groups. We can conclude from these data that there is a greater correlation between vasospasm and CBF than between CBF and clinical grade.

Correlations with Patients' Outcome (Table 3)

The 62 patients were divided into three groups according to clinical condition at the time of leaving hospital: 37 recovered with no disability; 16 showed motor, aphasic, or intellectual deficit; 9 of them died while in hospital. Initial CBF was greater in those patients having good recovery compared with those either suffering disabilities or dying (Fig. 2). Mean values were significantly different for F_1 cortical flow and for ISI. Taking the value of 60 ml/100 g/min as the critical level for F_1 and 35 as that for ISI, it appeared that below this level the risk of complications was much greater; of the 24 patients with cortical flow below this level, 17 experienced complications, and all the deaths had shown values below this critical level except one case where cortical flow was measured at 60.3 ml. Conversely, in every case but one outcome was good when cortical flow rose above 70 ml.

The initial clinical grade, evaluated at the time of the first CBF measurement, was poorly correlated with patients' outcome. We compared the prognostic values of CBF and clinical grading (Fig. 3): mean cortical flow values were significantly higher in patients with favorable outcome than in subjects with complications or who died, regardless of clinical grading at time of measurement.

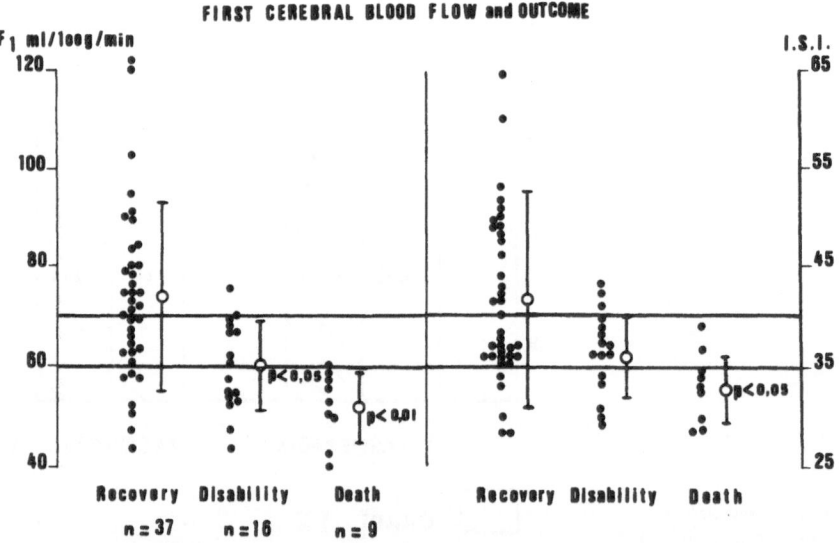

Fig. 2. Initial cerebral blood flow values for the 62 patients in terms of their outcome. F_1 represents cortical flow. ISI is an index of mean flow

Table 3. Correlations between patients' outcome and clinical, angiographic, and hemodynamic parameters

Outcome	n	Age (years)	Clinical grade I–II	Clinical grade III–IV	Vasospasm 0	Vasospasm +	Cortical flow (F_1) ml/100 g/min	ISI	Oligemic focus 0	Oligemic focus +
Recovery	37	46±13	68%	32%	68%	32%	74±19]*]*	42±11]*	73%	27%*
Disability	16	50±13	56%	44%	56%	44%	60± 9]*	36± 4	44%	56%
Death	9	49±13	33%	67%	22%	78%	52± 7	33± 4	22%	76%

* $P < 0.05$, ** $P < 0.01$

Table 4. Correlations between the 31 surgical patients' outcome and clinical, angiographic, and hemodynamic parameters

Outcome	n	Age (years)	Clinical grade I–II	Clinical grade III–IV	Vasospasm 0	Vasospasm +	Cortical flow (F_1) ml/100 g/min	ISI	Oligemic focus 0	Oligemic focus +
Favorable	20	46±14	55%	45%	65%	35%	72±19]*	40±8]*	80%	20%**
Unfavorable	11	51±11	64%	36%	73%	27%	56±10	34±4	18%	82%

* $P < 0.05$, ** $P < 0.01$

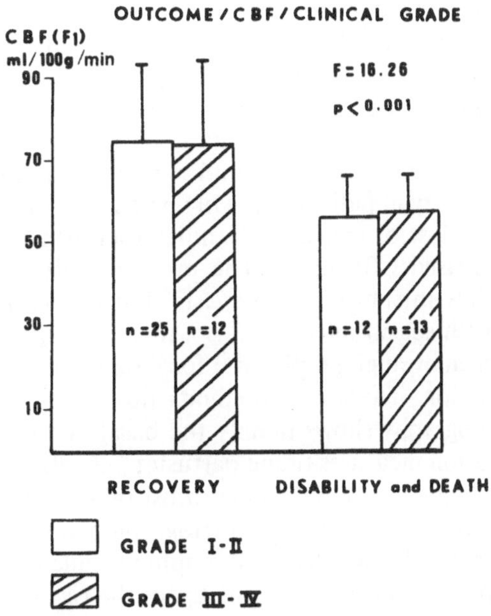

Fig. 3. Comparative prognostic value of clinical grade and initial CBF measurements evaluated in the early days following meningeal hemorrhage (mean 8.9 ± 8.9 days)

Focal oligemia occurred more frequently in cases with poor outcome than in those with good recovery, whereas the same type of comparison was not significant for vasospasm.

For statistical reasons, the 31 surgical patients were divided into two groups, one with good recovery (20 cases), the other with postoperative complications (11 cases, including 3 deaths). Neither age, clinical condition, nor absence or presence of vasospasm correlated significantly with outcome. However, marked decrease of initial cortical CBF was noted in those subjects experiencing complications (Table 4).

Discussion

As many other authors [2, 4, 5, 8] have found, a certain correlation was observed between CBF values and the clinical state of patients: those with both alteration of consciousness and motor deficit showed a mean CBF decrease of 30% compared with the others. However, it seemed more valuable to emphasize the discordances which may exist between clinical and hemodynamic parameters: Among grade I−II patients, certain subjects showed abnormally low CBF values − 28% had a cortical flow value less than 60 ml/100 g/min. The study of cerebral hemodynamics in these patients could furnish additional results and particularly, as Ferguson has noted (2), enable us to detect those risk-prone subjects with insufficient cerebral perfusion whose clinical condition gives no indication of the same.

Analysis of regional CBF can also provide valuable information: Each of the eight patients with localized clinical symptoms presented a concordant ischemic or oligemic territory. The advantage of this type of atraumatic measurement is that of following the evolution of such focal ischemia and comparing it with the patient's clinical outcome. However, discordances could also have existed between clinical and hemodynamic results: thus, in 34% of the grade I−II patients and 57% of those in grade III, the first CBF measurement showed clinically silent focal oligemia. In six patients this oligemic focus was seen a few days before the concordant clinical deficit appeared. The existence of such clinically silent oligemic foci could explain ischemic strokes following arteriography or surgery in patients showing no clinical or arteriographic symptoms of high risk.

Most of the earlier studies establish a relationship between spasm and decreased CBF [2, 8]. The present study also revealed decreased mean cortical flow in subjects with spasm, but it was not possible to correlate this decrease with spasm intensity. Also, there was no systematic correlation between spasm seen by angiography and focal oligemia revealed through hemodynamic examination. These two methods do not explore the same vascular territory: angiography shows mainly the basal and cortical arteries, while xenon 133 inhalation measures tissue perfusion, i.e., distal circulation. It is known that the narrowing of arterial lumen must be greater than 80% before distal circulation is affected. That is why angiographic vasospasm and focal ischemia may be dissociated, and why angiographic results do not provide an exact picture of tissue perfusion and, consequently, of the risk of ischemia.

Finally, the essential focus of this study remains the patients' outcome and the search for prognostic factors. It was observed that neither age, nor the presence of vasospasm, nor the extent of bleeding, nor the location of the aneurysm seemed to be determinant in the patients' evolution. Nor was significant correlation found between initial clinical grade and outcome. It is not to be inferred, nonetheless, that the clinical state has no predictive value, particularly for surgical results. To do so would be to oppose well-established ideas and daily experience. Like most other medical – surgical teams, ours does not consider operable those patients with alteration of consciousness or significant neurologic deficit. Spontaneous improvement must occur before surgery is indicated. Nonetheless, it must be stated that, in this study, a significant correlation appeared between CBF measured in the first 2 weeks of illness and the patient's outcome on leaving hospital. Such a correlation was not evident with initial clinical condition at the time of CBF measurement (Fig. 3). This observation stands when considering solely the 31 patients later undergoing surgery (Table 4).

The prognostic value of cerebral hemodynamic measurements in subarachnoid hemorrhage has been suggested by several authors [2, 4, 6, 7]. These studies record lower CBF values in those subjects having poor outcome. The decrease falls between 25% and 40% depending on the study, absolute values varying widely according to the method of CBF measurement. Ishii [4], by the intracarotid method, and Meyer et al. [6], by the inhalation method, place the critical level at 30 ml/100 g/min, this value corresponding to mean flow rather than gray matter flow. Our study seemed to indicate the critical level to be 60 ml/100 g/min for cortical flow and 35 for flow, as evaluated by ISI.

This study thus confirms the significant prognostic value of CBF measurement in subarachnoid hemorrhage. Thanks to its atraumatic nature, measurement by inhalation can be repeatedly performed on the same patient with no ill-effect. The approach we presently adopt in cases of spontaneous subarachnoid hemorrhage is performing the first CBF measurement as soon as possible. Mean cortical flow values less than 60 ml/100 g/min and/or focal oligemia temporarily contraindicate surgery. In these conditions, we prefer delaying angiography and repeating CBF measurements until hemodynamic and clinical improvement safety permit further investigation.

References

1. Brawanski A et al. (1982) Atraumatic rCBF measurement: an aid in the timing of surgery and the management of spasm following SAH. Acta Neurochir (Wien) 63:43 – 51
2. Ferguson GG et al. (1981) Serial measurements of CBF as a guide to surgery in patients which ruptured intracranial aneurysms. J Cereb Blood Flow Metabol 1, Suppl 1:s518 – s519
3. Hunt WE, Hess RM (1968) Surgical risks as related to time of intervention in the repair of intracranial aneurysms. *J Neurosurg* 28:14 – 20
4. Ishii R (1979) Regional cerebral blood flow in patients with ruptured intracranial aneurysms. J Neurosurg 50:587 – 594
5. Merory J et al. (1980) Cerebral blood flow after surgery for recent subarachnoid haemorrhage. J Neurol Neurosurg Psychiatry 43:214 – 221

 6. Meyer CHA et al. (1982) Subarachnoid haemorrhage: older patients have low cerebral blood flow. Br Med J ii:1149−1152
 7. Nilsson BW (1977) Cerebral flow in patients with subarachnoid haemorrhage studied with an intravenous isotope technique. Its clinical significance in the timing of surgery of cerebral arterial aneurysms. Acta Neurochir 37:33−48
 8. Pitts LH et al. (1977) Cerebral blood flow, angiographic cerebral vasospasm and subarachnoid haemorrhage. Acta Neuro Scand 56:334−335
 9. Risberg J et al. (1975) Regional cerebral blood flow by 133 xenon inhalation. Stroke 6:142−148
10. Weir B et al. (1978) Regional cerebral blood flow in patients with aneurysms: estimation by xenon 133 inhalation. Can J Neurol Sci 5:301−305

Xenon 133 CBF Measurements in the Clinical Management of Patients with Subarachnoid Hemorrhage

J. K. FARRAR, G. G. FERGUSON, C. G. DRAKE, and S. J. PEERLESS

Introduction

A large number of patients with subarachnoid hemorrhage (SAH) develop neurologic deficits as a result of critical reductions in cerebral blood flow (CBF) and it has been suggested that CBF measurements may be useful in the clinical management of these patients. Although a number of studies' have shown that flow is decreased following SAH, the relationship between CBF and clinical status is not well understood and often a poor correlation exists between these two parameters. On the other hand, factors which would tend to cause hemodynamic instability, such as arterial spasm or intracranial surgery, are frequently associated with neurologic deterioration suggesting a cause and effect relationship. In an attempt to clarify these issues, we have examined the progressive changes in CBF at various stages in the clinical course of patients with SAH and have related these changes to the corresponding alterations in neurologic status and angiographic and CT information.

Methods

The study group consisted of 120 patients with recent SAH confirmed by lumbar puncture, CT scan, and four-vessel angiography. An additional group of 43 patients with unruptured aneurysms served as an age-matched control group. CBF was measured by the xenon 133 inhalation technique using a 32-channel detector system (Novo Cerebrograph) with 16 detectors in a parallel array situated on each side of the head. Mean hemispheric CBF was calculated from the mean hemispheric clearance curve obtained by summing the individual data points from all 16 detectors. This method has been found to be more stable than simply taking the arithmetic mean of the individual CBF values. Arterial spasm was assessed from measurements of the middle cerebral artery following angiography and the data were corrected for magnification. These results were then compared to control values obtained from similar measurements in a group of 50 patients without vascular disease. Arterial narrowing was con-

Department of Clinical Neurological Sciences, University Hospital, P.O. Box 5339, Terminal "A", London, Ontario, N6A 5A5, Canada

Cerebral Blood Flow and
Metabolism Measurement.
Eds. Hartmann/Hoyer
© Springer-Verlag Berlin Heidelberg 1985

sidered significant if the measured diameter was less than 70% of the control value. Whenever possible, a CBF measurement was obtained immediately prior to angiography. Neurologic status was assessed regularly and recorded in relation to individual flow studies. For the purposes of this report, the mean CBF (Risberg initial slope index) of a single hemisphere was chosen to represent each study. If lateralizing clinical signs were present, the side responsible for the deficit was used and in the remainder, the value for the left hemisphere was chosen. A total of 250 CBF studies were done in the SAH patients and relevant angiographic data were available in 111.

Results

At an early stage in these experiments, we found that in patients who were alert and oriented but had a mild focal deficit, CBF did not differ from the values obtained in those without deficit. In terms of cerebral perfusion, these patients were considered to have a falsely low clinical grade in either the Botterell or Hunt and Hess classifications. Therefore, we have adopted the grading system shown in Table 1. The data were grouped according to neurologic grade and the results are presented in Table 2. There were no significant differences between groups in terms of age, mean arterial blood pressure, or arterial pCO_2. Only three patients were grade 4 when studied and CBF was between 20 and 32. In agreement with previous studies, we found that CBF generally decreased with worsening clinical grade; however there was considerable overlap of flow values between groups. In the absence of angiographic information, there was a poor correlation between CBF and neurologic status in approximately one-third of the studies. In the majority of these, CBF was far below the group mean for patients of that clinical grade. The addition of angiographic data provided a much clearer distinction between clinical grades and revealed the cause of unexpectedly low flows as shown in Table 3. Mean flow decreased by approximately 5 ml/100 g/min with each decline in neurologic status for those without significant vascular narrowing. These flow values were taken to define the "normal" limits of CBF for each grade and flow was considered "abnormal" if the observed value was more than one standard deviation from the group mean for that grade. The distribution of flow rates according to this definition is shown in Table 4. There was a marked increase in the incidence of abnormally low flows in all patients with SAH (grades $1-3$) when compared to those without hemorrhage (grade 0). As indicated in Table 3, arterial spasm causes a pronounced reduction in CBF in patients of all clinical grades and is the most likely cause of the increased incidence of unexpectedly low flows. When only those studies with angiographic data were considered, we found that approximately 80% of those with low CBF had moderate to severe arterial narrowing and only 25% of those with normal CBF had spasm (Table 5). The influence of arterial spasm on CBF was most pronounced in grade 1 patients. Only 4/18 (22%) of patients with spasm had flows within the normal range (≥ 40) and all patients with CBF ≤ 35 had spasm.

Table 1. Clinical grading system

Grade	Criteria
0	Asymptomatic, without subarachnoid hemorrhage
1	Alert/oriented, with or without minor deficit
2	Drowsy/disoriented, without deficit
3	Major deficit or drowsy with minor deficit
4	Stuporous or comatose

Table 2. Grouped data [a] for all studies

Grade	# patients	# studies	Age	MABP	pCO_2	CBF
0	43	43	50±14	95±14	37±5	49±7
1	62	110	47±13	92±11	37±4	41±7
2	35	63	50±14	94±13	37±4	38±6
3	36	74	51±11	94±11	35±4	33±6

[a] Mean ± SD

Table 3. Mean CBF and angiographic findings

Grade	Normal angiograms	> 30% narrowing
0	49±7 (43) [a]	–
1	45±5 (29)	34±6 (18)
2	40±5 (19)	34±4 (7)
3	34±6 (21)	29±5 (17)

[a] Mean CBF ± SD (# of studies)

Table 4. Distribution of flow rates (all studies)

Grade	# studies	Low CBF	Normal CBF	High CBF
0	43	9%	79%	12%
1	110	37%	51%	12%
2	63	33%	51%	16%
3	74	22%	66%	12%

Table 5. Incidence of arterial spasm

Grade	Low CBF	Normal CBF	High CBF
1	14/16 (88%)	4/27 (15%)	0/4 (0%)
2	4/4 (57%)	3/15 (20%)	0/4 (0%)
3	7/8 (88%)	10/27 (37%)	0/3 (0%)
Overall	25/31 (81%)	17/69 (25%)	0/11 (0%)

Discussion

These results indicate that there is a close coupling between CBF and neurologic status in patients with normal angiograms following subarachnoid hemorrhage and that there are well-defined limits of "normal" CBF values for each clinical grade. Individual CBF measurements may be interpreted with respect to these limits in order to provide clinically useful information – particularly as to the presence of unsuspected arterial spasm. We have found that the detection of arterial spasm may be of considerable importance in the timing of surgery for these patients. In a previous study (Farrar et al. 1981), we found that patients with arterial spasm preoperatively did not autoregulate to hypotension during surgery. They developed relatively severe flow disturbances at the site of retraction postoperatively and had a high incidence of late-onset ischemic deficits. In another study (Farrar et al. 1983), we examined the serial changes in CBF in a group of 26 patients with recent SAH. Those with no blood in the basal cisterns on early CT scan had a low incidence of arterial spasm (11%) and showed relatively stable flow values preoperatively (mean fluctuation of 5 ml/100 g/min). Surgery had very little effect on CBF or vessel caliber in this group and only one patient had a poor clinical outcome. In contrast, patients with blood in the basal cisterns had a high incidence of spasm (82%) resulting in unstable flows preoperatively (mean fluctuation of 13 ml/100 g/min). Both CBF and vessel caliber decreased by 25% – 30% during the second week after hemorrhage. Surgery resulted in a worsening of preoperative spasm and a marked depression of mean CBF. Eight patients (50%) sustained a cerebral infarction; four had a poor clinical outcome and two died. It is clear that arterial spasm was the major factor in producing unstable hemodynamics in both of the above studies and that this results in a very high risk of neurologic deterioration following surgery.

We would conclude that inhalation CBF measurements can provide information which is clinically useful in the management of patients with subarachnoid hemorrhage. Patients with low clinical grade but normal CBF usually have discrete focal lesions and are unlikely to develop ischemia during or following surgery. In contrast, CBF studies showing abnormally low flow or instability from one study to the next should alert one to the need for angiography and caution regarding surgery. It is important to note that over 30% of the SAH patients studied fell into this category.

Acknowledgments. This study was supported by grants from the Ontario Heart Foundation. We would also like to acknowledge the technical assistance provided by I. Morrison.

References

Farrar JK, Gamache FW, Ferguson GG et al. (1981) Effects of profound hypotension on cerebral blood flow during surgery for intracranial aneurysms. J Neurosurg 55:857 – 864
Farrar JK, Meguro K, Ferguson GG et al. (1983) Cerebrovascular instability following subarachnoid hemorrhage: relation to early CT scan and the effects of surgery. J Cereb Blood Flow Metab 3 (Suppl 1):59 – 60

Utility of Noninvasive rCBF Measurements for Assessment of Cerebral Function Following Subarachnoid Hemorrhage[1]

A. Brawanski and V. A. Maximilian[2]

Introduction

One of the main problems related to aneurysm surgery is the proper timing of the operative intervention. On the one hand, clipping of the aneurysm should not be performed whilst the patient is either in a poor clinical condition or shows signs of progressive neurologic deficits. A delay in surgery, on the other hand, increases the likelihood of rebleeding. Consequently, for the proper management of patients suffering from subarachnoid hemorrhage (SAH), it is highly desirable to have physiologic parameters that will yield quantitative information on cerebral function. In this context, we began to evaluate the applicability and relevance of xenon 133 clearance regional cerebral blood flow (rCBF) measurements in patients after SAH. The purpose of the present study was to assess the interdependence of CBF, clinical status, and further clinical course for this group.

Patients and Method

All 52 SAH patients had the typical history. Blood in the subarachnoid spaces, and often in the basal cisterns, was documented by CT. The lumbar puncture performed immediately after admission showed blood-tinged spinal fluid. The patients were independently graded and grouped according to the Hunt and Hess (1968) scale, accounting for the neurologic status on the day of the rCBF investigation. The number of male and female patients in each subgroup, their mean ages, and the time interval between the SAH and initial CBF measurement are listed in Table 1, whereas sites of angiographically localized aneurysms are presented in Table 2.

For control, we used the flow values of a group comprising 42 psychiatric patients (24 male, 18 female) with a mean age of 51. All were neurologically inconspicuous, had normal CT scans, and showed no evidence of organic deficiency.

The rCBF investigations were performed with a standard cerebrograph (Novo Diagnostic Systems, Denmark), measuring flow via 16 homologous

1 Supported by DFG-grants (Bra 843/2-1)
2 Neurochirurgische Universitätsklinik, Josef-Schneider-Strasse 11, D-8700 Würzburg

Cerebral Blood Flow and
Metabolism Measurement.
Eds. Hartmann/Hoyer
© Springer-Verlag Berlin Heidelberg 1985

Table 1. Number of patients (male and female), mean age, and mean time interval between SAH and rCBF study in each Hunt-Hess group

	n	♂	♀	Mean age (years)	CBF after SAH/days
HH 1	16	8	8	47.06	12.44
HH 2	12	4	8	54.92	8.92
HH 3	17	2	15	54.06	14.71
HH 4	7	5	2	49.43	14.14
Mean	52	19	33	51.00	12.34

Table 2. Location of the aneurysms in each Hunt-Hess group, as found by angiography

	A. comm. ant.	media	carotis	basilaris	None found
HH 1	4	2	1	1	8
HH 2	2	2	1	1	6
HH 3	6	3	2	0	6
HH 4	3	1	1	0	2
	15	8	5	2	22

probes placed in parallel over each cerebral hemisphere. All rCBF measurements were performed with the xenon 133 inhalation technique (for a more detailed description of the measurement procedure and the underlying methodology, see Obrist et al. 1975; Risberg 1980). Using a specially developed respirator at our laboratory connected to the cerebrograph, we were also able to measure CBF in three automatically ventilated patients of group HH$_4$. The CBF data were calculated using the Obrist et al. (1975) model, and are here reported in initial slope index (ISI) units. None of the original results were corrected for

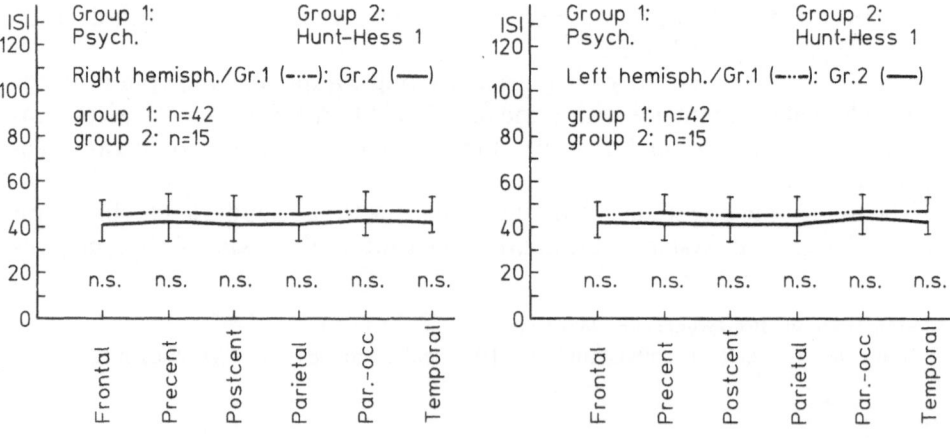

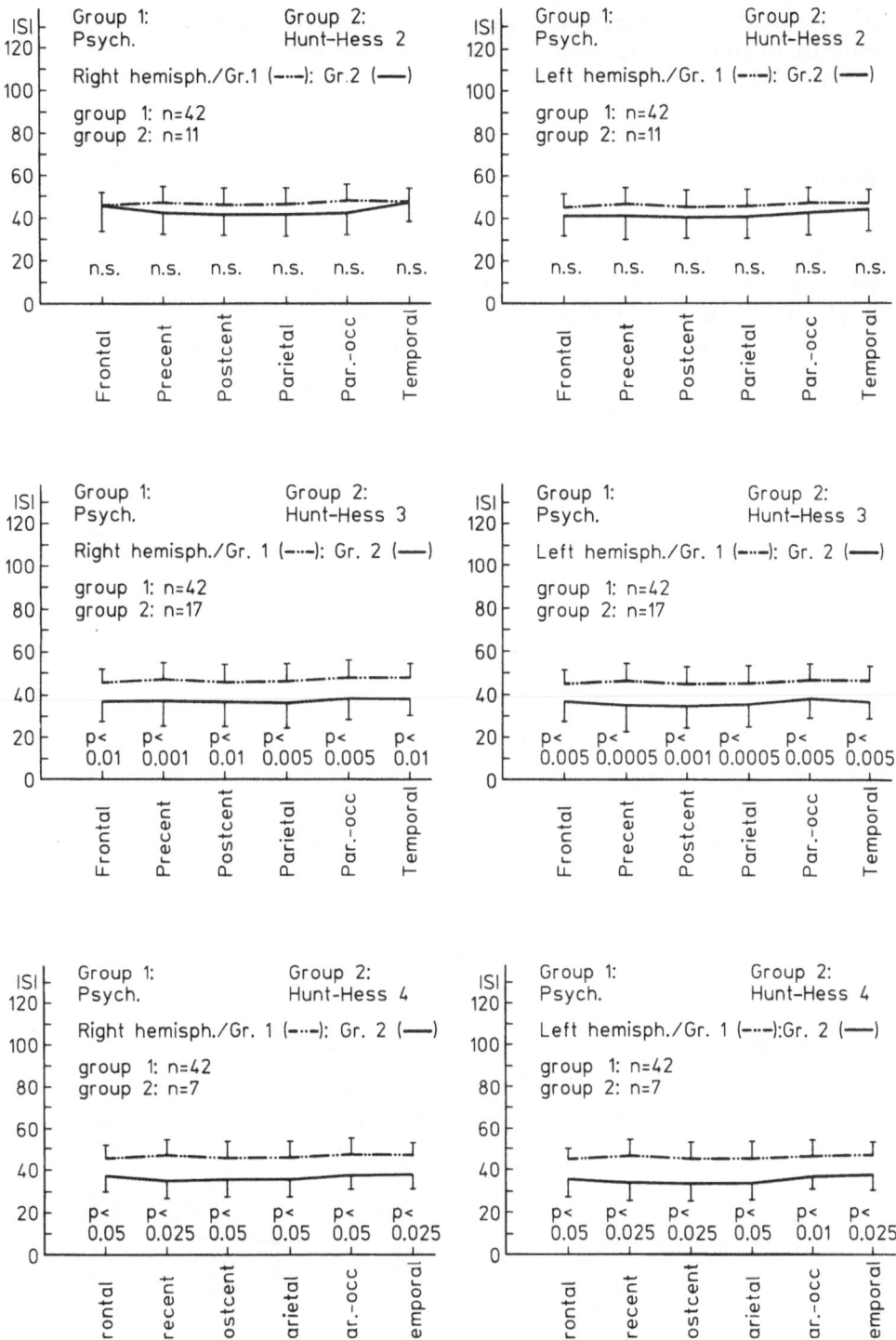

Fig. 1. ISI values of the Hunt-Hess groups (−··−) compared with the psychiatric group (—). Below the ROIs and the level of significance are given

pCO$_2$ variances, assuming that normal pCO$_2$ reactivity could be disregarded in patients following SAH. The statistical analysis consisted of t-tests for dependent and independent samples.

Results

As shown in Table 3, mean hemispheric ISI clearly decreases with clinical deterioration. The mean ISI levels of groups HH$_3$and HH$_4$ were significantly lower ($P < 0.05$) than those of groups HH$_1$ and HH$_2$ and psychiatric patients. In order to gain a more practical survey of regional flows we gathered detectors over specific regions of interest (ROI) and calculated their corresponding average ISI values. As can be seen in Fig. 1, significant flow differences between controls and SAH patient groups HH$_3$ and HH$_4$ were found in all ROIs.

A further subdivision of each HH group into patients either showing subsequent improvement or deterioration was performed pending a clinical reevaluation 1−2 weeks after the initial rCBF measurement. The results are summarized in Fig. 2. In group HH$_1$ the deteriorating patients had significantly

Table 3. Mean hemispheric flow (ISI) of the psychiatric group and the Hunt-Hess groups [+ indicates a significant difference ($p < 0.05$)]

Group	n	X̄ CBF (ISI units)	
		L. hem	R. hem
Controls	41	46.24 (7.4)	46.75 (7.9)
HH 1	16	42.93 (6.8)	42.08 (6.3)
HH 2	12	42.23 (6.7)	43.37 (7.2)
HH 3	17	36.56 (7.3)[+]	37.34 (6.9)[+]
HH 4	7	35.45 (7.1)[+]	36.75 (7.0)[+]

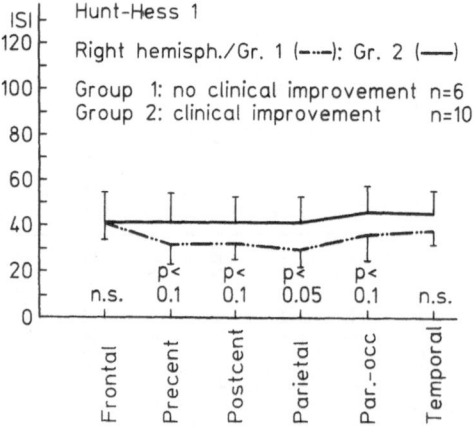

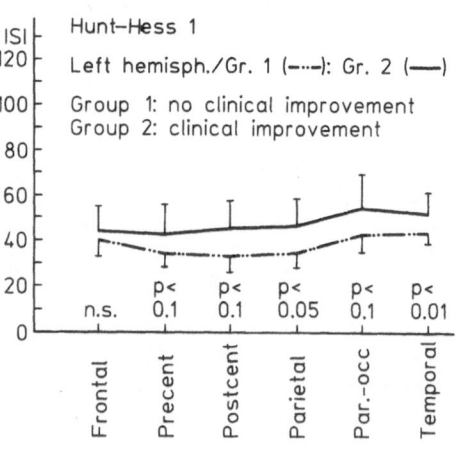

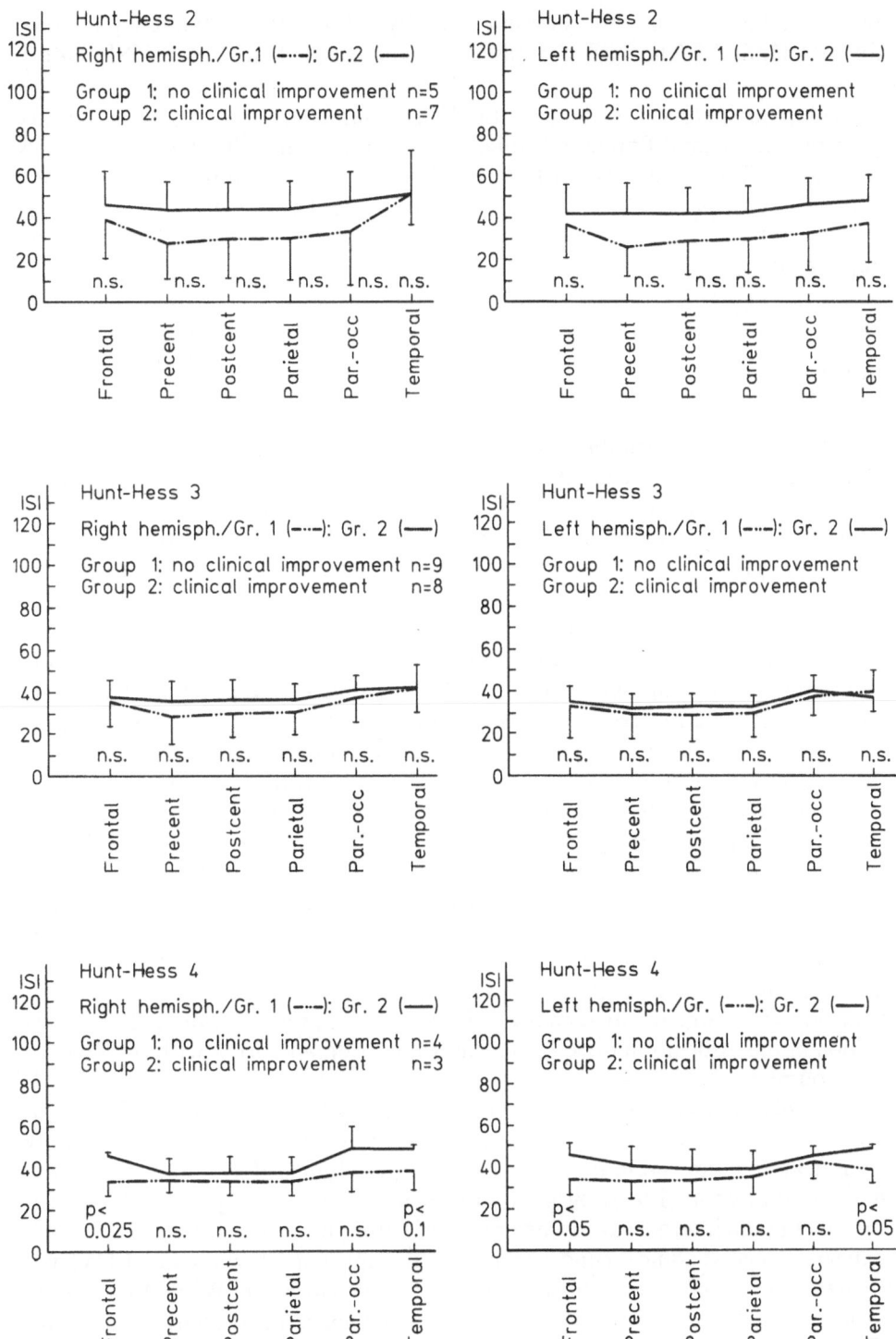

Fig. 2. ISI values of patients with clinical improvement (—) or deterioration (−··−) after the CBF study in each Hunt-Hess group. Below the ROIs and the level of significance are given

lower CBF levels in all regions except frontally, with a similar tendency observable in temporal ROIs. CBF decreased to around 30% below mean ISI control values.

A comparable trend was found in the HH_2 group as well, though with no statistically significant differences (probably due to the somewhat higher standard deviations). There were no distinctive changes between improving and deteriorating HH_3 patients, but in the HH_4 group significant differences were recorded in frontotemporal areas, whereby flows were nearly identical centroparietally.

Discussion

In the present study cerebral function of patients with aneurysmal SAH was estimated by rCBF measurements. Our SAH patient population had a preponderance of females (60%), a mean age of 51, and prevalence of aneurysm localization in the anterior part of the circle of Willis, closely corresponding to the SAH population profiles reported by others (e.g., Boullin, 1980). The choice of controls (restricted by local regulations) was not ideal, especially because regional flow distribution lacked a clear hyperfrontality. We found, however, very similar CBF levels and distribution patterns in a group of 24 patients suffering from low pressure glaucoma. The CBF results of the psychiatric patients could therefore be representative for a hospitalized population, possibly comprising a more adequate reference group than young and healthy volunteers (this aspect will be studied in more detail at our laboratory). Nevertheless, we found marked flow differences between the SAH patients and psychiatric controls. Moreover, patients of groups HH_1 and HH_2 with subsequent clinical deterioration showed a clear decrease in CBF parietally and centrally, whereby frontal and temporal regions remained relatively unaffected. Such centroparietal hypoperfusion was not found in deteriorating HH_3 and HH_4 patients, probably due to the already significantly reduced cerebral circulation. We observed, however, additional CBF decreases frontoparietally in deteriorating HH_4 patients, who were practically comatose. This aspect supports Ingvar's (1976) contention of the association of rCBF in the frontal human cortex with the level of consciousness.

Generally, we found that CBF decreases in conjunction with clinical deterioration first became evident in central and parietal ROIs representing the supply area of the middle cerebral artery; with further clinical deterioration, the hypoperfusion extended to frontotemporal locations, now comprising the perfusion area of the internal carotid artery. In addition, we repeatedly observed that patients in a good clinical condition having unexpectedly low mean CBF values showed a clear tendency to subsequent clinical deterioration. While this finding yields a valuable parameter for prognosis of further clinical course, it remains to be validated and statistically analyzed in a larger group of patients.

Finally it should be noted that the CBF measurements were performed within the high incidence time interval of vasospasm, generally between the 8th and 14th day after SAH (Grubb et al. 1977). Vasospasm following SAH is a life-threatening complication. The concepts of vasospasm treatment are still under discussion, but it is generally accepted that clipping of the aneurysm should not be attempted during a vasospastic condition. Consequently, one of the main aims during the clinical care of such patients is detecting and confirming the vasospasm as early as possible, whereby the investigation procedure should not include additional spasm-inducing components. Presently, angiography is still the method of choice, but it is associated with hazardous side-effects (Maurer 1980). At our laboratory, besides CT scanning as an initial diagnostic procedure, noninvasive measurements are performed as soon as is feasible. When there are any hints of a markedly reduced cerebral perfusion coupled with concurrent neurologic deficits, angiography is delayed until the patient shows signs of clear clinical improvement or follow-up CBF measurements indicate global flow increases (Maximilian and Brawanski 1984). In conclusion, we propose that noninvasive rCBF measurements providing a functional correlate to clinical condition serve as a propitious adjunct for clinical evaluation and management of candidates for aneurysmal surgery following SAH.

References

Boullin, D. J. (1980) Cerebral vasospasm. John Wiley and Sons, Chichester-New York-Brisbane-Toronto

Grubb, R. L., Raichle, M., Eichling, J., Gado, M. (1977) Effects of subarachnoid haemorrhage on cerebral blood volume, blood flow and oxygen utilization in humans. J. Neurosurg. *46*, 446–452

Hunt, W. E., Hess, R. M. (1968) Surgical risk as related to time of intervention in the repair of intracranial aneurysms. J. Neurosurg. *28*, 14–19

Ingvar, D. H. (1976) Functional landscape of the dominant hemisphere. Brain *107*, 181–197

Maurer, H. J. (1980) Risiken bei Kontrastmitteluntersuchungen. Dtsch. Ärzteblatt *77*, 1555–1564

Maximilian, V. A., Brawanski, A. (1984) Application of nuclear medicine procedures in neurosurgery: Advantages of 133-xenon measurements for assessment of functional changes in the brain. In: Schmidt, H. A. E., Adam, W. E. (eds): Imaging of metabolism and organ function, Schattauer Verlag, Stuttgart

Obrist, W. D., Thompson, H. K., Wang, H. W., Wilkinson W. E. (1975) Regional blood flow estimated by 133-xenon inhalation. Stroke *6*, 245–256

Risberg, J. (1980) Regional cerebral blood flow measurements by 133-xenon inhalation. Methodology and applications in psychiatry. Brain and Language *9*, 9–34

nrCBF for the Assessment of Cerebrovascular Spasm and the Timing of Angiography and Operation in Subarachnoid Hemorrhage

G. Meinig, P. Ulrich, R. Pesch, and K. Schürmann

Introduction

With the aid of atraumatic measurement of cerebral blood flow (xenon inhalation) we attempted to detect the development and course of cerebral vasospasm and the influence of treatment; moreover we tried to establish the consequences for the timing of the aneurysm operation and cerebral angiography. In the present paper, preliminary results on single observations will be presented because of the small number of cases.

Methods

The nrCBF measurements were performed under resting conditions in the supine position with a ^{133}Xe inhalation system (Novo diagnostic systems, Hadsund) simultaneously monitoring 16 homologous regions of both hemispheres. Calculations were performed with a Hewlett Packard (Model 9825/M) desk-top computer using the two-compartmental analysis of Obrist et al. [8] and Risberg et al. [9]. F_1 (the flow of rapidly perfused gray matter) and ISI (the initial slope index calculated between the initial 30 and 90 s after the start of the measurement) were the CBF observations used in this study. The arterial pCO_2 was determined in blood samples from the radial artery by an AVL gas check and controlled recordings of the end-tidal CO_2 concentrations (Novo capnograph).

Calcium antagonist was generally administered for treatment of cerebral vasospasm. In this study we gave nimodipine (Nimotop, Bayer, Germany) in a dose of $1-2$ mg/h in 40 patients. The influence on cerebral blood flow was evaluated by nrCBF follow-up studies in 18 patients.

Results

The correlation between cerebral blood flow and neurologic state described by various authors (e.g. Hunt and Hess [6]) appeared to be confirmed in 46 patients with subarachnoid hemorrhage when the measurement was performed

Universitätsklinik Mainz, Abt. für Neurochirurgie, Langenbeckstrasse 1, D-6500 Mainz

Cerebral Blood Flow and
Metabolism Measurement.
Eds. Hartmann/Hoyer
© Springer-Verlag Berlin Heidelberg 1985

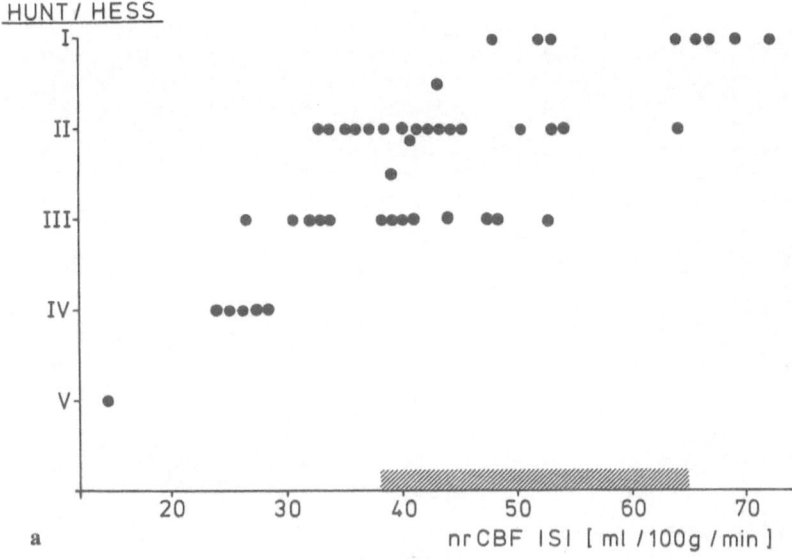

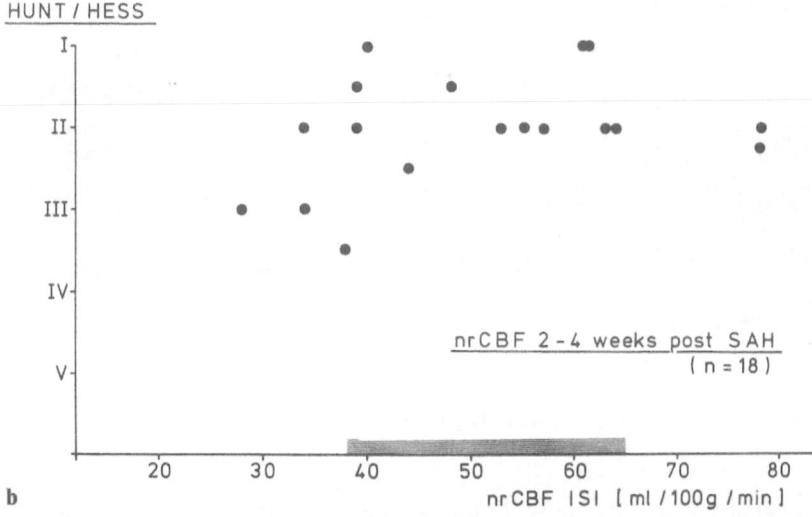

Fig. 1. a Correlation of nrCBF (ISI) and grading after Hunt and Hess in 46 patients with SAH (measurement during the first week after the SAH). **b** Correlation of nrCBF (ISI) and grading after Hunt and Hess in 18 patients with SAH (measurements 2−4 weeks after SAH)

within 1 week, as is to be seen in Fig. 1a. However, five patients in whom we found normal cerebral blood flow values in clinical stages IV − V immediately after the subarachnoid hemorrhage are not included in this figure. The relationship was less unequivocal in 18 patients (Fig. 1b) in whom the cerebral blood flow was measured 2−4 weeks after subarachnoid hemorrhage. Despite normal cerebral blood flow in the majority of these patients, grade I was

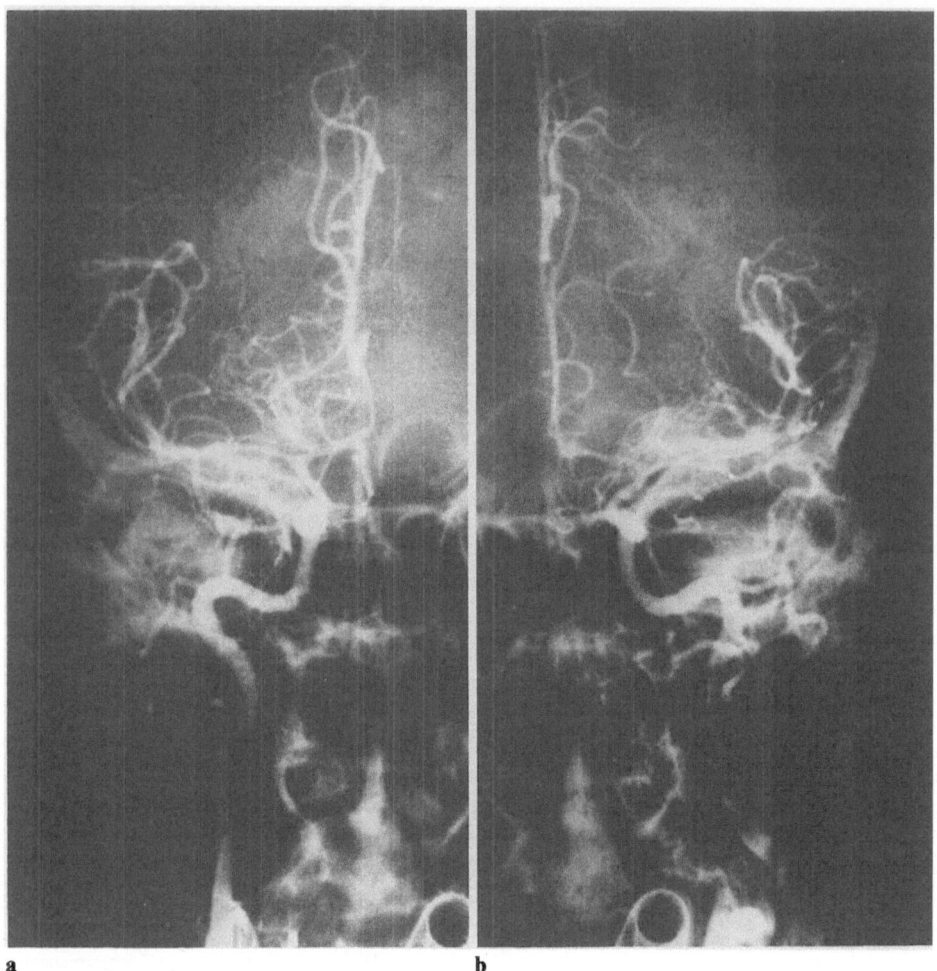

a   b

Fig. 2a, b. 47-year-old female with SAH. Angiography was performed on the sixth day after the bleeding and showed high-grade global bilateral vasospasm

reached only occasionally. In all patients in whom we had diagnosed a cerebral vasospasm angiographically, the result was confirmed by measurement of CBF.

Figure 2a, b shows the angiogram of a 47-year-old female patient which was performed on the sixth day after subarachnoid hemorrhage in stage I according to Hunt and Hess. A high-grade global bilateral vasospasm, which could not be suspected on the basis of clinical findings, was found. Over the following days the patient deteriorated clinically into stage IV (almost V). The measurement of cerebral blood flow finally performed showed a dramatically reduced perfusion in the blood supply on both sides. The ISI (Fig. 3) showed a value of 24 ml/100 g/min. After intravenous administration of 1 mg nimodipine within 1 h, the cerebral blood flow was raised to 30 ml/100 g/min; 2 days later, there was a rise to 39 ml/100 g/min with the continued infusion of nimodipine (1 – 2 mg

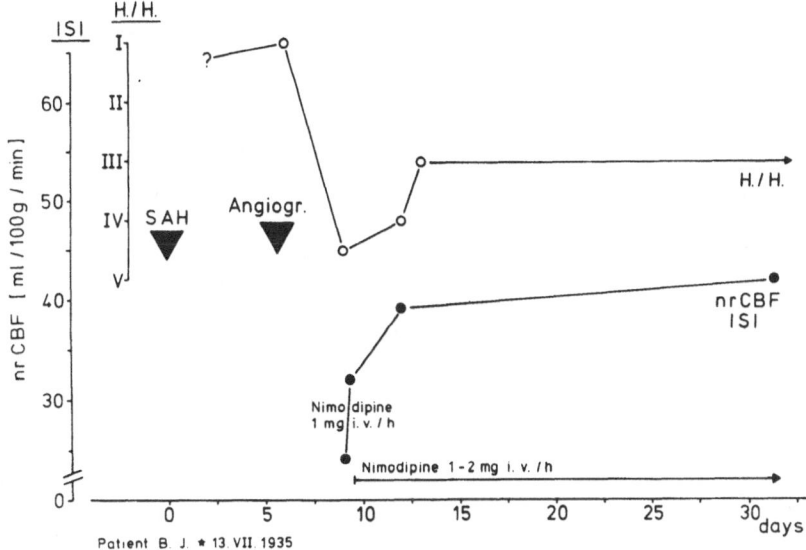

Fig. 3. 47-year-old female (see Fig. 2a, b) with SAH. Clinical course after Hunt and Hess (*open circles*) and nrCBF (*solid circles*) before and during administration of nimodipine

were administered intravenously per hour). In the course of a further 3 weeks (Fig. 3) an almost normal cerebral blood flow was reestablished. However, the severe neurologic symptoms regressed only slowly and incompletely.

In the majority of our 18 cases there was an improvement of cerebral blood flow with nimodipine. However, one can by no means expect dramatic improvement of blood flow in every case, especially when an improvement is impossible because of raised intracranial pressure (hydrocephalus, brain edema, or hematoma), cardiopulmonary disorders, hypoxia, disorders of autoregulation, etc.

Figure 4 shows the course of neurologic change and nrCBF in a 23-year-old man: decline and improvement of the neurologic state run parallel with nrCBF change.

Figure 5 shows the case of a 59-year-old female patient in whom administration of nimodipine resulted in an improvement of cerebral blood flow. After a secondary fall after withdrawal of the medication, nrCBF improved with delay. The discrepancy between CBF and clinical symptoms is striking. By the third day after the subarachnoid hemorrhage, a continuous improvement of clinical symptoms occurred, despite a drop in cerebral blood flow after the withdrawal of nimodipine.

Figure 6 shows the nrCBF of a 36-year-old female 7 days after SAH. This patient was the first (case number 15 of our nrCBF study during nimodipine infusion) in whom cerebral vasospasm occurred during continuous infusion of nimodipine (2 mg/h). First nrCBF performed 3 days after SAH was normal; angiography showed an aneurysm of the MCA and probably a second one of the basilar artery. Except for headache and meningism there was no neurologic deficit. Seven days after SAH, hemiparesis as well as severe aphasia occurred

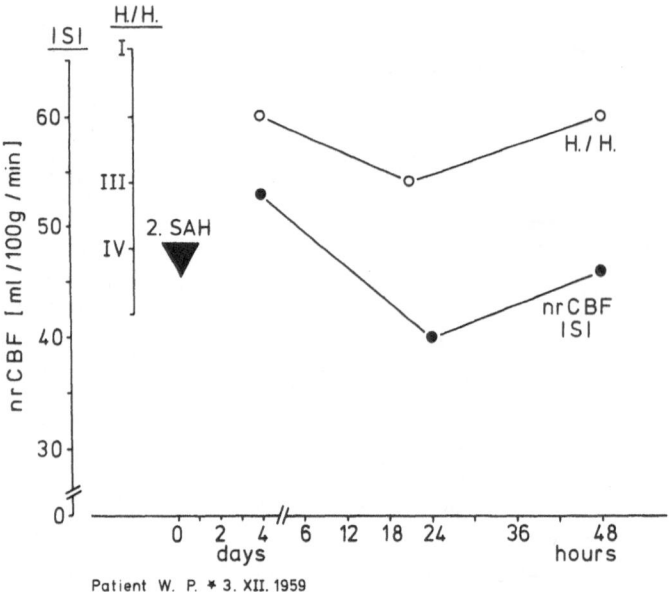

Patient W. P. ✳ 3. XII. 1959

Fig. 4. 23-year-old man with SAH. The neurologic situation corresponds to the nrCBF change

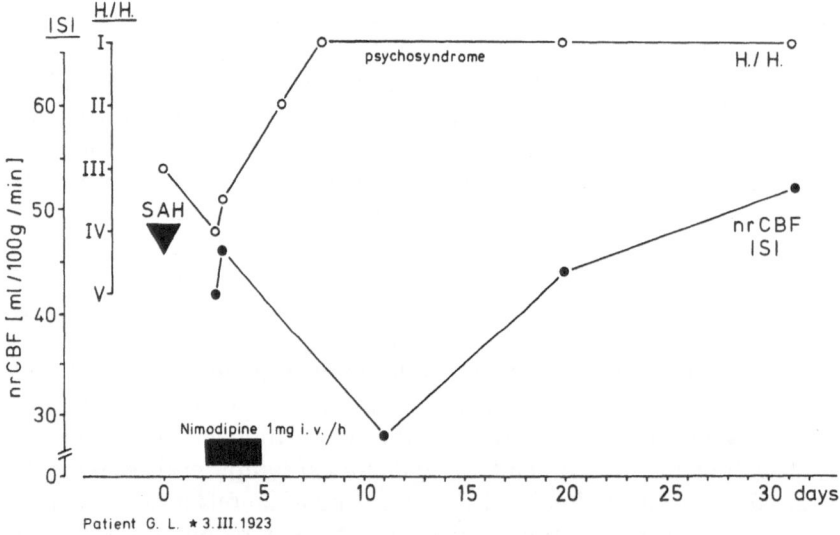

Patient G. L. ✳ 3. III. 1923

Fig. 5. 59-year-old female with SAH. Relationship between clinical course after Hunt and Hess (*white circles*) and nrCBF (*black points*)

(CT showing no rebleeding), disappearing within 4 days during continuation of nimodipine infusion.

In all patients disturbances of cortical blood flow (corresponding to the F_1 value) were evident after the subarachnoid hemorrhage. Regional disorders were likewise to be demonstrated in the distribution pattern, although regional

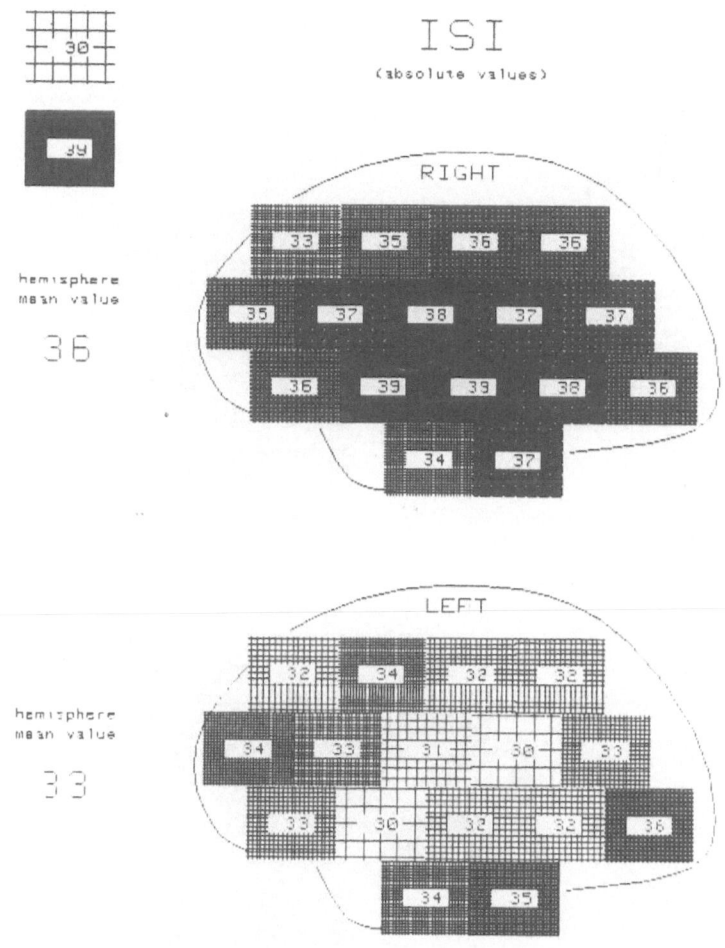

Fig. 6. nrCBF of a 36-year-old female suffering from an aneurysm of the left MCA and probably a second one of the basilar artery 7 days after SAH. The first nrCBF, performed 3 days after the bleeding, was normal; the neurologic situation was primarily normal except for headache and meningism. Seven days after SAH hemiparesis and severe aphasia occurred, nrCBF showed significant decrease of flow on the left side due to spasm since CT showed no rebleeding or signs of ICP increase. Nimodipine (2 mg/h i.v.) was given continuously from the 2nd day after SAH

reductions of the blood flow by 10% or less below the hemispheric mean indicate a trend but cannot be regarded as significant. In most cases the blood flow disorders persisted for weeks (or even months, as in Fig. 7). In our experience they were usually accompanied by mild and transient mental and affective disorders which could not be graded according to the Hunt and Hess scale.

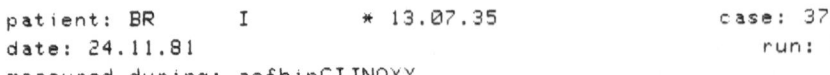

patient: BR   I   * 13.07.35   case: 372
date: 24.11.81   run: 5
measured during: acfhipCIJNOXY

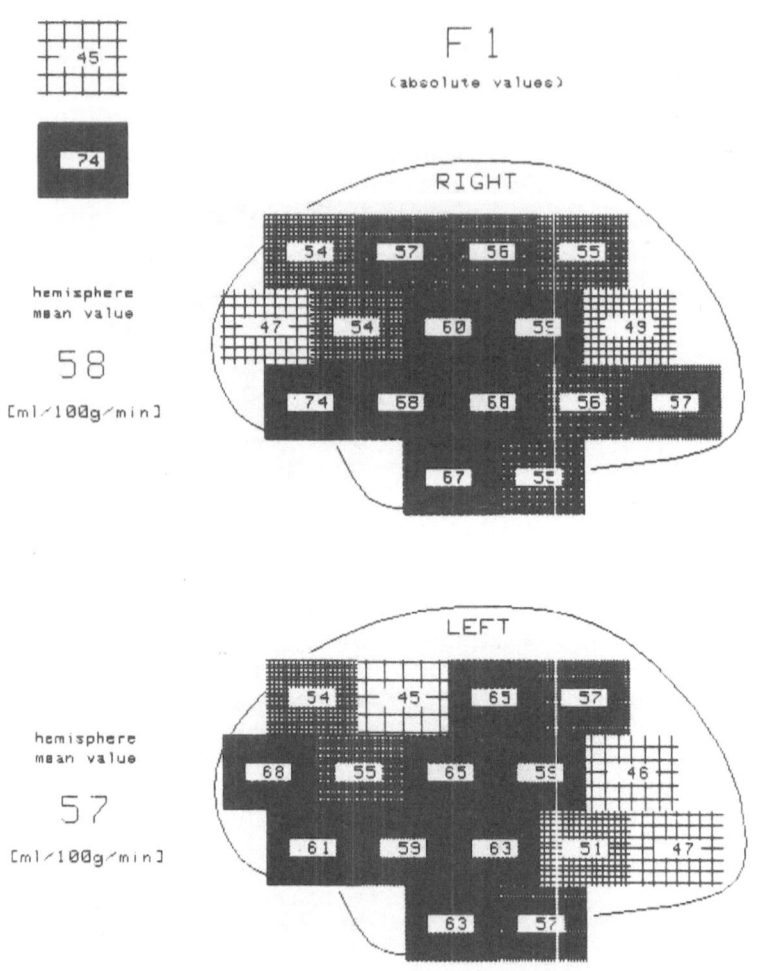

Fig. 7. Altered cortical flow pattern of the right and left hemispheres in a 47-year-old female (see Fig. 2) 5 weeks after SAH

Conclusions

In about 70 patients we carried out measurements of cerebral blood flow applying the xenon inhalation technique. In some cases we did only sporadic measurements, in others, serial investigations. We made the observation that angiographically demonstrated vasospasm could always be confirmed by cerebral blood flow measurement. Moreover, cerebral vasospasm may be detected by CBF measurement even in cases without corresponding neurologic symptoms. Serious damage to the patient by angiography or operation may be avoided by early diagnosis of vasospasm. Thus we perform CBF measurements instead of control angiographies for exclusion of vasospasm immediately before operation. Beyond this, serial nrCBF measurements provide important data for evaluating changes in cerebral blood flow during treatment of cerebral vasospasm. In a group of 18 patients in whom we administered the calcium antagonist Nimodipine intravenously, we observed a pronounced improvement of cerebral blood flow. The alteration of cortical blood flow, which frequently persists over several weeks even in patients whose neurologic deficits completely disappear, requires further observation and correlation to psychological monitoring.

Summary

The ^{133}Xe inhalation method for measurement of rCBF proved to be of some help in avoiding additional damage by angiography or operation in patients with SAH in the period of vasospasm. It may also provide indications of vasospasm in cases without clearly corresponding neurologic signs. Manifest vasospasm in angiography was confirmed by nrCBF in all cases. Alterations of cortical flow persisting for several months are detectable in many cases, correlating well with a longer lasting psycho-organic syndrome (not detectable with the Hunt and Hess scale). Thus nrCBF measurements seem to be suitable and reliable for observation of the efficiency of medical treatment.

Acknowledgment. We are grateful to Prof. Wende, head of the Dept. of Neuroradiology of the University Hospital Mainz, for making available the angiograms shown.

References

1. Drake, C. G. (1968) Surgical risk as related to time of intervention in the repair of intracranial aneurysms. J. Neurosurg. *28*, 19–20
2. Drake, C. G. (1975) Intracranial aneurysms. In: Tower, D. (ed.), The nervous system. Vol. 2. The clinical neurosciences. Raven Press, New York, 287–295
3. Grubb, R. L., Raichle, M. E., Eichling J. O., Gado. M. H. (1977) Effects of subarachnoid hemorrhage on cerebral blood volume, blood flow, and oxygen utilization in humans. J. Neurosurg. *46*, 446–453
4. Hamer, J. (1981) Häufigkeit und klinische Bedeutung des cerebralen Vasospasmus nach aneurysmatischer Subarachnoidalblutung. Nervenarzt *52*, 108–112
5. Hartmann, A., Menzel, J., Buttinger, C., Lange, D., Albert, E. (1981) Die regionale Gehirndurchblutung des Pavians beim ischämischen Hirninfarkt unter Dexamethasonbehandlung. Fortschr. Neurol. Psychiat. *49*, 380–392
6. Hunt, W. E., Hess, R. M. (1968) Surgical risk as related to time of intervention in the repair of intracranial aneurysms. J. Neurosurg. *28*, 14–20
7. Ishii, R. (1979) Regional cerebral blood flow in patients with ruptured intracranial aneurysms. J. Neurosurg. *50*, 587–594
8. Obrist, W. D., Thompson, H. K., Wang, H. S. (1975) Regional cerebral blood flow estimated by ^{133}xenon inhalation. Stroke *6*, 246–256
9. Risberg, J., Ali, Z., Wilson, E. M. (1975) Regional cerebral blood flow by ^{133}xenon inhalation. Stroke *6*, 142–148

Specific Calcium Antagonism: A New Therapeutic Principle in Cerebrovascular Diseases?

M. R. Gaab[1], C. P. Rode[4], I. Haubitz[3], J. Bockhorn[2], and A. Brawanski[2]

Introduction

"Calcium antagonism" now plays an accepted therapeutic role in cardiovascular diseases. Especially in coronary spasm, calcium antagonists have become the therapy of first choice. According to Allen et al. (1979), Hayashi et al. (1977), and Towart (1981), the cerebral vascular smooth muscle is still more sensitive to drugs blocking the calcium influx. Especially the dihydropyridine named nimodipine, a derivative of nifedipine, shows a predilective cerebrovascular action in animal experiments. Here, it was able to inhibit the vascular spasm following experimental subarachnoid hemorrhage as well as the postischemic impaired reperfusion in different models of cerebral ischemia. According to Kazda et al. (1979), Hoffmeister et al. (1979), and Takagi et al. (1979), nimodipine particularly inhibits the spasmogenic calcium influx following potassium depolarization of the vascular smooth muscle. This mechanism may also have great importance in vasospasm after SAH (Brawanski et al. 1982; Gaab et al. 1982).

Patients and Methods

We studied the acute effects of nimodipine on the CBF and related parameters in 42 patients, whose diagnoses are shown in Table 1. The group with vasospasm is very homogeneous, consisting entirely of patients who had aneurysms of the anterior cerebral artery causing the hemorrhage. The stroke patients had reversible, or only mildly irreversible ischemic attacks of the "minor stroke" type. The short interval between the onset of the disease and nimodipine administration is noteworthy.

We investigated the effect of a single oral dose of nimodipine on these parameters after 1 h. This interval of 60 min is suggested by pharmacologic data showing that nimodipine reaches its maximum plasma value at this time. The ISI was used as the CBF parameter; however, the results with F1 were exactly the same (Gaab et al. 1982).

1 Neurochirurgische Klinik der Universität, Alser Strasse 4, A-1090 Wien
2 Neurochirurgische Universitätsklinik, Josef-Schneider-Strasse 11, D-8700 Würzburg
3 Computer-Center der Universität, D-8700 Würzburg
4 Bayer AG, Wuppertal

Cerebral Blood Flow and
Metabolism Measurement.
Eds. Hartmann/Hoyer
© Springer-Verlag Berlin Heidelberg 1985

Table 1. Patients and therapy

Groups	n	Age (yrs)	Therapy	Diagnosis
Controls	12	38–62	None – "test–retest" after 60 min	"arteriosclerotic" ischemic stroke (TIA, PRIND, minor stroke) within 4 weeks following attack
Stroke	25	33–67	Nimodipine p.o. 40–80 mg	
Vasospasm	11	21–58	Nimodipine p.o. 40–80 mg	Vasospasm after SAH (within 8 days following attack)
Vasospasm	6	22–46	Nimodipine i.v. 0.5–2 mg/h	

Table 2. Methods and evaluation

Methods	Time (min) (0 = before)
rCBF by ^{133}Xe inhalation, *ISI* (Novo Cerebrograph, 32 detectors)	0/60
Art. pCO_2 (microprobe/earlobe)	0/60
Blood pressure (atraumatic)	0/30/60
Hb, HCT, blood cells	0/60
Plasma levels of nimodipine	0/60

Evaluation:
Paired values before/after nimodipine administration.
Wilcoxon rank and other parameter-free tests

Our methods of investigation and evaluation are shown on Table 2. All patients were investigated in the supine position in a dark room after at least 15 min resting. The washout curves were calculated with Obrist and Fourier software. The 32 detector regions were grouped into 14 anatomic areas and compared before and after therapy (Gaab et al. 1982).

Results

Initially, we tested the reliability of our inhalation machine and the stability of the CBF. These 12 control patients received a placebo (lactose) and were investigated after an interval of 1 h. Comparing these test – retest values (Figs. 1, 2), we find an almost identical regional and global cerebral blood flow in both investigations. This stability of the uninfluenced CBF, blood pressure, and pCO_2 with our investigation technique should allow a reliable demonstration of real drug effects (Figs. 1, 2; Gaab et al. 1982; Gaab 1983). Regarding only the mean CBF of all regions in our 36 patients, we find an increase of the flow of about

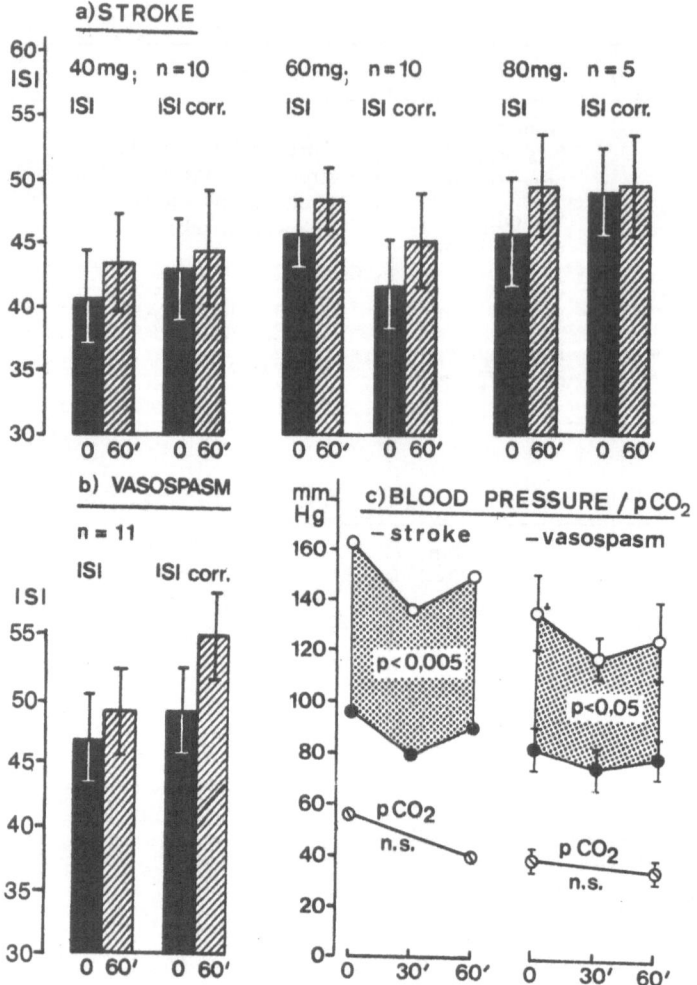

Fig. 1a–c. *CBF,* blood pressure, and arterial pCO_2 after nimodipine p.o.. The increase in CBF (measured as ISI) 60 min after nimodipine (*shadowed*) is more pronounced in vasospasm (**b**) than in stroke (**a**); the highest increase is achieved with 60 mg p.o. (**a**). pCO_2 is not significantly altered, whereas the drop in blood pressure (**c**) is significant

8% (Gaab 1983). This increase seems rather small; however, it is significant in Wilcoxon's rank test. No differences were found between actual and pCO_2-corrected values; the arterial CO_2 tension is not significantly altered by nimodipine. The blood pressure, however, is slightly decreased; the fall in systolic pressure is more pronounced, reaching (Fig. 1) its lowest value 30 min after administration of nimodipine (Gaab 1983, Gaab et al. 1982). If we compare the patients with spasm with those suffering from ischemic stroke with underlying arteriosclerosis, the effect of nimodipine is clearly greater in the *spastic* vessel (Figs. 1, 2). Here we see a mean CBF increase of 14%, whereas it remains as low as 6% in ischemic stroke, although the dosage of nimodipine was equal in both groups. And only the CBF increase in vasospasm is statistically significant.

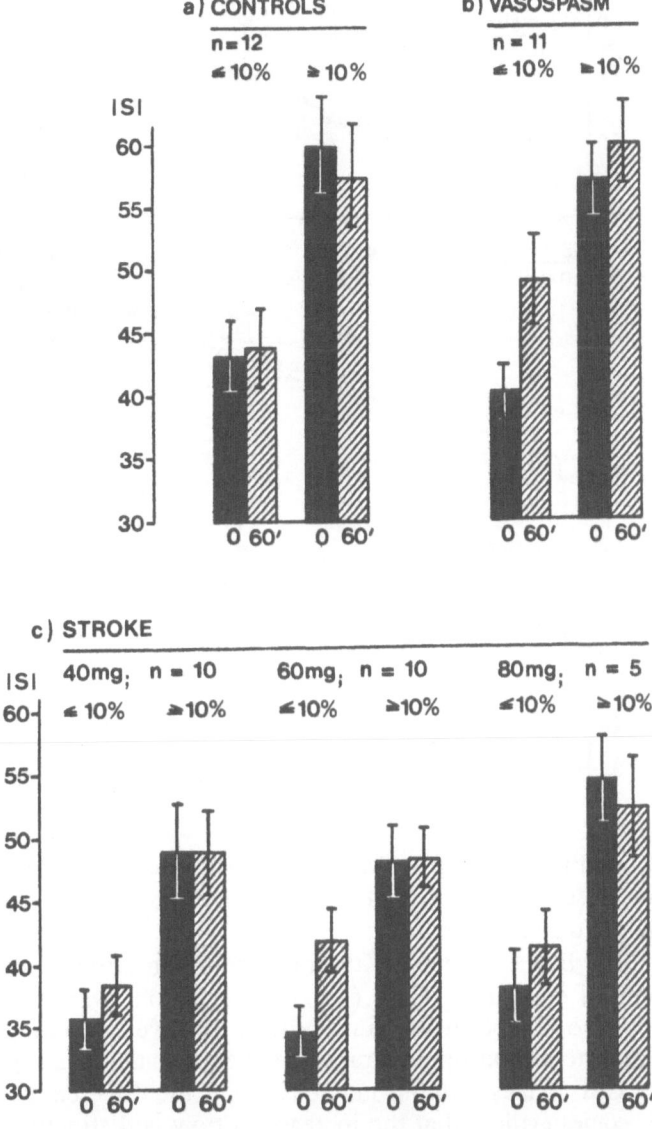

Fig. 2a–c. Reaction of the unequally perfused brain regions to nimodipine. Only the CBF in those regions with a perfusion of at least 10% below the mean hemispheric value markedly increase after nimodipine, whereas the "high-flow areas" with a rCBF of at least 10% more than the mean hemispheric flow remain unchanged. Identical values in the control group (**a**)

If we compare the reaction of the unequally perfused regions, no indication of a steal effect is seen: By comparing the reaction of the cerebral regions (Fig. 2) which have at least a 10% lower perfusion than the mean hemispheric value with those which have a perfusion of 10% or more above this mean hemispheric value, the effect of nimodipine in the low flow areas is significantly better than in the high flow regions. This desirable preference of the action of

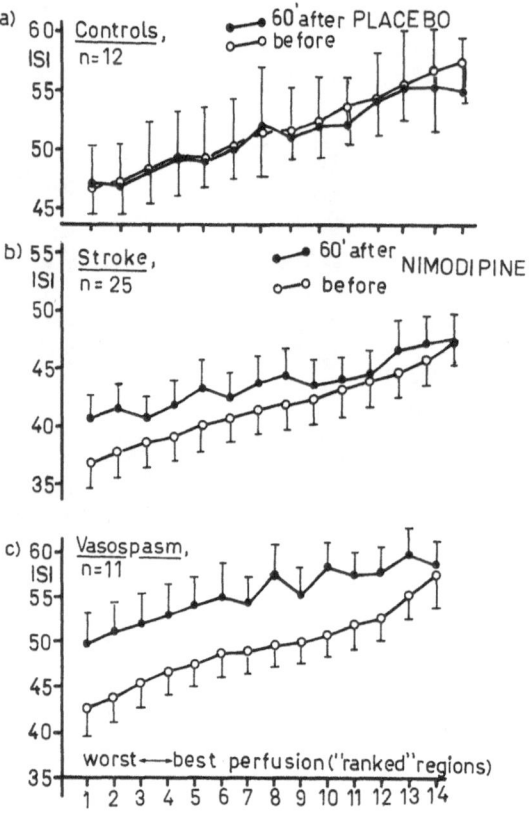

Fig. 3a – c. Different regional activity of nimodipine. By plotting the different brain regions at initial rCBF investigation from worst (*1*) to best (*14*) regional perfusion (*circles*), the regional activity of nimodipine can be evaluated by comparing the corresponding rCBF 60 min after nimodipine (*dots*). Nimodipine predominantly increases the rCBF in the low-flow regions; there is no indication of a steal phenomenon. Identical values in the control group (**a**)!

nimodipine for the impaired areas is more pronounced in stroke patients than in those with vasospasm, which is a more diffuse disturbance with less interregional differences.

The best evidence for the absence of an adverse steal effect is given by a ranked evaluation. Here (Fig. 3), the regions were ordered in sequence from worst to best perfusion during first CBF investigation. The reaction of these regions to nimodipine therapy was then calculated and plotted; thus, the increase of CBF in these regions may be seen as the difference between the two lines. It becomes evident that the increase in flow is distinctly higher in the worst perfused areas; the improvement here may exceed 20%, whereas the areas which are initially well perfused do not react further to Nimodipine. This preference of nimodipine activity for low perfused areas can again only be demonstrated in stroke patients; in vasospasm, there is an overall and relatively uniform rise of CBF (Fig. 3c).

Conclusions

According to our results, nimodipine (a) markedly and significantly improves the (r)CBF in cerebral vasospasm after SAH, (b) only moderately increases

CBF in "arteriosclerotic" *stroke*, and (c) produces no steal effect, but (d) predominantly increases rCBF of initially lower perfused areas and (e) significantly decreases blood pressure, especially in hypertensive patients.

Discussion

Our conclusions are open to discussion; one may especially question the validity of the relatively small overall increase in CBF after nimodipine. Kohlmeyer and Blessing (1978) call for at least a 7% CBF increase for reliability. Our results following nimodipine show a rise in CBF just above 7%, and when subjected to careful statistical evaluation our findings were significant at least in the vasospasm group. The reliability of our investigation method is demonstrated by the identical CBF values in the test − retest group.

The favorable effect of nimodipine on the CBF in patients with vasospasm suggests routine nimodipine administration in patients suffering from SAH after aneurysm rupture, especially if signs of cerebral ischemia are seen. Our initial clinical observations of a rapid clinical improvement of vasospasm (Gaab et al. 1982) after nimodipine have now been confirmed by a double-blind controlled study; according to these results reported by Allen et al. (1983), even routine prophylactic administration of nimodipine may be recommended after SAH.

This cerebrovascular action of nimodipine, however, may have negative effects in patients suffering from an already extended edema; here the rise in ICP, probably due to the increase in CBV, may become critical, as it is associated with an overproportional decrease in CPP (drop in blood pressure, Gaab et al. 1983).

Acknowledgment. This work was supported by grants from the DFG (Ga 273/1) and the VW Foundation.

References

Allen GS, Bahr AL, Banghart SB (1979) Cerebral arterial spasm, parts 9 and 10. Neurosurgery 4:37−46

Allen GS, Ahn HS, Preziosi TJ et al. (1983) Cerebral arterial spasm − a controlled trial of nimodipine in patients with subarachnoid hemorrhage. New Engl J Med 308:619−624

Brawanski A, Gaab MR, Bockhorn J, Haubitz I (1982) Atraumatic rCBF measurement: An aid in the timing of surgery and the management of spasm following SAH. In: Auer LM, Heppner F, Symon L (eds) Aneurysm surgery in the acute stage. Springer, Wien−New York, p 43−51

Gaab M (1983) CBF and ICP during nimodipine therapy in neurosurgical patients. Excerpta Medica, Amsterdam, in press

Gaab MR, Brawanski A, Bockhorn J et al. (1982) Calcium antagonism: a new therapeutic principle in stroke and cerebral vasospasm? rCBF Bulletin (Novo, Copenhagen) 3:47−51

Hayashi M, Marukawa S, Fuji H et al. (1977) Intracranial hypertension in patients with ruptured intracranial aneurysm. J Neurosurg 46:584−590

Hoffmeister F, Kazda S, Krause HP (1979) Influence of nimodipine (Bay e 9736) on the post-ischemic changes of brain function. Acta Neurol Scand 60, suppl 72:358 – 359

Kazda S, Hoffmeister F, Garthoff B, Towart R (1979) Prevention of the postischemic impaired reperfusion of the brain by nimodipine (Bay e 9736). Acta Neurol Scand 60, suppl 72:302 – 303

Kazda S, Garthoff B, Luckhaus G, Nash G (1982) Prevention of cerebrovascular lesions and mortality in stroke-prone spontaneously hypertensive rats by the calcium antagonist nimodipine. In: Godfraid T, Albertini A, Paoletti R (eds) Calcium modulators. Elsevier, Amsterdam, p 155 – 167

Kohlmeyer K, Blessing J (1978) Zur Wirkung von Dihydro-Ergocristin-Methansulfat auf den Hirnkreislauf des Menschen im Akutversuch. Arzneim Forsch/drug res 28 (II):1788 – 1797

Takagi I, Kamiya K, Fukuoka H et al. (1979) The effect of Ca-antagonists on experimental cerebral vasospasm. Acta Neuro Scand 60, suppl 72:486 – 487

Towart R (1981) The selective inhibition of serotonin-induced contractions of rabbit cerebral vascular smooth muscle by calcium antagonistic dihydropyridines. An investigation of the mechanism of action of nimodipine. Circ Res 48:560 – 657

Calcium Antagonists for the Treatment of Cerebral Ischemia?

L. M. AUER

In recent years, calcium entry blockers have been discussed as potentially beneficial drugs for the prevention and treatment of cerebral ischemia via a direct vascular as well as via a metabolic effect. Pathophysiologic circumstances mainly discussed in this context are the acute, the subacute, and the chronic state of ischemic cerebral infarction, the formation of ischemic infarcts following an episode of severe symptomatic vasospasm in patients after subarachnoid hemorrhage from a cerebral aneurysm, as well as migraine.

The basic functions attributed to such drugs are the improvement of cerebral blood flow and metabolism, the prevention of metabolic disturbances, and support of functional recovery following an ischemic episode. In more concrete terms, the improvement of cerebral blood flow should be achieved by dilatation of resistance vessels, especially the opening of collateral circulation; moreover, lowering of blood viscosity and reduction of platelet aggregability are of interest. The local metabolic disturbances in cerebral ischemia to be prevented or reversed are an elevated extracellular potassium, induced by the ischemia-dependent membrane depolarization, the normalization of elevated intracellular calcium, and the normalization of one of its consequences, namely the disturbed formation of energy-rich phosphates, where the phosphorylation of adenosine-diphosphate into adenosine-triphosphate is disturbed by an increased amount of intracellular calcium. Moreover, intracellular osmolarity is increased as a consequence of membrane depolarization and sodium influx. An increase of phospholipase activity causes destruction of membranes mainly via neutral proteases; primarily however, free fatty acids are accumulated, above all arachidonic acid. One of the reasons why incomplete ischemia ends up with a worse outcome than an episode of complete ischemia is probably the fact that reoxidation leads to the formation of free radicals from arachidonic acid and thereby causes delayed cellular death.

To counter these pathophysiologic processes by therapeutic measures, Siesjö (1981) suggested the use of three groups of compounds: calcium antagonists, phospholipase and lipoxygenase blockers, and free radical scavengers. The argument for the use of calcium antagonists was mainly derived from the observation that an increased influx of calcium is able to turn reversible ischemic neuronal dysfunction into irreversible damage (Farber 1981; Hass 1981; Siesjö 1981).

Universitätsklinik für Neurochirurgie, A-8036 Graz

Cerebral Blood Flow and
Metabolism Measurement.
Eds. Hartmann/Hoyer
© Springer-Verlag Berlin Heidelberg 1985

Calcium antagonists have been considered as potential candidates not only for their metabolic effect but also, if not predominantly, for their well-known direct vascular activity. It has to be stressed that such a vasodilatory effect of calcium antagonists varies considerably between different compounds; this circumstance forces us to analyze at least each group of compounds carefully for its cerebrovascular effect as compared with the systemic effect. The cerebrovascular dilatory effect would be welcome for brain regions with reversible cerebral ischemia, provided that flow is elevated to these areas, i.e., there is no intracerebral steal effect which elevates flow to healthy regions and further deprives the ischemic area. In the acute stage of cerebral ischemic stroke, a penumbra zone around irreversibly damaged tissue has been described with blood flows which are too low for neuronal function but are sufficient for cell survival by preservation of structural metabolism (Astrup et al. 1977). For the subacute and chronic states of cerebral infarction, it has not yet become clear whether and under which circumstances penumbra areas do exist around definitely infarcted zones. Low blood flow in the neighborhood of an infarcted area could in fact be a sign of graded ischemia around an infarct, causing a slow progression of neuronal death around the primary infarct. However, this hypoperfusion could be a relative one which reflects an adaptation of flow to metabolic demands which are reduced according to graded and/or selective neuronal death as a consequence of the initial ischemic event (Auer et al. 1982).

Recent observations in patients (Olsen et al. 1983) have shown that peri-infarct tissue not only shows a timely sequence of initial hypoperfusion followed by hyperemia again followed by hypoperfusion; rather in addition a regional dissociation of proximal hyperemia and distal oligaemia have been observed simultaneously, and the hypoperfused area was regularly several times larger than the definite infarct seen on computerized tomography.

In subarachnoid hemorrhage, the vascular effect of calcium antagonists has become of interest mainly for prevention of slowly progressing spastic contraction of main feeding arteries as well as resistance vessels in their periphery, caused by a multiplicity of substances accumulated in the perivascular space,

Table 1. Calcium entry blockers used in recent years

Substance	Authors
Verapamil	Bedford et al. 1983
	Edvinsson et al. 1983
	Golenhofen and Hermstein 1975
	Haeusler 1972
	Hayashi et al. 1977
	Kovach et al. 1983
	Latchaw et al. 1983
	Leblanc et al. 1983
	Reedy et al. 1983
	Roy et al. 1983
	Shimizu et al. 1980
	Van Nueten and Vanhoutte 1981

Table 1. (continued)

Substance	Authors
Flunarizine	Hossmann et al. 1983 Newberg et al. 1983 Van Nueten and Vanhoutte 1981 White et al. 1982
Nimodipine	Allen et al. 1983 Edvinsson et al. 1981 Gaab et al. 1982 Harper et al. 1981 Harris et al. 1981, 1982 a, b Haws and Heistad 1983 a, b Haws et al. 1983 Heller et al. 1983 Hoffmeister et al. 1979 Kaste et al. 1983 Kazda and Hoffmeister 1979 Kazda and Towart 1982 Kazda et al 1982 a, b; 1983 Mohamed et al. 1983 Müller-Schweinitzer and Neumann 1983 Steen et al. 1983 Symon et al. 1982 Tanaka et al. 1980 Towart and Kazda 1979, 1980 a, b Towart et al. 1982 Van Nueten and Vanhoutte 1981 White et al. 1983
Nifedipine	Allen et al. 1979 Boisvert 1983 Brandt et al. 1979, 1980 a, b, 1981 a, b, 1983 Edvinsson et al. 1981, 1982 Högestätt et al. 1982 Hosobuchi et al. 1982 Mikkelsen et al. 1978 Nowicki et al. 1982 Shimizu et al. 1980 Van Nueten and Vanhoutte 1981
Diltiazem	Harper et al. 1981 Nagao et al. 1978 Pearce et al. 1983 Shimizu et al. 1980
D-600	Hester et al. 1979 Golenhofen and Hermstein 1975 Van Nueten and Vanhoutte 1981
PN200-110	Müller-Schweinitzer and Neumann 1983
Lidoflazine	Van Nueten and Vanhoutte 1981
Nitrendipine	Kazda et al. 1983

especially blood degradation products (Auer 1983a; Boullin 1980; Towart 1982).

The effect of calcium antagonists on vascular smooth muscle cells is to inhibit an excess-influx of calcium ions, which are required for the actin-myosin coupling process. Three groups of membrane Ca^{2+}-channels have been described: (1) Receptor operated channels (ROCs) which open in the presence of agonists such as noradrenaline, serotonin, and other substances. (2) Potential sensitive channels (PSCs), also called ion-sensitive channels, which react to the state of membrane polarization and depolarization; thus, for example, they open in the presence of increased extracellular potassium. (3) Stretch-dependent calcium channels; these have been described only recently (Bevan 1983) and are a new aspect in addition to the well-known effect of an activation of intracellular calcium as a response to stretch (Nakayama 1982). The system is further complicated by a great number of further ion-sensitive channels, especially potassium channels which are presently on the verge of being better understood (Henry 1983).

Smooth muscle contraction is also effected via the intracellular calcium stores, which are mobilized not only by stretch but also as a second effect of agonist action at the receptor site.

The strongest calcium antagonistic effect is known from metals such as La^{3+}, followed by Cd^{2+}, Co^{2+}, Ni^{2+}, Mn^{2+}, and Mg^{2+} in that sequence; these substances also change the surface potential and are not of clinical relevance because of their side-effects (Bolton 1979).

In recent years, a great number of selective blockers of calcium entry to vertebrate heart and smooth muscle have been developed. A list of these compounds is given in Table 1. According to their different channel blocking properties, these substances show different cerebrovascular effects. However, none of these compounds has been investigated to answer all questions posed at the beginning of this contribution.

However, it is worth mentioning that different effects of different drugs derive in part from their predominant effect on ROCs or PSCs as well as the distribution of the one and the other type of channel on the vascular tree of the brain and the remaining peripheral circulation.

Vascular Effects

The most convincing vascular effect has been observed with substances mainly blocking the ROCs such as nifedipine and nimodipine; accordingly, these compounds have been best investigated until now (among the PSC-blockers, verapamil and D-600 have been mainly investigated). Nimodipine has turned out to be the strongest predominantly cerebroarterial dilator with the most long-lasting effect because of its lipophilia. In vitro, nimodipine has a stronger inhibitory effect on blood-induced constriction of helical strips of canine basilar arteries than do nifedipine and PN 200-110 (Müller-Schweinitzer and Neumann 1983). Nimodipine inhibited serotonin-induced contraction of basilar and middle cerebral arteries (Silverberg et al. 1979).

Verapamil showed a predominantly cerebrovascular effect in inhibiting calcium-induced contraction in vitro when compared with the effect on coronary and mesenteric vessels (Hayashi and Toda 1977). With 1.0 µmol verapamil, the effect was almost identical in all three types of vessels. Prostaglandin (PGF2)-induced spasm was more successfully reversed on cerebral than on mesenteric and coronary arteries; potassium-induced contraction was fully prevented in all three types of vessels (Shimizu et al. 1980). Nitroglycerin and sodium nitroprusside fully prevented prostaglandin-induced contraction of cerebral, coronary, and mesenteric arteries, which has important clinical implications regarding the blood pressure reducing effect of these compounds parallel with their cerebroarterial dilatation properties.

The predominantly cerebroarterial effect of nimodipine and nifedipine has also been demonstrated in vivo: intravenous infusion of nimodipine resulted in a dose-dependent pial arterial dilatation without a significant decrease in blood pressure up to doses of 1 µg/kg/min (Auer 1981; Tanaka et al. 1980).

In the acute stage of cerebral ischemia, a vascular effect might be of special interest when considering the observation that cerebral arteries constrict in the borderzone around a developing infarct (Teasdale et al. 1981). Ischemia-induced pial arterial constriction could be reversed into significant dilatation by perivascular administration of nifedipine (Brandt et al. 1983).

Perivascular application of nimodipine in patients during aneurysm surgery also resulted in arterial dilatation cross-correlated to vessel size (Auer et al. 1984). Likewise, intravenous infusion of nimodipine during EC-IC bypass surgery resulted in significant pial arterial dilatation when compared with infusion of the solute of nimodipine (Auer et al. 1983).

As a consequence of the cerebroarterial dilatory effect of nimodipine, cerebral blood flow was shown to be increased in a group of experiments (Harper et al. 1981; Haws et al. 1983; Kazda et al. 1982b) as well as by Harris et al. (1982a) in an open skull preparation; no effect was seen by Harris et al. in a closed skull preparation or by Ott et al. in patients (1980).

In patients with acute cerebral infarction, Gelmers (1982) observed an increase in CBF with nimodipine; an intracerebral steal effect did not occur. Nimodipine increased cerebral blood flow more than bencyclane and cinnarizine or a comparable dose of papaverine, whereas femoral blood flow was almost unaffected by nimodipine; by contrast bencyclane and cinnarizine had a stronger effect on femoral than on cerebral blood flow (Shimizu et al. 1980). Haws et al. (1983) measured blood flows with the microsphere technique and observed significant cerebral and coronary blood flow increases with nimodipine as opposed to unchanged muscle, bowel, and kidney flows; cerebral venous oxygen saturation and oxygen consumption remained stable. The permeation of nimodipine through the blood-brain barrier is limited, as became apparent from a comparison of flow increases during intravenous infusion and during intracarotid infusion together with osmotic opening of the blood-brain barrier; flow increases in the latter situation were about three-fold those during intravenous infusion (Harper et al. 1981).

Cerebral blood volume as estimated from surface reflectance was increased in a dose-dependent manner by verapamil and D-600 together with stable NADH fluorescence (Kovach et al. 1983).

Following an episode of complete ischemia in dogs, cerebral blood flow fell below baseline after a period of hyperemia in nimodipine-treated and untreated animals, but remained significantly higher in treated animals (Steen et al. 1983). Likewise, after global ischemia in cats, cerebral blood flow during recirculation was significantly less reduced in nimodipine-treated animals (Nowicki et al. 1982). Verapamil could not prevent acute ischemic changes by occlusion of the middle cerebral artery in cats (Reedy et al. 1983), caused inverse steal, and even increased the ischemic areas (Roy et al. 1983). Flunarizine, too, did not improve neurologic outcome, although postischemic hypoperfusion was slightly reduced (Newberg et al. 1983).

Metabolic Effect

Metabolic changes during treatment with calcium antagonists in acute ischemia are less well documented. Despite a significantly higher recirculation flow, CMRO2 was identical in nimodipine- and untreated animals; phosphocreatine and ATP also had the same levels in both groups (Steen et al. 1983). The flow thresholds at which extracellular calcium decreases and potassium rises were higher in nimodipine-treated animals than in untreated animals, and there was a disturbed autoregulatory and CO_2 response in nimodipine-treated animals (Harris et al. 1982a). Calcium metabolism after cerebral ischemia and a period of recirculation has been investigated during treatment with flunarizine: The calcium content of brain samples was the same in treated and untreated animals; electrocorticographic changes and evoked potentials were the same in treated and untreated animals, and recovery was even worse in the flunarizine-treated animals (Hossmann et al. 1983).

Measurements of brain water content following ischemia and recirculation showed a higher degree of edema formation in nimodipine-treated than in untreated animals (Harris et al. 1982a).

Conclusion

Viewed together from the clinical point of view, there are a number of papers indicating a beneficial effect of calcium antagonists in acute cerebral ischemia (Brandt et al. 1983; Kazda and Towart 1982; Kazda et al. 1982a; Karasawa et al. 1982; Mohamed et al. 1983; Nowicki et al. 1982; White et al. 1982). Similarly, the occurrence of delayed ischemic neurologic deficit from vasospasm in patients with subarachnoid hemorrhage could be significantly reduced by their administration (Allen et al. 1983; Auer 1983b, 1984). However, in none of the studies so far performed − neither on ischemic stroke nor on vasospasm − was normalization of elevated intracellular calcium shown. With the present state of the art, data lead to the conclusion that calcium antagonists with a predominantly cerebrovascular effect are probably of benefit via the cerebral blood flow in preventing irreversible ischemia under special circumstances such as subarachnoid hemorrhage, and in treating reversible ischemia (penumbra). A beneficial metabolic effect has not yet been proven.

Calcium antagonists might prove disadvantageous in recirculated areas of severe ischemia by supporting the formation of free radicals via improved flow. Moreover, the formation of brain edema might be increased; this latter concern could be less important in patients with circumscribed infarcts, but might become essential in patients with massive ischemic infarcts and in head-injured patients.

References

Allen GS, Banghart SB (1979) Cerebral arterial spasm: Part 9. In vitro effects of nifedipine on serotonin-, phenylephrine-, and potassium-induced contractions of canine basilar and femoral arteries. Neurosurgery 1:37−42

Allen GS, Ahn HS, Preziosi TJ et al. (1983) Cerebral arterial spasm − a controlled trial of nimodipine in patients with subarachnoid hemorrhage. N Engl J Med 308:619−624

Astrup J, Symon L, Branston NM, Lassen NA (1977) Cortical evoked potential and extracellular K^+ and H^+ at critical levels of brain ischaemia. Stroke 8:51−57

Auer LM (1981) Pial arterial vasodilatation by intravenous nimodipine in cats. Drug Res 31:1423−1425

Auer LM (1983a) Prophylaxis of cerebral ischemic damage from vasoqpasm after subarachnoid hemorrhage. In: Brain protection (Wiedemann K, Hoyer S, eds.) pp. 124−139, Springer Verlag, Berlin/Heidelberg

Auer LM (1983b) Acute surgery of cerebral aneurysms and prevention of symptomatic vasospasm. Acta Neurochir 69:273−281

Auer LM (1984) Acute operation and preventive nimodipine improve outcome in patients with ruptured cerebral aneurysms. Neurosurgery 15:57−66

Auer LM, Mies G, Ebhardt G, Traupe H, Heiss WD (1982) Reduced blood flow and neuronal density in cortex surrounding chronic infarcts in cats. Proc. 6th Int. Symp. on Microsurgical Anastomoses for Cerebral Ischemia, Kyoto

Auer LM, Oberbauer RW, Schalk HV (1983) Human pial vascular reactions to intravenous nimodipine infusion during EC-IC bypass surgery. Stroke 14:210−213

Auer LM, Suzuki A, Yasui N, Ito Z (1984) Intraoperative topical nimodipine after aneurysm clipping. Neurochirurgia 27:36−38

Bedford RF, Dacey R, Winn R, Lynch C III (1983) Adverse impact of a calcium entry-blocker (verapamil) on intracranial pressure in patients with brain tumors. J Neurosurg 59:800−802

Bevan JA (1983) Calcium and vascular tone. In: Proceedings of June 10−11, Workshop held in Dorado Beach, Puerto Rico, pp. 6−14

Boisvert D (1983) In vivo effect of nifedipine and naloxone on primate cerebral arterioles. J Cereb Blood Flow Metab 3, Suppl 1, S552−S553

Bolton TB (1979) Mechanisms of action of transmitters and other substances on smooth muscle. Physiol Rev 59:606−718

Boullin DJ (1980) Cerebral vasospasm. Chichester, J Wiley

Brandt L, Andersson KE, Bengtsson B, Edvinsson L, Ljunggren B, MacKenzie ET (1979) Effects of nifedipine on pial arteriolar calibre: An in vivo study. Surg Neurol 12:349−352

Brandt L, Andersson KE, Bengtsson B, Edvinsson L, Ljunggren B, MacKenzie ET (1980a) Effects of a calcium antagonist on cerebrovascular smooth muscle in vitro and in vivo. In: Cerebral arterial spasm (Wilkins RH, ed) pp. 604–607, Williams–Wilkins, Baltimore

Brandt L, Andersson KE, Edvinsson L, Hindfelt B, Ljunggren B, MacKenzie ET, Tamura A, Teasdale G (1980b) Effects of perivascular microapplication of nifedipine on cat pial arteriolar and venular caliber. In: Pathophysiology and pharmacotherapy of cerebrovascular disorders (Betz E, Grote J, Heuser D, Wüllenweber R, eds) pp. 19–23, Gerhard Witzstrock Verlag, Baden-Baden

Brandt L, Ljunggren B, Andersson KE, Hindfelt B, Teasdale G (1981a) Vasoconstrictive effects of human post-hemorrhagic cerebrospinal fluid on cat pial arterioles in situ. J Neurosurg 54:351–356

Brandt L, Ljunggren B, Andersson KE, Hindfelt B, Teasdale G (1981b) The effect of cerebrospinal fluid from patients with subarachnoid haemorrhage on cat pial arterioles in situ. Acta Neurochir 56:111–144

Brandt L, Ljunggren B, Andersson KE, Edvinsson L, MacKenzie ET, Tamura A, Teasdale G (1983) Effects of topical application of a calcium antagonist (nifedipine) on feline cortical pial microvasculature under normal conditions and in focal ischemia. J Cereb Blood Flow Metab 3:44–50

Edvinsson L, Andersson KE, Brandt L, Ljunggren B, MacKenzie ET, Skärby T, Young A (1981) Effects of Ca++ antagonists on cerebral blood vessels. J Cereb Blood Flow Metab 1, Suppl 1:S344–S345

Edvinsson L, McCulloch J, Tatemoto K, Uddman R (1983a) Neuropeptide Y: A unique system of perivascular nerves mediating contraction of cerebral vessels. J Cereb Blood Flow Metab 3, Suppl 1, S182–S183

Edvinsson L, Johansson BB, Larsson B, MacKenzie ET, Skärby T, Young AR (1983b) Calcium antagonists: Effects on cerebral blood flow and blood-brain barrier permeability in the rat. In press

Farber JL (1981) The role of calcium in cell death. Life Sci 29:1289–1295

Fleckenstein H (1977) Specific pharmacology of calcium in myocardium, cardiac pacemakers, and vascular smooth muscle. Annu Rev Pharmacol Toxicol 17:149–166

Gaab MR, Brawanski A, Bockhorn J, Haubitz I, Rode CHP, Maximilian VA (1982) Calcium antagonism: A new therapeutic principle in stroke and cerebral vasospasm? rCBF Bulletin 3:47–51

Gelmers HJ (1982) Effect of nimodipine (BAY e 9736) on postischaemic cerebrovascular reactivity, as revealed by measuring regional cerebral blood flow (rCBF) In: Auer LM, Heppner F, Symon L (eds) Proc. Symp. Aneurysm Surgery in the Acute Stage. Acta Neurochir 63:283–290

Golenhofen K, Hermstein N (1975) Differentiation of calcium activation mechanisms in vascular smooth muscle by selective suppression with verapamil and D-600. Blood Vessels 12:21–37

Haeusler G (1972) Differential effect of verapamil on excitation-concentration coupling in smooth muscle and on excitation-secretion coupling in adrenergic nerve terminals. J Pharmacol Exp Ther 180:672–682

Harper AM, Craigen L, Kazda S (1981) Effect of the calcium antagonist, nimodipine, on cerebral blood flow and metabolism in the primates. J Cereb Blood Flow Metab 1:349–356

Harris RJ, Symon L, Branston NM, Bayham M (1981) Changes in extracellular calcium activity in cerebral ischaemia. J Cereb Blood Flow Metab 1:203–209

Harris RJ, Branston NM, Symon L, Bayham M, Watson A (1982a) The effects of a calcium antagonist, nimodipine, upon physiological responses of the cerebral vasculature and its possible influence upon focal cerebral ischaemia. Stroke 13:759–766

Harris RJ, Branston NM, Symon L (1982b) The effect of a calcium antagonist on the formation of cerebral ischaemic oedema and ion homeostasis. In: Abstract Book 5th Int Symp on Brain Edema, Groningen, p 56

Hass WK (1981) Beyond cerebral blood flow, metabolism and ischemic thresholds: An examination of the role of calcium in the initiation of cerebral infarction. In: Proc. 10th Salzburg Conference on Cerebral Vascular Diseases, Vol. 3: Cerebral Vascular Disease (Meyer JS, Lechner H, Reivich M, Ott EO, Aranibar A, eds), Amsterdam, Excerpta Medica, pp. 3–17

Haws CW, Heistad DD (1983 a) Effects of nimodipine on cerebral vessels: Role of extracellular calcium in vasoconstrictor responses. Abstracts of the 8th Int. Conf. on Stroke and Cerebral Circulation, San Diego. Stroke 14:5

Haws CW, Heistad DD (1983 b) Cardiovascular effects of nimodipine: I. Effects on distribution of blood flow. II. Inhibition of cerebral vasoconstrictor responses. J Cereb Blood Flow Metab 3, Suppl 1:S524−S525

Haws CW, Gourley JK, Heistad DD (1983) Effects of nimodipine on cerebral blood flow. J Pharmacol exp Ther

Hayashi S, Toda N (1977) Inhibition by Cd^{2+}, verapamil and papaverine of Ca^{2+}-induced contractions in isolated cerebral and peripheral arteries of the dog. Br J Pharmac 60:35−43

Heller V, Poch B, Gaab MR, Sold M (1983) CBF, O_2 metabolism, and O_2 tension in cold-injured rat brain and pharmacological effects of dexamethasone and nimodipine. J Cereb Blood Flow Metab 3, Suppl 1:S536−S537

Henry PD (1983) Clinical pharmacology of calcium antagonists: effects on muscle and nonmuscle cells. In: Proceedings of June 10−11, Workshop held in Dorado Beach, Puerto Rico, pp. 15−23

Hester RK, Weiss GB, Fry WJ (1979) Differing actions of nitroprusside and D-600 on tension and 45Ca fluxes in canine renal arteries. J Pharmacol Exp Ther 208:155−160

Hoffmeister F, Kazda S, Krause HP (1979) Influence of nimodipine (BAY e 9736) on the postischaemic changes of brain function. Acta Neurol Scand 60, Suppl 72:358−359

Högestätt ED, Andersson KE, Edvinsson L (1982) Effects of nifedipine on potassium-induced contraction and noradrenaline release in cerebral and extracranial arteries from rabbit. Acta Physiol Scand 114:283−296

Hosobuchi Y, Baskin DS, Woo SK (1982) Reversal of neurological deficits by opiate antagonist naloxone after cerebral ischemia in animals and humans. J Cereb Blood Flow Metab 2, Suppl 1:S98−S100

Hossmann KA, Paschen W, Csiba L (1983) Relationship between calcium accumulation and recovery of cat brain after prolonged cerebral ischemia. J Cereb Blood Flow Metab 3:345−353

Karasawa A, Kumada Y, Yamada K, Shuto K, Nakamizo N (1982) Protective effect of flunarizine against cerebral hypoxia-anoxia in mice and rats. J Pharm Dyn 5:295−300

Kaste M, Sipponen JT, Sepponen RE (1983) Nuclear magnetic resonance (NMR) imaging and calcium antagonist (nimodipine) in ischemic penumbra in man. J Cereb Blood Flow Metab 3, Suppl 1:S141−S142

Kazda S, Hoffmeister F (1979) Effect of some cerebral vasodilators on the postischaemic impaired cerebral reperfusion in cats. Arch Pharmacol Suppl 307:R43

Kazda S, Towart R (1982 a) Nimodipine: A new calcium antagonistic drug with a preferential cerebrovascular action. In: Auer LM, Heppner F, Symon L (eds) Proc. Symp. Aneurysm Surgery in the Acute Stage. Acta Neurochir 63:259−265

Kazda S, Garthoff B, Luckhaus G, Nash G (1982 b) Prevention of cerebrovascular lesions and mortality in stroke-prone spontaneously hypertensive rats by the calcium antagonist nimodipine. In: Calcium Modulators (Godfraind T, Albertini A, Paoletti R, eds) pp. 155−167, Elsevier Biomedical Press, Amsterdam

Kazda S, Garthoff B, Krause HP, Schlossmann K (1982 c) Cerebrovascular effects of the calcium antagonistic dihydropyridine derivative nimodipine in animal experiments. Drug Res 32:331−338

Kazda S, Garthoff B, Luckhaus G (1983) Calcium antagonists prevent brain damage in stroke-prone spontaneously hypertensive rats (SHR-SP) independent of their effect on blood pressure: Nimodipine versus nitrendipine. J Cereb Blood Flow Metab 3, Suppl 1:S526−S527

Kovach AGB, Dora E, Szedlacsek S, Koller A (1983) Effect of the organic calcium antagonist D-600 on cerebrocortical vascular and redox responses evoked by adenosine, anoxia and epilepsy. J Cereb Blood Flow Metab 3:51−61

Latchaw JP, Capraro J, Moufarrij N, Redy DP, Lesser RP, Slugg RM, Cook AF, Stowe N, Skrinska V, Lucas F, Little JR (1983) The effects of verapamil, thromboxane synthetase inhibitor, chlorpromazine, and propranolol in acute focal cerebral ischemia. J Cereb Blood Flow Metab 3, Suppl 1:S367−S368

Leblanc R, Feindel W, Yamamoto L, Milton JG, Frojmovic MM (1983) Reversal of acute cerebral vasospasm by the calcium antagonist verapamil. In: Scientific Program − The American Association of Neurological Surgeons Annual Meeting April 24−28, Washington DC, pp 64−65

Mikkelsen E, Andersson KE, Lederballe Pedersen O (1978) The effect of nifedipine on isolated human peripheral vessels. Acta Pharmacol Toxicol 43:291−298

Mohamed AA, Shigeno T, Mendelow AD, McCulloch J, Harper AM, Teasdale GM (1983) The effect of the calcium antagonist, nimodipine, on local cerebral blood flow, glucose use, and focal cerebral ischaemia. J Cereb Blood Flow Metab 3, Suppl 1: S578−S579

Müller-Schweinitzer E, Neumann P (1983) In vitro effects of calcium antagonists PN 200-110, nifedipine and nimodipine on human and canine cerebral arteries. J Cereb Blood Flow Metab 3:354−361

Nagao T, Ikeo T, Sato M, Nakajima H, Kiyomoto A (1978) Effect of diltiazem on calcium- and noradrenaline-induced contractions in isolated rabbit aorta. In: Heart Function and Metabolism (Kobayashi T, Sano T, Dhalla NS, eds) pp. 437−440, University Park Press, Baltimore

Nakayama K (1982) Calcium-dependent contractile activation of cerebral artery produced by quick stretch. Am J Physiol 242:H760−H768

Newberg LA, Steen PA, Milde JH, Michenfelder JD (1983) Effect of flunarizine on cerebral blood flow and neurologic recovery after complete cerebral ischemia in the dog. J Cereb Blood Flow Metab 3, Suppl 1:S544−S545

Nowicki JP, MacKenzie ET, Young AR (1982) Brain ischaemia, calcium and calcium antagonists. Pathol Biol 30:282−288

Olsen TS, Larsen B, Herning M, Bech Skriver E, Lassen NA (1983) Blood flow and vascular reactivity in collaterally perfused brain tissue. Evidence of an ischemic penumbra in patients with acute stroke. Stroke 14:332−341

Ott E, Lechner H (1980) The influence of nimodipine on cerebral blood flow in patients with subacute cerebral infarction. (abstract) In: Pathophysiology and Pharmacotherapy of Cerebrovascular Disorders (Satellite Symposium of the XXVIII International Congress of Physiological Sciences) (Betz E, Grote J, Heuser D, Wüllenweber R, eds)

Pearce WJ, Bevan JA (1983) The cerebrovascular selectivity of diltiazem. J Cereb Blood Flow Metab 3, Suppl 1:S546−S547

Reedy DP, Little JR, Capraro J, Lesser RP (1983) Effects of verapamil on acute focal cerebral ischemia. In: Scientific Program − The American Association of Neurological Surgeons, Annual Meeting April 24−28, Washington DC, p 63

Roy MW, Tibbs PA, Holladay EP, Donaldson DL, Young AB (1983) Deleterious effects of verapamil following middle cerebral artery occlusion in cats. In: Scientific Manuscripts − The American Association of Neurological Surgeons Annual Meeting April 24−28, Washington DC, p 58−59

Shimizu K, Ohta T, Toda N (1980) Evidence for greater susceptibility of isolated dog cerebral arteries to Ca antagonists than peripheral arteries. Stroke 11:261−266

Siesjö BK (1981) Cell damage in the brain: A speculative synthesis. J Cereb Blood Flow Metab 1:155−185

Silverberg GD, Ross G, Corbin SD, New W (1979) Time course of serotonin-induced vasoconstriction. Neurosurgery 4:539−542

Steen PA, Newberg LA, Milde Jh, Michenfelder JD (1983) Nimodipine improves cerebral blood flow and neurologic recovery after complete cerebral ischemia in the dog. J Cereb Blood Flow Metab 3:38−43

Symon L, Harris RJ, Branston NM (1982) Calcium ions and calcium antagonists in ischaemia. In: Proc. Symp. Aneurysm Surgery in the Acute Stage (Auer LM, Heppner F, Symon L, eds) Acta Neurochir 63:267−275

Tanaka K, Gotoh F, Muramatsu F, Fukuuchi Y, Amano T, Okayasu H, Suzuki N (1980) Effects of nimodipine (Bay e 9736) on cerebral circulation in cats. Drug Res 30:1494−1497

Teasdale G, Tamura A, Graham D, Rabow L, MacKenzie ET (1981) Vasoconstriction in focal cerebral ischemia. In: Cerebral microcirculation and metabolism (Cervos-Navarro J, Fritschka E, eds) pp. 77−81, Raven Press, New York

Towart R (1982) The pathophysiology of cerebral vasospasm, and pharmacological approaches to its management. In: Proc. Symp. Aneurysm Surgery in the Acute Stage (Auer LM, Heppner F, Symon L, eds) Acta Neurochir 63:253 – 258

Towart R, Kazda S (1979) The cellular mechanism of action of nimodipine (BAY e 9736), a new calcium antagonist. Brit J Pharmacol 67:407

Towart R, Kazda S (1980 a) Selective inhibition of serotonin-induced contractions of rabbit basilar artery by nimodipine (BAY e 9736). IRCS Med Sci 8:206

Towart R, Kazda S (1980 b) The effect of the calcium antagonist nimodipine (BAY e 9736) on contractions of the rabbit basilar artery. Arch Pharmacol Suppl 311:169

Towart R, Wehinger E, Meyer H, Kazda S (1982) The effect of nimodipine, its optical isomers and metabolities on isolated vascular smooth muscle. Drug Res 32:338 – 346

Van Nueten JM, Vanhoutte PM (1981) Selectivity of calcium antagonism and serotonin antagonism with respect to venous and arterial tissues. Angiology 32:476 – 484

White BC, Gadzinski DS, Hoehner PJ, Krome C, Hoehner T, White JD, Trombley JH Jr (1982) Effect of flunarizine on canine cerebral cortical blood flow and vascular resistance post cardiac arrest. Ann Emerg Med 11:119 – 126

White RP, Shirasawa Y, Robertson JT (1983) Mechanisms of the cerebral arterial constriction caused by uridine triphosphate. Abstract of the 8th Int. Conf. on Stroke and Cerebral Circulation, San Diego. Stroke 14:9

The Lower Limit of Cerebral Autoregulation During Treatment of Severe Hypertension

G. Gulliksen, E. Højer-Pedersen, M. Møller, B. Harvald, and E. Enevoldsen

The lower mean blood pressure limit of CBF autoregulation is about 60 torr in normal man. In hypertensive patients the lower limit is shifted towards higher blood pressures.

Strandgaard (1976) showed that antihypertensive treatment might help to normalize the lower limit of the autoregulation. However, only occasionally did patients normalize completely. He considered this to reflect to varying functional and structural adaptation of the cerebral resistance veesels.

Little is known about the time course of the normalization of autoregulation in man during antihypertensive treatment. An estimation of this may make it possible to avoid cerebral ischemia due to blood pressure being lowered beneath the lower limit of autoregulation.

Methods

This study was performed to investigate the possibility of monitoring the lower limit of autoregulation during antihypertensive treatment using the ^{133}Xe inhalation technique. The CBF measurements were performed before the antihypertensive treatment was started and repeated several times during its course. CBF was measured at rest and during sodium nitroprusside-induced hypotension. The three patients to be presented received a standard antihypertensive medication, including the α-blocking agent prozozin and a β-blocker.

Results

The first patient (Fig. 1) was a 58-year-old man with hypertensive encephalopathy. The CBF measurements were performed on day one, at admission before the antihypertensive treatment was started, and repeated at day three, five, etc. Each study included a CBF measurement at rest and a measurement during induced hypotension. The blood pressure at rest and during the induced hypotension at each investigation is indicated in the upper part of the figure. In the lower part of the figure the difference between CBF at rest and during hy-

Neurological Department, Odense University Hospital, DK-5000 Odense C

Cerebral Blood Flow and
Metabolism Measurement.
Eds. Hartmann/Hoyer
© Springer-Verlag Berlin Heidelberg 1985

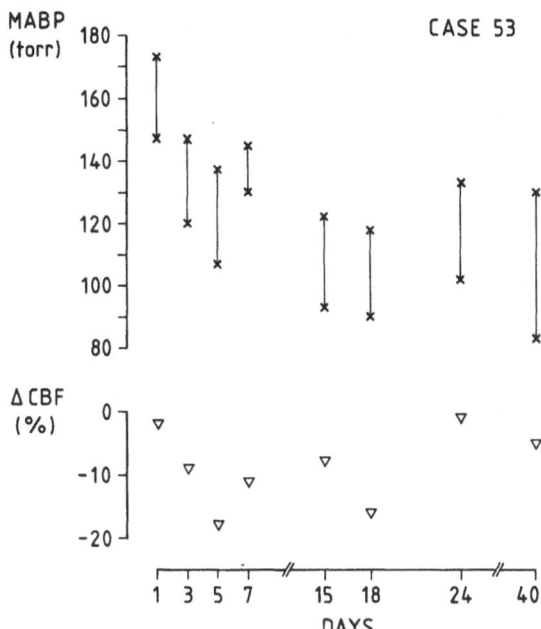

Fig. 1. The first patient, case 53. The reduction in cerebral blood flow during hypotension illustrated together with the mean arterial blood pressure during rest and hypotension at each investigation

potension is calculated as a percentage of the resting CBF. A decrease of more than 15% in CBF during hypotension is considered to indicate impaired autoregulation. At the first and third day the reduction in CBF during hypotension was less than 15%, indicating intact autoregulation. On the fifth day hypotension induced a reduction in CBF of 18%, indicating impaired autoregulation. Thus, the lower limit of autoregulation was estimated to be between the hypotensive blood pressure on the fifth day (107 torr) and the blood pressure during hypotension on the third day (120 torr). At the 15th and the 18th day the lower limit of the autoregulation was estimated in the same way to be below 93 torr and above 90 torr. Thus, the lower limit of autoregulation normalized from the fifth to the 15th day although this normalization was not complete. A further shift of the autoregulation towards a lower blood pressure is evident from the latest study. More than 1 year after the antihypertensive treatment was started the blood pressure could be reduced to 83 torr without reducing the CBF.

The results from the study of a 43-year-old man with severe hypertension are shown in Fig. 2 in the same way as in Fig. 1. The first two examinations localized the lower limit of the autoregulation within a narrow interval. The examination on the sixth day disclosed a quick shift in autoregulation towards lower blood pressures as blood pressure could now be reduced to 92 torr without influencing the CBF. Later examinations showed further normalization. However, the lower limit of the autoregulation was still abnormal 3 months after the start of antihypertensive treatment.

Figure 3 shows the results from the study of a 61-year-old severely arteriosclerotic and hypertensive man. The MABP observed during 3 weeks before admission was about 190 torr. Upon a tiny dose of antihypertensive medi-

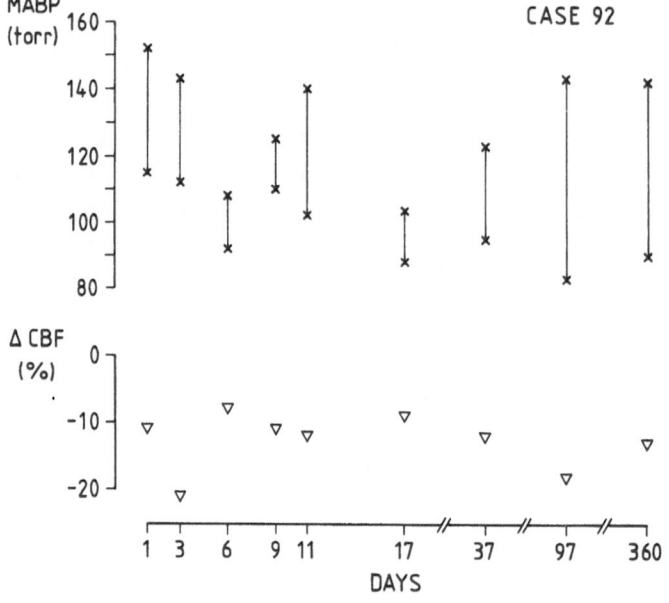

Fig. 2. The second patient, case 92. CBF changes and blood pressures illustrated as in Fig. 1

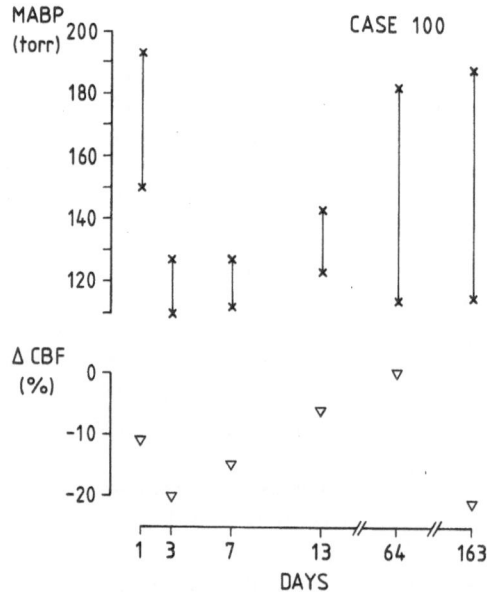

Fig. 3. The third patient, case 100. CBF changes and blood pressures illustrated as in Fig. 1

cation the blood pressure dropped rather abruptly and was recorded at below 112 torr several times during the third week. From the figure it is seen that this blood pressure is below the lower limit of the autoregulation. The patient eventually developed a lacunar infarct. In this case we were unable to demonstrate any shift in the lower limit of autoregulation.

Conclusion

In conclusion, we have found that the lower limit of CBF autoregulation can be estimated using the ^{133}Xe inhalation technique.

The demonstration of a shift towards lower MABP values of cerebral autoregulation is thought to implicate a reversal of the functional adaptation to high blood pressure of the cerebral resistance vessel. We believe that patients number 1 and 2 demonstrate this.

If no shift occurs this may be due to structural adaptation of the cerebral resistance vessels. Structural adaptation to high blood pressure in the cerebral resistance vessel indicates that the blood pressure reduction will have to be handled with great care. If high blood pressure is treated too vigorously and falls below the lower limit of the autoregulation, the patient may be exposed to the complication of cerebral ischemia. We believe that this is what might have happened in patient number 3.

Reference

Strandgaard S (1976) Autoregulation of cerebral blood flow in hypertensive patients. Circulation 53:720–727

Measurements of CBF in Patients with Epilepsy

K. Marguc, E. Ott, E. Flooh, E. Körner, F. Fazekas, G. Bertha, R. Wolf, B. Reinhart, J. Kordasch, and H. Lechner

It is well accepted that measurements of cerebral blood flow (CBF) under physiologic conditions may reflect the metabolic activity of the human brain. Hence, in patients with epilepsy, CBF has been found to be increased during seizure attacks coinciding with an increased metabolic rate for oxygen and glucose. However, little information exists on whether CBF varies in patients with epilepsy of various etiologies. It is therefore the purpose of the present paper to describe the results of CBF measurements in patients with epilepsy of different etiologies.

Patients and Methods

There were 31 patients (15 males, 16 females) with a mean age of 45 years (range 17–69) in whom on the basis of the history and clinical and electroencephalographic criteria epilepsy had been diagnosed and various etiologies established. Thus, 13 patients had been diagnosed as having idiopathic epilepsy, 10 as having symptomatic – but nonvascular – epilepsy (tumors, alcoholism, brain trauma), and 8 as having vascular epilepsy (Table 1).

Cerebral blood flow was measured using the noninvasive xenon 133 clearance technique and attaching ten collimators to each hemisphere. 15–20 mCi

Table 1. Etiology of epilepsy

Etiology	Number of patients	Mean age
Idiopathic	13	25
Symptomatic – nonvascular[a]	10	49
Vascular	8	62
Total	31	45

* $P < 0.001$
[a] Tumor, alcoholism, trauma

Universität Graz, Abteilung Psychiatrie und Neurologie, A-8036 Graz

Cerebral Blood Flow and
Metabolism Measurement.
Eds. Hartmann/Hoyer
© Springer-Verlag Berlin Heidelberg 1985

xenon 133 dissolved in physiologic saline solution was injected into an ante-cubital vein within 60 s. For this study the total CBF-gray (tCBFg) and the blood flow of the fast-clearing tissue compartments were calculated by a computer.

During each study the end-tidal pCO_2 and the blood pressure were monitored; in most of the patients investigated an eight-channel EEG was recorded simultaneously. In nine patients tCBFg was measured not only at rest but also during hyperventilation.

Results

As can be seen from Table 2, tCBFg was found to vary depending on the underlying pathology. There was no significant difference between the 13 patients with idiopathic epilepsy and a control group of 13 normals. However, tCBFg was found to be significantly different when compared with the groups of patients with symptomatic epilepsy or epilepsy of vascular origin.

Moreover, there was also evidence of abnormal cerebral vasomotor reserve in these patients, since hyperventilation revealed a paradoxical response of CBF relating to the changes in pCO_2 (Table 3). We are in agreement with previous reports indicating a 3%–4% change in CBF corresponding to a pCO_2 change of 1 mmHg in normals during hyperventilation.

In one case a seizure was provoked by 4 min hyperventilation characterized by 4/s spike-wave activity in the EEG. In this case end-tidal pCO_2 decreased from 42.3 mmHg to 31.9 mmHg, but the CBF remained unchanged.

Table 2. Global CBF (ml/100 g/min)

	CBF	pCO_2
Normal	78.2±11.5	37.3±1.9
Idiopathic	72.1±12.8	40.4±3.7
Symptomatic – nonvascular	58.3±10.9	38.0±3.9
Vascular	54.1± 8.8	36.8±1.8

* $P<0.05$,   ** $P<0.001$

Table 3. Response of CBF in relation to changes in pCO_2

	Mean CBF ml/100 g/min	pCO_2
CBF measurement at rest	70.38±12.76	40.99±3.23
CBF measurement during hyperventilation	73.65±17.55	32.68±4.92

* Nonsignificant,   ** $P<0.005$

Comments

The results set out in this paper indicate that CBF in patients with symptomatic epilepsy is significantly lower than in patients with idiopathic epilepsy. This finding would still be valid if the CBF values were to be corrected for pCO_2 differences between the groups.

Cerebral vasomotor reserve was abnormal in patients with epilepsy independent of the etiology.

It may be concluded that in our patients, the underlying etiopathologic process may have been responsible for the CBF reduction rather than the epilepsy itself.

References

Brodersen P, Paulson O, Bolwig T, Rogon E, Rafaelsen O, Lassen N (1973) Cerebral hyperemia in electrically induced epileptic seizures. Arch Neurol Vol 28

Hougaard K, Oikawa T, Sveinsdottir E, Skinhoj E, Ingvar D, Lassen N (1976) Regional CBF in focal cortical epilepsy. Arch Neurol Vol. 33

Lavy S, Melamed E, Portnoy Z, Carmon A (1976) Interictal regional CBF in patients with partial seizures. Neurology 26

Phleps M, Mazziotta J, Huang S (1982) Study of cerebral function with positron computer tomography. J CBF & M 2

Plum F, Duffy T (1975) The couple between cerebral metabolism and blood flow during seizures. Bentzon-S. VIII

Plum F, Howse D, Duffy T (1974) Metabolic effects of seizures. Brain dysfunction in metabolic disorders

Sakai F, Meyer J, Naritomi H, Hsu M (1978) Regional CBF and EEG in patients with epilepsy. Arch Neurol Vol 35

CBF Measurements in Parkinson's Disease

F. FAZEKAS, E. OTT, K. MARGUC, E. KÖRNER, H. VALETITSCH, H. LECHNER, and E. FLOOH

Introduction

It has been shown repeatedly that cerebral blood flow (CBF) measurements can be used to demonstrate functional changes in the brain. This is due to the fact that normally blood flow is controlled by the metabolism of the neuronal tissue. Measurements of CBF in Parkinson's disease (PD) revealed a reduction of total as well as of regional CBF, but little attention has so far been paid to the underlying etiology as a possible cause of the CBF decline. In this study CBF measurements have been performed in patients with idiopathic PD and the results compared with those obtained in patients with PD and with evidence of cerebrovascular disease (CVD), arbitrarily termed "increased vascular risk" (IVR).

Methods and Patients

Twenty-five patients with Parkinson's disease (PD) were investigated. On the basis of their history, of clinical, and of laboratory examinations as well as of CT findings, 15 patients have been referred to as idiopathic PD and 10 patients as PD with IVR. All but two patients received various drugs for treatment of PD.

CBF was measured by the xenon 133 technique following intravenous injection of 15–20 mCi xenon 133 dissolved in 2.5 ml saline. The clearance of the isotope was followed by a total of 20 collimators attached to both hemispheres. For this study only values of the fast-clearing tissue were considered (Fg). Mean arterial blood pressure (MABP) was registered before and after the examination. The PCO_2 was estimated from the CO_2 content in the expired air ($pECO_2$).

Results

As can be seen from Table 1, total CBF was lower in the group with IVR; however, the mean age of this group of patients was higher than in the idiopathic

Universität Graz, Abteilung Neurologie und Psychiatrie, Auenbruggerplatz 22, A-8036 Graz

Cerebral Blood Flow and
Metabolism Measurement.
Eds. Hartmann/Hoyer
© Springer-Verlag Berlin Heidelberg 1985

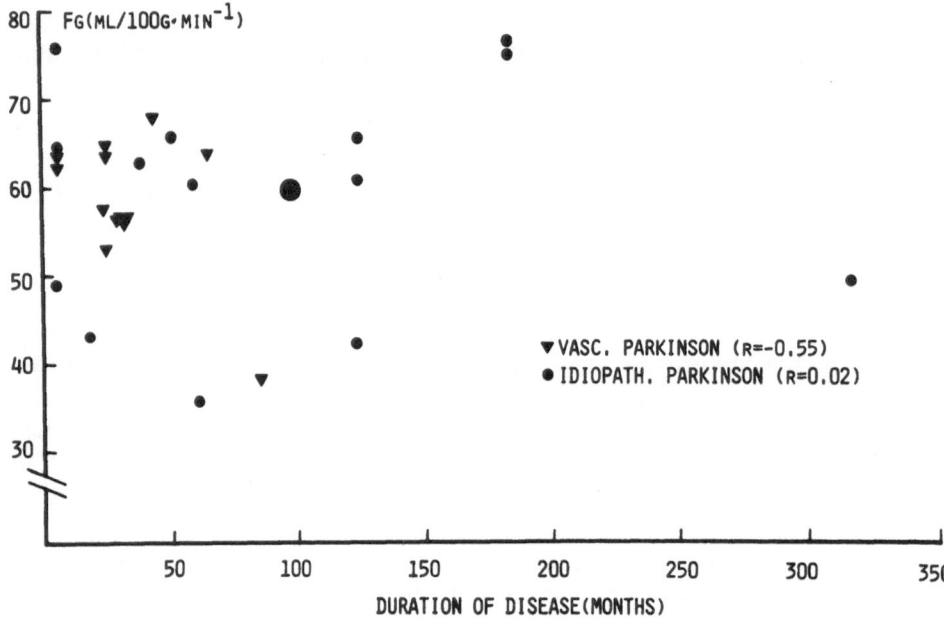

Fig. 1. tCBF (Fg) and duration of disease (months) correlated in patients with idiopathic PD and patients with PD and IVR

Table 1. Duration of PD, mean age, and tCBF

	Mean age (yrs)	Duration of disease (months)	tCBF (ml/100 g/min)	pECO$_2$ (mmHg)	MABP (mmHg)
Idiopathic PD	64±6	90.4±85.8	59.3±15.3	39.8±4.5	98.7±11.8
PD with IVR	69±6	33.4±24.6	55.8± 8.6	37.9±2.3	94.6±12.3

Table 2. Influence of treatment on tCBF in PD[a] patients

	tCBF (ml/100 g/min)	pECO$_2$ (mmHg)	MABP (mmHg)
Untreated ($n=2$)	55.9± 6.1	38.2±5.2	96.6± 7.3
Anticholinergic drugs ($n=7$)	57.4± 8.9	39.5±4.7	100.3±11.2
L-dopa and decarboxylase inhibitor ($n=6$)	63.6±11.9	40.5±3.2	97.3±12.3

[a] Only patients with idiopathic PD considered

PD group. Duration of the disease was longer in the group with idiopathic PD than in the group with PD and IVR.

The pattern of hyperfrontality was lost in both groups of patients with PD. This loss of frontal hyperperfusion, however, is more distinct in patients with PD and IVR than in patients with idiopathic PD.

While no correlation existed between duration of PD and tCBF in patients with idiopathic PD, a significant correlation was found in patients with PD and IVR (Fig. 1). Treatment with λ-dopa and decarboxylase inhibitor increased tCBF when compared with untreated PD patients as well as with PD patients treated with anticholinergic drugs (Table 2).

Discussion

The existence of biochemical as well as metabolic alterations in PD is well known, and they might in part be responsible for changes of CBF in patients with PD. Lavy et al. (1979) described a reduction of CBF in patients with idiopathic PD in comparison with age-matched controls, which was confirmed by our findings. However, it is a striking feature of the patients with PD reported in this paper that tCBF was found to be different in the presence of vascular risk factors. Since it is well known that cerebrovascular disease (CVD) affects tCBF, it may be concluded from the results that this additional factor led to a further decrease of tCBF when compared with patients with idiopathic PD. On the other hand the age difference might have further contributed to this finding.

Increased frontal regional CBF in normals has been described by several authors (Lassen 1977; Ingvar 1979). In patients with PD, however, a loss of hyperfrontality has been observed (Lavy 1979; Bes 1983) and degeneration of ascending dopaminergic pathways has been discussed as the possible cause. In addition, Mamo et al. (1979) pointed out that patients with CVD also show loss of hyperfrontality during the fifth and sixth decade, so that our results may also be discussed in the light of these findings.

Previous investigation did not find a correlation between duration of PD and decrease of CBF in patients with idiopathic PD. However, a significant correlation was established in patients with PD and IVR. In those cases we found a decline of CBF related to the duration of parkinsonian symptoms. We assume that cerebrovascular factors contributed to this otherwise age-dependent decline (Ott 1982).

In contrast to others, we saw some influence of λ-dopa treatment on CBF. Untreated patients and patients with anticholinergic treatment had lower perfusion rates. This is in accordance with the paper of Bes (1983); however, the real mechanism of CBF increase following λ-dopa treatment is not yet clear.

Summary

Cerebral blood flow (CBF) was studied in 25 patients with Parkinson's disease (PD) (15 patients with idiopathic PD, 10 patients with PD and increased vascular risk – IVR). Patients with PD and IVR were older than those with idiopathic PD, whereas the duration of disease was much shorter. Total cerebral blood flow (tCBF) was more reduced in patients with PD and IVR. In this group the loss of hyperfrontality was more marked and there was a close correlation to duration of disease in patients with PD and IVR whereas none existed in patients with idiopathic PD. PD patients treated with λ-dopa had higher tCBF values than untreated patients or those treated with anticholinergic drugs.

References

Bes A, Guell A, Fabre N, Geraud G, Larrue V (1983) Hyperfrontal distribution of grey matter flow studied by 133 xenon inhalation. Results in normal subjects, in aging and parkinsonism. In: Meyer JS, Lechner H, Reivich E (eds) Cerebral vascular disease 4, 11th Salzburg conference. Excerpta medica, Amsterdam, p 231–235

Ingvar DH (1979) "Hyperfrontal" distribution of the cerebral grey matter flow on resting wakefulness: on the functional anatomy of the conscious state. Acta Neurol Scand 60:12–25

Lassen NA, Roland PE, Larsen E et al. (1977) Mapping of human cerebral functions: A study of the regional cerebral blood flow pattern during rest, its reproductibility and activation seen during basic sensory and motor functions. Acta Neurol Scand 56, (suppl 64):262–263

Lavy S, Melamed E, Cooper G, Shlomo B, Rinot Y (1979) Regional cerebral blood flow in patients with Parkinson's disease. Arch Neurol 36:344–348

Mamo H, Meric Ph, Luft A, Seylaz J (1983) Hyperfrontal pattern of human cerebral circulation: Variations with age and atherosclerotic state. Arch Neurol 40:626–632

Ott E, Lechner H (1982) Hemorheologic and hemodynamic aspects of cerebrovascular disease. Path Biol 30:611–614

Quantitative Cerebral Blood Flow Mapping in Stroke and During Mental Stimulation After Intravenous Injection of ^{195m}Au

P. LINDNER and O. NICKEL

Methods

The new short life isotope ^{195m}Au has some suitable features for quantitative cerebral blood flow mapping. Its half-life is 30.5 s; therefore an injection can be repeated after 3 min (six half-lives) without any need for background subtraction and with the same specific activity. The calculated whole body radiation dose after three successive administrations of 25 mCi ^{195m}Au amounts to 50 mrad. In comparison to a ^{99m}Tc pertechnetate injection it is estimated that the dose to the patients is reduced by a factor of eight (Garcia et al. 1981).

The gamma-camera energy spectrum of a ^{195m}Au generator eluate shows two peaks, one at an energy level of 262 keV and a second at $68-70$ keV. The half-life times of both energy lines are nearly equal and lie in the range of 30 s. Both peaks can be used for imaging and perfusion studies. Details are described in Nickel et al. 1983. In the case of PA positions one should use the high energy peak of 262 keV, whereas for lateral views of the hemispheres the low energy peak (69 keV) is ideal because no significant "look-through" effect is detected. At least the possible "look-through" is diminished in comparison to the xenon photon energy of 85 keV. The necessary detection time lies in the range of $20-40$ s.

Clinical Investigations and Results

With the use of a multicrystal camera, which allows one to measure high count rates with high sensitivity, we made our investigations in the following way: The patient sits before the camera either in the dorsal position or in the lateral position. After eluating 2 ml we immediately inject into the cubital vein, followed by a hand-injected flush of 20 ml saline. A serial scintigram with 2 frames per second is generated over a detection time of 50 s. The data can then be evaluated by a computer program which computes quantitative cerebral blood flow values over selected regions of interest, following the mentioned theory.

Johannes-Gutenberg-Universität Mainz, Institut für Klinische Strahlenkunde, Langenbeckstrasse 1, D-6500 Mainz

Cerebral Blood Flow and
Metabolism Measurement.
Eds. Hartmann/Hoyer
© Springer-Verlag Berlin Heidelberg 1985

Fig. 1. Left lateral view; a patient after stroke in the area of the left middle cerebral artery. The area of the occluded vessel can be clearly seen

It is also possible to generate parametric images, which show the regional speed of inflow of the bolus, the distribution of reverse regional mean transit times, and a quantitative mapping of regional cerebral blood flow. The use of a multicrystal camera gives as high count rates with ^{195m}Au as with a bolus of 15−20 mCi ^{99m}Te.

Using this technique we routinely study all patients with cerebrovascular diseases, and especially all those with stroke. In order to demonstrate the effectiviness of the described method, Fig. 1 shows the quantitative mapping of cerebral blood flow from a patient after stroke in the area of the left middle cerebral artery. The region of the occluded vessel is clearly seen. No significant "look-through" effect can be recognized.

We have been very interested in examining the method in mental stimulation exercises with volunteers, in the same way as has described by many other authors.

Stimulation Exercise

A simple mental stimulation test can be performed as follows: A volunteer sits before the camera with closed eyes at rest for 2−3 min in the left lateral posi-

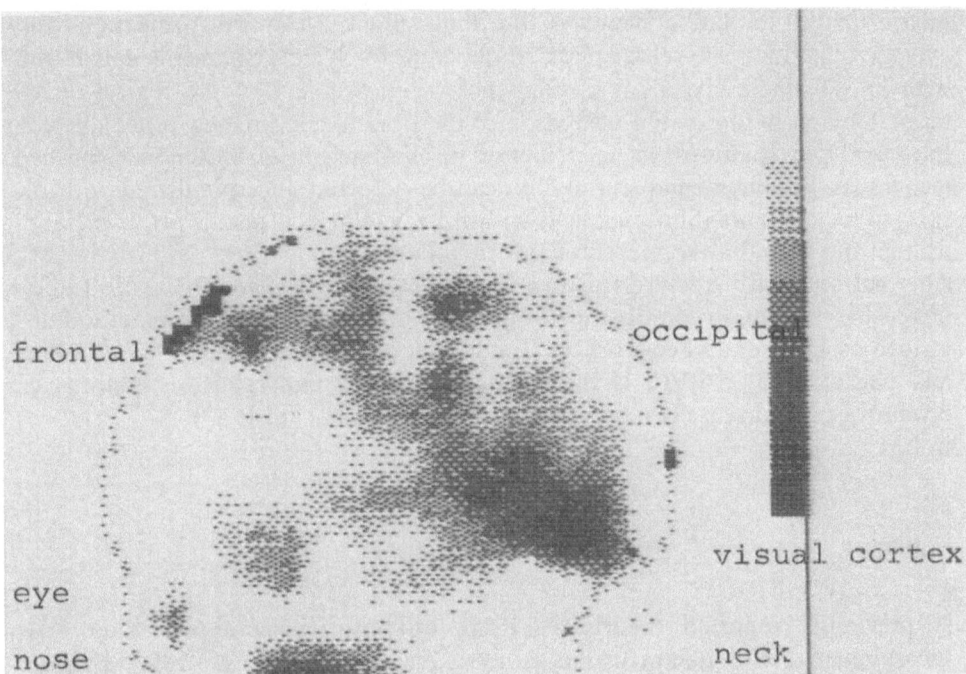

Fig. 2. Change in rCBF pattern during visual stimulation. The mean percentage change in rCBF is shown from the left lateral views of eight volunteers. The maximum change of about 10% is in the visual cortex

tion. A bolus injection is made as described when the volunteer has his eyes closed. In the same position the investigation is repeated after 3 min. The patient opens his eyes and fixes his gaze upon a moving object in front of his eyes during the second investigation.

For both investigations the quantitative mapping of regional cerebral blood flow is calculated. Then the study at rest is subtracted from the study during mental activation of the areas of optical perception. As seen in Fig. 2, there is a significant change in the cerebral blood flow pattern. The average increase during stimulation in the area of optical perception is 11.5%, and regionally it is as high as 30%.

Figure 2 shows the regional CBF differences with a regional maximum of increase over the visual perception center in the cortex of the occipital lobe. You can also see local changes in the motion activation.

Discussion

The results demonstrate the high sensitivity of the method. The spatial resolution is 1 – 2 cm approximately. Most xenon clearance techniques use 16 single detectors for each hemisphere. The presented technique using a multicrystal camera (Baird Atomic) uses about 80−100 single crystals for generating parametric images for lateral views of the hemispheres. For PA projections the average quantitative cerebral blood flow value for a hemisphere is calculated with 15−30 single crystals in dependence on the seizure of the region of interest. Obviously the spatial resolution of the parametric images, especially for the blood flow patterns, is much better in comparison with xenon clearance techniques. Simple mental stimulation can be detected and quantified with respect to regional blood flow changes within 5%−10%. The results prove primarily that the quantitative measurement of regional blood flow patterns is possible not only with freely diffusible xenon but also with nondiffusible radiotracers like [195m]Au. In combination with the quantitative evaluation method described by Lindner and co-workers (Lindner et al. 1980; Lindner 1983; Lindner and Nickel 1983), [195m]Au is the ideal isotope for clinical investigations of cerebrovascular disease (Lindner 1983; Lindner and Nickel 1983).

Summary

A previously reported theory regarding quantitative cerebral blood flow measurements with nondiffusible radiotracers (Lindner et al. 1980; Lindner 1983) has been applied to patients after stroke and to volunteers undergoing a mental stimulation exercise. Consecutive quantitative measurements of cerebral blood flow patterns not only in PA but also in lateral views of the brain are possible by the use of the short-lived (30 s) isotope [195m]Au (Nickel et al. 1983). The studies last less than 1 min and can be repeated after 3 min. Parametric images for quantitative regional cerebral blood flow can demonstrate the area of the occluded vessels. Quantitative activation patterns of cerebral blood flow during mental stimulation can be generated. The results prove that it is possible to measure quantitatively cerebral blood flow patterns not only with freely diffusible indicators like xenon but also with nondiffusible indicators.

References

Garcia E, Mena J, de Jong R (1981) Gold Au 195m, short lived migle photon emitter for haemodynamic studies. J Nucl Med 22, p 71
Lassen NA, Ingvar DH (1963) Regional cerebral blood flow measurements in man. Archs Neurol Psychia Chicago, pp 615−622
Lindner P (1983) Quantitative, noninvasive cerebral blood flow measurements with nondiffusible tracers using a heart rate dependent recirculation correction; application in carotid surgery. Accepted for publication in Europ J Nucl Med
Lindner P, Nickel O (1983) Quantitative activation patterns of cerebral blood flow during mental stimulation after intravenous injection of [195m]Au. Neuroradiology, Vol. 25, No 3
Lindner P, Wolf F, Schad N (1980) Assessment of regional blood flow by intravenous injection of [99m]technetium pertechnetate. Europ J Nucl Med, 5, pp 229−235

Meier P, Zierler KL (1954) On the theory of the indicator dilution method for measurement of blood flow and volume. J appl Physiol 6, pp 731 – 743

Nickel O, Lindner P, Schad N (1983) Parametric imaging of cerebral blood flow with the short lived isotope Au 195m. Accepted for publication in Europ J Nucl Med

Schad N, Schön H, Nickel O, Le-Thi HO, Lindner P (1983) Aurum 195m: Application in first pass cardiac examinations. To be published in J Nucl Med

Zierler KL (1965) Equations for measuring blood flow by external monitoring of radioisotopes. Circulation Res 16, pp 306 – 321

Measurement of the Intracerebral Blood Flow Distribution Using a Dispersion Model

C. C. NIMMON, K. E. BRITTON, M. GRANOWSKA, J. S. P. LUMLEY, B. MAHENDRA, J. DRINKWATER, M. CHARLESWORTH, L. A. HAWKINS, M. J. CARROLL, and D. P. E. KINGSLEY

In the study of the cerebral circulation, the use of a nondiffusible tracer such as ^{99m}Tc-labeled erythrocytes carries the inherent attraction that measurements are made directly of blood flow rather than of a secondary effect of washout, as is the case with diffusible indicators. In a noninvasive study utilizing a rapid intravenous injection of the labeled tracer, the transit time information contained in the cerebral activity time (A/T) curves was masked by the effects of the central circulation (Nilsson et al. 1977). Britton et al. (1979) introduced a method in which regional cerebral impulse retention functions, CRFs, are computed by the application of deconvolution analysis to an A/T curve recorded over the region of the aortic arch as input, and the regional cerebral A/T curves corresponding to selected regions of interest, ROI, on the vertex view from a gamma camera study. These CRFs correspond to the hypothetical A/T curves which would be expected following intra-arterial injection and without any recirculation component. The measurement solely of regional mean blood flow, which is proportional to the height of the corresponding CRF (Sapirstein 1961), disregards potentially valuable information contained in the variation of the shape of the CRF. This paper is concerned with a method for extraction of characteristics of the intracerebral flow distribution which influence the shape of the CRF, and with the investigation of their relevance to the patient study.

Methods

CRFs were obtained corresponding to ROIs defining (i) each cerebral hemisphere and (ii) anterior, middle, and posterior regions formed by division of each hemisphere into three parts, each with an equal length in the sagittal plane. A Fourier Transform method of deconvolution was used, in which the problem of noise amplification is overcome by the use of constraints imposed on the solution (Nimmon et al. 1981). In particular, minimum smoothing was applied sufficient to produce a curve with a single peak and with a monotonic rise and fall pre- and postpeak, respectively. This "monotonicity" constraint is equivalent to the assumption firstly, that all of the spike input originating at the aortic arch reaches the brain before the minimum cerebral transit time and secondly, that there is no backflow of tracer.

St. Bartholomew's Hospital, London EC1A 7BE, Great Britain

Cerebral Blood Flow and
Metabolism Measurement.
Eds. Hartmann/Hoyer
© Springer-Verlag Berlin Heidelberg 1985

The overall asymmetrical dispersive nature of the CRF arises from the summation of contributions from flow channels with differing transit times. Based on classical concepts of the cerebral blood supply (Bergh and Eecken 1968), the main arterial supply may be divided into (a) the superficial arteries, and (b) the deep penetrating arteries. The superficial consist of branches of the anterior, middle, and posterior cerebral arteries arising from the circle of Willis at the base of the brain. The deep intracerebral circulation is fed from these cortical vessels and also from deep penetrating vessels supplying the regions around the ventricles. In the normal cerebrovascular tree, the shortest transit times are observed from angiography to be associated with a fraction of the superficial flow. This is also supported by the fact that the fastest flow observed in both ^{133}Xe and positron measurements of rCBF is superficial.

A simple two parallel component model is introduced in which one component (A) is represented by a delay and without any dispersion. The second component (B) is dispersive and contains a continuum of transit times. The component (A) reflects the shortest transit times associated with a fraction f_1 of the superficial flow, and the component (B) contains the longest transit times associated with the deep circulation which is fed by the remaining fraction f_2 of the superficial flow ($f_1 + f_2 = 1$). The separation of these components is accomplished by an iterative algorithm in which the monotonicity criterion is preserved separately for each component and the peak values are constrained to occur within 0.6 s. A parameter Q is defined as the ratio of the dispersive to nondispersive components of flow and is calculated from the ratio of the heights of the components B and A, respectively, as illustrated in Fig. 1.

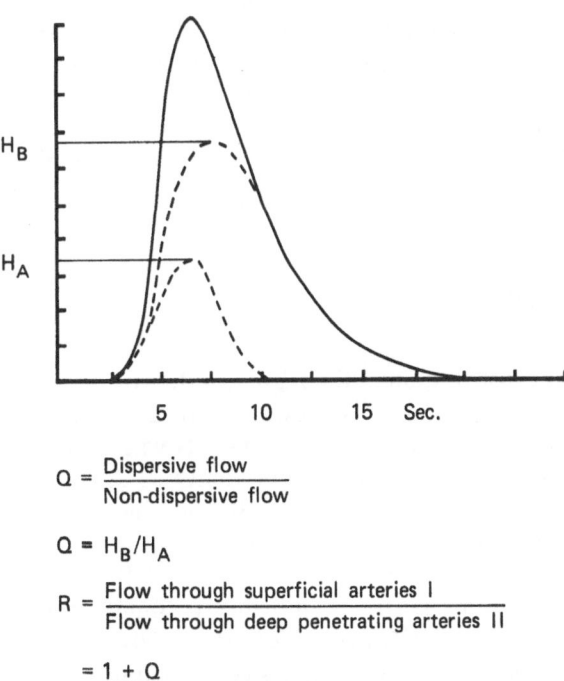

$$Q = \frac{\text{Dispersive flow}}{\text{Non-dispersive flow}}$$

$$Q = H_B / H_A$$

$$R = \frac{\text{Flow through superficial arteries I}}{\text{Flow through deep penetrating arteries II}}$$

$$= \frac{1 + Q}{Q}$$

Fig. 1. Separation of dispersive (B) and nondispersive (A) components of flow from the cerebral retention function, CRF

The behavior of the ratio Q has been studied in the following three groups of patients.

(I): 7 patients presenting with mild essential hypertension and 2 volunteers with normal blood pressure.

(II): 47 patients presenting with transient ischemic attack (TIA) or stroke.

(III): 26 patients presenting with some form of mental disorder: intellectual deterioration, memory impairment, or personality change.

Group (III) was divided into two subgroups (A) and (B).

(A): 9 patients subsequently selected as probably having a cerebral vascular deficit as assessed by past history of TIA or stroke, or from infarction seen on a computerized tomography (CT) scan.

(B): 17 patients without evidence of any vascular abnormality.

In order to assess the significance of the dispersive component of flow, Q values corresponding to patients in group III(B) were compared with the results of quantitative CT scanning and psychometry.

For each patient, a series of CT scans consisting of between 8 and 12 nonoverlapping axial slices were obtained using an EMI CT1010 scanner. A method has been introduced for the quantitative evaluation of the histograms of CT numbers formed from the scan data (Nimmon et al. 1984). A low attenuation mode was identified as consisting of voxels with CT numbers below a threshold value corresponding to $0.61 \times$ the dominant modal frequency. This mode was then separated into ventricular and nonventricular components by means of an irregular ROI outlining the ventricles, and correction was made for the partial volume effect. The resulting nonventricular low attenuation mode L_{NV} was expressed as a percentage of the total brain volume.

Patients in this group were psychometrically assessed using the Wechsler Adult Intelligence Scale, WAIS, which includes the subtests: Digit Span, Similarities, Arithmetic, Vocabulary, Digit Symbol, Picture Arrangement, Block Design, and Picture Completion (Wechsler 1965).

Results

The distribution of the higher of the two Q values corresponding to the cerebral hemisphere ROIs is shown in Fig. 2 for the groups (I), (II), (III), and (III)A. The mean value for Q in group (I) was 1.44. Taking this group of results to define an upper normal limit for Q as 2.5, in group (II) 23 values (49%) were above this limit. In group (III), 5 values were above this limit, 4 of which belonged to group (III)A (44%).

Comparison of Q values in group (II) with the corresponding values of mean transit time is shown in Table 1. Significantly increased transit times are associated with Q values > 2.5 ($P < 0.001$).

Comparisons of Q values corresponding to the left anterior and right middle ROIs with the scores obtained for the Digit Span and combined Picture Ar-

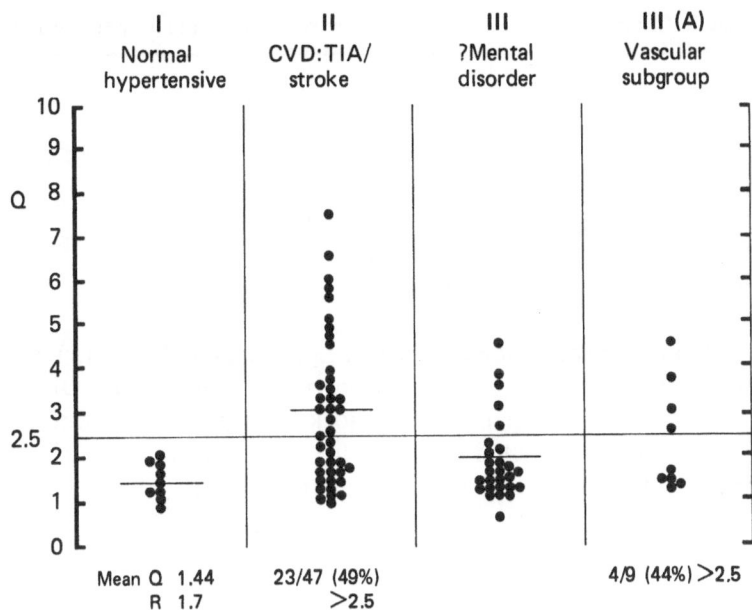

Fig. 2. The distribution of Q values obtained in the patient studies

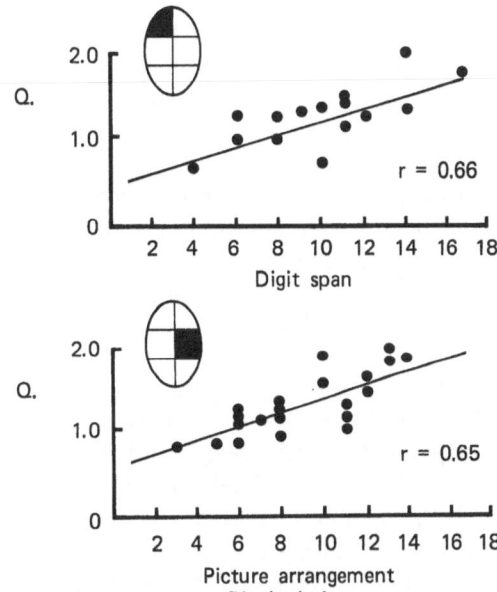

Fig. 3. Comparison between regional Q values and psychometry

rangement and Block Design WAIS subtests are shown in Fig. 3. The results show a linear correlation with values for the coefficient of linear correlation r of 0.66 ($P < 0.001$) and 0.65 ($P < 0.001$), respectively.

Comparison of the Q values and the nonventricular low attenuation volume L_{NV}, calculated for each cerebral hemisphere, yielded an inverse linear relationship with an r value of 0.56 ($P < 0.001$).

Table 1. Comparison of Q values and mean transit times (T) for patients in group (II)

Range	Q Mean ± (SE)	T Mean ± (SE) (s)
Q < 2.5	1.60 ± (0.09)	6.74 ± (0.28)
Q > 2.5	4.66 ± (0.42)	8.74 ± (0.44)
Significance of difference between means	P < 0.001	P < 0.001

Comparisons of regional values of L_{NV} calculated for the left anterior and the right hemisphere regions with the scores obtained in the Digit Span and combined Picture Arrangement and Block Design WAIS subtests using multiple linear regression resulted in values for the multiple correlation coefficient r of 0.69 ($P < 0.001$) and 0.65 ($P < 0.001$), respectively.

Discussion

The occurrence of high Q values in the groups with cerebrovascular disease (CVD), groups (II) and (III)A, is of interest. A raised Q value implies a decreased nondispersive fraction f_1 and an increased dispersive fraction f_2 of the total flow. From the range of Q values observed, up to 7.5, it is thought unlikely that these high values correspond to a real redistribution of flow between the superficial and deep components. A more plausible hypothesis is that there is dispersion with consequently increased transit times associated with the normally nondispersive superficial component of flow. Whether these findings have a common basis with the observations of decreased perfusion reserve (Gibbs et al. 1984) or with those of compartmental slippage in ^{133}Xe techniques (Obrist and Wilkinson 1984) reported in cerebrovascular disease remains speculative at present.

Although originally introduced as a means of assessing intelligence levels, subtests of the WAIS are sensitive to cerebral function in particular regions. For example, both Block Design and Picture Arrangement are nonverbal tests and involve the right parietal and right frontotemporal regions respectively. The Digit Span subtest involves the attention and coordinating centres in the left frontal and frontoparietal regions. The significant positive correlation which we have found between scores for these subtests and the Q values for the corresponding regions is of interest. Lower values of Q, corresponding to a decreased deep (long transit time) component of flow, are associated with lower scores on the WAIS subtests. No significant correlation was obtained when these subtest results were compared with Q values corresponding to adjacent or contralateral regions. In comparison with CT scanning, the Q values showed an inverse trend in which lower values of Q corresponded to increased values of the nonventricular low attenuation volume indicating loss of deep lying cerebral tissue. Finally, increased values of L_{NV} are associated with lower scores on the corresponding WAIS subtests. These results in combination suggest the hypothesis of an underlying process, either natural aging or pathologic, leading to loss of deep lying tissue, a decrease in the long transit time component of flow, and lower performance on the psychometric tests.

These preliminary results illustrate the potential value of the dispersion model in aiding clinical classification.

References

Bergh RVD, Eecken HV (1968) Anatomy and embryology of cerebral circulation. In: Luyendijk W (eds) Progress in brain research. Elsevier, Amsterdam, 30, p 1 – 25

Britton KE, Granowska M, Rutland M, Lee TY, Nimmon CC, Petrosino I, Lumley JSP (1979) Noninvasive measurement of regional cerebral flow before and after microvascular surgery. In: Greenlalgh RM, Rose FC (eds) Progress in stroke research I. Pitman Medical, London, p 307 – 318

Gibbs JM, Wise RJS, Leenders K, Jones T (1984) Measurements of regional cerebral blood flow, blood volume and oxygen metabolism in patients with extracranial vascular disease. In this symposium

Nilsson BW, Rikner G, Wolgast M (1977) On the theory of an intravenous isotope method for cerebral blood flow measurements. Scand J clin Lab Invest 37:195 – 200

Nimmon CC, Lee TY, Britton KE, Granowska M, Gruenewald S (1981) Practical application of deconvolution techniques to dynamic studies. In: Medical radionuclide imaging 1980. IAEA, Vienna, I, p 367 – 388

Nimmon CC, Charlesworth M, Lumley JSP, Mahendra B, Kingsley DPE (1984) Quantitative grading of ventricular volume and cortical atrophy using a histogram analysis of CT scan data. Submitted for publication

Obrist WD, Wilkinson WE (1984) ^{133}Xe methodology: stability and sensitivity of CBF indices. In this symposium

Sapirstein LA (1961) Measurement of the cephalic and cerebral blood flow fractions of the cardiac output in man. J clin Invest 41:1429 – 1438

Wechsler D (1955) Manual for the Wechsler Adult Intelligence Scale. The Psychological Corporation, New York

Methodological Problems in the Clinical Application of the Atraumatic Xenon Techniques

M. D. O'BRIEN

The purpose of this paper is to outline some of the problems in the clinical application of the noninvasive xenon 133 inhalation and intravenous injection methods of cerebral blood flow measurement, and any subsequent mention of methodology refers to these two techniques. There are a number of pitfalls for the unwary in the clinical use of these methods, and these can be divided into two main groups. Firstly, the inappropriate use of equipment and overinterpretation of data and, secondly, possible errors due to assumptions made in the method itself.

Of course, errors due to the inappropriate use of equipment are not confined to cerebral blood flow measurement, but this field has been, and is, particularly vulnerable to this type of problem. It was a problem in the early days and is again now a problem for quite different reasons. During the development of blood flow methods many departments devoted considerable time, effort, and resources to establishing a particular technique and quite understandably wished to use it for a wide range of clinical studies. There was therefore a tendency to ask questions of a particular method for which it might not be appropriate. The same problem now arises because of the availability of complete sets of equipment from manufacturers which print out a figure, and the question to be asked is: What does this figure mean? Once a biologic parameter can be quantified, all sorts of manipulations can be applied to it, but the result depends on the quality of the original numbers.

The use of an inappropriate method for the question asked leads to the overinterpretation or misinterpretation of the data produced. A good example of this was widespread in the early 1960s, when transit times were often measured and equated with blood flow. Many reports have equated blood flow with metabolism, which is the same error in reverse when measurements of changes in arteriovenous oxygen differences are equated with blood flow. Similarly, changes in blood volume have been used as an index of change in blood flow. Of course, all these indices are closely linked and in normal subjects they are to some extent coupled, but in pathologic states they frequently uncouple.

There is another way in which data is overinterpreted, and this is also the consequence of these blood flow techniques. They require both time and effort so that it is difficult to obtain sufficient patients who are similar enough to form a homogeneous series. As a result many reports in the literature are so heterogeneous as to be unanalyzable or consist of a single case or a very small

Department of Neurology, Guy's Hospital, London, SE1 9RT, Great Britain

Cerebral Blood Flow and
Metabolism Measurement.
Eds. Hartmann/Hoyer
© Springer-Verlag Berlin Heidelberg 1985

number of patients, and it is easy to find examples of hypotheses built on inadequate data.

It is important to remember that only blood flow is measured and this is to some extent an epiphenomenon: the result of the local perfusion pressure and the diameter of the blood vessels, and these are determined by many more factors, including systemic arterial pressure, viscosity, PO_2, PCO_2, local pH, and local metabolic demand. If only the blood flow is measured it is an easy error to assume that an increase is a good thing and a decrease a bad thing, but this may not be so since a low flow may be appropriate to a low metabolic state, and an increase in flow in these circumstances may not be beneficial and could even be harmful.

Several papers in this symposium have discussed some of the errors which might arise as a result of assumptions that are made in the technique and in the calculation of results. The principal advantage of these methods is the lack of trauma, which accounts for their current popularity, but to achieve this inevitably results in a number of sources of error. These can be divided into problems related to labeling and the problem of recirculation.

Problems of labeling concern both labeling of the brain and of the noncerebral cranial tissues. It may seem obvious, but clearance curves can only be recorded from tissue that the tracer has reached. Anything that is not labeled, either because it has no flow or because the labeling time is too short, is not seen; although in some circumstances it can be identified if the volume is sufficiently large because of the small intercept, spurious flow values may be obtained by counting scatter or look-through to more normal tissue. This is especially true if an infarcted area is a small part of the volume seen by the detector. Saturation of the brain with tracer takes at least 5 min in normal subjects and it may be 10 min or more in patients with slow flows (Kumer et al., this volume). The current practice of using 1 min inhalation is probably insufficient to label the slower components sufficiently for identification. In normal subjects this short inhalation time does not introduce a sizeable error; although very high apparent flow rates result from inhalation times of less than 1 min, this appears to be largely artifactual and due to scatter from the nasal sinuses which then becomes a significant part of the clearance curves. This is a source of error in frontal clearance curves with longer inhalation times (Crawley and Veall 1975).

Partial volume effects are a potential source of considerable error if attempts are made to record focal changes in blood flow, particularly if an infarcted area and hyperemic areas are included with normal brain in the volume of tissue seen by the counter. Serial measurements in these patients compound the problem, not only for positional reasons but also because of alteration in the relative volume of hyperemic and underperfused areas. I do not propose to discuss the errors associated with start-fit time, end-fit time, compartmental analysis, the problem of slippage in pathologic states, or the merits of the initial slope index because these have been fully discussed elsewhere in this volume (Herholz et al.; Obrist and Wilkinson; Seylaz; Prohovnic et al.; and Farrar); but we should be impressed by the very considerable errors that may arise in some of these calculations.

The partition coefficient of xenon used in the calculation may vary considerably from tissue to tissue. O'Brien and Veall (1974) showed wide variations in lambda in patients with different brain tumours and this data has been confirmed by the stable xenon technique (Meyer, personal communication).

There are, therefore, very definite problems in the labeling of the brain in pathologic states, and there are also major problems due to the labeling of noncerebral cranial tissue. This tends to produce artificially low flows (Obrist, this volume). It is worth mentioning that 15% of this compartment has a clearance rate which is very similar to that of gray matter; this cannot be easily identified and results in a systematic error, but it does produce a predictable deviation rather than a contamination of the results.

Crawley et al. (1971) described the gamma subtraction technique which largely eliminates the third component, but it introduces further error because of the very low count rates that result and the effects of this have been demonstrated by Herholz et al. (this volume).

The second major methodological problem with these techniques is due to recirculation of tracer. In order to maintain an atraumatic procedure it has become usual to record the end expired air radioactivity as an index of arterial activity and hence to make a correction for recirculation by deconvolution. Again this works reasonably well in normal subjects but in patients with respiratory disease and particularly those with ventilation/perfusion defects this correction cannot be applied and is a potential source of considerable error, particularly if the respiratory problem changes from time to time, as is likely to be the case in sick patients in intensive care wards and patients on ventilators. Such problems may preclude the use of these techniques.

The profound effect of PCO_2 on the cerebral circulation is well known. Any cerebral blood flow measurement without a record of the arterial PCO_2 is worthless. In experimental circumstances it may be possible to ensure that the PCO_2 is at normal levels and maintained there during the recording. However, in patients, particularly sick patients, it may be extremely difficult. Unfortunately it is not possible to make a correction for variation in arterial PCO_2 since the correction factor is unknown. Even in the middle of quite large infarcts some CO_2 responsivity is usually preserved, although impaired, and this impairment is likely to vary considerably throughout a lesion.

Conclusion

These techniques have very considerable advantages over many methods because they are relatively atraumatic, but this imposes definite limitations on their applicability. They give very good results in normal subjects but errors increase exponentially in pathologic states. It is important to use the appropriate methodology for the question asked and, if the appropriate method is not available, it may be necessary to modify the question. It is usually better to ask the question before doing the experiment rather than afterwards, as seems often to be the case. In order to do this, it is important to know what the method measures and what are its limitations together with the reliability and reproducibility of the data. It is also important to avoid overinterpretation of the data, that is to say, reading more into the results than is justified by the method. However, with care these methods can provide interesting and useful information.

References

Crawley JCW, Veall N (1975) Recent developments in the ^{133}xenon inhalation technique for cerebral flow, J Nucl Biol and Med *19*:205−212

Crawley JCW, O'Brien MD, Veall N (1971) The gamma spectrum subtraction technique applied to cerebral blood flow measurement by the inhalation of ^{133}xenon, in: Brain and blood flow, ed. R.W. Ross Russell. Pitman, London, p 54−56

O'Brien MD, Veall N (1974) Partition coefficients between blood and various cerebral tumours for Xenon-133. Phys Med Biol 19:472

Cerebral Blood Flow Tomography Using Xenon 133 Inhalation – Methodological Considerations

N. A. Lassen

Three tomographic techniques for measurement of cerebral blood flow (CBF) are currently being explored for clinical use. They are all founded upon the use of digital computers for image reconstruction based on a series of lateral projections. In 1978 Drayer et al. described the use of stable xenon combined with conventional CT scanning, viz. single-photon transmission tomography. Because the enhancement is small, the signal-to-noise ratio is very unfavorable. The high cost of stable xenon and the difficulty in obtaining a sequence of CT scans at several levels simultaneously also limit the clinical usefulness of the method (Gur et al. 1981). Positron emission tomography (PET) based on coincidence counting of the annihilation photons has been developed over the last decade. In 1979 Yamamoto et al. described the use of krypton 77 for obtaining CBF tomograms using the inert gas clearance principle. Other positron emitting tracers, notably oxygen 15 labeled CO_2 or H_2O, have also been used (see, e.g., the reports from the 10th and 11th International Symposia on CBF and Metabolism, 1981 and 1983). The clinical usefulness of these techniques is, however, limited by their cost and cumbersomeness, and they have so far been reserved for use in clinical research.

The third method, to be discussed in this paper, is based on single-photon tomography using a rotating gamma camera and xenon 133 inhalation. This isotope is commercially available at a relatively low cost and the signal-to-noise ratio is no major problem. The technique, described in 1980 by Stokely et al., is suited for routine clinical use. Its major drawback is the need for a very sensitive gamma camera. Recently two iodine 123-labeled amines have been introduced that can be used in conjunction with conventional rotating gamma cameras, albeit yielding a fairly gross spatial resolution mainly due to the low sensitivity of the systems. This approach will be commented on in the discussion.

Method

Xenon 133 tomography for CBF measurement is based on the principles pioneered by Kuhl et al. in 1976. The instrument used (Tomomatic 64, produced by Medimatic Inc., Hellerup, Denmark) was developed with the specific aim of

Department of Clinical Physiology, Bispebjerg Hospital, DK-2400 Copenhagen NV

Cerebral Blood Flow and Metabolism Measurement.
Eds. Hartmann/Hoyer
© Springer-Verlag Berlin Heidelberg 1985

allowing dynamic studies, viz. to record a series of 1-min images of the isotope distribution in slices of brain tissue. It consists of four gamma cameras rotating at a fairly rapid speed of 10 s/rotation close to the head. Each gamma camera consists of 16 thin sodium iodide crystals (13 mm wide). The instrument rotates at a constant speed and collects data in brief time intervals (1/8 of a second). Thus, in each time interval all four cameras record a projection of the isotope distribution in a 2 cm thick slice of brain tissue. Three slices of brain tissue are recorded simultaneously. Conventional filtering and back-projection techniques are used for reconstructing the isotope distribution in each slice.

One half turn of the tomograph − 5 s − allows sampling of a complete set of projections, 40 in all. This means that despite the dynamic aspect of the xenon 133 isotope, its arrival and washout, its concentration is practically constant during the collection of one complete set of projections. It is for this reason that the tomograph is designed to rotate so rapidly.

However, owing to the low count rate, a series of 5-s tomograms cannot be reconstructed with adequate resolution. Hence several sets of projections. typically 12 − i.e., covering 1 min − are added before reconstruction. Such a picture gives the *average* isotope concentration over that minute.

The xenon 133 is administered in a closed respiratory system of approximately 5 liters in volume. The gas ampoule is crushed inside the system and the xenon 133 gas is admixed to the air in the closed system in such a way that the concentration in the 5-liter system reaches about 20 mCi/liter. When the patient rebreathes the air in the system, the lung concentration will reach a maximum of about 10 mCi/liter towards the end of the first minute. The patient is connected to the system for a total of 1.5 min.

The radiation dose per study calculated for 10 mCi/liter lung air for 1.5 min is approximately 0.6 rad in the critical organ − the lung. The gonadal dose is much lower, approximately 0.06 rad per study (Stokely et al. 1980). These doses must be compared with the considerably higher doses obtained with iodine 123 or positron emitting isotopes as currently applied to CBF measurements in man. By inference this would mean that we could increase the dose of xenon 133 considerably, say by a factor of 5 (or use a less sensitive collimator system), without surpassing acceptable gonadal doses. Yet we decided not to do so because the reasonably low doses mentioned allow us freely to study normal subjects of all ages (we only exempt pregnant normal women) and even to repeat the measurements in up to approximately five studies over 1 year if so desired. In the context of studying patients with cerebral diseases the radiation dose is considered quite negligible. When it is relevant for following the disease, up to ca. ten studies per patient have been performed without this being taken to constitute a significant radiation hazard and hence also without it being considered necessary to require explicit permission from the patient in accordance with the Helsinki declaration: As each study gives an exposure approximately equal to a single conventional lung X-ray exposure (that is what we tell the patient) the radiation problem is considered small indeed.

In this context it may be mentioned that xenon 133 is also a very clean isotope for the personnel. Its radiation is so soft that it is easy to shield the hotter radioactive sources. And, since atmospheric air acts as a carrier, any leak of xenon 133 is rapidly diluted. We use the conventional X-ray film badges that do not exceed acceptable levels for personal exposure for the person (a medical doctor) operating the system daily. The instrument is placed in a conventional laboratory without special shielding. An exhaust hood is used for placing hotter sources that might leak some xenon 133 (e.g., phantoms for studying the resolution). It should be mentioned that for the following 3 min after disconnection from the rebreathing system, the expired air from the patient is led out through a tube connected to a *recycling* xenon 133 trap. This trap recovers ca. 80% of the xenon 133 and thus reduces the cost involved. We currently use 1 Ci of xenon 133 per week and usually perform 20 – 30 studies per week. The only other direct costs involved in a study are the Polaroid films taken as a permanent document of the final results and the floppy discs used to store the data.

Sensitivity of the Tomograph and Spatial Resolution

Using an energy window of ca. 30% around the peak energy for the 81 keV γ-ray energy of xenon-133 emitted in about 1/3 of disintegrations, the sensitivity is approximately 20 000 cps per slice using a water-filled phantom with a diameter of 20 cm and filled with water having a xenon 133 concentration of 1 μCi/ml. For technetium 99m, having a 100% incidence of γ-ray producing disintegration, the corresponding sensitivity is about 60 000 cps per slice, and the same pertains for iodine 123. (With three slices seen simultaneously the total sensitivity of the instrument is threefold higher).

These data are obtained using the high sensitivity – low resolution collimator designed for dynamic tomography of xenon 133. A clinical study lasting 4.5 min gives a maximum counting rate of about 10 000 cps per slice and a total number of counts of about 1 000 000 counts over the entire period. It is considered essential to accumulate about this number of counts to avoid undue random noise in the final CBF picture.

Using the xenon 133 collimator a spatial resolution of 2.0 cm axially and about 1.7 cm in the plane is achieved, being expressed as the full width at half maximum (FWHM) of a line source.

As mentioned above, the current version of the dynamic tomograph only allows measurement from three 2.0 cm wide slices simultaneously. The three slices are spaced so that the center-slice planes are 4.0 cm apart: each slice is thus separated from the next by a practically "unseen" in-between slice also 2.0 cm thick. This design with unseen interslices was chosen to enhance the instrument's sensitivity, as it allows 4.0 cm of crystal to look at each 2.0 cm thick slice of brain via the slightly converging collimator.

Experiments are now being made in which two of the four gamma cameras are offset by 2.0 cm in order to obtain six contiguous slices at the "expense" of doubling the xenon 133 dose. The advantage of routinely obtaining a complete set of slices must be weighed against the expense involved in doubling the dose and the possible loss in resolution due to the offsetting.

The Performance of a Study

The patient is connected to the valve allowing him to be connected to the re-breathing system. We most often use a mouthpiece fixed with adhesive tape and a nose clip. A mask can be used if the discomfort of the noseclip is considered unacceptable. Leaking of xenon 133 from the mouthpiece/nose or the mask is the most significant source of error. It is readily recognized in the final pictures as radioactivity over the skull: normally the xenon 133 concentration in the extracranial tissue tissues only reaches a very low level.

The patient's head is fixed in the aperture of the apparatus using the balloon of an ordinary arm blood pressure cuff. It is gently inflated, so that only minimal movements are possible. The head is positioned using light markers. Routinely we position the midslice planes 1.0, 5.0, and 9.0 cm above the plane going through the lateral canthus of the orbita and the external auditory meatus, the orbitomeatal plane.

A single nonmoving probe placed over the upper part of the right lung is used to record the shape of the lung xenon 133 radioactivity curve. This curve is used to represent the shape of the arterial input curve to the brain. A 10-s time delay is allowed for, i.e., it is assumed that the true arterial curve reaching the brain has the lung curves' shape after a 10-s delay.

Calculation of CBF

The raw data from a study consist of the lung curve and the three sequences of tomographic images taken. Each sequence comprises four images averaging the xenon 133 concentration over the time intervals $0-1.5$ min, $1.5-2.5$, $2.5-3.5$, and $3.5-4.5$.

The sum of the first and second pictures is called the *"early picture,"* as used in the calculation outlined below.

CBF is calculated as described by Celsis et al. in 1981. In essence the bolus distribution principle is applied to the early picture. The calculation is essentially the same as that used by Kety in developing the autoradiographic method for calculating CBF. However, because scaling factors relating the lung curve to arterial concentration and the head tomograms to brain concentrations are not available, the scaling is accomplished by using the entire sequence of four images: for all high-count rate pixels (from 50% to 100% of the maximal count rate) the four counts recorded are analyzed as a clearance curve using the clearance principle.

The principle involved in the final calculation essentially consists in a relatively minor modification of this early picture. In fact, in relative terms this early picture looks very much like the final CBF map. The essential raw data, i.e., the early picture, thus shows exactly the same low flow (low count rate) and high flow (high count rate) areas as the final CBF map. The spatial resolution is also the same. This is mentioned in order to stress that the rater involved calculations do not materially change the essential raw data.

One point deserving special mention concerns the partition coefficient (λ) for xenon 133 in the different areas. Contrary to what one might have thought, this unknown factor plays but a minor role in the calculation of CBF using the bolus distribution principle outlined above. This is so because the bolus distribution principle expresses that the early picture's count rate in a pixel $\bar{C}_i$ is calculated from the convolution integral of the input function to that pixel $f_i C_a(t)$ and the impulse response wash-out function exp. $(- f_i \lambda_i \cdot t)$:

$$\bar{C}_i \underset{0-2.5\,min}{} = B \int_0^{2.5} f_i\, C_a(t) * \exp(- f_i/\lambda_i \cdot t)\, dt$$

where B is the unknown scaling factor common to all pixels and the asterisk denotes the convolution integral. Hence it follows that in the ideal situation of an ultrabrief experiment, where essentially no wash-out has yet taken place, the convolution term vanishes and hence C_i becomes proportional to f_i: no partition coefficient correction is needed.

In this context it should be stressed that the algorithm used to find the absolute blood flow level applying the clearance principle to all four data points for high count rate pixels (= high flow) employs a partition coefficient of 0.85, i.e., a value typical of gray matter structures (when the hematocrit is normal). This means that it is assumed that hyperemic areas are gray matter, not myelinated white matter. On the basis of this assumption the scaling of the early picture is performed. Thus when performing the bolus distribution calculation outlined above an error is only made for myelinated white matter, i.e., for low flow regions where the correction for wash-out indicated by the convolution is a very small one. In practice, therefore, the use of the erroneous λ of 0.85 for these areas, where the correct value should be about 1.50, involves only a minor overestimation in the order of 10% in white matter flow levels: a value of 0.22 ml/g/min instead of 0.20 ml/g/min, as described in some detail by Holm in the study by Shirahata et al. (1983). As stated in that same study, the errors due to uncertainty of the value of the partition coefficient may be considered completely negligible compared with the rather massive errors caused by Compton scatter, which results in a massive overestimation of CBF in very low flow regions.

Some Results

In normal man the CBF tomograms are side-to-side symmetrical and outline the gray matter structures as high flow areas with the blurring inherent in the limited spatial resolution. The average CBF value of the slices is 0.60 ml/g/min (± 0.11) (Lauritzen et al. 1981). In the upper slice, with its midline 9 cm above the orbitomeatal (OM) plane, flow tends to be almost uniform − only in patients with a large cranium can the lower flows inside the hemispheres (the white matter of centrum semiovale) be discerned. The middle slice, at OM + 5 cm, shows a pattern with highest flow levels of about 0.80 ml/g/min in the

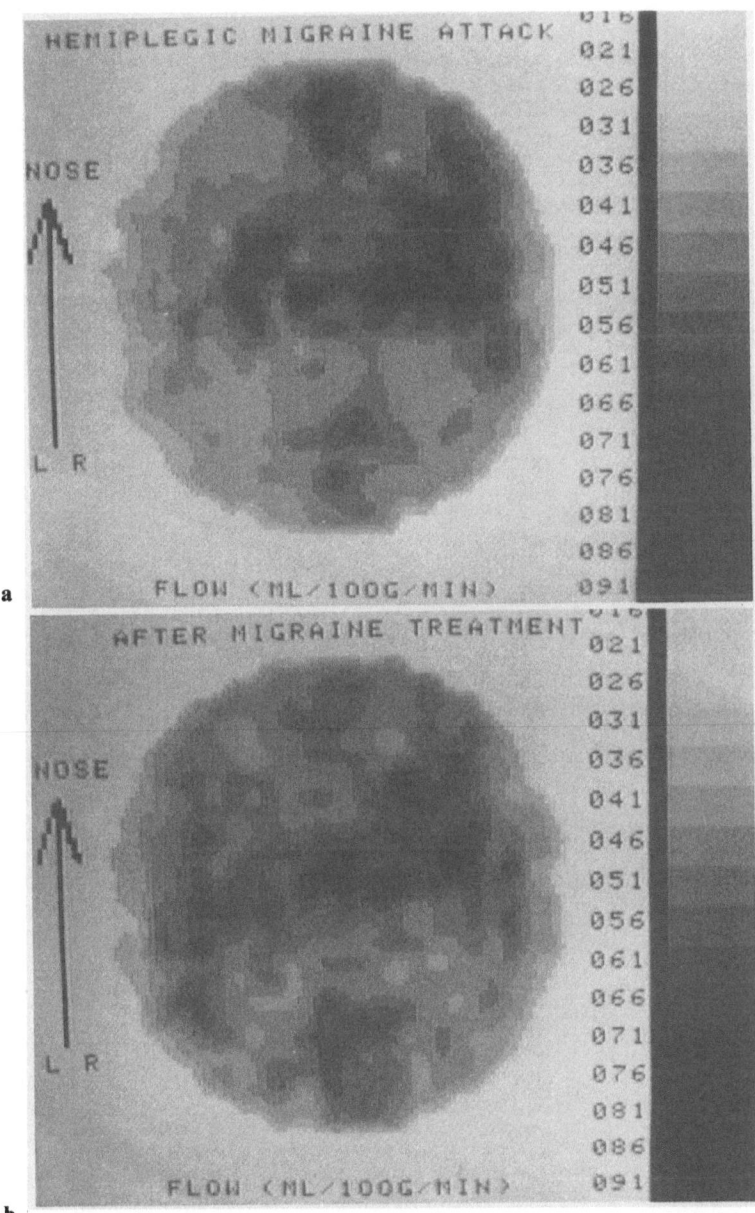

Fig. 1 a, b. Classical migraine with visual symptoms and right-sided hemiparesis and aphasia as prodromal symptoms persisting during headache. During attack there is low CBF in the left hemisphere cortex (but not in the subcortex). After migraine treatment with subsidence of symptoms the CBF map returns to the normal symmetrical pattern (Lauritzen et al. 1983). Xenon 133 CBF tomogram, middle slice, OM + 5 cm

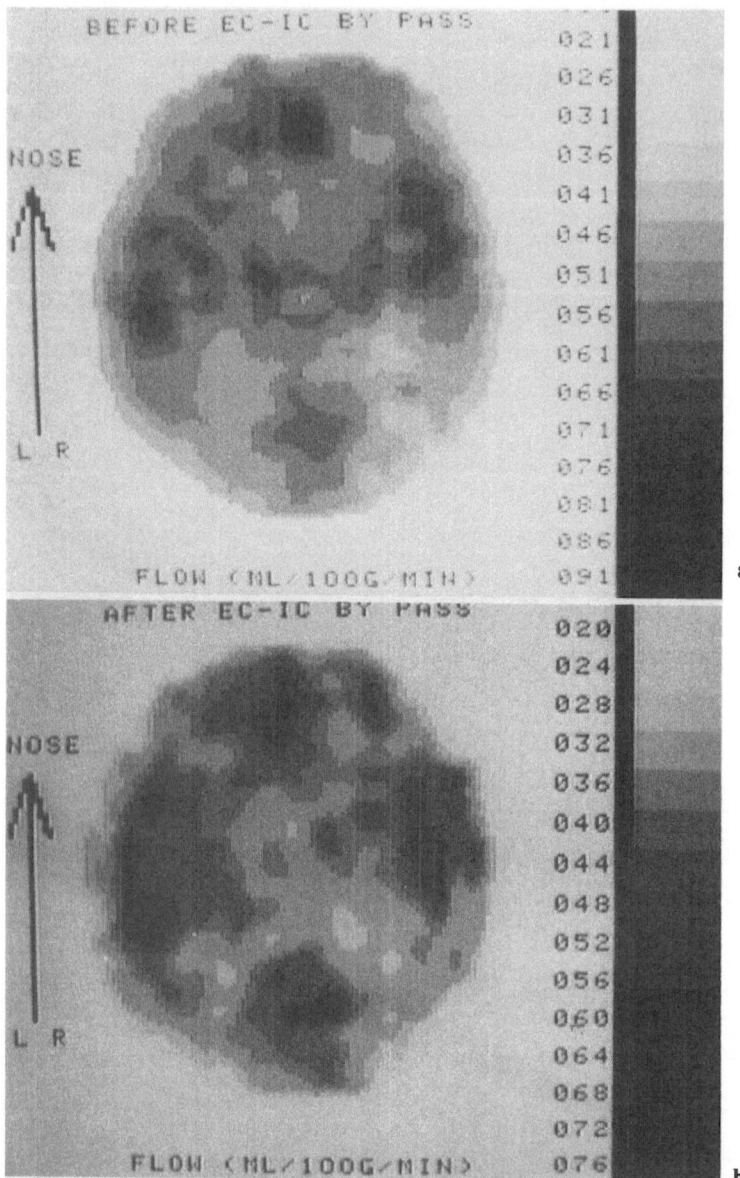

Fig. 2a, b. EC-IC bypass surgery in patients with amaurosis fugax and two TIAs from the right hemisphere; right internal carotid occlusion; normal CT scan. Before EC-IC bypass a slight but reproducible asymmetry of CBF map is noted, with the lowest flow in the watershed area between middle and posterior cerebral artery territory on the right side. After shunting the CBF map is unchanged, suggesting that the low flow is due not to ischemia ("misery perfusion") but to a slight degree of irreversible ischemic tissue damage, viz. selective neuronal loss, also called incomplete infarction (a lesion that cannot be seen on the CT scan) (Vorstrup 1983). Xenon 133 CBF tomogram, middle slice, OM + 5 cm midline anteriorly and posteriorly as well as laterally, corresponding to the Sylvian and insular cortex and underlying nuclei: The lowermost slice, OM + 1 cm, shows the cerebellum, the tip of the temporal lobes, and − to a variable degree − an artifact anteriorly in the midline corresponding to xenon 133 in the nasal sinuses.

Focal areas of increase in CBF are normally seen with *various types of brain activity.* In normal man a 30% increase in CBF is seen in the visual cortex − in the midline posteriorly in the middle slice (OM + 5 cm) during visual perception (Henriksen et al. 1981). Movements of the hand augment CBF in the contralateral hand area as well as in both supplementary motor areas in the midline anteriorly (Lauritzen et al. 1981). In a preliminary study of language functions an asymmetry has been found to develop, with highest flow in the left hemisphere laterally.

Among the pathologic states showing abnormalities of the CBF maps, classical migraine and apoplexy must above all be mentioned. In classical migraine an area of decreased flow in the cortex is seen corresponding to the focal neurologic symptoms (Lauritzen et al. 1983) (Fig. 1). Typically the flow is decreased in the posterior-lateral cortex, the visual cortex, during visual prodromal symptoms. The low flow area usually persists during the headache phase. In common migraine, where no focal neurologic symptoms develop, the CBF map is normal, displaying neither decreased our increased flow (neither focally nor globally).

In acute apoplexy due to large vessel occlusion the CBF shows low flow areas that typically are larger than the hypodense area seen on the CT scans (Lassen et al. 1981). Sequential studies show that the CBF map often changes spontaneously in that a previously low flow area all of a sudden may have the highest flow (be hyperemic). This is the effect of spontaneous lysis of the arterial obstruction − usually an embolus arising from the carotid bifurcation or further proximal in the circulation. We currently use the xenon CBF tomograms routinely to study all cerebrovascular cases considered for reconstructive or bypass vascular surgery (Fig. 2) (see chapter by Vorstrup and Lassen, this symposium). Another interesting group of patients studied by CBF tomography are patients with subarachnoid hemorrhage due to a spontaneous rupture of an arterial aneurysm. In such patients the flow tomograms carried out sequentially (five to eight per patient) allow one to time when the dreaded complication of severe arterial spasm sets in, causing the so-called delayed ischemic deficit, a slow onset stroke (B. Mickey et al. 1983).

Discussion

This easy and atraumatic CBF method lasting only 4.5 min is not yet in full clinical use. Many patient categories other than those mentioned above could be considered, yet it may already be surmised that the method's major application will be in specialized centers performing vascular surgery in stroke or aneurysm cases. The possibility of performing same setting, say with an interval

of 20 min, should be emphasized. Current experience with a cerebral vasodilator compound such as Diamox (acetazolamide, a carbon anhydrase inhibitor accumulating CO_2 in the brain and hence dilating the brain vessels) makes
use of this facility. The approach appears to allow one to discern areas of critically low flow caused by inflow obstruction ("misery perfusion") from areas of
low flow due to low metabolism due to tissue damage due to complete or incomplete (partial) infarction.

In the clinical routine setting the low cost of xenon 133 is attractive. Using the
recycling trap the cost per investigation is at the same level as conventional
X-ray studies and CT scans. This includes considerations of instrument acquisition and depreciation, as well as costs in respect of maintenance, space, and
personnel. As the instrument is programmed it can be run by a medical doctor
or a nurse after a brief training period. The short duration of the study, 4.5 min,
is also advantageous (in special cases we have reduced the time to 2.5 or
2.0 min, taking only the early picture as this closely represents the flow distribution in relative units).

The fact that I have stressed the advantages of xenon 133 should not be taken
to mean that this is the ideal isotope. Indeed, due to the soft primary radiation,
Compton scatter is a source of major error rendering the tomogram nonlinear:
an area with no radioactivity will typically show a count rate of ca. 40% of the
mean value of the slice (see chapter by S. Holm et al., this symposium). This
error, and not the errors due to the unknown solubility coefficient, is – as already stressed – the main technical limitation.

Other single-photon emitting radioactive tracers can be used to map CBF
tomographically. Iodine 123-labeled amines that cross the blood-brain barrier
freely and remain fixed to the brain have been developed. The most promising
are N-isopropyl-para-iodo-amphetamine (IMP) developed by Winchell et al.
(1980), Kuhl et al. (1982), and Hill et al. (1982), and N,N,N'-trimethyl-N'-(2-
hydroxy-3-methyl-5-iodobenzyl)-1,3-propanediamine (HIPDM) developed by
Kung et al. (1983). However, due to the costs, the short half-time of iodine 123,
and the rather high radiation dose, these tracers are not ideal either. A direct
comparison using our high sensitivity camera for xenon 133 and for IMP
showed fair agrrement between the CBF maps obtained (Lassen et al. 1983).
But, when used in conjunction with a conventional rotating gamma camera
having a much lower sensitivity and rotating at a greater distance from the
brain, the spatial resolution is not impressive and the counting time is long despite the use of fairly high doses (Ell et al. 1983). Technetium 99m, with its
mono-energetic medium-energy gamma radiation, would constitute a better
radionuclide. However, tracers have not yet been developed that are labeled
with this isotope and have the required properties displayed by the amines
mentioned: the chemical microembolus to be injected intravenously to enter the
brain freely in proportion to local blood flow and yet being retained for a sufficiently long time to allow imaging. Were such a compound to become available, advantage would still accrue from having a brain-dedicated four-faced
tomograph as high intrinsic sensitivity is a prerequisite for obtaining good spatial resolution.

References

Drayer BP, Wolfson JR SK, Reinmuth OM, Dujovny M, Boehnke M, Cook EE (1978) Xenon enhancement CT for analysis of cerebral integrity, perfusion and blood flow. Stroke 9:123–130

Ell PJ, Lui D, Cullum I, Jarritt PH, Donaghy M, Harrison MJG (1983) Cerebral blood flow studies with 123-iodine-labelled amines. The Lancet ii:1348–1352

Gur D, Yonas H, Herbert D, Wolfson SK, Kennedy WH, Drayer BP, Gray J (1981) Xenon enhanced dynamic computed tomography: Multilevel cerebral blood flow study. J Comput Ass Tomogr 5:334–340

Henriksen L, Paulson OB, Lassen NA (1981) Visual cortex activation recorded by dynamic emission computed tomography of inhaled xenon-133. Eur J Nucl Med 6:487–489

Hill TC, Holman BL, Lovett R, O'Leary DH, Front D, Magistretti Ph, Zimmerman RE, Moore S, Clouse ME, Wu JL, Lin TH, Baldwin RM (1982) Initial experience with SPECT (single-photon computerized tomography) of the brain using N-isopropyl 1-123-p-iodoamphetamine: Concise communication. J Nucl Med 23:191–195

Journal of Cerebral Blood Flow and Metabolism. Proc. of Tenth International Symposium on Cerebral Blood Flow and Metabolism. St. Louis, USA, 1981. 1: suppl. 1

Journal of Cerebral Blood Flow and Metabolism. Proc. of Eleventh International Symposium on Cerebral Blood Flow and Metabolism. Paris, France, 1983. 3: suppl. 1

Kuhl DE, Edwards RQ, Ricci AR (1976) The Mark IV system for radionuclide computed tomography of the brain. Radiology 121:405–413

Kuhl DE, Barrio JR, Huang SC, Selin C, Ackermann RF, Lear JL, Wu TH, Lin TH, Phelps ME (1982) Quantifying local cerebral blood flow by N-isopropyl-p-(123-I) iodoamphetamine (IMP) tomography. J Nucl Med 23:196–203

Kung HF, Tromposch KM, Blau M (1983) A new brain perfusion imaging agent: N,N,N'-trimethyl-N'-(2hydroxy-3 methyl-5-iodobenzyl)-1, 3-propanediamine (HIPDM). J Nucl Med 29:66–72

Lassen NA, Henriksen L, Paulson O (1981) Regional cerebral blood flow in stroke by [133]xenon inhalation and emission tomography. Stroke 12:284–288

Lassen NA, Henriksen L, Holm S, Barry DI, Paulson OB, Vorstrup S, Rapin J, Poncin-Lafitte M le, Moretti JL, Askienazy S, Raynaud C (1983) Cerebral blood-flow tomography: Xenon-133 compared with isopropyl-amphetamine-iodine-123: Concise communication. N Nucl Med 24:17–21

Lauritzen M, Olesen J (1983) Regional cerebral blood flow during migraine attacks by xenon-133 inhalation and emission tomography. Brain

Lauritzen M, Henriksen L, Lassen NA (1981) Regional cerebral blood flow during rest and skilled hand movements by xenon-133 inhalation and emission computerized tomography. J CBF & Metab 1:385–389

Shirahata N, Henriksen L, Vorstrup S, Lauritzen M, Paulson OB, Lassen NA. Tomographic cerebral blood flow in man by [133]xenon inhalation. Normal values and illustrative clinical cases. In preparation

Stokely EM, Sveinsdottir E, Lassen NA, Rommer P (1980) A single photon dynamic computer-assisted tomography (DCAT) for imaging brain function in multiple cross-sections. J Comput Assist Tomogr 4:230–240

Winchell HS, Horst WD, Brain L, Oldendorf WH, Hattner R, Parker H (1980) N-Isopropyl-(123-I)p-iodoamphetamine: Single-pass brain uptake and washout; binding to brain synaptosomes, and localization in dog and monkey brain. J Nucl Med 21:947–952

Yamamoto YL, Thompson E, Meyer E, Nukul H, Matsunaga M, Feindel W (1979) Three dimensional tomographical regional cerebral blood flow in man, measured with high efficiency mini-BGO two ring positron device using krypton-77. In: Cerebral Blood Flow and Metabolism. Acta Neurol Scand 60 (suppl. 72):186–187

Physical Factors Affecting Calculated Cerebral Blood Flow Values in Hypoperfused Areas in Single Photon Emission Computerized Tomography

S. Holm, S. Vorstrup, N. A. Lassen, and O. B. Paulson

Introduction

Single photon emission computed tomography (SPECT) is a recently introduced method for measurement of cerebral blood flow (CBF) of inhaled xenon 133 [1–3]. CBF has predominantly been measured in patients with cerebrovascular diseases. These studies have documented that several patients with TIA and a high percentage of patients with stroke have focal low flow areas. These hypoperfused areas are seen both in the acute and in the chronic phase. The evaluation of the tomographic CBF pictures has mostly been based on visual inspection. Calculation and interpretation of side-to-side asymmetries in selected regions require a more detailed analysis of the factors influencing the detection of hypoperfused areas. For this reason a brief description will be given of the tomograph, the algorithm used for calculation of CBF, and some results of studies of linear sources and of various phantoms.

Methods

The tomographic device (Tomomatic 64) has a detector arrangement consisting of four specially constructed gamma cameras rotating around the head at a uniform speed of 6 rev/min, recording data from three slices of brain simultaneously. Data may be transferred to the computer for each 1.5° of rotation but normally three intervals (4.5°) are added to form one projection. To obtain a sufficient number of counts, projections from subsequent turns are accumulated. The fast rotation ensures a true time average picture. The projections are corrected for detector sensitivity and background and the tomographic pictures are reconstructed using a filtered back-projection algorithm with an additive attenuation correction.

A series of four images is taken, one during a 1.5-min period of xenon 133 inhalation, and one during each of the following 3 min. After xenon 133 inhalation (10 mCi/liter, at equilibrium) the count rate reaches a maximum of 400 000 cpm per slice. The xenon 133 input curve is determined from a single stationary scintillation detector placed over the right lung. A linearization and

Department of Neurology, Rigshospitalet, Blegdamsvej 9, DK-2100 Copenhagen

Cerebral Blood Flow and
Metabolism Measurement.
Eds. Hartmann/Hoyer
© Springer-Verlag Berlin Heidelberg 1985

scaling of the sum of the first two images is used to calculate CBF for each pixel [3]. The absolute values are influenced by several factors. However, the process only minimally disturbs the relative values of the xenon 133 concentration among the pixels. Thus, detection of low flow areas is mainly determined by the quality of the "early picture."

In a technically perfect image (i.e., free from artifacts) the information is limited by the resolution of the collimator, the absorbed or scattered photons, and the statistical noise which is mainly caused by the limited number of registered photons. The resolution was measured using line sources. Twenty-centimeter long capillary tubes with an internal diameter of 1.2 mm were filled with approximately 100 µCi of xenon 133 dissolved in water and sealed. Uniformity of the lines was controlled with an extremely narrow collimated NaI-detector. Sensitivity, uniformity, attenuation correction, and statistical noise were examined using a 20 cm diameter cylindrical perspex phantom containing various amounts of xenon 133. The contrast in the pictures was evaluated using the same cylindrical phantom containing five tubes of varying diameter (1 – 5 cm).

Results

The resolution in the plane was calculated as a full width half maximum (FWHM) from the tomographic pictures of a line source parallel to the axis by changing its position, the sampling and reconstructive variables. Using the standard sampling and reconstructive variables, the resolution in air at the center was 16.3 mm. At a distance of 9 cm from the center, these values were 15.9 mm (radially) and 16.5 mm (tangentially). When water or paraffin was used as scattering media the corresponding values were 17.3 mm, 16.8 mm, and 14.4 mm, respectively. FWHM increased only slightly with increasing energy window width. Using an angular sampling of 9° instead of 4.5°, the resolution at 9 cm from the center increased by 1 mm in the tangential direction while the radial resolution remained unchanged. Reconstruction using different filter functions yielded minor changes (± 1 mm).

The resolution in the axial direction was measured by recording a sequence of 20 tomographic pictures of a line source, now positioned diagonally in the field at right angles to the axis, moving the line source 2 – 5 mm between recordings. The values corresponding to a fixed position in the 20 tomographic pictures form a curve describing the slice at the given distance from the center. Near the center these curves are triangles having a FWHM of 20 mm. The width at the 10% level (FWTM) is consequently 36 mm. The FWHM is independent of the position, but the FWTM increases to 44 mm near the edge of the picture, corresponding to a distortion of the triangular form of the curve. From the axial measurements the cross-talk between slices can be estimated to be 5%.

The sensitivity with the normal energy window setting around the 81 keV peak of xenon 133 is 20 000 cps/(mCi/liter) per slice. Recordings of the uniform cylinder source with different attenuation correction coefficients yield pictures with some ratio (< 1) between the center and a rim just inside the edge. With an optimized value of the attenuation coefficient this ratio is approximately 0.95.

Noise in the pictures was estimated from calculations performed on images of the uniform cylinder. Pictures were recorded with total number of counts ranging from 10^3 to $3 \cdot 10^7$. The relative standard deviation (coefficient of variation) was calculated from pixels placed at a constant distance from the center. Reconstructions were carried out using the different filter functions. Noise was in this way found to be independent of the distance from the center except for regions containing the four central pixels (central artifacts) or the edge of the picture (partial volume effect). The noise of a picture based on $0.5 \cdot 10^6$ counts was 8% when using the standard filter and ranged from 5% to 11% with application of extreme filter functions. As expected, noise varied almost in inverse proportion to the square root of the number of the counts over a wide range.

Contrast was compared in pictures obtained from water-filled or gas-filled cylinder phantoms containing five tubes (cold spots). Both visual inspection and calculation of the counting rate in the center of the largest tube (5 cm) relative to the surroundings were used. The energy threshold was increased stepwise from 15 keV to 80 keV. The quality of the pictures was unchanged up to approximately 54 keV, but a marked improvement was then seen in the contrast and the calculated ratio value decreased from 0.60 to about 0.30. The picture obtained with xenon 133 gas (air-filled tubes) showed a significantly higher contrast and a ratio value of only 0.15.

Discussion

The resolution of systems using single photon detection is predominantly determined by the collimation. The use of the FWHM concept for description of the resolution is simple but yields only a rough estimate. For example, two sources separated from each other by a distance of 1 FWHM do not necessarily appear separated on the picture, as separation depends not only on the FWHM but also on the form of the entire line spread function. Further, the exact position relative to the pixels and the precise number of counts relative to the selected color scale may determine whether a separation is or is not actually obtained. As a supplement, FWTM might be used. However, this value is not well defined in pictures of line sources as the finite number of projections results in a "star artifact." The more advanced concept of modulation transfer function (MTF) has not yet been applied. Using a cold spot phantom, visual inspection may allow for an estimate of the resolution. Under favorable conditions, i.e., a high counting rate, cold tubes of 1 cm diameter can be recognized. Under less favorable conditions with a lower counting rate or less contrast in the object (i.e., some radioactivity in the tubes), detection requires a larger diameter of up to approximately three times the FWHM (if the counting rate is not unacceptable low).

In order to detect low flow areas it is essential that the pictures are free from artifacts. Especially it should be noticed that angle-dependent detector failures not observable during the calibration may cause artificial low flow regions to appear. Such artifacts are easily recognized in a picture of the uniform cylinder, but may be misinterpreted in a (noisy) picture of the human brain.

Measurements of the sensitivity obtained from a uniform water-filled cylinder showed a decreasing, almost linear relationship with the energy threshold, a finding that was anticipated from the absence of a well-demarcated 81 keV photopeak. Lowering the energy threshold increases the sensitivity, but most of the photons gained will be false (scattered) and tend to blur the picture, as can be seen from the comparisons of spectra and images using xenon 133 gas or xenon 133 dissolved in water. On the other hand, using a too high threshold yields an unacceptably low count rate and corresponding high noise level. Therefore, a compromise must be made as the poor energy resolution and the minimal energy losses of the scattered photons preclude effective rejection of the scattered photons.

Using such a compromise (corresponding to normal conditions for patient studies) a count rate of 40% was found in the center of a water-filled tube of 5 cm diameter surrounded by xenon 133 in water. This finding indicates that due to Compton scatter true flow values cannot be obtained from ischemic lesions. It follows that flow values in all low flow areas, including white matter tissue areas, are overestimated.

Finally the effect of the blood-brain partition coefficient λ should be mentioned. Although this value has not been commented on in this report, calculations (unpublished, S. Holm) have documented that the influence of varying λ values in the brain tissues yields only a negligible effect compared with the Compton effect described above.

References

1. Stokely EM, Sveinsdottir E, Lassen NA, Rommer P. (1980) J Comput Assist Tomogr 4:230−40
2. Lassen NA, Henriksen L, Paulson OB. (1981) Stroke 12:284−288
3. Celcis P, Goldman T, Henriksen L, Lassen NA. (1981) J Comput Assist Tomogr 5:641−645

Xenon-133 Dynamic SPECT in Cerebrovascular Disease

U. Buell, G. Leinsinger, T. Kreisig, and P. Schmiedek

Inhalation of xenon-133 gas has reached the stage of useful clinical application in recent years. Besides employment of multiple probes, a dynamic SPECT procedure (DSPECT) was introduced by the group of Lassen (Stokely et al. 1980; Lassen et al. 1981). Since first reports have mainly been aimed at computation methods and technical considerations (Celsis et al. 1981; Bonte and Stokely 1981), clinical results are still the subject of discussion (Buell et al. 1983; Moser et al. 1983; Kanaya et al. 1983).

We therefore undertook a study to validate the DSPECT method in selected patients, referring to clinical (history) and morphological (transmission CT, radiographic angiography) findings. To determine the sensitivity of the method, patients were additionally seperated by angiographically confirmed alterations in the cranial vasculature (unilateral or bilateral).

Material and Methods

DSPECT apparatus and procedure have been described in detail elsewhere (Stockely et al. 1980; Lassen et al. 1981; Bonte et al. 1981). In short, inhalation of xenon-133 gas (10 mCi/liter) and a fast rotating, 64 crystal SPECT scanner for measurement of rCBF in ml/100 g/min in three axial slices were used. After inhalation for 1 min and exhalation for 3 min, calculation was done with the software provided by the manufacturer (Medimatic, Copenhagen). During the examination, patients had their eyes closed and were asked to breathe normally. End-tidal pCO_2 was monitored by a capnograph during the examination, and arterial pCO_2 was determined by a blood gas analyzer in blood samples taken from the radial artery at the end of exhalation. Patients with a $p_aCO_2 < 33$ or > 40 mmHg were excluded from the study. No correction of rCBF for p_aCO_2 was performed.

In addition, we developed a computer program (Moser et al. 1983) which divides each slice by midline section with further subdivision into 12 areas of interest. From these, areal flow (AF) values were derived (Figs. 2b, 3b, 4b). For data reduction, only slices two (6 cm above the canthomeatal line) and three (10 cm above the canthomeatal line) were used.

Universität München, Abteilung für Nuklearmedizin und Neurochirurgie, Klinikum Großhadern, Marchioninistraße 15, D-8000 München 70

Cerebral Blood Flow and
Metabolism Measurement.
Eds. Hartmann/Hoyer
© Springer-Verlag Berlin Heidelberg 1985

Table 1. Xenon-133 DSPECT: parameters for clinical employment

---

I. rCBF – absolute values (ml/100 g/min) = flow (F)
II. rCBF – relative values (right-to-left ratios) = R
III. regions of low flow, visually detected = VLF

    1. per pixel (32 × 32)   2. standardized area
    3. per half slice   4. individual areas

Areas: 12 per slice (2 cm) from
              9 cm^2 (frontal or occipital) to
              18 cm^2 (temporal or parietal)
Voxel size: 18 cm^3 to 36 cm^3

Furthermore, for clinical validation of xenon-133 DSPECT, the following parameters were employed per patient (Table 1): absolute values within 24 areas, and 12 ratios of such flow values (right-to-left, AR). In addition, differences (Δ) from thresholds of normal values (ΔAF, ΔAR) were formed (ΔAF: difference of actual flow from mean minus 2 SD of the normal value for the corresponding area; ΔAR: difference of the actual AR from the normal mean AR from corresponding areas). Moreover, visual evaluation of slices was used to compare the results of DSPECT to the findings of TCT, referring to the pattern of A (12 A per slice).

All examinations (xenon-133 DSPECT, TCT, radiographic angiography) were performed within 2 weeks. Results of invasive angiography, were used to subdivide the patients into groups having unilateral or bilateral CVD. Therefore, biplane radiographic angiography was performed in two projections for both carotid and vertebral arteries and evaluated by experienced observers. The evaluation included determination of stenosis or occlusion of the common carotid or internal carotid artery and of the main cerebral arteries or their branches. In the group of patients with unilateral CVD, the normal side was allowed to reveal only slight irregularities but no stenosis of > 30% of luminal diameter or plaques.

The patient's clinical presentation was used to subdivide the patients into three different groups: completed stroke (CS), prolonged reversible ischemic neurologic deficit (PRIND), and transient ischemic attack (TIA) (Schmiedek et al. 1977; Buell et al. 1981). Only patients with CS had considerable neurologic deficits at the time of examination. In all cases more than 1 week had elapsed since the ictus.

One hundred and eight patients are included in this study. CVD was excluded in 27 right-handed patients (18 male, 9 female: mean age 45 ± 13 years). This group represents "normals". The group of patients with unilateral CVD ($n = 47$) included 15 patients (age 52 ± 10 years) with CS, 16 (age 49 ± 14 years) with a history of PRIND, and 16 (age 58 ± 11 years) with a history of TIA. The bilateral CVD group consisted of 12 patients (age 54 ± 7 years) with CS, 8 (age 53 ± 9 years) with a history of PRIND, and 14 (age 62 ± 7 years) with a history of TIA.

To determine the repeatability of results with the areal method, in addition six patients were examined twice a day, at 6-h intervals (intraindividual comparison). To evaluate the interobserver variability, 22 patients of the normal group were evaluated by two observers. The correlation coefficients *r* were determined.

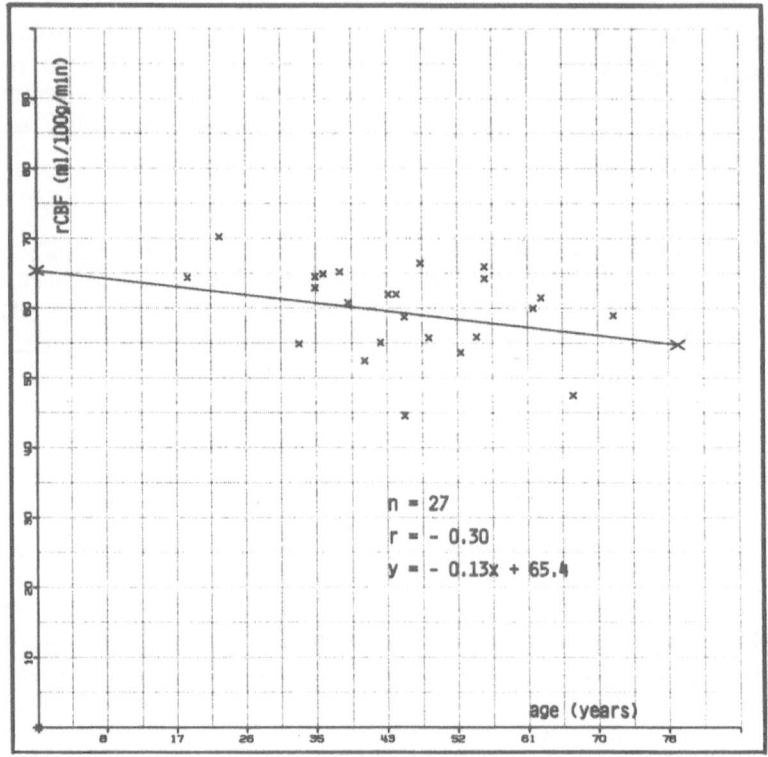

Fig. 1. rCBF (ml/100 g/min) determined in 27 "normals" as the mean from all areas versus age (years). Note slight decrease of flow with age

Table 2. Xenon-133 DSPECT: Normal rCBF values for areal flow (AF) (ml/100 g/min) and right-to-left ratios (AR) in slices 2 and 3 ($n=27$)

Slice 2	frontal		Slice 3	frontal	
Left AF	Right AF	AR	Left AF	Right AF	AR
30.4	29.8	0.80–1.32	17.2	18.8	0.75–1.35
48.1	50.8	0.86–1.27	38.8	40.9	0.79–1.23
57.9	59.8	0.86–1.14	51.9	54.4	0.88–1.16
53.3	54.9	0.89–1.09	54.5	54.4	0.90–1.10
49.0	47.4	0.87–1.11	44.6	43.9	0.86–1.14
39.1	37.2	0.81–1.13	27.2	28.3	0.80–1.24

AF: mean minus 2 SD; AR: from mean minus 2 SD to mean plus 2 SD

Results

1. Normal Values

Normal lower thresholds for the 24 areas (12 areal flow values for slice two, 12 areal flow values for slice three) and the normal range of area ratios (AR) are shown in Table 2. From these data, differences Δ areal flow and Δ areal ratios were computed. Correlation of mean areal flow (mean out of 24 areal flows per "normal") and age of the normals revealed a slight decrease of areal flow with age (Fig. 1).

2. Repeatability

Intraindividual comparison and interobserver variability revealed high correlation coefficients, with a minimum of 0.63 for intraindividual comparison and 0.67 for interobserver variability (for slice two, Table 3). The highest interobserver variability was found for the two frontal and occipital areas. Evaluating six areas for the left and six areas for the right slices or comparing the total slice mean values gave correlation coefficiens from 0.84 to 0.94 (Table 3).

3. Unilateral CVD

In patients with CS, radiographic angiography revealed alterations of the ICA in 12 patients, and the middle cerebral artery was involved in three. TCT revealed normal findings in three patients and unilateral low density areas in 12.

In patients with PRIND, radiographic angiography revealed alterations of the ICA in 11 patients and involvement of the middle cerebral artery in five. TCT showed low density areas in 11 patients (Fig. 2c).

In patients with TIA, radiographic angiography revealed ICA alterations in 14 patients and involvement of the middle cerebral artery in two. TCT detected low density areas in seven patients.

Table 3. Xenon DSPECT: repeatability of results from the areal method

I. Intraindividual comparison (6 pts/2 exams 6 h apart/one obs)
II. Interobserver variability (2 "normals"/one exam/two observers)

	I	II		I	II		I	II
A1	0.88	0.67	A12	0.88	0.72	A1–6	0.84	0.94
A2	0.94	0.93	A11	0.96	0.92	A7–12	0.92	0.89
A3	0.92	0.98	A10	0.90	0.96	A1–12	0.89	0.93
A4	0.63	0.96	A9	0.80	0.93			
A5	0.65	0.96	A8	0.90	0.93	Slice 2		
A6	0.80	0.84	A7	0.61	0.85	Correlation coeff, r		

A1–A12: areas of slice 2

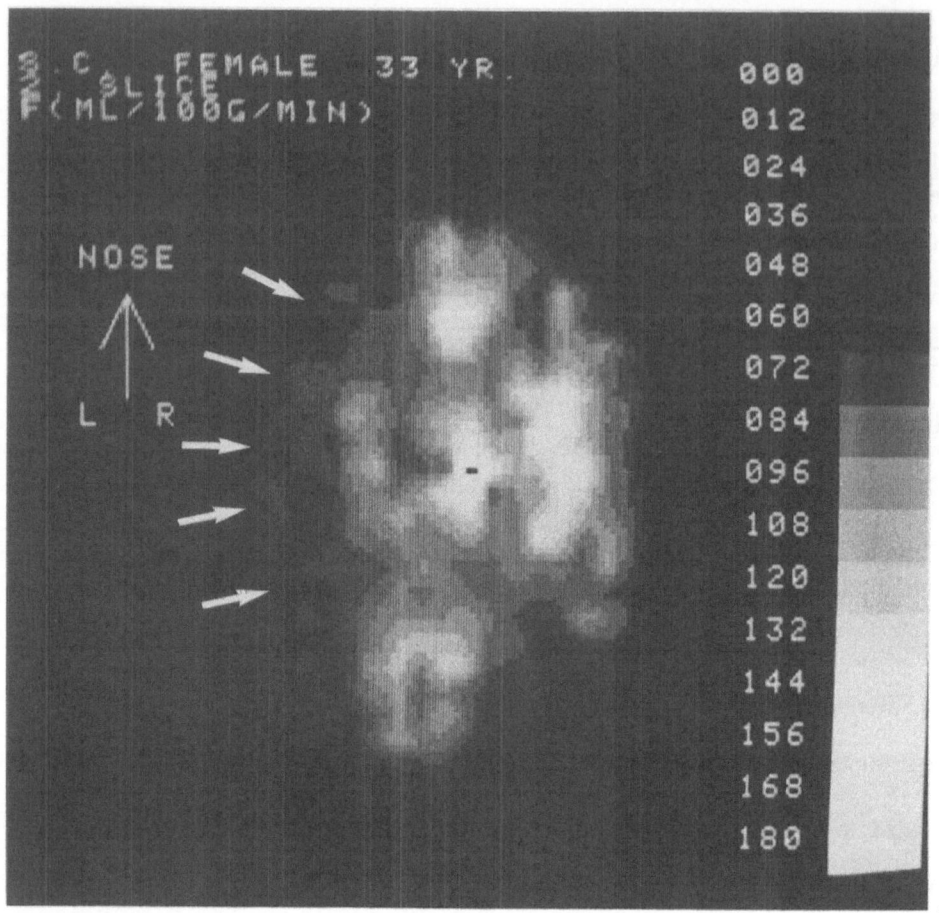

a

Fig. 2a–c. 33-year-old female patient with left-hemispheric PRIND. Stenosis of 75% of the left internal carotid artery. In the DSPECT flow map (**a**) there is considerable side difference, with a low flow pattern throughout the whole left hemisphere (*arrows*). The areal method (**b**) *revealed normal absolute values (L =* left, *R =* right) for both hemispheres. However, right-to-left ratios (AR = PQ, 1.30–1.45) exceeded the normal range. In TCT (**c**) only a small low density area was found in the parietal region (*arrow*). Absolute flow values (AF) were false-negative, area ratios (AR) were correct-positive (**b**), and visual inspection was correct-positive (**a**) and revealed considerably larger areas of low flow as compared with areas of low density in TCT (**c**)

Least areal flow, measured in the affected hemispheres, corresponding contralateral areal flow, Δ areal ratio, and Δ areal flow in patients with CS, PRIND, and TiA are shown in Table 4. Minimum areal flow decreased and Δ areal ratio and Δ areal flow increased with increasing severity of the clinical history. Contralateral areal flow values were nearly independent from these sequences.

Significantly low areal flow (AF below mean minus 2 SD of normals) in the affected hemisphere was found in 36% of areas. However, low areal flow was

S C FEMALE 33 YR.
08/18/83
X
L R
PQ*100
000
012
024
036
048
044 057 130 060
058 079 136 072
Y
084
071 103 145
096
067 092 137
108
071 071 2. SLICE 100 120
054 039 072 132
144
0608 0735 1200 156
168
180

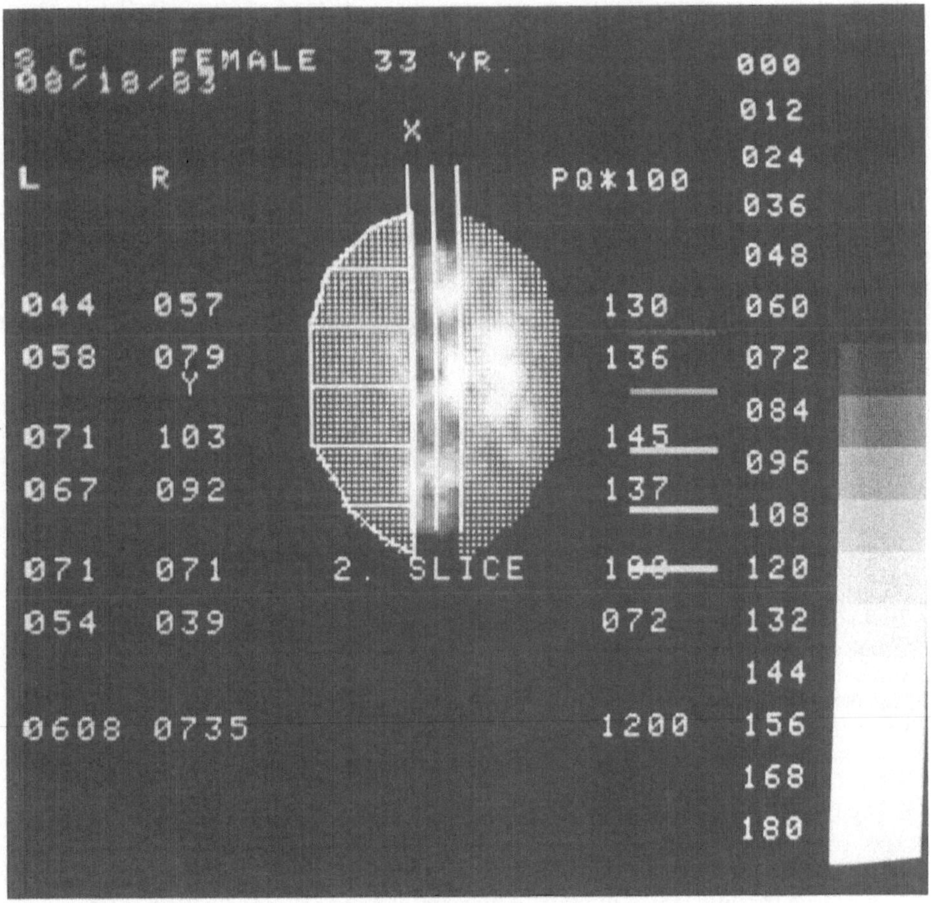

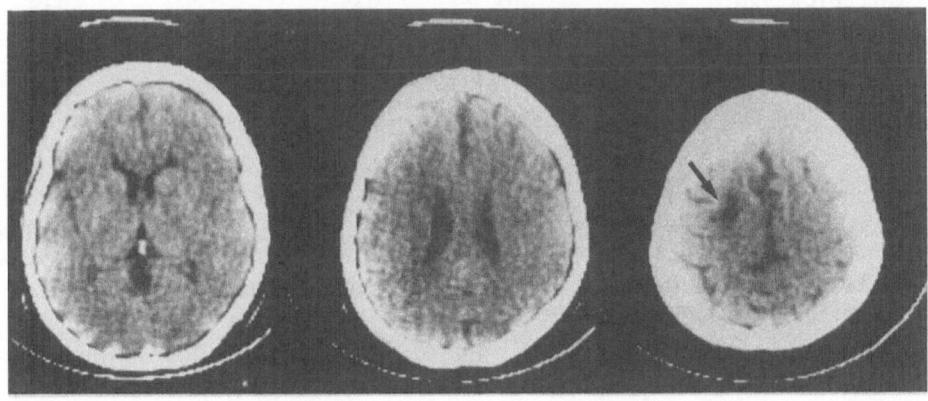

b

c

Table 4. Least AF (affected side), corresponding contralateral AF, ΔAR, ΔAF in patients with CS, PRIND, and TIA (unilateral CVD)

	n	Least AF[a] (ml/100 g/min)	Contralat. AF[a] (ml/100 g/min)	ΔAR[a]	ΔAF[a, b] (ml/100 g/min)
CS	15	43±17[c]	60±15	0.39±0.22	13.5±8.4
PRIND	16	46±14[c]	62±15	0.29±0.14	9.9±6.9
TIA	16	50±13[c]	60±16	0.20±0.15	8.3±6.8

AF: areal flow; AR: areal flow ratio; for Δ see "Material and Methods"
[a] All values mean±SD
[b] In patients with least AF below normal (mean minus 2SD)
[c] $P < 0.005$ (Student's t-test for paired data)

Table 5. Quantitative (rCBF) evaluation searching for low AF values in angiographically and/or clinically affected hemispheres

	Incidence of low AFs in one slice per patient
Affected side (unilat)	102/282 = 36%
Affected side (bilat)	96/270 = 36%
Nonaffected side (unilat)	45/282 = 16%
Nonaffected side (bilat)	34/138 = 25%

unilat or bilat.: uni- or bilateral CVD, confirmed by angiography

Table 6. Comparative results of visual evaluation of one corresponding D-SPECT and TCT slice in patients with CVD[a]

Clinical history of	CS	PRIND	TIA	Total
VLF in D-SPECT (uni)	34%	28%	12%	24%
VLF in D-SPECT (bil)	32%	30%	5%	20%
Defect in TCT (uni)	20%	13%	5%	13%
Defect in TCT (bil)	24%	13%	4%	13%

uni, bil: uni- or bilateral CVD, confirmed by RGA; VLF: visually detected low flow
[a] Employing pattern for A in both methods

also detected in the nonaffected hemisphere in 16% of areas (Table 5). Comparative visual evaluation of one corresponding DSPECT and TCT slice per patient revealed 24% of areas to have visually low flow in DSPECT and 13% ($P < 0.01$) to have low density in TCT, the percentages increasing with the severity of the disease (Table 6, Fig. 2).

Sensitivities, derived from employment of either low areal flow in the affected hemispheres or of areal ratio outside the normal range (high ΔAR, Table 2)

Table 7. Number of patients with true-positive low AF (affected side), AR outside normal, or a combination thereof

Clinical history of	CS	PRIND	TIA
Low AF[a]	12/15	9/16	8/16
Sensitivity	80%	56%[b]	50%[b]
AR $\gtrless$ mean $\pm$ 2SD	13/15	15/16	10/16
Sensitivity	87%	94%[c]	63%[c]
Combination	14/15	16/16	14/16
Sensitivity	93%	100%	88%

[a] Below mean minus 2 SD of normals
[b] NS
[c] $P < 0.01$

or of a combination thereof, are given in Table 7. Highest sensitivities were derived by combined evaluation, yielding 93% sensitivity in CS, 100% in PRIND, and 88% in TIA. A false-negative case is illustrated in Fig. 3.

4. Bilateral CVD

Radiographic angiography in the group of patients with CS revealed alterations of the ICA in all 12 patients (11 right-sided, 12 left-sided), and other arteries were involved in three patients. TCT revealed unilateral low density areas in eight patients and bilateral low density areas in three (Fig. 4c). One patient had no TCT.

In patients with PRIND, radiographic angiography revealed alterations of the ICA in all eight, and additional involvement of other cerebral arteries in two. TCT showed low density areas in five patients (one bi-, four unilateral).

In patients with TIA, radiographic angiography revealed ICA alterations in all 14 patients and additional involvement of other cerebral arteries in one patient. TCT detected low density areas in four patients (two bi-, two unilateral).

Least areal flow, corresponding contralateral areal flow, Δ areal ratio, and Δ areal flow in patients with CS, PRIND, and TIA are shown in Table 8. Least

Table 8. Least AF (per patient), corresponding contralateral AF, ΔAR, and ΔAF in patients with CS, PRIND, TIA, and double-sided vascular alterations

	n	Least AF (ml/100 g/min)	Contralat. AF (ml/100 g/min)	ΔAR	ΔAF[a] (ml/100 g/min)
CS	12	43 ± 10[b]	46 ± 9	0.10	12 ± 9
PRIND	8	51 ± 12[b]	53 ± 12	0.04	10 ± 7
TIA	14	54 ± 11[b]	55 ± 13	0.01	6 ± 3

AF: areal flow; AR: areal flow ratio; for Δ see "Material and Methods"
[a] In patients with least AF below normal
[b] not significant

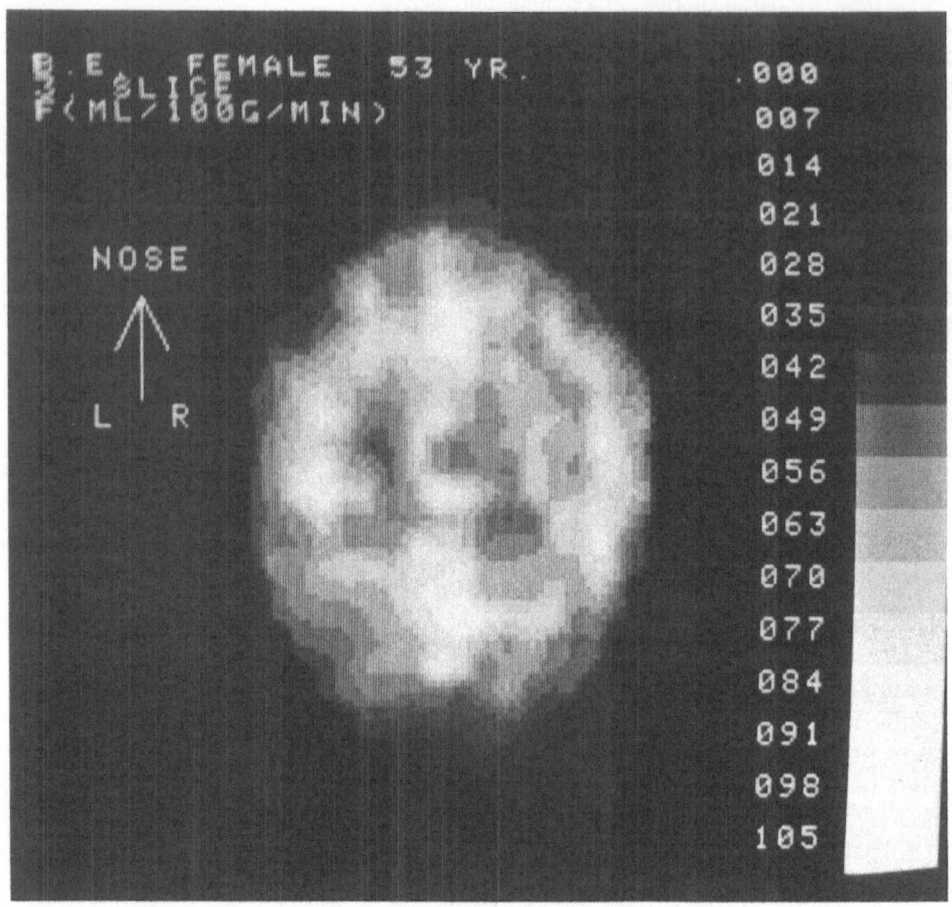

Fig. 3a–c. 53-year-old female patients with a right-hemispheric PRIND: High-grade stenosis of the right middle cerebral artery. DSPECT flow map (**a**) revealed a nearly symmetrical perfusion pattern with lower flow values in the ventricular areas. Areal flow (AF) and area ratios (AR) were completely normal (*b*); in TCT (**c**) enlarged ventricles and two small low density areas within the internal capsule (*arrows*) were found. The patient was completely false-negative in AF and AR and by visual inspection (only enlarged ventricles may be assumed)

areal flow decreased and Δ areal flow increased with increasing severity of clinical history. However, Δ areal ratio did not exactly discriminate (Fig. 4b), and contralateral areal flow in such patients, in contrast to unilateral CVD, seemed also to depend on the clinical history.

Significantly low areal flow in the clinically affected right and left hemispheres was found in 36% of the areas (Fig. 4). Low areal flow in the clinically nonaffected hemispheres was detected in 25% of areas (Table 5). Comparative visual evaluation of one corresponding DSPECT and TCT slice revealed 20% of areas to have visually low flow in DSPECT and 13% ($P < 0.01$) to have low density in TCT (Table 6).

B E FEMALE 53 YR. *27.2.000
08/11/83
X
L R PQ*100
Y
3. SLICE
034 033 097
053 054 102
059 061
062 063
057 057 100
038 037 097
0505 0508 1002
007
014
021
028
035
042
049
056
063
070
077
084
091
098
105

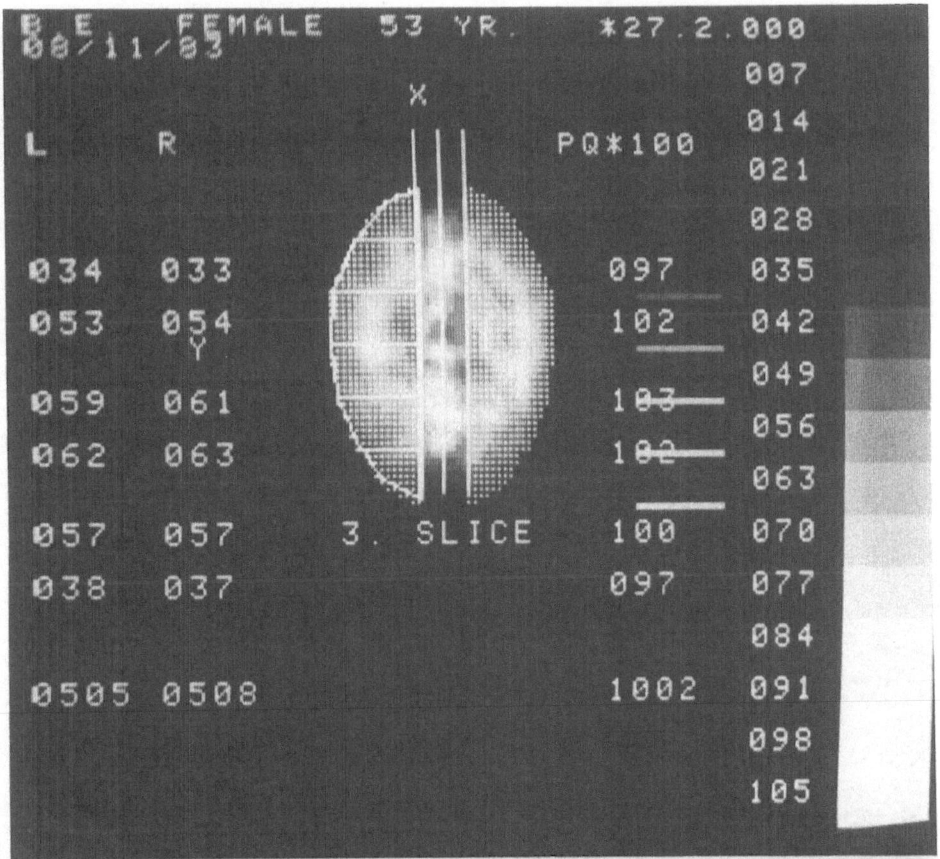

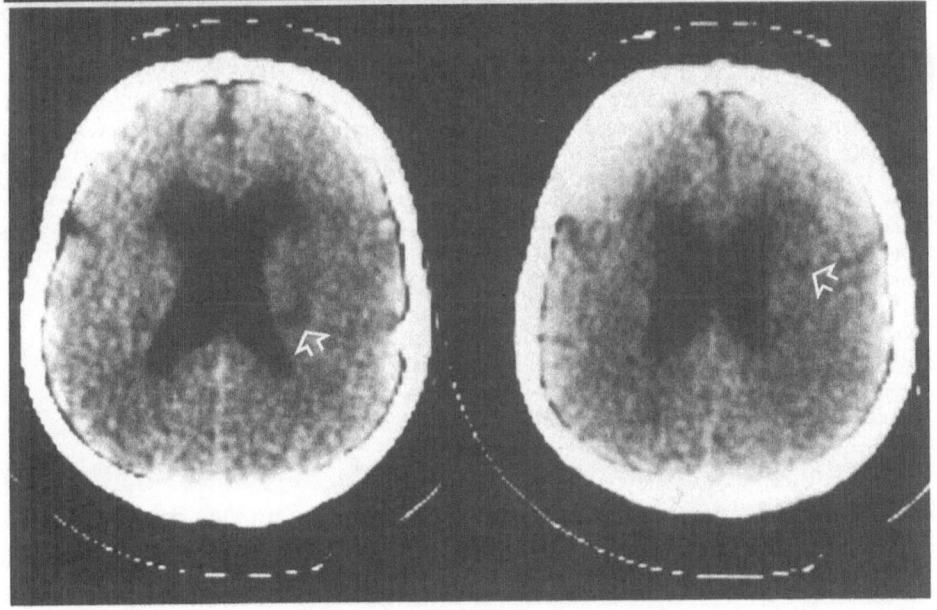

b

c

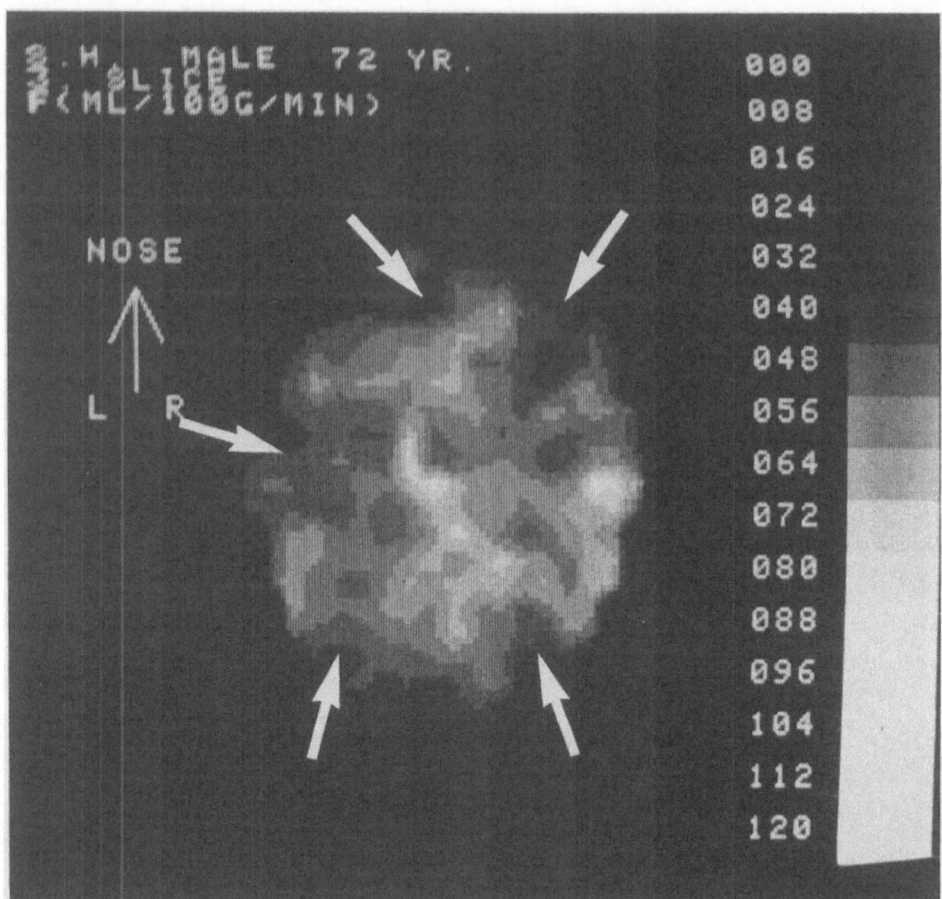

a

Fig. 4a–c. 72-year-old male patient with left-hemispheric PRIND and subsequent TIA in both hemispheres. The left internal carotid artery was occluded, and the right internal carotid artery showed high-grade stenosis. The DSPECT flow map (**a**) revealed numerous areas of low flow (*arrows*) in both hemispheres. Areal flow (AF) was significantly below normal in five areas (*arrows*, **b**). Areal ratios were normal with one exception (*arrow*, **b**). TCT (**c**) revealed large areas of low density in the right and two small areas of low density in the left hemisphere. This patient was true-positive in visual inspection and areal flow but only in one ratio (bilateral CVD!). Note the difference between low flow areas and TCT findings, in particular concerning the left hemisphere

Table 9. Number of patients (double-sided vascular changes) with true-positive low AF in either one or both hemispheres

Clinical history of	CS	PRIND	TIA
Total (*n*)	12	8	14
Both hemispheres	7 (58%)	2 (25%)	5 (36%)
One hemisphere	3 (25%)	4 (50%)	2 (14%)
Sensitivity (%)	92	75	50

AF: areal flow

MALE 72 YR.
X
L
R
PQ*100
023 022
036 037
Y
043 044
045 054
049 053
036 033
0387 0405
096
103
102
120
108
092
1035
3. SLICE
000
008
016
024
032
040
048
056
064
072
080
088
096
104
112
120

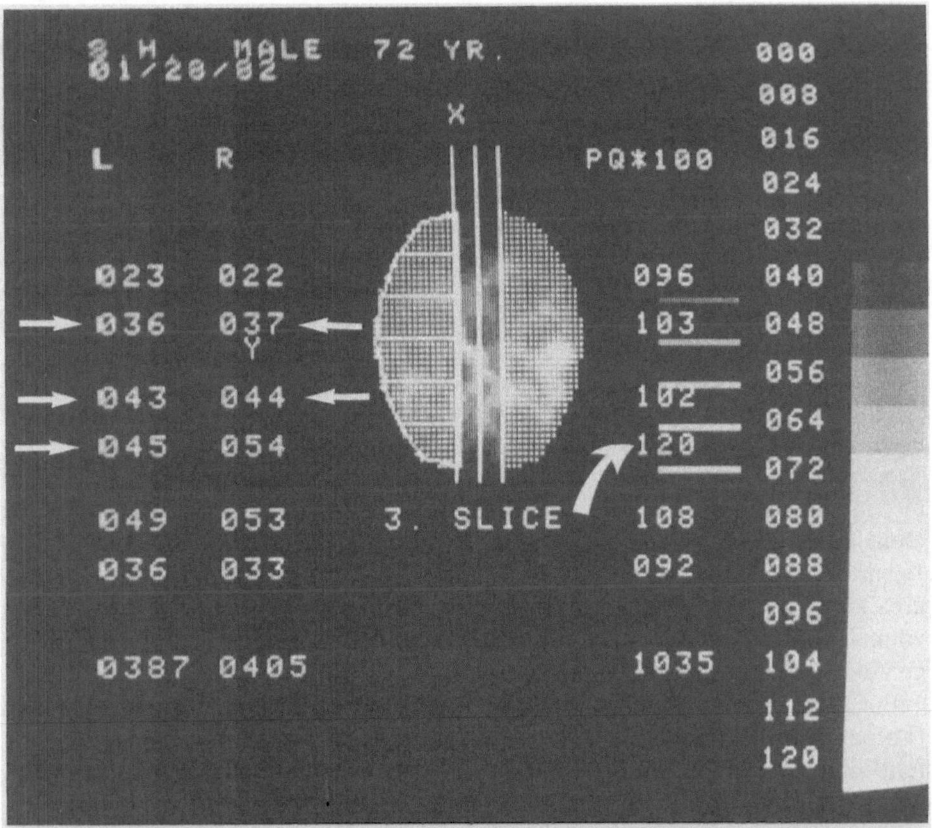

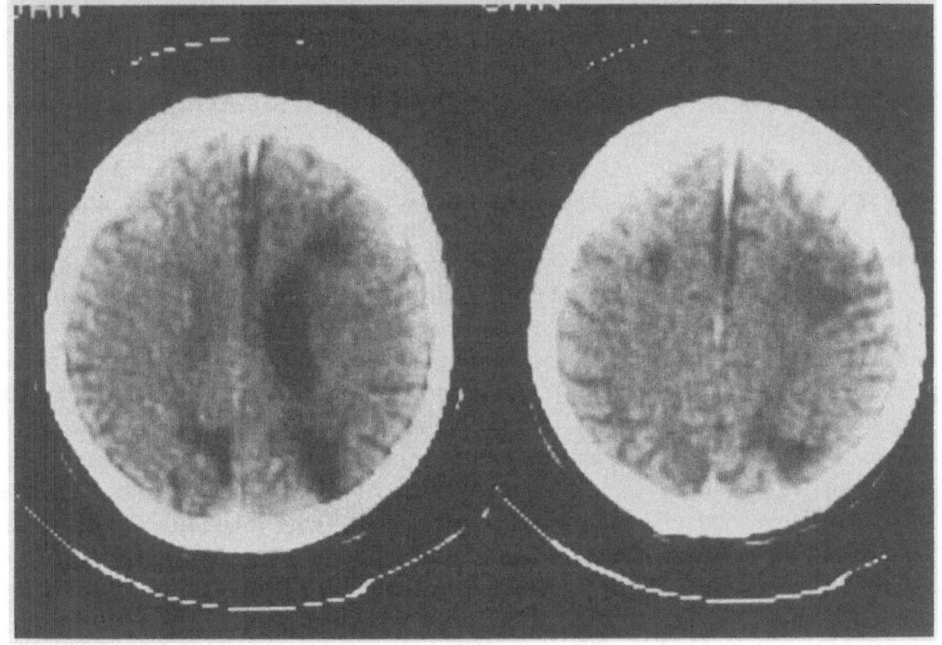

b

c

Sensitivities, derived from employment of low areal flow in such patients, are illustrated in Table 9. In comparison with radiographic angiography (bilateral alterations) only 36% (TIA) to 58% (CS) of patients were correctly identified. If positive findings in only one hemisphere were added to the correct positives, sensitivities increased. However, in patients with a history of reversible ischemia (PRIND, TIA), highest sensitivities were only 50% (TIA) or 75% (PRIND).

Discussion

Facing the relatively low spatial resolution of the xenon-133 DSPECT system used (> 1.7 cm FWHM; Stokely et al. 1980), we restricted the subdivision of slices into 12 areas each. To avoid contributions from large vessels or asymmetries, the middle strips of the slices were excluded [± 5 pixel (± 1.7 cm) from the midline; Figs. 2 b, 3 b, 4 b]. Thus, areas cover a volume (voxel) from 18 cm³ (frontal or occipital) to 36 cm³ (temporal or parietal; Table 1). These area sizes fit the recommendation of Todd-Pokropek and Jarritt (1982) that volume elements employed for quantitative assessment in SPECT should be greater than twice the spatial resolution of the SPECT system.

For clinical employment of such methods, various parameters may be used (Lassen et al. 1981; Buell et al.1981; Coleman et al. 1982). We have found the following to be helpful in explaining the patient's situation and clinical presentation: absolute flow values (ml/100 g/min), right-to-left ratios from areas, and visually detectable "lesions" in flow maps. Since the 81 patients with cerebrovascular disease (CVD) were carefully selected out of the xenon-133 DSPECT group of more than 300 patients by clinical presentation, radiographic angiography, and transmission CT (TCT), the results of the present study may reflect the clinical validity of xenon-133 DSPECT. Moreover, greater physiologic deviations in rCBF values were prevented by monitoring arterial and end-tidal pCO_2 and by keeping the patients at rest during the examinations.

Repeatability of results from the area method (Table 3) revealed relatively high correlation coefficients. In intraindividual comparison only the temporal, left hemispheric areas (A 4, A 5) gave low ($r = 0.63$, $r = 0.65$) coefficients. This may reflect variability in motor centers of the dominant left hemisphere since the corresponding areas in the right hemisphere (A 8, A 9) had coefficients of $r = 0.90$ and $r = 0.80$. It has to be emphasized that repetitive studies with xenon-133 DSPECT in the same patient should be done at least 1 h apart. We found that background problems caused by the first examination influenced the results of the second one. Interobserver variability was excellent. Coefficients below 0.90 were only found in the frontal and occipital areas. These areas are the smallest by voxel size. We did not experience problems from cross-talk by the scalp since the xenon-133 DSPECT calculation method uses only 1 min for inhalation (and saturation) (Kanno and Lassen 1979). Lower thresholds of rCBF and the range of normal areal (right-to-left) ratios (Table 2) therefore seemed to be determined as exactly as possible.

The usefulness of such normal areal flow values is illustrated by the results from the unilateral CVD group (Table 4). Xenon-133 DSPECT determined correctly involved hemispheric sides *and* hemispheric areas; furthermore it discriminated rCBF values in good correlation with the severity (CS, PRIND, TIA) of the disease. Angiographically and clinically affected hemispheres revealed low areal flow in 36% of the areas. As expected, in unilateral CVD, nonaffected hemispheres had less areas involved (16%) than in angiographically bilateral CVD (25%; Table 5).

In patients with double-sided vascular alterations, least areal flow values separated in good correlation with the patient's history (CS, PRIND, TIA), and the decrease in area flow versus normal was twice as high in CS (12 ml/100 g/min) as in TIA (6 ml/100 g/min). However, contralateral areal flow values, as expected in double-sided disease, also depended on the type of disease. Thus, areal ratios did not discriminate (Table 8) very well (Fig. 4) versus normal.

In comparison with morphologically aimed methods like TCT, xenon-133 DSPECT in both groups (unilateral and bilateral) revealed more low flow areas than morphological defects were detected (Table 6). In patients with bilateral CVD, visual evaluation of DSPECT flow maps was more difficult (Fig. 4a) than in patients with unilateral CVD (Fig. 2a). This finding and problems with small lesions (Fig. 3) were the only drawbacks derived from the relatively low spatial resolution of the system and, therefore, were not experienced with TCT (Table 6).

Employment of xenon-133 DSPECT as a screening procedure in patients with suspected CVD needs more sensitivity than could be derived from absolute flow values (Table 7, 9). Additional sensitivity may be gained from right-to-left ratios. However, such ratios are valid only in unilateral CVD (Fig. 2). In patients with bilateral CVD, combination of areal flow, areal ratio, and visual evaluation did not yield higher sensitivities than 50% in TIA patients. This group was the exception, for double-sided CVD with PRIND or CS (Fig. 4) and unilateral CVD in all three clinical groups were detected with high sensitivities.

Conclusion

1. To detect or to confirm CVD with xenon-133 DSPECT, both thresholds of absolute areal CBF (ml/100 g/min) and interhemispheric comparison of rCBF may be employed. This combination yielded 90% sensitivity in CS, > 75% in PRIND, and between 50% and 88% in TIA. This includes correct localization of low flow areas in patients some time after the ictus.

2. In comparison with methods aimed at morphological imaging (digital vascular imaging, invasive radiographic angiography, TCT), xenon-133 DSPECT provides significantly more functional data.

3. In clinical practice (to detect or to confirm CVD) doppler sonography, DVI, and TCT may be employed first. Xenon-133 DSPECT follows if one desires to gain data on:
   a) The hemodynamic relevance of vascular alterations
   b) Extension of regional low flow in correlation with defective brain tissue
   c) rCBF in morphological normal brain tissue in patients with CVD
   d) rCBF (absolute or relative) after treatment

4. Regional CBF (ml/100 g/min) is influenced by various physiologic parameters and by computational compromises inherent in the DSPECT methods. Despite good repeatability of areal flow values, a number of factors may lead to blurred thresholds of normal values. However, rCBF is the only absolute value derived from SPECT methods which reflects actual blood flow. Thus, "flow maps" classify the patient's situation voxel by voxel and hereby the individual severity of CVD.

References

Bonte FJ, Stokely EM (1981) Single-photon tomographic study of regional cerebral blood flow after stroke: concise communication. J. Nucl. Med. 22:1049−1053

Buell U, Scheid KF, Lanksch W, et al. (1981) Sensitivity of computer assisted radionuclide angiography in transient ischemic attack and prolonged reversible ischemic neurologic deficit. Comparison with findings in radiographic angiography and transmission computerized axial tomography. Stroke 12:829−834

Buell U, Moser E, Schmiedek P, et al. (1983) Evaluation of Xe-133 DSPECT in unilateral cerebrovascular disease. A comparative study to transmission CT and X-ray angiography. J. nucl. Med. 24:P6

Celsis P, Goldman T, Henriksen L, et al. (1981) A method for calculating regional cerebral blood flow from emission computed tomography of inert gas concentrations. J. com. ass. Tomogr. 5:641−645

Coleman RE, Drayer BP, Jaszczak RJ (1982) Studying regional brain function: a challenge for SPECT. J. Nucl. Med. 23:266−270

Kanaya H, Endo H, Suguyama T, et al. (1983) "Crossed cerebellar diaschisis" in patients with putaminal hemorrage. J. Cerebr. Blood Flow Metab. 3:S27−S28

Kanno I, Lassen NA (1979) Two methods for calculation regional cerebral blood flow from emission computed tomography of inert gas concentrations. J. comp ass Tomogr. 3:71−76

Lassen NA, Henriksen L, Paulson O (1981) Regional cerebral blood flow in stroke by 133-xenon inhalation and emission tomography. Stroke 12:284−288

Moser EA, Schmiedek P, Kirsch CM, et al. (1983) Xe-133 dynamic single photon emission computerized tomography (DSPECT): regional cerebral blood flow (rCBF) in normals and patients with cerebrovascular disease (CVD). J. Cerebr. Blood Flow Metab. 3:S25−S26

Schmiedek P, Lanksch W, Olteanu-Nerbe V, et al. (1977) Combined use of regional cerebral blood flow measurement and computerized tomography for the diagnosis of cerebral ischemia. In: Schmiedek P, Gratzl O, Spetzler RF (eds) Microsurgery for Stroke. New York, Springer, Chap. 8, p 67−78

Stokely EM, Sveinsdottier E, Lassen NA, et al. (1980) A single photon dynamic computer assisted tomograph (DCAT) for imaging brain function in multiple cross sections. J. comp ass Tomogr. 4:230−240

Todd-Pokropek AE, Jarritt PH (1982) The noise characteristics of SPECT systems. In Computed emission tomography, PJ Ell, BL Holaman, Eds. Oxford Univ. Press, New York, pp 390−398

Cerebellar Blood Flow in Ischemic Stroke Studied by Xenon 133 Inhalation and Single Photon Emission Computed Tomography

G. Meneghetti, S. Vorstrup, B. Mickey, H. Lindewald, and N. A. Lassen

Introduction

A parallel reduction of cerebral blood flow and oxygen uptake in the cerebellar hemisphere contralateral to the side of supratentorial ischemic infarction has recently been reported using positron emission tomography (Baron et al. 1980; Lenzi et al. 1982). This phenomenon was termed "crossed cerebellar diaschisis."

Single photon emission computed tomography following ^{133}Xe inhalation constitutes a technically simpler approach for a three-dimensional noninvasive assessment of CBF that greatly facilitates serial measurements. With this method we carried out a detailed evaluation of cerebellar blood flow in the early and late period following acute cerebral ischemia in order to decide whether crossed cerebellar diaschisis is a transient or a persistent phenomenon.

Material and Method

A consecutive series of 12 patients, 8 men and 4 women, ranging from 33 to 75 years in age – mean age 60 years – were included in this study as they developed sudden focal neurologic symptoms from the hemispheres. CT scan excluded the presence of hemorrhages or tumors and in most cases confirmed the presence of an ischemic infarction. CBF was measured as soon as possible after admission, usually at $1-2$ days after development of symptoms. In the days following admission CBF was measured every other day, yielding a total of four to six CBF measurements per patient in the acute phase. CBF was studied again in the chronic phase of the disease after 8 weeks and after 6 months. CT scan was performed in the acute phase of the disease in a period ranging from 1 to 6 days after the ischemic stroke. A further CT scan was repeated in the chronic phase of the disease at the 8-week follow-up. CBF measurements were obtained with the same technique in a control group of ten normal volunteers – mean age 43 years – each of whom was studied three times at 1-week intervals. CBF was measured by ^{133}Xe inhalation and a rapidly rotating single photon emission computed tomograph (Stokely et al. 1980). During and after 1½ min of ^{133}Xe inhalation a sequence of four tomographic pictures were recorded from three

Clinica Neurologica dell' Università di Padova, Via Giustiniani 5, I-35100 Padova

Cerebral Blood Flow and
Metabolism Measurement.
Eds. Hartmann/Hoyer
© Springer-Verlag Berlin Heidelberg 1985

slices simultaneously. These slices were positioned 1 cm, 5 cm, and 9 cm above the orbitomeatal line and were termed slices 1, 2, and 3 respectively. Mean hemispheric CBF was calculated from slices 2 and 3. Mean hemispheric cerebellar flow values were calculated from the posterior part of slice 1 in order to avoid the high flow values from nasal cavities which would have led to underestimation of the cerebellar flow values.

Results

In the control group the side-to-side cerebellar flow asymmetry was 3.4 ± 2.4 (± 1 SD). On this basis we called an asymmetry significant when side-to-side differences exceeded 8%.

In five patients a significant cerebellar side-to-side asymmetry, with the lower value contralateral to the side of cerebral ischemic infarction, was seen in the early as well as in the late flows (Table 1). In this group the average value for cerebellar asymmetry was 16% in the acute phase and 18% in the chronic phase. These five patients had large low flow areas in the contralateral hemisphere that corresponded to large cortical and subcortical infarcted areas on the CT scan (Fig. 1). These hypodense lesions were located in the distribution of the middle cerebral artery territory and involved the posterior frontal and parietotemporal regions. In one patient with a small deep CT lesion involving the internal capsule a significant (14%) cerebellar flow asymmetry was evident only in one CBF measurement during the acute phase (Fig. 2). In this patient the cerebellar flow values returned to normal in the chronic phase. The other six patients showed a significant cerebellar side-to-side asymmetry neither in the

Table 1. Comparative values of cerebellar blood flow during the acute phase (average of 4–6 flows) and the chronic phase (average of 1–2 flows)[a]

Case	Acute phase			Chronic phase		
	Left	Right	% asymmetry	Left	Right	% asymmetry
1	51 ± 4	56 ± 4	9%	46 ± 2	55 ± 1	16%
2	50 ± 4	61 ± 3	18%	58 ± 10	67 ± 1	13%
3	58 ± 7	49 ± 3	16%	65 ± 7	54 ± 2	17%
4	52 ± 4	59 ± 4	12%	57 ± 9	66 ± 7	14%
5	69 ± 11	53 ± 10	23%	63 ± 7	45 ± 2	28%
6	63 ± 6	67 ± 5	6%	62 ± 11	64 ± 13	3%
7	63 ± 6	60 ± 5	5%	67 ± 14	67 ± 13	0
8	56 ± 8	57 ± 6	2%	52 ± 6	53 ± 7	2%
9	50 ± 5	51 ± 6	2%	–	–	–
10	44 ± 7	46 ± 7	4%	34 ± 3	34 ± 4	0
11	51 ± 11	52 ± 11	2%	59 ± 1	61 ± 1	3%
12	50 ± 7	49 ± 4	2%	58 ± 10	54 ± 8	7%

[a] Flow: ml/100 g/min

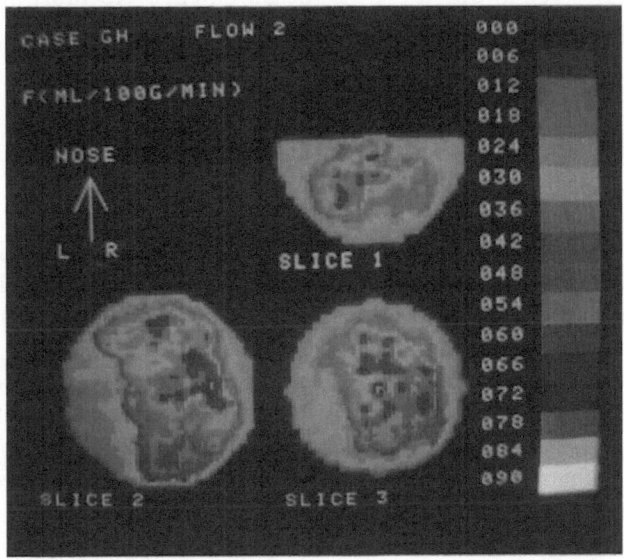

Fig. 1. Case 3. CBF tomography 2 days after stroke shows a large low flow area in the left m.c.a. territory (slices 2 and 3). On slice 1 a contralateral cerebellar flow depression is evident

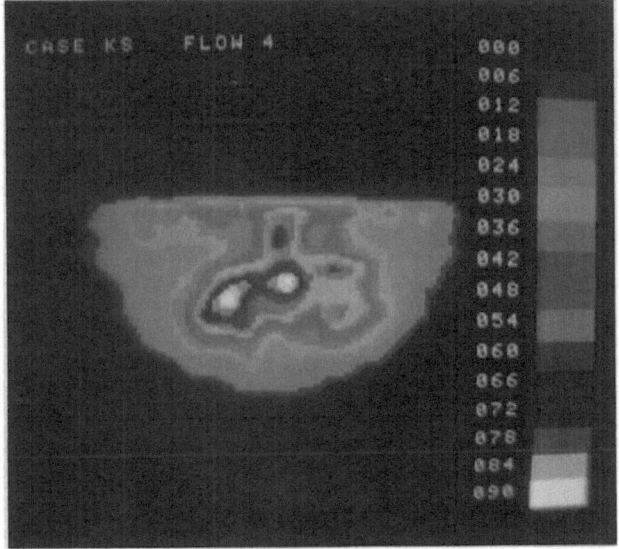

Fig. 2. Case 6. CBF tomography 4 days after stroke shows a significant cerebellar flow asymmetry

early nor in the late CBF measurements. No morphological alterations in the posterior fossa were ever observed on the CT scan.

Discussion

Our serial study of cerebellar blood flow confirmed the frequent appearance of crossed cerebellar diaschisis in unilateral ischemic infarction. From our observations it appeared that the phenomenon is persistent when the anatomic lesion is large while it can be transient when the damage affects more limited areas. In agreement with Baron et al. (1980) we think that supratentorial ischemic lesions can induce a functional transneural depression in the contralateral cerebellar hemisphere through a deactivation of cortico-ponto-cerebellar pathways, a system that, as stated by Brodal (1972), is particularly important in mediating cortical influences on the cerebellum. Recently Martin and Raichle (1983) observed that crossed cerebellar diaschisis was present in patients with frontal lobe infarction and not in patients with parieto-occipital infarcts. All our cases with crossed cerebellar diaschisis had lesions involving both pre- and postcentral cortex. Hence our data do not allow us to determine whether isolated damage in either area might suffice for eliciting the phenomenon. Experimental studies in animal models would be important to determine the relation between hemispheric lesions and cerebellar flow/metabolism.

References

Baron JC, Bousser MG, Comar D, Castaigne P (1980) Crossed cerebellar diaschisis in human supratentorial brain infarction. Ann Neurol 8:128
Brodal A (1972) Cerebrocerebellar pathways: anatomical data and some functional implications. Acta Neurol Scand, Suppl 51:153–195
Lenzi GL, Frackowiak RSJ, Jones T (1982) Cerebral oxygen metabolism and blood flow in human cerebral ischemic infarction. J Cereb Blood Flow Metab 2:321–335
Martin WRW, Raichle ME (1983) Cerebellar blood flow and metabolism in cerebral hemisphere infarction. Ann Neurol 14:168–176
Stokely EM, Sveinsdottir E. Lassen NA, Rommer P (1980) A single photon dynamic computer assisted tomograph (DCAT) for imaging brain function in multiple cross sections. J Comput Assist Tomogr 4:230–240

Xenon 133 Dynamic Single Photon Emission Computerized Tomography (D-SPECT) in Extra-intracranial Arterial Bypass Patients

P. Schmiedek, E. A. Moser, C. M. Kirsch, T. Kreisig, and U. Buell

With its supposedly direct effect on brain blood flow, extra-intracranial arterial bypass (EIAB) surgery represents a clinical situation where measurements of cerebral blood flow (CBF) are of particular interest. However, not least due to the well known limitations of the more conventional techniques for the measurement of CBF [5, 6], the hemodynamic effects of EIAB on brain ischemia remain a matter of controversy [2, 7, 8]. With the introduction of the dynamic single photon emission computerized tomography (D-SPECT) technique, a new method has now become available which promises to be more suitable for the clinical requirements of CBF studies and in addition provides a more accurate, namely a transaxial, image of brain blood flow (4). In the following report our preliminary experience with the D-SPECT technique in a selected group of EIAB patients is presented.

Clinical Material and Methods

The study population included 24 patients, 22 men and two women. The age range was 32−72 years with an averae age of 55 years. Clinically, six patients had a history of previous transient ischemic attacks (TIA), 12 presented with prolonged reversible ischemic neurologic deficits (PRIND), and six had a completed stroke, though of minor extent. Angiographically, 20 patients had unilateral and four bilateral carotid artery lesions. Occlusion of the internal carotid artery was demonstrated in 19 patients. Three patients had a distal stenosis of the internal carotid artery, one patient presented with an occlusion of his middle cerebral artery, and one had a common carotid artery occlusion. All patients of this group had postoperative angiography in order to demonstrate the patency of the newly established collateral and also to show the extent of the intracranial arterial filling via the bypass. Angiography was usually done at the end of the first postoperative week. In five patients repeat angiography was done 5 weeks to 6 months following the operation.

In addition to preoperative routine diagnostic studies, including cerebral angiography and computerized tomography, measurements of CBF were

Neurochirurgische Universitätsklinik, Klinikum Großhadern, Marchioninistraße 15, D-8000 München 70

Cerebral Blood Flow and
Metabolism Measurement.
Eds. Hartmann/Hoyer
© Springer-Verlag Berlin Heidelberg 1985

performed in all patients using the D-SPECT technique with inhalation of xenon 133.

The principle of the method and details of data analysis have been reported previously [1, 4]. Briefly, following a 1-min period of xenon 133 inhalation, three tomographic slices through the brain are obtained simultaneously from which CBF can be calculated by employing specially designed algorithms [3]. Using a computer program, each slice is then divided into two half slices corresponding to the right and left hemisphere. Each half slice is further divided into six equally distributed regions of interest. In addition to a direct quantitative evaluation of CBF, this approach allows an interhemispheric comparison of flow values which is expressed as flow ratio. Since at this stage of our investigation the quantitative analysis of flow was found to be still unsatisfactory, only flow ratios were included in the present study. Furthermore, due to the fact that bypass surgery is thought to affect CBF in only a limited area of one hemisphere, the middle two regions of the second tomographic slice which correspond primarily to the territory of the middle cerebral artery were taken into consideration. Any flow changes were expressed as the difference of flow ratios of the central two regions of slice two by comparing the affected over the nonaffected hemisphere.

In all patients CBF studies were performed preoperatively and then repeated following the operation. Nineteen patients had early postoperative CBF studies before they left the hospital and in 15 patients follow-up CBF studies were performed during the later postoperative course. Results of early and late postoperative studies were available from ten patients.

Results

Angiography

Postoperative angiography showed an occluded bypass in two patients. In one of them, however, repeat angiography 3 weeks later demonstrated that the initially occluded bypass was now patent, with intracranial filling of two branches of the middle cerebral artery. A secondary occlusion of an initially functioning bypass was found in one case. In nine patients angiography revealed a patent anastomosis with filling of one or two cortical arteries. In seven patients more than two intracranial arteries were filled via anastomosis and in six patients excellent results were found where the anastomosis supplied the entire territory of the middle cerebral artery.

Cerebral Blood Flow (Figs. 1 and 2)

The comparison of pre- and early postoperative CBF studies in 19 patients showed an improvement of interhemispheric flow ratios in 12 patients. A de-

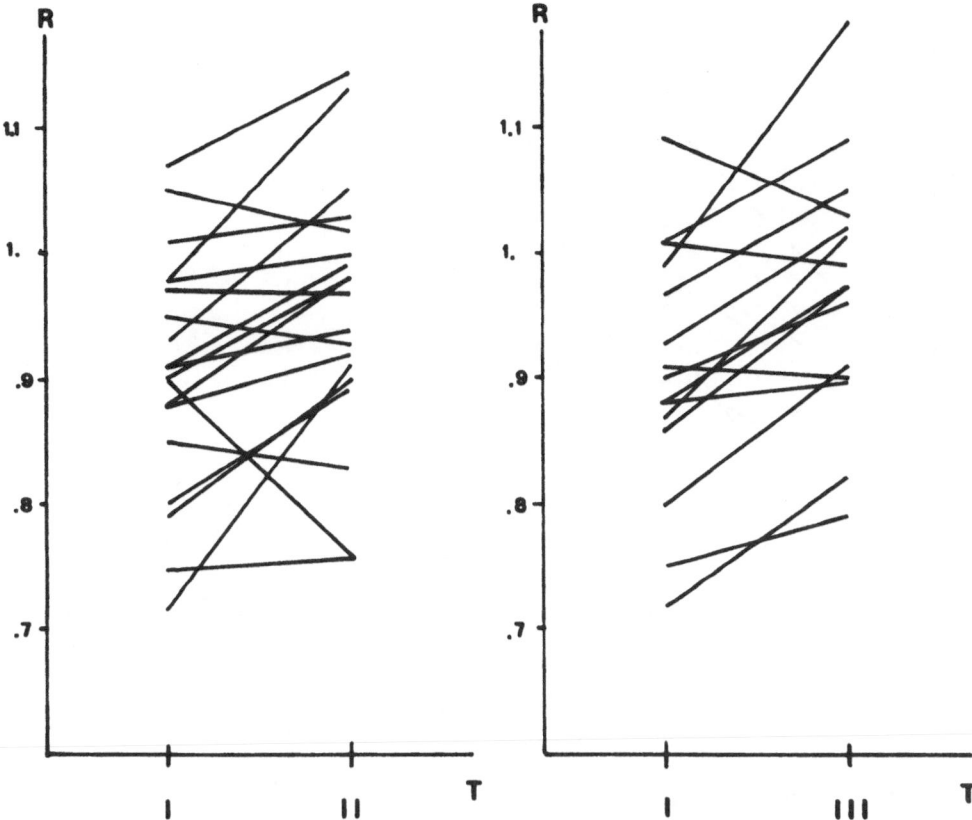

Fig. 1. Changes in flow ratios in 19 patients studied pre- and postoperatively (*left*) and in 15 patients with pre- and late postoperative CBF studies (*right*). *R*, flow ratio of the central two regions of slice two of the affected over the nonaffected hemisphere

crease in the flow ratio was found in three patients and in four cases no change in the interhemispheric flow ratio was seen. The mean value for all 19 patients demonstrated an increase in CBF of 6% soon after the operation. Out of 15 patients in whom CBF studies were done before surgery and 6 weeks to several months later, an increase in the flow ratio was noted in ten and a decrease in four, while in one case the ratio remained unchanged. The analysis of serial follow-up studies in ten patients showed an increase in flow ratios in eight when the preoperative and the early postoperative studies were compared. In six patients a further increase in flow ratios was found when comparing the early postoperative and the late postoperative result. Except for two cases with a marked increase in flow ratios during the later follow-up period, the initial flow increase which occurred soon after the operation was more pronounced than that observed over the subsequent period. It was not possible to establish a correlation between the postoperative angiographic findings and the results obtained from CBF studies.

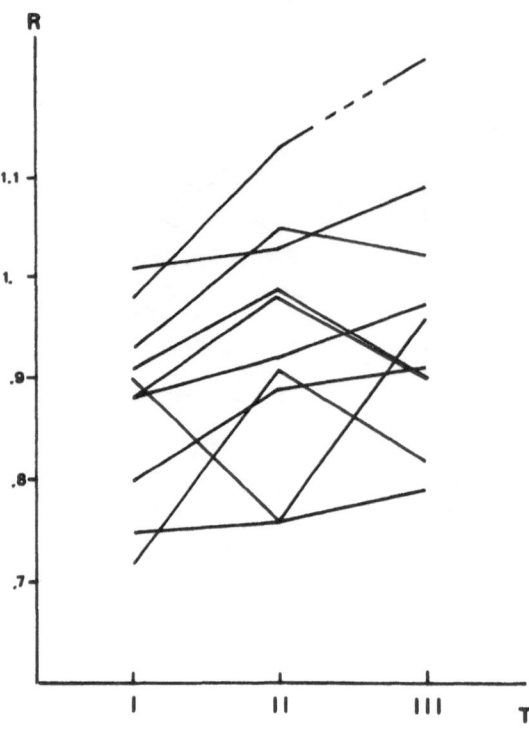

Fig. 2. Follow-up CBF studies in ten patients. *I,* preoperative; *II,* early postoperative; *III,* late postoperative; *R,* as in Fig. 1

Case Report (Figs. 3 and 4)

This 32-year-old patient was admitted with a 4-month history of recurrent right-sided TIAs. Cerebral angiography revealed an occlusion of his left internal carotid artery. The CBF study showed an area of hypoperfusion within his left hemisphere. An EIAB operation was done on the left side and the patient had an uneventful postoperative course. Selective external carotid artery angiography on the 7th postoperative day demonstrated a well functioning bypass. The CBF study was repeated on the following day, showing an improvement in the right-to-left flow ratio of 8%. When the patient was seen again 2 months later, the bypass was still functioning angiographically; however, the extent of the intracranial filling was significantly less than on the previous study. The result of the CBF study was in good agreement with this, revealing a secondary decrease in the flow ratio. In the meantime the patient had not experienced any further ischemic attacks. It is planned to repeat both angiography and the CBF study in 6 months.

Discussion

The D-SPECT technique with inhalation of xenon 133 for the measurement of CBF represents a new approach to the clinical study of cerebral ischemia. The

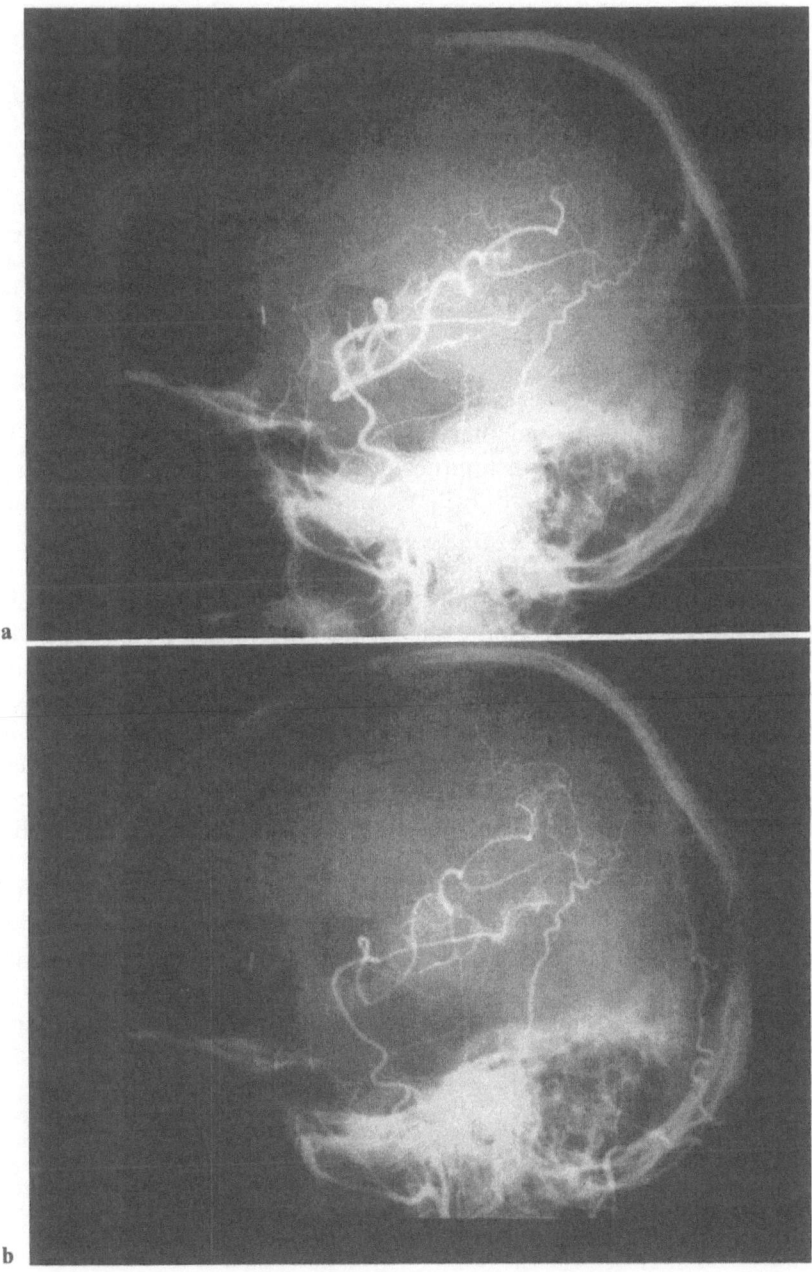

Fig. 3a, b. Early and late postoperative angiography. For details see "Case Report"

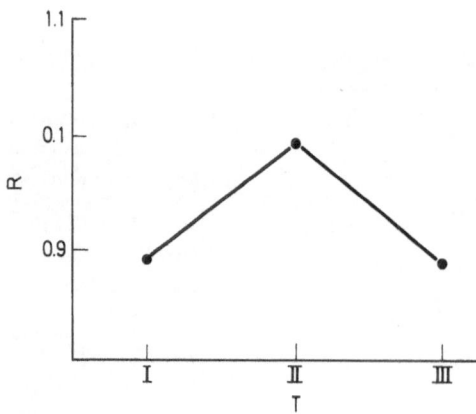

Fig. 4. Preoperative (*I*), early postoperative (*II*), and late postoperative (*III*) flow ratios. For details see "Case Report"

method is atraumatic and therefore easily repeatable. Moreover, when compared with conventional techniques using stationary multidetector systems, the major advantage of the D-SPECT technique is that it provides a transaxial image of brain blood flow. Owing to the limited experience with this new technique, however, its clinical usefulness needs to be further evaluated. In the present study, therefore, an attempt was made to investigate the effect of EIAB surgery on brain ischemia. The results of this preliminary study indicate that the D-SPECT technique provides valuable information by demonstrating areas of hypoperfusion in patients presenting with symptoms of cerebral ischemia due to occlusions or stenoses of cerebral arteries. We were also able to show an improvement of cerebral hemodynamics following surgery according to an increase in interhemispheric flow ratios on D-SPECT studies. The observed flow changes were in good agreement with those that have been reported previously by Yonekura et al. and Halsey et al. [2, 8].

Until now, however, a reliable quantitative estimation of CBF has been found to be unsatisfactory with the D-SPECT technique. This is probably due to methodologic reasons but may also be due to the fact that the scatter of flow values under various conditions exhibits a greater range than expected. There is some evidence that this limitation will at least partially be improved in the future by internal standardization of the technique and possibly also by the introduction of a modified algorithm for the calculation of CBF.

References

1. Buell U, Moser EA, Kirsch CM et al. (1983) 133-Xe-D-SPECT, Ergebnisse einer neuen nicht-invasiven Methode zur Messung der regionalen Hirndurchblutung (rCBF). Ein Vergleich mit kranialer Angiographie und Transmissions-CT. Fortschr. Röntgenstr. 139:4, 351−358.
2. Halsey JH, Morawetz RB, Blauenstein UW (1982) The hemodynamic effect of STA- MCA bypass. Stroke 13:2, 163−167
3. Kanno I, Lassen NA (1979) Two methods for calculating cerebral blood flow from emission computed tomography of inert gas concentrations. J. Comp. Assist. Tomogr. 3:71−76

4. Lassen NA, Henriksen L, Paulson O (1981) Regional cerebral blood flow in stroke by 133-xenon inhalation and emission tomography. Stroke 12:3, 284–288
5. O'Brien M (1985) Methodological problems in the clinical application of the xenon-133 methods. This volume, pp. 220–223
6. Olsen TS, Larsen B, Skriver EB et al. (1981) Focal cerebral ischemia measured by the intra-arterial [133]xenon method. Limitations of 2-dimensional blood flow measurements. Stroke 12:6, 736–744
7. Schmiedek P, Gratzl O, Steinhoff H et al. (1976) Blood flow and cerebral revascularization. Clin. Neurosurg. 23:270–286
8. Yonekura M, Austin G, Hayward W (1982) Long-term evaluation of cerebral blood flow, transient ischemic attacks and stroke after STA-MCA anastomosis. Surg. Neurol. 18:123–130

CBF in Patients with Ischemic Cerebrovascular Disease Studied with ^{133}Xe Inhalation and Single Photon Emission Tomography

S. Vorstrup and N. A. Lassen

Introduction

The recent development of a three-dimensional tomographic technique for measurements of cerebral blood flow (CBF) of inhaled ^{133}Xe enables accurate detection and visualization of focal low flow areas. Several measurements may be performed in individual patients and normal volunteers due to the noninvasive nature of the method and the low radiation exposure. It has therefore become possible to study the changes in regional CBF following cerebral ischemia in man, and also to evaluate the effects of medical or surgical therapy upon the cerebral circulation. Analyzing the findings obtained by this method in patients with cerebrovascular disease in conjunction with the clinical course, the angiographic lesions, and the structural findings as seen on CT scan has increased our knowledge of the sequence of pathophysiologic events occurring after cerebral ischemia. This paper aims to provide a brief description of the results in our series of patients with cerebrovascular disease, i.e., ischemic stroke and transient ischemic attacks (TIA), along with the findings following vascular surgery, studied with ^{133}Xe inhalation and single photon emission computer tomography (SPECT).

Methods

CBF was measured with a Tomomatic 64 (Stokely et al. 1980), which has a fast rotating detector array composed of 64 NaI crystals. Collimation of the crystals allows for three slices of brain tissue to be studied simultaneously with a resolution of 1.7 cm as measured at full width half maximum (FWHM) in the horizontal plane. The three slices are approximately 2.0 cm thick (FWHM), and are interspersed by 2 cm. ^{133}Xe is inhaled from a closed system with a CO_2 absorber during a period of 1½ min, yielding a maximum concentration of 10 mCi/liter at equilibrium. During the ^{133}Xe inhalation period and each of the subsequent three 1-min periods counts are summed, yielding a sequence of four tomographic pictures of the isotope distribution of each slice. These pictures, along with the input profile obtained by a single narrowly collimated detector po-

Department of Neurology, Rigshospitalet, Blegsdamvej 9, DK-2100 Copenhagen

Cerebral Blood Flow and
Metabolism Measurement.
Eds. Hartmann/Hoyer
© Springer-Verlag Berlin Heidelberg 1985

sitioned over the apical part of the right lung, permit the calculation of CBF by a conventional algorithm (Kanno and Lassen 1979; Celsis et al. 1981), and the final result is presented in a 32×32 pixel matrix.

Routinely, the midplanes of the three slices are placed 1 cm, 5 cm, and 9 cm above the orbitomeatal plane. Mean hemispheric blood flow values are calculated from the middle slice from all pixels representing brain tissue. Calculation of flow in a visually depicted region in one hemisphere is performed by encircling the region of interest by a cursor. This yields the average flow value of the encircled pixels as well as the value in the symmetrical region of the opposite hemisphere. A low flow value is considered significant when the side-to-side asymmetry exceeds 10%, as such values are not seen in normal man.

Results

TIA

In a small series of 14 patients with TIAs and arteriosclerotic neck vessel disease studied days to months after the most recent attack, focal low flow areas were seen in nine (Vorstrup et al. 1983). The localization of these hypoperfused areas agreed well with the clinical symptoms and angiographic lesions. In four of these nine cases, the CT scan showed a corresponding but smaller hypodense lesion, whereas in the remaining ten patients of this series the CT scan was without focal lesions.

A detailed analysis of the symptom-provoking factors and the angiographic lesions was undertaken in an attempt to classify the patients in embolic or hemodynamic cases on clinical grounds. A hemodynamic pathogenesis to the ischemic event was suspected when the symptoms were shortlasting (minutes) and occurred closely related to positional changes, with severely stenosing occlusive lesions being seen on the angiograms. As only a few patients could be classified as hemodynamic, it was presumed that the focal low flow area might in many cases represent a permanent ischemic tissue lesion invisible on CT scan.

In a later series now comprising 31 patients with TIAs, CBF measurements and CT scan were performed both before and 3 months after reconstructive neck vessel surgery (Vorstrup et al. 1983). Of the 13 patients showing a focal low flow preoperatively, only three in this series showed a focal improvement in regional CBF following surgery. These three patients all showed severely stenosing (> 90%) arteriosclerotic lesions before surgery. Thus the majority of the patients, ten in all, showed a completely unchanged flow map. The finding of a persistent low flow following surgery in patients with no CT lesion was taken as suggestive evidence of an "incomplete infarction," i.e., the low flow corresponds to the lowered metabolic demands caused by ischemic neuronal tissue damage, a lesion not visible on CT scan.

Stroke

More than 60 patients with ischemic stroke have been studied weeks to months after the ischemic event. Focal low flow areas were readily seen in agreement with the clinical symptoms and the hypodense lesions on the CT scan in the majority of cases. In most of these patients the size of the low flow area was considerably larger than the CT lesion. In several cases with persistent clinical symptoms without evidence of focal CT lesions, the low flow area was the only evidence of cerebral dysfunction. Three patients had a normal symmetrical flow map despite a focal CT lesion in the hemispheres: In two cases the CT lesions were very small (lacunar or just visible on CT), and in one case the hypodense lesion was located deep in the temporal pole, corresponding to a level of 3 cm above the orbitomeatal plane as judged by the CT scan and therefore not demonstrated by the routine positioning of the head yielding slices 1 cm, 5 cm, and 9 cm above the orbitomeatal plane. Two patients with clinical evidence of hemispheric infarction but persistently normal CBF maps deserve special mention: The first showed a pontine hypodense area on CT scan, probably explaining the symptoms. The second had normal CT scans on repeated examinations. We suspect a lacunar hemispheric or brain stem lesion to be the cause of the severe hemiparesis in this case. This latter patient is thus the only one in whom the combined imaging techniques, CT scan and CBF measurement, failed to show a relevant abnormality.

Serial measurements – up to ten CBF studies per patient – have also been performed in a smaller series of stroke patients with symptoms compatible with hemispheric lesions; they were studied in the acute phase within 48 h after onset and followed every other day during the first few weeks. This showed that during the first or second week following ischemia, a previously low flow area may transiently change into a region with a high flow level, i.e., the phenomenon of luxury perfusion. During this phase the clinical symptoms remain unchanged. We believe that spontaneous lysis of an embolic occlusion may explain the transient hyperemia. Repeated CBF measurement in these patients 8 weeks and ½ year later showed that the findings in the chronic phase resembled the early findings of a large area with quite low flow values. In addition, in the patients with minor flow defects corresponding to less severe clinical symptoms, the low flow areas persisted at the late follow-up studies.

Extracranial-Intracranial Bypass

Twenty-five patients with symptoms of ischemic cerebrovascular disease (predominantly patients with TIAs or minor strokes) and arteriosclerotic lesions not allowing conventional neck vessel surgery, i.e., occlusion of the internal carotid artery or lesions of the middle cerebral artery, were studied before and after the establishment of an extracranial-intracranial bypass (EC-IC) shunt. CBF measurement and CT scanning were included in the preoperative evaluation of these patients, and only patients with low flow areas exceeding the size of the

CT lesion or without CT lesions were selected for surgery. CBF was repeated 3 months after shunting (1 month in three patients for geographic reasons).

Also in this series an effort was made to use clinical criteria to classify the patients as hemodynamic or embolic TIA cases. Although it could a priori be expected that a high percentage of these patients with severe, often multiple, arteriosclerotic lesions suffered ischemic symptoms related to orthostatic changes, only one patient in this series showed such symptoms. This patient and three others were the only ones who showed a focal increase in CBF with a concomitant decrease in side-to-side asymmetry following shunting. The majority – 21 of 25 patients – showed an unchanged tomographic flow map with a persistent low flow area. During the observation period (at present averaging 14 months with a range of 3–45 months), only one of the 14 patients suffering TIAs prior to surgery has had new attacks. This patient had a well functioning shunt as assessed angiographically a few weeks after the most recent postoperative ischemic event. One patient suffered a minor stroke 16 months after surgery from the ipsilateral hemisphere, despite a well functioning shunt (verified by angiography after the stroke).

It may be assumed that the suggestive evidence of a decrease in ischemic events following surgery in our series is perhaps caused by an increased perfusion pressure due to the establishment of an additional collateral channel. An increase in the collateral capacity may be documented by studies on the cerebral vasoactive capacity. Preoperative measurements following the intravenous injection of 1 g acetazolamide were undertaken in the later part of the series. These results suggest that the patients in whom no or only a minor increase in flow – or even a decrease of flow – is seen in the region of interest, are those who will increase in perfusion pressure and/or CBF.

Comment

Focal low flow areas were detected in the majority of the patients studied in these series suffering symptoms of ischemic cerebrovascular disease. In the series of patients with TIA and arteriosclerotic neck vessel hypoperfused areas were seen in several cases both with and without evidence of focal CT lesions. The finding of a persistent low flow area following reconstructive neck vessel surgery implying the removal of possibly hemodynamically significant arteriosclerotic lesions in patients with no CT lesions *indicates that the low flow is in all likelihood a consequence of a lowered metabolic demand due to "incomplete infarction."* Transient ischemia may accordingly cause a selective neuronal cell loss without affecting the supporting tissue, the glia cells and the microvasculature (Marcoux et al. 1982) – a lesion not evidenced by CT scan. The finding in these series of a high proportion of patients with hypodense lesions on the CT scan (complete infarction) and/or areas of incomplete infarction indicates that TIAs cause irreversible cerebral damage despite the lack of permanent neurologic symptoms.

In patients with minor or severe strokes, somewhat larger hypoperfused areas which also reach lower flow values are typically seen. The low flow areas in most cases exceed the area of complete infarction seen on the CT scan, a finding which suggests that areas of incomplete infarction surround the hypodense lesions (Lassen et al. 1981). The term "ischemic penumbra," originally introduced by Astrup et al., describes a condition where the degree of tissue ischemia is such that the normal neuronal function is abolished, but may be restored if the blood flow is increased. In the experiment of Astrup et al. the reversibility was only tested after a short period of ischemia lasting $1-2$ h. Our finding of low flow areas weeks to months after an ischemic event in areas without CT lesions could theoretically represent a state of chronic ischemic penumbra. However, the unchanged flow-distribution in the majority of these patients after successful vascular surgery suggests that the phenomenon of a chronic ischemic penumbra does not exist, and that the low flow rather represents irreversible ischemic tissue damage or neuronal undercutting caused by remote infarction (Strong et al. 1983; Mies et al. 1983; Marcoux et al. 1982). However, although a chronic penumbra is unlikely to exist, a state of chronic "misery perfusion" (Baron et al. 1980) due to compromised collateral circulation may exist in rare cases with chronic cerebrovascular disease. In these patients with severely stenosing and/or occluding arteriosclerotic lesions, ischemic clinical symptoms may occur related to positional changes, further reducing the low local perfusion pressure. Even in the resting recumbent state such patients may be assumed to have regions in which maximal or close to maximal vasodilatation is required to maintain CBF at normal or almost normal levels despite low perfusion pressure. Recognition and selection of these patients, in whom reconstructive vascular surgery or EC-IC bypass may be considered the only rational procedure for preventing permanent ischemic damage, is important. This may be achieved, as already indicated, upon the occurrence of the clinical symptoms. These cases may be identified by studies on the cerebral vasoactive capacity – either by i.v. injection of acetazolamide (DIAMOX) or CO_2 inhalation (Norrving et al., 1982) – or by recently presented preliminary studies on cerebral blood volume with positron emission tomography (Gibbs et al. 1983; Powers et al. 1983). We consider undertaking such studies, which may be performed with the present equipment by labeling the red blood cells with [99m]Technetium.

References

Astrup J, Symon L, Branston NM, Lassen NA (1977) Cortical evoked potential and extracellular K[+] and H[+] at critical levels of brain ischemia. Stroke 8:51 – 57

Baron JC, Bousser MG, Rey A, Guillard A, Comar D, Castaigne P (1981) Reversal of focal "misery-perfusion syndrome" by extracranial-intracranial arterial bypass in hemodynamic cerebral ischemia. A case study with [15]O positron emission tomography. Stroke 12:454 – 459

Celsis P, Goldman T, Henriksen L, Lassen NA (1981) A method for calculating regional cerebral blood flow from emission computed tomography of inert gas concentrations. J Comput Assist Tomogr 5:641 – 645

Gibbs J, Wise R, Leenders K, Jones T (1983) The relationship of regional cerebral blood flow, blood volume, and oxygen metabolism in patients with carotid occlusion: Evaluation of perfusion reserve. J Cereb Blood Flow Metabol 3, suppl 1:S590–S591

Kanno I, Lassen NA (1979) Two methods for calculating cerebral blood flow from emission computed tomography of inert gas concentrations. J Comput Assist Tomogr 3:71–76

Lassen NA, Henriksen L, Paulson OB (1981) Regional cerebral blood flow in stroke by [133]Xenon inhalation and emission tomography. Stroke 12:284–288

Marcoux FW, Morawetz RB, Crowell RM, DeGirolami U, Halsey JH (1982) Differential regional vulnerability in transient focal cerebral ischemia. Stroke 13:339–346

Mies G, Auer LM, Ebhardt G, Traupe H, Heiss WD (1983) Flow and neuronal density in tissue surrounding chronic infarction. Stroke 14:22–27

Norrving B, Nilsson, Risberg J (1982) rCBF in patients with carotid occlusion. Resting and hypercapnic flow related to collateral pattern. Stroke 13:155–162

Powers W, Martin W, Herscovitch P, Raichle M, Grubb R (1983) The value of regional cerebral blood volume measurements in the diagnosis of cerebral ischemia. J Cereb Blood Flow Metabol 3, suppl 1:S598–599

Stokely EM, Sveinsdottir E, Lassen NA, Rommer P (1980) A single photon dynamic computer assisted tomograph (DCAT) for imaging brain function in multiple cross sections. J Comput Assist Tomogr 4:230–240

Strong AJ, Venables GS, Gibson G (1983) The cortical ischaemic penumbra associated with occlusion of the middle cerebral artery in the cat: 1. Topography of changes in blood flow, potassium ion activity, and EEG. J Cerebral Blood Flow and Metabolism 3:86–96

Strong AJ, Tomlinson BE, Venables GS, Gibson G, Hardy A (1983) The cortical ischaemic penumbra associated with occlusion of the middle cerebral artery in the cat: 2. Studies of histopathology, water content, and in vitro neurotransmitter uptake. J Cerebral Blood Flow and Metabolism 3:97–108

Vorstrup S, Hemmingsen R, Henriksen L, Lindewald H, Boysen G, Paulson OB, Lassen NA (1983) Cerebral blood flow in patients with transient ischemic attacks studied with [133]Xenon inhalation and emission tomography before and after reconstructive vascular surgery. J Cereb Blood Flow Metabol 3, suppl 1:S592–S593

Vorstrup S, Hemmingsen R, Henriksen L, Lindewald H, Engell HC, Lassen NA (1983) rCBF in TIA studied by [133]Xenon inhalation and emission tomography. Stroke 14:903–910

Asymmetric Cerebral Blood Flow in Hemiparkinsonian Patients: Tomography of Inhaled Xenon 133 Before and After Levodopa Treatment

L. Henriksen and J. Boas

Introduction

Parkinson's disease is an entity of symptoms mostly confined to tremor, rigidity, and akinesia, and often with tremor as the initial symptom. The organic substrate for the syndrome is both controversial and elusive, but structural and biochemical changes in the substantia nigra, the caudate, the putamen, and the globus pallidus seem to be constant findings (Marsden and Parkes 1977).

In the average parkinsonian population the response to L-dopa therapy is highly variable, with both dramatic responses and resistant cases. The response variability and the on/off symptoms are not fully understood, but it is most unlikely that one anatomic structure could be incriminated in the metabolic disturbances which underlie parkinsonism. The different course of the apparently same disease could be a multifactorial anatomic destruction per coincidense grouped under the same syndrome.

With the assumption that subcortical flow changes and asymmetries could be of pathogenetic significance, 18 parkinsonian patients with mostly unilateral symptoms were selected from our clinic and were studied with CBF tomography before and after L-dopa treatment.

Material

The 18 parkinsonian patients consisted of 11 men and 7 women, all with predominantly unilateral symptoms. Half of them had left-sided symptoms and the other half right-sided symptoms. Their mean age was 59 years, ranging from 38 to 72. The mean duration of their parkinsonian symptoms was 6.7 years (range 3 to 13), and they were treated with L-dopa for a mean of 4.0 years (range 1 to 11). One patient had a left-sided stereotactic operation 10 years previously which produced a moderate right-sided hemiparesis. The other patients had no symptoms or clinical signs of additional cerebrovascular diseases. All patients gave their informed consent prior to the study.

Department of Neurology, Rigshospitalet, Blegsdamvej 9, DK-2100 Copenhagen

Cerebral Blood Flow and
Metabolism Measurement.
Eds. Hartmann/Hoyer
© Springer-Verlag Berlin Heidelberg 1985

Methods

Regional cerebral blood flow (rCBF) was measured by emission computerized tomography of inhaled ^{133}Xe (Tomomatic-64, Medimatic Inc., Copenhagen). Detailed descriptions of the device have recently been published (Lassen et al. 1981; Celcis et al. 1981). Briefly, ^{133}Xe is inhaled for 1.5 min at a concentration of 10 mCi/liter, and the washout followed for three additional 1-min periods. The instrument records from three slices simultaneously, each with a resolution element of $1.7 \times 1.7 \times 2.0$ cm. From the reconstructed images of local isotope concentration in the four periods, rCBF is calculated by a two-step procedure (Celcis et al. 1981). It is assumed that the partition coefficient has the same value of 0.85 ml/g in all high flow areas.

On the day of investigation all patients were without their usual medicine for a minimum of 10 h. The CBF measurements were obtained with the patients in the supine position, eyes closed, and the noise was reduced to a minimum. After a baseline CBF measurement L-dopa was administered in fractionary doses until optimal clinical improvement was achieved, then the second CBF measurement was made. Regional cerebral blood flow was calculated in cortical, subcortical, and cerebellar regions on both the symptomatic and the asymptomatic side.

The radiation exposure from one study has been calculated to be approximately 0.08 rad of gonadal radiation and 0.6 rad for the lungs (target organ). Both beta and gamma radiation has been included in the calculations (Atkins et al. 1980).

The partial pressure of carbon dioxide in arterial blood was estimated by spectrophotometric analysis of end-expiratory CO_2 fraction ($FeCO_2$).

Statistical analysis was performed by Spearmen's and Wilcoxon's nonparametric tests. $P < 0.05$ was used as the significance level.

Results

Mean CBF and Mean Hemispheric CBF

Mean CBF in our group of parkinsonian patients was 54.9 and 56.1 ml/100 g/min before and after L-dopa, respectively. Mean CBF in the resting state was not different from a normal age- and sex-matched population, and the affected mean hemisphere flow was not different from the "healthy" contralateral hemisphere. L-dopa did not change mean CBF, but regional flow changes occurred.

Regional CBF

The cortical CBF was related to the duration of Parkinson's disease ($P < 0.05$), probably reflecting an increasing mental deterioration with time. The 18 pa-

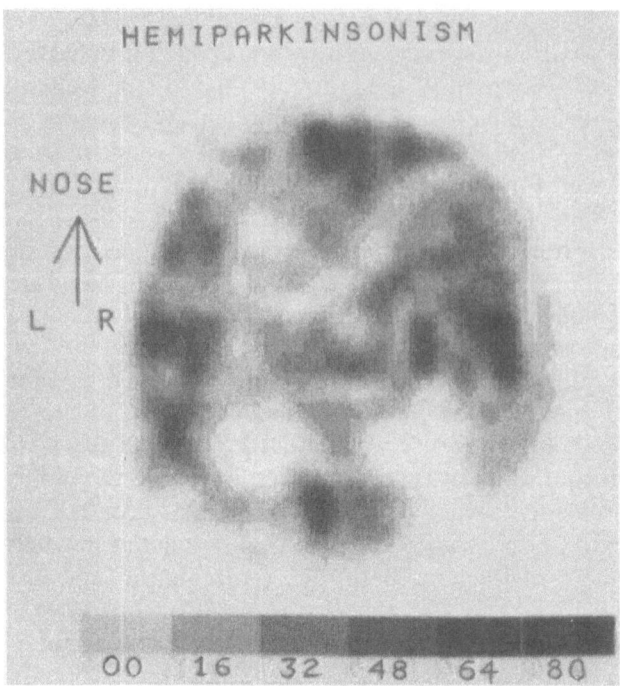

Fig. 1. This 46-year-old man had had hemi-parkinsonian symptoms from his right extremities during the previous 12 years. CBF tomography shows asymmetric rCBF corresponding to the basal ganglia, with a reduced flow in left-sided subcortical structures. His symptoms primarily consisted in tremor and rigidity

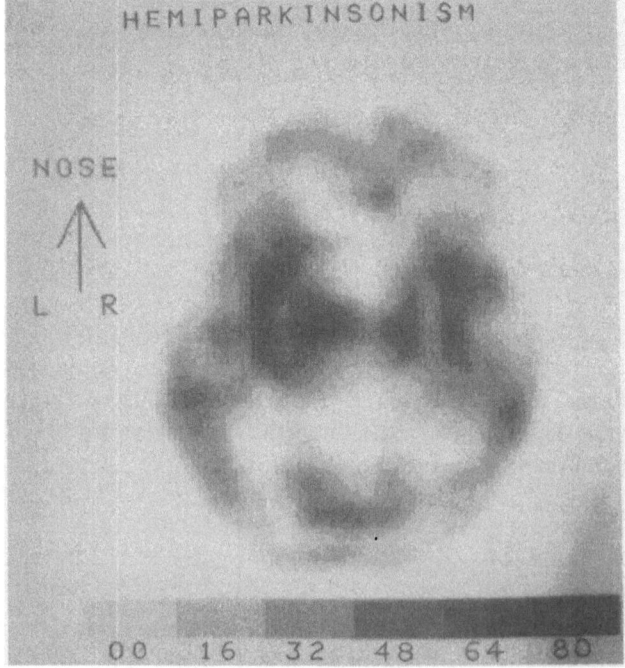

Fig. 2. This 66-year-old woman had left-sided extremity symptoms mostly consisting of rigidity, some akinesia, and minimum tremor. In this paient the new brain seeking compound iodo-123-amphetamine was used, and its imaging properties with a good delineation of gray and white matter tissue must be emphasized. Reduced tracer uptake is noted in areas anatomically related to the right globus pallidus and the anteroventral thalamic nuclei

tients were divided into three groups according to whether or not L-dopa elicited adverse reactions. From these results it appeared that a good response to L-dopa therapy was associated with a reduced flow in the contralateral striatum ($P < 0.05$), whereas an increased flow was observed in association with adverse reactions ($P < 0.05$).

Subcortical flow asymmetries were noted in all 18 patients with a reduced flow in the contralateral basal ganglia, as illustrated in Fig. 1. One patient was investigated with iodo-123-amphetamine and is presented to show its good delineation of gray and white matter tissue (Fig. 2). According to the flow maps the subgroup with mainly akinesia could represent pallidal degeneration, and those with mainly tremor and rigidity could have a nigrostriatal disconnection syndrome.

A flow asymmetry in the contralateral cerebellar hemisphere was noted in only the patient who had a stereotactic operation 10 years previously. The other patients showed only minor insignificant flow asymmetries with no relation to right or left extremity symptoms.

Discussion

Anatomic Considerations

The basal ganglia is usually synonymous with the combination of caudate nucleus, putamen, globus pallidus, subthalamic nucleus, and substantia nigra. Phylogenetically the caudate nucleus and putamen have a common origin and are often both together called the striatum. The substantia nigra projects to all parts of the striatum, and destruction of this pathway is pathogenetic for Parkinson's disease. The striatum receives input from the cerebral cortex, the substantia nigra, and the intralaminar nuclei of the thalamus. Thus the basal ganglia are constantly informed about most aspects of cortical function. The major efferents from the striatum funnel through the globus pallidus, into the ventral lateral and anterior nuclei of thalamus, and back to the motor and premotor cortex. With such interconnections the striatum depends on the globus pallidus for most of the influence it exerts on the control of movements. Globus pallidus is thought to be excitatory, and destruction in some parts of the loop could release globus pallidus from some of its restraints. Damage to the inhibitory nigrostriatal pathway leads to excessive excitatory output from the striatum with increased activity in ventrolateral thalamic nuclei. The parkinsonian symptoms could be classified as a dyscoordination of subcortical excitation and inhibition, producing an excessive and abnormal activity.

The brains of parkinsonian subjects are dopamine deficient (Hornykiewicz 1966). The progressive disability of parkinsonian patients is associated with a progressive loss of nigrostriatal neurons and probably associated with a central dopaminergic receptor supersensitivity (Marsden 1975; Pycock and Marsden 1977).

Cerebral Blood Flow Findings

Melamed and co-workers (1978) found with a stationary detector system an unchanged average CBF after L-dopa, a concept supported by our findings, where
CBF was 55 and 56 ml/100 g/min before and after L-dopa, respectively. It is
generally believed that progressive neurologic symptoms in parkinsonian patients are associated with gradually increasing mental impairment. A reduced
neuronal activity due to either neuronal dysfunction or neuronal cell loss will be
reflected in the CBF measurements. A mean CBF of 55 ml/100 g/min is within
the normal limits of CBF in this age group, but mean CBF and cortical CBF
(avoiding the subcortical structures) showed the expected tendency of a reduced CBF with an increasing duration of the disease.

Marked subcortical asymmetries were noted in all cases, despite Compton
scatter, partial volume effects, and a resolution element of $1.7 \times 1.7 \times 2.0$ cm.
X-ray tomography was performed in four of our parkinsonian patients, and the
apparently normal structure despite marked subcortical flow asymmetries must
be emphasized. The same pattern of normal structure despite marked functional deficits was also noted in a group of children with attentional deficit disorder (Henriksen et al. 1985), and both studies demonstrate that CNS dysfunction may precede X-ray indications of cell loss. The pilot study made with
iodo-123-amphetamine showed good imaging properties (Fig. 2), and this
radioisotope or maybe other new brain seeking compounds will hopefully more
accurately delineate the anatomic regions involved in the various parkinsonian
symptoms.

Following L-dopa, regional CBF decreased in the contralateral striatal area.
However, in the patients in whom adverse reactions appeared an increased
striatal flow was noted, often bilaterally, and involving more central parts of the
basal ganglia, i.e., the pallidus and thalamic-hypothalamic areas. In the average
parkinsonian population the therapeutic response to L-dopa declines after a few
years of treatment. As already stated, some are dramatic responders, others are
resistant to treatment, and within these extremes there is a highly varied response. The anatomic basis for the different symptoms is probably multifactorial. Tremor and rigidity could be explained on the basis of an increased activity
in the globus pallidus (Pakkenberg 1963) and represent a nigrostriatal disconnection syndrome, whereas patients with mainly akinesia could represent
pallidal degeneration. The postural instability and in some cases mental changes could be from other sites, e.g., the cerebellum and cortex.

In conclusion, an asymmetric subcortical blood flow was found in the basal
ganglia contralateral to the extremity symptoms in all 18 patients with hemi-
parkinsonism. Mean CBF was unaffected by L-dopa treatment, but alleviation
of symptoms was associated with a decreased rCBF in the contralateral striatal
area. Patients with a reduced or nor effect of L-dopa, often combined with adverse reactions, showed an increased flow in the contralateral and the ipsilateral
striatal areas.

Acknowledgment. This study was supported by a grant from Svend Aa. Nielsen
Wackerhausen's Foundation.

References

Atkins HL, Robertson JS, Croft BY, Tsui B, Susskind H, Ellis KJ, Loken MK, Treves S. (1980) Estimates of radiation absorbed doses from radioxenons in lung imaging. J Nucl Med 21:459−465

Celcis P, Goldman T, Henriksen L, Lassen NA. (1981) A method for calculating regional cerebral blood flow from emission computed tomography of inert gas concentrations. J Comput Assist Tomogr 5:641−645

Henriksen L, Lou H, Bruhn P. (1985) Focal frontal hypoperfusion in children with attentional deficit disorder. This book, pp. 278−282

Hornykiewicz O. (1966) Dopamine (3-hydroxytyramine) and brain function. Pharmacol Rev 18:925−964

Lassen NA, Henriksen L, Paulson OB. (1981) Regional cerebral blood flow in stroke by 133 xenon inhalation and emission tomography. Stroke 12:284−288

Marsden CD. (1975) The neuropharmacology of abnormal involuntary movement disorders (the dyskinesias). In: D. Williams (ed.), Modern trends in neurology, vol 6, Butterworths. London, 141−166

Marsden CD, Parkes JD. (1977) Success and problems of long-term levodopa therapy in Parkinson's disease. Lancet 1:345−349

Melamed E, Lavy S, Cooper G, Bentin S. (1978) Regional cerebral blood flow in parkinsonism. J Neurol Sci 38:391−397

Pakkenberg H. (1963) Globus pallidus in parkinsonism. Acta Neurol Scand suppl 4, 139−144

Pycock CJ, Marsden CD. (1977) Central dopaminergic receptor supersensitivity and its relevance to Parkinson's disease. J Neurol Sci 31:113−121

Measurement of Regional Cerebral Blood Flow in Psychiatry

J. P. HEDDE, F. M. REISCHIES, H. GUTZMANN, R. FELIX, and H. HELMCHEN

In the past, description of cerebral blood flow disturbances in psychiatric diseases has mainly been based on use of the conventional multidetector system. The introduction of dynamic single photon emission tomography by Lassen and co-workers for the special purpose of cerebral blood flow measurement reduced disturbance factors such as cross-talk, look-through, noise artifacts etc. compared with the multidetector system. It therefore appears interesting to carry out a search for characteristic perfusion disturbances which are specific to certain psychopathologic disorders by means of this new system. It should be taken into account, however, that a lesser amount of perfusion deviation can be expected than in the case of primary vascular disease. This fact should be stressed even more because conventional scintigraphy in this context did not exhibit significant pathologic findings later discovered with the help of the ^{133}Xe clearance.

Method and Patients

Since 1982 we apply the dynamic SPECT, developed by Lassen and co-workers. The indications for examination in 61 psychiatric patients were as follows: 27 cases of nonspecific organic brain syndrome, 6 cases of advanced dementia, 18 cases of a depressive syndrome, and 10 cases of schizophrenia or paranoid hallucinal psychosis. The aim of our investigation was to determine regional hypoperfusion under resting conditions. Examination took place in a dimly lit room. The patient was requested to keep his eyes closed during the examination. Patients with miscellaneous psychiatric diagnoses as well as patients with circumscribed lesions in CT (with the exception of multi-infarct dementia) were excluded from the evaluation.

Results

In 23 of the 27 patients with nonspecific organic brain psychosyndrome without neurologic deficits we found one or several regions of regional reduced

---

Psychiatrische und Radiologische Klinik, Klinikum Charlottenburg, D-1000 Berlin 19

Cerebral Blood Flow and
Metabolism Measurement.
Eds. Hartmann/Hoyer
© Springer-Verlag Berlin Heidelberg 1985

perfusion. Up until now there has been no evidence to support a preference for either hemisphere.

The six cases of dementia (three patients with multi-infarct dementia and three patients with Alzheimer dementia) all showed a clear deviation from the standard perfusion pattern, without preference for one hemisphere. Two of the patients with Alzheimer dementia had a from frontal to occipital decreasing blood flow, which permits differentiation from multi-infarct dementia.

In six of the ten patients with schizophrenia or paranoid hallucinal psychosis we found a regional decreased perfusion, which in four cases involved the frontal brain, as was first described by Ingvar and co-workers and subsequently replicated by positron emission tomography.

Up until now we have not been able to discover any characteristic changes in the perfusion pattern of patients with a depressive syndrome. But in a part of them there seems to be a tendency of regionally diminished blood flow in the right hemisphere.

Discussion and Summary

The following conclusions seem possible:

1. In a high percentage of cases the nonspecific organic brain syndrome shows one or several regions of diminished perfusion. The homogeneous reduction of the perfusion mainly temporally may be characteristic for severe Alzheimer dementia, whereas multi-infarct dementia exhibits multiple regions of circumscribed hypoperfusion associated with hypodense zones in CT. A difficulty exists in distinguishing between multi-infarct dementia and degenerative dementia because the diagnosis of both is often arrived at from multiple regions of reduced perfusion. At least in the early phase of disease the extent of the perfusion reduction may be a diagnostic sign, a higher degree occurring in the case of multi-infarct dementia. Frontal hypoperfusion was confirmed in four of the ten patients with paranoid psychosis. In patients with depression we found sometimes a mild hypoperfusion on the right. In a lot of cases in agreement with J. Risberg, differentiation between advanced dementia and depression seems possible.

2. Our results indicate that tomographic ^{133}Xe clearance contributes to the pathogenetic explanation and nosologic classification of mental disorders.

3. Our preliminary experience leads us to suppose that use of activation tests and consideration of the absolute flow values enable us to estimate more exactly the degree of the disease and its nosologic classification.

The development of new radionuclides to investigate functions of the brain other than CBF by means of SPECT is in progress and no more the aim only for the much more complicated positron emission tomography.

We hope to have shown that the tomographic measurement of rCBF by means of ^{133}Xe clearance can be a helpful biologic parameter in clinical psychiatry.

Focal Frontal Hypoperfusion in Children with Attentional Deficit Disorder

L. Henriksen, H. Lou, and P. Bruhn

Attentional deficit disorder (ADD) is a major cause of learning disability and behavioral disturbance in childhood. It is a controversial and poorly defined entity which has been associated with brain damage in early childhood. To date there has, however, been no information available on the cerebral structures involved in its pathogenesis. The cardinal symptom in ADD is distractability and short attention span. The child appears restless, with incessant motor response to stimuli in a haphazard fashion — hence the traditional eponym of the disorder: "hyperkinetic syndrome." The behavioral disturbance is often well controlled by central stimulants such as amphetamine and methylphenidate, but such treatment is still controversial as there is even disagreement of the organic nature of the syndrome.

There has been much hypothesizing about the etiology, but few attempts to understand the pathogenetic mechanisms (Kinsbourne 1973). Lesions of the mesial frontal cortex are known to severely impair the capacity for voluntary attention, and such patients are incessantly distracted by irrelevant and extraneous stimuli. Furthermore, lesions of the neighboring basal (orbital) zones of the frontal cortex give rise to generalized disinhibition and gross changes in the affective processes (Luria 1973). Based on these facts we propose the hypothesis that ADD in childhood is essentially a mesial frontal lobe syndrome, and the associated symptoms are due to dysfunction of neighboring or connected structures. This hypothesis has been tested in 12 children studied consecutively with tomography of inhaled xenon 133.

Material

Patient Group

The clinical diagnosis was confirmed by neuropsychological testing; a typical ADD should exhibit attentional deficit, impulsiveness, hyperactivity, and temper, severely hampering scholastic achievement. Twelve such children with ADD were admitted to the study consecutively over a 2-year period (1980–1981). One patient was unable to cooperate during the CBF measurement. The remaining 11 ranged between 6.5 and 15 years in age (median 9).

Department of Neurology, Rigshospitalet, Blegdamsvej 9, DK-2100 Copenhagen

Cerebral Blood Flow and
Metabolism Measurement.
Eds. Hartmann/Hoyer
© Springer-Verlag Berlin Heidelberg 1985

There was only one girl, and two patients were brothers. Transmission computerized tomography (TCT) was performed in all patients. Six patients were treated with methylphenidate 10–30 mg daily, but this was discontinued 1 week before the investigation. The patients who were treated with methylphenidate were studied twice, before and after medication, to visualize the effect produced by the drug. Informed consent was obtained in all cases, and the parents were encouraged to help in the examination.

Normal Material

Nine normal children of median age 12 years (range 7–15), mostly consisting of siblings of the study group, were used as controls. We felt that we were not able to select a completely independent control group where the parents had no motivation for the study. Therefore complete age and sex matching was not obtained. However, comparing the three girls with the six boys in the group, no systematic difference was noted. Similarly, comparing the three youngest with the six oldest, no difference was noted. Informed consent was obtained in all cases and the parents were encouraged to help in the examination.

CBF Method

Regional CBF was measured by xenon 133 inhalation and emission computerized tomography. The instrument has 64 sodium iodide crystals arranged in four banks around the head. The CBF study lasts 4.5 min. Xenon 133 is inhaled for 1.5 min at an equilibrated concentration of 10 mCi/liter. The washout of the radioisotope is followed for three additional 1-min periods. These four integral pictures are used for the flow calculations, which have been described in detail (Celcis et al. 1981). The air curve monitored over the upper part of the right lung with a narrow collimated scintillation detector is taken to represent the arterial input of ^{133}Xe. The tomograph measures from three slices simultaneously positioned routinely 1, 5, and 9 cm above the orbitomeatal (OM) line. The resolution element (full width at half maximum, FWHM) is $17 \times 17 \times 20$ mm, and the slice thickness 2 cm, FWHM. The middle slice containing the central part of the mesial frontal cortex, perisylvian regions, the upper part of basal ganglia, the parietotemporal cortex, and the central part of the occipital cortex was used for calculations.

The test-retest variability was found to be small. In 30 unselected patients from our clinic in test-retest variability of the mean CBF + 1 SD was 0.2 ml + 6.4 ml/100 g/min. The mean regional test-retest variability was 7%. The mean CBF + 1 SD calculated for our control group was 71 + 9.9 ml/100 g/min. In order to recognize regions with relative hypoperfusion in the patients, the mean flow distribution pattern in the control group was calculated (regional CBF/mean CBF) and its standard deviation in all regions calculated. With these control flow maps it was possible in the patients to identify areas with a regional flow below 1 SD or 2 SD.

The six children who received treatment with methylphenidate were examined immediately before and 30–60 min after medication (10–30 mg). The rCBF pattern before methylphenidate was subtracted from the rCBF after medication in each patient to identify regions with changes in flow, and hence neuronal activity, following treatment.

Results

A CBF tomogram of a normal child is presented to show the symmetrical flow pattern from the anterior to the posterior and from the left to the right side (Fig. 1). The flow pattern was identical in boys and girls, and dividing the normal control group into two groups according to age showed no difference except for a smaller size of the heads in the youngest.

Transmission computerized tomography by X-ray failed to show abnormalities in any of the patients, whereas regions with hypoperfusion were found in all cases (Figs. 2 and 3). The hypoperfused regions tended to be located symmetrically in the central parts of the brain (Fig. 2) but always periventricularly in the frontal lobes (Fig. 3).

After methylphenidate all six patients showed an increased flow in areas corresponding to the basal ganglia and mesencephalon, and the cortical sensory and motor regions decreased in flow.

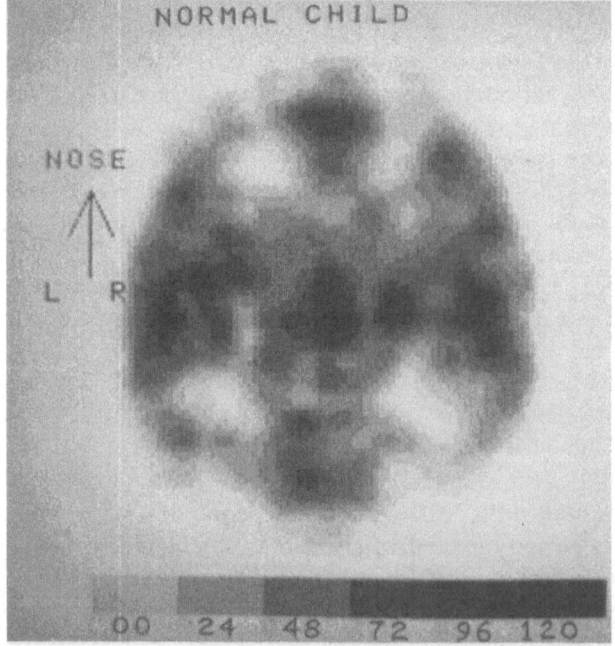

Fig. 1. Regional cerebral blood flow by emission tomography of inhaled ^{133}Xe in a normal 7-year-old boy. Note a symmetrical cerebral blood flow from the anterior to the posterior, and from the left side to the right side

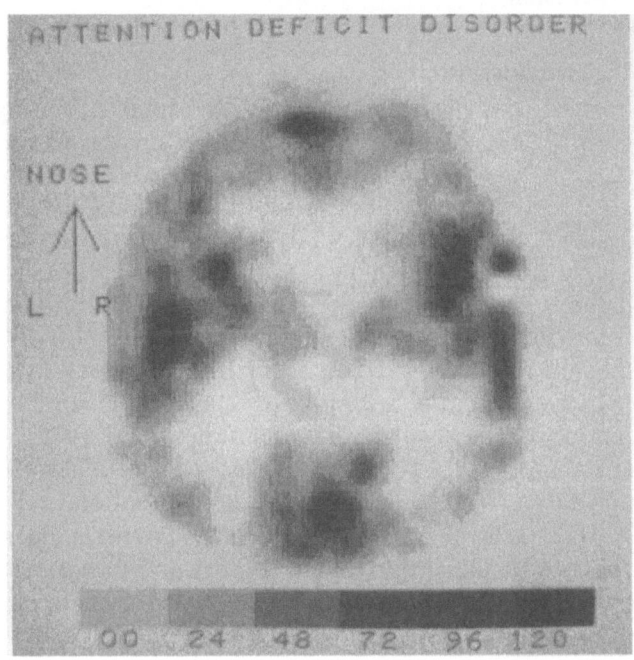

Fig. 2. Regional cerebral blood flow by emission tomography of inhaled ^{133}Xe in a 7-year-old boy with ADD. Note a markedly reduced flow periventricularly in artery boundary zones (the watershed areas) between cortical and subcortical brain areas. A normal structure (attenuation) was noted on transmission computerized tomography (X-ray), but apparently a dysfunction exists clinically and in terms of low periventricular blood flow

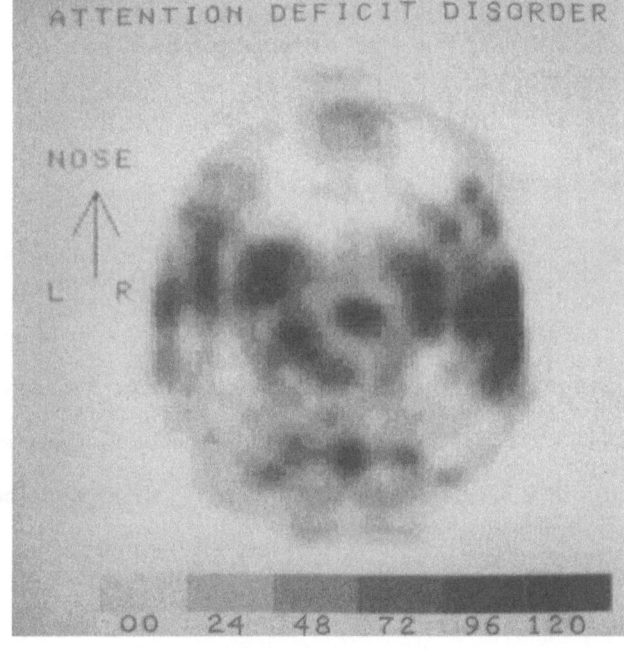

Fig. 3. Regional cerebral blood flow by emission tomography of inhaled ^{133}Xe in a 6-year-old boy with ADD. Note a reduced flow symmetrically in both frontal lobes around the anterior ventricles involving the corpus callosum and the mesial frontal cortex. This child, just like all the other children with ADD, had a normal transmission computerized tomography (X-ray)

Discussion

The hypoperfusion in our 11 ADD patients was typically located periventricularly but always centrally in the frontal lobes, suggesting a relationship to border zones between major arterial territories supplying the phylogenetic older parts of the brain with cortical areas. Hypoperfusion and hence low metabolic activity may be due to subtle morphological abnormalities (white matter lesions?) not detectable with TCT but with important pathogenetic implications. This interpretation is supported by the fact that axons from the dopaminergic neurons originating in the mesencephalon pass through the central frontal lobes to reach the prefrontal cortex (Lindwall et al. 1974), which is thought to be involved in regulation of attention (Fuster 1980). These dopaminergic neurons are probably activated by methylphenidate, which blocks the membrane reuptake of dopamine (Ross 1979). They may constitute the anatomic basis for the interaction between reticular formation of the brain stem and the prefrontal lobe in regulation of attention.

After methylphenidate we observed increased perfusion of central regions and a concomitant decrease of sensory and motor cortex. In recent experiments on rats the same pattern was noted (Bell 1982). These findings suggest an inhibition of function of these structures, noted clinically as less distractability and decreased motor activity during treatment.

The finding that hypoperfused regions in ADD seem to be located symmetrically in both hemispheres around the ventricles in arterial border zones is consistent with an etiologic role for early hypoxic-ischemic lesions and supports the notion of a critical role of mesial frontal cortex dysfunction in the pathogenesis of ADD.

Acknowledgment. This work was supported by a grant from Svend Aa. Wackerhausen's Foundation.

References

Bell RD, Alexander GM, Schwartman RJ, Yu J (1981) The methylphenidate-induced stereotype in the awake rat: Local cerebral metabolism. Neurology 32:377 – 381.

Celcis P, Goldman T, Henriksen L, Lassen NA (1981) A method for calculating regional cerebral blood flow from emission computed tomography of inert gas concentrations. J Comput Assist Tomogr 5:641 – 645

Fuster JM (1980) The prefrontal cortex. Anatomy, physiology and neuropsychology of the frontal lobe. Raven Press, New York

Kinsbourne M (1973) Minimal brain dysfunction as a neurodevelopmental lag. Ann. NY, Acad-Sci 205:268 – 273

Lindwall O, Björklund A, Moore RY, Stenevi U (1974) Mesencephalic dopamine neurons projecting to neocortex. Brain Res 81:325 – 331

Luria AR (1973) The working brain. Penguin, London

Ross SB (1979) The central stimulatory action of inhibitors of dopamine uptake. Life Sci 24:159 – 168

The Usefulness of Brain SPECT with ^{123}I-IAMP and HIPDM

C. Raynaud, G. Rancurel, E. Kieffer, F. Soussaline, S. Ricard,
S. Askienazy, J. L. Moretti, M. Bourdoiseau, and J. Rapin

As soon as iodoamphetamine labeled with ^{123}I was available, brain lesions such as tumors and infarcts were clearly visible on single photon emission computerized tomography (SPECT) images with excellent accuracy (Lafrance et al. 1981; Kuhl et al. 1981; Moretti et al. 1982; Hill et al. 1982). These initial results were very encouraging. However, it soon became evident that this costly procedure could not compete with techniques such as computerized tomography (CT) or scintiscans, the reliability of which is excellent in diagnosing tumors and infarcts. The usefulness of brain SPECT was therefore questioned. In this paper we would like to evaluate the usefulness of this technique at the present time.

Methodological Considerations

SPECT systems now marketed can be classified into two groups, detector array systems and rotating gamma cameras. Detector array systems have a high sensitivity but provide only a few transverse slices. Due to their large area detector, rotating gamma cameras image the entire brain, which can be studied on complete sets of transverse, coronal, and sagittal slices, but have a low sensitivity. Details concerning individual system characteristics, which do not have their place here, will be found in several excellent reviews (Budinger 1981; Soussaline 1982). We insist only on the necessity of careful and frequent quality control of the gantry, detector, and computer, which should be carried out weekly because these devices are highly sophisticated. Immobilization of the patient's head also requires full attention, especially when rotation gamma cameras are used.

Indicators used in brain SPECT must penetrate the normal blood-brain barrier (BBB) and be labeled with gamma emitter. Two have so far been proposed, ^{123}I-isopropyl amphetamine (IAMP) and ^{123}I trimethyl propane diamine (HIPDM). Their specific activity varies between 1 and 10 mCi/mg depending on the isotope-producing companies. Labeling is done by these companies except for IAMP, which can be labeled with commercial kits (Oris, France). Injected amounts vary from 2 to 8 mCi depending on the type of device used.

Service Hospitalier Frédéric Joliot CEA, Département de Biologie, F-91406 Orsay

Cerebral Blood Flow and
Metabolism Measurement.
Eds. Hartmann/Hoyer
© Springer-Verlag Berlin Heidelberg 1985

Biodistribution Considerations

After i.v. injection with IAMP, the brain activity curve increases rather slowly; it reaches a maximum value of approximately 5% of the injected dose after 30 min and remains at a plateau for several hours (Kuhl et al. 1982; Moretti et - al. 1982). With HIPDM the maximum brain activity value seems slightly lower (Holman et al. 1982).

Kuhl et al. (1982), Lassen et al. (1983), Devous et al. (1983), Holman et al. (1983), and Drayer et al. (1983) demonstrated that during the first 10 min after injection the results obtained with IAMP and HIPDM correlate significantly with regional cerebral blood flow (rCBF), and proposed the use of these indicators in measuring rCBF.

After 30 min the brain distribution of IAMP is modified, the results obtained by Lassen et al. (1983) suggesting a redistribution. Rapin et al. and Devous et - al. (1983) studying mice with IAMP and dogs with HIPDM, respectively, confirmed this redistribution and the poor correlation with rCBF. The new distribution of IAMP and HIPDM may reflect the metabolic activity of the brain more accurately than the rCBF. In this paper we will only consider these late SPECT images obtained after 30 min.

Applications

When the head of a patient is placed in the orbitomeatal position and maintained there throughout acquisition, the images obtained may be compared with those found in anatomic atlases using the same head position (Salamon 1971). It is then possible to propose schematic diagrams of eight transverse, ten frontal, and seven or eight sagittal SPECT slices of the normal brain (Figs. 1 – 3). For transverse sections, the brain is sliced from top to bottom, for frontal sections from the forehead to the occipital region, and for sagittal sections from the left to the right ear. The presentation of the latter differs from the other two because the median line of the brain may be used as an axis of symmetry, the images placed on the left and right side of the diagram sheet are then symmetrical compared to the median line of the brain. The top couple of images represent the most external slices. When the slice number is odd, the lower (S6) represents the middle line section; when the slice number is even, the two lowest (S6 and S7) represent the two sections around the median line. Depending on an even or odd slice number one or the other sagittal diagram should be used. This schematic representation should help to localize a lesion. For example, the caudate nucleus is well delineated on the slices T5 and F5, the thalamus on T6, F6, F7, and F8, and the hippocampal gyrus on T7, F6, F7, and F8. The proposed diagrams were established for a slice thickness of 12.5 mm; they will of course be different if the slice thickness is different.

At present, applications of brain SPECT using IAMP or HIPDM are limited to focal epilepsy and Alzheimer disease. However, results obtained in

Transverse sections

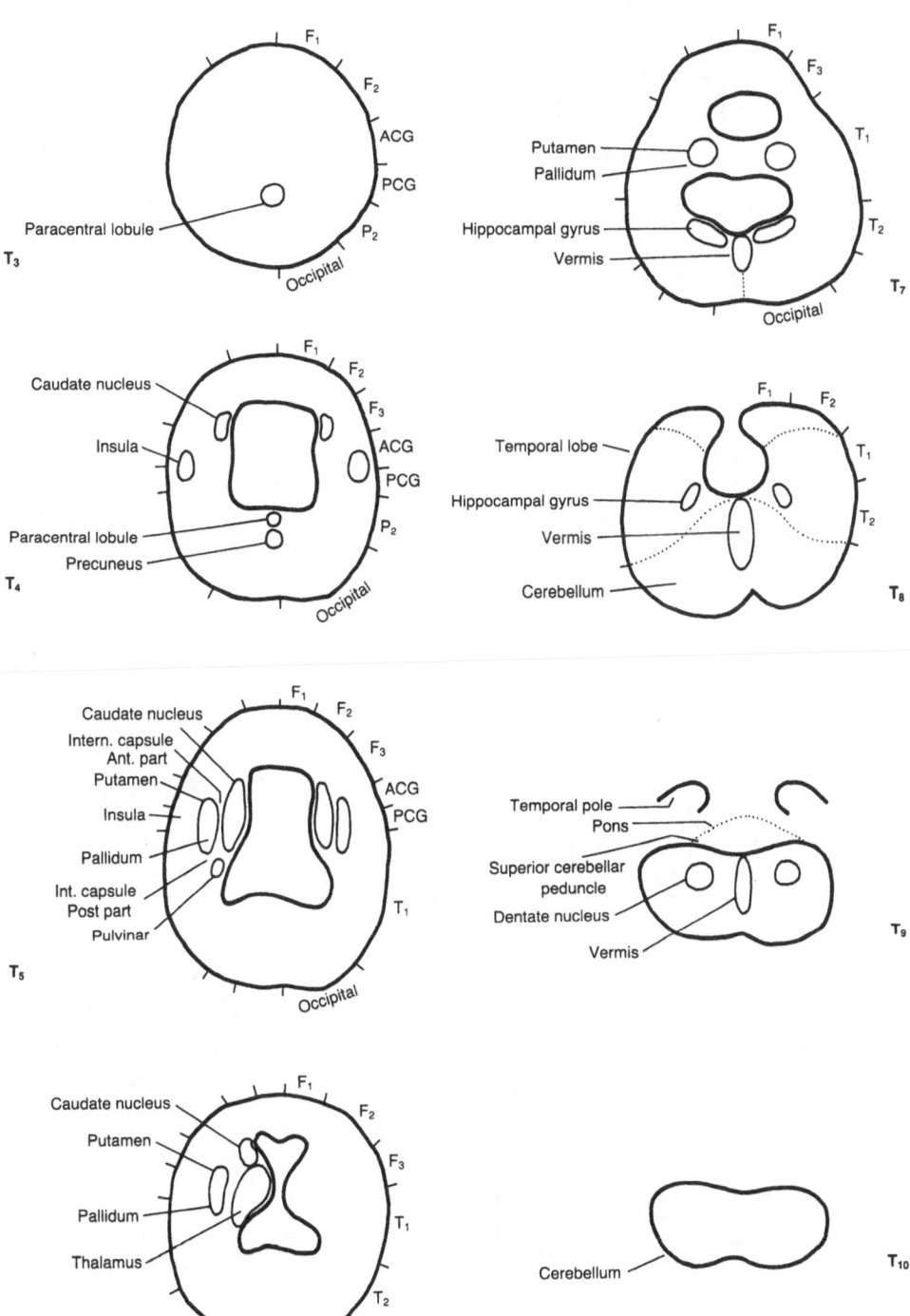

Fig. 1. Diagram of transverse sections

Frontal sections

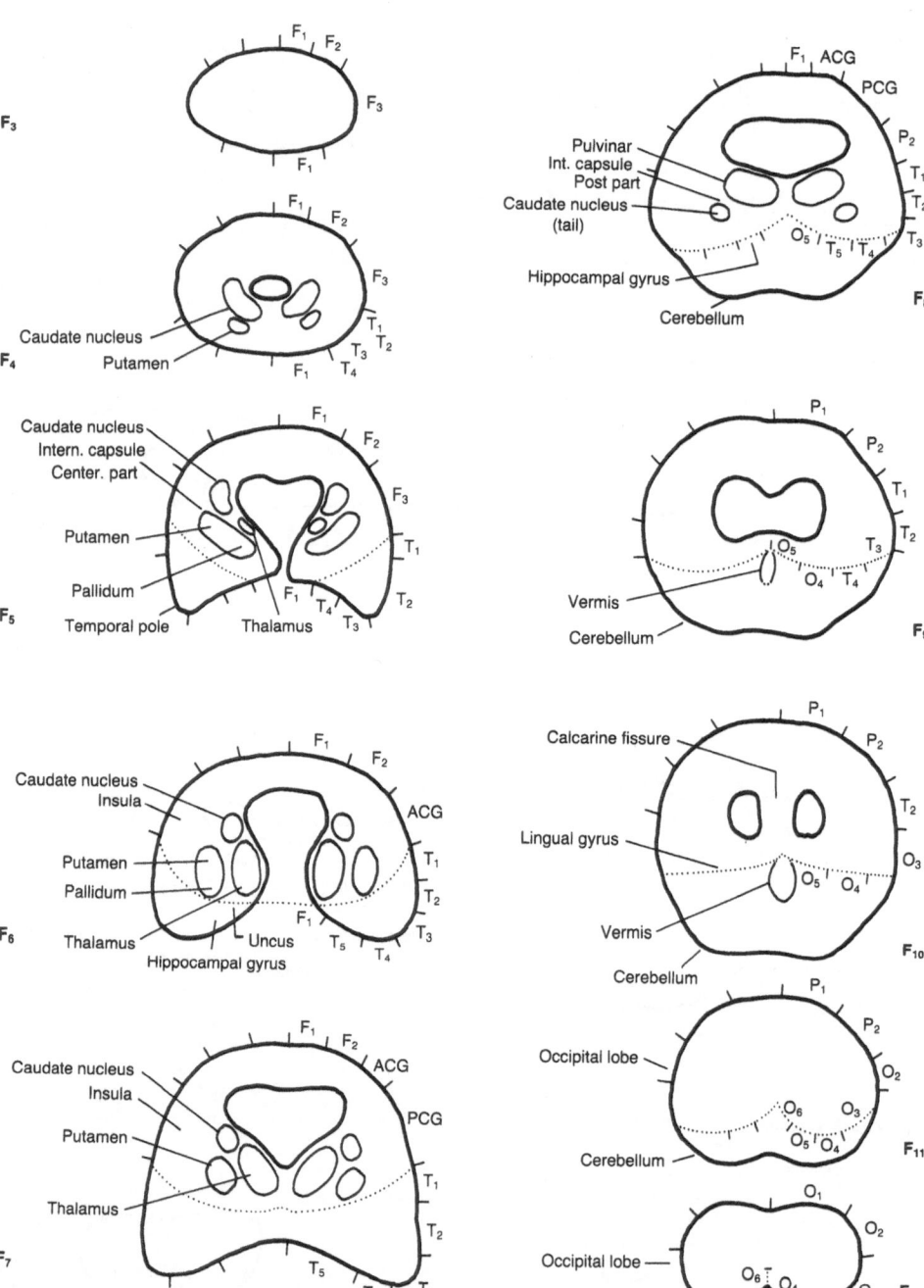

Fig. 2. Diagram of frontal sections

Sagittal sections

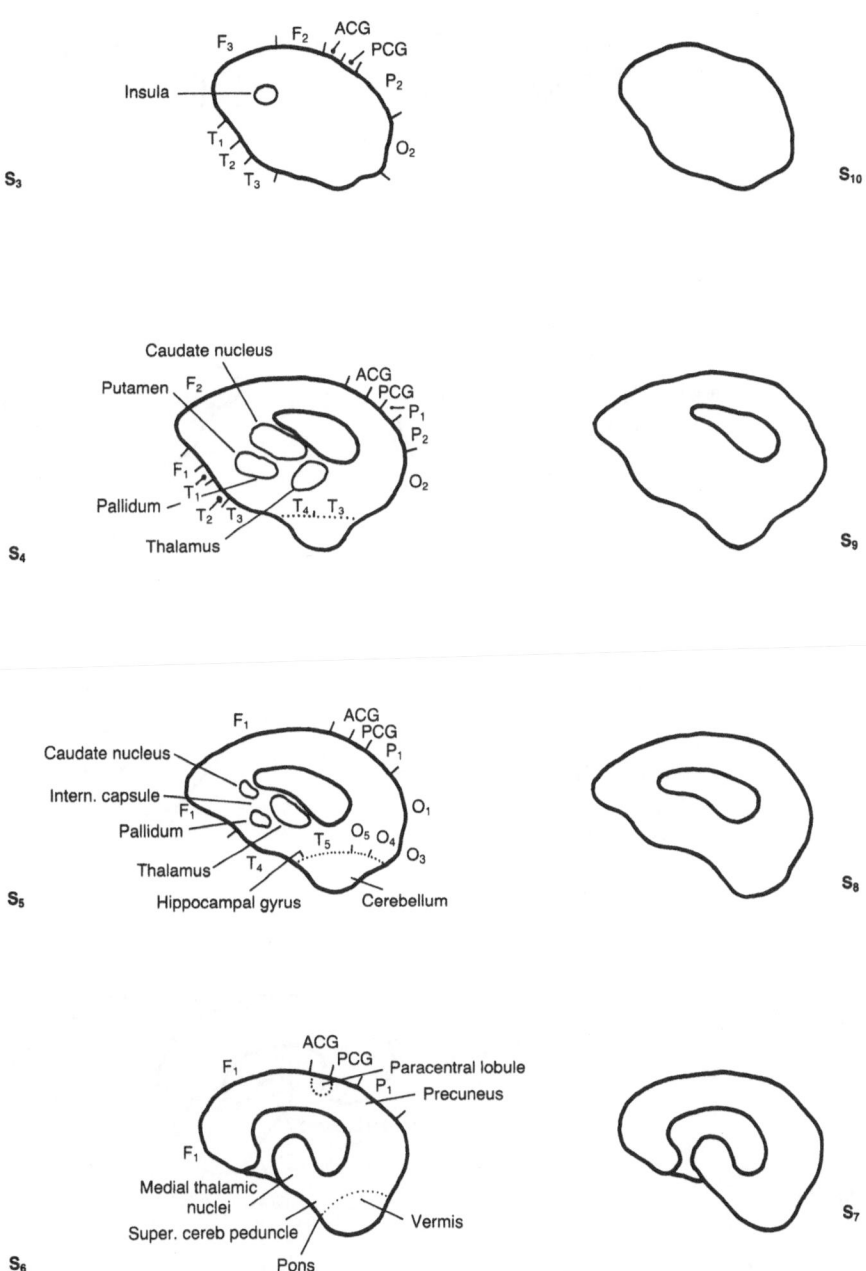

Fig. 3a, b. Diagram of sagittal sections. Diagram (**a**) should be used when the slice number is even, diagram (**b**) when it is odd (see next page)

Sagittal sections

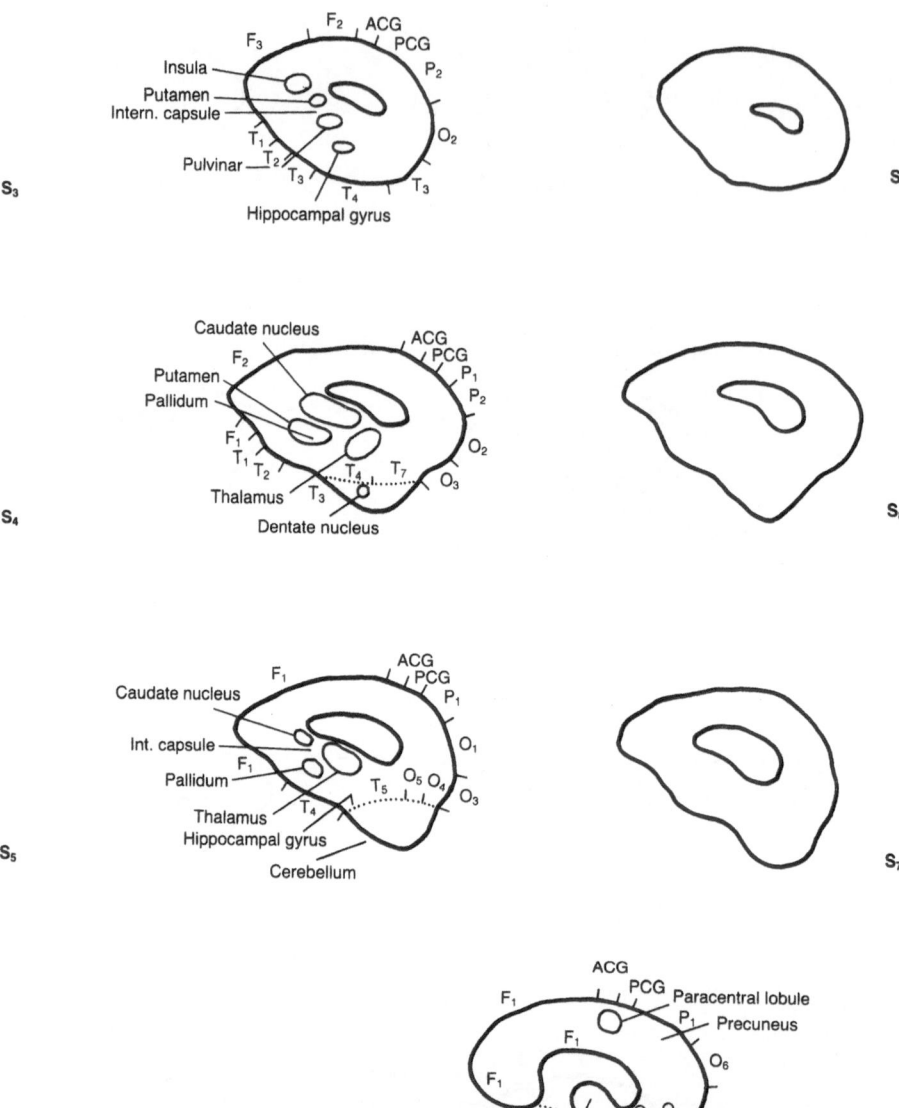

Fig. 3b.

cerebrovascular patients are encouraging although not confirmed. In addition potential applications to be tested will be discussed.

Focal Epilepsy

Results obtained by practically all groups working with IAMP or HIPDM indicate that lesional and epileptogenic areas are hypoactive during interictal periods on SPECT image, as they are with positron emission tomography (PET) when fluorodeoxy-glucose is used (Kuhl et al. 1981; Kuhl et al. 1982). Askienazy recently reported in a group of 32 patients with focal epilepsy (those with brain tumor were excluded) that lesional and epileptogenic areas were de-

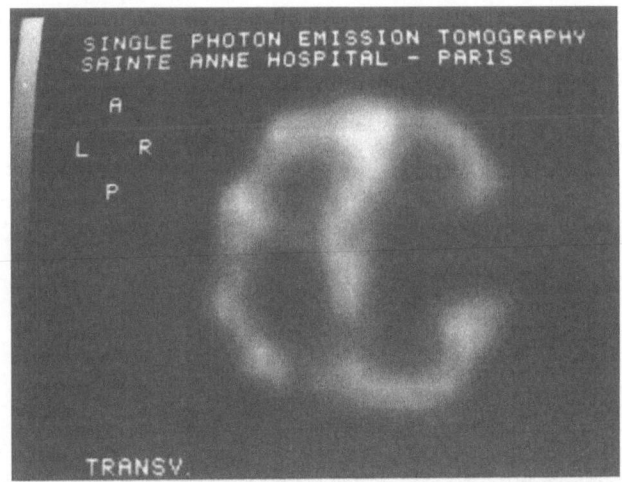

Fig. 4. 24-year-old patient with frontotemporal epilepsy. CT scan showed right frontotemporal porencephalia. On IAMP SPECT a large right frontotemporal hypoactive area is visible

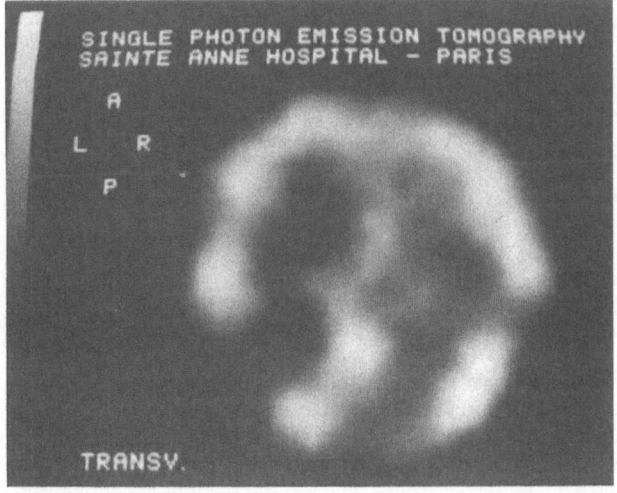

Fig. 5. 25-year-old woman with left parietotemporal epilepsy. On CT scan the occipital horn of the left ventricle was slightly enlarged. A large left parieto-temporo-occipital hypoactive area is visible on IAMP SPECT

tected in approximately 70% of the cases with CT and in all cases with IAMP SPECT. Localization of these regions (Figs. 4 and 5) correlates well with more accurate neuroradiologic procedures or stereotactic techniques such as stereotactic electroencephalography. At present, in partial epilepsy IAMP has proved to be an irreplaceable atraumatic method whose sensitivity and specificity must be studied.

The mechanism of the observed hypofixation is still not clarified but it may not always be due to an rCBF modification.

Alzheimer Syndrome

It was recently reported that with IAMP, multi-infarct dementia could be differentiated from Alzheimer syndrome (Cohen et al. 1983). In multi-infarct dementia, cortical defects are discrete, multiple, and asymmetric; in Alzheimer syndrome they are extensive and symmetric. In our view IAMP SPECT can potentially diagnose early cases of Alzheimer syndrome, a disease for which there is at present no diagnostic test.

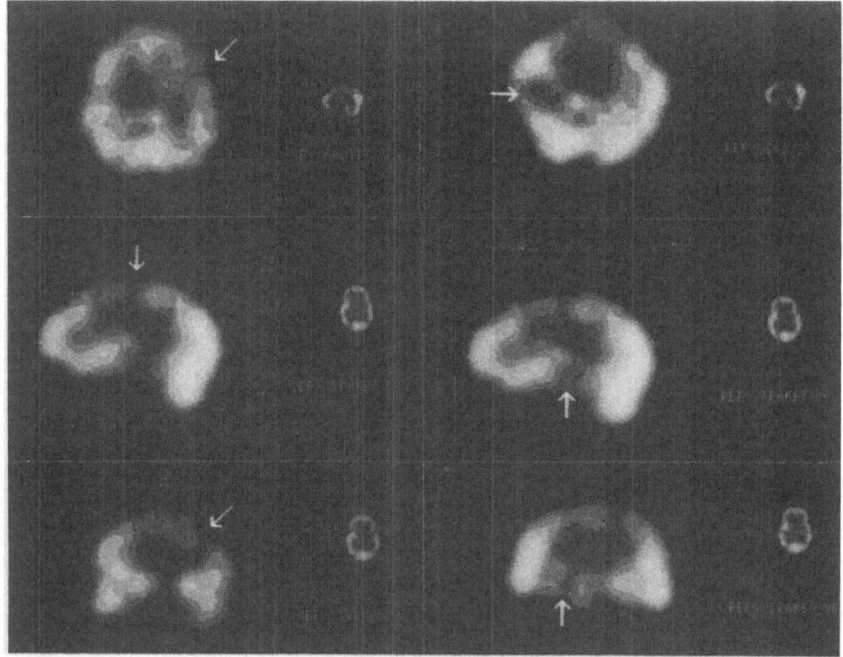

Fig. 6. 50-year-old patient with asymptomatic bilateral carotid occlusion. CT scan showed moderate cortical atrophy without parenchymal lesions. On IAMP SPECT sections, four areas are hypoactive, of which two are shown here; one is the left prerolandic (*left images*), the other is located in the deep right temporal region (*right images*). The latter appears to extend to T4 and T5. Transverse, frontal, and sagittal sections are shown from top to bottom respectively

Cerebrovascular Patients

Infarcts are clearly visible on SPECT images as on CT scans, but they are generally larger. Other hypoactive areas not visible on CT scan can also be observed on SPECT images. The significance of these areas is still not completely understood. However, it can be proposed – at least as a working hypothesis – that they correspond to hypofunctional areas. In this respect knowledge of their size, number, and volume should be useful in establishing a functional evaluation of the cerebral parenchyma. We studied eight patients with bilateral carotid artery occlusion who were clinically asymptomatic or paucisymptomatic. In all of these patients, hypoactive areas were found, without corresponding modification of the CT scan (Figs. 6 and 7). Can the rCBF be decreased in these areas? Can hypoactivity be improved by extracranial-intracranial bypass shunting? These questions are still unanswered. In transient ischemic attack (TIA) several hypoactive areas are also found, and IAMP brain SPECT should be useful in determining the functional state of whole brain (Fig. 8). In some infarcts,

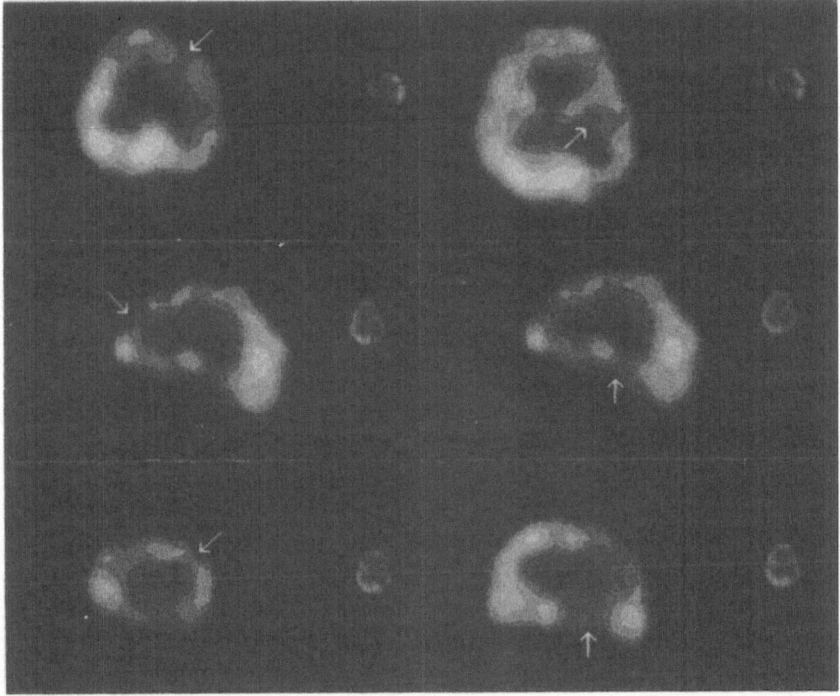

Fig. 7. 72-year-old patient with asymptomatic bilateral carotid occlusion. In fact transitory visual symptoms were noticed on the left eye 3 years earlier. CT scan showed a scar from a posterior sylvian ischemic accident on the left side. On IAMP SPECT a large hypoactive area corresponds to the CT image (*images on the left of the figure*). Other hypoactive areas are visible which cannot be seen on CT scan, and especially a left temporal cortical area extended to subcortical territories, including the left thalamus and probably the hippocampal gyrus (*images on the right of the figure*)

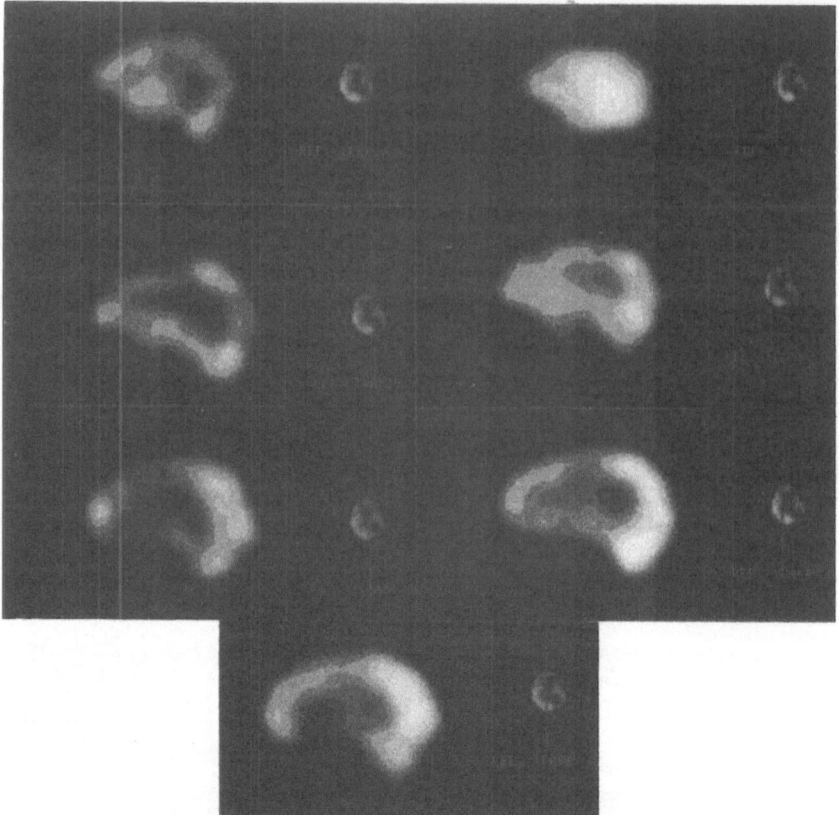

Fig. 8. 71-year-old woman with right TIA for 20 years (hemiparesia, focal epilepsy, aphasia, and hyperthermia). The last attack occurred 30 days before examination. On CT scan, two scars of old infarcts (left temporo-occipital and right cerebellar) are visible. On IAMP SPECT most of the left cortex is hypoactive, as are the left subcortical areas and the left cerebellum, extending to the anterosuperior part of the right cerebellum. This is easily seen when comparing sagittal images of the left brain (*left side of the figure*) and the symmetric saggital images of the right brain (*right side of the figure*)

IAMP brain SPECT may be necessary to appreciate the functional state of the brain with so-called ischemic penumbra for therapeutic purposes. It should also be useful in patients with impairment of the cognitive function whether or not it is associated with ischemic hemiplegia, in establishing the prognosis of stroke rehabilitation, and, above all, in long-term amnesic syndrome of hippocampal origin.

Brain Tumor

Brain tumor can be detected easily by competitive methods and brain SPECT is certainly not the most convenient method. However, in some tumors obtaining

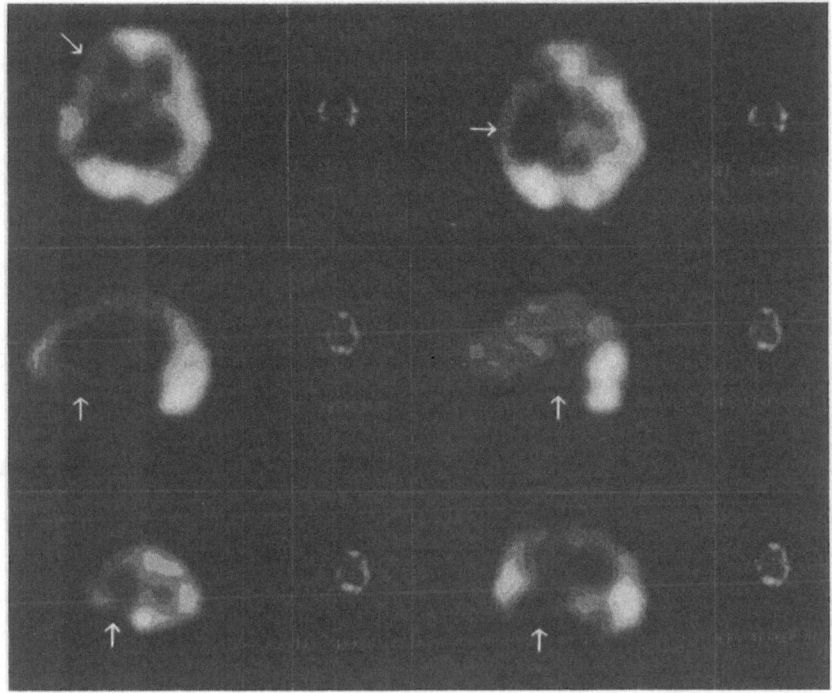

Fig. 9. 47-year-old woman with a right frontotemporal tumor visible on CT scan with a thalamic extension and displacement of the right ventricle. A hypoactive frontotemporal area corresponds to the CT image, but is larger, extending to the lower part of the parietal lobe, the right thalamus, and the hippocampal gyrus, and seems to displace the right cerebellum and the left thalamus

an image of the whole brain in a complete set of transverse, frontal, and sagittal sections may be very useful in determining the extent of a tumoral lesion in deep structures such as thalamic nuclei or basal ganglia. This is the case, for example, with tumors the evolution of which is very slow, such as low-grade astrocytomas, and which are not visible on CT scan for a long time (Fig. 9).

In conclusion, the usefulness of IAMP or HIPDM brain SPECT have already been shown for focal epilepsy and Alzheimer syndrome. Several other applications are now being carefully studied, including use in cerebrovascular patients, which could become the most important application of the method in the near future. Such encouraging results should stimulate research into other brain parenchyma indicators labeled with a gamma-emitter; we may hope that IAMP and HIPDM are the first in a series of BBB penetrating indicators. However, the future of brain SPECT will certainly depend on the discovery of new BBB indicators labeled with less expensive radioisotopes which can be more easily used.

References

Budinger TF (1981) Revival of clinical medicine brain imaging. J. Nucl. Med. 22:1094

Cohen MB, Metter EJ, Graham LS, Wasterlain C, Spolter L, Lake RR, Rose G, Yamada L and Chang CC (1983) Differential diagnosis of dementia with "pure" I^{123} iodoamphetamine and a clinical camera. J. Nucl. Med. 24:P.106

Devous MD, Lewis SE, Kulkarni PV and Bonte FJ (1983) A comparison of regional cerebral blood flow measured with I-123-labelled diamine of amphetamine and radioactive tracer microspheres. J. Nucl. Med. 24:P.6

Drayer B, Albright R, Jaszczac R, King H, Coleman E, Friedman A and Greer K (1983) Quantitative rCBF in man: The SPECT-HIPDM method. J. of Cerebral Blood Flow and Metabolism. Proceed. Congress Paris, Raven Press, N.Y. P.156

Hill TC, Holman BL, Lovett R, O'Leary D, Front D, Magistretti P, Zimmerman RE, Moore S, Clouse ME, Wu JL, Lin TH and Baldwin RM (1982) Initial experience with SPECT (single photon computerized tomography) of the brain using N-isopropyl-I-123-p-iodoamphetamine: concise communication. J. Nucl. Med. 23:191

Holman BL, Lee RGL, Hill TC, Lovett RD and Lister-james J (1983) A comparison of two cerebral blood flow tracers, N-isopropyl I^{123}-p-iodoamphetamine and I^{123}HIPDM. Jucl. Med. 24:P.6

Kuhl DE, Wu JL, Lin TH, Selin C and Phelps M (1981) Mapping local cerebral blood flow by means of emission computed tomography of N-isopropyl-p-(^{123}I)-iodoamphetamine (IMP). J. Nucl. Med. 22:P.16

Kuhl DE, Barrio JR, Huang S, Selin C, Ackerman RF, Lear JL, Wu JL, Lin TH and Phelps ME (1982) Quantifying local cerebral blood flow by N-isopropyl-p-(^{123}I)-iodoamphetamine (IMP) tomography. J. Nucl. Med. 23:196

Lafrance ND, Wagner HN Jr, Whitehouse P, Corley E and Duelfer T (1981) Decreased accumulation of isopropyl-iodoamphetamine (I-123) in brain tumors. J. Nucl. Med. 22:1081

Lassen NB, Henriksen L, Holm S, Barry DI, Vorstrup S, Rapin J, Le Poncin-Lafitte M, Moretti JL, Askienazy S and Raynaud C (1983) Cerebral blood-flow tomography: xenon-133 compared with isopropyl-amphetamine-iodine-123: concise communication. J. Nucl. Med. 24:17

Moretti JL, Askienazy S, Raynaud C, Mathieu E, Sanabria E, Cianci G, Bardy A and Le Poncin-Lafitte M (1982) Brain single photon emission tomography with isopropyl-amphetamine I-123: preliminary results. Proceedings of third World Congress of nuclear medicine and biology, Ed. C. Raynaud, Pergamon Press, Paris Vol I:135

Moretti JL, Askienazy S, Cesaro P, Chauvel P, Sanabria E, Raynaud C, Bardy A and Rapin J (1983) I^{123}N-isopropyl amphetamine (I-123 AMP) SPECT in epilepsy and cerebral ischemia. J. Nucl. Med. 24:P.108

Rapin J and Le Poncin-Laffite M. Unpublished observations

Salamon G (1971) Atlas of the arteries of the human brain. Sandoz editions

Soussaline F (1982) The single photon tomograph. In: Kuhl DE (ed) Principles of radionuclide emission imaging, Pergamon Press, Paris, p. 153−185

SPECT of the Brain Using Radiolabeled Amphetamines and a Rotating Gamma Camera

H. J. BIERSACK[1], A. HARTMANN, W. FRÖSCHER, K. REICHMANN, S. N. RESKE, R. LEDDA, and C. WINKLER

The noninvasive measurement of regional cerebral blood flow and metabolism in man was originally limited to departments where positron emission tomography, including an on-site cyclotron or special xenon devices, was available. In contrast, single photon emission computed tomography (SPECT) can be performed with gamma-emitting radiopharmaceuticals which are routinely in use and may therefore gain wider importance in clinical practice.

Radiolabeled amphetamines, such as I-123 N-isopropyl amphetamine [11], are lipophilic compounds which are exracted by the brain proportional to the blood flow [7]. Images representing the regional blood flow can therefore be obtained using commercially available radionuclides and standard SPECT devices such as a rotating gamma camera computer system. The purpose of this paper is to give an overview of our own clinical results in epilepsy, cerebrovascular disease, migraine, and tumors using SPECT and radiolabeled amphetamine as a tracer.

Methods and Material

SPECT of the brain was performed 60 min after injection of 6.5 mCi ^{123}I-labeled N-isopropyl-iodoamphetamine – provided by Amersham Buchler – using a rotating gamma camera (Gammatome T9000/CGR) equipped with a high resolution low energy collimator. During one 360° rotation, 64 frames with 4k matrix were acquired within 20 min (2). Transversal, sagittal, and coronal slices were reconstructed within short processing time using an array processor. ^{123}I was produced according to the reaction:

$$^{124}_{52}\text{Te} \, (p, 2n) \, ^{123}_{53}\text{I}$$

^{124}I contamination was less than 2% as the investigations were carried out within 12–13 h after cyclotron production of the radionuclide.

A total of 28 patients were investigated: 14 patients suffered from partial epilepsy, 10 from cerebrovascular diseases, 2 from brain tumors, and 2 from migraine. In two patients with cerebrovascular disease, SPECT was repeated within 1 week. One patient with prolonged reversible ischemic neurologic deficit (PRIND) underwent three studies within 2 weeks.

---

1 Institut für Nuklearmedizin der Universität Bonn, D-5300 Bonn-Venusberg

Cerebral Blood Flow and
Metabolism Measurement.
Eds. Hartmann/Hoyer
© Springer-Verlag Berlin Heidelberg 1985

Results

Epilepsy

The results of our investigations in epileptic patients ($n = 14$) are summarized in Tables 1 and 2. In three patients, CT and SPECT were negative; in eight, CT and SPECT revealed concordant lesions. One patient with cerebral atrophy had a CT-ascertained lesion but a negative SPECT study whereas in two cases, positive SPECT results were obtained despite negative CT. Of the nine cases wherein the extent of lesions determined with both techniques was compared, three showed the SPECT lesion to be larger than that of the CT study, one revealed a CT lesion larger than the SPECT defect, four presented with similar CT and SPECT results, and one patient with cerebral atrophy had a negative amphetamine SPECT.

Cerebrovascular Disease

Our results in this group of patients ($n = 10$) are summarized in Tables 3 and 4. In one patient both CT and SPECT were negative, while in seven cases both techniques showed defects. In two patients, negative CT lesions were found with SPECT. No patient was found to have an abnormal CT and normal SPECT. Table 4 shows the comparative results with respect to the extent of lesions. In two patients, the SPECT lesions were larger than those shown by the CT; in five cases, concordant results were obtained. There was only one patient with a long history of cerebrovascular disease and extra/intracranial bypass who showed a crossed cerebellar diaschisis. This phenomenon could not be observed in other patients even under weekly follow-up.

Table 1. Correlation of CT and SPECT results in epilepsy

CT + SPECT negative	CT + SPECT positive
3	8
CT positive/ SPECT negative	CT negative/ SPECT positive
1	2

Table 3. Correlation of CT and SPECT results in cerebrovascular disease

CT + SPECT negative	CT + SPECT positive
1	7
CT positive/ SPECT negative	CT negative/ SPECT positive
0	2

Table 2. Extent of lesions ($n = 9$) as estimated by SPECT and CT (epilepsy)

SPECT > CT	$n = 3$
CT > SPECT	$n = 1$
CT = SPECT	$n = 5$

Table 4. Extent of lesions ($n = 7$) as estimated by SPECT and CT (cerebrovascular disease)

SPECT > CT	$n = 2$
CT > SPECT	$n = 0$
CT = SPECT	$n = 5$

Migraine

Two patients with migraine were included in our study. In one case regional hyperperfusion consistent with the neurologic findings was established. In another patient with right-sided hemisyndrome and clonic convulsions, regional left-sided hypoperfusion was observed despite negative CT.

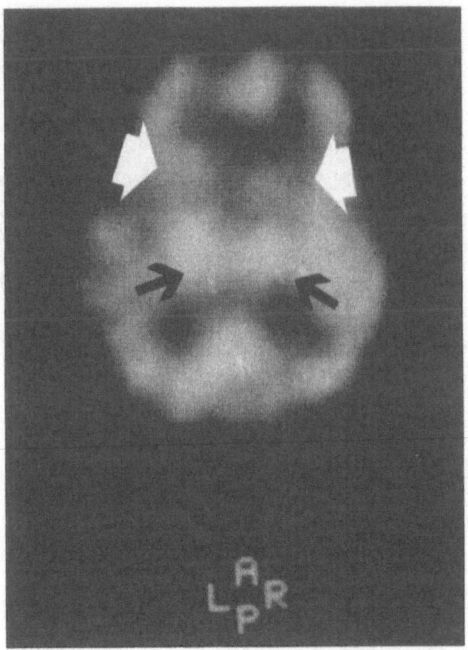

Fig. 1. Normal brain SPECT, the *arrows* indicating the thalamic region and the caudate nucleus

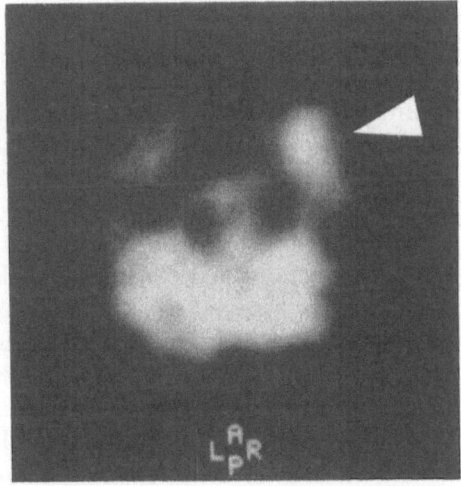

Fig. 2. Brain SPECT in epilepsy, the *arrow* indicating the focus in the right temporal lobe

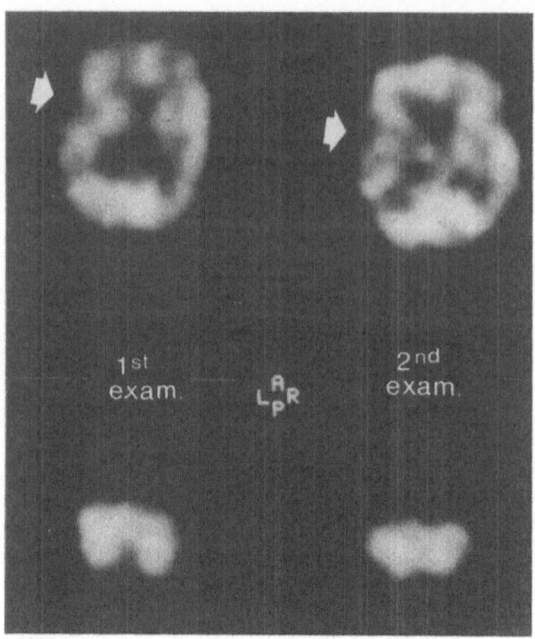

Fig. 3. Brain SPECT in PRIND
(left-sided) during follow-up over
8 days; there was constantly reduced perfusion of the left hemisphere but no crossed cerebellar
diaschisis at any time

Tumors

In our two cases with brain tumors (metastasis, glioblastoma), the respective
lesions presented with diminished IMP accumulation.

Discussion

Epilepsy

PET scans obtained *interictally* were able to detect dysfunctional brain zones
considered most likely to be responsible for seizures in patients with epilepsy.
These areas usually appeared unaltered on CT scan [1]. In a study of Alvi et al
[1] 12 of 15 patients with focal or unilateral EEG abnormalities, broad regions
of cortical hypometabolism, and hypoperfusion were demonstrated on PET
scans. Our results using SPECT and I-123 IMP as a tracer are comparable with
the above-mentioned PET findings. However, in our group of patients there
were no cases with negative CT and hypoperfusion. In the *ictal* state, PET investigations revealed increased perfusion and metabolism in cortical epileptic
foci. The same finding was also achieved using I-123 IMP [3, 9]. Two of our patients had locally increased amphetamine uptake consistent with the EEG findings despite negative CT. During the SPECT study both patients presented with
EEG spike and wave complexes but had no clinically evident seizure. A similar
case was described by Holman et al. [6]. Thus, increased perfusion can occur at

a time when there is focal electrical discharge even in the absence of clinically recognizable seizures.

In approximately 20% of patients with partial epilepsy uncontrolled by medication, additional information is usually required if surgery (temporal lobectomy) is being contemplated [1, 4]. Depth electrodes, although valuable in these patients, may be not successful in localizing the primary epileptic focus in each case [4]. Moreover, radiographic techniques usually fail to reveal foci. The data from PET scans may help to distinguish the exact localization of a focus, especially when a secondary onset is propagated at a site distant from the recording electrode [1].

However, PET scans, though highly informative, can only be obtained in centers with a cyclotron or reactor on hand. The clinical use of amphetamine SPECT may therefore gain widespread application in the management of epilepsy as this investigation can be performed in every nuclear medicine department where SPECT facilities are available. It seems obvious that with a combination of a surface EEG and ictal or interictal amphetamine SPECT, the need for depth electrode studies [1] may be reduced in patients in whom surgery is planned.

Cerebrovascular Disease

In patients with cerebral infarction the abnormality on SPECT with IMP is immediately detected, unlike on TCT wherein the defect appears only 3 or 4 days after onset of the disease. In a study reported by Hill et al. [5], 21 of 22 patients with acute cerebral infarction had perfusion defects on IMP study. Eight of these patients presented with normal TCT findings at the time of hospital admission. In our series of patients with vascular disease, we observed two cases (one with PRIND) with positive SPECT consistent with the neurologic findings but negative CT. In addition, two patients presented with larger SPECT lesions compared with that on TCT. However, Holman et al. [6] found that ⅔ of their patients revealed SPECT lesions considerably larger than those of TCT. In our two patients with transient ischemic attacks without focal neurologic symptoms, both the IMP study and the TCT scans were normal, which is in accordance with Holman et al. [6].

From these results it can be concluded that IMP SPECT studies are useful in assessing the extent of tissue involvement in stroke patients. In patients with acute infarction (first 3−4 days), functional iodoamphetamine studies show lack of perfusion to areas that subsequently develop TCT changes consistent with infarction. The perfusion abnormalities may also be larger than the TCT changes. In patients with PRIND, IMP-SPECT offers the opportunity to visualize diminished regional perfusion in the absence of morphological lesions.

Migraine, Tumors

To date there are no reports dealing with migraine and amphetamine SPECT which might confirm our inconsistent observations of regional hyper- or hypoperfused areas. In respect to brain tumors, Lafrance et al. [8] reported focal lesions in grade III and IV astrocytoma, whereas Ell et al. [3] observed focal increased IMP uptake in one case of astrocytoma III, a finding also described by Moretti et al. [10] in a patient suffering from astrocytoma grade II. Our two patients with glioblastoma and metastases presented with accumulation defects. However, further investigations will have to prove the usefulness of IMP-SPECT in differentiating the various brain tumors.

References

1. Alavi, A., M. Reivich, S. C. Jones, J. H. Greenberg, A. P. Wolf (1982) Functional imaging of the brain with positron emission tomography, in: Nucl. Med. Ann., ed. by L. M. Freeman and H. Weissmann, Raven Press 1982, p. 319
2. Biersack, H. J., W. Fröscher, H. Klünenberg, S. N. Reske, A. Rasche, K. Reichmann, C. Winkler (1983) SPECT des Hirns mit ^{123}J-Isopropylamphetamin bei Epilepsie. Nuc Compact 14:62
3. Ell, P. J., I. Cullum, M. Donaghy, D. Lui, P. H. Jarritt, M. J. G. Harrison (1983) Cerebral blood flow studies with ^{123}iodine-labelled amines. Lancet 1 348
4. Engel, J., R. Rausch, J. Leib, D. E. Kuhl, P. M. Crandall (1980) Reevaluation of criteria for localizing the epileptic focus in patients considered for surgical therapy of epilepsy. Epilepsia 21:184
5. Hill, T. C., B. L. Holman, R. Lovett, D. H. O'Leary, D. Front, P. Magistretti, R. E. Zimmerman, S. Moore, M. E. Clouse, J. L. Wu, T. H. Lin, R. M. Baldwin (1982) Initial experience with SPECT (single-photon computerized tomography) of the brain using N-isopropyl I-123 p-iodoamphetamine: Concise communication. J. Nucl. Med. 23:191
6. Holman, B. L., T. C. Hill, P. L. Magistretti (1982) Brain imaging with emission computed tomography and radiolabeled amines. Invest. Radiol. 17:206
7. Kuhl, D. E., J. R. Barrio, S. C. Huang (1982) Tomographic mapping of local cerebral blood flow using N-isopropyl-p(^{123}I)-iodoamphetamine (IMP), in: Proc. III. World Congr. Nucl. Med. & Biol., ed. by C. Raynaud, Pergamon Press, p. 1731
8. Lafrance, N. D., H. N. Wagner, P. Whitehouse, E. Corley, T. Duelfer (1981) Decreased accumulation of isopropyl-iodoamphetamine(I-123) in brain tumors. J. Nucl. Med. 22:1081
9. Magistretti, P., R. Uren, D. Shomer, H. Blume, B. L. Holman, T. Hill (1982) Emission tomographic scans of cerebral blood flow using (^{123}I)iodoamphetamine in epilepsy, in: Proc. III. World Congr. Nucl. Med. & Biol., ed. by C. Raynaud, Pergamon Press, p. 139
10. Moretti, J. L., S. Askienazy, C. Raynaud, E. Mathieu, E. Sanabria, G. Cianci, A. Bardy, M. Leponcin-Lafitte: Brain single photon emission tomography with isopropyl-ampethamine I-123 (1982) Preliminary results, in: Proc. III. World Congr. Nucl. Med. & Biol., ed. by C. Raynaud, Pergamon Press, p. 135
11. Winchell, H. S., R. M. Baldwin, T. H. Lin (1980) Development of I-123-labeled amines for brain studies: Localization of I-123 iodophenyl-alkyl amines in rat brain. J. Nucl. Med. 21:940

Evaluation of [125I]HIPDM as a Potential Tracer for Quantitative Measurement of Regional Cerebral Blood Flow

G. Lucignani[1], F. Fazio[2], A. Nehlig[1], R. Blasberg[3], C. S. Patlak[4], L. Anderson[5], H. F. Kung[6], C. Fieschi[7], and L. Sokoloff[1]

External measurement of radioisotope concentration in brain tissue by positron emission tomography (PET) makes it possible to measure physiologic processes such as regional cerebral blood flow (rCBF) and cerebral metabolism in human subjects. The ability to achieve the same result with single photon emission computed tomography (SPECT), a technique which is much simpler and less costly than PET, is presently limited by the lack of suitable gamma-emitting labeled compounds and methods appropriate to the specific requirements of SPECT. While the development of methods for the measurement of rCBF by means of SPECT is presently the object of studies by several groups, this paper deals with the development of a specific tracer molecule and a kinetic model that might allow its use for the measurement of rCBF in man.

The indicator fractionation method has been proposed for quantitative measurement of rCBF with SPECT. It is based on a two-compartment model which assumes complete first pass indicator extraction and no significant backflux from tissue to blood during the experimental period. A popular application of the indicator fractionation model is the use of microspheres to measure regional cerebral blood flow in animals. Validity of these assumptions allows the measurement of tissue concentration of radioisotope over the long period of time required by SPECT carried out with a rotating gamma camera. The operational equation for measurement of rCBF with such a method is the following:

$$rCBF = \frac{Ci\,(T)}{\int_0^T Ca\,(t)\,dt} \tag{1}$$

---

1 Laboratory of Cerebral Metabolism, National Institute of Mental Health, U.S. Public Health Service, Department of Health and Human Services, Bethesda, Maryland 20205, USA
2 Institute S. Raffaele, University of Milano, I-Milano
3 Department of Nuclear Medicine, Clinical Center, National Institutes of Health, U.S. Public Health Service, Department of Health and Human Services, Bethesda, Maryland 20205, USA
4 Theoretical Statistics and Mathematics Branch, Division of Biometry and Epidemiology, National Institute of Mental Health, U.S. Public Health Service, Department of Health and Human Services, Bethesda, Maryland 20205, USA
5 Laboratory of Medicinal Chemistry and Pharmacology, DTP, DCT, National Cancer Institute, NIH. U.S Public Health Service, Department of Health and Human Services, Bethesda, Maryland 20205, USA.
6 SUNY Department of Nuclear Medicine at Buffalo, Buffalo, NY, USA
7 Department of Neurological Sciences, University of Rome, I-Rome

Cerebral Blood Flow and
Metabolism Measurement.
Eds. Hartmann/Hoyer
© Springer-Verlag Berlin Heidelberg 1985

where Ca is the changing tracer concentration in the arterial blood and Ci(T) is the tracer concentration in the tissue at any time, T, following its i.v. administration.

Various efforts have been made in the synthesis and study of radio-pharmaceuticals with the properties of "chemical microspheres." Winchell et al. (1980) developed several gamma-labeled amines, and among these N-iso-propyl [^{123}I]p-iodoamphetamine has been extensively studied in animal models as well as in man (Kuhl et al. 1982). The mechanism of trapping of this compound in the cerebral tissue, although not established, has been attributed to receptor binding, high blood-brain partition coefficient, and pH gradient across the blood-brain barrier (BBB) (Winchell et al. 1980).

Kung and Blau (1980) have developed a series of compounds, neutral and lipid soluble in blood, which can freely diffuse into the tissue but allegedly become charged at the lower pH in tissue and are, therefore, trapped in the tissue. One of the molecules synthesized by Kung and Blau, N,N,N'-trimethyl-N'-(2-hydroxy-3-methyl-5-[^{123}I]-iodobenzyl)-1,3-propanediamine (HIPDM) has been extensively studied, and its regional distribution in the brain tissue has been reported to correlate with the regional blood flow in rats (Kung et al. 1983).

The purpose of this study was: (1) to measure the capillary first pass cerebral extraction of this compound; (2) to determine the extent of the backflux of the tracer from tissue to blood during the experimental period; and (3) to measure the rate of HIPDM metabolism in vivo. The clarification of these properties was considered to be essential before HIPDM could be used to measure rCBF by application of the indicator fractionation model and Equation 1.

Male adult Sprague-Dawley rats were used for all the experiments. The first pass extraction fraction of [^{125}I]HIPDM was determined experimentally by the indicator diffusion method. The [^{125}I]HIPDM was found to be incompletely extracted by the brain tissue on one pass, the extraction fraction ranging between 75% and 85%. The results were the same whether the tracer was dissolved in saline, plasma, or blood.

The values of rCBF estimated with [^{125}I]HIPDM were compared with concurrent measurements obtained with [^{14}C]iodoantipyrine (IAP) in 12 animals. Studies with [^{125}I]HIPDM were performed over different experimental times up to 60 min after i.v. injection. [^{14}C]IAP rCBF measurements were always performed over the last minute of the experimental period in the standard manner (Sakurada et al. 1978). Our results confirmed the correlation between rCBF, as measured with [^{14}C]IAP, and [^{125}I]HIPDM distribution in different regions of the brain.

We then tried to apply the indicator fractionation model, assuming complete trapping of the tracer after 80% extraction and negligible metabolism, for quantitative assessment of rCBF with [^{125}I]HIPDM. The values of rCBF estimated from the [^{125}I]HIPDM data and Equation 1 were lower than those measured with [^{14}C]IAP, and the underestimation increased with longer experimental periods. One possible explanation for the time-dependent underestimation of rCBF calculated by Equation 1 from the [^{125}I]HIPDM uptake data was the metabolism of the molecule with distribution of the labeled metabolites in blood and brain. Tissue and arterial blood samples were, therefore, extracted with ethylacetate and analyzed by high performance liquid chromatography (HPLC) for unaltered [^{125}I]HIPDM and labeled breakdown products.

The [^{125}I]HIPDM concentration drops rapidly in the arterial blood after an i.v. pulse and represents only 30% of the total blood radioactivity 60 min after the i.v. injection. Radioactive HIPDM metabolites in the brain tissue were minimal; more than 92% of the radioactivity in brain tissue 60 min after i.v. administration was in unaltered [^{125}I]HIPDM.

Corrections for partial first pass extraction and for the presence of metabolites in the arterial blood were applied to the experimental data. Despite these corrections the values of rCBF calculated from [^{125}I]HIPDM uptake by Equation 1 continued to result in a time-dependent underestimation of rCBF. A possible explanation for this time-dependent error could be the presence of substantial backflux of tracer from the brain tissue to blood.

Previous observations have shown a constant level of radioactivity in brain tissue over a long time following i.v. administration of a pulse of iodine-labeled HIPDM and a low but constant level of radioactivity in the arterial blood at the same time (Fazio et al. 1983; Drayer et al. 1983). In our experiments the tracer's concentration in the tissue reaches a plateau within a few minutes after the i.v. pulse and remains fairly constant for up to 60 min. The concentration of [^{125}I]HIPDM in the arterial blood falls to very low but still measurable levels. Inasmuch as the integral of the concentration of the tracer in the arterial blood is slowly but continuously increasing with increasing experimental time, an increase in the tissue concentration of the tracer should be expected if HIPDM is trapped in brain tissue. An increase in brain activity over the $10-60$ min time period was not, however, observed in our experiments. These observations suggest that a steady state bidirectional flux of [^{125}I]HIPDM between blood and brain tissue is achieved, rather than a complete irreversible trapping of this compound in the brain.

The experimental data obtained with HIPDM were then analyzed on the basis of a distribution model different from the indicator fractionation model. In this model bidirectional exchange between tissue and blood compartment is assumed instead of unidirectional flux as required by the indicator fractionation model.

The rate constant for [^{125}I]HIPDM transport across the BBB was calculated by a nonlinear best fitting least squares routine on the basis of the following equation:

$$Ci(T) = k_1 e^{-k_2 T} \int_0^T Ca(t) e^{k_2 t} dt \tag{2}$$

where k_1 and k_2 are influx and efflux constants, respectively. The overall average values for the whole brain of k_1 and k_2, respectively, are 0.915 ml/g/min and 0.021 min^{-1}. K_1 is several-fold bigger than k_2, but k_2 is clearly not negligible. The present study reveals serious limitations to the application of the indicator fractionation method for the measurement of rCBF with [^{125}I]HIPDM. The effect of a low efflux is negligible in a short experimental period, but it becomes progressively more important with increasing duration of the experimental period.

A kinetic model and an operational equation can be derived for measurements of regional CBF with this molecule only by taking into account at least the three following variables: (a) incomplete first pass extraction; (b) HIPDM metabolism and labeled metabolites in the arterial blood; (c) efflux of tracer from the brain tissue.

These preliminary observations suggest that efforts toward the synthesis of new molecules for CBF measurements with SPECT are worthwhile, provided that the validation of their use moves beyond the goal imaging techniques to the goal of quantitative measurements.

References

Drayer B, Albright R, Jaszczak R, Kung H, Coleman E, Friedman A, Greer K (1983) Quantitative rCBF in man: the SPECT-HIPDM method. J Cereb Blood Flow Metab 3 (Suppl. 1):S156–S157

Fazio F, Lenzi GL, Gerundini P, Fieschi C, Collice M, Pozzilli C, Gilardi MP (1983) Regional cerebral perfusion with SPECT and I-123-trimethyl-propanediamine (HIPDM). J Cereb Blood Flow Metab 3 (Suppl. 1):S158–S159

Kuhl DE, Barrio JR, Huang S-C, Selin C, Ackermann RF, Lear JL, Wu JL, Lin TH, Phelps ME (1982) Quantifying local cerebral blood flow by N-isopropyl-p-[^{123}I]iodoamphetamine (IMP) tomography. J Nucl Med 23:196–203

Kung HF, Blau M (1980) Regional intracellular pH shift: a proposed new mechanism for radiopharmaceutical uptake in brain and other tissues. J Nucl Med 21:147–152

Kung HF, Tramposch KM, Blau M (1983) A new brain perfusion imaging agent: [I-123]HIPDM: N,N,N'-trimethyl-N'-[2-hydroxy-3-methyl-5-iodobenzyl]-1,3-propanediamine. J Nucl Med 24:66–72

Sakurada O, Kennedy C, Jehle J, Brown JD, Carbin GL, Sokoloff L (1978) Measurement of local cerebral blood flow with iodo[^{14}C]antipyrine. Am J Physiol 234(1):H59–H66

Winchell HS, Baldwin RM, Lin TH (1980) Development of I-123-labeled amines for brain studies: localization of I-123 iodophenylalkyl amines in rat brain. J Nucl Med 21:940–946

Kinetics of Radioiodinated Isopropylamphetamine in Blood and Brain

S. N. Reske[1], A. Rasche[1], W. Eichelkraut[2], H. J. Biersack[1], and C. Winkler[1]

Radioiodinated alkylated sympathicomimetics have gained considerable interest for the noninvasive evaluation of regional cerebral blood flow (rCBF) [1]. One of the most promising substances is N-isopropyl-p-^{123}I-amphetamine (IMP) due to its nearly 100% cerebral extraction fraction, its slow wash-out from brain tissue, the resulting high cerebral tracer concentration, and the most favorable detection characteristics of ^{123}I for single photon tomography (SPECT) [2]. The use of IMP for quantitative assessment of rCBF, however, is hampered by its largely unknown metabolism. Tissue perfusion, lipophilicity, and affinity of IMP for high-capacity low-specificity adrenergic receptor sites have been suggested to be major determinants of regional IMP uptake in the brain [1]. In addition, reutilization of IMP and/or its metabolites, conceivably released from lung and liver, may influence local tracer tissue concentrations [3]. Thus complex tracer tissue concentrations in the brain can be expected. We have therefore investigated the kinetics of IMP brain uptake with a radioactivity residue detection technique in dogs after i.v. tracer application. In addition, the relation of microsphere-determined rCBF and late regional IMP uptake in the brain (i.e., 60 min p.i.) was examined.

Material and Methods

Iodine 123 or 131 labeled IMP, supplied by Amersham Buchler, Braunschweig, was used in this study. Specific activity was $0.3-1$ mCi/mg.

In total three pentobarbital-anesthetized (25 mg/kg) mongrel dogs (mean weight 17.3 kg) were investigated. In all dogs the abdominal aorta, the left femoral vein, and the left V. antebrachialis were cannulated. Subsequently the animals were thoracotomized during assisted ventilation with room air and a polyethylene tubing was inserted into the left atrium. ^{103}Ru-labeled microspheres (15 ± 1 mm diameter, NEN, Boston, Mass.) were injected into the left atrium and IMP i.v. Simultaneously arterial blood was withdrawn from the aorta by means of a calibrated syringe. rCBF was determined by the arterial reference technique (4). Brain tissue was sampled 60 min after tracer injection. rCBF was calculated from ^{131}I/^{123}I or ^{103}Ru radioactivity in brain tissue and in the refer-

1 Institut für Nuklearmedizin der Universität Bonn, D-5300 Bonn
2 Abteilung für experimentelle Chirurgie der Universität Bonn, D-5300 Bonn

Cerebral Blood Flow and
Metabolism Measurement.
Eds. Hartmann/Hoyer
© Springer-Verlag Berlin Heidelberg 1985

ence blood sample. Background and cross-contamination correction was performed in all tissue samples. Immediately after IMP injection, the time course of $^{131}I/^{123}I$ radioactivity was registered over the head by means of a gamma-camera equipped with a dedicated collimator. Sequential data ($\Delta t = 20$ s, T = 60 min) were stored in a minicomputer for later display and analysis. In addition, macroautoradiographys of coronal brain slices were performed by means of AGFA Scopix CR 3-ray films, exposed for 48–72 h. IMP blood clearance was determined by repeated blood sampling from the aorta and femoral vein.

Results and Discussion

After i.v. tracer application IMP is rapidly accumulated in brain tissue. The time course of radioactivity registered over the brain is characterized by a rapid radioactivity inflow (about 50% of maximal brain uptake observed 60 min p.i.). Subsequently a protracted multicomponent tracer accumulation was observed (Fig. 1). The maximal count-rate ratio registered over brain and neighboring neck tissue was about 2.5.

The tracer was rapidly cleared from arterial and venous blood until 15 to 20 min p.i. Thereafter a plateau phase of radioactivity concentration of about 5%–30% of maximal arterial concentration was established (Fig. 2). A positive

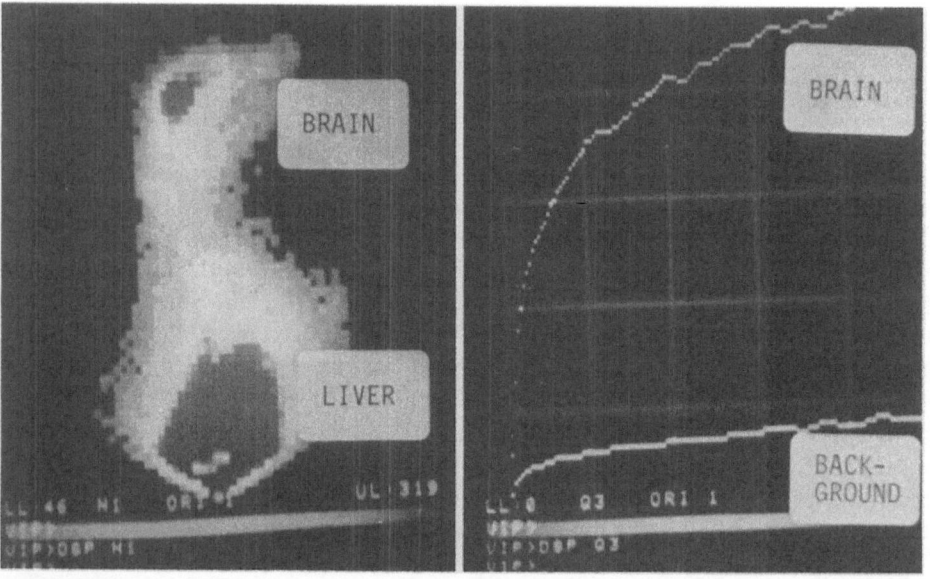

Fig. 1. Scinti-image of a dog, recorded in the left lateral projection 20–30 min after i.v. injection of 0.5 mCi IMP. Note significant tracer uptake in the brain (*left*). Time activity curves recorded over the brain and neck ("background") are shown on the *right*. Recording time was 60 min

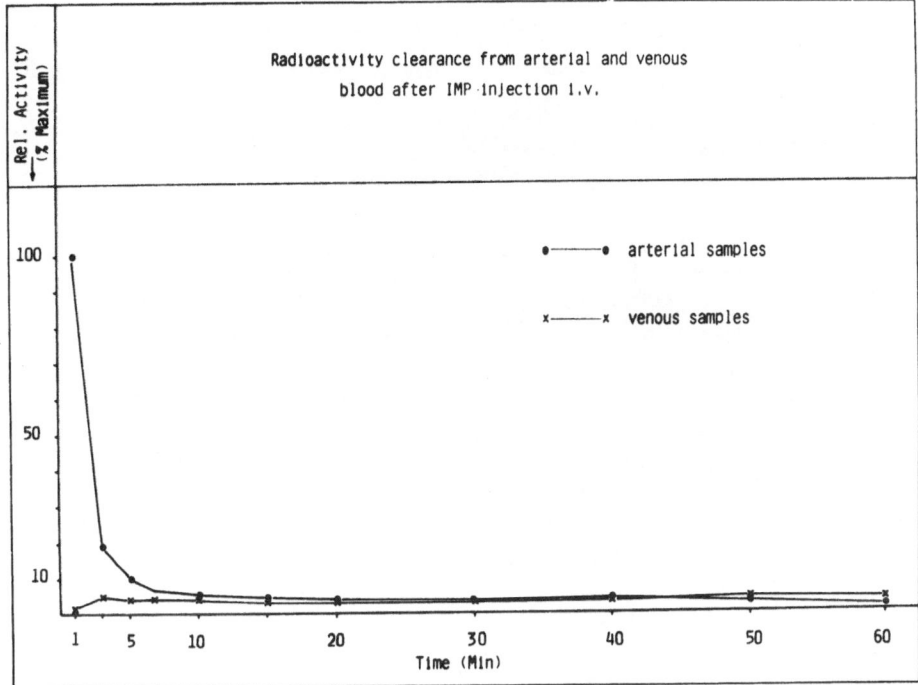

Fig. 2. Time course of radioactivity in arterial and venous blood after i.v. application of IMP

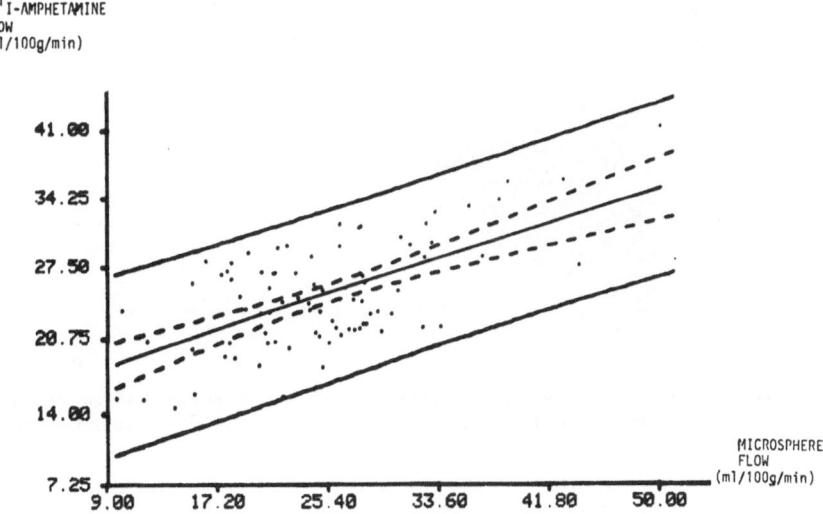

Fig. 3. Significant correlation of IMP brain uptake and rCBF at 60 min p.i. ($r = 0.61$; $n = 85$; $P < 0.05$)

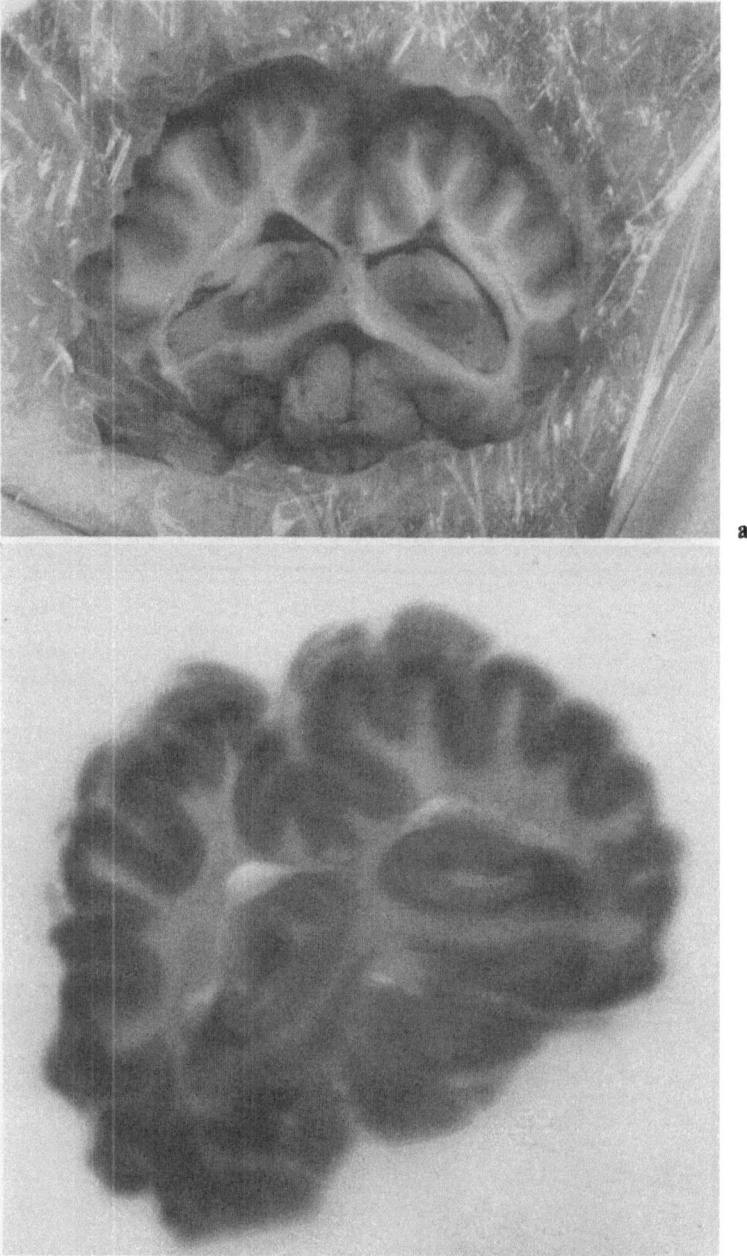

Fig. 4a, b. Coronal slice of a dogs brain (**a**) and corresponding macroautoradiography (**b**) Note fine delineation of brain cortex and ganglia due to increased IMP uptake in high flow areas

arteriovenous concentration gradient was observed immediately after injection until $10-15$ min p.i. Afterwards concentrations in both arterial and venous blood had equilibrated.

The observed kinetics of IMP in brain and blood shows marked deviations from that of an ideal rCBF tracer [5]. Especially progressive radioactivity increase in brain tissue, which was observed in patients with quantitative SPECT [2] as well as in the animal model initially evaluated [1], may imply reutilization of IMP and/or its lipophilic metabolits in the brain.

Regional cerebral blood flow determined by the microsphere technique correlates significantly with that assessed by means of IMP ($r = 0.61$, $P < 0.05$, $n = 85$) (Fig. 3). A closer correlation of microsphere- and IMP-determined rCBF has been reported by Kuhl and collaborators (2). Progressive redistribution of IMP after initial, probably flow-dependent, tracer uptake in the cerebrum may explain the looser correlation obtained at later tissue sampling times (60 min p.i. 5 min p.i.).

In macroautoradiographies of the brain the cerebral cortex, basal ganglia, white matter, and ventricular system were clearly delineated (Fig. 4). Intracerebral structures of gray and white matter were very clearly visualized. Therefore high quality tomographic imaging of normal and pathologic brain structures is feasible with a dedicated imaging device [2]. Quantitative measurements of rCBF by means of IMP brain tissue concentrations will, however, depend heavily on data sampling duration and time after tracer application. Thus a proper correction procedure for the changing IMP brain tissue concentration is required.

References

1 Winchell, H. S., Baldwin, R. M., Lin, T. H. (1980) Development of I-123 labeled amines for brain studies: Localization of I-123 iodophenylalkyl amines in rat brain. J. Nucl. Med. 21:940−946
2. Kuhl, P. E., Barrio, J. R., Huang, S. C., Selin, C., Ackerman, R. F., Lear, J. L., Wu, J. L., Liu, T. H., Phelps, M. E. (1982) Quantifying local cerebral blood flow by N-isopropyl-p-(I-123) iodoamphetamine (IMP) tomography. J. Nucl. Med. 23:196−203
3. Caldwell, J. (1980) Amphetamines and related stimulants: chemical, clinical, and sociological aspects. CRC-press, Boca Raton, Florida
4. Heyman, M. A., Payne, B. D., Hofman, J. I. E., Rudolph, A. M. (1977) Blood flow measurements with radiolabeled particles. Progress in Cardiovascular Diseases, Vol. XX, No. 1, 55−79
5. Shipley, R. A., Clark, R. E. (1977) Tracer methods for in vivo kinetics. Academic press, New York

New Approaches to the Study of Cerebral Ischemia in Man Using Single Photon Labeled Indicators

G. L. Lenzi[1], F. Fazio[3], C. Pozzilli[1], M. Negri[2], L. Bozzao[1], L. M. Fantozzi[1], P. Gerundini[3], F. Colombo[3], F. Cacace[4], M. Attina[4], P. Pozzilli[2], B. Guidetti[1], and C. Fieschi[1]

Our understanding of the pathophysiology of brain ischemia has greatly increased over the past decade. Recently the development of emission tomography has stimulated new investigative work in this area designed to evaluate in vivo important physiologic parameters such as cerebral blood flow and glucose and oxygen metabolism (Kuhl et al. 1980; Lenzi et al. 1982).

Our initial investigations have been devoted to some of these parameters in patients with cerebrovascular ischemia using a commercial rotating gamma camera (C.G.E. 400 T) connected to a dedicated computer. In particular we have utilized: (a) a new brain imaging agent, I-123-trimethyl-propanediamine (HIPDM), to assess the regional cerebral perfusion; (b) indium 111-labeled leukocytes to investigate the inflammatory process following brain ischemia; and (c) Tc 99 pyrophosphate-labeled red cells according to Callahan et al. (1982) to assess cerebral blood volume. In this report we discuss some methodological aspects of these techniques and their clinical application to the study of brain ischemia.

Regional Cerebral Blood Volume with Tc 99 Pyrophosphate-Labeled Red Cells

Assessment of rCBV was made using Tc 99 pyrophosphate-labeled red cells in eight patients. Such data, obtained with a simple and noninvasive technique, are particularly useful for the evaluation of critical perfusion situations (Gibbs et al. 1983) and of postaneurysmal spasms. In fact, instances of local CBV increase with normal TCT scan and neurologic examination were detected in some TIA patients, thus indirectly supporting the hypothesis of a critical perfusion.

Regional Cerebral Perfusion and I-123-HIPDM

Recently, organic compounds of the amphetamine-diamine group have been synthesized and recommended for use as tracers of cerebral perfusion. Among

---

1 Department of Neurological Science, University of Rome, I-Rome
2 II Medical Clinic, University of Rome, I-Rome
3 University of Milan, S. Raffaele Hospital, Division of Nuclear Medicine, I-Milano
4 Institute of Pharmaceutical Chemistry, University of Rome, I-Rome

Cerebral Blood Flow and
Metabolism Measurement.
Eds. Hartmann/Hoyer
© Springer-Verlag Berlin Heidelberg 1985

these compounds a new perfusion imaging agent of the brain, I-123-HIPDM, has been developed (Kung et al. 1983) and preliminarily used both in experimental and in clinical studies (Lucignani et al. 1983; Fazio et al. 1983).

After intravenous injection of I-123-HIPDM in man, brain activity rapidly increases and remains stable over a 240-min period. This steady state allows one to obtain high quality tomographic images of the brain using a rotating gamma camera (SPECT).

The present report concerns preliminary data from a study performed on ten patients with a clinical history of reversible ischemic attack (RIA) within the last 12 months. The selection of patients was based on the presence of an occlusion of the internal carotid or middle cerebral artery previously documented by angiography. Transmission computed tomography (TCT) brain scan was performed early before each I-123-HIPDM study.

TCT scans were normal in four cases and abnormal in six in respect of the presence of a limited hypodensity area reflecting irreversible tissue damage; by contrast HIPDM images of these patients always showed a focal or diffuse brain perfusion abnormality in the territory corresponding to the arterial occlusion. Even when TCT showed small hypodense regions, larger areas of perfusion deficit were detected with SPECT. These findings could be referred both to regional ischemia and to related acidosis that may affect regional HIPDM extraction by pH shift. An example of the discordance between TCT scan and HIPDM image is presented in Figs. 1 and 2.

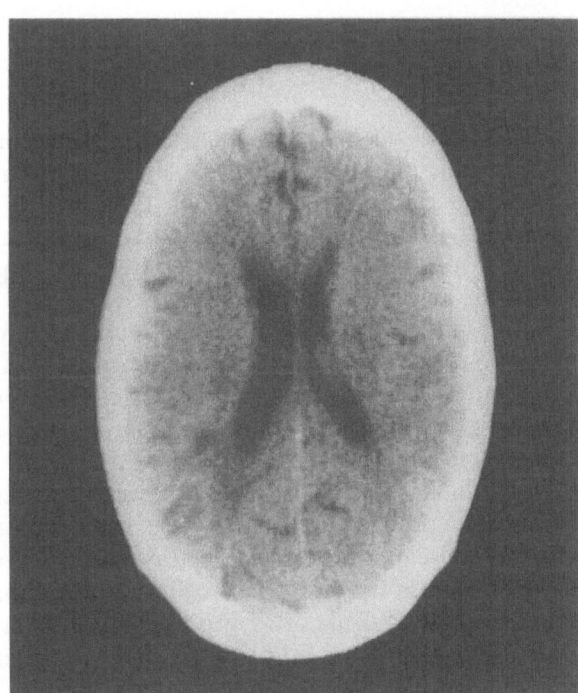

Fig. 1. Cerebral TCT scan of a TIA patient showing a small hypodense region in the left cerebral hemisphere

Fig. 2. I-123-HIPDM brain SPECT of the same patient as in Fig. 1, showing diffuse reduction of activity over the left cerebral hemisphere

The assessment of regional cerebral perfusion with HIPDM in patients with RIA before and after EC-IC bypass has also been performed by our group (Fazio et al. 1983), indicating the potential of this technique for evaluating the degree of cerebral reperfusion after bypass surgery.

Indium 111-Labeled Leukocytes in Cerebral Ischemia

Recent studies on the physiopathology of cerebral ischemia performed with positron emission tomography revealed the presence of an uncoupling between oxygen and glucose metabolism in the ischemic area. This increased glucose uptake may reflect an anaerobic macrophagic metabolism (Wise et al. 1983). The in vivo assessment of the inflammatory response to brain ischemia may clarify this physiopathologic aspect.

The in vitro labeling of peripheral leukocytes with ^{111}In and the study of cell circulation after their reinjection by means of gamma camera imaging is a technique widely used for detecting abscesses (Segal et al. 1976; Peters et al. 1980) and more recently for imaging the inflammatory response to acute myocardial infarction in man (Davies et al. 1981).

Using this technique we have examined six patients affected by acute cerebral ischemia during the first 2 weeks after the onset of neurologic symptoms. The procedure of labeling leukocytes was done according to Peters et al. (1983) using "mixed buffy coat white cells." Approximately 300 micro-Ci of In 111 tropolone-labeled leukocytes was injected into a cubital vein. Venous blood samples were collected immediately and after 2 and 20 h for the measurement of whole blood, leukocytes, and plasma activities.

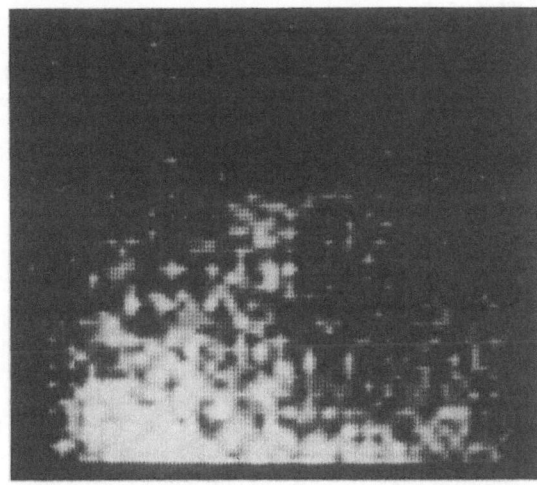

Fig. 3. Indium 111-labeled leucocyte activity in a patient with right cerebral infarct. The affected hemisphere shows a higher uptake than the contralateral hemisphere

At the same time intervals, anterior and lateral brain scans were recorded. Unfortunately, the activity achievable with In 111 tropolone was insufficient for emission tomographic studies. Despite this limitation, conventional scintigraphy showed in all cases a clear increase of the radioactivity in the ischemic area, indicating the active migration and tracking of labeled leukocytes in cerebral infarct.

Figure 3 shows an example of asymmetric brain uptake of labeled cells in a patient with right cerebral infarction studied 10 days after the onset of symptoms.

These preliminary findings suggest that this technique is able to demonstrate leukocyte infiltration in infarcted cerebral tissue (Pozzilli et al. 1985).

Conclusion

Single photon emission computerized tomography is a technique achievable by every nuclear medicine division, and its potential in neurology appears quite rewarding. It is relevant to underline that a more profound knowledge of the models and very careful consideration of the technology are necessary before filling the gap between qualitative and quantitative measurements, but evidence is being presented in this symposium (Lucignani et al. 1983) of the potential for quantitative evaluation of rCBF with particular radiopharmaceuticals.

References

Callahan R. J., Froelich J. W., McKusick K. A., Leppo J. and H. W. Strauss (1982) A modified method for the in vivo labeling of red blood cells with Tc-99m: concise comunication. J. Nucl. Med. 23:315–318

Davies R. A., Thakur M. L., Berger H. J., Wackers F. J., Gottschalk A., Zaret B. L. (1981) Imaging the inflammatory response to acute myocardial infarction in man using indium-111 labelled autologous leukocytes. Circulation 60:297

Fazio F., Lenzi G. L., Gerundini P., Fieschi C, Collice M., Pozzilli C., Gilardi M. P. (1983) Regional cerebral perfusion with SPECT and I-123 trimethyl-propanediamine (HIPDM). J.C.B.F. Metabol. S158–159

Fazio F., Gerundini P., Lenzi G. L., Collice M., Gilardi M. C., Taddei G., Piacentini M., Colombo F., Colombo R., Kung H. F. and M. Blau (1983) Evaluation of cerebrovascular disorders using the brain imaging agent (I-123) HIPDM and single photon emission computerized tomography (SPECT). J. Nucl. Med., 24, 5 P 5–6

Gibbs J., Wise R., Leenders K. and T. Jones (1983) The relationship of regional cerebral blood flow, blood volume and oxygen metabolism in patients with carotid occlusion: evaluation of perfusion reserve. J.C.B.F. Metabol. 1983, Vol. 3, Suppl. 1, S590–591

Kuhl D. E., Phelps M. E., Kowell A. P., Metter E. J., Selin C., Winter J. (1980) Effects of stroke on local cerebral metabolism and perfusion: Mapping by emission computed tomography of 18-FDG and 13-NH. Ann Neurol. 8:47–60

Kung H. F., Tramposch K. M. and Blau M. (1983) A new brain perfusion imaging agent: (I-123) HIPDM: N,N,N'-trimethyl-N'-(2-hydroxy-3-methyl-5-iodobenzyl)-1,3-propanediamine. J. Nucl. Med. 24:66–72.

Lenzi G. L., Frackowiak R. S. J. and Jones T. (1982) Cerebral oxygen metabolism and blood flow in human cerebral ischemic infarction. J.C.B.F. Metabol. 2:321–335

Lucignani G., Fazio F., Nehlig A., Blasberg R., Patlak C. S., Fieschi C., Kung H. F., Sokoloff L. (1985) Evaluation of 123-I-HIPDM as a potential tracer for the measurement of regional blood flow. In: Methods of cerebral blood flow and metabolism measurement in man. Heidelberg, pp. 301–304, Springer-Verlag Berlin, Heidelberg, New York, Tokyo

Peters A. M., Lavander J. P., MDermott J. (1980) Diagnosing cerebral abscess with indium-111 labelled leukocytes. Lancet 2:309–310

Peters A. M., Saverymuttu S. H., Reavy H. J., Danpure H. J., Osman S., Lavander J. P., (1983) Imaging of inflammation with indium-111 tropolonate labelled leukocytes. J. Nucl. Med. 24:39–44

Pozzilli C., Lenzi G. L., Signore A., Argentino C., Rasura M., Carolei A., Bozzao L. and P. Pozzilli. Imaging of leucocyte infiltration in cerebral ischemic infarct in man. Stroke 1985:162

Segal A. W., Arnot R. N., Thakur M. L., Lavander J. P. (1976) Indium-111 labelled leukocytes for localization of abscesses. Lancet 2:1056–1058

Wise R., Rhodes C., Gibbs J., Frackowiak R. and T. Jones (1983) The relationship between oxygen metabolism and glucose utilization in early cerebral infarcts. J.C.B.F. Metabol. 1983, Vol. 3, Suppl. 1, S 580–581

Measurement of Cerebral Blood Flow by Stable Xenon Contrast Computerized Tomography

J. S. MEYER

Introduction

Computerized reconstruction techniques have made remarkable progress, providing anatomic resolution in three dimensions by X-ray transmission and positron emission tomography [1−3]. If advantage is taken of these technical advances, localized measurements of cerebral blood flow (LCBF), local tissue: blood partition coefficients (Lλ), and local metabolism are possible with excellent resolution [4−7]. The present communication describes improved methods for measuring LCBF and Lλ, utilizing the CT scanner during 35% stable xenon (Xe^s) inhalation [6, 7], and their application in normal aged volunteers. Computerized emission tomography has also been used for measuring LCBF during ^{77}Kr or ^{133}Xe inhalation [4, 5]. The latter emission methods have poorer resolution [3] than Xe^s methods because of technical problems with absorption of isotope emissions, partial volume effects, and Compton scatter.

Plain CT scanning has sufficient resolution for differentiating gray and white matter, although gray and white matter differentiation becomes less well defined with advancing age and dementia. The reason for this loss of differentiation is a matter of contention − it has been attributed to reductions in gray matter flow or changes in the physical properties of brain tissue that may occur with advancing age and dementia [8, 9].

Apart from the excellent resolution afforded by CT scanners, they are widely available in the majority of medical centers and the method is cost-effective. Thus, development of satisfactory CT CBF methods avoids the high cost and practical limitations of positron emission tomography and makes widely accessible correlation of CBF values, which provide a functional and metabolic index to be correlated with regional anatomy or pathology since tight coupling of LCBF with regional neuronal activity and metabolism is well established.

The possibility of utilizing the CT scanner for measuring LCBF and Lλ during inhalation of Xe^s was first conceived by Kelcz et al. [10]. There are several possible approaches to calculation of LCBF from serial CT scans which record changes in Hounsfield Units (ΔH) which, in turn, are proportional to regional diffusion of inhaled XE^s gas when it is used as the indicator: LCBF may be calculated either by analysis of clearance during desaturation [11, 12] or by the in

Cerebrovascular Research Laboratories, VA Medical Center and Department of Neurology, Baylor College of Medicine, Houston, Texas 77211, USA

Cerebral Blood Flow and
Metabolism Measurement.
Eds. Hartmann/Hoyer
© Springer-Verlag Berlin Heidelberg 1985

vivo autoradiographic method during saturation or by the monoexponential analysis method during saturation or desaturation [7, 12].

Early experience with Xe^s CT methods in this laboratory has shown that clearance curves for brain tissue obtained during spontaneous breathing in human subjects in the desaturation phase are extremely rapid compared with slower saturation curves, so that multiple, rapid exposures are necessary to provide sufficient data points to plot their time course. In this respect, utilization of either autoradiographic or monoexponential analysis during saturation was found to be more satisfactory, since optimal scanning times could be determined and utilized [7, 12].

In order to measure $L\lambda$ values for small regions of interest in gray and white matter, multiple serial scans are necessary to plot the points at which equilibrium of Xe^s between blood and brain is achieved [10, 12]. Unless reliable computer programs are available to estimate accurately the saturation points at infinity, inhalation intervals of longer than 10 min become necessary in order to saturate regions of low perfusion, so that even when low concentrations of Xe^s (such as 35%) are inhaled, restlessness with movement of the head may result due to eventual subanesthetic effects of the gas [12]. This produces movement artifacts in CT images, with loss of accuracy of calculated LCBF and $L\lambda$ values [12, 13].

Suitable computer programs have been investigated by the use of double integration techniques [14] whereby the XE^s saturation curves for arterial blood and brain tissue were optimally matched against a time base computed as 6-s intervals. The computer program thereby adjusted for slower scanning times (1-min scans) of the EMI 1010 instrument used, versus faster changes of arterial concentrations.

Since the program automatically estimates saturation at infinity for both blood and tissue saturation curves, $L\lambda$ values are provided without prolonged inhalation of 35% Xe^s beyond 8 min, thereby minimizing subanesthetic effects and restlessness. After computing $L\lambda$ values in regions of interest, our program in the final step used these derived $L\lambda$ values to compute LCBF by the use of a single compartment analysis model [12].

In order to determine normative values for LCBF and $L\lambda$ and to define any effects of normal aging, 13 normal healthy volunteers aged between 20 and 80 years were selected for measurements by use of the 35% stable xenon inhalation method during CT scanning utilizing this new computer program [12].

ICBF and $L\lambda$ measurements were carried out in 13 normal right-handed healthy volunteers after explaining the procedure to them and obtaining informed consent. (The informed consent forms and protocol for these measurements were approved by the Institutional Review Boards of VA Medical Center and Baylor College of Medicine, Houston, Texas.) Since advanced cardiopulmonary disease might distort normal relationships between end-tidal xenon and arterial xenon inhalation curves, patients with these conditions are excluded from LCBF and $L\lambda$ measurements when the inhalation method is used. The mean age of the normal, healthy volunteers without risk factors and with normal mentation was 49.3 years, and there were ten males and three females. These healthy volunteers were free of risk factors for cerebral arteriosclerosis, including hypertension, diabetes mellitus, hyperlipidemia, heart disease, or evidence of arteriosclerosis elsewhere in the body. They were all normal by general medical and neurologic examination and mentation was estimated to be normal according to complete or abbreviated neuropsychological test batteries, which included the Wechsler Adult Intelligence Scale (Wais).

On the same day as the CT CBF measurements, ^{133}Xe inhalation rCBF measurements were recorded for purposes of comparison, using collimated sodium iodide crystal detectors which were mounted over the scalp by means of a lead-lined helmet [15, 16].

Statistical comparisons, made for comparative analysis of pooled data, were performed by means of the t-test.

All patients were fasted for at least 6 h prior to Xe[s] CT CBF measurements and were then placed in the EMI 1010 CT scanner. The particular EMI 1010 scanner used in these measurements has excellent reproducibility and signal-to-noise ratio [12], and is calibrated and adjusted at monthly intervals to maintain optimal performance. The machine was adjusted and calibrated using the standard EMI zero (water) phantom to derive zero Hounsfield units with standard deviations. (Housfield units were always obtained by doubling the values provided by the EMI scanner, since each of the units of density on the EMI machine equals 2 HU.) The mean and standard deviations obtained with this particular scanner when cursoring the entire Plastic phantom were circa 0 ± 2.9 HU.

Each subject's head was restrained with three velcor belts placed around the head, face, and chin to minimize movement artifact and a rubber mask was mounted over the mouth and nose and was sealed with grease to minimize leaks. The EEG and EKG were recorded from suitable electrodes placed outside the CT field. The mask was fitted with ports to permit continuous sampling of $PECO_2$, PEO_2, and PEXe[s][12]. The end-tidal Xe[s] was monitored with a Gow-Mac Thermoconductivity analyzer.

Figure 1 shows a typical calibration curve for stable xenon and oxygen mixtures. The changing oxygen concentrations were monitored with a Beckman oxygen analyzer, and the xenon concentrations with thermoconductivity analyzers. Zero for the thermoconductivity instruments war first calibrated by flushing into the anesthesia gas bag made of Tedlar plastic, which is impermeable to xenon. The oxygen was then gradually mixed with xenon by adding aliquots from a tank of 100% xenon gas. Consequent changes in oxygen concentration were measured with the Beckman OM-15 oxygen analyzer, which has a linear response between zero and 100% oxygen. The concentrations of xenon in oxygen mixture were estimated by subtraction, and the voltage outputs from the thermoconductivity analyzer were thereby calibrated as percentage xenon concentration. There was always excellent linearity between zero and 35% xenon in oxygen for all thermoconductivity analyzers tested. This was the range used.

Before beginning 35% Xe[s] inhalation in 65% oxygen, body nitrogen was displaced by inhalation of 100% oxygen for 10 min. This ensured accuracy of zero calibrations for the xenon detectors in vivo and improved the efficiency of xenon saturation of body tissues when inhaling the gas [12].

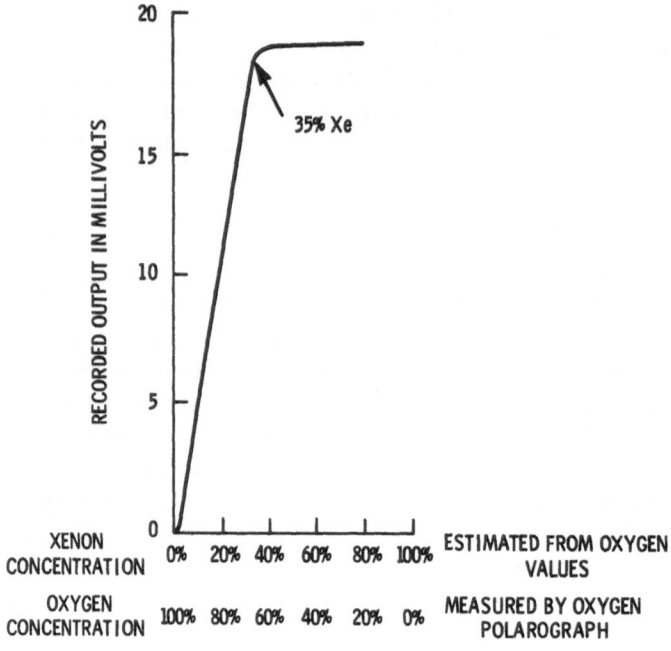

TYPICAL EXAMPLE OF CALIBRATION CURVE FOR STABLE XENON
CONCENTRATIONS MEASURED BY THERMOCONDUCTIVITY GAS ANALYZER

Fig. 1. A representative calibration curve is illustrated for stable xenon gas mixtures plotted against electrical output of one of several thermoconductivity analyzers used. The response is linear between concentrations of 0% and 35% xenon although at concentrations above 40% the sensitivity of this particular instrument decreased

Three control scans (zero enhancement scans) were made prior to inhalation of 35% Xe[s], which was administered through a semiclosed, partial rebreathing system. The semiclosed system ensured that the inhaled mixture was gradually replenished with 35% xenon as the body tissues were saturated, so that gradual build-up was assured, thereby providing optimal curve fitting and minimizing subanesthetic effects.

After starting 35% Xe[s] inhalation, scanning was begun immediately in three normal volunteers and was continued thereafter for the ensuing 8 min. Apart from these three cases, CT scanning was started in all subjects 2 min after the Xe[s] inhalation was begun, which takes advantage of optimum scanning times to be discussed later and reduces X-ray exposure. Usually four or five scans were performed during the 8-min xenon inhalation interval. After this, the delivery system was switched from 35% Xe[s] to 100% oxygen. The 35% Xe[s] mixture inhaled via the semiclosed system described induced a 30.0% ± 3.0% ($n = 21$) mixture measured in the end-tidal air after 8 − 10 min of inhalation, which confirms that subanesthetic mixtures were used, in terms of measured alveolar concentrations of the gas. [For illustrations of typical end-tidal xenon curves during CT scanning, together with discussion of the anesthetic effects of mixtures of xenon gas in oxygen above 35% (i.e., their pharmacologic effects), which may include a loss of consciousness, EEG slowing, and depression of cerebral blood flow, the reader is referred to references [7, 12, and 13].] The 5% difference between inhaled and exhaled xenon concentrations is due to dilution and loss by solution of the gas in all body tissues.

For measurement of LCBF and $L\lambda$ values in the present series of adult human subjects the EMI 1010 CT scanner was fitted with an 8-mm collimator. The instrument also has a 4-mm collimator for infants which may be used for Xe^s CBF measurements if greater resolution is desired [7]. The X-ray beam was adjusted to 120.2 kVp (mean) and 32.6 mamp (mean). Continuous records were obtained of $PEXe^s$ on a polygraph, which were converted to ΔHU by [1] scanning saturated cubital venous blood samples at the end of the saturation interval to obtain 0 and maximal ΔHU or [2] cursoring the sinus blood in the final CT scan, or [3] using of Kelcz's formula [7, 10, 12, 13]:

$$\Delta H = \frac{5.15 \times Xe^s \times C}{\mu\varrho/\mu\varrho^{Xe^s} \times 100}$$

where $\Theta Xe^s = 0.001 \times Hct\,(\%)$ and $C = \%$ mixture of Xe^s used (35%).

$\mu\varrho^\omega/\mu\varrho Xe^s$ values were empirically determined to be 4.64×10^{-2} at the 120 kVp setting. This made possible estimation of saturated ΔH values for arterial blood, at equilibrium, after Xe^s gas inhalation.

Single Compartment Computer Program Used for LCBF Measurements

Originally all LCBF values were derived by using the inert, freely diffusible, tracer modification of Kety's formula [17]: $C_i = \frac{mi\,Fi}{Vi}\,e\,Ca^k\,i^t\,dt$. This formula was originally applied for use with the autoradiographic method, after sacrifice, for measuring LCBF in animals [18]. Later this mathematical model was modified for in vivo CBF measurements in man by Obrist et al. and has been widely used for deriving the fast and slow components [15, 19].

The Xe^s CT CBF method has sufficient resolution (as small as 80 mm³) for cursoring gray or white matter regions of interest, such as the cortex, regional white matter, basal ganglia, thalami, and individual nuclei of the brainstem [12]. If the small volumes of tissue cursored are homogeneous then Kety's formula, cited above, can be used for calculating LCBF utilizing the single compartmental analysis model [12].

Although CT scanning makes possible measurement of concentrations of tracer in brain tissue with high resolution, the EMI 1010 scanner has a relatively slow scanning interval (1 min) which may be a disadvantage for single compartmental analysis of saturation curves of brain tissue if tissue saturations for xenon are changing rapidly during each 1-min interval. To correct for this disadvantage of slow scanning intervals, double integration of the end-tidal Xe^s curves (which is in equilibrium with arterial blood) is carried out by the computer program in order to correlate as exactly as possible the ΔHU changes for arterial blood with those ΔHU changes measured during each corresponding CT scan.

The concept for this computer program was originally developed in collaboration with Professor Walter Obrist of the University of Pennsylvania as a single compartment-double integration modification of the formula as first applied to ^{133}Xe inhalation [19, 20]. It has now been successfully modified, so that raw data points drawn at 6-s intervals from the PEXe^s curve on the polygraph can be keyed into the computer terminal, along with the ΔHU derived from each CT scan of the brain. The program then prints out Lλ and LCBF values for each region of interest.

The formula used for solving changes in Xe^s concentrations in both tissue and artery between the time intervals tn$_1$ and tn$_2$ may be expressed as follows:

$$C tn_1 \rightarrow tn_2 = P \int_{tn_1}^{tn_2} \int_0^t \alpha CA(u)\, e^{-k(t-u)}\, du\, dt, \text{ where: } C tn_1 \rightarrow tn_2 = \text{concentrations}$$

of tracer in brain tissue over the number of time intervals scanned, CA = the concentration of tracer in the arterial blood and α = the proportionality constant between thermoconductivity output (CA) and ΔHU changes occurring in arterial blood, tn$_1$ = the starting time for each scan, tn$_2$ = the finishing time for each scan, and P = λk.

In order to estimate P and k, the computer program requires information regarding (a) the changing Xe^s concentrations in the artery, estimated from end-tidal air, multiplied by the proportionality constant at every 6-s interval, and (b) the Xe^s concentration in the brain region of interest during each of the scanning times tn$_1$ to tn$_2$. The time base for the computer program is based on 6-s interval points.

The computer program was designed to start each calculation by using arbitrarily assigned k values for each scan time. The final k and P values were computed from each scanning interval by best fitting using the least squares approach to minimize: $C tn_1 \rightarrow tn_2 = P \int_{tn_1}^{tn_2} \int_0^t CA(u)\, e^{-k(t-u)}\, du\, dt^2$. The solution was designed to accomplish the final k and P values according to the variable metric algorithm by Fletcher [15, 21].

Local λ values were derived from the formula P/k = λ and LCBF was analyzed by the product of the computed λ value $\times$ k. Thus, computed L½ values are mathematically estimated at infinity.

Optimal Scanning Time

In three normal healthy volunteers CT scanning was started immediately after 35% stable Xe inhalation was begun and was continued for the ensuing 8–10 min. Five to six consecutive 1-min scans were thus derived during the build-up of Xe gas after its inhalation. It has been reported previously [12], using the method of in vivo autoradiographic analysis, that early scans obtained during the first 2 min of inhalation give unreliable values due to mismatching of brain and end-tidal build up, as both are changing too rapidly. This problem was now reevaluated by taking advantage of the newly developed computer

Table 1. Optimal CT scanning times for LCBF and Lλ determined for gray and white matter

		Regions examined (n)	Calculations made from scans 1 to 5 or 6	Calculations made from scans 2 to 5 or 6	Calculations made from scans 3 to 5 or 6
Gray matter	LCBF	35	103.2 ±26.7	82.9 ±12.5	78.1 ±17.0
	Lλ		0.84± 0.07	0.85± 0.07	0.86± 0.08
White matter	LCBF	16	27.6 ± 5.5	27.2 ± 5.3	24.9 ± 5.1
	Lλ		1.20± 0.13	1.22± 0.13	1.39± 0.16

Scans $\begin{cases} 1: & 0.0 \text{ to } 1.1 \\ 2: & 1.8 \text{ to } 2.9 \\ 3: & 3.5 \text{ to } 4.6 \\ 4: & 5.2 \text{ to } 6.3 \\ 5: & 6.9 \text{ to } 8.0 \\ 6: & 8.8 \text{ to } 9.9 \end{cases}$ Time intervals in minutes from beginning of Xe inhalation

program and its double integration methods, so that P and k values could be derived using data points sampled from multiple scans.

Table 1 illustrates the effects of different scanning times on LCBF and Lλ values estimated for both gray and white matter. Mean gray matter flow values calculated from scans 1 through 5 or 6 were 103.2 ± 26.7 ml/100 g brain/min but 82.9 ± 12.5 when calculated from scans 2 through 5 or 6 and 78.1 ± 17.0 when calculated from scans 3 through 5−6. Optimal scanning times were for scans 2 through 5 or 6, which, as stated above, gave mean values of 82.9 ± 12.5 ml/100 g brain/min and provided the lowest standard deviations. Local λ values for gray matter, however, were not significantly altered by differences in scanning times.

Overestimation effects on LCBF values when using early scans were even greater when LCBF was estimated for white matter because of its slower rate of perfusion. Lλ values for white matter were underestimated if early scans measured during the first 2 min after inhalation were used. Hence, it is recommended that serial scanning be deferred until the end of the second minute of Xe[s]

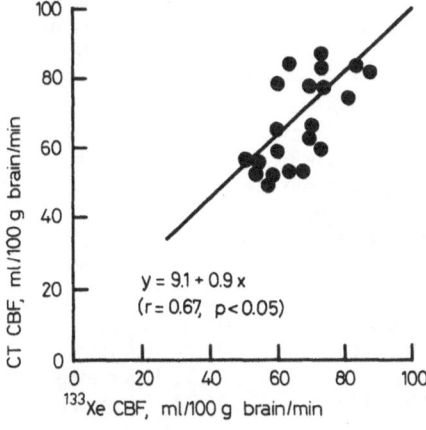

Fig. 2. Correlation between mean Fg values measured by the ^{133}Xe inhalation method and mean cortical gray matter LCBF values measured by the stable xenon CT CBF method in the same subjects ($n = 21$). There was also some correlation between mean white matter flow values, but ^{133}Xe values were consistently lower for reasons discussed in the text

inhalation and white matter vaules for Lλ and LCBF be measured between scans 3 through 5 or 6. This also reduces X-ray exposure.

When used in the manner described, the new double integration program improved the fitting of arterial Xe^s changes (estimated from $PEXe^s$ changes) with corresponding changes in Xe concentration for brain tissue for the entire series of CT scans, utilizing the least square method of curve fitting. When this recommended method was used (i.e., scans 2 or 3 through 5–6, as shown in Fig. 2), LCBF values obtained for gray matter were in good agreement with ^{133}Xe values measured ($n = 21$) on the same day ($r = 0.67$, $P < 0.05$) in this population of subjects, and with values reported in the literature using the ^{133}Xe inhalation method [16–18].

Results

Comparison of [133]Xe Inhalation rCBF Measurements with CT LCBF Measurements

Figure 2 compares data for cortical gray matter flow values measured by the CT CBF method with those measured as F_1 values utilizing the ^{133}Xe method in 13 normals ($n = 13$). There was excellent agreement. However, for the same groups mean white matter flow values measured by the ^{133}Xe inhalation method were consistently lower (18.1 ± 2.3 ml/100 g brain/min) than those measured by the CT CBF method (27.2 ± 5 ml/100 g brain/min).

Normative Values Calculated by Single Compartment Analysis Using the Double Integration Program

Table 2 displays normative values for the cerebral cortex, basal ganglia, thalamus, and white matter calculated by means of the CT CBF method utilizing the

Table 2. Normative values compiled from measurements in normal healthy and active volunteers without risk factors

	Regions examined (n)	LCBF (ml/100 g brain/min)	Lλ
Frontal cortex	51	75.5 ± 11.9	0.87 ± 0.10
Parietal cortex	14	71.2 ± 10.8	0.86 ± 0.10
Temporal cortex	44	76.2 ± 12.7	0.86 ± 0.07
Occipital cortex	44	77.9 ± 11.5	0.85 ± 0.08
Basal ganglia	21	73.9 ± 12.4	0.87 ± 0.09
Thalamus	37	70.7 ± 11.6	0.92 ± 0.07
White matter	69	27.1 ± 5.7	1.36 ± 0.11

Subjects (n)	13 (10 m., 3f)
Mean age	49.3 ± 19.4 years
MABP	95.3 ± 12.8 mmHg
Mean PECO₂	36.9 ± 2.8 mmHg

double integration program. Values were obtained from 13 normal, healthy volunteers aged 20 to 83 years with a mean age of 49.3 years. $L\lambda$ values for cortical and basal ganglia gray matter were $0.85-0.87$, for thalamus 0.02, and for white matter 1.36. Highest values for LCBF were seen in the occipital cortex (77.9 ml/100 g brain/min), with intermediate values of 72.6 ml/100 g brain/min for the temporal cortex and 75.5 ml/100 g brain/min for the frontal cortex and lowest values of 71,2 ml/100 g brain/min for the parietal cortex. Basal ganglia and thamalic flow values were 73.9 and 70.7 ml/100 g brain/min respectively, while white matter values were 27.1 ml/100 g brain/min.

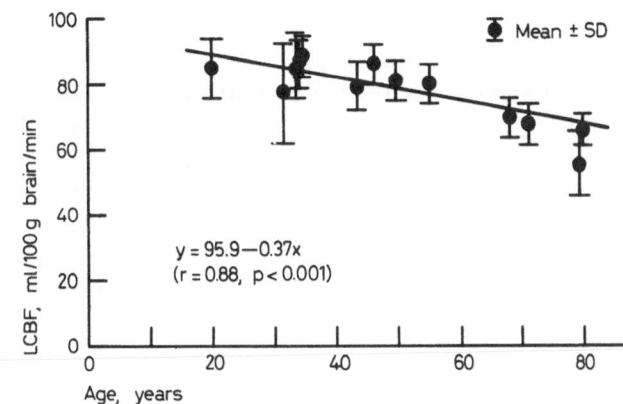

Fig. 3. The decline of cortical gray matter LCBF values with advancing age between 20 and 80 years among normal healthy volunteers

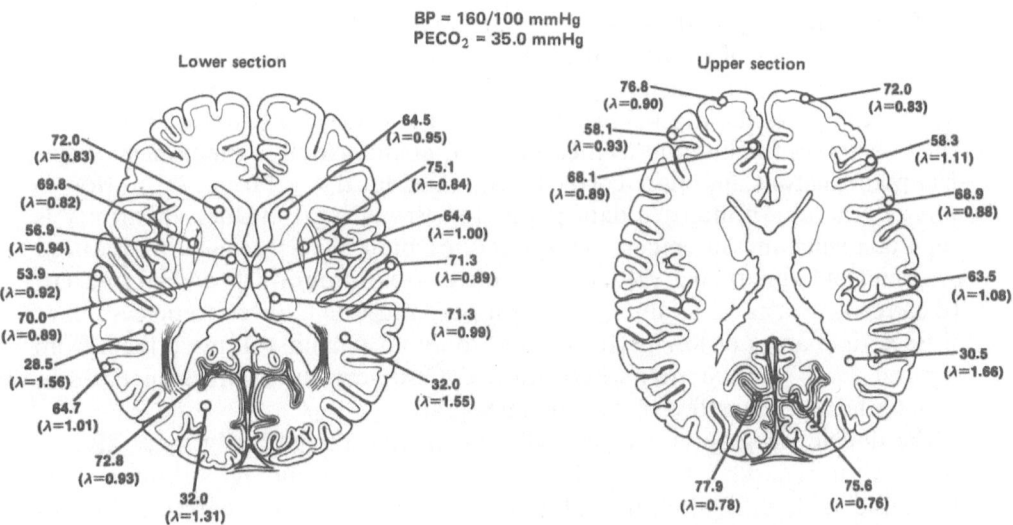

Fig. 4. Stable xenon CT CBF values measured in 72-year-old normal healthy male volunteer. Flow values are given as ml/100 g brain/min above $L\lambda$ values in parenthesis. All $L\lambda$ values are within normal limits

Effect of Advancing Age on LCBF and Lλ Values in Normal Volunteers

The influence of advancing age on resting LCBF values measured in normal volunteers between 20 and 80 years of age by the stable xenon CT method are shown in Fig. 3. There were no significant changes in Lλ values with advancing age (Fig. 4), but cerebral cortex, basal ganglia, thalamus, and white matter LCBF values all showed progressive declines with advancing age. Cerebral cortical flow showed remarkable decreases with advancing age ($y = 5.0 - 0.37$ x, $r = 0.88$, $P < 0.01$), as did flow values for basal ganglia and thalamus.

Discussion

The double integration-single compartment model for estimation of LCBF and Lλ values utilizing serial CT scanning during 35% stable xenon inhalation offers technical advantages over the in vivo autoradiographic method, which requires precise matching of arterial and tissue concentration at a single point in time; this is technically difficult (a) as both variables are rapidly changing within the first 2 min of inhalation and (b) if a slow scanner is used, such as the EMI 1010 instrument. Previous data have been confirmed that LCBF estimated for gray and white matter derived during the first 2 min of Xe^s inhalation tend to be overestimated [12]. One reason for this error is that the early rapid build-up of xenon in brain measured with a slow scanner and the use of the midpoint of the scanning interval may not accurately approximate the time base for ΔHU values with those occurring in blood. After 2 min the build-up is less abrupt and this source of error becomes negligible. In practice, we have found that the in vivo autoradiographic model, which uses a single data point derived from one CT exposure, is vulnerable to artifacts caused by slight head movement.

The double integration-single compartment model utilizes multiple data points (as a series of in vivo autoradiographic measurements) for accurately predicting the rapidly changing saturation curves in blood and tissue and thereby computes the most satisfactory P and k values by least square fitting. This overcomes the liability to measurement errors which is inherent in single data point analysis and more correctly estimates local λ values at saturation to infinity. The use of multiple data points for estimating Lλvalues at infinity is important when measurements are applied to abnormal tissues having reduced perfusion and abnormal λ values, as occurs in brain tumor [21] and in infarcts. To correctly measure Lλ values, sufficient time must be allowed to permit entry of tracer into areas of low perfusion and to avoid partial volume effects from adjacent zones of infarction where limited tissue-capillary diffusion equilibrium (m factor) may affect tracer exchange [22].

The double integration program, described here theoretically, may also be used with tracers other than stable xenon during CT scanning, although there are considerable difficulties in finding the ideal tracer for use with CT, including iodine − labeled radionuclides such as iodantipyrine, as reviewed by Drayer [23]. The model first calculates Lλ values for the region of interest at saturation of tissue and blood [10] and then calculates LCBF.

Based on practical experience, recommended optimum scanning times are 2 through 8 min for gray matter and 3 through 8 min for white matter. At least three scans are recommended for obtaining optimal $L\lambda$ and LCBF values. 35% xenon gas in 65% oxygen is purchased in 1140-liter volumes and shipped in small storage tanks. The Tedlar sampling bags used conserve the gas when administered. The current price is $ 2000 per tank, a quantity sufficient for at least 15 patients for a series of measurements.

Local λ values were not significantly altered by advancing age so that the use of assumed normal λ values for calculation of rCBF values in this condition, by the ^{133}Xe method, appears justified. In practice, excellent agreement was found between mean F_1 or cortical gray matter flow values measured by both ^{133}Xe and stable xenon CT methods in normals, and the decline of CBF with normal aging was confirmed [16, 24, 25].

White matter flow values were consistently higher when measured by the CT CBF method than by the ^{133}Xe inhalation method, and $L\lambda$ values were slightly lower than postmortem estimates [26]. These differences are attributed to three important differences between ^{133}Xe and CT CBF methods: (1) ^{133}Xe uses a bicompartmental analysis model whereby all low flow values are arbitrarily assigned to the F_2 or white matter compartment, even though they may not represent the flow of anatomic white matter and may include scalp flow, (2) due to tissue overlap or partial volume effects when LCBF is measured by the CT scanner, some higher flows derived from gray matter may be included in white matter scans, (3) consistent overestimates of the rate of white matter saturation by the CR CBF method may occur, since white matter may not be fully saturated at $8-10$ min and this may not be entirely corrected by the computer program. These three factors, alone or together, may account for the relatively lower $L\lambda$ values and higher LCBF values measured for white matter by the CT CBF method.

The CT CBF method, with its improved dimensional resolution, confirms the decline of gray matter flow with advancing age [25, 27 − 29] and has also shown accompanying age-related declines in local blood flow of basal ganglia, thalamus, and white matter.

The CT CBF method has certain advantages and disadvantages compared with radioisotopic tracers using external counters or single emission or positron emission tomographic (PET) detectors [13]. The resolution (80 mm³) for measuring LCBF in $L\lambda$ in three dimensions is unsurpassed, and CT scanners are widely available. Although the cost of commercially available xenon gas is considerable, the cost per patient is modest. The method can define regions in three dimensions having zero flow or low flow in any part of the brain with better precision than can any other currently available method. Changes in $L\lambda$ values due to changes of tissue solubility for xenon brought about by disease can be defined. Apart from the intrinsic usefulness of $L\lambda$ information, such knowledge is essential for accurate measurement of LCBF values in abnormal tissues. The method is reproducible, can be repeated at intervals, and is totally devoid of contamination of flow values from extracranial sources.

Limitations of Xe[s] CT CBF measurements are (1) irradiation exposure, which amounts to $6-8$ rad to the center of the brain for a series of measurements, and (2) the requirement for immobility of the head during scanning. Patients with advanced pulmonary disease cannot be studied satisfactorily by the method. Future improvements may permit the possibility of automatic computer printouts of maps defining LCBF and Lλ values in color, particularly if enhanced signal-to-noise ratios are achieved by digital subtraction methods.

References

1. Hounsfield GN (1973) Computerized transverse axial scanning (tomography): Part 1. Description of system. British Journal of Radiology 46:1016–1022
2. Ambrose J (1973) Computerized transverse axial scanning (tomography): Part 2. Clinical application. British Journal of Radiology 46:1023–1047
3. Adair T, Karp P, Stein A, Bajesy R and Reivich M (1981) Computed assisted analysis of tomographic images of the brain. Journal of Computed Assisted Tomography 5:929–932
4. Yamamoto YL, Thompson C, Meyr E and Feindel W (1981) Positron emission tomography for measurement of regional cerebral blood flow. In: Advances of neurology Vol. 30: Diagnosis and treatment of brain ischemia. Ed. by Carney AL and Anderson EM, Raven Press, NY
5. Lassen NA, Henriksen L and Paulson O (1981) Regional cerebral blood flow in stroke by [133]xenon inhalation and emission tomography. Stroke 12:284–288
6. Drayer BP, Gur D, Wolfson SK and Cook EE (1980) Experimental xenon enhancement with CT imaging: Cerebral applications. American Journal of Roentgenology 134:39–44
7. Meyer JS, Haymann LA, Yamamoto M, Sakai F, Nakajima S (1980) Local cerebral blood flow measured by CT after stable xenon inhalation. American Journal of Neuroradiology 1:213–225
8. George AE, deLeon MJ, Ferris SH, Kricheff II (1981) Paenchymal Ct correlates of senile dementia (Alzheimer Disease: loss of gray white matter discriminability. AJNR 2:205–213
9. Neaser MA, Gebhardt C, Levine HL (1980) Decreased computerized tomography numbers in patients with presenile dementia. Detection in patient with otherwise normal scans. Arch Neurol 37:401–409
10. Kelcz F, Hilal SK, Hartwell P and Joseph PM (1978) Computed tomographic measurement of the xenon brain-blood partition coefficient and implications for regional cerebral blood flow: A preliminary report. Radiology 127:385–393
11. Drayer BP, Wolfson SK, Reinmuth OM, Dujovny M, Boehnke M and Cook EE (1978) Weno enhanced CT for analysis of cerebral integrity, perfusion and blood flow. Stroke 9:123–130
12. Meyer JS, Hayman LA, Amano T, Nakajima S, Shaw T, Lauzon P, Derman S, Karacan I, Harati Y (1981) Mapping local blood flow of human brain by CT scanning during stable xenon inhalation. Stroke 12:426–436
13. Meyer JS: Applications of studies of cerebral blood flow (1981) Ischemic cerebrovascular disease. In: Cerebrovascular disease. Ed by Moosy J and Reinmuth OM, Raven Press, New York, 125–141
14. Kanno I and Lassen NA (1979) Two methods for calculating regional cerebral blood flow from emission computed tomography of inert gas concentrations. Journal of Computer Assisted Tomography 3:71–76
15. Obriat WD, Thompson KH, Wang HE, Wilkinson WE (1975) Regional cerebral blood flow estimated by ^{133}Xe inhalation. Stroke 6:245–256
16. Meyer JS, Ishihara N, Deshmukh VD, Naritiomi H, Sakai F, Hsu MC, Pollack P: Improved method for non-invasive measurement of regional cerebral blood flow by ^{133}Xe inhalation. Part I (1980) Description of method and normal values obtained in healthy volunteers. Stroke, 1978, Radiology 135:501–505
17. Kety SS (1981) The theory and applications of the exchange of inert gas at the lungs and tissues. Pharmacological Review 3:1–41

18. Landau WM, Freygang WH, Roland LP, Sokoloff L and Kety SS (1955) The local circulation of the living brain: values in the unanesthetized and anesthetized cat. Trans. Amer. Neurol. Assoc. 80:125−129
19. Obrist WD, Thompson HK, King CH and Wang HS (1967) Determination of regional cerebral blood flow by inhalation of [133]xenon. Circulation Research 20:124−135
20. Fletcher R (1967) A new approach to variable metric algorithms. Computer Journal 13:317−335
21. O'Brien MD and Veall N (1974) Partition coefficients between various brain tumors and blood for ^{133}Xe physics. In Medicine and Biology 4:472−475
22. Tomita M and Gotoh F: Local cerebral blood flow values as estimated with diffusible tracers: validity of assumptions in normal and ischemic tissue. Journal of Cerebral Blood Flow and Metabolism
23. Drayer B, Coleman E, Bates M, Hedlund L and Petry N (1980) Nonradioactive iodantipyrine enhanced cranial computed tomography: preliminary observations. Journal of Computer Assisted Tomography 4:186−190
24. Sakai F, Meyer JS, Karacan I, Yamaguchi F and Yamamoto M (1979) Narcolepsy: Regional cerebral blood flow during sleep and wakefulness. Neurology 29:61−67
25. Shaw T, Meyer JS (1982) Aging and cerebrovascular disease. In: Diagnosis and management of stroke and TIAs. Addison-Wesley Publishing Co., Menlo Park, California, 1−24
26. Veall N, Mallett BL (1965) The partition of trace amounts of xenon between human blood and brain tissues at 37 °C. Phy. Med. Biol. 10:375−380
27. Melamed E, Lavy S, Benton S, Cooper G, Ruist Y (1981) Reduction in regional cerebral blood flow during normal aging in men. Stroke 11:31−35
28. Kety SS (1956) Human cerebral blood flow and oxygen consumption as related to aging. J. Chronic. Dis. 3:478−486
29. Naritomi H, Meyer JS, Sakai F, Yamaguchi F and Shaw T (1979) Effects of advancing age on regional cerebral blood flow. Studies in normal subjects and subjects with risk factors for atherothrombotic stroke. Arch. Neurol. 36:410−416

Regional Cerebral Blood Flow: The Xenon CT Method

B. P. DRAYER and R. E. ALBRIGHT

Introduction

In addition to its superb anatomic specificity, cranial computed tomography (CT) can be used to delineate brain function. An analysis of the kinetics of the blood-brain barrier is performed on a daily basis in the CT laboratory by intravenously injecting iodinated contrast material followed by the immediate and delayed performance of CT scanning to define whether there is an abnormal accumulation denoting a brain lesion. It has been suggested that the rapid infusion of iodinated contrast material followed by rapid serially performed CT scans might give us information concerning the cerebral transit time as well as the cerebral blood volume (Ladurner et al. 1976; Axel 1980; Drayer et al. 1979). The results obtained using this dynamic CT technique following iodinated intravenous contrast material infusion have been limited by various factors, including the limited enhancement which is obtained in the brain-capillary bed, alterations in autoregulation, blood pressure, blood volume, and blood flow which occur following the rapid intravenous infusion of iodinated contrast material, the fact that approximately 70% of iodinated contrast material resides in the venules rather than the capillary bed, the decrease in hematocrit which occurs following the infusion of iodinated contrast material, the problems of a dispersed bolus when injecting intravenously rather than rapidly into the carotid artery, the variable corrections necessary for volume calculation to convert cerebral to large vessel hematocrit, and finally quantitative difficulties which occur when there is leakage of contrast material across an abnormally permeable blood-brain barrier (Phelps and Kuhl 1976; Drayer 1981).

In order to overcome some of these limitations and quantitate regional cerebral blood flow in absolute terms, the inhalation of a diffusible indicator, nonradioactive xenon, followed by CT scans obtained at approximately 1-min intervals has been utilized. Nonradioactive (stable) xenon is an inert inorganic gas which freely moves across the blood-brain barrier. There is a rapid buildup and clearance of xenon from the brain which has been well characterized over the past 20 years using radioactive xenon 133 (Obrist et al. 1975). The atomic number of stable xenon is 54, very close to that of iodine, and therefore it will increase the density (enhancement) of the brain substance as monitored by CT scanning if inhaled in a sufficient concentration. The xenon CT method has

Departments of Radiology and Neurology, Duke University Medical Center, Durham, North Carolina 27710, USA

Cerebral Blood Flow and
Metabolism Measurement.
Eds. Hartmann/Hoyer
© Springer-Verlag Berlin Heidelberg 1985

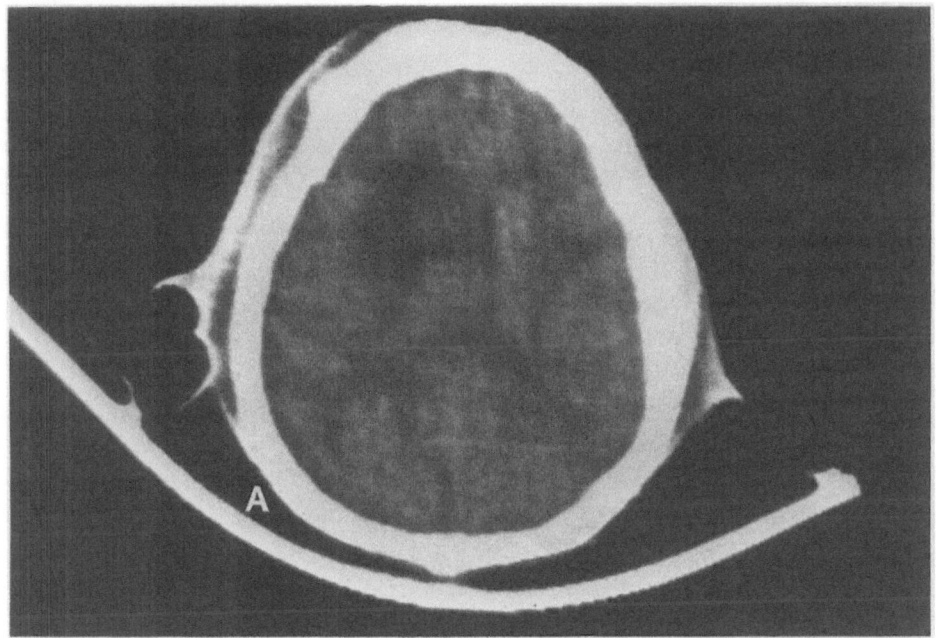

a

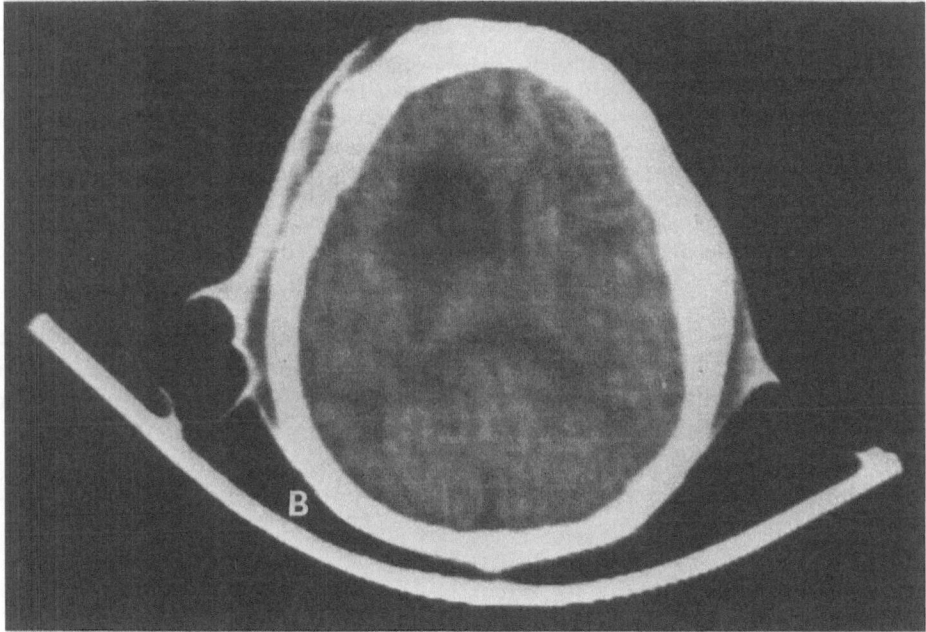

b

Fig. 1a, b. *Cerebral infarction, nonhuman primate model.* The region of infarction in the lenticulostriate distribution of the right middle cerebral artery is far better delineated on the xenon-enhanced CT scan (**B**) than on the nonenhanced CT scan (**A**)

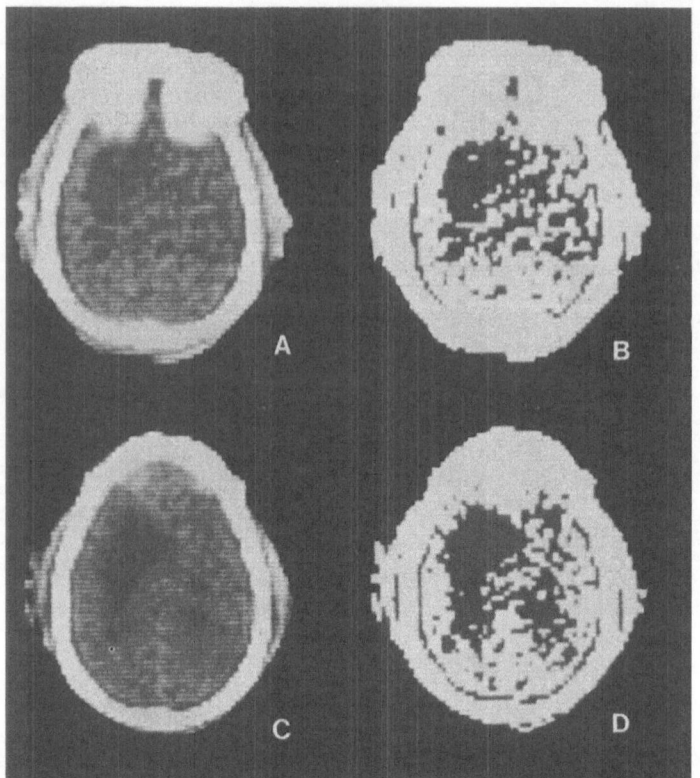

Fig. 2A–E. *Cerebral infarction, nonhuman primate model.* Xenon-enhanced scans performed 2 weeks after the lodging of a silastic tantulum embolus in the horizontal portion of the right middle cerebral artery. The resultant infarction in the lenticulostriate distribution of the right middle cerebral artery is well visualized following the inhalation of stable xenon in a 35% concentration. In addition, areas of decreased perfusion are noted in the contralateral hemisphere consistent with diaschisis although no pathologic change is seen in these regions (**E**). The CT scans were performed at two different brain levels (**A, C**) and the areas of decreased perfusion with infarction are highlighted using the measure mode (**B, D**) filming technique

various potential advantages compared with the xenon 133 technique (Drayer 1980). Particularly, because CT scanning is used, there is excellent anatomic specificity. In addition, the CT technique permits the calculation of direct brain-blood partition coefficient for any given region of brain to be studied. The technique is also rapid and easily repeated and, perhaps most importantly, CT equipment is widely available even at community hospitals.

Historical Background

The concept of using nonradioactive xenon to enhance the CT scan was first developed by Winkler and associates (Winkler et al. 1977). This concept was then extended to the estimation of regional cerebral blood flow (Drayer et al. 1980;

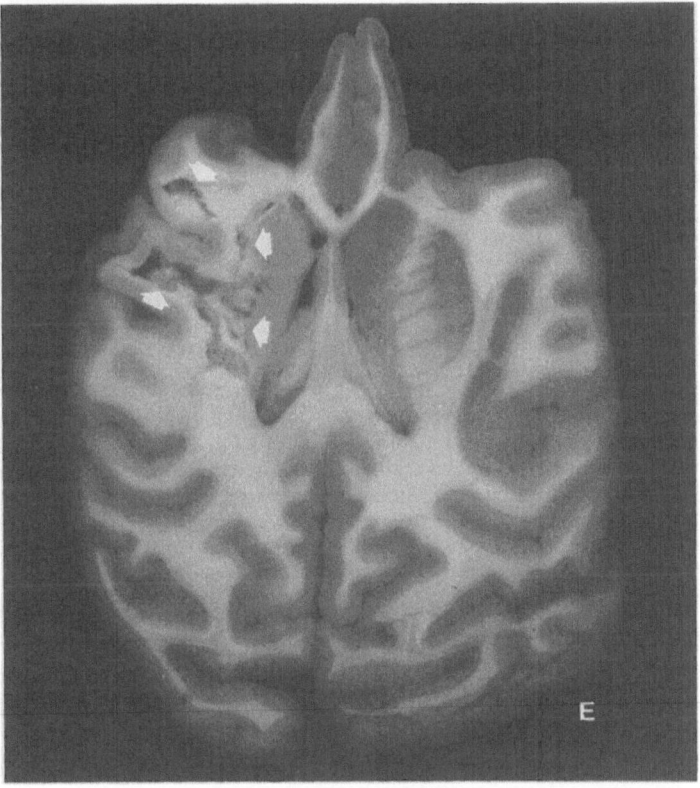

Fig. 2 E.

Drayer et al. 1978) using not only the brain enhancement seen by CT but also the increase in density within the arterial blood as measured directly by CT or indirectly using end-tidal oxygen and then later xenon levels. Kelcz et al. (1978) defined a method for estimating the brain-blood partition coefficient using the inhaled xenon concentration. Over the past few years various groups have been active in clinical applications as well as in further refining the precision and reproducibility of the xenon CT cerebral blood flow method (Meyer et al. 1980; Meyer et al. 1981; Gur et al. 1982; Rottenberg et al. 1982; Drayer 1983; Segawa et al. 1981).

Our initial studies with xenon showed that the visualization of brain abnormalities such as cerebral infarction was greatly improved by inhalation of stable xenon (Drayer et al. 1980) (Fig. 1). Not only were we able to see decreases in xenon distribution within the known area of infarction in a baboon model using a silastic tantulum embolus occlusion of the middle cerebral artery, but we also were able to see areas of decreased distribution in the contralateral hemisphere. Using the data obtained from analyzing the progressive increase of xenon concentration in the brain with CT scanning and in the arterial blood, we were able to determine that there were not only significant declines in the regional cerebral blood flow (ml/100 g/min), but also in the brain-blood partition coefficient (Drayer et al. 1980). In addition, we showed areas of

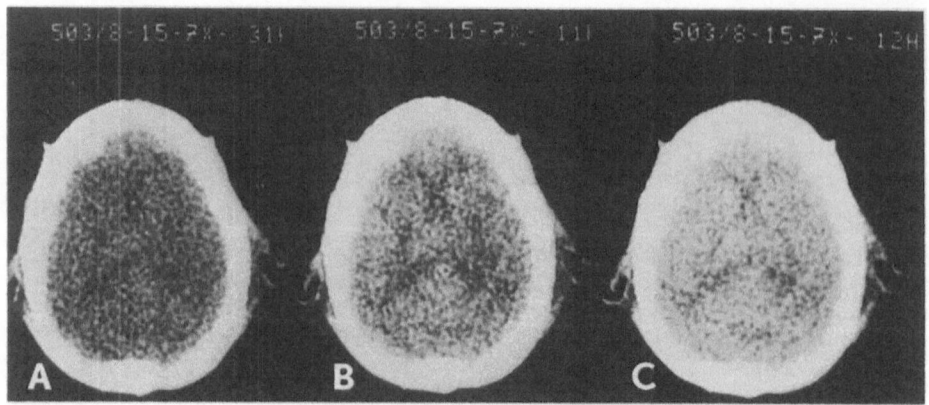

Fig. 3A−C. *Normal xenon-enhanced CT, nonhuman primate.* Initial study from our laboratory comparing a nonenhanced CT scan (**A**) with CT scans performed at 3 min (**B**) and 6 min (**C**) during the inhalation of 40% stable xenon. There is an obvious increase in density in the brain substance which can be used in conjunction with information concerning the arterial concentration of xenon to calculate regional cerebral blood flow in absolute terms

decreased flow with normal partition coefficient in the contralateral hemisphere consistent with diaschisis (Fig. 2). A large number of normal nonhuman primates were also studied and gray matter flow rates were consistently in the range of 60−90 ml/100 g/min while white matter flow rates were in the 10−30 range (Fig. 3). The calculated partition coefficient ranged from 0.85 to 1.05 in gray matter and from 1.30 to 1.60 inwhite matter. In later studies we were also able to accurately define cerebrocirculatory arrest (Drayer et al. 1980) in a group of baboons with large intracerebral hematomas and clinical findings suggestive of brain death by the absence of any xenon enhancement within the brain substance even upon prolonged inhalation of 70% xenon.

Methodologic Considerations

The measurement of *cerebral blood flow* using a diffusible indicator such as xenon is based on the Fick principle, which in the simplest terms states that the brain uptake of an indicator is equal to the amount supplied to the brain from the arterial blood minus the amount taken away from the brain by the venous blood. Using these principles of inert gas exchange, Kety (1951) determined that regional cerebral blood flow in a single compartmental tissue volume (Fi) may be derived from the time-dependent concentration of gas (e.g., xenon) in the arterial blood (Ca) and specified brain region (Ci) if the brain-blood partition coefficient (λi) is known:

$$Ci\,(T) = \lambda i \; Ki \int_0^T Ca \; e^{-Ki(T-u)} \, du \tag{1}$$

$$Fi = \frac{\lambda i \cdot Ki}{m} \tag{2}$$

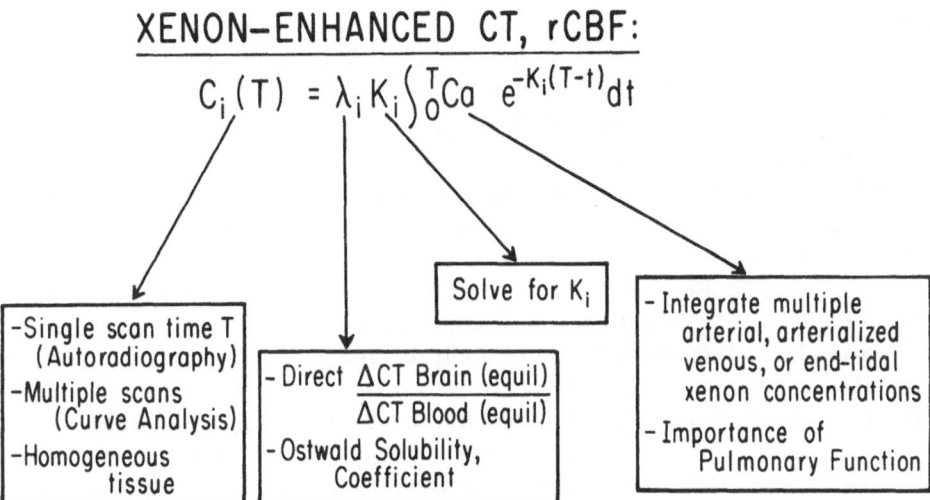

Fig. 4. Simplification of the *working equation* used in estimating regional cerebral blood flow using xenon-enhanced CT scanning

where Ci (T) equals the concentration of diffusible indicator in a given tissue, i, at time T; Ki represents the flow rate constant; and Fi is cerebral blood flow. The same assumptions may be used whether analyzing the buildup into or clearance from a tissue compartment and are dependent on free diffusion of the inert gas across the blood-brain barrier (m = 1). It has been generally assumed that except at extremely fast flows in the greater than 100 ml/100 g/min range that xenon achieves complete diffusional equilibrium, i.e., m = 1 (Tomita and Gotoh 1981).

The time-dependent *xenon concentration in the arterial blood* can be measured using various methods (Fig. 4). The arterial blood may be withdrawn in a wide plastic syringe which is immediately sealed and placed in the CT scanner to get a direct measurement of CT number (Drayer et al. 1980; Drayer et al. 1980). As the patient is often in the scanner at the same time, it is difficult to scan the syringes. With delay of scanning the syringes, there will probably be a loss of xenon from the blood, making the number inaccurate. If the patient is denitrogenated,followed by the drawing of arterial blood, the arterial concentration of xenon can be estimated by subtracting the partial pressures of $PaCO_2$ and PaO_2 and the water vapor tension from the atmospheric pressure (Drayer et al. 1978). Another method for measuring the arterial concentration of xenon is to directly scan the carotid artery. This method also has some difficulties because it requires the quantitation of enhancement from a small anatomic region (i.e., the lumen of the carotid artery) and in addition requires repeatedly moving the patient from the head to the neck level. Even with the most accurate CT scanner tables, achieving precisely the same brain and carotid level after repeated moves of the table is difficult.

Due to the above difficulties of measuring the arterial concentration of xenon as well as the invasiveness of drawing direct arterial blood, various methods have been utilized to use the end-tidal xenon concentration which is presumably in equilibrium with the arterial blood. However, when there is severely impaired gas exchange in the lung this direct relationship between the end-tidal xenon gas and arterial blood is unreliable (Obrist et al. 1975). The most accurat method for measuring the end-tidal xenon concentration requires the use of a mass spectrometer. This will also give you the end-tidal oxygen and carbon dioxide levels. Unfortunately, a mass spectrometer is an expensive piece of equipment which is technically difficult to run. For this reason, other techniques have been developed to measure end-tidal xenon gas. The most successful of these has been the use of a modified thermoconductivity analyzer (Meyer et al. 1981; Gur et al. 1982). There is a close correlation between the results obtained with a modified thermoconductivity analyzer and a mass spectrometer (Gur et al. 1984), and for this reason most centers are now using the far less expensive and complex thermoconductivity analyzer. When using end-tidal xenon measurements to estimate time-dependent concentrations of xenon in the arterial blood, these concentrations must be converted to CT units for installation into the Kety formula. This is generally done using a relationship developed by Kelcz et al. (1978):

$$Ca \ (\text{in CT units}) = \frac{5.15 \times \Theta \, Xe \times C \, Xe\%}{U_p^w / u_p^{Xe} \times 100} \tag{3}$$

$$\Theta \, Xe = 0.0011 \times HCT \, (\%) + 0.10. \tag{4}$$

Up^{Xe} is a scanning factor (mA, kVp) dependent conversion factor established by scanning various concentrations of either xenon or iodine in a phantom.

The *brain-blood partition coefficient (λ)* can then be calculated from our knowledge of the brain and arterial blood concentration of xenon if equilibrium has been achieved:

$$\lambda i = \frac{Ci \ (\text{at equilibrium}) - Ci \ (\text{baseline})}{Ca \ (\text{at equilibrium}) - Ca \ (\text{baseline})} . \tag{5}$$

One of the major difficulties in accurately estimating the partition coefficient is related to partial volume averaging. It is extremely difficult if not impossible to choose any region which is pure gray matter. It is far simpler to choose a pure white matter location to estimate the partition coefficient; however, equilibration of arterial blood and white matter xenon concentration may require from 30–60 min of inhalation. In clinical practice, this prolonged inhalation is not practical due to patient cooperation and expense. For this reason, the acquired white matter data over a 5–8 min period is generally extrapolated to equilibrium, leading to some degree of inaccuracy.

Various methods have been employed for the final *calculation of regional cerebral blood flow* from the time-dependent xenon concentrations in the brain and arterial blood (Drayer et al. 1980; Meyer et al. 1981; Gur et al. 1982; Rottenberg et al. 1982; Segawa et al. 1981). Most calculations assume a single compartmental model when calculating flow from either gray or white matter regions even though most investigators believe that flow in ghe gray matter is probably multicompartmental. It should be noted that, as stated above, a pure area of gray matter is extraordinarily difficult to find within the brain substance and therefore the multicompartmental nature of flow in the gray may be related to partial volume averaging rather than a mixture of fast and slower flows. In general, a nonlinear least-squares curve fitting of the brain data is used. Various methods have been applied to address the noise in the brain CT data. Methods have been used to correct for patient movement, which is easily done in the x,y (horizontal) axis but is extremely difficult to do when movement has occurred in the z (vertical) axis. Techniques have been developed to eliminate outlying pixels which are beyond 2 or 3 standard deviations from their surrounding pixels. Additional smoothing may be done by comparing any given pixel to a matrix of a 3×3 area of pixels or $3 \times 3 \times 3$ volume of surrounding pixels. It is generally assumed that the arterial data (Ca) is accurate.

The partition coefficient from each given region of interest may be determined independently followed by estimation of the flow rate constant (Ki) from a single enhanced scan during the buildup phase of xenon inhalation or from a group of scans from nonlinear least-square curve fitting of the brain data. Another method involves a multivariable analysis using a series of enhanced scans performed at sequential time intervals and then simultaneously solving the equation for both the partition coefficient and the flow rate constant (Gur et al. 1982; Rottenberg et al. 1982; Good et al. 1982; Gur et al. 1983). This latter method seems superior for the accurate calculation of regional cerebral blood flow. Of interest, however, is that even though the blood flow data is most accurate by this method, the partition coefficient estimations may be extremely inaccurate. This error in partition coefficient is compensated for by the estimations of the flow rate constant and therefore the method is extremely insensitive to errors in partition coefficient (Gur et al. 1982). However, if the goal of the study is to accurately calculate the partition coefficient from a given region of interest, an independent determination using a late (equilibrium) CT scan and the equilibrium arterial blood (or end-tidal xenon) concentration seems the method of choice (Drayer et al. 1980).

There have been two basic methods of *data presentation* used for xenon enhanced blood flow studies. The first involves the selection of multiple regions of interest in different anatomic locales and deriving a mean blood flow and partition coefficient with standard deviation (Drayer 1981; Drayer et al. 1980; Meyer et al. 1980; Gur et al. 1982; Rottenberg et al. 1982; Segawa et al. 1981) (Fig. 5). The second involves the generation of a flow map for the entire brain at the scanned level. Using this technique, each pixel on the flow map represents regional cerebral blood flow in ml/100 g/min rather than attenuation coefficient in CT units (Fig. 6). As there is pixel-to-pixel variation in CT images and there is generally some minimal patient movement over a 5-min scanning sequence, some smoothing of the data is necessary to produce a visually pleasant flow map. Even with this smoothing, the generated map of regional cerebral blood flow has a spatial resolution of approximately $3-4$ mm (Gur et al. 1982). These maps, which have the subjective appearance of flow, oxygen consumption, or glucose utilization maps derived using positron emission tomography, compare extremely favorably with emission tomography in terms of anatomic specificity.

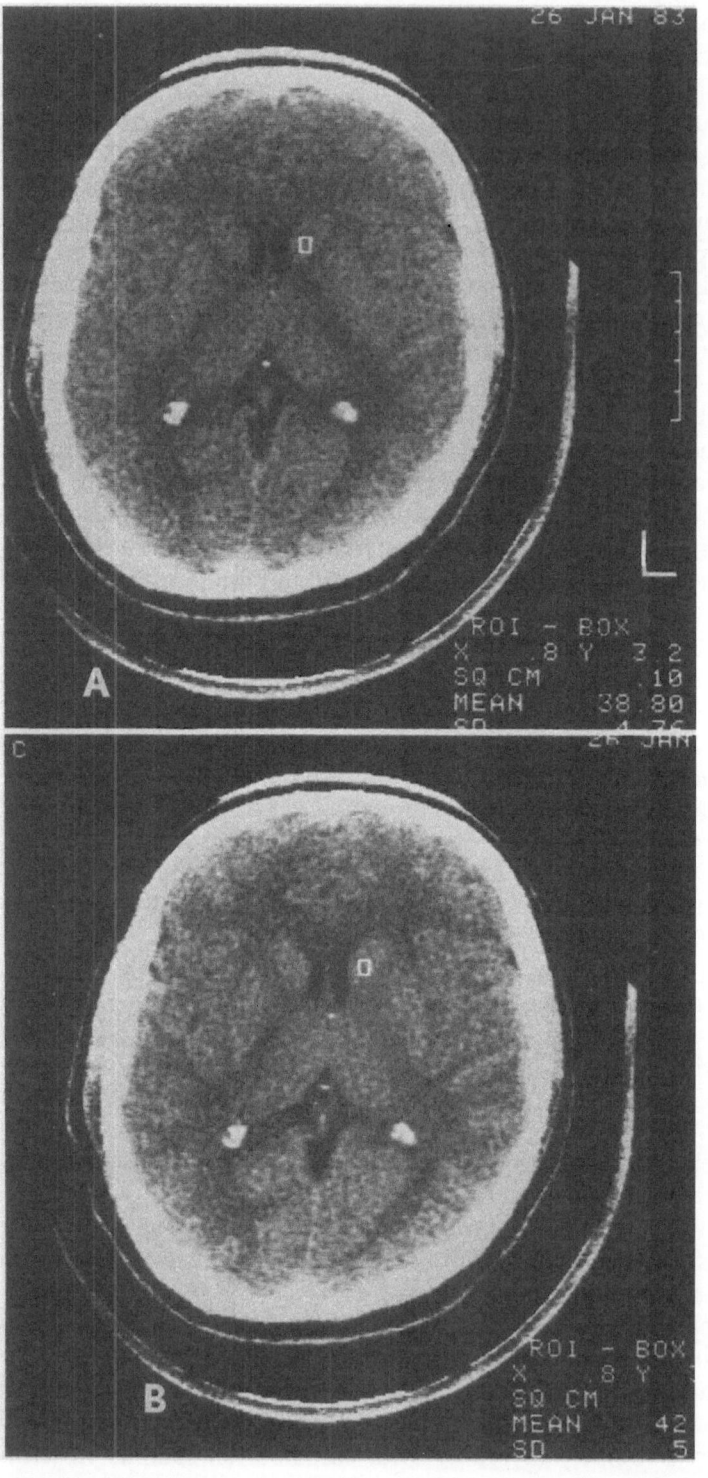

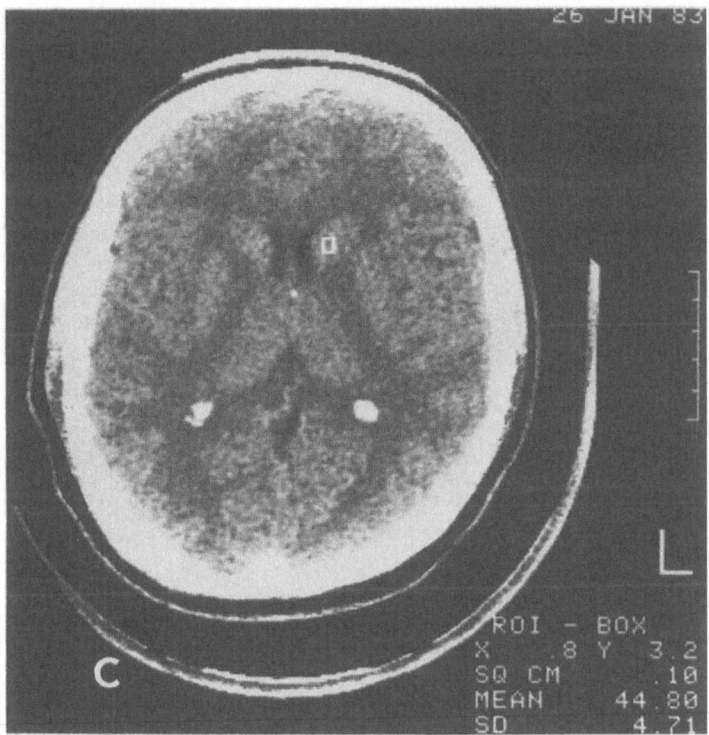

Fig. 5A – C. *Xenon-enhanced CT, normal man.* Xenon-enhanced study in a normal 28-year-old man showing the progressive increase in density in the brain substance when comparing a baseline (**A**) CT scan with one performed at 2 (**B**) and 4 (**C**) min during the inhalation of 32% stable xenon. The region of interest cursor within the caudate nucleus region shows the progressive increase in brain density that can be quantitated using the CT scanner. This increase in density is also apparent on visual analysis as the faster flow gray matter increases in density more rapidly than the white matter and thereby the caudate nucleus, thalamus, and lenticular nuclei are highlighted against the slower flow internal capsule white matter

Applications

The major applications of xenon-enhanced CT scanning to measure cerebral blood flow have involved the evaluation of cerebrovascular disease. In general terms, an area of decreased blood flow often has extended beyond the structural lesions seen on nonenhanced CT. In addition, there have been various instances where no abnormality is seen on the nonenhanced CT scan while large areas of ischemia are noted both in the symptomatic areas of brain as well as in the contralateral brain.

The CT scan is generally normal with transient ischemic attacks (TIA). Nevertheless, on xenon-enhanced CT scans an area of decreased cerebral blood flow generally in the distribution of the middle cerebral artery has been noted with a normal partition coefficient. With completed cerebral infarction, areas of decreased flow and partition coefficient are well delineated in a vascular dis-

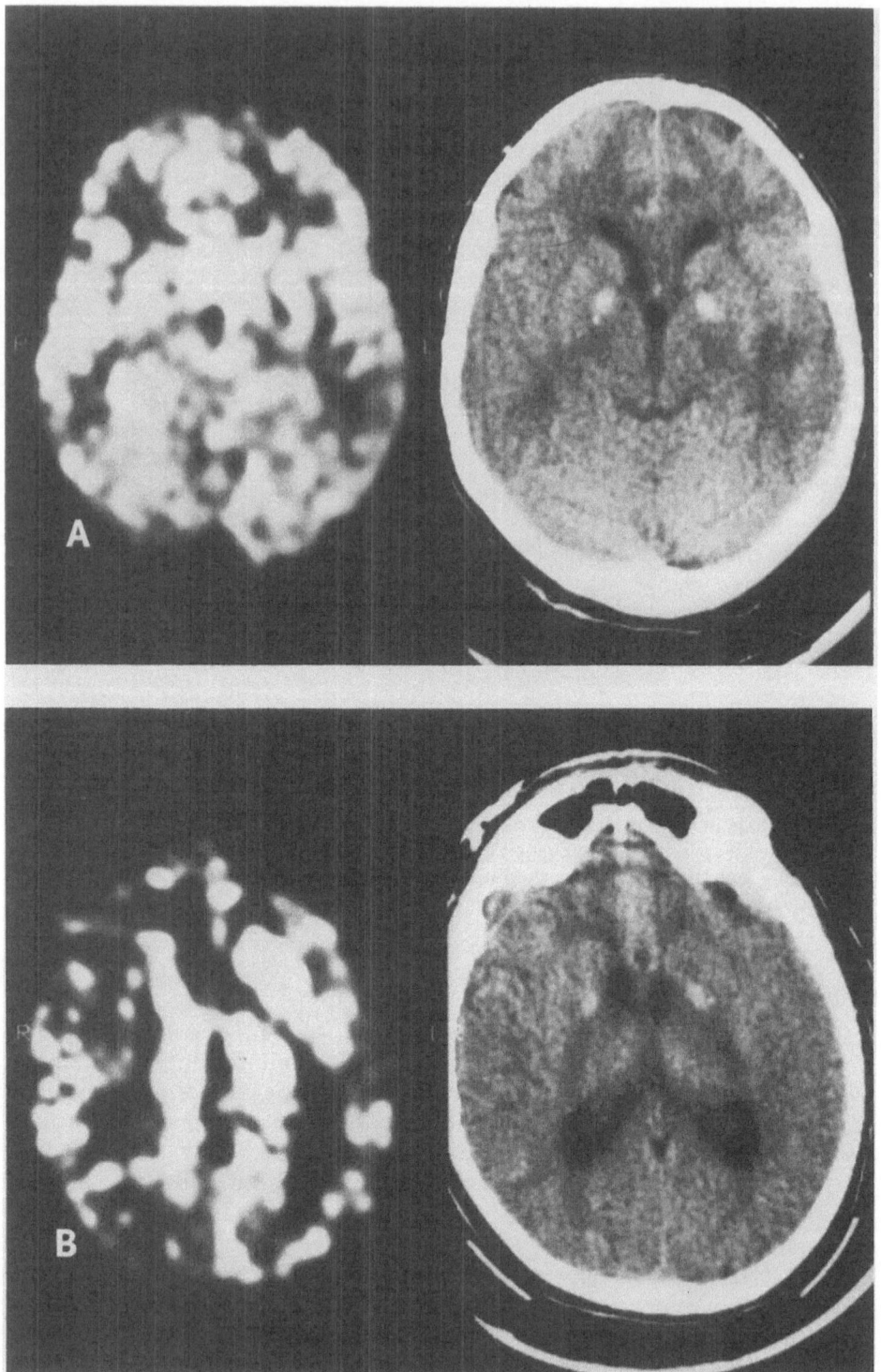

tribution (Drayer 1981; Meyer et al. 1981; Drayer et al. 1980; Tomita and Gotoh 1981;Gur et al. 1981). In many instances there is also decreased flow in homologous regions of the contralateral hemisphere without an associated abnormality in partition coefficient (Meyer et al. 1981; Drayer et al. 1980). Even without calculating flow in absolute terms, areas of infarction which are often not well seen on an unenhanced scan may be highlighted due to the normal xenon enhancement of the nonaffected brain and decreased enhancement in the infarction (Drayer 1981; Drayer 1983; Radue and Kendall 1978). Decreased flow has also been seen with intracerebral hematomas (Radue and Kendall 1978) with either a more normal or surrounding rim of relatively increased flow. A potential major application of xenon-enhanced CT scanning is in the evaluation of ischemia secondary to vasospasm following subarachnoid hemorrhage. In the few patients that have been investigated to date, decreased flow in the distribution of the vasospasm has been noted (Fig. 6). It is hoped that xenon blood flow studies may provide an important tool for evaluating vasospasm without the need for repeated cerebral angiography. Finally, in the ultimate case of cerebrovascular disease, cerebrocirculatory arrest (Drayer et al. 1980), essentially zero flow is noted within the brain even with prolonged xenon inhalation. A xenon CT study may prove an additional criteria in the evaluation of brain death.

Xenon CT scanning has also been used to better evaluate the flow characteristics of nonvascular brain abnormalities. Applications include the evaluation of Alzheimer's disease, Parkinson's disease, multiple sclerosis, glioma, metastasis, metabolic disorders, and trauma. In addition, the expected increase in regional cerebral blood flow with an increase in $PaCO_2$ has been well defined. Studies of the nonvascular applications of xenon CT await further confirmation from larger series of patients. As the xenon method is gaining widespread application, this type of information should be available in the next few years.

Limitations and Pitfalls

Although many of the limitations have been overcome, a review of the potential difficulties (Drayer 1981; Drayer et al. 1980; Gur et al. 1982; Rottenberg et al. 1982; Gur et al. 1983) involving this technique is important.

Fig. 6A, B. *Xenon-enhanced CT, Vasospasm.* A 30-year-old woman with rupture of an anterior communicating artery aneurysm. The initial xenon flow map (**A**) was performed 3 days following the subarachnoid hemorrhage and shows essentially normal flow. A second xenon blood flow study (**B**) was performed 15 days following the subarachnoid hemorrhage and shows markedly decreased flow, particularly in the anterior cerebral artery distribution bilaterally. An angiogram in this patient showed severe vasospasm at this time. The density of each pixel in this flow map is directly proportional to absolute regional cerebral blood flow in ml/100 g/min rather than merely the xenon concentration in the brain as seen in the previous figures. (Courtesy of Roger Bird, Stanford University Medical Center)

Practical Limitations

The most important practical limitation is the avoidance of patient motion. During the inhalation of xenon for a period of approximately 5 min slight patient motion is to be expected. This motion can easily be corrected in the x,y plane (horizontal), but cannot be corrected for in the z plane (vertical). Most centers are now using some type of moldable headholder as has been used for emission computed tomography. If patients are appropriately chosen, successful studies can be obtained over 90% of the time without patient motion that will significantly affect quantitative information. Another practical problem is the use of a respiratory apparatus for xenon inhalation and end-tidal xenon analysis in the CT scanning room. This can be done simply and cheaply at present if a thermoconductivity analyzer is used to estimate end-tidal xenon concentration (Gur et al. 1984). $PaCO_2$ may be obtained by adding a capnograph to the breathing circuit.

Quantitative Limits of CT

The major limitation of the xenon CT technique for accurately quantitating regional cerebral blood flow is the relatively limited signal-to-noise ratio obtained during the buildup phase of xenon enhancement as the brain may increase only a few CT units initially. Even with the marked improvements that have occurred in CT technology in terms of noise reduction, an overall error of approximately 10% − 15% is probably all that can be achieved at subanesthetic xenon doses with an anatomic specificity of the acquired absolute blood flow data of less than 4 mm. Certain trade-offs must be understood; a very high dose, obtaining additional scans which also will greatly increase the dose, or inhaling a high concentration of xenon will improve numerical accuracy. In addition, region of interest compromises must be made. A larger region of interest will have a smaller error; however, the degree of anatomic specificity is by definition decreased when a larger region of interest is used. It is also important to remember that with all quantitative imaging techniques, including CT, emission tomography, and magnetic resonance, partial volume averaging is a major limitation. Using 5 or 10 mm slice thickness, it is extremely difficult if not impossible to choose an area of pure gray matter without any partial volume averaging from white matter. Regions of pure white matter, particularly in the centrum semiovale, are far easier to select. However, accurate quantitative information from the white matter is limited by the length of xenon inhalation in that far longer inhalation times would be needed to extrapolate flow information from the white matter more accurately.

Theoretical Limitations

Theoretical limitations are intertwined with those of quantitative CT accuracy. If larger numbers of scans are performed using both multiple baseline scans and multiple CT scans at, for example, 30-s intervals, the accuracy of the obtained information will be greatly improved but the radiation dose increases significantly. This is particularly a problem if repeated scans are to be performed at the same scan level for comparison at a later date. Using a high resolution mode, the radiation dose is approximately 4 – 5 rads/scan within the scanning field. The dose outside the scanning field is far lower. When selecting the scan level, particular attention should be directed to avoid including the lens or thyroid in the scanning field. The use of more than one CT scan at the selected brain level during the xenon buildup phase, nonlinear least-squares curve fitting of this brain data, and a multivariant, multiexponential analysis simultaneously varying the partition coefficient and flow rate constant have increased the quantitative accuracy of measurements of regional cerebral blood flow. Additional methods have been developed to address the noise in the brain CT data by eliminating outlying points and smoothing the information derived from each pixel. These types of data analysis become even more important when developing flow maps. In further trying to understand the error involved in blood flow measurement, it is important to also realize that an assumption has been made that the arterial concentration or end-tidal xenon data is accurate. Finally, tissue inhomogeneity is the rule rather than the exception within areas of brain often considered "pure" gray matter. This inhomogeneity is magnified by problems of partial volume averaging, particularly when using 5 or 10 mm slice thicknesses.

Biological Limitations

When inhaling xenon in concentrations of greater than 35%, many individuals will experience paresthesiae and some may become quite agitated. At even-higher concentrations the dose becomes anesthetic. For this reason, most gropus have suggested a limit of 35% xenon concentration in the individual who is not intubated. In addition, issues arise concerning the effects that xenon may have on regional cerebral blood flow. Although EEG slowing (Meyer et al. 1981; Morris et al. 1955) occurs with xenon inhalation, suggesting that the brain may become hypometabolic and therefore have decreased blood flow, the final answer is not yet established. The best compromise at present seems an inhalation of xenon in a 32% – 35% concentration for approximately 5 min, fully recognizing that the signal-to-noise particularly on CT scans performed during the initial 2 min of inhalation will be somewhat limited. The typical study that is now used by most centers involves obtaining one or two baseline CT scans followed by scans at approximately 1.5, 2.5, 4.0, and 5.0 min after the initiation of xenon inhalation.

Conclusion

Xenon-enhanced CT scanning is a useful technique for the in vivo measurement of regional cerebral blood flow and brain-blood partition coefficient in animal studies and man. The method is highly competitive with any other technique that has been used to date to measure absolute regional cerebral blood flow in living animals. Further studies concerning the utility and quantitative accuracy of the technique in man are now underway. Initial data suggest that xenon CT may become a routine procedure for measuring regional cerebral blood flow in any institution which has a high resolution CT scanner. Although initial studies have predominantly involved the use of xenon CT studies in vascular disease, future applications in the analysis of degenerative and demyelinating diseases, neoplasms, metabolic disorders, and trauma will hopefully be developed owing to the widespread availability of CT scanners.

References

Axel L (1980) Cerebral blood flow determination by rapid-sequence computed tomography. Radiology 137:679–686

Drayer BP (1981) Functional applications of CT of the central nervous system. AJNR 2:495–510

Drayer BP, Wolfson SK, Reinmuth OM, Dujovny M, Boehnke M, Cook EE (1978) Xenon enhanced CT for analysis of cerebral integrity, perfusion and blood flow. Stroke 9:123–130

Drayer BP, Heinz ER, Dujovny M, Wolfson SK, Gur D (1979) Patterns of brain perfusion: dynamic computed tomography using intravenous contrast enhancement. J Comput Assist Tomogr 3:633–640

Drayer BP, Gur D, Wolfson SK, Cook EE (1980) Experimental xenon enhancement with CT imaging: cerebral applications. AJR 134:39–44

Drayer BP, Gur D, Yonas H, Wolfson SK, Cook EE (1980) Abnormality of the xenon brain: blood partition coefficient and blood flow in cerebral infarction. Radiology 135:349–354

Drayer BP, Dujovny M, Wolfson SK, et al. (1980) Xenon and iodine enhanced CT of diffuse cerebral circulatory arrest. AJNR 1:227–232

Drayer BP, Friedman A, Osborne D, Albright R, Bates M (1983) Anatomic applications of xenon-enhanced CT scanning: Visual image analysis and brain-blood partition coefficient studies in man. AJNR 4:577–582

Good WF, Gur D, Shabason L, Wolfson SK, Yonas H, Latchaw RE, Herbert DL, Kennedy WH (1982) Errors associated with single-scan determinations of regional cerebral blood flow by xenon enhanced CT. Phys Med Biol 27:531–537

Gur D, Yonas H, Herbert D, Wolfson SK, Kennedy WH, Drayer BP, Gray J (1981) Xenon enhanced dynamic computed tomography: Multilevel cerebral blood flow studies. J Comput Assist Tomogr 5:334–340

Gur D, Wolfson SK, Yonas H, Good WF, Shabason L, Latchaw RE, Milar DM, Cook EE (1982) Progress in cerebrovascular disease: Local cerebral blood flow by xenon enhanced CT. Stroke 13:750–758

Gur D, Shabason L, Wolfson SK, Yonas H, Good WF (1983) Measurement of local cerebral blood flow by xenon-enhanced computerized tomography imaging: A critique of an error assessment. J Cerebr Flow Metab 3:133–135

Gur D, Heron JM, Molter BS, Good BC, Albright RE, Miller JN, Drayer BP (1984) Simultaneous mass spectrometry and thermoconductivity measurements of end-tidal xenon concentrations: A comparison. Med Phys 11:209–212

Kelcz F, Hilal SK, Hartwell P, Joseph PM (1978) Computed tomographic measurement of xenon brain-blood partition coefficient and implications for regional cerebral blood flow: A preliminary report. Radiology 127:385–392

Kety SS (1951) The theory and applications of the exchange of inert gas at the lungs and tissues. Pharmacol Rev 3:1–41

Ladurner G, Zilkha E, Iliff LD, DuBoulay GH, Marshall J (1976) Measurement of regional cerebral blood volume by computerized axial tomography. J Neurol Neurosurg Psychiatry 39:152−158

Meyer JS, Hayman LA, Yamomoto M, et al. (1980) Local cerebral blood flow measure by CT after stable xenon inhalation. AJNR 1:213−225

Meyer JS, Hayman LA, Takahira A, Nakajima S, Shaw T, Lauzon P, Derman S, Karacan I, Harati Y (1981) Mapping local blood flow of human brain by CT scanning during stable xenon inhalation. Stroke 12:426−436

Morris LE, Knott JR, Pittinger CB (1955) Electroencephalographic and blood gas observations in human surgical patients during xenon anesthesia. Anesthesiol 16:312−319

Obrist WD, Thompson HK, Wang HS, Wilkinson WE (1975) Regional cerebral blood flow estimation by xenon-133 inhalation. Stroke 6:245−256

Phelps ME, Kuhl DE (1976) Pitfalls in the measurement of cerebral blood volume with computed tomography. Radiology 121:375−377

Radue EW, Kendall BE (1978) Xenon enhancement in tumors and infarcts. Neuroradiology 16:224−227

Rottenberg DA, Lu HC, Kearfott KJ (1982) The in vivo autoradiographic measurement of regional cerebral blood flow using stable xenon and computerized tomography: The effect of tissue heterogeneity and computerized tomography noise. J Cereb Blood Flow and Metab 2:173−178

Segawa H, Wakai S, Tamura A, Sano K, Ueda Y, Ohshima M (1981) CSF study by CT with Xe enhancement − experience in 30 cases. J Cereb Blood Flow and Metab 1:52−53

Tomita M, Gotoh F (1981) Local cerebral blood flow values as estimated with diffusible tracers: validity of assumptions in normal and ischemic tissue. J Cereb Blood Flow and Metab 1:403−411

Winkler SS, Sackett JF, Holden EE, et al (1977) Xenon inhalation as an adjunct to computerized tomography of the brain: preliminary study. Invest Radiol 12:15−18

Multiple Parameter Estimation from Tomographic Inert Gas Clearance Curves: A Modification on the Double Integral Method

E. M. Stokely, M. D. Devous, Sr., and F. J. Bonte

Introduction

This paper describes a method for measuring relative perfusion (uncoupled from partition coefficient), relative "pseudo" partition coefficient, and input function delay using data from a dynamic single photon emission tomograph (DSPECT). The method produces regional transverse section images of these three parameters. While the examples presented here will use data from the TOMOMATIC 64*, the method is valid for application to data from inert gas clearance studies where (a) the input function is known, and (b) the observed tissue can be assumed to be dominated by a single compartment washout.

Kanno and Lassen (1979) proposed a method for processing tomographic data from diffusible tracer studies. The technique was implemented on the TOMOMATIC 64* by Goldman (1980), who added a number of innovations, and tested and described in the literature by Celsius (1981) and Smith (1983). The implementation will be referred to as the Kanno-Lassen-Goldman-Celsius (KLGC) method in this presentation.

It has been recognized by those who work with inert gas techniques for measuring blood perfusion that methods which use only the derivative, or curve shape, to calculate perfusion often overestimate perfusion values in areas of ischemia. This problem arises because little tracer infuses into the ischemic region, and scintillation detectors which monitor radioactive tracers are sensitive to photon flux from well-perfused regions near the ischemic volume (e.g., scatter or region cross talk). Thus, the shape of presumed clearance curves from the ischemic volume becomes contaminated by perfusion effects from other tissues. .

The KLGC algorithm deals with this problem by generating a mapping between activity in the tissue and the relative perfusion index, k. The algorithm is very sensitive to ischemic areas and does not suffer from problems in this regard, as do curve shape methods. The algorithm is based on the following double integral Kety (1951) equation.

$$C_i = K\lambda k \int_{(i-1)\Delta}^{i\Delta} \int_0^t C_a(u) \, e^{-k(t-u)} \, du \, dt, \quad i = 1, 2, 3, 4, \tag{1}$$

Radiology Department, The University of Texas Health Science Center at Dallas, 5323 Harry Hines Blvd., Dallas, Texas 75235, USA

Cerebral Blood Flow and
Metabolism Measurement.
Eds. Hartmann/Hoyer
© Springer-Verlag Berlin Heidelberg 1985

where C_i are the counts in a voxel during the ith time interval, K is an unknown scaling constant which must be found, λ is the partition coefficient for xenon, Δ is fixed at 60 s for the DSPECT system, $C_a(t)$ is the input function, and k is the relative perfusion index ($F/\lambda V$ where F is flow and V is voxel volume) calculated by the algorithm. The KLGC implementation assumes that λ is unity over the image, and that the arrival of the input function in each voxel is fixed at 10 s following the inhalation of the first breath of xenon 133 mixture. An error study (Smith 1983) has shown that air curve arrival errors caused an overestimation of the average per-slice k value by as much as 23% in a series of 11 patients whose input curves were monitored by a separate head detector. In addition, a simulation study showed (Smith 1983) that the assumption of unity λ caused extreme overestimation of the value of p, the perfusion (F/V) uncoupled from λ.

Because of these problems a different algorithm was developed for handling inert gas clearance data from tomographic studies. Equation 1 was modified to include a delay term for the arrival of the input function, and an attempt was made to estimate the partition coefficient.

Method of Estimation

Steps in the estimation procedure will be presented chronologically.

1. The first step is the estimation of the unknown scaling constant, K. Note that there are three unknowns in Eq. $1 - K$, λ, and k. The proposed method adds another unknown $- \delta$, the error in input function arrival time. The observed counts can be normalized to the total:

$$\overset{o}{C_i^*} = \frac{\overset{o}{C_i}}{\sum_i \overset{o}{C_i}} \tag{2}$$

where $\overset{o}{C_i}$ is the observed number of counts in the voxel in the ith time interval. By normalizing the model in this way, two of the unknown parameters cancel out:

$$C_i^* = \frac{C_i}{\sum_i C_i} = \frac{g_i(\delta, k)}{\sum_i g_i(\delta, k)}$$

where:

$$g_i(\delta, k) = \int_{(i-1)\Delta+\delta}^{i\Delta+\delta} \int_0^t C_a(u) \, e^{-k(t-u)} \, du \, dt \tag{3}$$

where C_i is the model output for the ith time interval. By using a nonlinear estimator (Marquardt 1963), one can optimally choose parameters k and δ to match the shape of C_i^* to $\overset{o}{C_i^*}$. Denoting estimated parameters with a "^"

superscript, once $\hat{k}$ and $\hat{\delta}$ are known, one can solve the following equation:

$$\hat{K} = \frac{\sum_i \overset{o}{C}_i}{\lambda^* \hat{k} \sum_i g_i(\hat{\delta}, \hat{k})} \tag{4}$$

which is derived from Eq. 1 after summing both sides over i. The only remaining unknown in Eq. 4 is the relative "pseudo" partition coefficient term, λ^*.

A major assumption is introduced at this point. A region of reconstructed image is chosen which contains gray matter (e.g., the visual cortex). It is assumed that the average λ^* for all voxels inside the gray matter regions is 0.8. The value for $\hat{K}$ can then be estimated from Eq. 4. An average scaling factor, $\hat{\bar{K}}$, is found by averaging $\hat{K}$ over all selected gray matter voxels. This scaling factor is assumed to apply to all voxels in the brain slice.

2. Armed with the scaling factor, voxels in the four reconstructed images from one slice of a DSPECT study are examined one at a time. First, the normalization described above is performed and $\hat{k}$ and $\hat{\delta}$ are found for each voxel. Since $\hat{\bar{K}}$ is assumed to be known, λ^* can now be found:

$$\lambda^* = \frac{\sum_i \overset{o}{C}_i}{\hat{\bar{K}} \hat{k} g_i(\hat{\delta}, \hat{k})} . \tag{5}$$

Finally, the relative perfusion term can be calculated, $\hat{p} = \hat{k} \lambda^*$. Three functional images have been produced from the data for each slice — relative perfusion, relative "pseudo" partition coefficient, and input function delay.

Results Using the New Method

A field of p-λ values were generated for testing the algorithm. Figure 1 shows the distribution of p-λ combinations. The entire field could also be assigned an input function delay error, and Poisson noise could be introduced to simulate statistical fluctuations. Four simulated perfusion images were generated from this field and submitted to the KLGC algorithm, and to the new three-param-

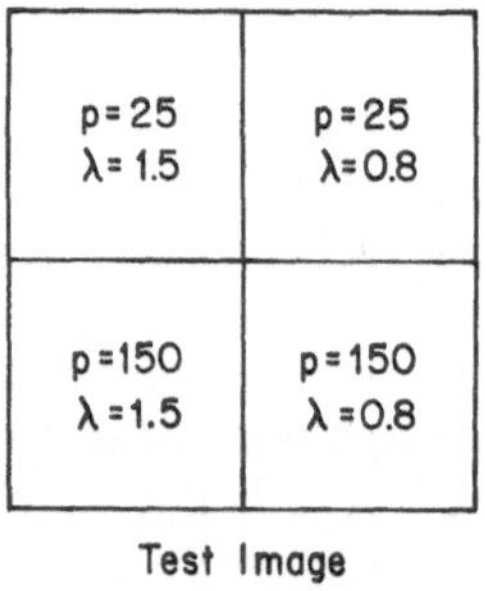

p = 25 λ = 1.5	p = 25 λ = 0.8
p = 150 λ = 1.5	p = 150 λ = 0.8

Test Image

Fig. 1. Simulated 32×32 field of perfusion values (p) and a partition coefficient (λ) used to generate test images

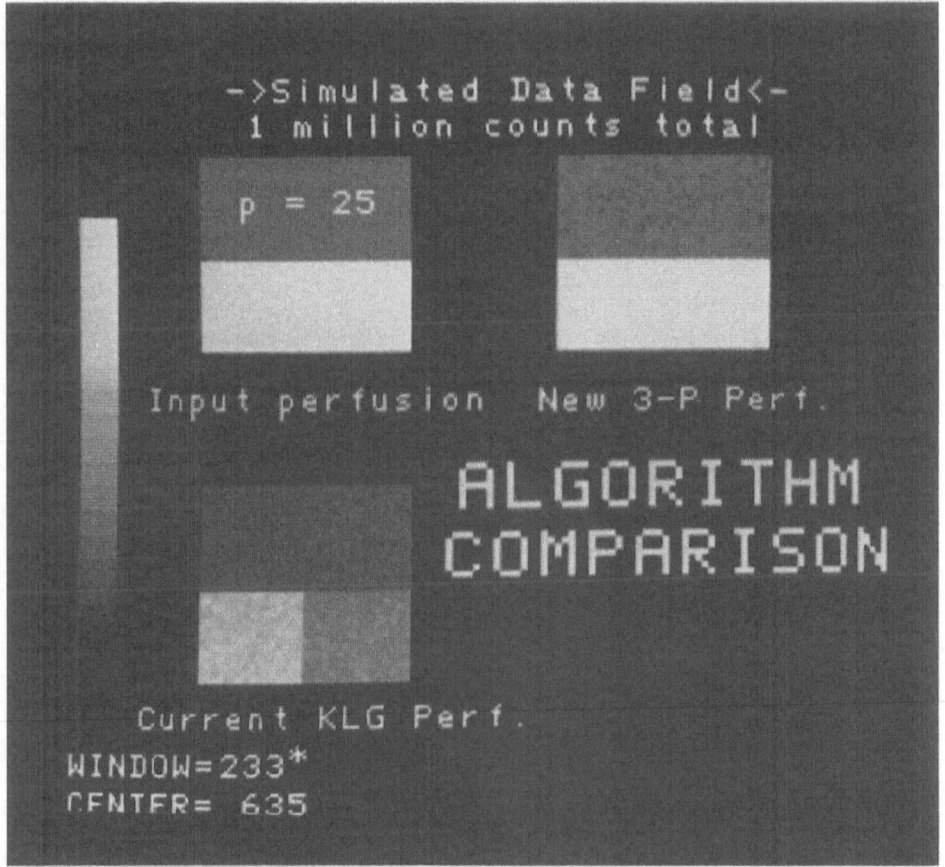

Fig. 2. Comparison of three-parameter relative perfusion (p) and KLGC relative perfusion index (k) values for simulated data from the field shown in Fig. 1. Note the sensitivity of the KLGC k value to changes in λ

eter method. The results are shown in Fig. 2. Note that the KLGC k map was influenced by the λ value, and also the fixed input curve error of -3 s. The p map from the three-parameter method gave results quite close to the input map, and all three parameters were essentially exactly estimated when no noise was introduced. In addition, the three-parameter method provides a λ-like map which will be referred to as λ^*, and a distribution of input function delay values. The delay term is extremely sensitive to noise fluctuations, although the estimation of this term was very important in providing input function correction for calculation of p.

When a study from a stroke patient is processed with both the three-parameter algorithm and the KLGC method, the three-parameter p map retains the sensitivity to low values of perfusion, its distribution of perfusions seems to be more uniform than the KLGC method, and a λ^* map is provided (Fig. 3).

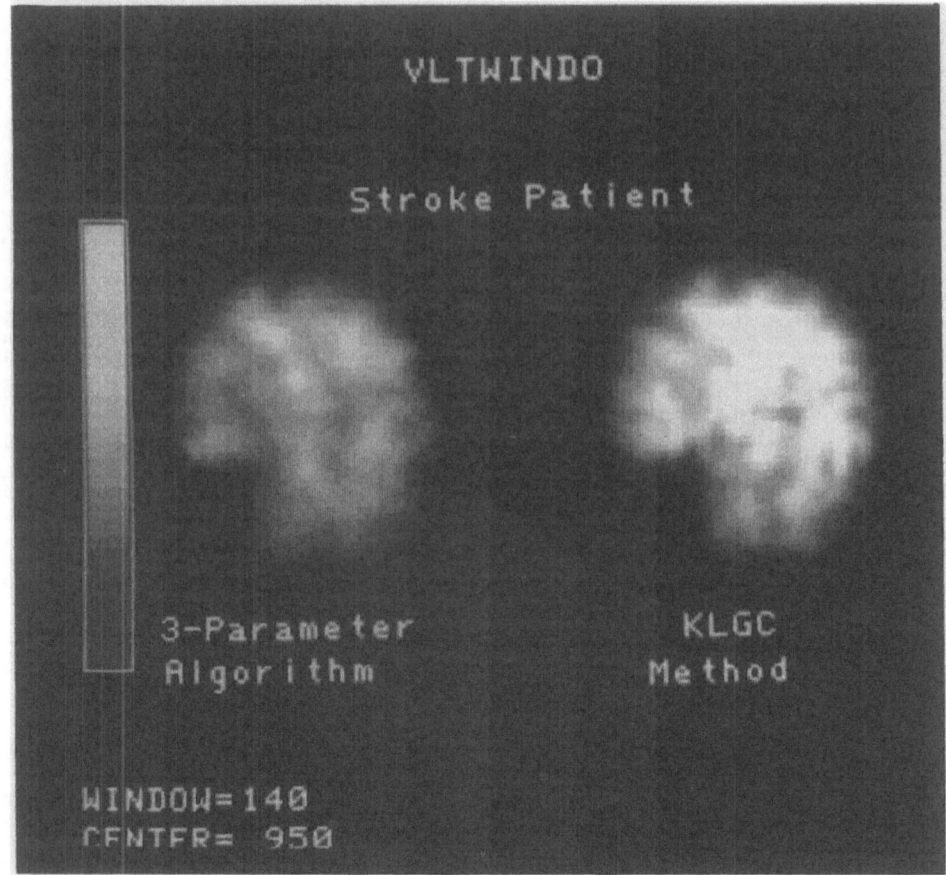

Fig. 3. Slice 2 of TOMOMATIC 64 study of stroke patient. Image on the *left* is the p-image from the three-parameter method; image on *right* is KLGC k image. Values of p image are lower and provide a generally smoother presentation

Discussion

There is reason to argue that the resolution and scatter fraction of the DSPECT do not support the use of a single compartment model. For a 4-min study, however, it is assumed that the clearance curve is dominated by the fastest compartment. It may be that this phenomenon contributes to the overestimation of white matter perfusion values (found to have a mean of 59 ml/m/100 g for 39 normals) using the KLGC method.

While the new algorithm parallels many of the steps of the KLGC method, it differs in several important aspects: (a) all 4 min of data are used at each step, whereas the KLGC method uses only 2 min of data during the final lookup process, (b) curve shape methods are used to estimate k in the new algorithm, whereas the KLGC technique uses only activity for the final calculation, (c) the new method attempts to estimate relative regional λ, whereas the KLGC technique assumes that it has a global value of unity, (d) the new method attempts to estimate voxel input function delay, which is ignored by the KLGC implementation, and (e) the new nethod uses the λ^* estimate to break out the relative perfusion, p.

It is proper to speak of the λ^* and p parameters as relative values because of the assumption that λ in the selected gray matter region is 0.8. All λ^* and p values will be in error proportional to the error in this assumption; however, there is no nonlinear treatment of error with regard to perfusion values, as in the case of the KLGC method. It is strictly speaking improper to call λ^* partition coefficient since for very low levels of perfusion little xenon 133 will enter the ischemic region, and the clearance curve will be small in amplitude but dominated in shape by extraregional activity. The effect of ischemia on λ^* can be seen by examining Eq. 5. The numerator of the right-hand side of Eq. 5 will be small, and the estimated k will be too large, for reasons already given. Thus, λ^* will tend to be severely underestimated, and will no longer bear a reasonable resemblance to the partition coefficient. Its role in determining p is nonetheless important – it acts as a scaling constant to correct for the overestimation of k, and p will correctly reflect the state of low perfusion in ischemic regions. In regions of good perfusion λ^* serves as a reasonable estimate of partition coefficient.

The method is programmed in FORTRAN VII on a Perkin-Elmer 3241 computer, and three slices of data can be processed in under 3 min, including three-dimensional table generation. The clinical utility of the three-parameter images is yet to be determined.

References

Celsis P, Goldman T, Henriksen L, Lassen NA (1981) A method for calculating regional cerebral blood flow from emission computerized tomography of inert gas concentrations. J Comput Asst Tomogr 5:641–645

Goldman TJ: Image reconstruction for emission tomography. Institute of Computer Science, Univ. of Copenhagen, Denmark, Report 80/11

Kanno I, Lassen NA (1979) Two methods for calculating regional cerebral blood flow from emission computed tomography of inert gas concentrations. J Comput Asst Tomogr 3:71–76

Kety SS (1951) The theory and applications of the exchange of inert gas at the lungs and tissue. Pharm Rev 3:1–41

Marquardt DW (1963) An algorithm for least squares estimation of nonlinear parameters. J Soc Indust Appl Math 11:431–437

Smith GT, Stokely EM, Lewis MH, Devous MD Sr, Bonte FJ: An error analysis of the double integral method for calculating brain blood perfusion from inert gas clearance data. J Cereb Blood Flow Metabol 4 (1):61–67, 1984

Cerebral Blood Flow Study in Patients with Intracranial Tumors by CT with Stable Xenon Enhancement

K. Aritake[3], H. Segawa[1], N. Yoshimasu[1], O. Nakamura[1], K. Kimura[2], K. Takakura[1], and M. Brock[3]

There have been various reports describing the changes in cerebral blood flow (CBF) in patients with brain tumors using radioactive xenon (Brock et al. 1971; Reulen et al. 1972). However, one main disadvantage of this method is that blood flow of deeply seated tumors may be incorrectly estimated. Computed tomographic measurement of local CBF by xenon enhancement (Xe^s-CT) was developed to overcome this disadvantage of the conventional isotope method. Xe^s-CT not only allows quantitative estimation of flow rates as related to CT images, but also permits determination of local partition coefficient (λ) (Drayer et al. 1978; Meyer et al. 1981; Segawa et al. 1983). In the present study, CBF was measured by Xe^s-CT in patients with malignant tumors. This paper reports our preliminary experience with this method and discusses its clinical usefulness.

Material and Method

Ten patients with malignant tumors were studied (Table 1). Diagnosis was confirmed histologically in all cases. Computed tomography, with and without contrast medium, and serial angiography were performed in each case. Forty to 60% nonradioactive Xenon was inhaled for 25 min, during which time serial CT scanning was performed every 3–5 min. End-tidal xenon, oxygen, and carbon dioxide concentrations were measured by mass spectrometry. The saturation curves were analyzed according to Fick's principle (Ketty 1951). Local flow rate constant (K), local λ, and flow values were calculated from the changes in Housefield units and end-tidal air concentration curves. These values were displayed separately on CRT either in black and white or in color images (Fig. 1). Resolution was 4×4 mm. These techniques have been reported in detail elsewhere (Ueda et al. 1981; Segawa et al. 1983).

Results

The results are summarized in Table 1. Blood flow was 82 ± 11 ml/100 g/min in gray matter and $24 \pm$ ml/100 g/min in white matter. The value of λ was

---

1 Department of Neurosurgery, University of Tokyo Hospital, Tokyo, Japan
2 Dokkyo University of Medicine
3 Neurochirurgische Klinik, Universitätsklinikum Steglitz, Freie Universität Berlin, Hindenburgdamm 30, D-1000 Berlin 45

Cerebral Blood Flow and
Metabolism Measurement.
Eds. Hartmann/Hoyer
© Springer-Verlag Berlin Heidelberg 1985

Table 1. Summary of results in ten patients with malignant tumors

Case no.	Age, sex	Diagnosis	Abnormal vascularity on angiogram	Enhancement with contrast on CT	Findings on flow map
1	37 f	Glioblastoma multiforme	+	+	Ring-like HFA
2	68 f	Astrocytoma grade III	+	+	Ring-like HFA
3	43 f	Astrocytoma grade II	−	−	Ring-linke HFA
4	47 m	Astrocytoma grade II	−	−	HFA
5[a]	32 m	Astrocytoma grade II	−	−	HFA
6[a]	30 m	Astrocytoma grade II	−	−	HFA
7[a]	22 f	Astrocytoma grade I	−	−	HFA
8	65 m	Metastasis adenocarcinoma	+	+	Ring-like HFA
9	39 m	Metastasis adenocarcinoma	+	+	Ring-like HFA
10	62 m	Metastasis adenocarcinoma	+	+	Ring-like HFA

[a] Postoperative XE^s–CT study
+ = present; − = absent; HFA = high blood flow area

0.9 ± 0.1 for gray matter, and 1.4 ± 0.2 for white matter (Segawa et al. 1983). Angiographic evidence of tumor vascularity usually corresponded to an enhancement on CT. Contrast enhancement was associated with high flow on flow maps. The λ value of tumors was reduced to mean values of 0.8 ± 0.2, as compared with the local λ of normal white matter, and flow rates were increased to mean values of 87 ± 27 ml/100 g/min. It was characteristic that these values were almost similar to those of normal gray matter. In highly malignant tumors, such as glioblastoma multiforme, high-grade astrocytoma, and metastasis, flow maps revealed high-flow areas corresponding to the "ring structures", as well as central areas of low flow (Fig. 2). In the center, both λ and flow values were almost 0. However, at the margin, both λ and flow values were almost equal to or higher than those of normal gray matter. It seems that the central low flows on Xe^s-CT are due to necrotic tissue or cysts and that the high-flow areas at the tumor margin are due to viable neoplastic tissue. Rather high flow values with low λ values were generally observed throughout low-grade astrocytomas (Fig. 3). In the edematous tissue surrounding the tumors, both λ and flow values were extremely reduced to mean values of 0.5 ± 0.3 and 9 ± 5 ml/100 g/min, respectively. Edema appeared to spread preferably within the subcortical white

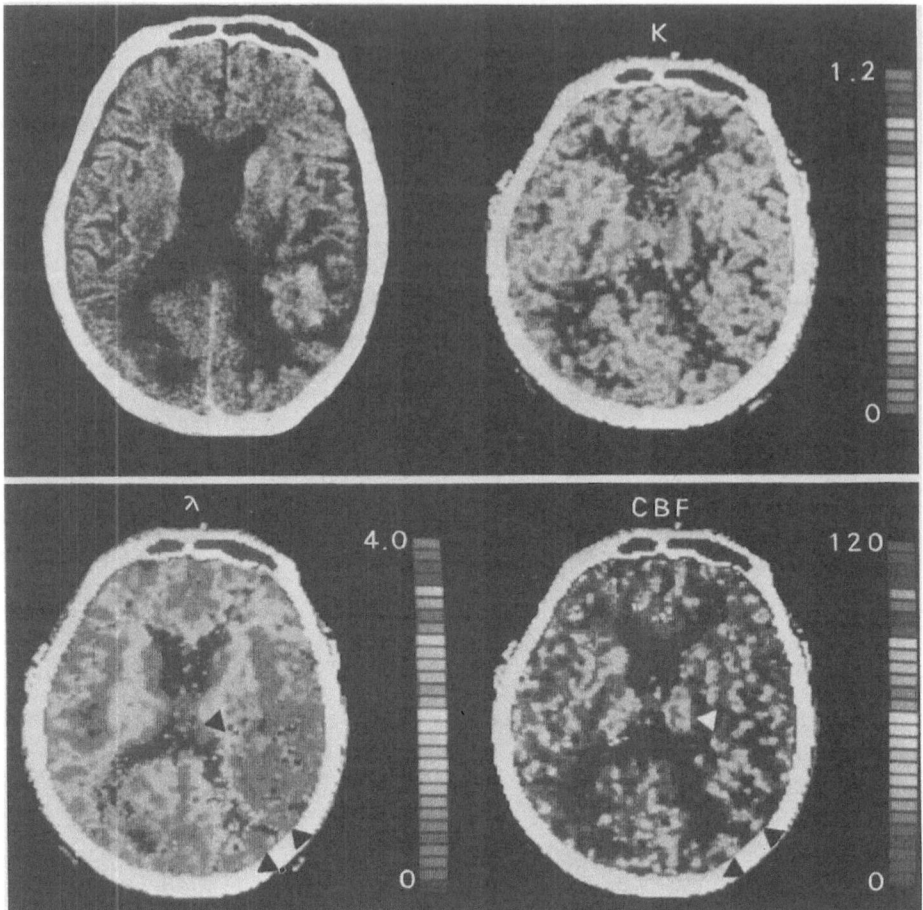

Fig. 1. Case 10. Conventional CT, K, λ, and flow maps. Either cortex or basal ganglia are visualized on the K or flow maps, while the white matter is recognized on the λ map because of its high λ value. Both the λ and flow values within the tumor are almost the same as or higher than those of cortex. In the surrounding edema, more clearly noted on the flow map, both the λ and flow values are very reduced. There are, however, no alterations in the λ and flow values of the cortex and the thalamus (*arrowheads*) adjacent to the edema

matter, sparing the gray matter of the overlying cortex and of the basal ganglia (Figs. 1, 2).

Discussion

As mentioned above, CBF measurement by CT has several advantages over conventional isotope methods. High spatial resolution achieved by CT scanning permits the measurement of CBF even in deep structures. Another major advantage is the capability of obtaining local λ, which has been demonstrated to vary considerably in pathologic conditions, such as tumors, edema, and infarct

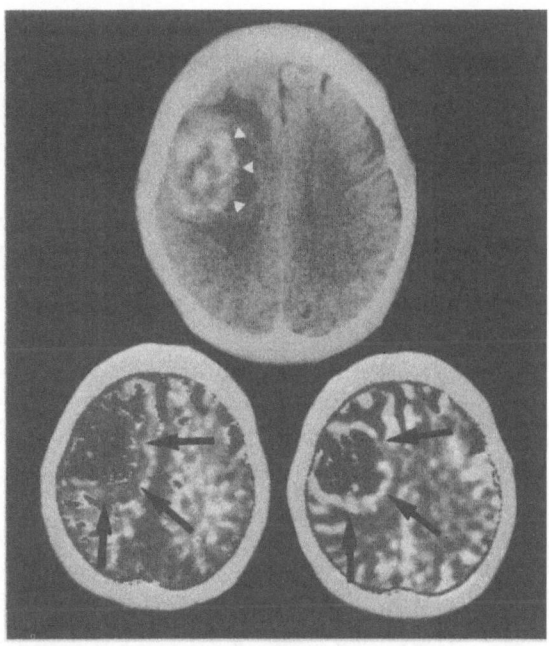

Fig. 2. Case 2. *Top:* Conventional enhanced CT; the tumor is enhanced as a ring structure. An area of low attenuation around the tumor was considered to be edematous. *Bottom left:* λ map; the λ values of the central parts of the tumor are almost 0, while those at its margin (*arrows*) are similar to those of the cortex. The λ values in edematous tissue are diminished. *Bottom right:* Flow map; an area of high flow rates at the margin of the tumor appears to extend beyond the enhancing ring on CT. Flow rates of edematous tissues are clearly reduced, while there are no obvious changes in the λ and flow values of the cortex overlying the edema

(Meyer et al. 1981; Radue et al. 1978; Segawa et al. 1983). Therefore, this method allows more accurate determination of flow rates in various tissues. Furthermore, xenon enhancement has proved useful in distinguishing abnormal from normal tissues through differences in λ values. In addition, the present method is helpful not only in relating flow rates to an anatomic specificity but also in observing the flow distribution at one brain slice.

It became apparent that both λ and flow values in tumor tissue were similar to those of gray matter. Contrary to these changes, both λ and flow values were remarkable reduced in edematous tissue. Therefore, it seems possible to differentiate an tumor from the surrounding edema more precisely based on λ and flow maps. At the same time, Xe^s-CT is helpful in defining the extent of peritumoral edema. The ability of the Xe^s-CT to ascertain the extent of tumors is of particular interest when surgery, chemotherapy, or radiation therapy is contemplated. Particularly, the combination of a λ map with a flow map made possible some exact statement about the growth of low- or iso-attenuating tumors in white matter.

Although our experimence of Xe^s-CT study in patients with brain tumors is small, it appears that this method could provide useful clinical and scientific information.

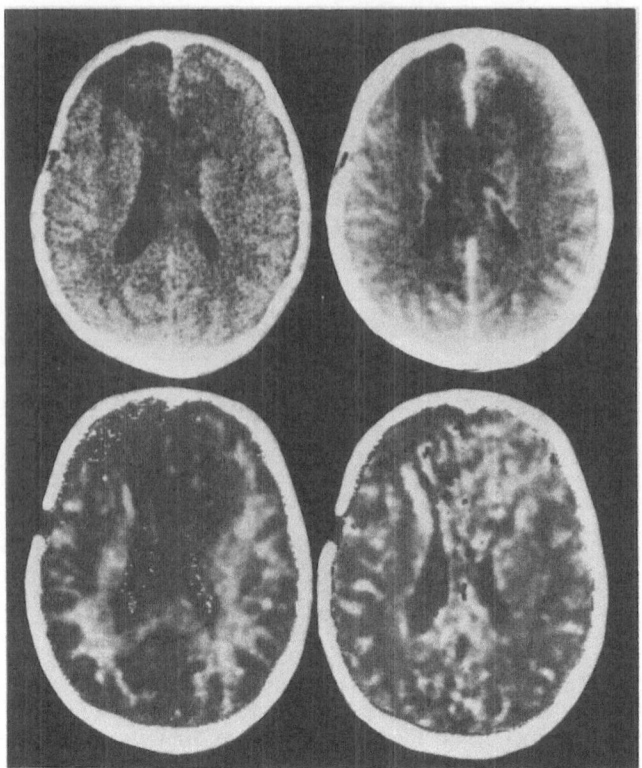

Fig. 3. Case 6. *Top:* CT with and without enhancement shows an unenhanced mass of low attenuation value. It is difficult to determine the extension of the tumor. *Bottom left:* λ map; reduced λ values either on both frontal regions or in the corpus callosum. *Bottom right:* flow map; as compared with the flow values in the white matter, a high flow area is observed in the regions corresponding to those with a reduced λ value. The tumor is shown to invade the corpus callosum

References

Brock M, Hadjikimos A, Derüaz JP, Schürmann K (1971) Regional cerebral blood flow and vascular reactivity in cases of brain tumor. In: Ross Russell RW (eds) Brain and blood flow. Pittman, London, pp 281−284

Drayer BP, Wolfson SK, Reinmuth OM, Dujovny M, Cook EE (1978) Xenon-enhanced CT for the analysis of cerebral integrity, perfusion and blood flow. Stroke 9:123−130

Kety SS (1951) Theory and applications of the exchange of inert gas at the lungs and tissues. Pharmacol Rev 3:1−41

Meyer JS, Hayman LA, Amano T, Nakajima S, Shaw T, Lauzon P, Derman S, Karacan I, Harati Y (1981) Mapping local blood flow of human brain by CT scanning during stable xenon inhalation. Stroke 12:426−436

Radue EW, Kendall BE (1978) Xenon enhancement in tumors and infarcts. Neuroradiol 16:224−227

Reulen HJ, Hadjidimos A, Schürmann K (1972) The effect of dexamethasone on water and electrolyte content and on rCBF in perifocal brain edema in man. In: Reulen HJ, Schürmann K (eds) Steroids and brain edema. Springer, Berlin Heidelberg New York, p 239–252

Segawa H, Wakai S, Tamura A, Yoshimasu N, Nakamura O, Ohta M (1983) Computed tomographic measurement of local cerebral blood flow by xenon enhancement. Stroke 14:356–362

Ueda Y, Kimura K, Nagai M, Segawa H, Sano K, Yamazaki T (1981) rCBF measurement by CT and its imaging. J Cereb Blood Flow Metabol 1 (Suppl 1):54–55

Xenon Effects on CNS Control of Respiratory Rate and Tidal Volume – The Danger of Apnea

S. Winkler[1], P. Turski[2], J. Holden[2], R. Koeppe[2], B. Rusy[2], and E. Garber[2]

Introduction

In the course of an investigation into the possibility of using a single breath technique with stable xenon (Xe) and CT for rCBF determination, a volunteer human subject developed apnea and had to be resuscitated. This episode led us to reevaluate our data and to discover some startling effects of xenon on respiratory rate and tidal volume. A review of central nervous system control of respiration (Mitchell and Berger 1981; Murray 1976) suggests that the site of both xenon effects (apnea and change in respiratory rate and tidal volume) is at the level of the pontine centers that control respiratory rhythm.

Material and Method

The procedure began with the subject inhaling 100% O_2 for several minutes to wash out N_2. He then inhaled one vital capacity of 100% Xe, held his breath for a fixed period of time, and then continued to breath O_2 as before. An open breathing circuit with mouthpiece was used. O_2, CO_2, and N_2 were monitored continuously at the mouth by Beckman OMII, Beckman LBII, and Nitralizer (Med. Sci.), respectively. Tidal volume was monitored by Hans-Rudolph Pneumotach on the expiratory limb. Two 10-s breathholds, six 20-s breathholds, and one 30-s breathhold of 100% Xe were performed, all on the same volunteer subject. An equal number of air control breathholds for each duration were performed. After acquisition of these preliminary data, a 30-s breathhold of 100% xenon was obtained on the CT scanner. Control and Xe scans were obtained.

Results

The air and 10-s Xe breathholds showed only transient change in respiratory pattern, which soon returned to normal. Respiratory patterns following 20-s and

---

1 Department of Radiology William S. Middleton Memorial Veterans Hospital 2500 Overlook Terrace, Madison, Wisconsin 53705, USA
2 University of Wisconsin Clinical Sciences Center, 600 Highland Avenue, Madison, Wisconsin 53792, USA

Cerebral Blood Flow and
Metabolism Measurement.
Eds. Hartmann/Hoyer
© Springer-Verlag Berlin Heidelberg 1985

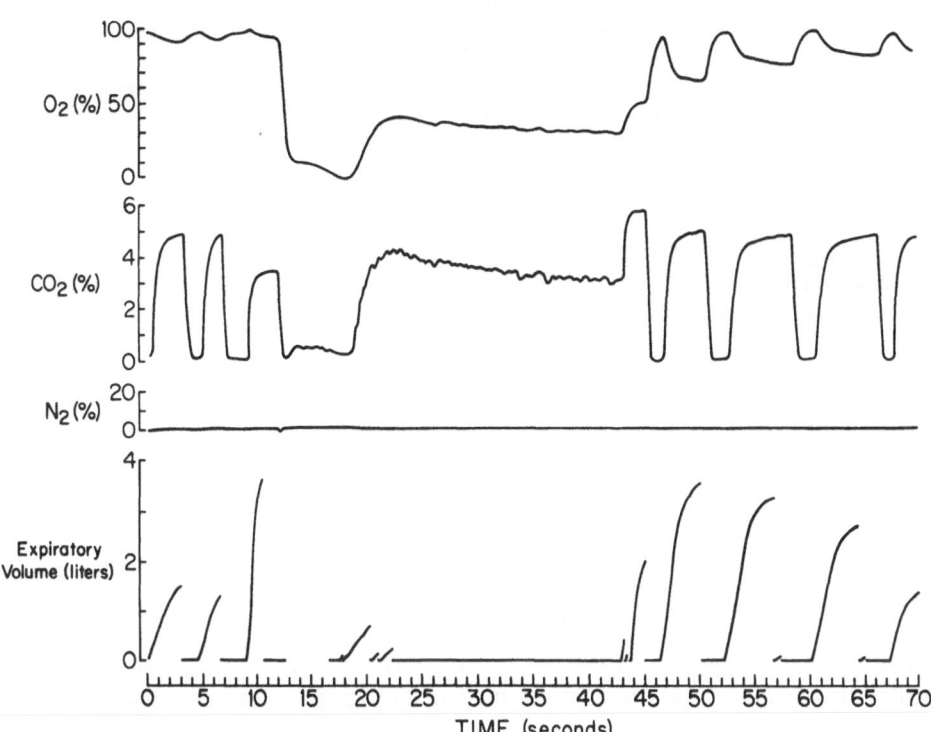

Fig. 1. Thirty-second breathhold of 100% Xe. The volume stroke at 10 s is the vital capacity maneuver prior to breathhold. Note large tidal volumes after breathhold almost as large as vital capacity

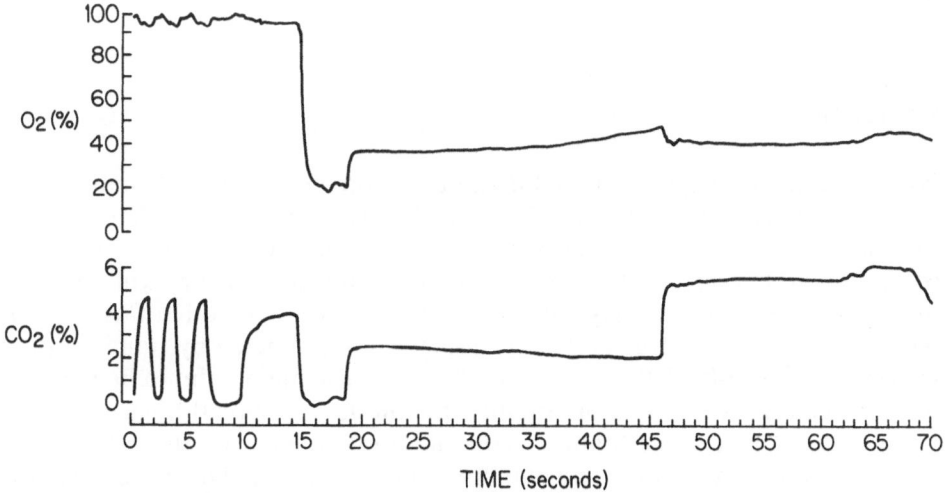

Fig. 2. Thirty-second breathhold of 100% Xe during which the subject became apneic

Table 1. Effects of xenon on tidal volume and respiratory rate

	Control period (1 min on 100% O_2)	1-min period after 20s, 100% Xe breathhold	1-min period after 30s, 100% Xe breathhold
Minute vent. liters min^{-1}	22.4	22.0	23.6
Respiratory rate min^{-1}	15	9	9
Average tidal volume (liters)	1.5	2.4	2.6

30-s Xe breathholds were abnormal. The washout pattern after the first 30-s breathhold (Fig. 1) is typical. There is a slowing of respiratory rate from a normal of 15 min^{-1} to 9 min^{-1}. At the same time, tidal volumes become very large, approaching the vital capacity in some instances. The effect lasts as long as 3 min. The effects of Xe on tidal volume and respiratory rate are summarized in Table 1. In the following discussion we call the increased tidal volume and slowed respiratory rate the "xenon effect".

The subject developed apnea during the 30-s Xe breathhold experiment in the CT scanner. Figure 2 shows the record of the apneic period. The subject recalls being awake at the time he was instructed to exhale and he exhaled exactly on cue. He was in no distress and was not aware that he then stopped breathing. He soon became unconscious, and toward the end of the record (Fig. 2) he became disconnected from the mouthpiece and no further record was made. When extracted from the scanner by his co-workers he was apneic, ashen in complexion, and completely relaxed. A strong carotid pulse was palpated, however, and after a few breaths of assisted ventilation, he began breathing on his own and soon regained consciousness. A CT scan obtained 68 s after start of breathhold showed a 4 HU increase in CT number.

Discussion

We constructed a computer simulation model of the expected CT enhancement for a region of grey matter having rCBF of 65 cc/100 g/min at 120 kVp (Fig. 3). This model is specific for the individual volunteer, because initial alveolar Xe concentration depends on the ratio of vital capacity to total lung volume and this ratio varies with age, body habitus, smoking history, and other factors. For our subject, the initial alveolar xenon concentration predicted by standard pulmonary function tests (Petty 1975) was 61% ± 3%. The main point to be noted is that the brain Xe concentration attained for the longest breathhold proposed (30 s) does not exceed the brain Xe concentration attained by the 35% Xe rebreathing protocol used by many authors (Gur et al. 1982). The 4 HU increase in CT observed on the one 30-s breathhold during which a scan was obtained

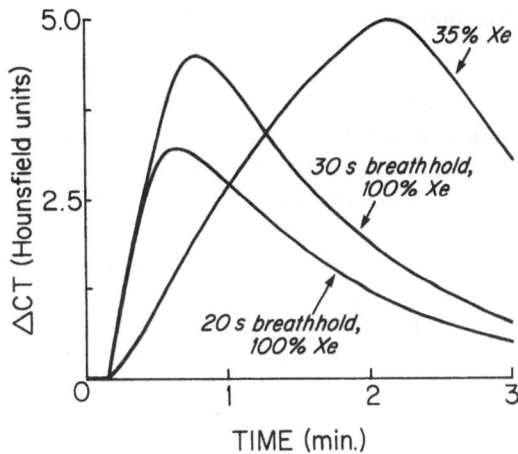

Fig. 3. Computer-simulated comparison of brain change in CT number for 20-s and 30-s breathhold of 100% Xe and 35% Xe rebreathing protocol for 2 min, all at 120 kVp

Assumptions:
1. Gray matter flow = 65 ml./min./100 g.
2. Initial Xe concentration = 60%.

support the general correctness of the computer simulation. Thus, we do not believe the effects we observed resulted from overdosage.

Could the xenon effect be an idiosyncrasy or somehow be dependent on our breathholding technique? While the literature makes no specific mention of respiratory rate, two authors who used 35% Xe in a rebreathing circuit included strip chart recordings. In one, the respiratory rate was 5 min^{-1} (Ip et al. 1982); in the other it was 9 min^{-1} (Gur et al. 1982). Thus the xenon effect can probably be observed in all subjects once it is looked for.

Unlike xenon, other inhalation anesthetics typically increase respiratory rate and decrease tidal volume (Hickey and Severinghous 1981). The xenon effect can be explained as a depression of the pontine centers that control respiratory rate and volume. These centers may be considered separate "on" and "off" switches. Stretch receptors in the lungs send signals along afferent vagal fibers, telling the off switch or pneumotaxic center in the upper pons what the state of inflation of the lung is and when to cut off inspiration. The very large tidal volumes observed after Xe breathholds may represent failure of the off switch. An effect identical to the xenon effect can be produced in the decerebrate cat by performing either a bilateral vagotomy or a mid-pons transection of the brain — in other words, disconnecting the pneumotaxic center from either its afferent or efferent fibers (Mitchell and Berger 1981; Murray 1976).

The most likely mechanism of the xenon effect is either a depression of the pontine centers or a direct action on the stretch receptors in the lung. We favor depression of the pontine centers as the xenon effect was not seen with 10-s breathhold of 100% Xe. Had we realized earlier the significance of the xenon effect, we might have anticipated that breathholding or voluntary apnea could proceed to a dangerous state of involuntary apnea, as it did in our subject.

Conclusion

Several authors have reported using stable Xe for clinical studies. Much of this work is summarized in a recent review (Gur et al. 1982). While some work has been done to evaluate the effect of Xe on cerebral blood flow, there have been virtually no studies on other aspects of its pharmacology. Our experience showing extreme sensitivity of the respiratory centers to Xe indicates the need for much more work to evaluate Xe effects not only on respiration but on other autonomic functions. In discussing the stable Xe technique one should not lose sight of the fact that Xe is a powerful anesthetic agent and not an inert tracer substance. The habit of referring to it as "mildly anesthetic" should be avoided.

References

Gur D, Wolfson SK Jr, Yonas H, Good WF, Shabason L, Latchaw RE, Miller DM, Cook EE (1982) Local cerebral blood flow by xenon enhanced CT. Stroke 12:750−758

Hickey RF and Severinghous JW (1981) Regulation of breathing: Drug effects. In: Hornbein TF (ed) Regulation of breathing, part II. Marcel Dekker, New York, Basel, p 1251−1312

Ip WR, Holden JE, Winkler SS (1982) Breath-by-breath radioabsorptometric assay of stable xenon in expired air. Clin Phys Physiol Meas 3:151−154

Mitchell RA and Berger AJ (1981) Neural regulation of respiration. In: Hornbein TF (ed) Regulation of breathing, part I. Marcel Dekker, New York, Basel, p 541−620

Murray JF (1976) Regulation of respiration. In: The normal lung. WB Saunders Company, Philadelphia, London, Toronto, p 226−228

Petty TL (1975) Pulmonary diagnostic techniques. Lea and Febiger, Philadelphia, p 42−46

Clinical Applications
of Xenon/CT Blood Flow Mapping

H. YONAS[1], S. K. WOLFSON, JR.[1,2], D. GUR[3,4], R. E. LATCHAW[3], and
J. V. SNYDER[5]

Introduction

Cerebral blood flow mapping utilizing the stable xenon/CT methodology has
evolved rapidly over the past few years [1, 2, 3]. Currently because of refinements in the mathematical analysis of flow, as well as improvements in the
stability and resolution of CT equipment, flow maps with a relatively high degree of spatial and anatomic resolution are being obtained on a routine basis.
Examples of the type of information being obtained are demonstrated in the
following cases.

Methodology

This methodology has been described in detail recently [1, 2]. In brief, patients
breathe room air until xenon delivery is initiated through either a face mask or
a mouthpiece accompanied by firm nasal compression. Patients are reminded
to remain without head movement before the baseline CT scans and then
throughout the 5-min period of xenon inhalation. During the examination, patient tolerance is monitored by predetermined foot movements. A chart recorder is used to record both the output from a thermoconductivity analyzer recording the xenon concentration and a caprograph measuring the pCO_2. After
two baseline scans are obtained, the patient begins to breathe from a 60-liter
plastic bag containing a premixed blend of 33% xenon and 67% oxygen, delivered as an open system for 5−6 min. After the first minute, four second scans
are then obtained at approximately 30-s intervals. For multiple level studies,
incremental table movements coupled with dynamic scanning are utilized. The
number of levels are currently limited to two or three due to tube heat loading.

Blood flow determination involves the averaging of two baseline scans before
xenon inhalation in order to reduce noise levels. The three to six scan images
obtained at each level are then subtracted from the average baseline image and

1 Department of Neurological Surgery, University of Pittsburgh, Pittsburgh, PA 15261, USA
2 Laboratory of Surgical Research, Montefiore Hospital, Pittsburgh, PA 15213, USA
3 Department of Radiology, University of Pittsburgh, Pittsburgh, PA 15261, USA
4 Department of Radiation Health, University of Pittsburgh, Pittsburgh, PA 15261, USA
5 Department of Anesthesiology, University of Pittsburgh, Pittsburgh, PA 15261, USA

Cerebral Blood Flow and
Metabolism Measurement.
Eds. Hartmann/Hoyer
© Springer-Verlag Berlin Heidelberg 1985

each voxel is subsequently defined by a series of enhancement values as a function of time $\Delta CT(t)$.

This series is used in conjunction with the end-tidal measurements, which are assumed to be proportional to the xenon concentration in the arterial blood, to solve for a monocompartmental Kety equation, in which $\Delta Ca(t)$ and $\Delta CT(t)$ are used as input data:

$$\Delta CT(t) = f \int_0^t \Delta Ca(u)\, e^{-f/\lambda(t-u)}\, du.$$

A weighted and/or nonweighted least square fit routine is used to derive the estimates of two parameters, λ and k or λ and f. Pre- and postanalysis smoothing routines can be used to reduce pixel-to-pixel variation. We presently use a 3×3 pixel bell-shape filter.

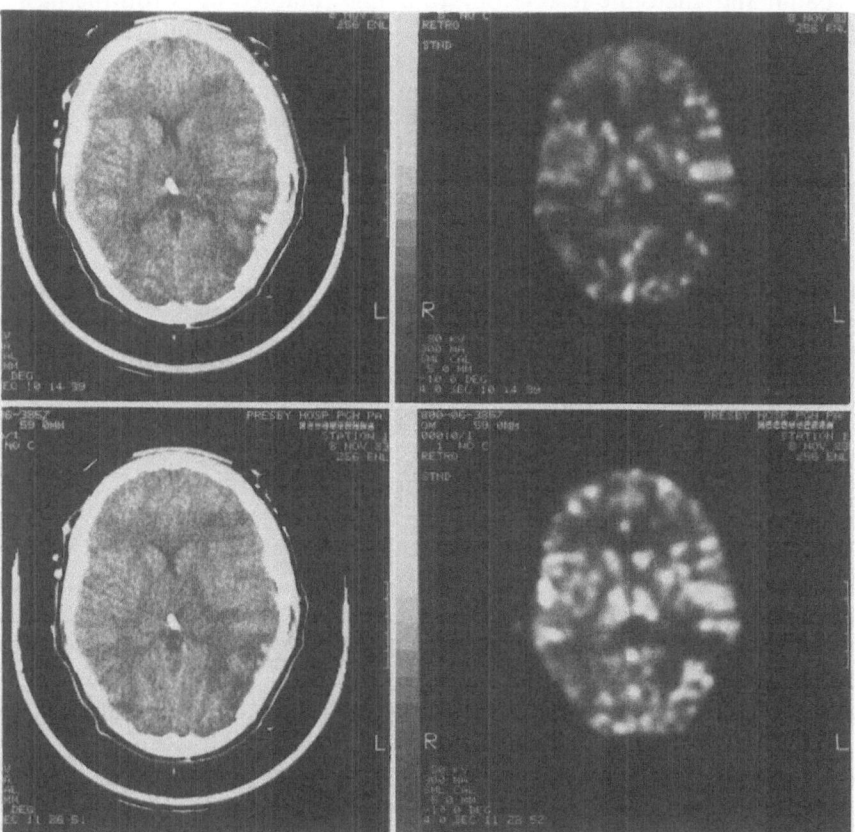

Fig. 1. The four images are the baseline CT images on the *left* for the blood flow determinations on the *right*. The scale for each blood flow determination is on the *bottom right* of the flow map. The window setting is the scale in cc/100 g/min, which is then displayed as a gray scale on the left of the flow map. The window level is the mean value of the scale. The upper pair of studies were obtained at a pCO_2 of 23 and the lower set of studies at a pCO_2 of 36 mmHg

Case Studies

Case 1. A 19-year-old female was involved in an automobile accident and brought to the hospital unresponsive. Following intubation and fluid replacement, she began to move spontaneously, but she remained plegic on the right side with nonpurposeful movements on the left. The CT scan demonstrated a small left-sided subdural hematoma with contusion of the lateral aspect of the left hemisphere. An external ventriculostomy was placed for intracranial pressure monitoring. After being elevated slightly for 48 h, the intracranial pressure returned to normal values. The accompanying blood flow studies were obtained on the third posttrauma day, initially at a pCO_2 of 23 and then at 36 mmHg (Fig. 1). A relative hyperemia was apparent over the lateral aspect of the left hemisphere. A loss of autoregulation in this area is apparent with a normal response of flow to pCO_2 manipulation occurring elsewhere in the hemisphere.

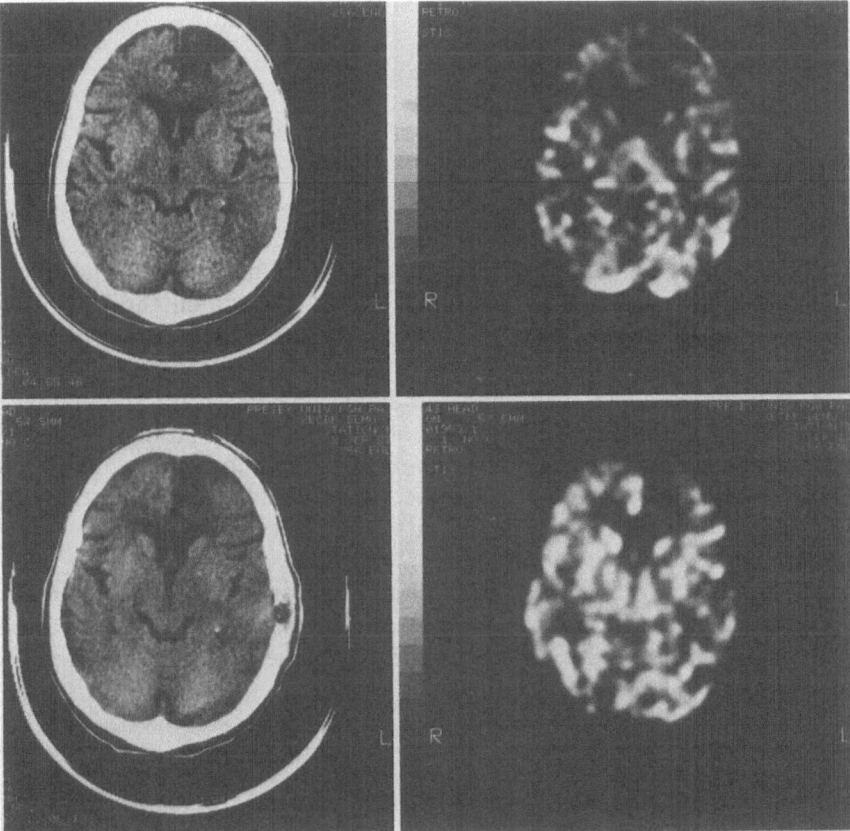

Fig. 2. The baseline CT images on the *left* for the flow maps on the *right* are displayed for studies obtained prior to STA bypass (*above*) and following bypass surgery (*below*). A marked improvement in blood flow values throughout both hemispheres is evident in the postoperative studies

Subsequent to this study, the patient began to improve neurologically and by the fourth posttrauma week was discharged to a rehabilitation facility with a spastic right hemiparesis and expressive aphasia.

Case 2. This 65-year-old right-handed female presented with the recent history of vertigo, dysarthriga, and confusion. She had been neurologically asymptomatic until 1 year previously when bilateral carotid bruits were identified and carotid endarterectomies performed. Four months following the carotid procedures, the patient had the onset of a right hemiparesis and was subsequently shown to have an infarction of the right anterior cerebral artery distribution. Vascular studies then demonstrated that both internal carotid arteries had occluded and that cerebral perfusion was being provided by a single vertebral artery. Subsequently the blood flow study (Fig. 2-above) demonstrated a marked reduction of blood flow values progressively as one moved anteriorly within both hemispheres. Subsequently the patient underwent a left STA/MCA bypass. She tolerated the procedure well and had an apparent stabilization of her neurologic course without further episodes of confusion or imbalance. The blood flow study 3 weeks following surgery (Fig. 2-below) demonstrated a marked improvement of blood flow values in both hemispheres.

Discussion

Cerebral blood flow mapping with the xenon/CT method has become a clinically useful tool in the past few years [1, 2, 3]. Refinements in both computerized transmission tomography and in computational methodologies used in the calculation of l-CBF have resulted in a substantial improvement in the information being obtained. The increased signal-to-noise ratio and improved stability of current scanners have made possible the quantitation of flow throughout the brain with acceptably low xenon concentrations in awake individuals. Shorter scanning times and programmed incremental table movements have also made possible the acquisition of data from more than one brain level during a single inhalation study.

The blood flow information that is being obtained by the xenon/CT method is consistent with that derived by other methodologies [4]. This information is particularly useful because it can be directly correlated with baseline CT anatomy.

These xenon/CT blood flow studies have been tolerated by nearly all individuals with a success rate of over 90%. The most common reason for a "poor" study has been movement, due to laughter. Studies are now being accomplished on a routine basis in a prototype unit incorporated totally within the GE/9800 scanning facility. The illustrations for this article were obtained directly from this unit. This study requires about 40 min and it can be performed on in-patients as well as out-patients without difficulty. The cost of the xenon for an average study is less than $ 60.00.

We believe that the xenon/CT method of blood flow determination provides a valuable new means of noninvasively obtaining tomographic blood flow information. Application of this type of information should have direct application to the management of a broad spectrum of clinical disorders.

References

1. Gur D, Wolfson SK, Yonas H, Good WF, Shabason L, Latchaw RE, Miller DM, Cook EE (1982) Progress in cerebrovascular disease: Local cerebral blood flow by xenon enhanced CT. Stroke 13 (6):750−758
2. Yonas H, Wolfson SK, Gur D, Latchaw RE, Good WF, Leanza R, Jackson DL, Jannetta PJ, Reinmuth OM: Clinical experience with the use of xenon-enhanced CT blood flow mapping in cerebrovascular disease. Stroke 15 (3):443−449, 1984
3. Meyer JS, Hayman LA, Sakai F, Yamamoto M, Nakajima S, Armstrong D (1979) High resolution three-dimensional measurement of localized cerebral blood flow by CT scanning and stable xenon clearance: Effect of cerebral infarction and ischemia trans. Ann Neurol 104:85−89
4. Pasztor E, Symon L, Dorsch NWC, Branston NM (1973) The hydrogen clearance method in assessment of blood flow in cortex, white matter and deep nuclei of baboons. Stroke 4:556−557

Cerebral Blood Flow in Patients with Extracranial Cerebrovascular Disease as Measured by Stable Xenon CT

F. J. Schuier[1], C. Härtel[1], A. Hartmann[2], D. Göbel[2], and A. Aulich[2]

Introduction

Advanced atherosclerotic disease of extracranial cerebral arteries in patients with no or negligible neurologic deficit and no brain damage on CT have been identified by new noninvasive test batteries [1]. The correct therapy for these patients sometimes poses a problem, because even an optimal assessment of their vascular disease does not predict their hemodynamic effect on brain perfusion. For such cases a clinically applicable, noninvasive, local, and quantitative method to measure brain blood flow is desirable.

Many patients with extracranial cerebrovascular disease are referred to our department and examined with noninvasive Doppler ultrasound techniques [4]. Stable xenon CT scanning [2, 3, 8, 9] therefore seemed to us a suitable method to complement the vascular diagnostic procedures.

This is a preliminary account of our initial experience with the stable xenon CT technique in patients with extracranial cerebrovascular disease.

Material and Methods

Scanner

We use a single slice ND 8000 scanner (CGR, France). Slice thickness is 6 or 9 mm, pixel size 5 by 5 mm. The scan time is 40 s and the scan-to-scan interval 100 s. Dynamic phantom studies showed that a 40-s scan gives a time weighted average of an attenuation changing during the scan interval (Fig. 1). The scan time used for the blood flow calculations therefore was defined as the midpoint of the scan interval.

Variation of energies is limited with the ND 8000 scanner. The 120 keV and 20 mAmp setting was found to be optimal with a Hounsfield enhancement of 120–150 for a 100% xenon phantom. Normal gray matter showed a Hounsfield

1 Neurologische Klinik, Medizinische Einrichtungen der Universität, Moorenstr. 5, D-4000 Düsseldorf 1
2 Universitäts-Nervenklinik, Abteilung Neurologie, Sigmund-Freud-Straße 25, D-5300 Bonn

Cerebral Blood Flow and
Metabolism Measurement.
Eds. Hartmann/Hoyer
© Springer-Verlag Berlin Heidelberg 1985

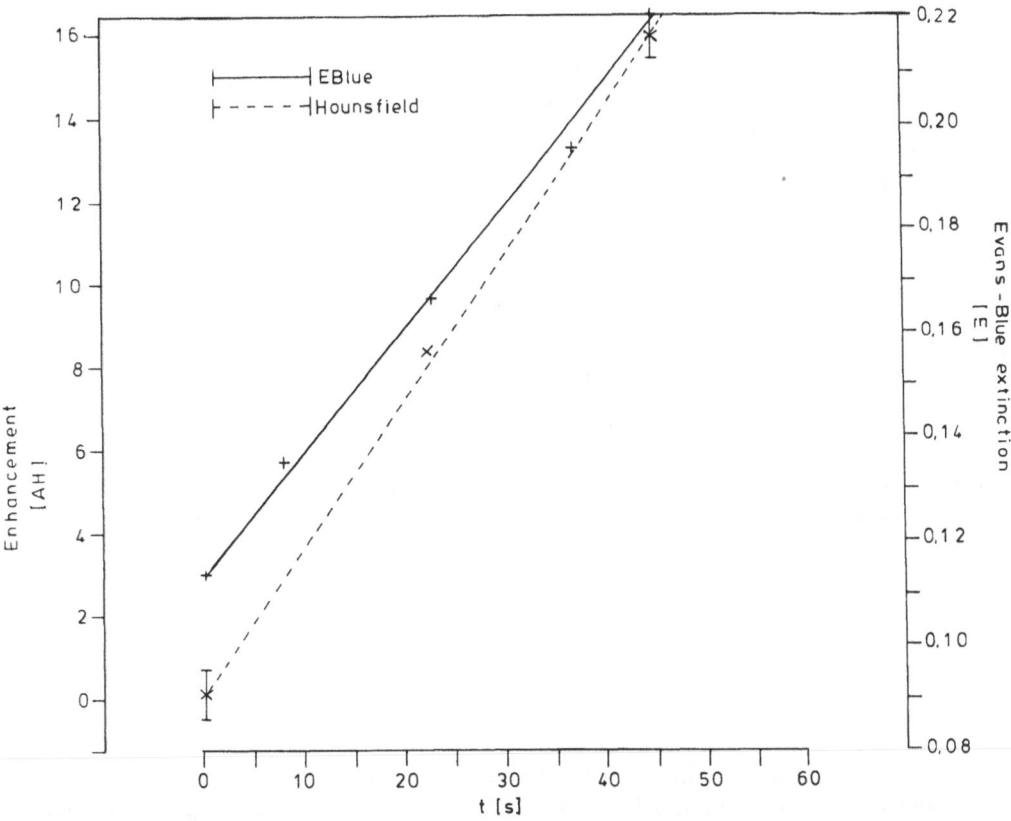

Fig. 1. Dynamic phantom study with iodine concentration changing from 0 to 20 Hounsfield units during the 40-s scan interval. Baseline scans were made before a linear increase in iodine concentration was produced with a ramp infusion into the phantom. Linearity of changing concentration was verified with additional Evans blue in the iodine solution. Repeat scans were made after the iodine infusion was stopped. The scan during infusion shows a mean Hounsfield value between the scans before and after infusion

value of 10 at that energy. Stability of the scanner was $\pm$ 0.7 Hounsfield units (SD) for an iodine phantom with a mean value of 9 Hounsfield units.

Gas Application

Prepared mixtures of xenon in oxygen (Messer-Griesheim, FRG) were inhaled through a face mask and a low resistance breathing system (Jaeger, FRG). Thirty-five percent xenon in 65% oxygen was used. Lower concentrations of xenon were found to yield too low enhancement. Higher concentrations considerably increased the side-effects of xenon, complicating the study. Patients were breathing room air during baseline scans. Inhalation of pure oxygen prior to xenon was abandoned because of a prolonged study time with increased discomfort and patient movement. With 5−8 min xenon-oxygen inhalation and three washin scans at 2, 4.5 and 7 min, negative side-effects were minimal with reasonable contrast enhancement [11].

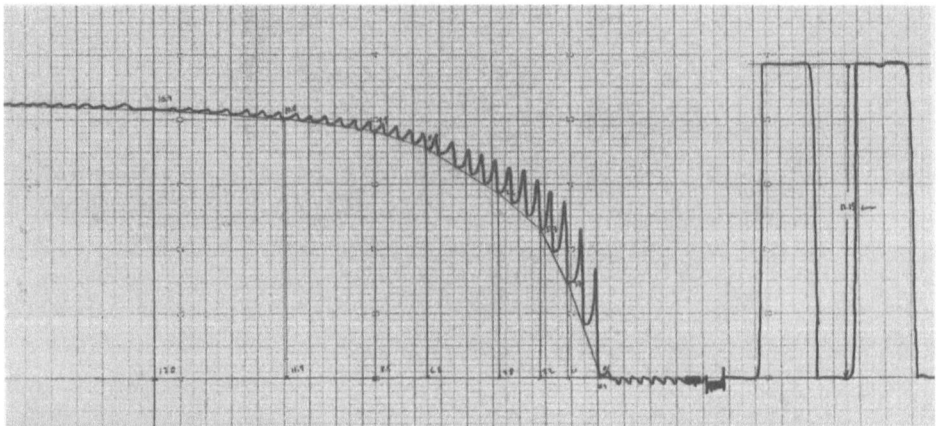

Fig. 2. Typical gas leak detector signal during calibration with 35% xenon in oxygen and xenon inhalation. The time axis is from right to left; the time base is 6 cm per minute

Arterial Xenon Concentration

End-tidal xenon concentration was measured with a gas leak detector (GowMac, USA, model 21–150). This detector gives a linear response to xenon from 0 to 35% [8]. The 100% response time to a step function of 35% xenon in oxygen was modified to 1 s. The detector was found to be uninfluenced by the temperatures in the face mask. The effect of humidity produced by patient respiration was also studied. A slight decrease in signal intensity was observed when gas saturated with water vapor at 32 °C was introduced into the sampling system. No further change was found, however, during the subsequent 30 min. Since patients saturated the breathing system with humidity during the baseline scans before xenon inhalations and a whole study lasted no longer than 20 min, water vapor should not distort the shape of the end-tidal xenon curve used for calculations. A typical example of such a curve is given in Fig. 2.

End-tidal xenon concentration was converted to arterial xenon enhancement in Hounsfield units using the patient's hematocrit and the specific absorption coefficients for water and xenon [5, 9]. The coefficients were determined before each study with a water and a xenon phantom rather than with iodine phantoms [5].

Blood Flow Calculations

Flow calculations were made from the converted end-tidal xenon concentration, three or four baseline scans, and three washin scans. Enhancement scans were obtained with a computer program (Fig. 3).

From these enhancement scans 2 by 2 cm regions of interest were selected with the ND 8000 software in frontal, temporal, parieto-occipital, and basal

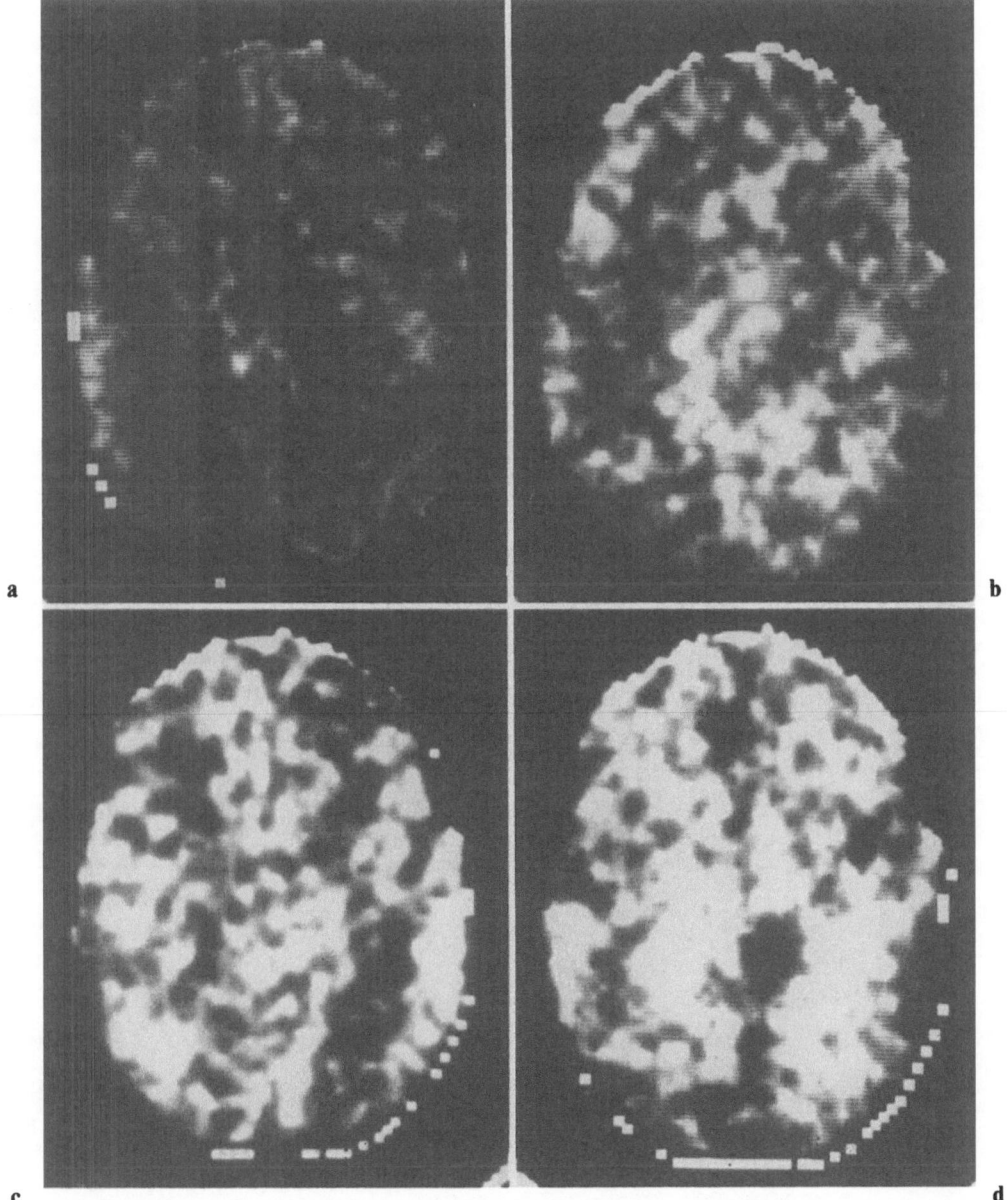

Fig. 3a–d. Enhancement scans of a monkey study at 1, 3, and 6 min (washin) and at 10 min (washout)

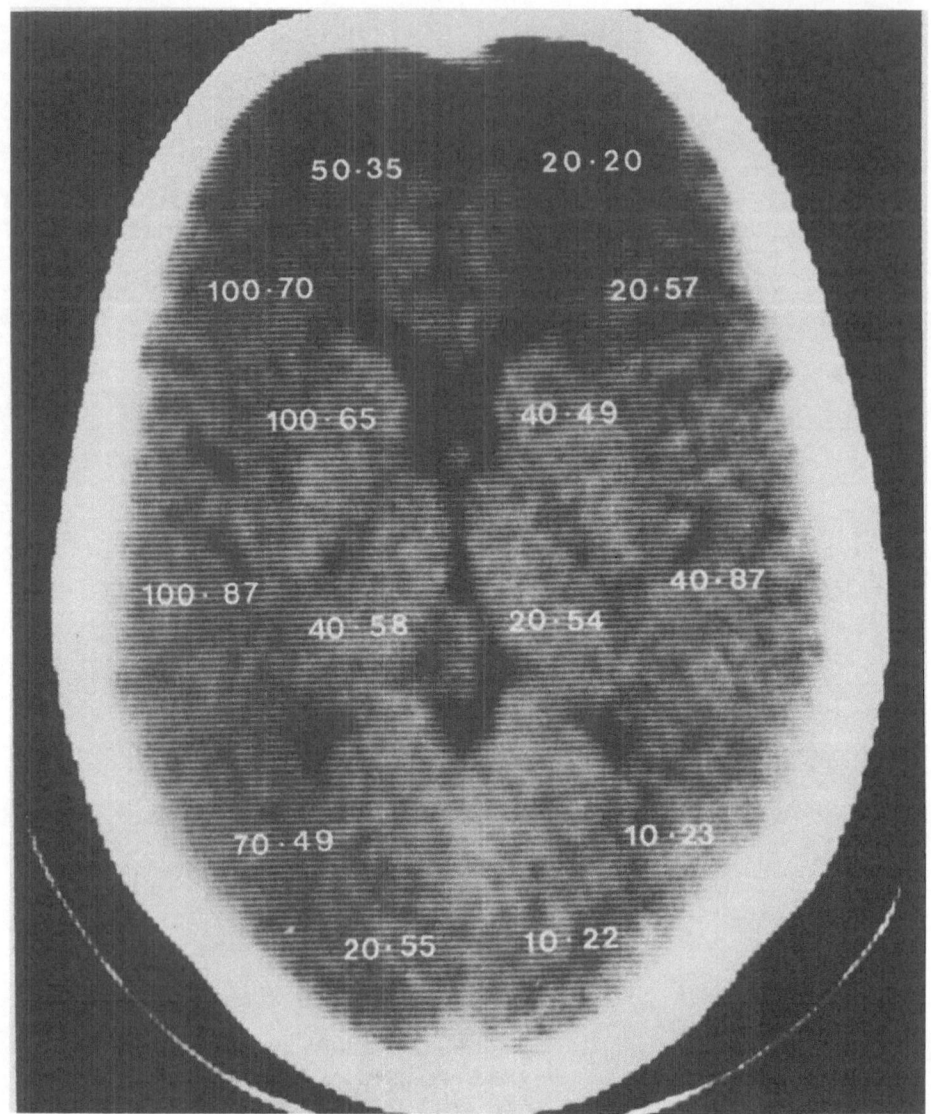

Fig. 4a, b. Cases 1 (**a**) and 2 (**b**). Right-left hemisphere. Flow values in ml/100 g/min. Values at 1st and 2nd study

ganglia gray matter regions. The mean enhancement values were recorded for each region of interest and used for further calculation. A common partition co-efficient, lambda, of 0.9 was used for all structures. Flow was calculated according to Kety [6] using a computer program:

$$Ci\,(T) = k * lambda * \int_0^T Ca\,(t) * exp - k\,(T - t) * dt. \qquad (1)$$

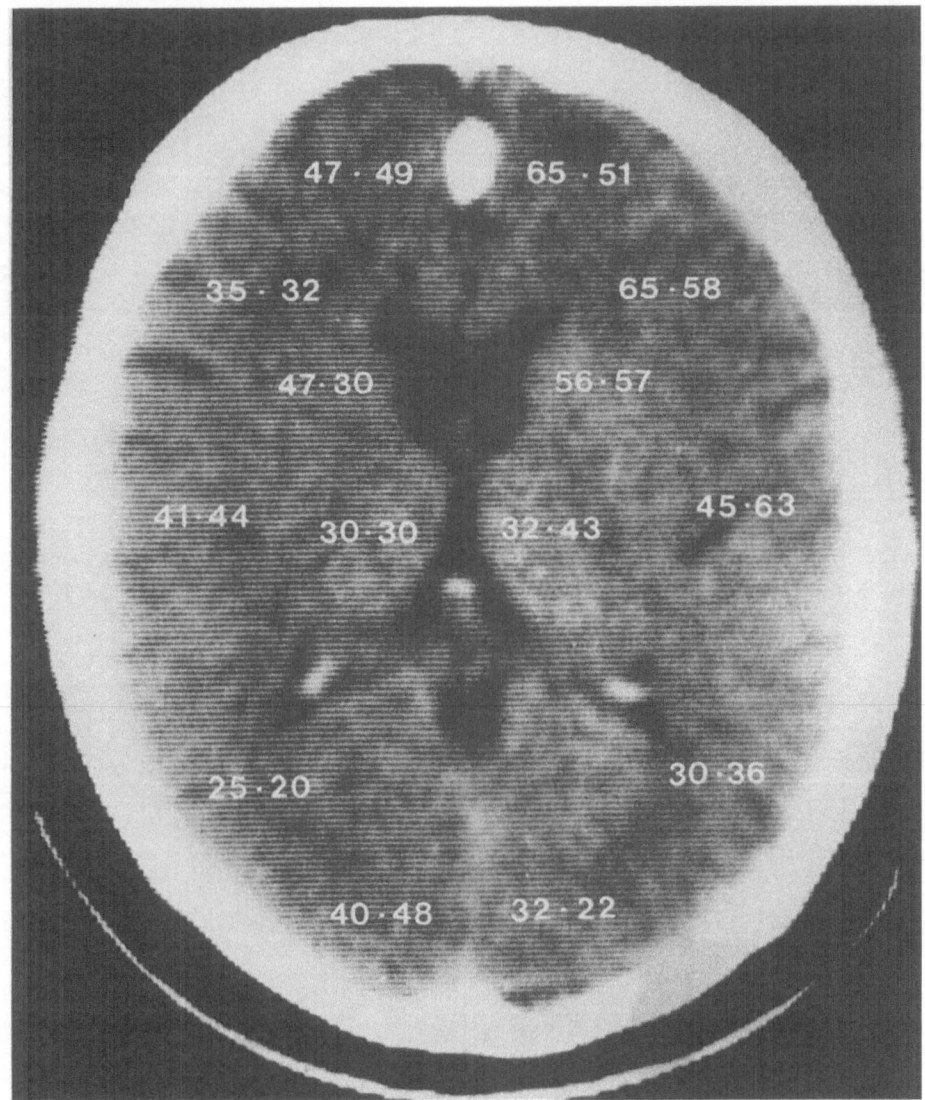

b

Fig. 4b.

The last enhancement scan with best contrast was mainly used for calculations. Typical enhancement values ranged from 2 to 6 delta Hounsfield units. In some cases with high blood flow the second washin scan could also be used. In such cases flow values of both scans were averaged. Flow values so obtained were then mapped by hand onto photographs of baseline scans.

We do not consider this method quantitative but qualitative. We therefore made no efforts to correct the flow data for arterial pCO_2.

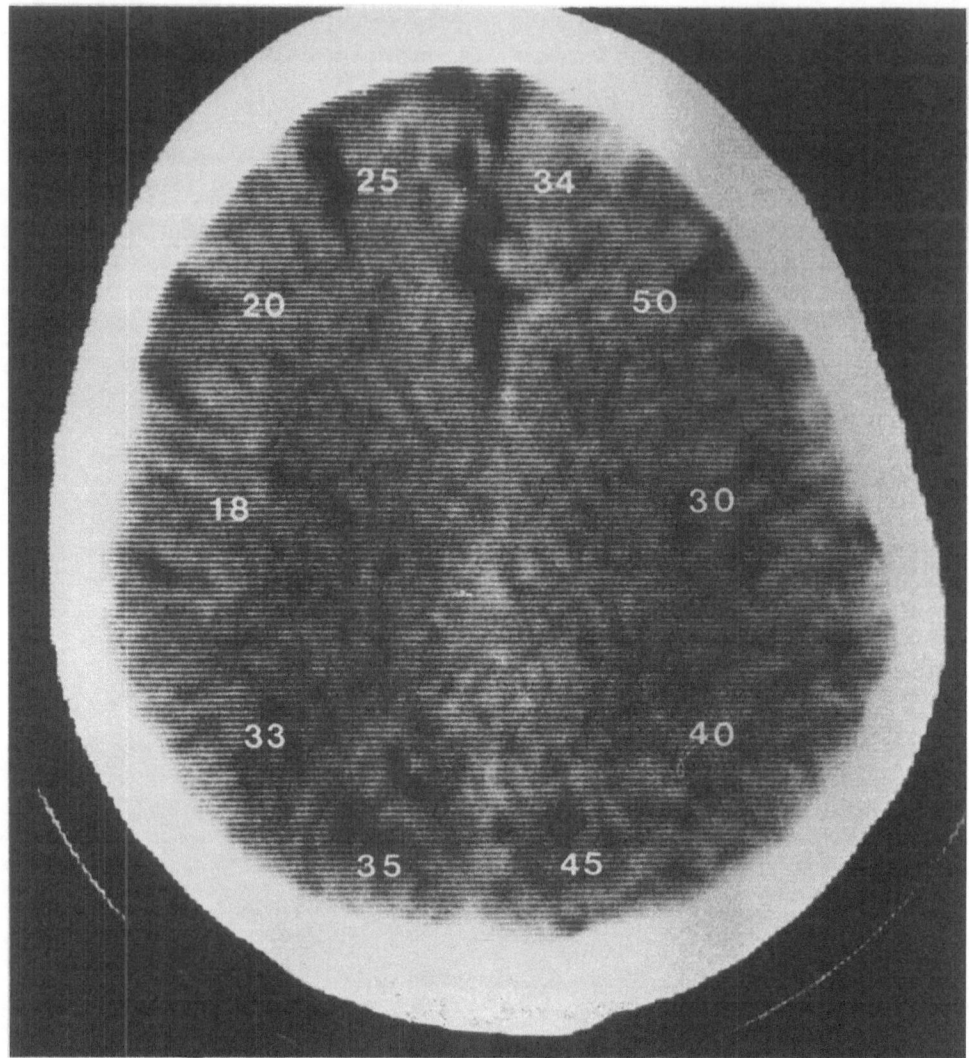

Fig. 5a, b. Cases 3 (**a**) and 4 (**b**). Right-left hemisphere. Flow values in ml/100 g/min

Results

Five typical examples of our studies in patients with severe extracranial cerebrovascular disease and no or minor neurologic deficit and no or minor abnormality on CT are shown in Table 1 and Figs. 4–6.

Case 1 (Fig. 4a) demonstrates that the stable xenon CT method is able to identify a hemodynamic TIA. Partial clinical improvement after EC-IC bypass surgery correlates well with a lower frequency of TIAs. These did not cease completely but occurred only in conjunction with severe hypotension.

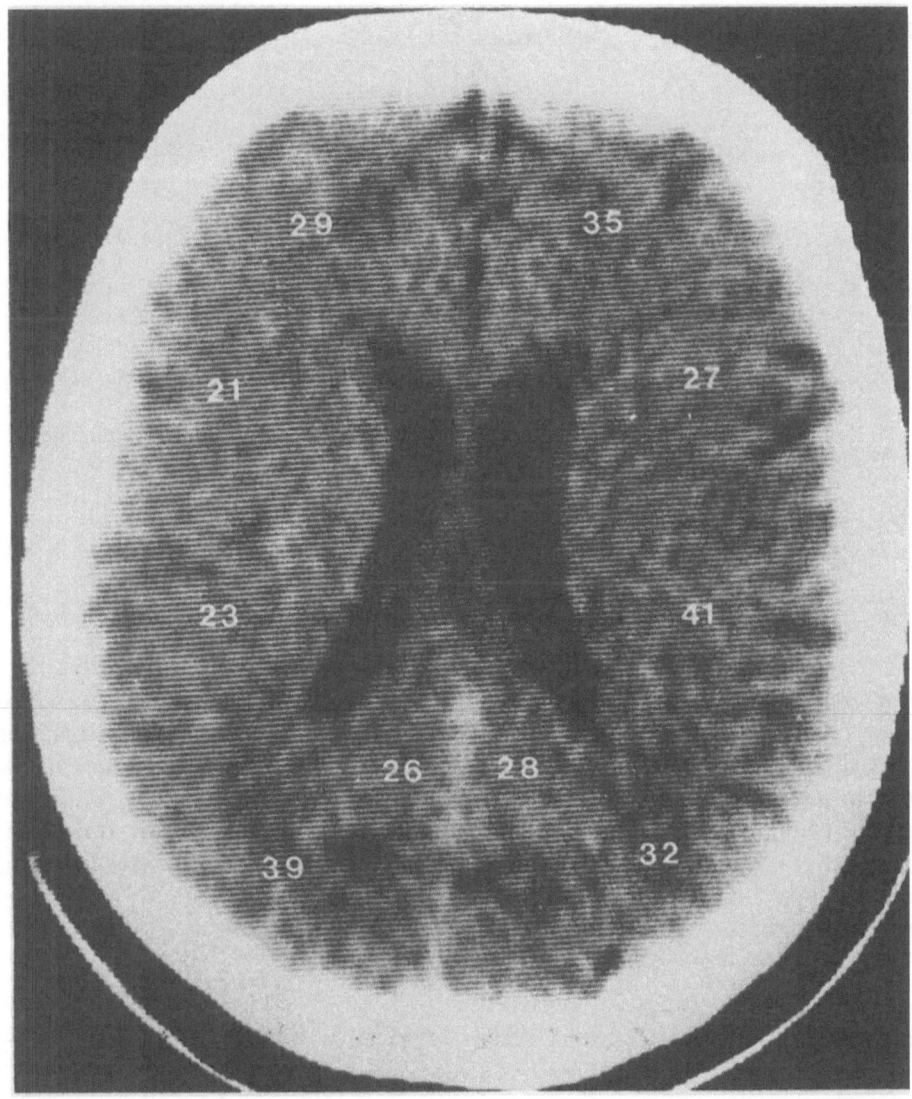

b

Fig. 5b.

Case 2 (Fig. 4b) shows that the 80% stenosis, which from a "vascular" point of view might or might not be hemodynamically significant, is indeed coupled to a low flow area ipsilateral to the stenosis. This, however, does not prove a causal relationship, since a TIA can be followed by neuronal loss undetectable on CT and subsequent matched low flow demand [7].

Cases 3 and 4 are good examples that unilateral ICA occlusion combined with contralateral stenosis has no predictable effect upon the degree and lateralization of flow deficit in the brain. Knowledge of the vascular disease therefore does not allow a complete description of its effect on the brain circulation.

Table 1. Details of five of our patients

Patient	Symptoms	Vasc. dis.	Neur. def.	CT	CBF
# 1: H.E., 45 y,	L. hem. TIA	L. ICA occl.	0	0	Gen. flow decr. l. hem.
# 1: H.E., 45 y, fem.	L. hem. TIA with hypotension	After EC–IC bypass	0	0	Focal decr. fro. and occ. lobe
# 2: A.J., 46 y male	R. hem. TIA	R. ICA sten. 80%	0	0	Focal decr. l. MCA terr. at 2wk and at 2mo
# 3: F.R., 63 y, male	R. hem. TIA r. amaur. f.	R. ICA occl. l. ICA sten. 50–60%	0	0	Bilat. flow decr., $r > 1$
# 4: K.W., 59 y, male	Interm. claudic.	L. ICA occl. r. CCA sten. bil. SA sten.	0	0	Bilat. flow decr., $r > 1$
# 5: G.B., 74 y, male	L. hemi. fac-brach. paresis	R. ICA occl. l. ICA sten. 50–60%	Minimal R. fac-brach. paresis	L. tempor. plaque	Flow decr. L. MCA terr.

Case 3 (Fig. 5a) showed the lowest flow values on the right side, ipsilateral to the complete occlusion, whereas case 4 (Fig. 5b) showed the most pronounced flow deficit on the right side, which is contralateral to the complete obstruction.

A similar situation is encountered in case 5 (Fig. 6): low flow values ipsilateral to the incomplete left ICA obstruction, but higher flow values on the side of the complete right ICA occlusion. In this case, however, low flow is not necessarily of hemodynamic origin, since it is confined to an area of minor morphological deficit on CT. Low flow in that area most probably reflects low flow demand.

Comment

Our preliminary data show no consistent relation between severity and laterality of extracranial cerebrovascular disease and the degree or side of flow changes in the brain. This is no surprise, because of the brain's unique collateral circulation.

The small number of observations and the nonquantitative nature of our data do not allow us to draw conclusions regarding brain perfusion patterns related to vascular disease. In addition, the incomplete assessment of brain perfusion by a single "flow slice" makes a correlation between vascular and flow changes difficult. Clearly, at least two or three scans at representative brain levels are needed for a sensible correlation.

Our data are not quantitative for three reasons: [1] low contrast with regard to the flow algorithm based on a single enhancement scan, [2] severe partial vol-

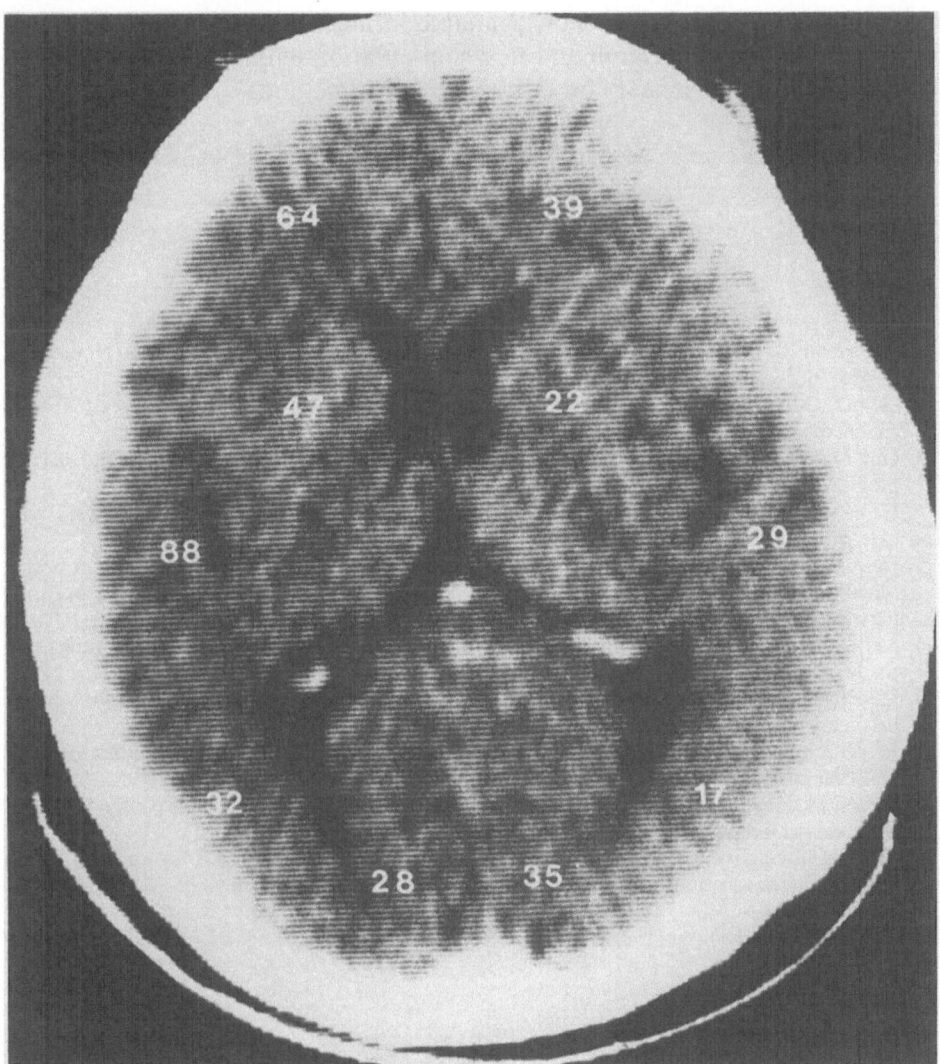

Fig. 6. Case 5. Right-left hemisphere. Flow values in ml/100 g/min

ume effect in large inhomogeneous regions of interest, and [3] unknown blood-brain partition coefficient. Low contrast can be counteracted by longer inhalation schedules with multiple washin-washout scans [2, 10]. Partial volume effect can be minimized by compartmental analysis. The partition coefficient, finally, can be estimated from multiple scans if longer inhalation schedules are used, as in the case of ongoing studies.

Before such improvements are made and the accuracy and precision of the method are known, no conclusions will be drawn from our patient data. It seems to us, however, that the stable xenon CT method should be well suited for the measurement of cerebral blood flow under clinical routine conditions.

The method is noninvasive, potentially quantitative, and does not need radioactive isotopes. CT scanners are now available in most hospitals, and additional equipment for gas application and flow computation is inexpensive compared with the equipment necessary for other blood flow methods.

Acknowledgment. This work was supported in part by grants from the DFG, SFB 200, DI and the MWF-NW, grant IV B 5FA 8375 (F.J.S.)

References

1. Ackerman R (1979) A perspective of noninvasive diagnosis of carotid disease. Neurology 29:615–622
2. Gur D, Good WF, et al. (1982) In vivo mapping of local cerebral blood flow by xenon-enhanced computed tomography. Science 215:1267–1268
3. Gur D, Wolfson KS, et al. (1982) Local cerebral blood flow by xenon enhanced CT: A progress review. Stroke 13:750–758
4. Hennerici M, Aulich A, et al. (1982) Incidence of asymptomatic extracranial arterial disease. Stroke 12:750–758
5. Kelcz F, Sadek K, et al. (1978) Computed tomographic measurement of the xenon brain-blood partition coefficient and implications for regional cerebral blood flow: a preliminary study. Radiology 127:385–392
6. Kety SS (1951) The theory and application of the exchange of inert gases at the lungs and tissue. Pharmacol Rev 3:1–41
7. Lassen NA (1982) Incomplete cerebral infarction – Focal incomplete ischemic tissue necroses not leading to emollition. Stroke 13:522–523
8. Meyer JS, Hayman LA, et al. (1981) Mapping local cerebral blood flow of human brain by CT scanning during stable xenon inhalation. Stroke 12:426–436
9. Meyer JS, Hayman LA, et al. (1980) Local cerebral blood flow measured by CT after stable xenon inhalation. AJNR 1:213–225
10. Thaler HT, Baglivo JA, et al. (1982) Repeated least squares analysis of stimulated xenon computed tomographic measurements of regional cerebral blood flow. J Cer Blood Flow Metabol 2:408–414
11. Yonas H, Grundy B, et al. (1981) Side effects of xenon inhalation. J Comp Ass Tomogr 5:591–592

Variability of the Partition Coefficient in Relation to Cerebral Blood Flow with a Freely Diffusible Tracer

R. M. Lambrecht[1], A. Rescigno[1,2], C. C. Duncan[2], C.-Y. Shiue[1], and L. R. Ment[2]

Cerebral blood flow (CBF) measurements with freely diffusible tracers have depended upon separate partition coefficients (lambda) for gray and white matter. As well, certain tracers have been considered inadequate in species with very high CBF as the lambda diminishes at high flow rates. Similarly, the relation of CBF in normal and pathologic circumstances related to the reliability of the measurement remains poorly understood.

As we have shown (Duncan et al. 1983; Lambrecht et al. 1983), the cerebral blood concentration (C_b (t) of a tracer can be described by the differential equation

$$dC_b(t)/dt + f/L \cdot C_b(t) = f \cdot C_a(t), \tag{1}$$

where $C_a(t)$ is the concentration of the tracer in the arterial blood, f is the CBF, and L the partition coefficient between brain and blood.

A servo controlled infusion pump (Lambrecht et al. 1983) was used to produce a linearly increasing arterial concentration

$$C_a(t) = S \cdot (t - t_0) \tag{2}$$

of the freely diffusible tracer, ^{18}F-4-fluoroantipyrine (^{18}FAP), in the baboon. At the same time we measured the concentration $C_b(t)$ of the tracer in different parts of the brain with a positron emission tomograph (PETT VI).

If f is constant at least in the interval from T_0 to t, then equation (1) can be integrated and with condition (2) it becomes

$$C_b(t) = LS \cdot (t - t_0 - L/f) + [C_b(t_0) + L_2S/f] \cdot \exp [- (t - t_0)f/L], \tag{3}$$

where $C_b(t_0)$ is the value of $C_b(t)$ at time t_0.

After a short interval of time, the exponential term in the expression above becomes negligible, and we have simply

$$C_b(t) = LS \cdot (t - t_0 - L/f), \tag{4}$$

i.e., the equation of a straight line of slope LS and intersecting the time-axis at the point $t = t_0 + L/f$.

As shown by equation (2), S is by definition the slope of $C_a(t)$ versus t; therefore the partition coefficient L is the ratio of the slopes of $C_b(t)$ and $C_a(t)$. Once L is known, the value of f can be computed from L/f.

1 Chemistry Department, Brookhaven National Laboratory, Upton, NY 11973, USA
2 Yale University

Cerebral Blood Flow and
Metabolism Measurement.
Eds. Hartmann/Hoyer
© Springer-Verlag Berlin Heidelberg 1985

Table 1.

ft/L	dC_b/C_b
1.0	1.000
1.2	0.601
1.4	0.381
1.6	0.252
1.8	0.171
2.0	0.119
2.4	0.061
2.8	0.033
3.2	0.018
3.6	0.010

Table 1 shows the relative errors in C_b by using equation (4) instead of (3), i.e., by ignoring the exponential term in equation (3):

With typical values of $f = 0.3$ min^{-1} and $L = 1$, the error in C_b is approximately 6% after 8 min.

With the PET, the acquisition time of a typical experiment is of the order of 2 min. Therefore the tracer activity is not measured as a continuous function of time, but rather by discrete intervals; instead of the instantaneous value $C_b(t)$, its average value

$$1/(t_2 - t_1) \cdot \int_{t_1}^{t_2} C_b(\tau) \, d\tau$$

over the interval t_1, t_2 is measured. The resulting values should be corrected for the radioactive decay of the nuclide used, as outlined in a previous paper (Rescigno et al. 1983).

Suppose that $C_b(t)$ has been measured at two different times such that the error due to the exponential term is negligible; call A and B the values found and depicted in Fig. 1:

t	$C_b(t)$
t_1	A
t_2	B

The slope of the straight line extrapolating $C_b(t)$ is given by

$$LS = (B - A)/(t_2 - t_1);$$

the intercept D of this straight line with the t-axis is

$$D = (t_1 B - t_2 A)/(B - A),$$

with

$$D = t_0 + L/f;$$

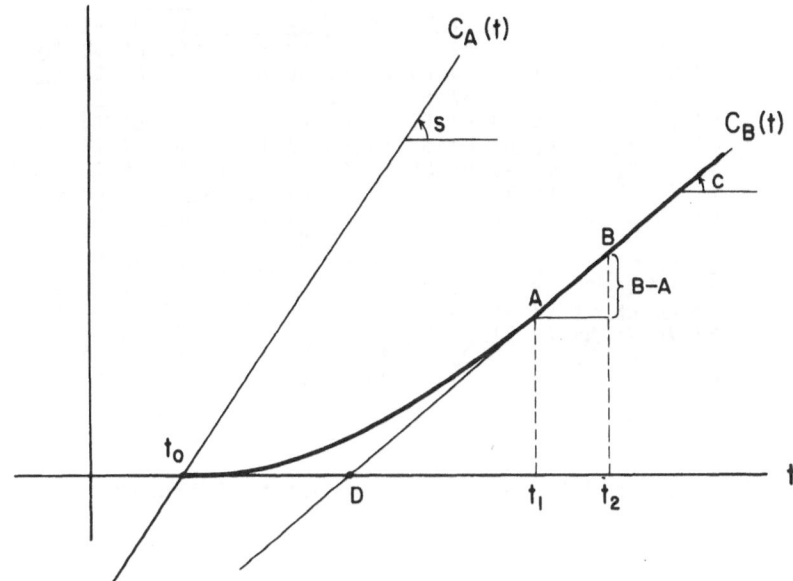

Fig. 1.

thence

$$L = 1/S \cdot (B - A)/(t_2 - t_1)$$

and

$$f = 1/S \cdot (B - A)/(t_2 - t_1) \cdot 1/(D - t_0).$$

Taking the total differentials of the last two expressions,

$$dL/L = (t_2 - D)/(t_2 - t_1) \cdot dB/B - (t_1 - D)/(t_2 - t_1) \cdot dA/A - dS/S, \tag{5}$$

$$
\begin{aligned}
df/f &= (t_2 - D)/(t_2 - t_1) \cdot [1 - (t_1 - D)/(D - t_0) \cdot dB/B \\
&\quad - (t_1 - D)/(t_2 - t_1) \cdot [1 - (t_2 - D)/(D - t_0)] \cdot dA/A \\
&\quad + t_0/(D - t_0) \cdot dt_0/t_0 - dS/S.
\end{aligned} \tag{6}
$$

These two equations show how the errors in the measurements of A, B, D, S, and t_0 propagate in the computation of L and f. Of course, all equations above ignore the infinitesimals of order higher than one; therefore these results are valid only for small errors.

Figure 2 shows the values of lambda computed for a particular region of the brain of a baboon, and Fig. 3 shows the corresponding values of f, in ml·min⁻¹ per 100 g of tissue. The iso-lines of these figures connect points of equal lambda.

The values shown are computed using equation (4), i.e., by ignoring the exponential term in equation (3); to give an idea of the precision obtained we can analyze a particular pixel, say the one in row 48, column 66, slice 5, as shown by the small arrows in Figs. 2 and 3. Table 2 summarizes the results.

```
64 66 69 71 73 75 77 79 79 79 79 78 79 81 82 81 77
64 69 73 76 79 81 82 83 83 83 82 82 82 83 84 83 80
62 69 75 80 83 84 85 85 85 85 84 84 84 84 85 84 80
60 68 75 81 85 87 87 86 86 85 85 84 84 84 84 82 78
60 68 76 82 86 88 87 86 84 84 83 83 83 82 81 78 74
63 71 79 85 88 88 87 85 83 82 81 81 81 80 77 74 69
69 77 84 89 91 90 88 85 83 81 80 80 79 77 75 70 65
76 84 91 94 95 94 91 87 84 82 81 80 79 77 74 69 62
82 90 96 99 99 98 94 91 88 85 84 82 81 79 75 70 62
85 94 99 99 99 99 98 95 92 89 87 85 83 81 77 71 63
84 93 99 99 99 99 99 98 95 92 90 87 85 82 78 72 63
75 88 96 99 99 99 99 99 97 94 90 88 84 81 77 70 61
72 80 88 94 97 98 98 97 95 92 88 85 81 77 72 65 56
62 70 78 85 89 91 92 91 90 87 83 79 75 70 64 57 48
52 59 66 73 77 80 82 82 81 79 76 71 66 60 54 47 39
40 46 53 59 64 67 69 70 70 68 65 61 56 50 44 37 30
```

Fig. 2. Values of L · 10² for measurements obtained 35 and 55 min into the ramp injection

```
24 32 39 42 39 34 30 28 29 32 37 40 35 27 21 17 15
27 31 32 32 31 30 29 28 28 30 32 33 30 25 20 17 14
41 36 32 30 29 29 29 29 30 31 31 30 28 24 20 17 15
   50 35 30 29 30 32 34 36 37 35 32 29 25 21 18 15
   59 37 31 30 33 38 46 53 54 47 39 33 28 24 20 17
   43 32 30 31 36 48 71       82 56 41 34 29 25 20
34 28 26 26 29 36 52 96          78 52 41 36 32 26
22 21 21 23 26 32 45 75          72 49 40 37 34 30
19 18 19 20 22 26 34 47 64 73 63 48 37 32 29 27 25
18 17 18 18 20 23 27 33 38 41 39 33 28 25 22 21 19
18 17 17 18 19 21 23 26 29 30 29 27 24 21 19 17 15
19 18 17 17 18 19 21 23 25 26 25 23 21 19 16 14 12
21 19 18 18 18 19 21 22 23 23 23 21 19 17 15 13 11
25 21 19 18 19 20 21 22 22 22 21 19 18 15 13 11  9
37 28 23 21 21 21 21 21 21 19 18 17 15 13 11  9  7
   79 36 27 25 24 23 21 19 17 15 14 12 10  8  7  6
```

Fig. 3. Values of f in ml · min⁻¹ per 100 g of tissue for measurements obtained 35 and 55 min into the ramp injection

Table 2.

	Measured value (nCi/gr)	Error due to exponential term
A	2245.44	1.8%
B	2712.04	0.88%

L = 0.997; f = 17.65

Equations (4) and (5) become

$$dL/L = 5.8 \, dB/B - 4.8 \, dA/A - dS/S,$$
$$df/f = -9.0 \, dB/B + 10.0 \, dA/A - dS/S + 4.9 \, dt_0/t_0;$$

and show (1) that the error in f is about twice as large as the error in L; (2) that the errors caused by A and by B act in opposite directions and could partly compensate each other; and finally (3) that the determination of the time t_0 of the beginning of the ramp injection is critical in the computation of f.

Acknowledgment. This research was performed at Brookhaven National Laboratory under contract DE-AC02-76CH0016 with the U.S. Department of Energy, and was supported by its Office of Health and Environmental Research; and by grant NINCDS RO1 NS-16801-03.

References

Duncan CC, Lambrecht RM, Rescigno A, Shiue C-Y, Bennett GW, Ment LR (1983) The ramp injection of radiotracers for blood flow measurements by emission tomography. Phys. Med. Biol. 28:963–972

Lambrecht RM, Duncan CC, Bennett GW, Ducote (1983) Method and Apparatus for injecting a substance into the bloodstream of a subject. U.S. Patent 4,409,966. October 18, 1983

Rescigno A, Lambrecht RM, Duncan CC (1983) Stochastic modelling of physiologic processes with radiotracers and positron emission tomography. In *Applications of Physics to medicine and biology* (Alberi, Bajzer and Baxa, editors). World Scientific Publ. Co., Singapore, Pages 303–318

Metabolism of the Human Brain: The Principle and Limitation of Global Measurements

S. HOYER

Global measurements of substrate consumption of any organ are based on the principle of Fick: the amount of a substrate which is used by that organ is represented by the difference in concentrations as measured supplying arterial and releasing venous blood, i.e., the arteriovenous difference. However, some conditions have to be fulfilled for this principle to hold: The arterial and venous concentrations of the substrate and blood flow through the organ have to be constant, and one main vein should drain the blood from the organ. The consumption or release of the substrate may then be calculated from the arteriovenous substrate difference and blood flow.

Arterial blood and substrate supply to the human brain takes place via large paired vessels. It is assumed that substrate concentration is identical in all large arterial vessels so that the substrate concentration in any large artery may be assumed to be representative for the arterial blood passing to the brain.

The main venous drainage from the brain takes place via the internal jugular vein. This vein is the continuation of the sigmoid sinus, which collects blood from the transverse sinus and both the superior and inferior petrosal sinus. The right transverse sinus is generally the direct continuation of the superior sagittal sinus, the left transverse sinus of the straight sinus. In 60% – 70%, the right transverse sinus is larger than the left one, and this also holds true for the jugular foramen and the internal jugular vein for phylogenetic reasons [20, 23]. The superior sagittal sinus receives its blood from superior cerebral veins but also from the pericranium via emissary veins. The inferior sagittal sinus, which ends in the straight sinus, leads blood from the falx cerebri and from the medial surface of the cerebrum. Blood from the cerebellum enters the inferior sagittal, the straight, and the transverse sinuses, and blood from the brain stem drains into the petrosal and transverse sinuses. The former are the continuation of the sinus cavernosus, into which the superior and inferior ophthalmic veins, the central vein of the retina, and the frontal tributary of the middle meningeal vein enter beside other cerebral veins. It thus becomes clear that the blood in the internal jugular vein mostly contains blood from the cerebrum, brain stem, and cerebellum but also from extracerebral structures. This venous blood has been mixed in the confluence of the sinuses, which usually directs the greater part of the flow from the superior sagittal sinus into the right transverse sinus and the greater part from the straight sinus into the left transverse sinus [5]. Mixed cere-

Institut für Pathochemie und Allgemeine Neurochemie im Zentrum Pathologie der Universität Heidelberg, Im Neuenheimer Feld 220–221, D-6900 Heidelberg 1

Cerebral Blood Flow and
Metabolism Measurement.
Eds. Hartmann/Hoyer
© Springer-Verlag Berlin Heidelberg 1985

Table 1. Cerebral arteriovenous differences of oxygen and glucose during blood flow measurements (saturation) in arterial
normotension, normocapnia, and normothermia in six young
adult dogs: Comparison of measurements in the same animal

	Oxygen (mmol/l)	Glucose (mmol/l)
1 a	4.90	0.63
b	4.96	0.68
c	4.83	0.59
mean	4.90	0.63
2 a	6.28	0.64
b	6.35	0.65
c	6.19	0.60
mean	6.27	0.63
3 a	4.42	0.43
b	4.38	0.46
c	4.32	0.43
mean	4.37	0.44
4 a	4.83	0.61
b	4.77	0.63
c	4.14	0.56
mean	4.58	0.60
5 a	4.28	0.40
b	4.31	0.43
c	4.21	0.37
mean	4.27	0.40
6 a	4.01	0.60
b	4.24	0.64
c	4.21	0.59
mean	4.15	0.61

a: Beginning of the measurements: t_0 min
b: Integrated blood sample: $t_0 - t_{10}$ min
c: End of the measurement: t_{10} min

bral venous blood can be collected in the superior bulb of the internal jugular
vein, which is situated in the posterior part of the jugular foramen. Distal from
the superior bulb, the internal jugular vein is joined by the facial, lingual,
pharyngeal, and thyroid veins, so that the extracerebral contamination increases
in this part of the internal jugular vein.

With respect to dynamic flow mechanisms, two-thirds of the blood supplied
to one hemisphere by means of an internal carotid artery is actually drained by
the internal jugular vein. As was shown by Shenkin, Harmel, and Kety [21], the
venous blood in the jugular bulb was found to be contaminated to only 3% by
blood from extracerebral structures. It thus may be stated that blood from the

superior bulb of the internal jugular vein represents greatly mixed blood from all brain structures.

The constancy of cerebral arteriovenous differences of oxygen and glucose was investigated in six young adult beagle dogs (10 months of age on average) in slight anesthesia with 0.5 vol% halothane under steady-state conditions of arterial normotension, normocapnia, normoxemia, and normothermia. Blood samples from one femoral artery and the superior sagittal sinus were taken at time t_0, t_{10} and as an integrated sample t_0 to t_{10} min. The concentrations of oxygen were measured by means of a sensitive gas chromatographic technique (22), those of glucose by means of a standard enzymatic method. As is demonstrated in Tables 1 and 2, no differences could be found either within the three measurements of each dog (Table 1) or within the time-dependent measurements of all dogs (Table 2). From these results, one may tentatively conclude that the cerebral arteriovenous substrate differences of oxygen and glucose are constant, so that one of the above conditions may be fulfilled under the steady-state conditions mentioned.

With respect to global cerebral blood flow measurements, some additional conditions have to be taken into account. The tracer which is used to calculate blood flow has to be nontoxic, may not produce any side-effects, may not be varied metabolically or induce by itself variations in metabolism, should be insoluble or only slightly soluble in blood and adipose tissue, should rapidly diffuse into the brain, should be able to be easily analyzed, should of course have a known brain-blood partition coefficient, and should be economic. Such an ideal gas is, however, not available. Indeed, only a small number of candidates fulfill a majority of the above-mentioned conditions: nitrous oxide, argon, and both labeled krypton and xenon. The properties of the labeled gases will not be discussed here in further detail since that has been done in preceding sessions of this symposium.

Table 2. Cerebral arteriovenous differences of oxygen and glucose during blood flow measurements (saturation) in arterial normotension, normocapnia, normoxemia, and normothermia in six young adult dogs: Time-dependent comparison between all dogs

	Oxygen (mmol/l)			Glucose (mmol/l)		
	a	b	c	a	b	c
1	4.90	4.96	4.83	0.63	0.68	0.59
2	6.28	6.35	6.19	0.64	0.65	0.60
3	4.42	4.38	4.32	0.43	0.46	0.43
4	4.83	4.77	4.14	0.61	0.63	0.56
5	4.28	4.31	4.21	0.40	0.43	0.37
6	4.01	4.24	4.21	0.60	0.64	0.59
$\bar{X}$	4.79	4.84	4.65	0.55	0.58	0.52

a: Beginning of the measurements: t_0 min
b: Integrated blood sample: t_0-t_{10} min
c: End of the measurement: t_{10} min

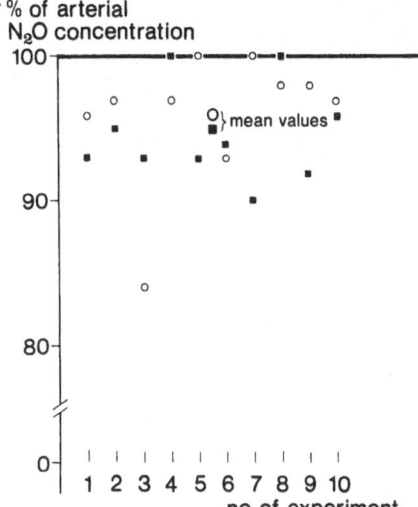

Fig. 1. Concentration of N_2O in superior sagittal sinus in percent of arterial N_2O concentration after a 60-min inhalation period of 25% N_2O in room air in dogs (○). ■ measurement after preceding desaturation and fresh saturation of 30-min duration

Nitrous oxide was originally used by Kety and Schmidt [15, 16]. It is quite well soluble in blood. This may facilitate manometric detection. On the other hand, saturation of arterial blood and brain tissue takes place rather slowly. Although Kety et al. [14] described a complete equilibration of nitrous oxide between brain and mixed cerebral venous blood after approximately 10 min of inhalation, such an equilibrium may be doubted somewhat.

In ten young adult dogs, the concentrations of N_2O were measured in arterial and venous blood from the superior sagittal sinus. The animals were ventilated by 0.5 vol% halothane and a gas mixture of nitrous oxide/room air 25:75 for 60 min under steady-state conditions of arterial normotension, normocapnia, normoxemia, and normothermia. After 60 min, the mean N_2O concentration in the sinus was 96% as related to the arterial concentration. After desaturation and fresh saturation of 30-min duration, the mean N_2O concentration in the sinus blood was 95% as related to the arterial concentration. In only two animals in each procedure could a 100% N_2O concentration in the sinus blood be found (Fig. 1).

After a 10-min saturation period, the mean concentration of N_2O in the superior sagittal sinus blood was found to be 96%. Thirty minutes later, after desaturation and another 10-min saturation, the mean N_2O concentration in the sinus blood was 93% in relation to the arterial concentration (Fig. 2).

From these findings it may be tentatively concluded that complete equilibration of N_2O between arterial blood, brain tissue, and cerebral venous blood does not occur − not after 10 min, the time usually used to perform measurements by means of the Kety-Schmidt technique, and not even after 60 min. That complete brain equilibration is never achieved either for ^{85}Kr or for N_2O can also be inferred from the earlier studies of Alexander et al. [1]. They found that brain saturation lags behind venous saturation at any level of cerebral perfusion and at any given time. Thus, cerebral blood flow as calculated from a 10-min saturation period at normal and low flow rates is overesti-

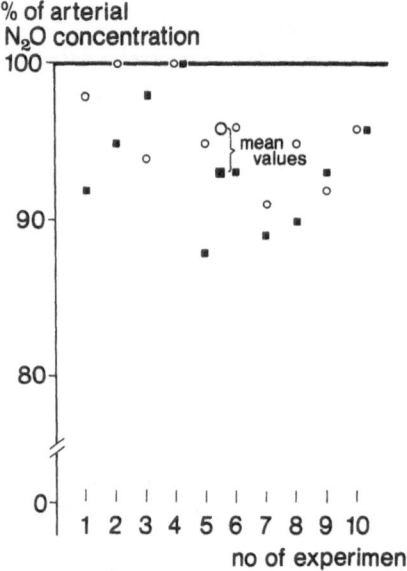

Fig. 2. Concentration of N_2O in superior sagittal sinus in percent of arterial N_2O concentration after a first (○) and a second (■) 10-min saturation period

mated by around 10% − 15%, as was demonstrated by Lassen and Klee [18]. This error can be minimized by prolonging the saturation period to 14 min and extrapolating to infinity [19]. The error may also be reduced by the use of a gas chromatographic technique which analyzes nitrous oxide and oxygen very sensitively. The standard error for both compounds was found to be below 1% [22]. Since the standard error of the glucose measurements is around 1% − 2%, it may be concluded that investigation of metabolic parameters is more accurate than blood flow measurement, provided that the Kety-Schmidt technique is used.

When we compared the results of cerebral blood flow after saturation and desaturation with nitrous oxide as a tracer in repeated measurements in the same animal, we found no statistically significant differences between saturation and desaturation although cerebral blood flow was around 5% lower in the first saturation measurement as compared with the corresponding desaturation study (Table 3).

Whether the use of the less soluble argon instead of nitrous oxide will be of advantage is still not clear. A drawback to argon is that it is hard to separate from oxygen by means of gas chromatography (3), so that at least two measurements would be necessary: one for cerebral blood flow and another for oxygen volume. If the Kety-Schmidt technique is used as mentioned, it should be borne in mind that cerebral blood flow can be measured only approximately. This indirect global method may be applied only in pathologic conditions which are not characterized by any focal cerebral disorders. With these limitations, the application of the Kety-Schmidt technique in man and in experimental animals is justified for the following reasons:

1. It is well suited for investigating cerebral metabolism, all the more so since the mixed cerebral venous blood in the jugular bulb derives from brain areas included in the blood flow study.

Table 3. Cerebral blood flow in dogs (mean values and standard deviations in ml/100 g/min) during saturation and desaturation with N_2O as a tracer under steady state conditions of arterial normotension, normocapnia, normoxemia, and normothermia

Saturation ($n = 10$)		Desaturation ($n = 10$)	
I	II	I	II
59.5	63.3	62.9	63.3
±7.8	±2.8	±5.3	±4.5

I: 1st measurement in arterial normotension, normocapnia, and normoxemia

II: 2nd measurement 30 min later under the same conditions

There are no statistically significant differences between I and II, or between saturation and desaturation

2. Several very sensitive analytic techniques are available for measuring cerebral arteriovenous differences, i.e., oxygen, glucose, lactate, pyruvate, ketone bodies, amino acids, free fatty acids, etc., and thus for calculating cerebral uptake or release under normal and abnormal conditions.

3. The duration of one study amounts to around $30-40$ min from the puncture of the vessels until removal of the needles. During this period it may be more easy to maintain a steady state than in longer lasting investigations, e.g., in those with labeled tracers and PET, even in mentally deranged people.

4. Although the technique is invasive, it may be applied almost daily if necessary without burdening the patient excessively.

5. The costs are quite low, amounting to around \$ 250 for one standard investigation of cerebral blood flow and the cerebral metabolic rates of oxygen, CO_2, glucose, lactate, and pyruvate.

Which cerebral metabolic pathways are of interest for investigation under normal and abnormal conditions?

It is well documented by many studies that under physiologic conditions the brain oxidizes only glucose to obtain energy. When glucose has passed the blood-brain barrier, it is glycolytically metabolized to form pyruvate, 7% of which is normally converted to lactate, which is released into the venous blood. Pyruvate is one of two precursors for acetylcholine synthesis [9]. Thus, the total amount of glucose taken up by the brain will yield some information on the activity of glycolysis. In addition, the amount of lactate and pyruvate formation and the lactate/glucose index will provide more information on pyruvate oxidation. This information can be completed by the calculation of the glucose oxidation ratio, GOR, which may be calculated by $\dfrac{\text{Glucose} - \text{Lactate}}{\text{Oxygen}}$ [11].

If the uptake of glucose is reduced but the cerebral metabolic rate of oxygen is still normal, it would be of great interest to investigate which substrates other than glucose are used oxidatively by the brain. Ketone bodies, amino acids, free fatty acids, and endogenous brain substrates are under discussion. Such studies might yield information on the pathobiochemical quality of an underlying abnormal process and will hopefully suggest therapeutic measures.

What results were obtained by the use of the Kety-Schmidt technique in human beings by our group? Are the results in agreement with the data of other studies performed with the same or another technique?

The findings in cerebral blood flow and the cerebral metabolic rates of oxygen, CO_2, glucose, and lactate in young healthy volunteers are listed in Table 4. These data are in good agreement with those published by Kety and Schmidt [16], Bernsmeier and Siemons [2], Lassen et al. [17], and Gottstein et al. [6]. Investigations in dementia yielded evidence that brain blood flow and oxidative metabolism are not uniformly varied in the two main types of primary dementia and with respect to its course [8, 10]. In the beginning of dementia of Alzheimer type and dementia of vascular type, the metabolic changes are found to be characteristically different, so that differential diagnosis becomes possible. In chronic dementias, these differences disappear. Brain blood flow and metabolism decrease to a functional low level [10]. With respect to the beginning of the disease, Gottstein et al. [7] have previously described the reduced CMR-glucose, as has also been found in several studies using deoxyglucose and PET.

First studies in schizophrenia showed a different pattern of brain blood flow and oxidative metabolism which was assumed to be related to psychiatric status. In productive schizophrenia, the biologic brain parameters were found to be abnormally increased. In schizophrenia simplex or paranoid schizophrenia, they did not vary from a healthy control group, and in nonproductive schizophrenia (hebephrenia or schizophrenic defects), they were abnormally decreased [12]. With respect to cerebral blood flow, Ingvar and Franzen [13]

Table 4. Mean values and standard deviations of cerebral blood flow, the cerebral metabolism rates of oxygen, CO_2, glucose, and lactate, cerebral RQ, lactate/glucose index, and GOR in 15 young adult healthy volunteers (mean age 25 yrs)

Cerebral blood flow	52.9 ± 4.9 ml/100 g/min
Cerebral metabolic rate of oxygen	3.54 ± 0.42 ml/100 g/min
Cerebral metabolic rate of CO_2	3.77 ± 0.51 ml/100 g/min
Cerebral metabolic rate of glucose	4.97 ± 0.75 mg/100 g/min
Cerebral metabolic rate of lactate	0.36 ± 0.22 mg/100 g/min
Cerebral RQ	1.06 ± 0.09
Lactate/glucose index	0.07 ± 0.01
Glucose/oxidation ratio	1.34 ± 0.17

found it to be reduced in frontal areas in patients with chronic schizophrenia. Buchsbaum et al. [4] described a lower glucose consumption in the frontal cortex in schizophrenic patients with a longer lasting history and obviously without any productive symptoms.

These few examples may be suffice to demonstrate that the Kety-Schmidt technique is still a useful tool in experimental brain research provided its limitations are thoroughly considered.

References

1. Alexander SC, Wollman H, Cohen PJ, Chase PE, Melman E, Behar M (1964) Krypton[85] and nitrous oxide uptake of the human brain during anesthesia. Anesthesiology 25:37−42
2. Bernsmeier A, Siemons K (1953) Die Messung der Hirndurchblutung mit der Stick-oxydulmethode. Pflügers Arch Ges Physiol 258:149−162
3. Bretschneider HJ, Cott L, Hilgert G, Probst R, Rau G (1966) Gaschromatographische Trennung und Analyse von Argon als Basis einer neuen Fremdgasmethode zur Durch-blutungsmessung von Organen. Verh dtsch Ges Kreislaufforschg 32:267−273
4. Buchsbaum MS, Ingvar DH, Kessler R, Waters RN, Cappelletti J, van Kammen DP, King AC, Johnson JL, Manning RG, Flynn RW, Mann LS, Bunney WE, Sokoloff L (1982) Cerebral glucography with position tomography. Use in normal subjects and in patients with schizophrenia. Arch Gen Psychiatry 39:251−259
5. Gibbs EL, Gibbs FA (1934) The cross section areas of the vessels that form the torcular and the manner in which flow is distributed to the right and to the left lateral sinus. Anat Rec 59:419−426
6. Gottstein U, Bernsmeier A, Sedlmeyer I (1963) Der Kohlenhydratestoffwechsel des menschlichen Gehirns. I. Untersuchungen mit substratspezifischen enzymatischen Methoden bei normaler Hirndurchblutung. Klin Wschr 41:943−948
7. Gottstein U, Bernsmeier A, Sedlmeyer I (1964) Der Kohlenhydratestoffwechsel des menschlichen Gehirns. II. Untersuchungen mit substratspezifischen enzymatischen Methoden bei Kranken mit verminderter Hirndurchblutung auf dem Boden einer Ar-teriosklerose der Hirngefäße. Klin Wschr 42:310−313
8. Hachinski VC, Iliff LD, Zilkha E, DuBoulay GH, McAllister VL, Marshall J, Ross-Russell RW, Symon L (1975) Cerebral blood flow in dementia. Arch Neurol 32:632−637
9. Hoyer S (1982) The young-adult and normally aged brain. Its blood flow and oxidative metabolism. A review-part I. Arch Gerontol Geriatr 1:101−116
10. Hoyer S (1982) The abnormally aged brain. Its blood flow and oxidative metabolism. A review − part II. Arch Geront Geriatr 1:195−207
11. Hoyer S, Becker K (1966) Hirndurchblutung und Hirnstoffwechselbefunde bei neuro-psychiatrisch Kranken. Nervenarzt 37:322−324
12. Hoyer S, Oesterreich K (1975) Blood flow and oxidative metabolism of the brain in patients with schizophrenia. Psychiatria Clin 8:304−313
13. Ingvar DH, Franzen G (1974) Abnormalities of cerebral blood flow distribution in patients with chronic schizophrenia. Acta Psychiat Scand 50:425−462
14. Kety SS, Harmel MH, Broomell HT, Rhode CB (1948) The solubility of nitrous oxide in blood and brain. J Biol Chem 173:487−496
15. Kety SS, Schmidt CF (1945) The determination of cerebral blood flow in man by the use of nitrous oxide in low concentrations. Am J Physiol 143:53−66
16. Kety SS, Schmidt CF (1948) The nitrous oxide method for the quantitative determination of cerebral blood flow in man: Theory, procedure and normal values. J Clin Invest 27:476−483

17. Lassen NA, Feinberg I, Lane MH (1960) Bilateral studies of cerebral oxygen uptake in young and aged normal subjects and in patients with organic dementia. J Clin Invest 39:491–500
18. Lassen NA, Klee A (1965) Cerebral blood flow determined by saturation and desaturation with krypton[85]. Circulation Res 16:26–32
19. Lassen NA, Munck O (1955) The cerebral blood flow in man determined by the use of radioactive krypton. Acta Physiol Scand 33:30–49
20. Padget DH (1956) The cranial venous system in man in reference to development, adult configuration, and relation to the arteries. Amer J Anat 98:307–355
21. Schenkin HA, Harmel MH, Kety SS (1948) Dynamic anatomy of the cerebral circulation. Arch Neurol Psychiat 60:240–252
22. Weinhardt F, Quadbeck G, Hoyer S (1972) Quantitative Bestimmung von Blutgasvolumina mit Hilfe der Gaschromatographie. Z Prakt Anaesth 6:337–347
23. Zeiger K (1923) Über die Ursachen der Asymetrie der Sinus transversi und sigmoidei beim Menschen. Beitr Anat Physiol Path Therap d Ohres d Nase u d Halses 19:184–208

Determination of Regional Glucose Metabolism in the Brain by FDG and PET

W.-D. Heiss, G. Pawlik, K. Herholz, R. Wagner, and K. Wienhard

Use of (^{18}F)-2-fluoro-2-deoxyglucose (FDG) (Reivich et al. 1979) made the deoxyglucose method developed by Sokoloff et al. (1977) for autoradiographic determination of regional cerebral glucose metabolism applicable to studies in humans by means of positron emission tomography (PET). A substantial number of investigations have been performed with this method in normal volunteers during physiologic stimulation and in patients suffering from various diseases (review in Phelps et al. 1982; Heiss and Phelps 1983). Since some of the major topics are dealt with in the following papers, this contribution focuses primarily on the kinetic aspects of glucose metabolism in normal subjects as well as in stroke patients, and on the topographic distribution of functional cerebral deactivation in patients with focal ischemic lesions. Distant effects of local brain lesions on glucose metabolism or oxygen uptake in areas without structural damage in X-ray computed tomography (XCT) have been repeatedly described (Kuhl et al. 1980; Lenzi et al. 1981; Baron et al. 1981; Heiss et al. 1983), but a definite relation either to localization and extent of the morphological lesion or to the neurologic deficit has not been established and differential effects on the kinetics of tracer accumulation were not determined.

Principles of Metabolic Model and Tracer Detection

According to the principles of the model (Sokoloff et al. 1977), glucose and deoxyglucose use the same carrier system for facilitated diffusion into the brain tissue and the same enzymatic reaction (hexokinase) for the first metabolic step of phosphorylation. However, due to the alteration at position 2 of the molecule, DG cannot be converted into fructose-6-phosphate and then further metabolized to CO_2 and H_2O, but is trapped in the cell because phosphatase activity in brain tissue is very low. The resulting accumulation makes this compound well suited for tracer kinetic studies if an appropriate radiolabel is used.

Individual steps of DG turnover can be represented in a three-compartment model (Fig. 1), with kinetic constants K_1 and k_2 for transport into and out of the cell and rate constant k_3 for phosphorylation (Sokoloff et al. 1977). A fourth constant, k_4, is required only for recordings of longer duration (more than 45 to 60 min after tracer injection) because only after that period of time does de-

Max-Planck-Institut für neurologische Forschung, Ostmerheimer Straße 200, D-5000 Köln 91

Cerebral Blood Flow and
Metabolism Measurement.
Eds. Hartmann/Hoyer
© Springer-Verlag Berlin Heidelberg 1985

FDG - MODEL (SOKOLOFF)

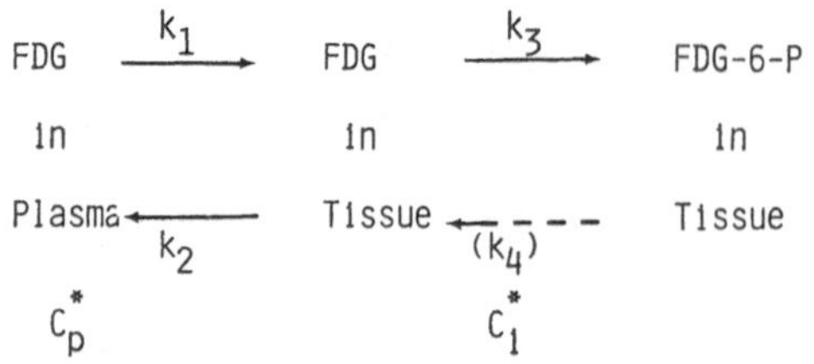

$$FDG \xrightarrow{k_1} FDG \xrightarrow{k_3} FDG\text{-}6\text{-}P$$

in   in   in

$$Plasma \xleftarrow{\quad} Tissue \xleftarrow{\;-\;-\;-} Tissue$$
$$\qquad k_2 \qquad\qquad (k_4)$$

$$C_p^* \qquad\qquad\qquad C_1^*$$

Static (autoradiographic) Study

$$CMRGlc = \frac{C_p}{LC} \cdot \frac{C_1^*(T) - k_1 \cdot e^{-(k_2+k_3)T} \cdot \int_0^T C_p^* \cdot e^{(k_2+k_3)t} dt}{\int_0^T C_p^* dt - e^{-(k_2+k_3)T} \cdot \int_0^T C_p^* \cdot e^{(k_2+k_3) \cdot t} dt} \qquad (1)$$

Dynamic Study

$$CMRGlc = \frac{C_p}{LC} \cdot \frac{k_1 \cdot k_3}{k_2 + k_3} \qquad\qquad (2)$$

$$C_1^*(t) = \frac{k_1 \cdot k_3}{k_2 + k_3} \int_0^t C_p^* dt + \frac{k_1 \cdot k_2}{k_2 + k_3} \cdot e^{-(k_2+k_3)t} \cdot \int_0^t C_p^* \cdot e^{(k_2+k_3)t'} \cdot dt' \qquad (3)$$

Note: $$\lim_{t \to 0} \frac{C_1^*(t)}{\int_0^t C_p^* dt'} = k_1 \qquad (4)$$

Fig. 1. Three-compartment model for determination of cerebral metabolic rate of glucose (CMRGl, Sokoloff et al. 1977) and equations for calculation of CMRGl by the autoradiographic and dynamic approach. For details see text

phosphorylation begin to play a significant role (Phelps et al. 1979). The mathematical model can be described by a complex integral equation in which individual expressions represent various measurable quantities or pools of metabolic constituents [Eq. (1), Fig. 1]:

- The total activity of label in the tissue, which is measured directly
- The precursor pool of unmetabolized DG in the tissue, which is calculated from the individual plasma curve and from kinetic constants previously determined in standard experiments; the difference between these two expressions yields the amount of metabolite, i.e., the DG-6-phosphate produced from time 0 to T
- The integrated precursor activity supplied to the brain, which equals the integrated plasma specific activity corrected for the lag in equilibration between plasma and tissue, the quotient of metabolized product and precursor supplied being a measure of the relative amount of phosphorylated DG, i.e., the rate of DG metabolism.

The latter value multiplied by plasma glucose concentration yields the metabolic rate of glucose. However, DG is not extracted at exactly the same rate as glucose and the value must therefore, be corrected by a lumped constant (LC) which is defined by a complex expression and has also been determined in standard experiments.

Errors caused by strongly deviating kinetic constants in pathologic tissue can be minimized by a dynamic approach: tissue activity is measured at short intervals from tracer injection until a steady state is reached between tissue and plasma DG. From the measured time course of plasma DG and tissue-label activity (Fig. 2), the best fitting set of parameters K_1, k_2, and k_3 is determined using equation (3) (Fig. 1). Metabolic rates for glucose are then calculated from these fitted rate constants and from the plasma glucose concentration Cp according to equation (2) (Fig. 1). K_1 is additionally defined by the initial activity accumulation in the tissue [equation (4), Fig. 1].

While the original method (Sokoloff et al. 1977) used (^{14}C)-2-deoxyglucose for autoradiographic determination of regional glucose metabolism in animal experiments, use of a positron-emitting isotope, e.g., ^{18}F, as the label offers several important advantages, also compared with tracers decaying by single photon emission. Positrons travel only a short distance in tissue, then they com-

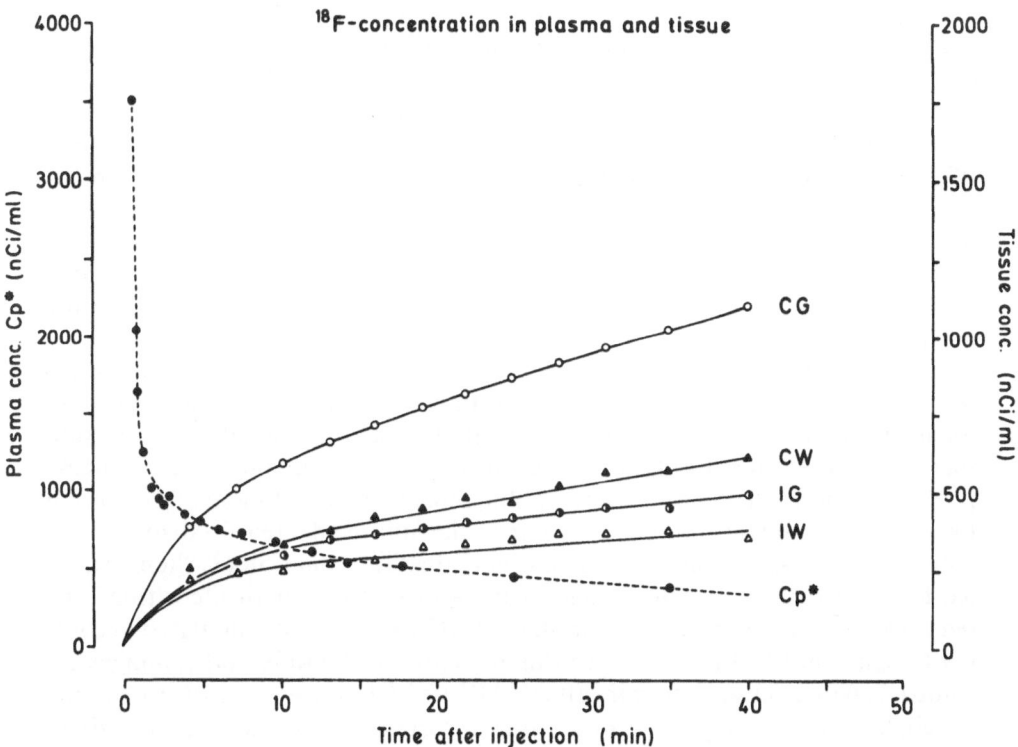

Fig. 2. Measured (^{18}F) activity and corresponding fitted curves in plasma (*CP**), in infarcted gray (*IG*) and white (*IW*) matter, and in the corresponding contralateral gray (*CG*) and white (*CW*) matter regions of a stroke patient, as a function of time after FDG injection

bine with an electron and both annihilate, emitting photons in opposite directions, both of which have the same energy of 511 keV. Recording of these coincident gamma quanta with two connected detectors permits one to assign the event to the region between the two detectors. Due to the high and identical energy of the two photons, detection probability is almost independent from where in between any paired detectors the event occurred, and electronic collimation can be used in the design of detector systems, providing high detection efficiency and accurate attenuation correction. From moving or steady multidetector arrays or rings, large numbers of projections are obtained at different angles, and local activity distribution is reconstructed in cross-sectional images.

Procedure of Investigations

The studies reported here were performed in six young male volunteers and in 22 patients with a recent ischemic stroke documented by XCT. All subjects rested on a reclining chair in a room with low ambient light and noise, with their eyes closed and ears unplugged. Approximately 15 min before the start of recordings, short catheters were placed into one cubital vein for injection and into a vein of the contralateral heated hand for blood sampling.

FDG was synthesized according to a modification of the method of Ido et al. (1977). Approximately 5 mCi FDG in normal saline solution were injected intravenously. Blood was sampled and plasma glucose as well as FDG concentrations were determined as described by Phelps et al. (1979). Seven equally spaced parallel planes, from the canthomeatal line (CML) up to 81 mm above, were simultaneously scanned with a four-ring positron camera (Scanditronix PC 384) at a spatial resolution of approximately 8 mm full width at half-maximum in 11-mm slices (Eriksson et al. 1982). Recordings were taken at consecutive intervals increasing from 1 to 5 min during a period of 40 min, starting at FDG injection. Data from the tomographic device and from a sample changer used for plasma counting, as well as plasma glucose values, were stored in the memory of a VAX 11/780 (DEC) computer for later processing.

Following decay correction the activity distribution in the scanned slices was reconstructed using an edge finding algorithm to determine the skull contour for attenuation correction (Bergström et al. 1982), a deconvolution for subtraction of the scattered radiation (Bergström et al. 1983), and a filtered backprojection algorithm. On each tomographic image between 10 and 30 regions of interest were outlined, representing either the infarcted area or distinct anatomic structures as demonstrated by individual XCT. Equation (3) then was fitted to the regional time-activity data using a fast nonlinear fit algorithm. The regional cerebral metabolic rate for glucose (rCMRGlc) was calculated according to equation (2). In addition to this dynamic evaluation, static images recorded between 30 and 50 min after FDG injection were transformed into metabolic maps employing a model that incorporates the slow hydrolysis of FDG-6-phosphate to FDG (Phelps et al. 1979). From the latter images slices could be reconstructed at any thickness and angle of cut for optimum visualization of points of interest.

Table 1. Average rate constants and metabolic rate for young normal adults (mean $\pm$ SD)

	K_1 min^{-1}	k_2 min^{-1}	k_3 min^{-1}	k_4 min^{-1}	MR µmol/ 100 g/min
This work:					
Cortex	0.090$\pm$0.009	0.133$\pm$0.006	0.075$\pm$0.016		40.06$\pm$4.19
White matter	0.050$\pm$0.009	0.117$\pm$0.007	0.046$\pm$0.009		17.09$\pm$1.95
Basal ganglia	0.093$\pm$0.012	0.134$\pm$0.010	0.078$\pm$0.014		42.09$\pm$4.71
Cerebellum	0.088$\pm$0.013	0.133$\pm$0.008	0.047$\pm$0.005		28.78$\pm$3.41
Plasma glucose concentration Cp = 95.6$\pm$7.4 mg/100 ml					
Huang et al					
Gray matter	0.102$\pm$0.028	0.130$\pm$0.066	0.062$\pm$0.019	0.0068$\pm$0.0014	40.56$\pm$6.56
White matter	0.054$\pm$0.014	0.109$\pm$0.044	0.045$\pm$0.019	0.0058$\pm$0.0017	18.94$\pm$3.56
Plasma glucose concentration Cp = 91.9$\pm$9.6 mg/100 ml					

rCMRGlc and Rate Constants in Healthy Volunteers

Table 1 shows average values of rCMRGlc and of rate constants K_1 through k_3 of six normal volunteers determined by dynamic curve fitting in comparison with the values given by Huang et al. (1980), in which the effect of hydrolysis of FDG-6-phosphate was accounted for by addition of another rate constant to the Sokoloff model. In our patient studies dynamic scanning was not extended beyond 40 min and, therefore, only the unmodified Sokoloff model was employed in both the patient and the control group. The variance of K_1, k_2, and k_3 in the present study is much smaller than that of the k-values presented by Huang et al. (1980). This may be due to our homogeneous group of volunteers who were neither apprehensive nor very excited about the investigation. Metabolic rates from the two studies, however, are almost identical. When the four-rate constants model was fitted to our data, metabolic rates calculated according to equation (2) (Fig. 1) turned out 5% – 10% higher. Remarkably enough, K_1 and k_2 in the cerebellum were comparable to the values found for the cerebral cortex, while k_3, and consequently also the metabolic rate, was significant lower. In comparisons among regions, k_2 ranging from 0.1 to 0.2 min^{-1} had the least total variation, showed little scatter within individual cases, and was comparable to the values given by Huang et al. (1980). Discrepancies with k_2- and k_3-values reported by Reivich et al. (1982) for (^{11}C)-deoxyglucose might be explained by slightly different affinities of the two glucose analogs for the facilitated diffusion system and for hexokinase.

In one of our volunteers we had the opportunity to perform FDG studies not only in the normal resting state, but also during sleep (Fig. 3). Compared with the waking state rCMRGlc was decreased in all gray matter structures. Even without a more precise definition of sleep stages by continuous EEG monitoring it would appear reasonable to conclude that gray matter structures were functionally inactivated and, consequently, a reduction in metabolic rate was found.

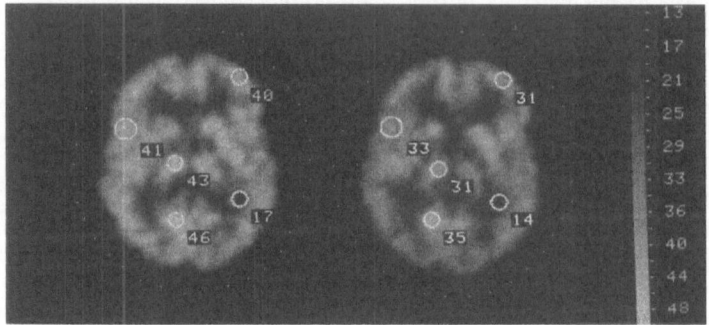

Fig. 3. FDG-PET images at 4 cm above CM line in a normal volunteer in the waking state (*left*) and during sleep (*right*): As can be read by comparing the gray of a region with the reference scale on the right, and as directly written in a few regions, CMRGl is markedly reduced during sleep in all gray matter structures of the brain

Table 2. rCMRGlc (µmol/100 g/min) in various brain regions of patients with an ischemic stroke

	n	rCMRGlc (µmol/100 g/min, mean ± SD)		Sign test
		Ipsilateral	Contralateral	P <
Infarction	17	8.2 ± 7.03	29.6 ± 11.84	0.00001
Frontal cortex	16	24.6 ± 10.49	30.0 ± 13.74	0.01
Temporal cortex	18	24.6 ± 7.89	31.5 ± 12.07	0.001
Parietal cortex	15	26.2 ± 9.27	31.9 ± 11.93	0.02
Occipital cortex	19	26.9 ± 10.68	33.2 ± 13.56	0.005
Thalamus	14	25.2 ± 7.36	31.9 ± 7.77	0.002
Striatum	14	25.6 ± 9.79	35.0 ± 11.72	0.001
Cerebellum	8	27.8 ± 9.65	21.7 ± 10.62	0.005

Metabolic Studies in Ischemic Stroke

Topographic Relationships Between Infarcts and Deactivated Areas

Characteristic distant effects of focal ischemic lesions on selected nonischemic brain regions in patients with typical infarction in the supply territory of the middle cerebral artery (Fig. 4) are demonstrated in Table 2. Metabolic changes were most severe in the infarct proper, but glucose consumption also was significantly decreased in other ipsilateral cortical and subcortical gray matter structures as well as in the contralateral cerebellum.

Detailed analysis of another series of stroke patients with rather small infarcts in various brain regions provided some insight into the topographical nature of remote deactivation. Patients with cortical or subcortical lesions and a neurologic deficit consisting exclusively in neuropsychological disturbances exhibited hypometabolism not only in the infarcted but also in related ipsilat-

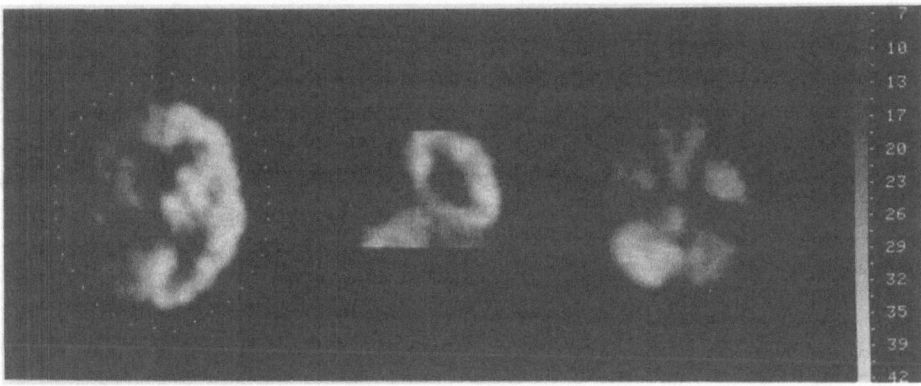

Fig. 4. Metabolic maps of a patient with an ischemic infarction in the right middle cerebral artery territory. Values on the reference gray scale are in μmol/100 g/min. *From left to right:* horizontal (bottom view) and coronal (front view) sections across the infarct, horizontal (bottom view) section across the cerebellum

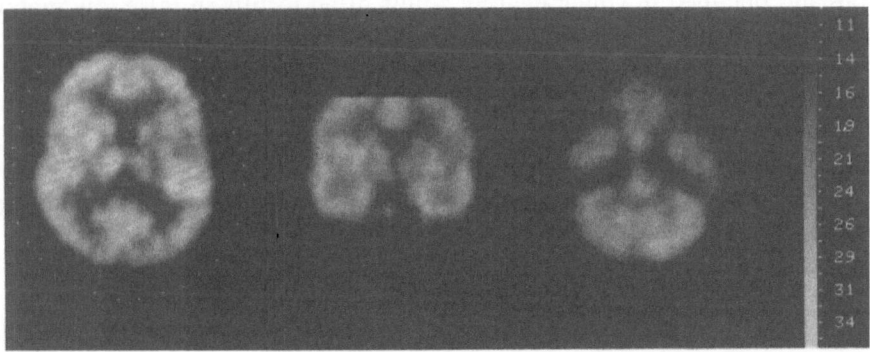

Fig. 5. Metabolic maps of a patient with a small ischemic infarction subcortical to the Broca area. Sections as in Fig. 4

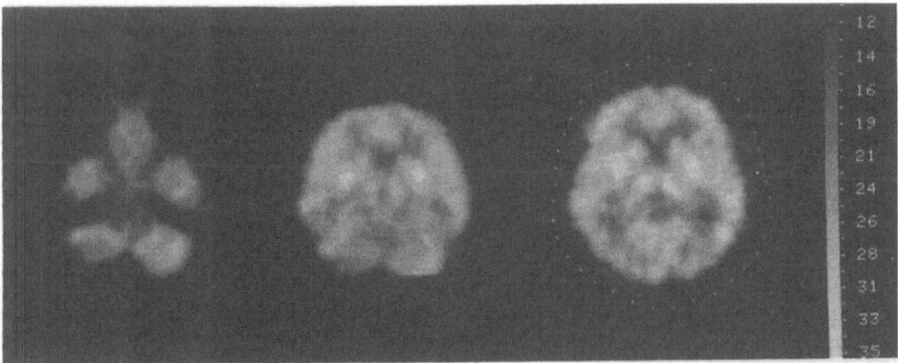

Fig. 6. Metabolic maps of a patient with an ischemic infarction in the right cerebellar hemisphere. Values on the reference gray scale are in μmol/100 g/min. *From left to right:* horizontal and 45° angular sections across the infarct, horizontal brain section at the level of the basal ganglia, all bottom view

eral cortical areas, basal ganglia, the thalamus, and the contralateral cerebellum, suggesting ischemic disruption of connecting fiber systems in the subcortical white matter. Figure 5 gives an example of such an aphasic patient. A lesion in the cerebral peduncle damaged the pyramidal tract and corticopontine fibers, and only the contralateral cerebellum was deactivated. However, in patients with severe impairment of motor function from an ischemic brainstem lesion leaving the corticopontine projections intact, no decrease in cerebellar glucose metabolism was found. Conversely, unilateral cerebellar infarction, as demonstrated in Fig. 6, caused no focal remote deactivation. From these examples it may be concluded that distant effects on cerebral glucose metabolism are brought about primarily by intracerebral deafferentation, and not by infarct size per se or by severe neurologic disturbances.

Regional Kinetics of Glucose Metabolism in Ischemic and Deactivated Tissue

As first described by Hawkins et al. (1981), indiscriminate use of average rate constants obtained in young healthy adults often results in false low metabolic values, particularly in ischemic tissue. The basis of this phenomenon becomes obvious when dynamic data on regional tracer accumulation are available. Under pathologic conditions leading to changes in enzyme kinetics and transport mechanisms, time-activity curves assume characteristic shapes that are quite different from the normal state (Fig. 2). Therefore, it may be expected that metabolic rates based on individually fitted rate constants approximate the un-

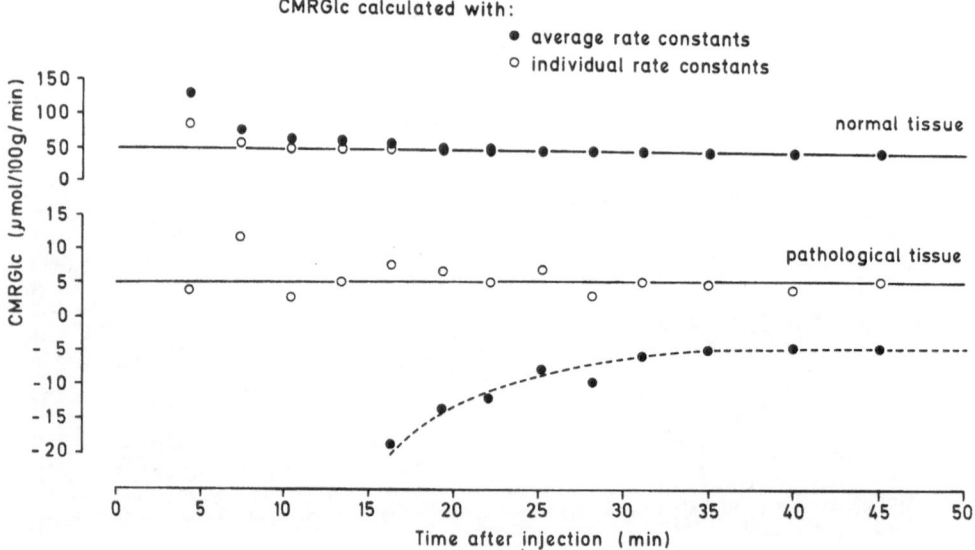

Fig. 7. Cerebral metabolic rate for glucose (CMRGlc) as computed from regional activity at various points in time after FDG injection, using either individually fitted or standard rate constants determined in healthy young adults. Data were obtained from a normal (*upper graph*) and infarcted (*lower graph*) cortex region of a typical stroke patient

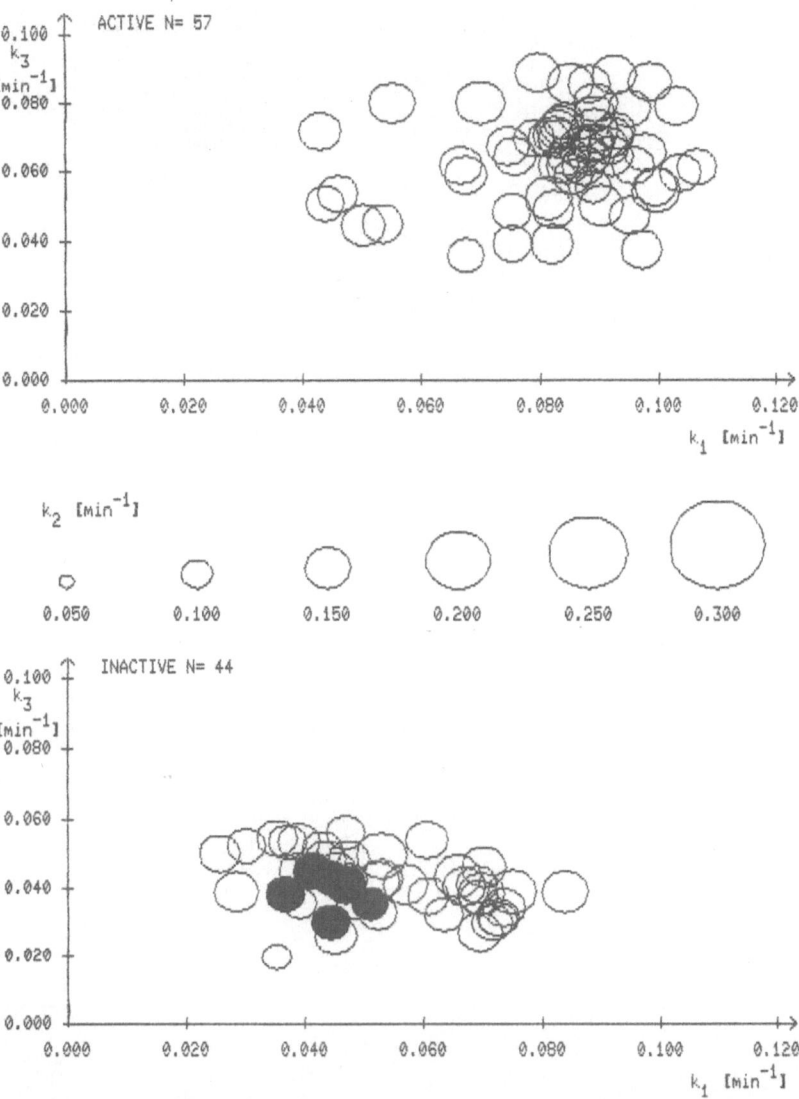

Fig. 8. Rate constants K_1, k_2, and k_3 of the Sokoloff model, determined by individual data fitting in 101 distinct brain regions of a stroke patient. k_2 is indicated by *circle size*, infarcted regions are represented by *full black circles*. Rate constants in normal (*upper graph*) and deactivated tissue (*lower graph*) are clearly separated

known true value more closely, although another possible source of errors, the lumped constant, cannot be determined regionally in man because this would require knowledge of regional brain glucose concentrations (Crane et al. 1983). Despite the fact that the operational equation of the Sokoloff model tends to minimize the influence of rate constants on the resulting metabolic value, it is not entirely insensitive to violations of the basic assumption of the model, as demonstrated in Fig. 7: in normal tissue metabolic rates computed from region-

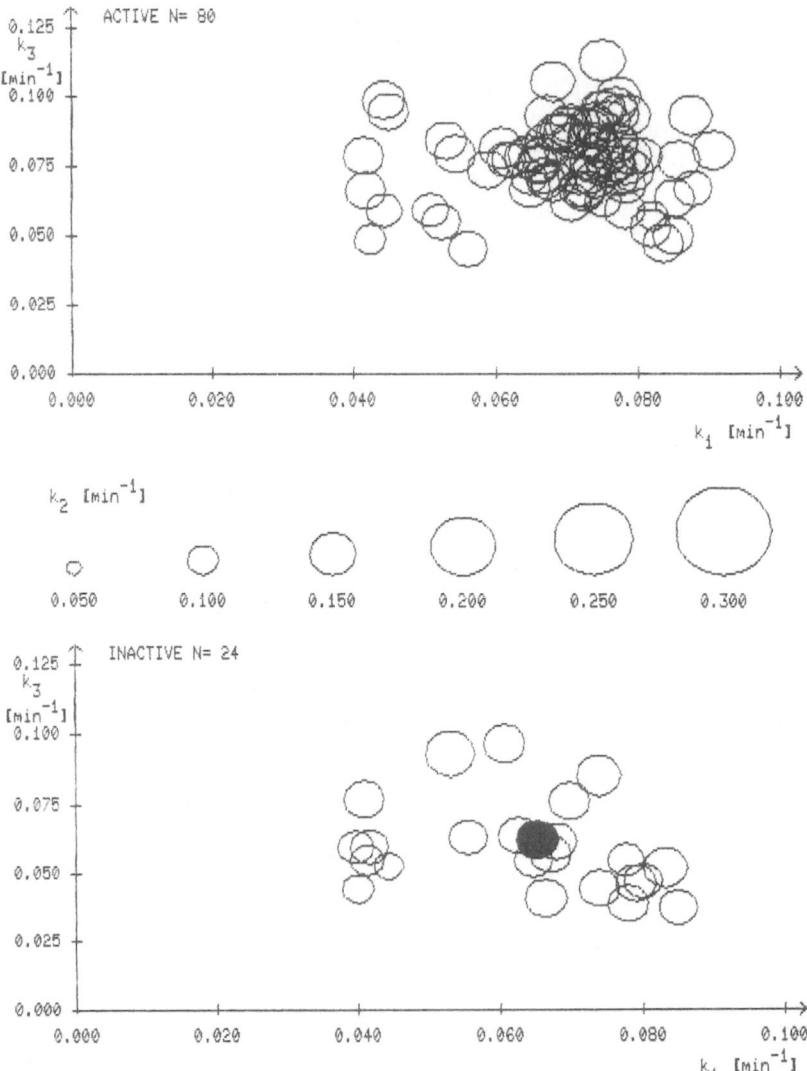

Fig. 9. Rate constants K_1, k_2, and k_3 of the Sokoloff model, determined by individual data fitting in 104 distinct brain regions of a stroke patient. k_2 is indicated by *circle size;* the infarcted region is represented by a *full black circle.* Rate constants in normal (*upper graph*) and deactivated tissue (*lower graph*) are poorly separated

al tracer activity after 20–30 min and standard rate constants approach the values obtained by the dynamic fit procedure, while in infarcted tissue even negative values may be found, no matter for how long regional activity is recorded.

Correction for biased metabolic rates, however, is not the only advantage of following the time course of tracer uptake. Individually determined rate constants may yield additional insight into the mechanism primarily affected in tissue exhibiting decreased glucose metabolism: transport, phosphorylation, or both. In the present series of stroke patients, within the infarcted area average reductions of K_1 by 36% and of k_3 by 38%, as compared with the corresponding contralateral region, were found, while k_2 showed only a slight decrease. In deactivated regions k_2 was similar to the value in ischemic and normal tissue, but K_1 was reduced by 11% and k_3 by 21% of the individual control values. In deactivated cerebellum even larger decreases were observed for K_1 (by 22%) and k_3 (by 32%). These results suggest that changes in the rate of phosphorylation in general are not exactly paralleled by changes in bidirectional transmembrane transport. Figure 8 shows predominant reduction of k_3 in 44 deactivated out of a total of 101 brain regions of a stroke patient during the early stage of the disease, who had a poor outcome. Figure 9, by contrast, demonstrates a quite proportional decrease of K_1 and k_3 in deactivated brain regions of a patient who later recovered almost fully from his ischemic stroke.

Conclusions

Dynamic PET scanning for 40 min and K_1-controlled curve-fitting yield least biased estimates of regional kinetic constants and metabolic rates in anatomically well-defined brain structures. Glucose consumption of cerebral gray matter seems to be significantly reduced during physiologic sleep. Deafferentation by ischemic disruption of connecting fiber tracts is apparently the major cause of deactivation of morphologically intact remote brain regions in stroke. In both infarcted and deactivated areas the reliability of metabolic measurements can be substantially improved by dynamic determinations. Regional rate constants provide a sound basis for multivariate analyses of the kinetics of cerebral glucose metabolism.

References

Baron JC, Bousser MG, Comar D, Castaigne P (1981) "Crossed cerebellar diaschisis" in human supratentorial brain infarction. Trans Am Neurol Ass 105:459–461

Bergström M, Litton J, Eriksson L, Bohm C, Blomqvist G (1982) Determination of object contour from projections for attenuation correction in cranial positron emission tomography. J Comput Assist Tomogr 6:365–371

Bergström M, Eriksson L, Bohm C, Blomqvist G, Litton J (1983) Correction for scattered radiation in a ring detector positron camera by integral transformation of the projections. J Comput Assist Tomogr 7:42–50

Crane PD, Pardridge WM, Braun LD, Oldendorf WH (1983) Kinetics of transport and phosphorylation of 2-fluoro-2-deoxy-D-glucose in rat brain. J Neurochem 40:160–167

Eriksson L, Bohm C, Kesselberg M, Blomqvist G, Litton J, Widén L, Bergström M, Ericson K, Greitz T (1982) A four ring positron camera system for emission tomography of the brain. IEEE Trans Nucl Sci 29:539–543

Hawkins RA, Phelps ME, Hung SC, Kuhl DE (1981) Effect of ischemia on quantification of local cerebral glucose metabolic rate in man. J Cereb Blood Flow Metabol 1:37–51

Heiss WD, Phelps ME (1983) Positron emission tomography of the brain. Springer-Verlag, Berlin Heidelberg New York

Heiss WD, Pawlik G, Herholz K, Wagner R, Wienhard K (1983) Determination of regional glucose metabolism in the brain by FDG and PET. Int Symp of Methods of Cerebral Blood Flow and Metabolism Measurements in Man, Sept 29th – Oct 1st, Heidelberg

Huang SC, Phelps ME, Hoffman EJ, Sideris K, Selin CJ, Kuhl DE (1980) Noninvasive determination of local cerebral metabolic rate of glucose in man. Am J Physiol 238:E69–E82

Ido T, Wan CN, Fowler JS, Wolf AP (1977) Fluorination with F 2, a convenient synthesis of 2-deoxy-2-fluoro-D-glucose. J Org Chem 42:2341–2342

Kuhl DE, Phelps ME, Kowell AP, Metter EJ, Selin C, Winter J (1980) Effects of stroke on local cerebral metabolism and perfusion: Mapping by emission computed tomography of ^{18}FDG and ^{13}NH$_3$. Ann Neurol 8:47–60

Lenzi GL, Frackowiak RS, Jones T (1981) Regional cerebral blood flow (CBF), oxygen utilization (CMRO$_2$), and oxygen extraction ratio (OER) in acute hemispheric stroke. J Cereb Blood Flow Metabol 1, Suppl 1:S504–S505

Phelps ME, Huang SC, Hoffman EJ, Selin C, Sokoloff L, Kuhl DE (1979) Tomographic measurement of local cerebral glucose metabolic rate in humans with (F-18)2-fluoro-2-deoxy-D-glucose: validation of method. Ann Neurol 6:371–388

Phelps ME, Mazziotta JC, Huang SC (1982) Study of cerebral function with positron computed tomography. J Cereb Blood Flow Metabol 2:113–162

Reivich M, Kuhl D, Wolf A, Greenberg J, Phelps M, Ido T, Casella V, Fowler J, Hoffman E, Alavi A, Som P, Sokoloff L (1979) The (^{18}F)fluorodeoxy-glucose method for the measurement of local cerebral glucose utilization in man. Circ Res 44:127–137

Reivich M, Alavi A, Wolf A, Greenberg JH, Fowler J, Christman D, Max Gregor R, Jones SC, London J, Shiue C, Yonekura Y (1982) Use of 2-deoxy-D(1-^{11}C)glucose for the determination of local cerebral glucose metabolism in humans: Variation within and between subjects. J Cereb Blood Flow Metabol 2:307–319

Sokoloff L, Reivich M, Kennedy C, DesRosiers MH, Patlak CS, Pettigrew KD, Sakurada O, Shinohara M (1977) The ^{14}C-deoxyglucose method for the measurement of local cerebral glucose utilization: Theory, procedure, and normal values in the conscious and anesthetized albino rat. J Neurochem 28:897–916

Regional Analysis of Steady-State Clearance of Fluor-Deoxyglucose into the Human Brain

A. Gjedde[1], W.-D. Heiss[2], and K. Wienhard[2]

Introduction

When labeled 2-fluoro-deoxyglucose (FDG) is given to a human, the compound enters brain tissue by the same mechanism that transports glucose. It is phosphorylated to 2-fluoro-deoxyglucose-6-phosphate (FDG-6-P) but remains trapped in this form because no other enzyme of the glycolytic chain accepts the substrate. To a still unknown extent, FDG-6-P may be reconverted to FDG by a phosphatase reaction.

If FDG-6-P were permanently trapped in the tissue, and if the FDG concentration in the blood quickly fell to zero following injection, the net rate of FDG-phosphorylation would equal:

$$K = \frac{\text{FDG-6-P in brain}}{\text{Concentration-time integral of FDG in blood}} . \tag{1}$$

Since neither is the case, one of two major controversies about the 2DG method focuses on the possibility of estimating K accurately when significant levels of unphosphorylated FD remain in blood and brain, and when trapping of FDG-6-P is incomplete (the other major controversy focuses on the "lumped constant").

Sokoloff et al. (1977) estimated the amount of FDG in brain by a simple three-compartment model. According to this approach, the measurement of K is based on a single measurement of the total radioactivity in brain:

$$K = \frac{\text{Total radioactivity in brain} - \text{FDG in brain}}{\text{Integral of FDG in blood} - \dfrac{\text{FDG in brain}}{K_1}} . \tag{2}$$

In this equation, the brain content of FDG-6-P is the difference between the total radioactivity and the estimated content of FDG in brain. Although the FDG content is normally quite low in the rat brain, compared with the total radioactivity, accumulation of FDG-6-P in the human brain is sufficiently slow to cause the estimate of FDG in brain, based on published average rate constants, to be so inaccurate that the difference in the nominator may occasionally be

---

1 Medicinsk-fysiologisk afdeling A, Panum-instituttet, Blegdamsvej 3, DK-2200 København
2 Max Planck-Institut für neurologische Forschung, Ostmerheimer Straße 200, D-5000 Köln-Merheim

Cerebral Blood Flow and
Metabolism Measurement.
Eds. Hartmann/Hoyer
© Springer-Verlag Berlin Heidelberg 1985

negative, particularly in brain regions with a low rate of accumulation of FDG-6-P (white matter).

A new approach (Heiss et al., this symposium) is to follow the accumulation of FDG and FDG-6-P during the entire period of uptake and use the information hidden in the uptake curve to fit individual sets of rate constants to each regional uptake curve. Surprisingly, this approach has yielded much more accurate values of the net rate of FDG accumulation, despite rate constant estimates that do not appear to be substantially different from the published, average rate constants.

The purpose of the present discussion is to show why accurate values for the net rate of FDG accumulation result from the so-called "kinetic" approach. The discussion is based on the regional analysis of FDG and FDG-6-P accumulation in 33 regions of the human brain.

Methods

Sokoloff's equation can be modified to express the net rate of FDG accumulation as a function of three variables rather than four:

$$K = \frac{\text{Total radioactivity in brain} - \dfrac{\text{FDG in brain}}{1 + \dfrac{k_3}{k_2}}}{\text{Integral of FDG in blood}}. \tag{3}$$

By division with the final FDG concentration in blood, the radioactivities are converted to apparent volumes of distribution (Gjedde 1982; Patlak et al. 1983):

$$K = \frac{V_{total} - \dfrac{V_{fdg}}{1 + \dfrac{k_3}{k_2}}}{\text{Theta}}. \tag{4}$$

In the steady-state, the apparent volume of distribution of FDG in brain equals $K_1/k_2 + k_3)$, and the equation further reduces to:

$$K = \frac{V - K_1 * k_2/(k_2 + k_3)^2}{\text{Theta}} = \frac{V - V_g}{\text{Theta}} \tag{5}$$

where Theta is the integral divided by the final concentration (unit of time), and V is the total radioactivity in the brain divided by the final radioactivity in blood. This equation shows that the apparent volume of distribution of the label in brain (V) is a linear function of Theta with a slope of K. The y-intercept is the "precursor pool" (V_g). Thus, when steady-state has been obtained, V rises linearly with Theta with the slope of K. The estimate of K is independent of the rate constants K_1, k_2, and k_3 (Patlak et al. 1983).

In the present analysis, all brain radioactivities recorded as functions of time were converted to apparent volumes of distribution as functions of Theta. Equation (4) was subsequently fitted to the volumes by nonlinear least-squares computer optimization. The fitting procedure yielded estimates of K and V_g and of their standard deviations. For comparison, the rate constants K_1, k_2, and k_3 were also fitted to the uptake curves in the conventional manner.

Results

A representative plot of the uptake of labeled FDG into a region of the human brain (ROI 11; visual cortex) is shown in Fig. 1, together with the curve of the arterial concentration of FDG. As shown in the insert, the accumulated radioactivity approached a plateau. In Fig. 2, the plasma curve was integrated, and pairs of V (radioactivity in brain, divided by activity in blood) and Theta (integral, divided by activity in blood) plotted as a graph of apparent volume versus time. The apparent volume rose with time. After an initial nonlinear phase (non-steady-state), the rise entered a linear phase that signified the onset of steady-state. The kinetic description of the slope and y-intercept is shown on the graph.

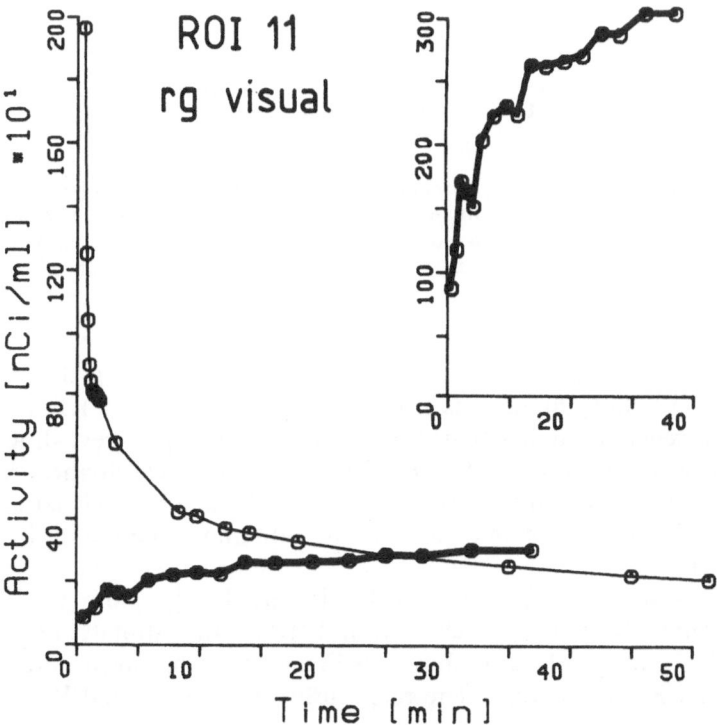

Fig. 1. Typical recording of fluor-deoxyglucose uptake by a region of the human brain (right hemisphere visual cortex). *Insert* shows uptake at greater magnification. *Abscissa:* Time of recording after injection of fluor-18-labeled deoxyglucose (min). *Ordinate:* Radioactivity in brain tissue (*heavy curve*) and plasma (*fine curve*) nCi/ml

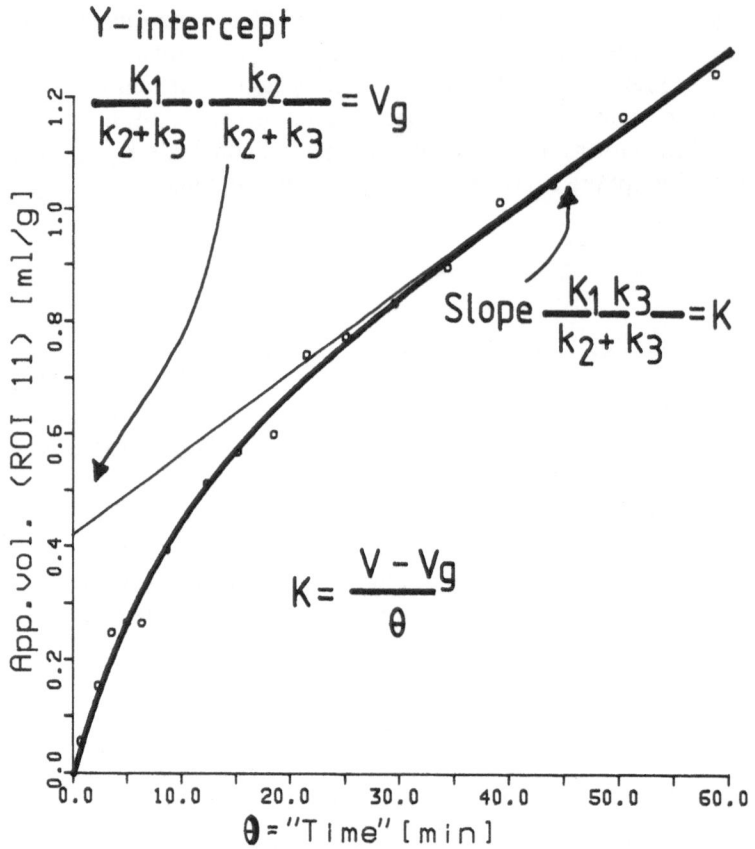

Fig. 2. Transformation of recording of region 11 according to equation (4) (see text). *Abscissa:* Apparent time of uptake, equal to time-concentration integral of radioactivity in blood plasma, divided by final concentration in blood plasma (min). *Ordinate:* Apparent volume of distribution, equal to radioactivity in brain, divided by final concentration in plasma (ml/g)

A total of 33 regions were plotted this way, as shown in Fig. 3. Gray matter slopes were generally higher than white matter slopes, and left hemisphere slopes were generally higher than right hemisphere slopes, revealing a global right hemisphere reduction of FDG-6-P accumulation. The highest rate of accumulation was estimated in region 12, the left hemisphere visual cortex, the lowest rate of accumulation in region 21, the right hemisphere occipitotemporal white matter.

The values of K, V_g, V_e ($= K_1/k_2$), K_1, k_2, and k_3 of regions 11, 12, 21, and 22 are shown in Table 1. The relationship between the estimates of V_g, V_e, and K_1, and the estimate of K, is shown in Fig. 4. A strong linear correlation between K and K_1 is apparent. The relationship between K and V_e is uncertain but the values of V_e did rise above the water volume of the brain for high values of K, indicating that the estimates of k_2 must be inaccurate. The relationship between K and V_g was practically horizontal, indicating a constant precursor pool volume at all rates of FDG phosphorylation.

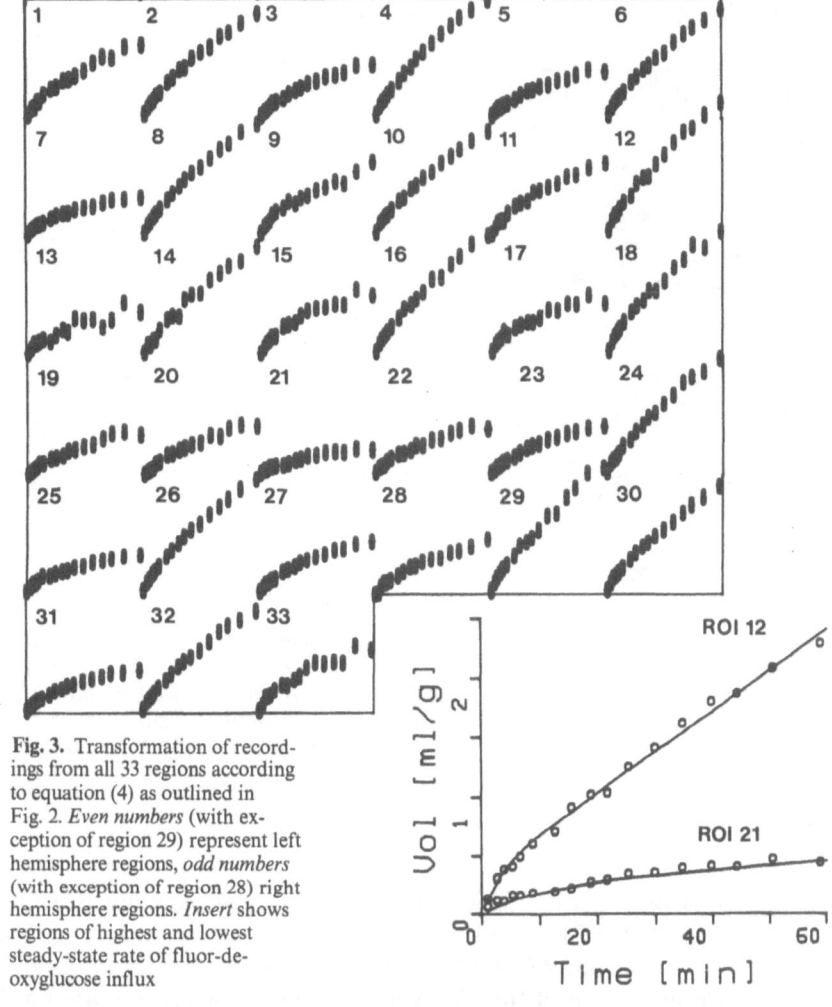

Fig. 3. Transformation of recordings from all 33 regions according to equation (4) as outlined in Fig. 2. *Even numbers* (with exception of region 29) represent left hemisphere regions, *odd numbers* (with exception of region 28) right hemisphere regions. *Insert* shows regions of highest and lowest steady-state rate of fluor-deoxyglucose influx

Table 1. Values of K, V_g, K_1, k_2, k_3 and V_e in regions 11, 12, 21, and 22

Region	Visual cortex		Occipitotemporal white matter	
Hemisphere	Left	Right	Left	Right
ROI	12	11	22	21
K (ml/hg/min)	3.40 (0.1)	1.40 (0.1)	0.90 (0.1)	0.40 (0.1)
V_g (ml/hg)	35 (4)	42 (5)	30 (6)	22 (6)
K_1 (ml/g/min)	0.14 (0.03)	0.07 (0.01)	0.05 (0.01)	0.03 (0.004)
k_2 (/min)	0.23 (0.09)	0.10 (0.02)	0.11 (0.03)	0.09 (0.03)
k_3 (/min)	0.07 (0.01)	0.03 (0.01)	0.02 (0.01)	0.01 (0.01)
V_e (ml/hg)	61 (12)	68 (8)	46 (8)	31 (6)

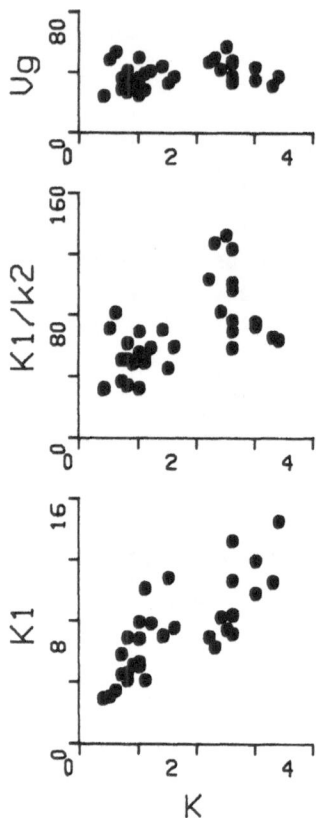

Fig. 4. Relationships between steady-state rate of fluor-deoxyglucose influx (*K*) (abscissa) and initial, unidirectional rate of fluor-deoxyglucose influx (K_1), permeability ratio ($K_1/k_2 = V_e$), and apparent precursor pool volume (V_g). *Abscissa:* Rate of influx (ml/hg/min). *Ordinate:* Rate of influx (k_1; ml/hg/min), apparent volumes of distribution (K_1/k_2 and V_g; ml/hg)

Discussion

The present analysis proved it possible to record measurements of the brain accumulation of labeled FDG and labeled FDG-6-P in such a way that the salient information, the net rate of FDG phosphorylation, is readily apparent. In fact, simple visual inspection of the "V" versus "Theta" curves revealed a dramatic difference between the left and right hemispheres of the brain.

The slope of the uptake curve in the linear phase is a model-independent estimate of the net rate of FDG phosphorylation (Patlak et al. 1983). This means that K is independent of the accuracy of the estimate of the three rate constants K_1, k_2, and k_3. However, the analysis requires the presence of a linear phase in the uptake curve. If such a phase does not exist, the estimate of K is no longer model independent, and such a phase would not exist if phosphatase activity of brain were significant. In that case, the uptake curve would have to be fitted to four rather than three rate constants. Also in that case the unidirectional rate of phosphorylation of FDG would bear no relation to the net rate of glucose phosphorylation, which would depend on all the subsequent reactions.

Estimates of four rate constants have much higher uncertainty than estimates of three rate constants, and the fits generally are no better than those obtained with three rate constants. Thus, from plots such as those obtained in the present analysis, there is no compelling reason to suspect significant FDG-6-phosphatase activity in the human brain.

Why are the estimates of K_1 and k_2 inaccurate? The symbol K_1 represents the initial (i.e., "unidirectional") clearance of the tracer. It is simply the ratio between the accumulated FDG in brain and the integral of FDG in blood at the very earliest times. When integration is based on arterial samples obtained at discrete times, the early values of the integral are very much in doubt. Integration of discrete time points assumes a certain initial shape of the curve which may be far from the actual, unknown shape. At later times, the uncertainty of the first very brief period makes no difference, but at early times, an independent measure of the integral is required, for example by continuous withdrawal of an arterial sample.

The inaccuracy of K_1 and k_2 makes it impossible to utilize the K_1/k_2 ratio to estimate the human brain glucose content, and thus to estimate the magnitude of the "lumped constant" in the various regions of the brain. The common measurement of the "lumped constant" as the ratio between the net extraction fractions of FDG and glucose in the steady-state is, of course, model independent, but this fact is of no help when regional changes are considered. These changes are associated with changes of the relationship between plasma and brain glucose and must be assessed by an "autoradiographic" index of the brain glucose. Such an index exists in the distribution of labeled 3-O-methylglucose (OMG) in the steady-state. It would be of enormous interest to perform the present analysis for a case of OMG uptake by the human brain, and thus to obtain regional estimates of the "lumped constant" of the human brain.

References

Gjedde A (1982) Calculation of cerebral glucose phosphorylation from brain uptake of glucose analogs in vivo: A re-examination. Brain Res Rev 4:237−274

Patlak C, Blasberg RG, Fenstermacher JD (1983) Graphical evaluation of blood-to-brain transfer constants from multiple-time uptake data. J Cerebr Blood Flow Metab 3:1−7

Sokoloff L, Reivich M, Kennedy C, Des Rosiers MH, Patlak CS, Pettigrew KD, Sakurada O, Shinohara M (1977) The (14C)deoxyglucose method for the measurement of local cerebral glucose utilization: Theory, procedure, and normal values in the conscious and anesthetized albino rat. J Neurochem 28:897−916

Measurement of Local Cerebral Glucose Metabolism: Effect of Pathology and Functional Stimulation

M. Reivich, M. Kushner, A. Alavi, and J. Greenberg

Introduction

The scientific rationale for associating specific neurologic functions with discrete regions of the brain arose from clinocopathologic correlations which began in the nineteenth century (Broca 1861). By the early twentieth century neurohistologic mapping had demonstrated the inhomogeneity of brain tissue (Brodmann 1907). Ablative experiments and early electrophysiologic studies also served to refine the view of regional neural specialization in lower mammals. These efforts reached a culmination of sorts when Penfield and Jasper (1954) demonstrated the inhomogeneous and specialized results of focal electrocortical stimulation. Despite these advances the technical ability to perform in vivo measurements of neurophysiologic parameters in man lagged far behind.

Modern cerebral metabolic studies in man began with the development of the Kety-Schmidt technique (1945). This method allowed the measurement of average brain blood flow. Early results using this technique demonstrated that cerebral blood flow varied according to the state of consciousness. Later, the development of techniques using the clearance rates of gamma-emitting isotopes allowed the measurement of brain blood flow in discrete regions (Lassen and Ingvar 1963). Both invasive and noninvasive modifications of these techniques permitted reliable study of diffuse and focal neuropathologic processes (Obrist et al. 1967). These methods, especially those involving xenon-133, provided for the first time a means to study higher cortical function in man under physiologic conditions (Lassen et al. 1977).

Early work with tomographic imaging devices offered the promise of adding the third dimension to brain imaging (Kuhl et al. 1976). These technical advances coupled with progress in basic science (Sokoloff et al. 1977) combined to lead to the successful imaging of cerebral metabolism in man (Reivich 1979). This new technique of tomographic imaging using specific positron emitting isotopes has already provided a means to measure local cerebral glucose metabolism, oxygen consumption, blood flow, blood volume, pH, and neurotransmitter distribution (Reivich et al. 1981; Frackowiak et al. 1980; Phelps et al. 1982). Many other aspects of neural function have the potential to be imaged.

The technique of positron emission tomography (PET) has already found application in the study of stroke, brain tumor, and dementia as well as the study

The Cerebrovascular Research Center, Department of Neurology, University of Pennsylvania, 429 Johnson Pavilion (G2) 36th & Hamilton Walk, Philadelphia, PA 19104, USA

Cerebral Blood Flow and
Metabolism Measurement.
Eds. Hartmann/Hoyer
© Springer-Verlag Berlin Heidelberg 1985

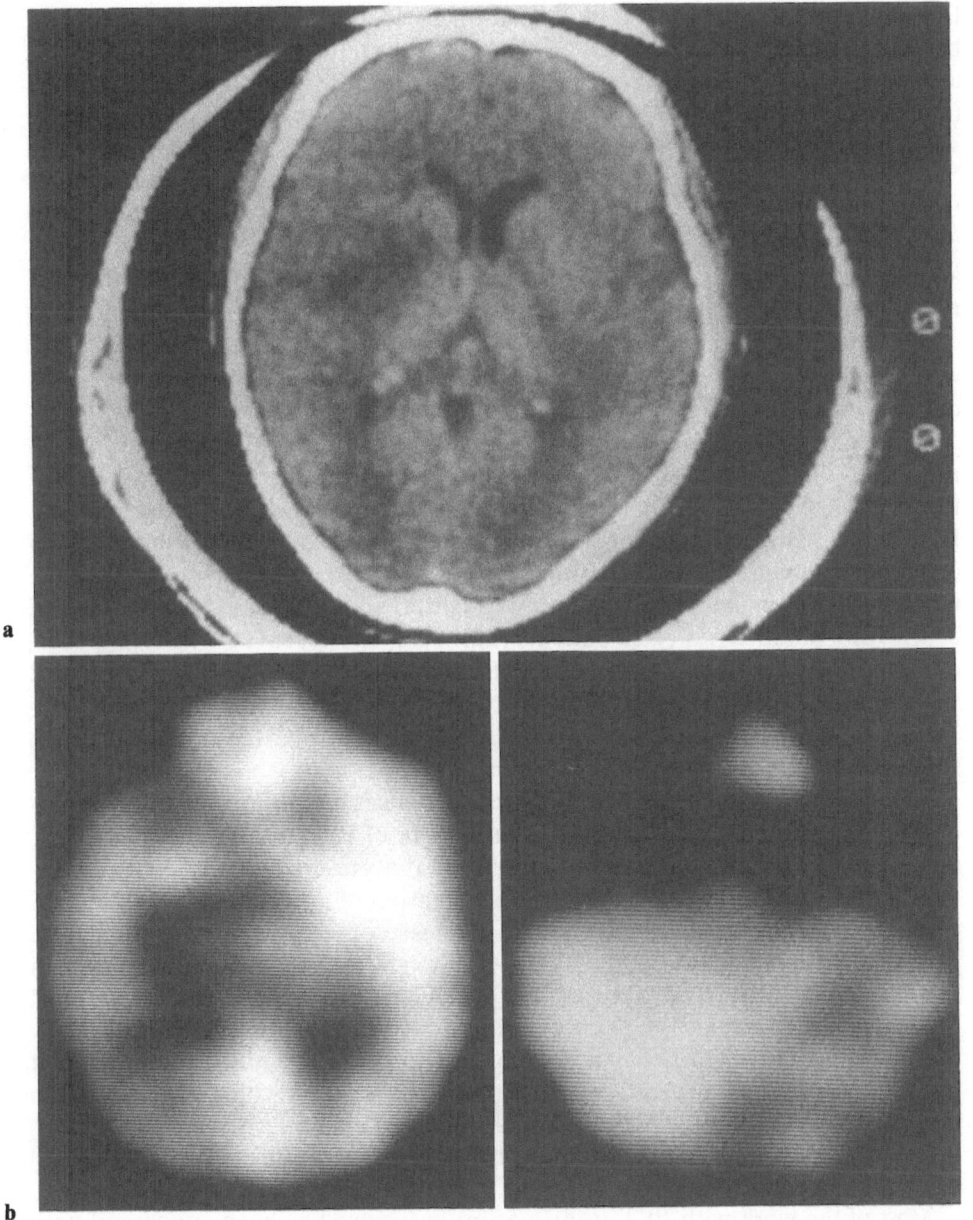

Fig. 1a–c. PET and CT images from a patient who had suffered an acute left-sided cerebral infarction 2 days previously (Kushner et al. 1984)

of normal states (Kuhl et al. 1980, 1982; Phelps et al. 1980; DiChiro et al. 1982; Frackowiak et al. 1980; Greenberg et al. 1981). Cerebral ischemia in general has lent itself to productive study using PET. Areas of ischemia are readily detected on images of glucose consumption, oxygen consumption, and blood flow (Kuhl et al. 1982; Wise et al. 1983). PET abnormalities are often present ictally when the CT shows no abnormality. Studies in stroke patients also have demonstrated unanticipated metabolic abnormalities. PET and CT images taken from a patient who suffered an acute left-sided cerebral infarction 2 days beforehand are shown in Fig. 1. The clinical presentation was one of global aphasia, right homonymous hemianopia, mild right hemiparesis, and mild right hemisensory loss. The CT scan showed some local edema and compression of the lateral ventricle. Notable in this case was the PET scan, which demonstrated widespread and profound metabolic depression throughout the left hemisphere. The extent of this hypometabolism was more widespread than expected given the clinical and CT findings. It should be remembered that this patient was not hemiplegic nor was the sensory deficit profound. Also shown is an image at the level of the cerebellum in this patient. Here a 40% depression of metabolism is present in the contralateral (right) cerebellar hemisphere. This finding of contralateral cerebellar metabolic disturbances has been noted by several investigators (e.g., Lenzi et al. 1982; Baron et al. 1981). In our series of 16 patients (nine with stroke, seven with brain tumor) cerebellar hypometabolism contralateral to the cerebral lesion was present in 80% of cases (Kushner et al. 1984). The presence of this phenomena resolved over time in the case of stroke and was not apparent in chronic cases studies more than 3 months after the ictus.

The significance of these metabolic disturbances in the cerebrum and cerebellum that seemingly are remote from the site of ischemia is uncertain. Several etiologic explanations can be offered for the presence of these remote effects. These include: (1) transneuronal deactivation of adjacent and distant (but functionally connected) neural tissue), (2) release of neurotransmitters or false neurotransmitters from ischemic tissue, (3) the effects of edema formation and local pressure shifts, (4) diaschisis of a purely vascular nature, and (5) a local effect secondary to an ischemic penumbra (Kushner 1984; Slater et al. 1977; Siesjo 1981). While further study is needed in order to resolve these issues it becomes clear that isolated cerebral ischemia triggers complex changes in wide areas of the brain. These complex effects may underly the disorders in higher cortical or integrative functions that can occur in association with focal cerebral ischemia.

Most of the work with PET studies of brain lesions has been concerned with the resting or inactive state. A next step would be to examine the patterns of activation in the lesioned brain in response to specific stimuli or tasks. Such studies could shed light on the relationship between fixed anatomic lesions and their possible interference with the normal pattern of brain activation. We have performed some studies in patients suffering from focal brain lesions using stimuli directed towards their sphere of disability. Normal individuals and patients with homonymous visual field defects were studied during patterned visual stimulation of the entire visual field (Reinvich et al. 1977). In normals

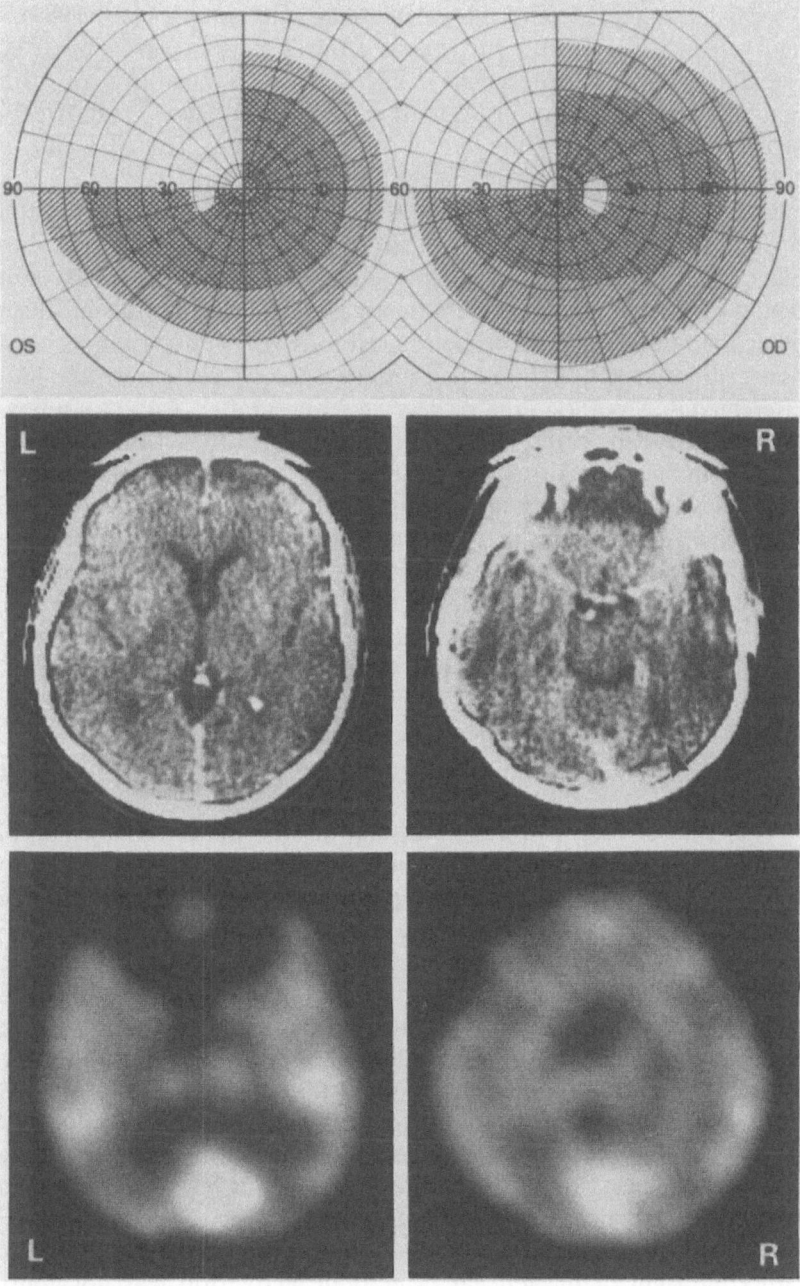

Fig. 2. Goldmann perimetry, CT scans and PET scans in a patient suffering from a homogenous visual field defect (by courtesy of Dr. A. Rosenquist)

such stimulation produces symmetric activation of the visual cortex (Greenberg et al. 1981). Stimulation confined to one or the other visual field produces only contralateral activation of the visual field (Greenberg et al. 1981). The Goldman perimetry, CT, and PET scans from a patient suffering a homogenous visual field defect are presented in Fig. 2. This 63-year-old man suffered a cerebral infarction 2 weeks prior to the PET scan. The CT showed an area of lucency in the region of the lateral optic radiations. The PET showed a unilateral area of hypometabolism in the inferior calcarine region on the right; this lateralization was appropriate for the clinical presence of a homonymous superior quadrantanopsia. Metabolism in other regions was unaffected and no other abnormal metabolic asymmetries were noted. In this case studied under nonresting conditions the metabolic abnormality was most apparent in a region other than the site of the ischemia. Here the hypometabolism was most pronounced in structurally intact primary sensory cortex which had been deafferented due to a lesion in projecting white matter. Future studies employing directed activation could expand our knowledge of the functional reserve in patients with disorders of higher cortical function such as aphasia, amnesia, and dementia. Paradigms using measurements of both the resting and activated states could aid in delineating the degree of disability and in prognosis.

It should be said that activation studies in pathologic states in themselves are somewhat incomplete. Complementary information about the response of the normal brain to such stimuli is needed. It has been demonstrated already that the effect of simple stimuli (visual, tactile, auditory) upon primary sensory cortex can be visualized with PET methodology (Greenberg et al. 1981). We have undertaken stimulation studies in man using more complex stimuli. These complex patterned stimuli may approximate more closely the input received by the CNS in the physiologic state.

The effect of verbal auditory stimulation was studied in a group of seventeen normal right-handed men. A discourse was presented monaurally to either the right or left ear during an uptake period with FDG. The same discourse was presented in an English ($n = 7$) and a non-English version ($n = 10$). No subject had any knowledge of the non-English language. During presentation the subjects were blindfolded and the contralateral ear was plugged. The vigilance of each subject was assessed by monitoring their recognition of a randomly presented meaningless target word. Bilateral handheld push-buttons were used by the subject to indicate recognition of the target word. The results of the studies were compared with those obtained from sensory deprived normal volunteers who were blindfolded with their ears plugged during the FDG uptake period. The metabolic activity in the auditory cortex was not symmetric. A relative increase in metabolism occurred contralateral to the stimulated ear (Fig. 3). The magnitude of the metabolic activation was the same for the stimulation in either the right or the left ear. Metabolic activation was not confined to the auditory cortex however. Regardless of the ear of stimulation, the left perisylvian cortex showed relatively greater metabolic activity than the right. This left-sided effect held true for both the English and non-English stimulations, although metabolic changes were more widespread in the case of the English language input.

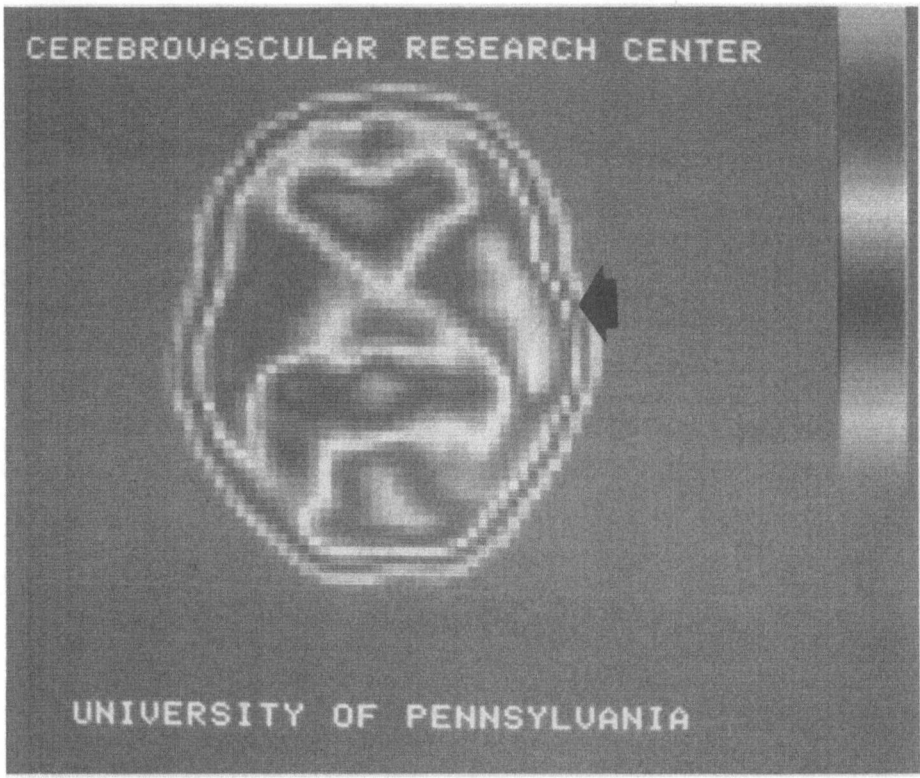

Fig. 3. Left ear stimulation produces significant right temporoparietal metabolic activation (*arrow*) (Reivich 1982)

Electrophysiologic studies have long established that in mammals the auditory pathways are predominantly unilateral (Phillips and Gates 1982). Nevertheless, it has been commonly assumed in clinical practice that auditory input achieves bilateral, and functionally equal, representation in the cerebrum. Despite this view Penfield and Jasper (1954) noted a contralateral activation in their electrocortical stimulation studies. An increasing number of neurobehavioral observations in man have supported the notion that auditory processing is inhomogeneous at the level of the cerebrum. Current views hold that the left hemisphere has primacy for processing of verbal meaning while the right hemisphere contributes to recognition of prosody and emotional content.

Our results suggest that deafferentation of the auditory cortex, either by interruption of the primary auditory pathway or of commissural connections, also could lead to dysfunction of auditory processing by way of cortical isolation. These results demonstrate the physiologic substrate whereby monaural auditory paradigms produce asymmetric responses in patients with brain lesions.

We have also studied the effect of patterned visual stimulation in normal man. This was done by using the method employed for the study of full-field visual stimulation in patients with homonymous field defects (Kushner et al.

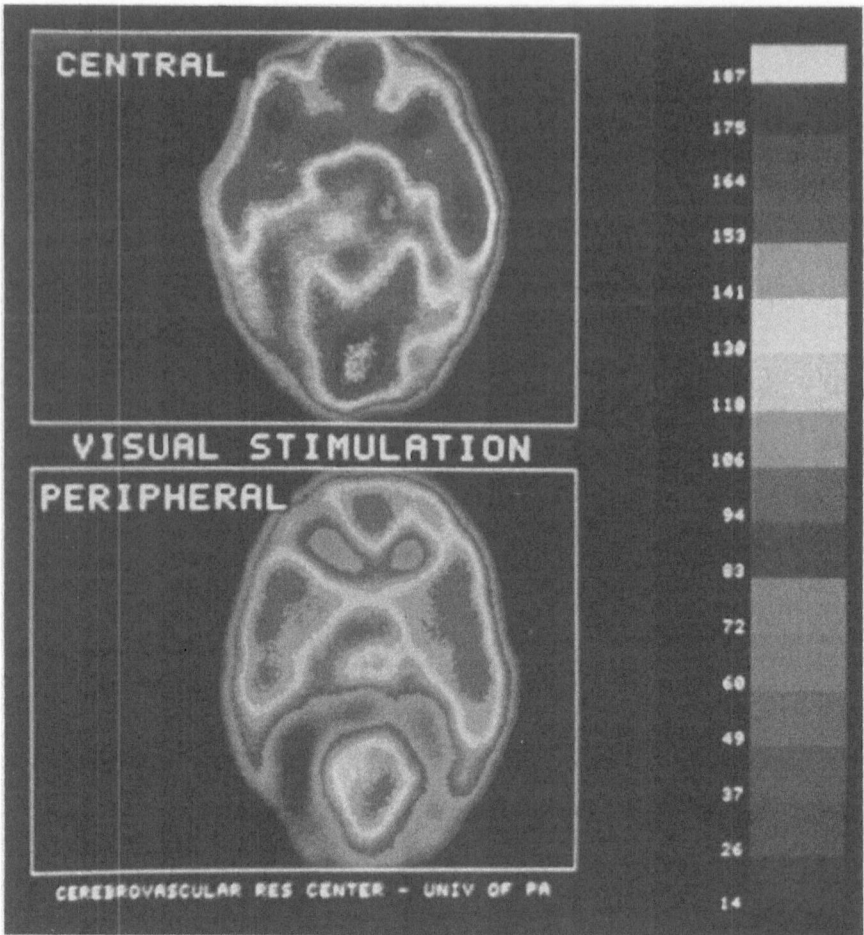

Fig. 4. *Top Panel* shows left calcarine posterior hypermetabolism (macular region) in response to a visual stimulus limited to the right central hemifield. *Bottom panel,* taken from a higher level, shows relatively greater metabolism in the right anterior calcarine area produced by a peripheral stimulation in the left hemifield. Neither panel contains the frontal eye fields. (By courtesy of Dr. A. Rosenquist)

1982). In these experiments the stimulus was confined to discrete regions of the visual field. While receiving the stimulus the subjects were required to fixate on a stationary point in order to assure a constant retinal projection. Exposure to a stimulus confined to the central region of the hemifield (within 20° from the point of fixation) caused vigorous metabolic activation of the contralateral posterior calcarine cortex (Fig. 4). When the stimulus was confined to the peripheral region of the hemifield (more than 60° from fixation) a consistent but less vigorous metabolic increase occurred in the contralateral anterior calcarine cortex (Fig. 4). Our previous results had shown that full hemifield stimulation produced contralateral activation of the whole contralateral calcarine cortex

(Greenberg et al. 1981). These results confirm long-held views of the retino-topic organization of the visual cortex (Holmes 1931; Teuber et al. 1960). In these studies cerebral metabolic activation was not confined to the primary visual (sensory) cortex. Nearby visual association areas (including Brodmann's areas 18 and 19) showed an increase in metabolism that was stimulus dependent. This effect was more vigorous in the right hemisphere.

Several subjects were studied during light exposure but without patterned stimulation. These individuals were required to perform the fixation task while receiving only uniform soft white illumination. In these subjects metabolic activity in the primary visual areas did not show any significant asymmetry. Notable in this fixated but nonstimulated group, and in the stimulated group as well was the presence of relative metabolic activation in the right midfrontal region. This focus of frontal hypermetabolism included Brodmann's area 8, or the "frontal eye fields." The frontal eye fields are part of the neural circuitry which subserves conjugate gaze in mammals. Hypermetabolism here was most dependent on the performance of the visual fixation task rather than the type of visual stimulation. No similar right frontal activation was present in the sensory deprived group nor in the auditory group. This finding might be anticipated given that visual fixation requires continuous correction of small ocular drifts and continuous input to the extraocular muscles. In this set of experiments PET proved useful in studying the response of the sensory association areas, and other brain regions remote from the primary sensory cortex, to complex stimuli and task performance.

References

Ackerman RH, Correia JA, Alpert NM, Baron JC, Goulianos A, Grotta JC, Brownell GL, Taveras JM (1981) Positron imaging in ischemic stroke disease using compounds labeled with oxygen-15. Arch Neurol 38:537−543

Baron JC, Bousser MG, Comar D (1981) "Cross cerebellar diaschisis": a remote functional depression secondary to supratentorial infarction in man. J Cereb Blood Flow Metabol [Suppl 1] S 500−S 501

Broca PP (1861) Remarques sur le siege de la faalte du language articule, suivie d'une observation de'aphemie (perte de la parche). Bull Soc Anat (Paris) 36:330−357

Brodmann K (1907) Die kortex gliederung des menschen. J Psychol Neurol 231

DiChiro G, Dela Paz RL, Brooks RA, Sokoloff L, Kornblith PL (1982) Glucose utilization of cerebral gliomas measured by (18F) fluorodeoxyglucose and positron emission tomography. Neurology 32:1323−1329

Frackowiak RSJ, Lenzi GL, Jones T, Heather JD (1980) Quantitative measurement of regional cerebral, blood flow and oxygen metabolism in man using 150 and positron emission tomography: theory, procedure, and normal values. J Comput Assist Tomogr 4:727−736

Greenberg JH, Reivich M, Alavi, et al (1981) Metabolic mapping of functional activity in human subjects with the 18F-fluorodeoxyglucose technique. Science 212:678−680

Holmes G (1931) A contribution to the cortical representation of vision. Brain 59:470−479

Kety SS, Schmidt CT (1945) Determination of cerebral blood flow in man by use of nitrous oxide in low concentrations. Am J Physiol 143:53

Kuhl DE, Edwards RQ, Ricci A, Yacob RT, Mich TJ, Alavi A (1976) The Mark IV system for radionuclide computed tomography of the brain. Radiology 121:405−413

Kuhl DE, Phelps ME, Kowell AP (1980) Effect of stroke on local cerebral metabolism and perfusion. Ann Neurol 8:47−60

Kuhl DE, Metter EJ, Riege WH, Phelps ME (1982) Effects of human aging on patterns of local cerebral glucose utilization determined by the (18F)fluorodeoxyglucose method. J Cereb Blood Flow Metab 2:163–171

Kushner M, Rosenquist A, Alavi A, Reivich M, Greenberg J, Cobb W (1982) Macular and peripheral visual field representation inthe striate cortex demonstrated by positron emission tomography. Ann Neurol 12:89

Kushner MJ, Alavi A, Reivich M, Dann R, Burke A, Robinson G (1984) Contralateral cerebellar hypometabolism following cerebral insult. A Pet study. Ann Neurol 5:425–434

Lassen NA, Ingvar DH (1963) Regional cerebral blood flow measurement in man. Arch Neurol 9:615–622

Lassen NA, Roland PE, Larsen B, Melared E, Suh K (1977) Mapping of human cerebral functions. Acta Neurol Scand [Suppl 64] 56:262–263

Lenzi GL, Frackowiak RSJ, Jones T (1982) Cerebral oxygen metabolism and blood flow in human cerebral ischemic infarction. J Cereb Blood Flow Metabol 2:321–339

Obrist WD, Thompson HK, Wang HS, Wilkinson WE (1967) Determination of regional cerebral blood flow by inhalation of 133-xenon. Circ Res 22:124–135

Penfield W, Jasper H (1954) Epilepsy and the functional anatomy of the human brain. Little, Brown Co, Boston

Phelps ME, Mazziotta JC, Kuhl DE (1981) Tomographic mapping of human cerebral metabolism: visual stimulation and deprivation. Neurology 31:517–529

Phelps ME, Mazziotta JC, Huang SC (1982) Study of cerebral function with positron computed tomography. J Cereb Blood Flow Metabol 2:113–162

Phillips DP, Gates GR (1982) Representation of the two ears in the auditory cortex: a re-examination. Intern J Neuro Science 16:41–46

Reivich M (1982) The use of cerebral blood flow and metabolic studies in cerebral localization. In Thompson RA, Green JR (eds), New perspectives in cerebral localization. Raven Press: New York 115–144

Reivich M, Cobbs W, Rosenquist A, Stein A, Schatz N, Savino P, Alarvi A, Greenberg J (1981) Abnormalities in local cerebral glucose metabolism in patients with visual field defects. J Cereb Blood Flow Metabol [Suppl 1] S 471–S 472

Reivich M, Kuhl DE, Wolf A, Greenberg J, Phelps M, et al. (1979) The 18F-fluorodeoxyglucose method for the measurement of local cerebral glucose utilixation in man. Circ Res 44:127–137

Siesjo BK (1981) Cell damage in the brain: a speculative synthesis. J Cereb Blood Flow and Metabol 1:155–185

Slater R, Reivich M, Goldberg H, Banka R, Greenberg J (1977) Diaschisis with cerebral infarction. Stroke 8:689–690

Sokoloff L, Reivich M, Kennedy C, Des Rosiers MH, Patlak CS, Pettigrew KD, Sakurada O, Shinoharo M (1977) The (14C) deoxyglucose method for the measurement of local cerebral glucose utilization: theory, procedure, and normal values in the conscious and anesthetized albino rat. J Neurochem 28:897–916

Teuber H, Battersby W, Bender M (1960) Visual field deficits after penetrating missile wounds. Harvard University Press: Cambridge

Wise RJS, Bernardi S, Frackowiak RSJ, Legg NJ, Jones T (1983) Serial observations on the pathophysiology of acute stroke. Brain 106:197–222

The Current State of rCBF, rCMRO$_2$, rCBV, and rCMRGlu Studies at the Hammersmith Hospital

J. M. Gibbs, R. S. J. Frackowiak, R. J. S. Wise, A. A. Lammertsma,
C. G. Rhodes, and T. Jones

Introduction

The development of positron emission tomography (PET) represents an important advance in the study of cerebral tissue function in man. By virtue of the characteristic double photon signal from positron emitting isotopes and tomographic reconstruction using paired coincidence detectors, precise regional measurements can be made of cerebral isotope concentration throughout a tomographic slice of brain. PET is thus a noninvasive, in vivo equivalent of the autoradiographic technique applied to experimental animals. Furthermore, the development of a variety of tracer models has led to the measurement not only of regional cerebral blood flow and blood volume, but also of local metabolic activity and its relationship to the available blood supply. Labeling of drugs and neurotransmitter precursors is further expanding the range of applications for PET in the study of both normal and abnormal cerebral physiology.

PET studies in a variety of neurologic disorders have produced some important advances in our understanding of pathophysiologic mechanisms, particularly in the field of cerebrovascular disease. However, the technique is still predominantly a research tool and relatively few observations of practical clinical value have so far emerged from PET studies of cerebral disease. The technique is extremely expensive, relatively time consuming, and demands considerable commitment from a large number of technical, scientific, and clinical staff. For these reasons its use will inevitably remain confined to a small number of specialized centers. One of the important roles of such centers should therefore include the identification of potentially rewarding areas of research or types of clinical measurement which might be pursued with less elaborate and more widely available techniques.

The emphasis of our PET program has been directed towards answering relatively simple questions about common neurologic disorders. Some illustrative examples of these studies are briefly reviewed below, but the main purpose of this paper is to discuss the quantitative methods involved. All our studies to date have included the measurement of regional cerebral blood flow (CBF), oxygen utilization (CMRO$_2$), and fractional oxygen extraction (OER) by means of the ^{15}O steady-state technique. Since early 1981, regional cerebral blood volume (CBV) has also been routinely measured by the ^{11}C-carboxyhemoglobin

MRC Cyclotron Unit and Department of Neurology, Hammersmith Hospital, Ducane Road, London W12 OHS, Great Britain

Cerebral Blood Flow and
Metabolism Measurement.
Eds. Hartmann/Hoyer
© Springer-Verlag Berlin Heidelberg 1985

method. Using ¹⁸FDG and a modified Sokoloff model, additional measurement of regional cerebral glucose metabolism has been carried out in certain groups of patients. Some theoretical and practical aspects of these techniques will be discussed, as well as their possible limitations in the study of neurologic disease.

Methodology

Tomographic Camera

All studies so far have been carried out with an ECAT II tomograph (EG & G Ortec), a single slice whole body scanner which in the medium resolution mode has a spatial resolution of $16.7 \times 16.7 \times 16$ mm at full width half maximum. Uniform resolution and a low scatter fraction are achieved at some expense to sensitivity, so relatively long scanning times are necessary and the use of traditional dynamic tracer techniques is impractical. We have therefore concentrated on measurements of the steady state uptake of short lived isotopes as a means of calculating the rates of physiologic processes. The sensitivity of the camera is such that during steady state inhalation of $^{15}O_2$ or $C^{15}O_2$, adequate coincidence counts are attained with a scanning period of 5 min for each tomographic plane. An external germanium 68 ring source is used for measurement of regional tissue attenuation, which is routinely used for correction of all emission data.

Oxygen-15 Steady-State Technique

Measurement of regional CBF is derived from the relationship of cerebral and arterial isotope activity during continuous inhalation of a tracer quantity of oxygen 15 labeled carbon dioxide. Under the influence of carbonic anhydrase in the lung, the ^{15}O label is rapidly transferred to circulating pulmonary capillary water. The effect of $C^{15}O_2$ inhalation is therefore equivalent to a continuous arterial infusion of $H_2^{15}O$. Because of the very short half-life of ^{15}O (2.1 min), a dynamic equilibrium is established in the brain after $8-10$ min, when the rate of arterial delivery of $H_2^{15}O$ and its diffusion into the tissues is balanced by the rate of washout and radioactive decay. The tissue concentration of $H_2^{15}O$ at this stage (measured by PET) is a function of regional CBF, which can be calculated in absolute units from the PET data and $H_2^{15}O$ concentration in arterial blood at the time of the scan.

A second scan is carried out during steady-state inhalation of molecular $^{15}O_2$, which is transferred in the lung to the hemoglobin of circulating red cells, and again transported to the brain according to blood flow. A proportion of the delivered arterial oxygen ($35\% - 50\%$ in the normal brain) is extracted by the tissues and rapidly consumed in the process of aerobic metabolism. The rate of consumption of extracted oxygen is such that the measured ^{15}O in the tissues is almost entirely in the form of the end product of oxidative metabolism, which again is labeled water. The PET signal during steady state $^{15}O_2$ inhalation is therefore largely due to this "water of metabolism," the concentration of which reflects both the rate of delivery of $^{15}O_2$ (CBF) and its fractional extraction by the tissues (OER). The combined data from both C $^{15}O_2$ and $^{15}O_2$ scans thus include sufficient information to calculate regional OER as well as CBF.

A proportion of the emission data collected during $^{15}O_2$ inhalation arises from two sources other than the labeled water produced locally by cerebral metabolism. Firstly, some $H_2^{15}O$ (produced by both cerebral and noncerebral tissues) is recirculating to the brain in the arterial blood and being distributed according to blood flow. Secondly there is unextracted hemoglobin-bound $^{15}O_2$ within all cerebral vessels, the vast majority of this signal being accounted for by the proportionally higher volume of the venous compartment. Both of these additional forms of ^{15}O contribute to the intracranial isotope activity and result in overestimation of OER unless appropriate corrections are made. The recirculating water component can be calculated from the plasma $H_2^{15}O$ concentration during $^{15}O_2$ inhalation, the ratio of plasma to whole blood $H_2^{15}O$ during $C^{15}O_2$ inhalation, and the regional CBF (Frackowiak et al. 1980). The amount of hemoglobin-bound $^{15}O_2$ in the vascular compartment is a function of cerebral blood volume and can therefore be calculated from an independent measurement of regional CBV (Lammertsma et al. 1981). This is achieved by scanning the brain after inhalation of a trace amount of ^{11}C-carbon monoxide, a red cell marker which remains confined to the intravascular compartment (Phelps et al. 1978).

The PET data collected during the three phases of gas inhalation and their relationship to regional CBF, OER, and CBV are summarized schematically in Fig. 1. Note that regional CMRO$_2$ is equivalent to the rate of oxygen extraction and can therefore be calculated from the rate of delivery of oxygen (CBF × arterial oxygen content) and its fractional extraction by the tissues (OER):

$$CMRO_2 = CBF \times Arterial\ [O_2] \times OER$$

The theoretical concepts of the ^{15}O steady-state model have been discussed in earlier papers by Jones et al. (1976), Subramanyam et al. (1978), and Lenzi et al. (1978). The final development of a quantitative technique was made possible by the arrival of PET, as described originally by Frackowiak et al. (1980). Details of the method of blood volume correction of the OER data have recently been reported in two papers by Lammertsma et al. (1983 a, b).

Two basic assumptions inherent in the steady-state oxygen model are firstly that the blood/tissue partition coefficient for water is equal to 1, and secondly that the permeability × surface area (PS) product is sufficiently high for the tissue extraction of water to be regarded as 100% under all conditions. The potential sources of error arising from these assumptions have been analyzed in detail by Lammertsma et al. (1981). It was concluded that within and below the expected physiologic range of blood flow, the effect on CBF and CMRO$_2$ measurements of variations in the partition coefficient or PS product are small. The greatest sources of error were found to be the overestimation of OER due

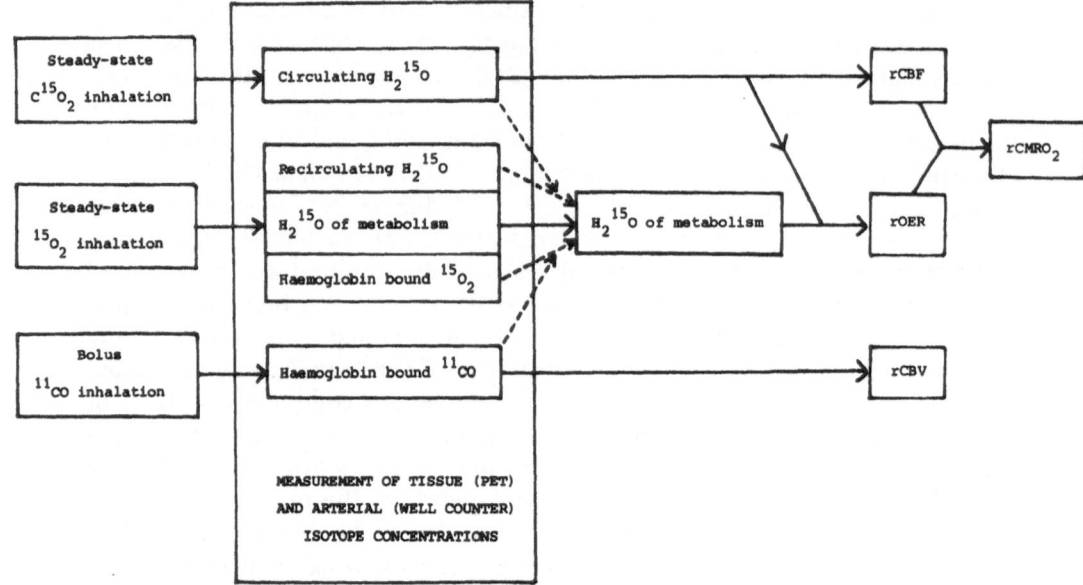

Fig. 1. Summary of tracer methodology for the measurement of regional CBF, OER, CMRO$_2$, and CBV (see text for details)

to unextracted intravascular $^{15}O_2$ during O_2 inhalation – now routinely corrected for from the CBV data – and the underestimation of both CBF and CMRO$_2$ which may arise from inclusion of nonexchanging tissue such as bone or CSF within the region of interest used for quantitation. It should be stressed that this latter effect does not apply to the OER measurement, or to any other quantitative expression produced by division of one set of scan data by another. To a large extent such physiologic ratios are also unaffected by the loss of recovery of counts which inevitably results from the finite spatial resolution of the PET scanner when sampling heterogeneous mixtures of gray and white matter (partial volume effect).

Measurement of Cerebral Glucose Metabolism

CMRGlu is measured in our unit by the deoxyglucose technique, originally described by Sokoloff et al. (1977) and adapted for use with PET and ^{18}F-fluorodeoxyglucose by Phelps et al. (1979). The theoretical and practical aspects of the technique are discussed elsewhere in this volume (Heiss W-D et al.). Much has been written about the importance of variation in the individual rate constants when applying the FDG model to the study of abnormal brain tissue. Although we routinely perform a series of uptake scans during the early stages after FDG injection, there is a limit to the dose of tracer which can be used in these multitracer studies, and activity within the brain is therefore low in the early stages. In conbination with the relative insensitivity of the ECAT II camera, this limitation of head counts precludes accurate regional measurement of the rate constants in each patient studied. However, as indicated by Sokoloff in his original description of the deoxyglucose model, the values of k_1, k_2, and k_3 become largely irrelevant if the tissue isotope concentration is measured at a sufficiently late stage after administration of the tracer (50 − 70 min). If it is assumed that the tissue ^{18}FDG at this stage is entirely in the phosphorylated form and that there has been no loss of ^{18}FDG-6-phosphate from the tissue, the exponential functions incorporating the rate constants can be excluded from the operational equation for calculating CMRGlu. This simplification of the model has been discussed in detail by Rhodes et al. (in press), who also examined the potential errors arising from such assumptions when measuring CMRGlu in ischemic brain. A test comparison was made using data published from normal subjects (Huang et al. 1980) and reported results from patients with stroke (Hawkins et al. 1981). CMRGlu was calculated for both groups of subjects in three different ways: using the individually measured k values for each subject, the mean k values taken from both populations, and finally applying the simplified equation which excludes the rate constants. There was close agreement between the CMRGlu values obtained by all three methods for normal brain and gray matter regions in ischemic brain. However, there was a tendency to overestimate CMRGlu in ischemic white matter when applying either the simplified equation or the correction with mean k values. There is considerable subject-to-subject variation of individual k values for both normal and abnormal brain. Hence the use of the simplified equation is at least as acceptable as the incorporation of 'standard' k values into the operational equation. When technically feasible, regional measurement of the k values in each patient studied is clearly the ideal.

However, it should be stressed that a further source of error (inherent in any technique which depends on a labeled analogue of glucose) arises from variations in the lumped constant, the expression relating the differential handling by the cerebral tissues of glucose and FDG. An indirect measurement of this constant for the whole brain of man has produced a value of 0.42 (Huang et al. 1980). This global value is now generally used for calculation of CMRGlu in both gray and white matter of normal brain as well as in various types of cerebral disease. The glucose/FDG lumped constant in the rat brain has been calculated by Crane et al. (1983) to range between 0.55 and 1.67 under extreme conditions, with a normal value of 0.8 − 0.9. A method for making regional measurements of the lumped constant has yet to be applied in man, but this major uncertainty about use of the FDG model in pathologic tissue clearly requires further elucidation.

Practical Aspects of the Steady-State Procedure

In order to achieve and maintain a steady state during $C^{15}O_2$ and $^{15}O_2$ inhalation, a constant supply of the tracer is essential. Both the flow rate and the concentration of supplied gas are monitored continuously and an electromechanical servo device is used to adjust the concentration in response to any fluctuations of the output from the cyclotron. A light, perforated face mask is used for administration of the gas, which is supplied to the patient at about 500 ml/min and diluted by room air inspired through the side holes in the mask. Approximately half the labeled gas is expelled from the mask during expiration, and this is removed to waste by a vacuum exhaust system via a perspex hood which covers the patient's face. This arrangement has been found to be preferable to either a mouthpiece or a closed anesthetic-type face mask. The valves in such systems introduce some resistance to both inspiration and expiration which often results in variation of the patient's respiratory pattern and hence loss of the steady state. It must be added that even with a low resistance system, achievement of a steady state may still be difficult in patients who are anxious, uncooperative, or restless as a result of their cerebral disease.

Arterial blood sampling is carried out by means of a fine gauge radial artery cannula, which is inserted under local anesthesia before final positioning of the patient in the scanner. Repeated samples can then be taken without any risk of disturbing the patient's position or physiologic steady state. Along with a continuous monitor of the head counts, serial measurements of arterial isotope concentration provide an additional check on the quality of equilibrium achieved. Loss of the steady state may necessitate restarting a set of scans and occasionally such instability makes an entire study nonquantitative. However, analysis of the serial blood counts (three or four specimens for each period of gas inhalation) from 60 consecutive studies revealed very little instability of the steady state. The percentage standard deviation from the mean in these 60 cases was 4.0, 4.2, 4.3, and 5% for the $C^{15}O_2$ whole blood, $C^{15}O_2$ plasma, $^{15}O_2$ whole blood, and $^{15}O_2$ plasma concentrations respectively.

Maintenance of the patient's head position is a crucial factor for any PET study in which the mathematical reconstruction of quantitative data depends on the relationship of one set of scans to another. In practice this applies to all quantitative measurements, since correction for tissue attenuation by means of a ring source transmission scan is now considered essential and is widely practised. However, the calculation of OER and $CMRO_2$, including the correction for blood volume effect, requires precise congruence of all four sets of scan data. The three emission scans during inhalation of $C^{15}O_2$, $^{15}O_2$, and ^{11}CO and the attenuation correction transmission scan must be all recorded from identical tomographic slices of brain. Of course this principle applies equally to the measurement of CBF, OER, and CBV using dynamic tracer techniques. Although the duration of each data collection period will clearly be much shorter with such methods, the time taken up by preparation, administration, and decay of each successive isotope still results in a total scanning period of around 1.5 h. A typical steady-state procedure with measurement of CBF, OER, $CMRO_2$, and CBV takes about 2 h. The relatively long duration of these multitracer studies requires the head restraint system to be extremely comfortable as well as effective. If glucose metabolism is measured in addition to CBF, OER, and CBV, the total duration of the study is extended to slightly more than 3 h. In practice, we have found that head movement of only a millimeter or two can be seen and corrected immediately by projecting a grid of light onto the patient's forehead and marking the position of a number of the intersections on the skin with a felt-tipped pen. The alignment of these marks and the light grid is constantly monitored during the scan on a closed circuit television image of the patient's face. The need for strict and prolonged immobility highlights again the difficulty of studying restless and uncooperative patients with this technique.

Clinical Studies

Since the introduction of the PET program at the Hammersmith Hospital in 1979, over 850 brain studies have been carried out in patients with a variety of neurologic disorders. Listed in Table 1 are those conditions which have been the subject of major projects, many of which are still in progress. A substantial proportion of our experience has been in the field of cerebrovascular disease. Studies have been carried out in multi-infarct dementia (Frackowiak et al. 1981), acute stroke (Lenzi et al. 1982; Wise et al. 1983), and occlusive carotid artery disease (Gibbs et al. 1983). Combined studies of CBF, CMRO$_2$, and CMRGlu in patients with stroke have recently been reported· by Wise et al. (in press).

The most significant pathophysiologic findings in these cerebrovascular studies have depended on analysis of the relationships between different physiologic variables, rather than on their absolute values considered in isolation. For example, serial measurements in acute stroke (Wise et al. 1983) have revealed a consistent evolution from high to low OER in regions of cerebral infarction, illustrating the changing relationship between oxygen supply and demand. The finding of low OER indicates that whatever the absolute value of CBF, regional blood flow is more than adequate to meet residual cerebral oxygen demands. This pattern of relative luxury perfusion is usually already present in patients studied between 24 and 48 h after the onset of stroke, which has an important bearing on any proposed treatment of such cases. Either medical or surgical manoeuvres designed to increase CBF at this stage are clearly irrational. Such

Table 1. Neurologic disorders studied in the Hammersmith Hospital PET program since 1979

Cerebrovascular disease:	Acute stroke
	Multi-infarct dementia
	Carotid artery occlusion
Degenerative disorders:	Alzheimer dementia
	Parkinson's disease
	Huntington's chorea
Miscellaneous:	Primary cerebral tumors
	Epilepsy
	Schizophrenia
	Multiple sclerosis
	Hydrocephalus

intervention might theoretically be of value during the early phase of true ischemia, when inappropriately low CBF is indicated by raised OER in the affected region. The available evidence suggests that any therapeutic measures in acute cerebral infarction should be carried out within the first few hours after the onset of symptoms.

In patients with multi-infarct dementia (Frackowiak et al. 1981), the finding of closely matched depression of CBF and $CMRO_2$ (and hence normal OER) supported the clinical and pathologic evidence that the basic lesion in this disorder is one of established cerebral infarction rather than a continuing state of true chronic ischemia. A similar situation is found in most patients with occlusive carotid artery disease, in whom reduced CBF is usually found to be appropriately matched to reduced cerebral metabolic demands (Gibbs et al. 1983). An important finding in these patients has been a focal increase of cerebral blood volume in regions distal to occluded carotid arteries. This is likely to reflect the compensatory vasodilatation which maintains CBF in the face of diminished cerebral perfusion pressure. Analysis of the relationship between regional CBF, CBV, and OER suggests that the ratio of CBF/CBV provides the most sensitive index of diminished perfusion pressure in these cases. The theoretical basis and evidence for this conclusion is discussed in more detail elsewhere in this volume (Gibbs et al., p. 452).

Comparison of $CMRO_2$ and CMRGlu in patients with stroke (Wise et al., in press) has revealed an altered but consistent relationship between oxygen and glucose consumption in the infarcted cerebral regions. In eight patients studied in this way there was relative preservation of CMRGlu in comparison with $CMRO_2$, despite a more than adequate oxygen supply within the infarct (low OER). For reasons discussed above, CMRGlu measurements based on FDG uptake and phosphorylation in regions of grossly damaged brain should be interpreted with some caution. However, these findings suggest that there is preferential nonoxidative metabolism of glucose (aerobic glycolysis to lactate) within the infarcted cerebral regions. This is not a recognized phenomenon in neuronal tissue studied during the period following severe cerebral ischemia and may represent the metabolic activity of the numerous infiltrating macrophages and neutrophils which are known to be present in regions of extensive cerebral infarction (Yates 1976).

Studies of nonvascular cerebral disorders have included degenerative dementia (Frackowiak et al. 1981), Parkinson's disease (see Leenders et al. in this volume, p. 459), and primary cerebral tumours (Ito et al. 1982; Rhodes et al. 1983). More recently we have examined the effects of dexamethasone treatment on regional CBF, CBV, and $CMRO_2$ in patients with cerebral tumors. Some preliminary results of this work are also discussed elsewhere in this volume (Leenders et al., p. 465). In studies of patients with temporal lobe epilepsy and a unilateral EEG focus, regional CBF and $CMRO_2$ were found to be reduced in the affected hemisphere in the interictal state (Bernardi et al. 1983). These patients were also noted to have reduced blood flow and oxygen consumption in the cerebellum. Although this could be a manifestation of the disease itself, it is more likely to represent an effect of chronic anticonvulsant medication. Regional CBF and $CMRO_2$ have been measured in patients with recently diagnosed and untreated schizophrenia (Shepherd et al., in press). In contrast to the results of some previously reported studies, no significant regional abnormalities were found when comparing these patients with age-matched controls. The only observed difference was a slightly more marked degree of asymmetry between left and right hemispheres amongst the normal population.

Summary

The greatest value of PET lies in the capacity of this technique to make regional measurements of multiple physiologic variables in the human brain. The ^{15}O-steady state technique is a well-established method for measuring regional CBF, OER, and CMRO₂ by PET. Reliable quantitation requires attenuation correction by means of a ring source transmission scan and correction of the OER data for the effects of unextracted ^{15}O₂ in the blood volume compartment. The technique is not difficult to apply but requires cooperative patients and considerable attention to detail at all stages of data collection.

Analysis of the relationships between different variables has consistently provided more valuable information than their absolute values considered in isolation. The use of physiologic ratios (OER, CBF/CBV,CMRO₂/CMRGlu) has the additional advantage that their quantitative values are largely unaffected by some of the major sources of error inherent in all absolute measurements with PET. Combined studies of regional CBF, CMRO₂ , OER, CBV, and CMRGlu have produced some stimulating results in patients with hemispheric stroke and cerebral tumours. However, interpretation of these findings is complicated by uncertainties about the FDG tracer model when applied to pathologic brain tissue. Measurement of the rate constants may eliminate one source of error but the possibility of variation in the lumped constant remains a major problem.

It should be pointed out that the measurement of CBF alone by rapid, serially repeatable, and much less cumbersome techniques can never be supplanted by PET. In many clinical and research situations such techniques will always be more appropriate. This applies, for instance, to rapidly evolving conditions such as migraine and TIAs, to very ill patients, and to peroperative measurements of CBF. However, CBF findings must clearly be interpreted with some caution in those conditions where PET studies have shown dissociated variations between flow and other physiologic variables. The independend disturbance of CBF and CBV in cerebrovascular disease is one area of investigation which can be explored with techniques other than PET.

Within the PET field, the considerable value of multitracer studies highlights the need for further technical refinement. The development of reliable dynamic tracer models and more sophisticated tomographs should reduce the duration and improve the quality of such studies in the future.

References

Bernhardi S, Trimble MR, Frackowiak RSJ, Wise RJS, Jones T (1983) An interietal study of partial epilepsy using positron emission tomography and the oxygen-15 inhalation technique. J Neurol Neurosurg Psychiatry 46:473–477

Crane PD, Pardridge WM, Braun LD, Oldendorf WH (1983) Kinetics of transport and phosphorylation of 2-fluoro-2-deoxy-D-glucose in rat brain. J Neurochem 40:160–176

Frackowiak RSJ, Lenzi GL, Jones T, Heather JD (1980) Quantitative measurement of regional cerebral blood flow and oxygen metabolism in man using ^{15}O and positron emission tomography: Theory, procedure, and normal values. J Comput Assist Tomogr 4:727–736

Frackowiak RSJ, Pozzilli C, Legg NJ, Du Boulay GH, Marshall J, Lenzi GL, Jones T (1981) Regional cerebral oxygen supply and utilisation in dementia: A clinical and physiological study with oxygen-15 and positron tomography. Brain 104:753–778

Gibbs JM, Wise RJS, Leenders KL, Jones T (1983) The relationship of regional cerebral blood flow, blood volume, and oxygen metabolism in patients with carotid occlusion: evaluation of perfusion reserve. J Cereb Blood Flow Metabol 3:S 590

Hawkins RA, Phelps ME, Huang S-C, Kuhl DE (1981) Effect of ischemia on quantification of local cerebral glucose metabolic rate in man. J Cereb Blood Flow Metabol 1:37–51

Huang S-C, Phelps ME, Hoffman EJ, Sideris K, Selin CJ, Kuhl DE (1980) Noninvasive determination of local cerebral metabolic rate of glucose in man. Am J Physiol 238:E69–E82

Ito M, Lammertsma AA, Wise RJS, Bernardi S, Frackowiak RSJ, Heather JD, McKenzie CG, Thomas DGT, Jones T (1982) Measurement of regional cerebral blood flow and oxygen utilisation in patients with cerebral tumours using ^{15}O and positron emission tomography: Analytical techniques and preliminary results. Neuroradiology 23:63–74

Jones T, Chesler DA, Ter-Pogossian MM (1976) The continuous inhalation of oxygen-15 for assessing regional oxygen extraction in the brain of man. Br J Radiol 49:339–343

Lammertsma AA, Jones T (1983a) The correction for the presence of intravascular oxygen-15 in the steady state technique for measuring regional oxygen extraction ratio in the brain: 1. Description of the method. J Cereb Blood Flow Metabol 3:416–424

Lammertsma AA, Jones T, Frackowiak RSJ, Lenzi GL (1981) A theoretical study of the steady-state model for measuring regional cerebral blood flow and oxygen utilization using oxygen-15. J Comput Assist Tomogr 5:544–550

Lammertsma AA, Wise RJS, Heather JD, Gibbs JM, Leenders KL, Frackowiak RSJ, Rhodes CG, Jones T (1983b) The correction for the presence of intravascular oxygen-15 in the steady-state technique for measuring regional oxygen extraction ratio in the brain: 2. Results in normal subjects and brain tumour and stroke patients. J Cereb Blood Flow Metabol 3:425–431

Lenzi GL, Jones T, McKenzie CG, Buckingham PD, Clark JC, Moss S (1978) Study of regional cerebral metabolism and blood flow relationships in man using the method of continuously inhaling oxygen-15 and oxygen-15 labelled carbon dioxide. J Neurol Neurosurg Psychiatry 41:1–10

Lenzi GL, Frackowiak RSJ, Jones T (1982) Cerebral oxygen metabolism and blood flow in human cerebral ischemic infarction. J Cereb Blood Flow Metabol 2:321–335

Phelps ME, Huang S-C, Hoffman EJ, Kuhl DE (1978) Validation of tomographic measurement of cerebral blood volume with C-11 labeled carboxy-hemoglobin. J Nucl Med 20:328–334

Phelps ME, Huang S-C, Hoffman EJ, Selin C, Sokoloff L, Kuhl DE (1979) Tomographic measurement of local cerebral glucose metabolic rate in humans with (F-18) 2-fluoro-2-deoxy-D-glucose: validation of method. Ann Neurol 6:371–388

Rhodes CG, Wise RJS, Gibbs JM, Frackowiak RSJ, Hatazawa J, Palmer T, Thomas DGT, Jones T (in press) In vivo disturbance of the oxidative metabolism of glucose in human cerebral gliomas. Annals of Neurology

Shepherd G, Gruzelier J, Manchanda R, Hirsch SR, Wise RJS, Jones T (in press) ^{15}O PET scanning in a predominantly first admission drug-naive acute schizophrenia population. The Lancet

Sokoloff L, Reivich M, Kennedy C, Des Rosiers MH, Patlak CS, Pettigrew KD, Sakaurada O, Shinohara M (1977) The [^{14}C] deoxyglucose method for the measurement of local cerebral glucose utilization: theory, procedure, and normal values in the conscious and anesthetized albino rat. N Neurochem 28:897–916

Subramanyam R, Alpert NM, Hoop B et al. (1978) A model for regional cerebral oxygen distribution during continuous inhalation of $^{15}O_2$, and $C^{15}O$ and $C^{15}O_2$. J Nucl Med 19:48–53

Wise RJS, Bernardi S, Frackowiak RSJ, Legg NJ, Jones T (1983) Serial observations on the pathophysiology of acute stroke: the transition from ischaemia to infarction as reflected in regional oxygen extraction. Braïn 106:197–222

Wise RJS, Rhodes CG, Gibbs JM, Hatazawa J, Palmer T, Frackowiak RSJ, Jones T (in press) Disturbance of oxidative metabolism of glucose in recent human cerebral infarcts. Annals of Neurology

Yates PO (1976) Vascular diseases of the nervous system. In Blackwood W, Corselius JAN (eds) Greenfields neuropathology 3rd edn, London, Arnold 86–147

Clinical Interpretation of Dynamic Positron Emission Tomography with C-11-methyl-D-glucose Before and After STA-MCA Anastomosis

H.-U. THAL and W. J. BOCK

Introduction

Neurologic examination, case history, and paraclinical investigations should help in diagnosing and providing the most beneficial therapy for cerebrovascular diseases. Doppler measurements, angiography, serial radionuclide scintigrams, and rCBF measurements are the methods of choice. Positron emission tomography and nuclear magnetic resonance will become future methods for evaluating the extent of a cerebral ischemic event; at present, however, these procedures have only minor importance in clinical practise because of our lack of knowledge about surgically treated stroke patients. Thus it was our goal to sample a series of patients suffering from different grades of cerebral ischemia and undergoing surgery. From the neurosurgical point of view we examined the patients before and after an extra-intracranial arterial bypass (EIAB) operation. Some patients were not operated upon and considered as a medically treated control group. Like the surgically treated group, they were given ASS and, in accordance with Heiss (1983), piracetam (Nootrop).

In this multidisciplinary study we investigated, among other patients with cerebrovascular diseases, 12 STA-MCA candidates. We studied eight of these patients before and after an STA-MCA bypass procedure in close cooperation with the Nuclear Research Center, Juelich. Besides using the above-mentioned methods, we determined the local perfusion rate (LPR) and the local unidirectional glucose transport rate (LUGTR) simultaneously using C-11-methyl-D-glucose (CMG) and dynamic positron emission tomography (dPET).

Results

As we know from 104 STA-MCA arterial bypass operations carried out between January 1979 and December 1981, the best clinical results are to be seen in grade II patients, i.e., in the TIA patients. In this group of 25 patients an improvement was observed in two-thirds of the reexamined cases. In 10 out of 16 PRIND patients (grade III) we saw an improvement. Minor improvement could be seen in only 32% of 26 grade IV a patients with a mild completed stroke. The

Neurochirurgische Klinik, Medizinische Einrichtungen der Universität Düsseldorf, Moorenstraße 5, D-4000 Düsseldorf 1

Cerebral Blood Flow and
Metabolism Measurement.
Eds. Hartmann/Hoyer
© Springer-Verlag Berlin Heidelberg 1985

worst results were to be seen in the 21 patients with a severe completed stroke grade IVb; in this group only three patients improved clearly. The mean age of the 53 men and 46 women was 54 years. The patency rate was 91%. Since according to the literature as summarized by Marshal (1964) the probability that patients with TIA will have a completed stroke within 1 year ranges from 18% to 60%, with, according to Reisner (1961) and Dorndorf (1969), a mortality from 21% to 54%, we decided to advise surgery — the mortality from cerebral bypass surgery ranges from 3% to 6% and the morbidity from 1% to 20% in general and from 1% to 7% neurologically. Our mortality of 3% and morbidity of 2% (two patients who developed epileptic seizures) were at the lower end of the statistics. The aim of the dPET studies was to improve prognosis, especially in grades IVa and IVb. In our investigation the LPR and LUGTR were calculated. In four adult volunteers the LPR was $0.8-0.98$ ml/g/min in the gray and $0.2-0.4$ ml/g/min in the white matter. The data for gray matter are in close agreement with those measured by Obrist and Ingvar with ^{133}Xe. The values for white matter are higher than those determined by Obrist et al. using ^{133}Xe. The glucose influx rate (LUGTR) was $0.43-0.6$ μmol/g/min in normal cortex and $0.09-0.12$ μmol/g/min in white matter. As expected, these values are higher than those reported by Phelps et al. with ^{18}F-deoxyglucose (FDG). Unexpected was the finding that in CMG-scintigrams the accumulation defects were larger than the hypodense zones in the corresponding slices of the computer tomogram. The Time-activity curves over areas of accumulation defects in dPET corresponding to morphologically intact cortex in CT showed changes of LPR which were equal, higher, or lower than the changes of LUGTR. Both parallel and non parallel changes of LPR and LUGTR could be observed. Overall we saw six types of reaction in our STA-MCA candidates.

— Normal or minimally lowered LPR; moderate decrease of LUGTR. These patients had a normal CT, alterations in the arteriogram, and TIA clinically. Very good result of the EIAB.

— Normal of minimally lowered LPR; more pronounced decrease in LUGTR. Normal CT, Minor or moderate alterations in the arteriogram, and TIA or RIND/PRIND. Very good or good result of the EIAB.

— Minor decreases in LPR; Relatively raised LUGTR. Normal or minimal altered CT, large and chronic abnormalities in the arteriogram. Difficult grouping (TIA-PRIND-CS) with psychic abnormalities as in a case of Moya-Moya-disease. Good Result of the EIAB.

— Greater or lesser reduction in LPR; major depression of LUGTR. Normal or minimal altered CT, abnormalities in the arteriogram. Clinically TIA — PRIND — mild CS. Different results from very good to moderate.

— Multiple accumulation defects. Here it is always necessary to examine whether in areas of parallel changes of LPR and LUGTR one is not dealing with the consequences of impairment of neural pathways and/or with the effects usually described as diaschisis. This means that it should be established that the accumulation defects (or concentration) do not reflect the physiologic alterations in physiologic inactivated cortex (or activated cortex).

– Major depression of LPR and LUGTR. Mostly hypodense zones in the CT and major stenoses or occlusions in the arteriogram. Clinically mild or (more usually) severe CS with spasticity. Poor result of EIAB.

The most important finding in these studies was that the increase in perfusion increased the transport ability for the glucose. Therefore we favor the EIAB procedure. The results of the postoperative dPET studies are in agreement with our clinical findings. We interpret the results as showing that simultaneous determination of LPR and LUGTR gives more sensitive information than rCBF measurement alone and may help us to select patients for surgery more precisely.

References

Dorndorf W (1968) Verlauf und Prognose bei spontanen zerebralen Arterienverschlüssen. Dr. A. Hüthig, Heidelberg

Heiss WD (1983) Remote functional depression of glucose metabolism in stroke and its alteration by activating drugs. In: W.-D. Heiss and M. E. Phelps (eds) Positron emission tomography of the brain. Springer, Berlin, Heidelberg, New York, p 162– 168

Marshall J (1971) Angiography in the investigation of ischemic episodes in the territory of the internal carotid artery. Lancet 1:719–721

Obrist WD, Thompson HK, King CH (1967) Determination of regional cerebral blood flow by inhalation of Xenon-133. Circulation res. 20:124

Phelps ME, Huang SC, Hoffmann EJ (1979) Tomographic measurement of local cerebral glucose metabolic rate in humans with (F-18)-fluoro-2-deoxy-D-glucose: Validation of Method. Ann Neurol 6:371–388

Reisner H et al (1971) Das weitere Schicksal von 1000 zerebralen Insulten. Wien klin Wschr 73:397–402

Vyska K et al. (1983) The use of 11 C-methyl-D-glucose for assessment of glucose transport in the human brain; Theory and Application. In: S. Levin (ed) Lecture Notes in Biomathematics. No. 48: Tracer Kinetics and Physiologic Modeling. R. M. Lambrecht and A. Rescigno (eds), Springer, Berlin, Heidelberg, New York, Tokyo

Measurement of Local Oxygen Consumption and Cerebral Blood Flow with Positron Emission Tomography in Patients with Cerebrovascular Disease

W. J. POWERS and M. E. RAICHLE

The development of positron emission tomography (PET) has provided a practical method for studying regional cerebral physiology in man. In patients with cerebrovascular disease, regional measurements of cerebral blood flow (rCBF), cerebral blood volume (rCBV), cerebral metabolic rate of oxygen (rCMRO$_2$), the fractional extraction of oxygen from blood (rOEF), and the cerebral metabolism rate for glucose (rCMRGlu) have all been successfully performed. This article will review the results of these studies and discuss how they have added to our knowledge of the pathophysiology of stroke in man.

Transient Ischemic Attacks

Only a few patients with transient ischemic attacks (TIA) have been studied with PET. Baron and co-workers reported three cases with TIA and carotid occlusion (Baron et al. 1981 d, e). One had a normal brain CT scan and two had watershed infarcts. PET demonstrated a region of relatively decreased CBF ipsilateral to the carotid occlusion in all three that was outside the area of infarction in the latter two. CMRO$_2$ in these areas was normal in one case and decreased in two. Regional OEF was normal in one (with decreased CMRO$_2$) and elevated in the other two. This combination of increased rOEF with decreased rCBF has been termed the "misery perfusion" syndrome (Baron et al. 1981 b, d, e). Whether or not patients with misery perfusion are at increased risk for cerebral infarction due to a decrease in perfusion reserve remains to be determined (Baron et al. 1981 e; Ackerman et al. 1981 a; Frackowiak and Wise 1983). Donnan and colleagues (1983) studied ten patients with hemispheric TIA and reported a wide variety of changes in CBF, CMRO$_2$, and OEF both ipsilaterally and contralaterally. We have investigated a number of patients with TIA as part of a study to assess the hemodynamic and metabolic effects of extracranial-intracranial bypass surgery. Many have had normal CBF, CBV, CMRO$_2$, and OEF in spite of severe carotid occlusive disease. We have, however, observed some patients with decreased rCBF ipsilateral to complete carotid occlusion with a concomitant increase in rCBV (Martin et al. 1982; Powers

The Department of Neurology and Neurological Surgery, Mallinckrodt Institute of Radiology and the McDonnell Center for Studies of Higher Brain Function, Washington University School of Medicine, St. Louis, MO 63110, USA

Cerebral Blood Flow and
Metabolism Measurement.
Eds. Hartmann/Hoyer
© Springer-Verlag Berlin Heidelberg 1985

et al. 1983 a, b). A similar increase in CBV as a response to decreased cerebral perfusion pressure has been consistently demonstrated in experimental animals and most likely represents compensatory vasodilation (Grubb et al. 1973, 1975). These few studies of transient cerebral ischemia indicate that it is a heterogeneous disorder with variable changes in both cerebral blood flow and metabolism. Further investigations with PET are needed to define different groups of patients according to physiologic criteria and to determine if these criteria have prognostic value.

Cerebral Infarction

PET studies of patients with stable cerebral infarcts greater than 31 days old have been in general agreement. The area of the brain corresponding to the infarct on CT scan demonstrates a decrease in both rCBF and $rCMRO_2$. Regional OEF is either normal or slightly decreased, suggesting that the blood supply is adequate to meet the oxidative requirements of the residual tissue and that normal coupling of blood flow to metabolism is present (Lenzi et al. 1982; Ackerman et al. 1981a; Baron et al. 1981b, 1982). Regional CMRGlu measured by ^{18}F-deoxyglucose (^{18}F-DG) in these old infarcts is also reported to be decreased (Baron et al. 1982; Kuhl et al. 1980; Metter et al. 1981). Several authors have looked at the relationship between CMRGlu and either rCBF or $rCMRO_2$. Kuhl et al. (1980) found that rCBF and rCMRGlu were equally reduced in old infarcts. Baron found that the $rCMRO_2$/rCMRGlu ratio was lower in the infarcted hemisphere than in the contralateral hemisphere, suggesting anaerobic glycolysis in the infarcted tissue (Baron et al. 1983b). Interpretation of these findings must be tempered by knowledge of the inaccuracies of the deoxyglucose method when applied to ischemic or infarcted tissue (Ginsberg and Reivich 1979; Hawkins et al. 1981; Baron et al. 1983b; Weinhard et al. 1983; Choki et al. 1983).

Most PET studies of ischemic cerebrovascular disease have been devoted to the study of acute stroke. The impetus behind this work comes from clinical and experimental evidence of recovery of cerebral function following variable periods of ischemia. PET provides a method to study changes in cerebral hemodynamics and metabolism in the early stages of stroke and to see if these changes can help to identify reversibly ischemic tissue. Early investigations by Ackerman and his colleagues (Ackerman et al. 1981a; Ackerman 1981b) demonstrated a variety of changes in rCBF, $rCMRO_2$, and rOEF during the first week following ischemic stroke. PET abnormalities often preceded findings on CT scan. Areas with marked depression of $rCMRO_2$ seen early on PET scan correlated well with the development of CT changes of infarction later on. Measurements of rCBF were less useful, tending to be decreased early and then to increase 10–20 days after the initial event. This increase was of no clinical or prognostic significance. One patient who showed subsequent clinical recovery was studied on day 9. PET studies demonstrated a hemispheric decrease in rCBF with normal oxygen metabolism and an increase in rOEF, suggesting that this combination is predictive of neurologic recovery. In a study of 45 infarcts less than 1 month old, Baron et al. (1981 b) reported that areas with increased rOEF were more likely to be seen early (three of three patients studied before day 3 and no patients studied after day 14). In contrast, patients studied after 4 days tended to have decreased rOEF. Regional CBF could be decreased, normal, or increased. As found by Ackerman and colleagues, CBF was highest between days 5 and 20. After this, it tended to decrease once more. Two patients with areas of increased rOEF on initial scans went on to develop infarction, indicating that this finding is not always predictive of recovery. Regional $CMRO_2$ values were not calculated in this study.

Lenzi and colleagues from Hammersmith (Lenzi et al. 1982) were the first to provide quantitative physiologic data on acute cerebral infarctions, the previous two studies having been based on PET counts alone. In the area of infarction they found that $rCMRO_2$ was decreased immediately and changed little with time. Regional CBF was decreased early and tended to rise somewhat from day 10 to 20. Regional OEF was low during the first week after stroke and became more normal as time went by. Several patients studied within the first few days showed relative elevations of rOEF. The borders of the infarct, in general, showed changes similar to those described for the infarct itself. A subsequent report from the Hammersmith group (Wise et al. 1983 a) again presenting quantitative data, specifically addressed the pathophysiologic significance of the early elevation in rOEF. This report confirmed that elevations in rOEF are an early phenomenon being present in 8 of 14 studies performed less than 24 h after acute stroke, but in only 2 of 14 studies performed from 24 to 72 h. Of nine patients with initial elevations in rOEF studied sequentially, eight showed a decrease when restudied within 7 days. All had low values for $rCMRO_2$ (less than 2.5 ml per min $\cdot$ 100 ml) initially, and $rCMRO_2$ fell further in all but one. This fall in $rCMRO_2$ was not accompanied by any clinical worsening. In fact, two patients showed clinical improvement with no change in $rCMRO_2$. Regional CBF was also decreased initially in all nine patients (less than 20 ml per min $\cdot$ 100 ml) but showed more variable changes with time, increasing in five and decreasing in four.

Based on these data the authors offered the following hypothesis. During the early stages of stroke, mitochondrial function remains intact, resulting in an increase in the extraction of oxygen from the residual cerebral blood flow. $CMRO_2$, previously determined by the energy requirements of the tissue, becomes dependent on CBF (the stage of "critical perfusion"). After a variable period of time this precarious state can no longer be maintained. Irreversible infarction occurs with a fall in both $rCMRO_2$ and rOEF regardless of any further decrease in rCBF. During the stage of critical perfusion, restoration of blood flow should lead to restoration of oxidative metabolism (Wise et al. 1983 a). The authors attempted this in one patient who had a persistent elevation of rOEF at day 4. Systemic blood pressure was raised by angiotensin infusion, resulting in increased blood flow to the ischemic area. Regional OEF decreased but $rCMRO_2$ did not change and there was no clinical improvement (Wise et al. 1983 a).

Some further insight into the pathophysiologic significance of the OEF is provided in a recent paper by Baron et al. (1983 a) correlating early PET findings with later CT evidence of infarction. These investigators found that, regardless of the rOEF, infarction occurred when rCMRO$_2$ was less than 1.5–1.7 ml per min · 100 g.

The picture of the evolution of acute cerebral infarction that emerges from these studies is remarkably consistent. During the first few days, both rCMRO$_2$ and rCBF are decreased, but disproportionately, such that rOEF is elevated. Within the first week rCMRO$_2$ remains the same or decreases slightly. Changes in rCBF during the first week are more variable. Flow may increase or decrease but, in general, becomes higher relative to rCMRO$_2$, resulting in a fall in rOEF. During the second and third weeks, rCBF increases with no concomitant change in rCMRO$_2$. This change most likely corresponds to the phenomenon of luxury perfusion described previously by Lassen (Lassen 1966). Finally, rCBF falls back to levels that more closely match the metabolic rate of the tissue, suggesting that the coupling between flow and metabolism has been reestablished (although at a somewhat lower level than normal as evidenced by persistently low values of rOEF in some patients). The pathophysiologic significance of these changes in cerebral blood flow and metabolism is still uncertain. Alterations in rCBF bear little relationship either to clinical status or to the metabolic rate of the tissue. Given the dependence of the brain on oxidative metabolism, one might expect rCMRO$_2$ to be the most important determinant of the functional and structural integrity of cerebral tissue. The available data on this point are confusing. Local decreases in rCMRO$_2$ are the best predictors of ultimate infarction regardless of rOEF or rCBF. On the other hand, declines in rCMRO$_2$ with no clinical worsening and clinical improvement without concomitant increases rCMRO$_2$ also occur. The interpretation of rOEF is also unclear. Most of these PET studies were performed without correction for residual intravascular ^{15}O-oxyhemoglobin, resulting in overestimation of OEF. The magnitude of this error increases as the local blood volume increases (Lammertsma et al. 1981, 1983; Wise et al. 1983 a). Since local increases in CBV may occur during cerebral ischemia (Gibbs et al. 1983; Martin et al. 1982; Powers et al. 1983 a, b), the reported elevations in rOEF may be, at least in part, artifactual. Furthermore, the use of rOEF as an index of tissue oxygenation presupposes that the factors which control OEF in ischemic or infarcted tissue are the same as in normal brain. In fact, the diffusion of oxygen in diseased tissue may be limited by edema and necrotic debris (Silver 1977), making a normal OEF a poor indicator of adequate cellular oxygenation. Measurements of arteriovenous oxygen gradients across the brain during reversible global ischemia have demonstrated an increase in OEF as a compensatory mechanism to maintain oxidative metabolism in the face of decreased flow (Finnerty et al. 1954). However, the usefulness of PET measurements of rOEF in the identification of reversibly ischemic tissue following acute stroke remains to be proven.

Gibbs et al. (1983) have reported elevations of rCBV concomitant with decreases in rCBF ipsilateral to carotid artery occlusions in 20 patients. CMRO$_2$ was decreased in most cases. Regional OEF was increased only in those four patients with the highest CBV/CBF ratio, suggesting that vasodilation is an earlier compensatory response to ischemia than increased rOEF. These findings of increased CBV with decreased CBF are similar to those already mentioned in patients with TIA. It is important to note, however, that we have also observed increases in the CBV/CBF ratio in patients with stable, completed stroke and that this finding does not necessarily represent a state of reversible ischemia.

Studies of glucose metabolism during the acute stages of cerebral infarction using ^{18}F-DG have, in general, shown local decreases in CMRGlu (Hawkins et al. 1981; Kuhl et al. 1980; Baron et al. 1982; Wise et al. 1983b). There has been a great deal of interest in determining the relative magnitude of this decrease compared with changes in rCBF and rCMRO$_2$. Kuhl et al. (1980), using ^{13}NH$_3$ as a CBF tracer, demonstrated two types of mismatch between rCBF and rCMRGlu. In two patients studied at less than 2 days, rCBF was decreased to a greater degree than rCMRGlu. In three patients studied during the second week, rCBF was increased relative to rCMRGlu. These findings were interpreted as evidence of a transient period of anaerobic glycolysis during the first few days following acute ischemic stroke. Baron et al. (1983b) have reported a low CMRO$_2$/CMRGlu ratio in the core of a 5-day-old infarct (again suggesting anaerobic glycolysis) but found high ratios in the borders of three infarcts less than 12 days old, suggesting utilization of an alternate energy substrate in these areas. Wise and colleagues (1983b) have studied eight recent large cerebral infarcts and also found a decrease in the CMRO$_2$/CMRGlu ratio. Since rOEF was also low, they postulate that tissue O$_2$ delivery must be adequate and that these findings are due to infiltrating phagocytic cells that rely mainly on anaerobic glycolysis for their energy demands. As already mentioned, this interpretation of rOEF as an indicator of cellular oxygenation may not be valid in pathologic states.

All of these studies with ^{18}F-DG report a common finding. In the early days following cerebral infarction, rCMRGlu is depressed less than rCBF or rCMRO$_2$. This mismatch is usually interpreted as indicating anaerobic glycolysis. The validity of this interpretation rests on the assumption that the measurements of glucose metabolism are accurate. Unfortunately, values for rCMRGlu in ischemic or infarcted tissue determined with the deoxyglucose method are subject to error because of variations in both the rate constants and the lumped constant (Ginsberg and Reivich 1979; Hawkins et al. 1981; Baron et al. 1983b; Weinhard et al. 1983; Choki et al. 1983). Even when rate constants have been measured individually (Hawkins et al. 1981; Baron et al. 1983b; Weinhard et al. 1983), uncertainty about the value of the lumped constant makes it hazardous to draw any conclusions from these studies.

Diaschisis

The discovery of areas with decreased CBF or metabolism in structurally normal brain distant from the site of a hemispheric infarction has been a common finding in PET studies. This phenomenon of diaschisis has been demonstrated previously in the hemisphere contralateral to infarction using other techniques to measure CBF (Hoedt-Rasumssen and Skinhoj 1964; Meyer et al. 1970; Lavy et al. 1975; Slater et al. 1977). Regions of diaschisis detected by PET have been reported in both the ipsilateral hemisphere and the posterior fossa as well as the contralateral hemisphere.

In the ipsilateral hemisphere, Kuhl et al. (1980), Phelps et al. (1981), and Reivich et al. (1981) have reported reductions in rCMRGlu in structurally normal visual cortex of patients with infarction involving the optic radiations anteriorly. Metter et al. (1981) has analyzed five patients with aphasia and found decreased rCMRGlu in cortical areas overlying deeper infarctions. Neither the CT nor the PET findings correlated will with the observed clinical language deficits in this study. A left frontal decrease in rCMRGlu has been reported in a man who was found to have a small lacunar infarct in the anterior limb of the left internal capsule at autopsy (Metter et al. 1983). Heiss has reported a decrease in rCMRGlu in uninfarcted cortex ipsilateral to cerebral infarction (Heiss et al. 1983). Baron et al. (1982) has reported two patients in whom rCBF, rCMRO$_2$, and rCMRGlu were all measured. Matched reductions of all three were seen at cortical sites ipsilateral but distant from the site of infarction. By the far the most common area of diaschisis reported in the isilateral hemisphere of patients with cerebral infarction has been the thalamus. These patients have all had infarction within the middle cerebral artery territory with sparing of the thalamus by CT scan. Decreases in rCMRGlu in this structure have been reported by several investigators (Kuhl et al. 1980; Metter et al. 1981; Baron et al. 1982; Heiss et al. 1983). Similar findings of reduced glucose metabolism in the basal ganglia have been reported in patients in whom these structures were spared (Metter et al. 1981; Heiss et al. 1983). Decreases in both rCBF and rCMRO$_2$ with little change in rOEF occurring in both thalamus and basal ganglia have also been reported (Wise et al. 1983a; Baron et al. 1982).

Reductions in CBF and metabolism contralateral to cerebral hemispheric infarction have been reported for both the homologous cortical area (Lenzi et al. 1982; Wise et al. 1983a; Heiss et al. 1983) and for the whole hemisphere (Baron et al. 1982; Kuhl et al. 1980; Lenzi et al. 1982). Regional OEF may be normal or decreased (Lenzi et al. 1982; Wise et al. 1983a). These changes of crossed hemispheric diaschisis are most pronounced within the first days following infarction (Lenzi et al. 1982; Wise et al. 1983a).

Crossed cerebellar diaschisis was first reported by Baron in 1981 (Baron et al. 1981a, c). He observed matched depression of rCBF and rCMRO$_2$ (normal rOEF) in the contralateral cerebellar hemisphere of patients with supratentorial infarction less than 2 months old. These changes were seen only in patients with significant hemiparesis, but were not related to the size of the infarct on CT. Seven patients studied greater than 2 months post ictus did not show similar changes. Lenzi et al. (1982) confirmed the presence of this phenomenon and also reported a lesser decrease in rCBF and rCMRO$_2$ in the ipsilateral cerebellar hemisphere of patients with a depressed level of consciousness. In this series, crossed cerebellar diaschisis was not seen with small infarcts and was more pronounced with parietal than with frontal lesions. Martin et al. (1983) have recently reported results from 13 patients. Ten with frontal infarctions showed marked depression of rCMRO$_2$ and rCBF in the contralateral cerebellar hemisphere with lesser depressions in the ipsilateral cerebellar hemisphere also.

Three patients with infarction restricted to the parietal-occipital region showed mild symmetric depressions in both cerebellar hemispheres. Crossed cerebellar diaschisis was not related to infarct size, neurologic deficit, or the time since infarction and was not seen in two patients with unilateral hemispheric decreases in rCBF and rCMRO$_2$ who had no evidence of infarction by CT. All these reports are in general agreement on the common occurrence of crossed cerebellar diaschisis with hemispheric infarction, but there remains some controversy over the influence of infarct size, location, and age.

The occurrence of diaschisis in multiple sites following hemispheric infarction is now well established by PET. The best explanation for these findings is a local functional depression of neuronal activity due to interruption of afferent fiber pathways. Regional CBF and rCMRO$_2$ are generally decreased to the same degree, demonstrating the normal coupling of flow and metabolism expected with functional depression. A primary disturbance in cerebral vascular regulation that produced diaschisis by causing ischemia of distant areas might, on the other hand, be expected to produce mismatched reductions of rCBF and rCMRO$_2$ with consequent elevation of rOEF. The clinical significance of diaschisis remains unknown. Determination of its relationship, if any, to the development and resolution of neurologic deficits in ischemic stroke will require further study.

Studies of cerebrovascular disease with PET have, even at this early stage, provided a great deal of new information about the pathophysiology of stroke. However, the clinical significance of most of these findings remains unclear at the present time. Further investigations are necessary to correlate the results of PET with both clinical and radiologic data in order to increase our understanding of the complicated pathophysiology of stroke.

Acknowledgments. This work was supported by NIH grants NS 06833 and HL 13851, The Retirement Research Foundation, and Teacher Investigator Development Award NS 00647 from the NINCDS (Dr. Powers).

References

Ackerman RH (1981 b) Positron imaging in stroke disease. In: Moosy J, Reinmuth DM, (eds), Cerebrovascular Disease: Twelfth Research Conference. Raven Press. New York, p 67–72

Ackerman RH, Correia JA, Alpert NM, Baron JC, Gouliamos A, Grotta JC, Brownell GL, Taveras JM (1981 a) Positron imaging in ischemic stroke disease using compounds labeled with oxygen 15. Arch Neurol 38:537–543

Baron JC, Bousser MG, Comar D, Duquesnoy N, Sastre J, Castaigne P (1981 a) Crossed cerebellar diaschisis: A remote functional depression secondary to supratentorial infarction in man. J Cereb Blood Flow Metabol 1 (Suppl 1) S 500–S 501

Baron JC, Bousser MG, Comar D, Soussaline F, Castaigne P (1981 b) Noninvasive tomographic study of cerebral blood flow and oxygen metabolism in vivo. Eur Neurol 20:273–284

Baron JC, Bousser MG, Comar D, Castaigne P (1981 c) Crossed cerebellar diaschisis in human supratentorial brain infarction. Trans Am Neurol Assoc 105:459–461

Baron JC, Bousser MG, Rey A, Guillard A, Comar D, Castaigne P (1981 d) Reversal of focal "misery-perfusion syndrome" by extra-intracranial arterial bypass in hemodynamic cerebral ischemia. Stroke 12:454–459

Baron JC, Rey A, Guillard A, Bousser MG, Comar D, Castaigne P (1981 e) Non-invasive tomographic imaging of cerebral blood flow (CBF) and oxygen extraction fraction (OEF) in superficial temporal artery to middle cerebral artery (STA-MCA) anastomosis. In: Meyer JS, Lechner H, Reivich M, Ott EO, Aranibar A (eds), Cerebral vascular disease 3. Excerpta Medica, Amsterdam-Oxford-Princeton, p 58 – 64

Baron JC, Lebrun-Grandie PH, Collard PH, Crouzel C, Mestelan G, Bousser MG (1982) Noninvasive measurement of blood flow, oxygen consumption and glucose utilization in the same brain regions in man by positron emission tomography: Concise communication. J Nucl Med 23:391 – 399

Baron JC, Rougemont D, Bousser MG, Lebrun-Grandie P, Iba-Zizen MT, Chivas JC (1983 a) Local CBF, oxygen extraction fraction and $CMRO_2$: Prognostic value in recent supratentorial infarction. J Cereb Blood Flow Metabol 3 (Suppl 1): S 1 – S 2

Baron JC, Rougemont D, Soussaline F, Crouzel C, Bousser MG, Comar D (1983 b) Positron tomography investigation of local coupling among CBF, oxygen consumption, and glucose utilization. J Cereb Blood Flow Metabol 3 (Suppl 1): S 242 – S 243

Choki J, Greenberg J, Reivich M (1983) Regional cerebral glucose metabolism during and after bilateral cerebral ischemia in the gerbil. Stroke 14:568 – 574

Donnan GA, D'Alton JG, Chang JY, Correia JA, Alpert NM, Ackerman RH, Taveras JM (1983) Alterations in cerebral blood flow (CBF) and metabolism ($CMRO_2$) after transient ischemic attacks. Neurology 33 (Suppl 2): 115

Finnerty FA, Wilkins L, Fazekas JF (1954) Cerebral hemodynamics during cerebral ischemia induced by acute hypotension. J Clin Invest 33:1227 – 1232

Frackowiak RSJ, Wise RJS (1983) Positron tomography in ischemic cerebrovascular disease. Neurologic Clinics 1:183 – 200

Gibbs J, Wise R, Leenders K, Jones T (1983) The relationship of regional cerebral blood flow, blood volume and oxygen metabolism in patients with carotid occlusion: Evaluation of perfusion reserve. J Cereb Blood Flow Metabol 3 (Suppl 1): S 590 – S 591

Ginsberg MD, Reivich M (1979) Use of the 2-deoxyglucose method of local cerebral glucose utilization in the abnormal brain: Evaluation of the lumped constant during ischemia. Acta Neurol Scand 60 (Suppl 72): 226 – 227

Grubb RL, Phelps ME, Raichle ME (1973) The effects of arterial blood pressure on the regional cerebral blood volume by X-ray fluorescence. Stroke 4:390 – 399

Grubb RL Jr, Raichle ME, Phelps ME, Ratcheson RA (1975) Effects of increased intracranial pressure on cerebral blood volume, blood flow and oxygen utilization in monkeys. J Neurosurg 43:385 – 398

Hawkins R, Phelps ME, Huang SC, Kuhl DE (1981) Effect of ischemia on quantification of local cerebral glucose metabolic rate in man. J Cereb Blood Flow Metabol 1:37 – 52

Heiss WD, Pawlik G, Wagner R, Ilsen HW, Herholz K, Wienhard K (1983) Functional hypometabolism of noninfarcted brain regions in ischemic stroke. J Cereb Blood Flow Metab 3 (Suppl 1): S 582 – S 583

Hoedt-Rasumssen K, Skinhoj E (1964) Transneural depression of the cerebral hemispheric metabolism in man. Acta Neurol Scand 40:41 – 46

Kuhl DE, Phelps ME, Kowell AP, Metter EJ, Selin C, Winter J (1980) Effects of stroke on local cerebral metabolism and perfusion: Mapping by emission computed tomography of ^{18}FDG and $^{13}NH_3$. Ann Neurol 8:47 – 60

Lammertsma AA, Jones T, Frackowiak RSJ, Lenzi GL (1981) A theoretical study of the steady-state model for measuring regional cerebral blood flow and oxygen utilization using oxygen-15. J Comput Assist Tomogr 5:544 – 550

Lammertsma AA, Wise R, Heather J, Rhodes C, Gibbs J, Jones T (1983) The correction for intravascular oxygen-15 in the steady state technique for measuring regional oxygen extraction ratio. J Cereb Blood Flow Metabol 3 (Suppl 1): S 19 – S 20

Lassen NA (1966) The luxury-perfusion syndrome and its possible relation to acute metabolic acidosis localized within the brain. Lancet ii:1113 – 1115

Lavy S, Melamed E, Portnoy Z (1975) The effect of cerebral infarction on the regional cerebral blood flow of the contralateral hemisphere. Stroke 6:160 – 163

Lenzi GL, Frackowiak RSJ, Jones T (1982) Cerebral oxygen metabolism and blood flow in human cerebral infarction. J Cereb Blood Flow Metabol 2:321 – 335

Martin WRW, Raichle ME (1983) Cerebellar blood flow and metabolism in cerebral hemisphere infarction. Ann Neurol 14:168–176

Martin WRW, Baker RP, Herscovitch P, Zeiger HE, Grubb RL, Raichle ME (1982) The selection of patients for extracranial–intracranial bypass surgery: Hemodynamic and metabolic criteria. Neurology (NY) 32: A89

Metter EJ, Wasterlain CG, Kuhl DE, Hanson WR, Phelps ME (1981) ^{18}FDG-positron emission computed tomography: A study of aphasia. Ann Neurol 10:173–183

Metter EJ, Mazziotta J, Itabashi H, Kuhl DE, Phelps ME (1983) Pathologic comparisons to FDG positron computed tomography in stroke: Case study. Neurology 33 (Suppl 2): 115

Meyer JS, Shinohara Y, Kanada T, et al (1970) Diaschisis resulting from acute unilateral cerebral infarction. Arch Neurol 23:241–247

Phelps ME, Mazziotta JC, Kuhl DE, Nuwer M, Packwood J, Metter J, Engel J Jr (1981) Tomographic mapping of human cerebral metabolism: Visual stimulation and deprivation. Neurology 31:517–529

Powers WJ, Martin WRW, Herscovitch P, Raichle ME, Grubb RL Jr (1983a) Hemodynamic results of cerebral bypass surgery measured by positron emission tomography. J Nucl Med 24:P108

Powers W, Martin W, Herscovitch P, Raichle M, Grubb R (1983b) The value of regional cerebral blood volume measurements in the diagnosis of cerebral ischemia. J Cereb Blood Flow Metabol 3 (Suppl 1) S598–S599

Reivich M, Cobbs W, Rosenquist A, Stein A, Schatz N, Savino P, Alavi A, Greenberg J (1981) Abnormalities in local cerebral glucose metabolism in patients with visual field defects. J Cereb Blood Flow Metabol 1 (Suppl 1): S471–S472

Silver IA (1977) Local factors in tissue oxygenation. J Clin Path 30 (Suppl 11) 7–13

Slater R, Reivich M, Goldberg H, Banka R, Greenberg J (1977) Diaschisis with cerebral infarction. Stroke 8:684–690

Weinhard K, Pawlik G, Eriksson L, Wagner R, Ilsen HW, Herholz K, Heiss WD (1983) Kinetic constants of cerebral glucose metabolism in pathological conditions. J Cereb Blood Flow Metabol 3 (Suppl 1): S474–S475

Wise RJS, Bernardi S, Frackowiak RSJ, Legg NJ, Jones T (1983a) Serial observations on the pathophysiology of acute stroke. Brain 106:197–222

Wise R, Rhodes C, Gibbs J, Frackowiak R, Jones T (1938b) The relationship between oxygen metabolism and glucose utilization in early cerebral infarcts. J Cereb Blood Flow Metabol 3 (Suppl 1): S580–S581

Theory for Noninvasive Measurement of Oxygen Extraction Fraction by Intravenous Bolus Injection or Bolus Inhalation of 15-Oxygen

S. Takagi[2], K. Ehara[1], P. J. Kenny[1], and A. J. Gilson[1]

Oxygen extraction fraction (OEF) in human subjects has been measured by the continuous inhalation (Jones et al. 1976) or bolus inhalation method (Mintun et al. 1983). However, arterial sampling is necessary to measure the arterial tracer concentration. We have developed a theory and a mathematical procedure by which OEF can be measured noninvasively in humans by positron emission tomography (PET). Our model has not only a possibility of being noninvasive, but also many advantages over the previously established methods. Three tracers are administered by intravenous bolus injection or bolus inhalation: (1) O-15 labeled O_2 as an OEF tracer, (2) a diffusible tracer, as a reference tracer for the OEF study, and also as a cerebral blood flow (CBF) tracer, (3) carbon monoxide, as a cerebral blood volume (CBV) tracer.

Model and equations for OEF: Symbols used are defined as follows.

F: Cerebral blood flow (ml/100 g/min)

C_{ao}^*, C_{aw}^*, C_a: Arterial concentrations of oxygen, recirculated water, and reference tracer, respectively (μCi/ml)

c_{ao}^*, c_{aw}^*, c_a: Measured count rates of oxygen, recirculated water, and reference tracer from arterial samples, respectively (count/min)

Y: Calibration factor between arterial concentrations and measured count rates [(count/min)/(μCi/ml)], then $c_{ao}^* = Y\,C_{ao}^*$, $c_{aw}^* = Y\,C_{aw}^*$, and $c_a = Y\,C_a$

$Q_{O_2}^*$, $Q_{H_2O}^*$, Q: The amounts of oxygen, water, and reference tracer in the brain, respectively (μCi)

$C_{O_2}^*$, $C_{H_2O}^*$, C: Brain concentrations of oxygen, water, and reference tracer, respectively (μCi/ml)

N^*, N: Measured count rates (PET) from the brain (count/min)

X: Calibration factor between the concentration in the brain and measured count rate (PET) [(count/min)/(μCi/ml)], then $N^* = X\,(C_{O_2}^* + C_{H_2O}^*)$ and $N = X\,C$

V: Brain volume (ml)

V_{O_2}: Distribution volume of nonextracted oxygen (ml), which is CBV multiplied by the ratio between large and small vessel hematocrit

1 Baumritter Institute of Nuclear Medicine, Mount Sinai Medical Center, 4300 Alton Road, Miami Beach, FL 33140, USA
2 Department of Neurology, Tokai University, School of Medicine, Bohseidai, Isehara, Kanagawa, 259-II, Japan

Cerebral Blood Flow and
Metabolism Measurement.
Eds. Hartmann/Hoyer
© Springer-Verlag Berlin Heidelberg 1985

$\lambda*,\lambda$:   Partition coefficients of the water and reference tracer, respectively

$\alpha*,\alpha$:   Physical decay constants (/min)

Symbols with and without * indicate the parameters in the oxygen study and in the reference study, respectively.

We assume the model developed by Subramanyam et al. (1978) for kinetics of the O-15-label in the brain. The system is modeled as two compartments, one being the brain's blood pool, the other the tissue water pool. It is assumed that the $^{15}O_2$ present in the tissue as dissolved O_2 is negligible compared with that labeling the red cells.

The basic equations under the usual assumptions for compartmental modeling are as follows.

$$\frac{dQ^*_{O_2}}{d\tau} = F\,C^*_{ao}(\tau) - B - \left(\frac{F}{V_{O_2}} + \alpha*\right) Q^*_{O_2},$$

$$\frac{dQ^*_{H_2O}}{d\tau} = F\,C^*_{aw}(\tau) + B - \left(\frac{F}{\lambda*V} + \alpha*\right) Q^*_{H_2O}.$$

The quantity B represents the rate at which $^{15}O_2$ is metabolized to H_2O, and can be expressed as $B = OEF\ F\ C^*_{ao}(\tau)$.

Substituting, dividing by V, and integrating results in the following equation

$$\frac{N^*(\varphi)}{X} = \frac{1}{Y}[(1-OEF)\,fc^*_{ao}(\varphi) \otimes e^{-k_1\varphi} + OEF\,fc^*_{ao}(\varphi) \otimes e^{-k_2\varphi} + fc^*_{aw}(\varphi) \otimes e^{-k_2\varphi}]$$

$$\left(k_1 = \frac{f}{V_{O_2}/V} + \alpha*,\ k_2 = \frac{f}{\lambda*} + \alpha*,\ f = \frac{F}{V}\right)$$

where $\otimes$ indicates the operation of convolution.

Suppose that injection of a reference tracer was performed. The equation becomes as follows:

$$\frac{N(\varphi)}{X} = \frac{1}{Y}\,fc_a(\varphi) \otimes e^{-k_3\varphi}\qquad \left(k_3 = \frac{f}{\lambda} + \alpha\right).$$

The next step is the most important point in our theory. We take the ratio of the measured count rate in the oxygen study, $N^*(\varphi)$, to that in the reference study, $N(\varphi)$. Calibration factors X and Y cancel out and the f also cancels out partially. An integration from time t_1 to t_2 results in the final equation.

$$\frac{\int_{t_1}^{t_2} N^*(\varphi)\,d\varphi}{\int_{t_1}^{t_2} N(\varphi)\,d\varphi} =$$

$$= \frac{(1-OEF)\int_{t_1}^{t_2} c^*_{ao}(\varphi) \otimes e^{-k_1\varphi}\,d\varphi + OEF\int_{t_1}^{t_2} c^*_{ao}(\varphi) \otimes e^{-k_2\varphi}\,d\varphi + \int_{t_1}^{t_2} c^*_{aw}(\varphi) \otimes e^{-k_2\varphi}\,d\varphi}{\int_{t_1}^{t_2} c_a(\varphi) \otimes e^{-k_3\varphi}\,d\varphi}$$

OEF can be calculated by this equation. The left side is a ratio of the total head counts of the oxygen and the reference tracer from t_1 to t_2, and can be measured easily by the PET scanner. Arterial concentration curves can be measured by direct arterial sampling or exhaled air counting. In the latter case, recirculated water must be neglected, resulting in minimal overestimation of OEF.

Simulation Studies

Two typical arterial curves of oxygen and ^{11}C-butanol in rats were used in the simulation study. These curves were obtained by an external coincidence detector system positioned over a catheter inserted in the femoral artery. Figure 1 shows the effect of measurement error in CBV on calculated OEF in a study where CBF was 70 ml/100 g min, $t_1 = 0$ s, and $t_2 = 40$ s. The effect of CBV on OEF is very small. For instance, if the true CBV is 0.033 and the assumed CBV is 0.030, we will obtain an OEF of 0.506 instead of a true OEF of 0.500. It means that 10% error in CBV results in only 1.2% error in OEF. The error increases if CBF is low or measurement time is short.

Figure 2 shows the effect of measurement error in CBF on OEF in a study where CBV = 0.030, $t_1 = 0$ s, and $t_2 = 40$ s. This error is also small. If the true CBF is 63 and the assumed CBF is 70, and OEF value of 0.502 will be obtained instead of a true value of 0.500. The error is only 0.4%. This error increases with increasing CBV or shorter measurement time.

The effect of changing measurement time on OEF is negligible.

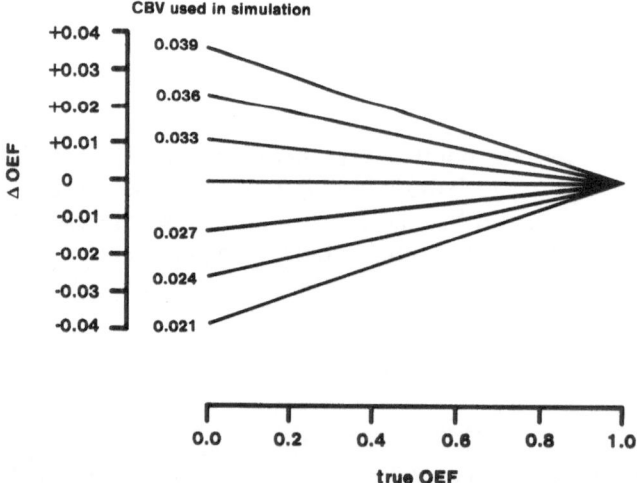

Fig. 1. The *horizontal axis* shows true OEF values, and the *vertical axis* shows the difference between calculated and true OEF values (ΔOEF = calculated OEF − true OEF). ΔOEF values in situations where various CBV values used in simulation and assumed to be 0.030 in calculation are plotted against the true OEF values

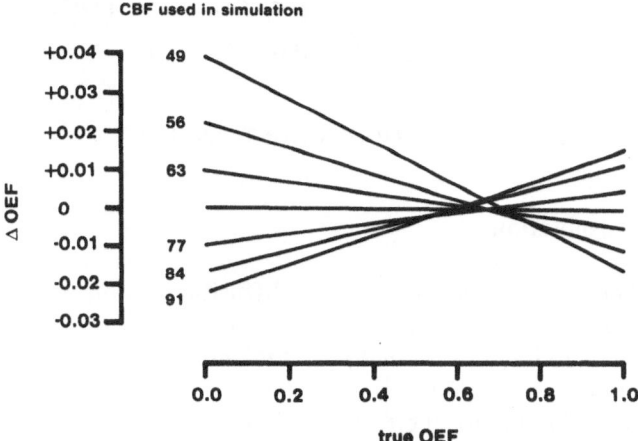

Fig. 2. Axes are same as in Figure 1. ΔOEF values in various situations where various CBF values used in simulation and assumed to be 70 ml/100 g min in calculation are plotted against the true OEF values

Discussion and Conclusion

Our theory seems to provide a basis for applying these methods to human studies. The most prominent advantage of our model is that it provides a basis for noninvasive measurement, if the arterial curves are measured noninvasively. Even if this cannot be done noninvasively, the other important advantages compared with the continuous inhalation method include (1) low radiation dose, (2) short measurement time, (3) no requirement for maintaining constant $^{15}O_2$ (mCi/liter) in administered gas, (4) low error sensitivity, (5) no calibration necessary between the arterial and head count. The advantages compared with the bolus inhalation method (Mintun et al.) include (1) low error sensitivity of OEF to CBF, (2) low error sensitivity to the time shift effect between the head and arterial curves, (3) no calibration necessary between the arterial and head count.

Acknowledgment. This work was supported by PHS grants NINCDS 1RO1 NS 15639.

References

Jones T, Chesler DA, Ter-Pogossian MM (1976) The continuous inhalation of oxygen-15 for assessing regional oxygen extraction in the brain of man. Brit J Radiol 49:339–343

Mintun MA, Raichle ME, Martin WRW, Herscovitch P (1983) Oxygen utilization measured by O-15 radiotracers and PET. J Nucl Med 24:p63

Subramanyam R, Alpert NM, Hoop B Jr, Brownell GL, Taveras JM (1978) A model for regional cerebral oxygen distribution during continuous inhalation of $^{15}O_2$, $C^{15}O$, and $C^{15}O_2$. J Nucl Med 19:48–53

Measurement of Cerebral Blood Flow Using ^{15}O-Labeled Water and Positron Emission Tomography with Special Attention to the Volume of Distribution of Radiowater

J. C. Depresseux, J. Hodiaumont, and R. Czichosz

Introduction

The methods using inert and diffusible radiotracers for evaluation of cerebral blood flow all rest upon equations derived from the model initially designed by Kety and Schmidt (1948). The formulation of this one-compartmental, two-parameter model describes transfer function of the tracer in terms of local blood flow (Fi) and local effective volume of distribution (Vi).

In most approaches, the insufficient quantity or the difficulty of calibration of data only allow computation of Fi, a predetermined value having to be allotted to the corresponding parameter Vi.

One of the interesting features of positron emission tomography (PET) is its ability to noninvasively realize three-dimensional autoradiograms of the distribution of positron emitting radiopharmaceuticals within the human brain.

Exercising caution to accurately correct data for auto-attenuation (Huang et al. 1979) and to suitably size regions of interest (Hoffmann et al. 1979), it is possible to calibrate each other input and transfer functions of the tracer (Eichling et al. 1977) an to perform a two-parameter analysis of the data (Huang et al. 1982; Depresseux et al. 1982).

The purpose of this study was to compare two different computational approaches to the evaluation of blood flow and volume of distribution of water, using bolus administration of $H_2^{15}O$ and PET.

Material and Methods

The data were obtained in five normal individuals. The subject received a bolus of $C^{15}O_2$ by taking a single breath from a 1.5-liter airbag containing 80 mCi of $C^{15}O_2$; this procedure is equivalent to the injection of a bolus of $H_2^{15}O$ within the pulmonary capillary bed (West and Dollery 1962).

Sequential PET detections (8 × 53 s) were performed from time zero of inhalation with an ECAT II tomography (EG & G Ortec). The tomographic plane was adjusted parallel to the orbitomeatal reference plane, 5 cm above it.

The time course of radioactive concentration within the arterial blood was measured by catheterization of one humeral artery and by sequential counting of blood samples. Arterial blood gases were monitored. Well counter and PET

Université de Liège, Cyclotron Research Center, Sart Tilman, B.30, B-4000 Liège

Cerebral Blood Flow and
Metabolism Measurement.
Eds. Hartmann/Hoyer
© Springer-Verlag Berlin Heidelberg 1985

measurements were calibrated by imaging and counting radioactive concentration in a classical pie phantom 20 cm in diameter (Eichling et al. 1977).

Equations

The differential equation expressing Qi(t), the local concentration of $H_2^{15}O$ within the detection element i at time t, expressed in radioactivity units per gram of organ is, quite generally:

$$dQi(t) = FiCa(t)dt - (Fi/Vi + \lambda) \, Qi(t)dt \tag{1}$$

with Ca(t), the concentration of $H_2^{15}O$ within the arterial blood at the input point to compartment i, expressed in radioactivity calibrated units per cm^3 of blood; λ, the radioactive decay constant of 15-oxygen, in min^{-1}; Fi, local blood flow, in $cm^3/min \cdot g$ and Vi, the local volume of distribution of radiowater, in cm^3/g.

Detected data and simulated values were analyzed using two different computational approaches.

The first uses numerical integration of data from $t = 0$ (Huang et al. 1982) and derives values for Fi and Vi as follows:

$$Fi = \frac{\int_0^t Qi^*(t)\,dt \left[\lambda \int_0^t Qi(t)\,dt + Qi(t)\right] - Q^*(t)\int_0^t Qi(t)\,dt}{\int_0^t Ca(t)\,dt \int_0^t Qi^*(t)\,dt - \int_0^t Qi(t)\,dt \int_0^t Ca^*(t)\,dt} \tag{2}$$

$$Vi = \frac{\int_0^t Qi^*(t)\,dt \left[\lambda \int_0^t Qi(t)\,dt + Qi(t)\right] - Qi^*(t)\int_0^t Q(t)\,dt}{\int_0^t Ca^*(t)\,dt \left[\lambda \int_0^t Qi(t)\,dt + Qi(t)\right] - Qi^*(t)\int_0^t Ca(t)\,dt} \tag{3}$$

with superscript * indicating values corrected for radioactive decay from time zero of input of radiowater within the brain.

As Qi(t) becomes negligible in front of $\lambda \int_0^t Qi(t)\,dt$ for sufficiently high values of t, equations (2) and (3) were simplified for speeding up data processing (Huang et al. 1982):

$$Fi = \frac{\lambda \int_0^t Qi^*(t)\,dt \int_0^t Qi(t)\,dt - Qi^*(t)\int_0^t Qi(t)\,dt}{\int_0^t Qi^*(t)\,dt \int_0^t Ca(t)\,dt - \int_0^t Qi(t)\,dt \int_0^t Ca^*(t)\,dt} \tag{4}$$

$$Vi = \frac{\lambda \int_0^t Qi^*(t)\,dt - Qi^*(t)}{\lambda \int_0^t Ca^*(t)\,dt - Qi^*(t)\int_0^t Ca(t)\,dt \Big/ \int_0^t Qi(t)\,dt} \tag{5}$$

The second method uses an analytical approach and a sequence of time integrated detection data. The input function is fitted as:

$$Ca(t) = Ca(o)\, e^{-\mu t} + A(t) \tag{6}$$

$A(t)$ being the error between one-exponential fit and observed values.

The transfer function is then expressed (Depresseux et al. 1983) as

$$\int_{tn}^{tn+1} Qi(t)\, dt = \frac{Ca(o)\, Fi}{Fi/Vi + \lambda - \mu} \int_{tn}^{tn+1} [e^{-\mu t} - e^{-(Fi/Vi+\lambda)t}]\, dt$$

$$+ Fi \int_{tn}^{tn+1} [e^{-(Fi/Vi+\lambda)t} * A(t)]\, dt + \varepsilon_i(t) \tag{7}$$

Values for Fi and Vi are derived using a least-squares procedure based on the first term of equation (7).

Results

Equations (2) to (7) were applied on simulated data, as generated from the first term of equation (7), with addition of a noise computed for resolution cells $(14.3\ mm)^2$ in size (Budinger et al. 1977), with a peak arterial activity of $1.9\ \mu Ci/cm^3$.

Systematic and aleatory errors were estimated by processing 100 independently noised simulated curves.

Table 1 shows that the use of equations (4) and (5) introduces bias: blood flow-dependent underestimation of Fi and slight overestimation of Vi. Both methods result in comparable error propagation.

Computed values of Fi and Vi using equations (2) and (3) fit integrated data from time o to time t. *Do these values fit the whole time course of the transfer function Qi(t)?* We examined this question using sequential PET data collected from 0 to 8 min after inhalation of the tracer. Values for Fi and Vi are comput-

Table 1. Systematic and aleatory errors on Fi and Vi, estimated by applying "integration" [equations (4) and (5)] and "iteration" methods to simulated PET sequential data after bolus inhalation of $C^{15}O_2$

Actual V (cm^3/hg)	Actual F $(cm^3/min \cdot hg)$	Computed F $(cm^3/min \cdot hg)$ Integration		Iterations		Computed V (cm^3/hg) Integration		Iterations	
		$\bar{x}$	σ	$\bar{x}$	σ	$\bar{x}$	σ	$\bar{x}$	σ
	20	16.5	0.5	20.3	0.3	80.4	3.8	68.2	1.8
70	40	35.6	0.8	40.8	1.3	73.8	1.2	70.0	1.5
	60	55.7	1.5	60.5	1.5	71.4	0.8	70.0	0.8
	80	77.5	2.2	80.3	2.6	70.8	0.7	70.1	0.6
	100	100.3	3.0	100.8	4.2	70.7	0.5	70.0	0.5

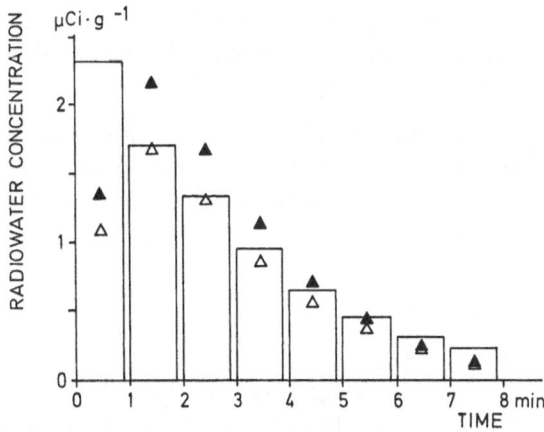

Fig. 1. Illustration of sequential PET data obtained after bolus inhalation of $C^{15}O_2$ (*rectangles*); *tops of black triangle* indicate back-simulation with F and V obtained from Eqs. (2) and (3); *tops of open triangles* indicate back-simulation with F and V obtained by the iterative method on data detected from 2 to 8 min after administration of the tracer

ed by applying equations (2) and (3) to observed data, being cautious to take into account dead times occurring between successive 53-s sampling periods. Back-simulation of sequential PET data using these values for Fi and Vi shows a systematic underestimation of the image content integrated from $t = 0$ to $t = 1$ min; although the individual data are not reproduced by back-simulation, the $t = 0$ to $t = 8$ min integrates of observed and back-simulation values are equal (Fig. 1).

On the other hand, results obtained from the iteration method [equations (6) and (7)], when back-simulated, fit observed data, when the interval of analysis of data is set from $t = 1$ min to $t = 8$ min (Fig. 1).

Several hypotheses could be evoked to explain this difficulty in numerically applying the Kety model to the initial part of the transfer function obtained after bolus administration of $H_2^{15}O$.

− An error in the determination of time zero, which precisely is the time of onset of input of tracer within the region of interest: This selective error could not explain the selective unfitting of the first value, as illustrated in Fig. 1.
− If A(t) reaches nonnegligible values at early times after injection, back-simulation using the first term of equation (7) is not valid in the whole time interval under consideration: Even if blood multisampling procedure for assessing Ca(t) gives poor definition of the early parts of the input function, A(t) has generally negative values and could not explain a systematically positive difference between observed and backsimulated values.
− The transfer function suffer distortions due to the rotation-translation sampling made of ECAT II (Williams et al. 1979): While requiring consideration, this distortion seems not to explain the observation of a first PET image content higher than could be foreseen from the Kety model.
− The value Vi could vary during the invasion period of the region by the tracer: this hypothesis is in accordance with the widely validated "single capil-

lary transit extraction" model (Crone 1963), whose conception implicates Vi being assimilable to infinity during the first transit of the tracer through the organ.

On the other hand, we demonstrated that the last two terms of equation (7) vanish for $t > 1$ min, with good stationarity in the estimated Vi for subsequent intervals of analysis of data (Depresseux 1983).

Conclusions

The present investigation used PET detected data and simulated values to describe the cerebral kinetics of bolus-administered radiowater following the one-compartmental two-parameter model of Kety. The results allow us to infer

- that, with suitable detection and computation methodology, PET has the potential to yield data for determining both blood flow and volume of distribution of the tracer;
- that equations using time-integrated data from time 0 of input of the tracer within the region of interest, while being mathematically correct, underestimate Fi and overestimate Vi;
- that a different approach, based on iterative procedures, is feasible, although implying longer computation times.

Table 2 compares published normal values obtained with both approaches, with differences in results being in the same direction as for the simulated values reported in Table 1.

Acknowledgment. This research was supported by grant 1.5.647.84 F from the Belgian National Foundation for Scientific Research.

Table 2. Published normal human values for cerebral blood flow and cerebral distribution volume of radiowater, using PET and bolus inhalation of $C^{15}O_2$; ref. (1): Huang et al. (1983); ref. (2): Depresseux (1983)

		Blood flow (cm³/min·hg ± SD)	Distribution volume of radiowater (cm³/hg ± SD)
Predominantly gray matter	(1)	58.5 ± 11.4	85 ± 3
	(2)	72.5 ± 8.0	69.6 ± 5.4
Predominantly while matter	(1)	20.0 ± 3.7	76 ± 3
	(2)	31.4 ± 6.3	59.3 ± 5.0

References

Budinger TF, Derenzo SE, Gullberg GT, Greenberg WL, Huesman RG (1977) Emission computer assisted tomography with single-photon and positron annihilation photon emitters. J. Comput. Assist. Tomogr., 1:131–145

Crone C (1963) The permeability of capillaries in various organs as determined by use of the "indicator diffusion" method. Acta Physiol Scand 58:292–305

Depresseux J (1983) A method for local evaluation of the volume of rapidly exchangeable water in the human brain. In: Heiss WD, Phelps ME (eds) Positron emission tomography of the brain. Springer, Berlin, p 93–102

Depresseux J, Cheslet JP, Hodiaumont J (1982) Evaluation tomographique chez l'homme du débit sanguin cérébral et du volume d'eau échangeable. J Biophys Med Nucl 6:167–171

Eichling JO, Higgins CS, Ter-Pogossian MM (1977) Determination of radionuclide concentrations with positron CT scanning (PET). J Nucl Med 18:845–847

Hoffman EJ, Huang SC, Phelps ME (1979) Quantitation in positron emission computed tomography: 1. Effect and object size. J.4 Comput Assist Tomogr 2:299–308

Huang SC, Hoffman EJ, Phelps ME, Kuhl DE (1979) Quantitation in positron emission computed tomography: 2. Effect and inaccurate attenuation correction. J Comput Assist Tomogr 3:804–814

Huang SC, Carson R, Phelps ME (1982) Measurement of local blood flow and distribution volume with short-lived isotopes: a general input technique. J Cereb Blood Flow Metab 2:99–108

Huang SC, Carson RE, Hoffman EJ, Carson J, Mac Donald N, Barrio JR, Phelps ME (1983) Quantitative measurement of local cerebral blood flow in humans by positron computed tomography and ^{15}O-water. J Cereb Blood Flow Metab 3:141–153

Kety SS, Schmidt CF (1948) The nitrous oxide method for the quantitative determination of cerebral blood flow in man: theory, procedure and normal values. J Clin Invest 27:476–403

West JB, Dollery CT (1962) Uptake of oxygen 15-labeled CO_2 compared with carbon 11-labeled CO_2 in the lung. J Appl Physiol 17:9–13

Williams CW, Crabtree MC, Burgiss SG (1979) Design and performance characteristics of a positron emission computed axial tomograph – ECAT II. IEEE Trans Nucl Sc Ns – 26:619–627

Measurement of Cerebral Blood Flow, Blood Volume, and Oxygen Utilization in Patients with Extracranial Vascular Disease

J. M. Gibbs, R. J. S. Wise, and T. Jones

Introduction

There is continuing controversy about the management of patients presenting with a TIA or minor stroke who are found to have occlusive disease of the carotid or vertebral arteries. Even the widespread belief in the value of carotid endarterectomy for accessible stenotic lesions has yet to be supported by a satisfactory clinical trial. There is still more uncertainty about the role of extracranial-intracranial (EC-IC) bypass surgery, although this procedure is now being widely practised in a variety of clinical circumstances. A large proportion of patients considered for EC-IC bypass have complete occlusion of an internal carotid artery (ICA), often associated with continuing symptoms referable to the ipsilateral cerebral hemisphere. These delayed ischemic episodes beyond an occluded ICA may still be explained by embolic mechanisms in many cases (Barnett 1978; Countee et al. 1980; Finklestein et al. 1980). However, in a minority of patients the clinical features are highly suggestive of a hemodynamic rather than an embolic circulatory disturbance (Shanbrom and Levy 1957; Baron et al. 1981; Stark and Wodak 1983). Ischemic events of this type, resulting from a state of critically reduced cerebral perfusion pressure (CPP), provide one of the few convincing arguments in favor of EC-IC bypass surgery. Since the majority of patients' symptoms cannot be conveniently classified as being either embolic or hemodynamic in origin, an indirect assessment of regional CPP, and hence residual circulatory reserve, would clearly be of value.

Baron et al. (1981) have reported a single patient with ICA occlusion in whom regional CBF was inappropriately low in relation to cerebral oxygen demands, $CMRO_2$ being maintained only by a compensatory increase of fractional oxygen extraction (OER). This pathophysiologic disturbance provides clear evidence that regional CPP has fallen below the range of effective autoregulation. Both experimental findings in the primate (Grubb et al. 1973) and studies of patients with cerebrovascular disease (Martin et al., in press) have suggested that elevation of cerebral blood volume (CBV), reflecting compensatory vasodilatation, may also provide evidence of reduced CPP.

With a view to more rational selection of patients for EC-IC bypass surgery, we have measured regional CBF, CBV, and OER in patients with ICA occlu-

MRC Cyclotron Unit and Department of Neurology, Hammersmith Hospital, Ducane Road, London W 12, Great Britain

Cerebral Blood Flow and
Metabolism Measurement.
Eds. Hartmann/Hoyer
© Springer-Verlag Berlin Heidelberg 1985

sion in an attempt to identify those most at risk of cerebral ischemia on hemodynamic grounds.

Patients

Over 50 patients with occlusive carotid disease have been studied during the last 2 years. A small number of cases with severe carotid stenosis have been studied before and after endarterectomy, but the majority of patients have had complete occlusion of one or both ICAs. The data presented here are from the first 32 of these patients, 29 of whom had experienced recent ischemic symptoms in the carotid territory ipsilateral to an occluded vessel. The symptoms consisted of a combination of minor stroke and TIAs in 13 cases, TIAs alone in 7, and minor stroke alone in 3. These 23 patients had no significant neurologic disability at the time of their PET studies. Six patients had suffered more substantial strokes and had a persisting neurologic deficit as well as clear-cut evidence of cerebral infarction on their CT scans. Of the remaining three patients, one had presented with a minor brainstem infarct and the other two had experienced no symptoms of cerebral ischemia at any stage.

The patients were divided into two groups on the basis of the angiographic findings. The first group (24 cases) had unilateral ICA occlusion and minimal or no stenosis of the contralateral ICA. The second group (eight cases) included six with complete bilateral occlusion and two with complete occlusion on one side and greater than 90% stenosis of the contralateral vessel. Mean age of the patients was 58 ± 9 (± SD) and there was no significant age difference between the "unilateral" and "bilateral" groups. The normal data included below are from a somewhat younger group of subjects (mean age 38 ± 12).

Methods

The patients were studied by positron emission tomography, using an ECAT II camera with a spatial resolution of 16.7 × 16.7 × 16 mm at full width half maximum. Regional cerebral blood flow (CBF), fractional oxygen extraction (OER), and oxygen utilization (CMRO$_2$) were measured by the oxygen-15 steady state technique (Frackowiak et al. 1980), and regional cerebral blood volume (CBV) by the ^{11}C-carboxyhemoglobin method (Phelps et al. 1978). The theoretical and practical aspects of these techniques are discussed in some detail elsewhere in this volume (Gibbs et al., p. 419).

Each patient was scanned in two transaxial tomographic planes, 4.5 and 6.5 cm above the orbitomeatal line, the primary area of interest being the territory of the middle cerebral artery (MCA). All physiologic values were taken from standardized cerebral regions corresponding to the superficial distribution of the MCA in each hemisphere. These regions are illustrated schematically in Fig. 1. In practice they were delineated by an objective method of "cortical

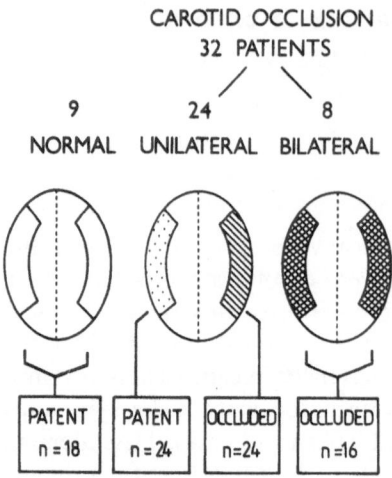

Fig. 1. Breakdown of subjects studied and classification of their MCA territories into four categories (see text)

plotting" previously described by Wise (1983). The values quoted for each MCA territory represent a mean from the standard regions of both tomographic planes studied in each patient. In the six patients with visible areas of cerebral infarction encroaching on the standard regions, this infarcted tissue (clearly demarcated on both X-ray CT and PET images) was excluded from the region of analysis.

In view of the hemodynamic continuity between the two carotid systems via the circle of Willis, MCA regions from the patients and normal subjects were classified into four categories according to the patency or otherwise of both their ipsilateral ICA and the contralateral vessel. The territories of patent and occluded ICAs in cases of unilateral occlusion therefore represent two intermediate categories between the two extremes of bilateral patency (normals) and bilateral occlusion of the ICAs. This classification is summarized diagramatically in Fig. 1.

Results and Discussion

Mean values of CBF, CMRO$_2$, OER, and CBV in the four categories of MCA territory are shown in Table 1. Note that although there was a successive reduction of CBF with increasingly severe occlusive disease, this was matched to a very similar pattern of decline in CMRO$_2$. Regional OER was within the normal range (less than 0.50) in 26 of the 32 patients, confirming that cerebral oxygen supply was well in excess of metabolic demands. OER was elevated to a modest degree (between 0.51 and 0.57) in five patients and to a greater extent (0.73) in only one. Note that even in this latter case, 27% of the available arterial oxygen was still left unextracted by the tissues, so the substantial oxygen

Table 1. Mean values ($\pm$ SD) of regional CBF, CMRO$_2$ OER, and CBV in the four categories of MCA territory (see Fig. 1 and text)

	Normal	Unilateral occlusion		Bilateral occlusion
		Patent	Occluded	
	($n=18$)	($n=24$)	($n=24$)	($n=16$)
CBF (ml/100 ml/min)	44 $\pm$4	38 $\pm$4	33 $\pm$5	32 $\pm$7
CMRO$_2$ (ml O$_2$/100 ml/min)	3.3 $\pm$0.3	2.9 $\pm$0.4	2.7 $\pm$0.5	2.6 $\pm$0.3
OER	0.43$\pm$0.04	0.43$\pm$0.04	0.46$\pm$0.04	0.50$\pm$0.08
CBV (ml/100 ml)	4.3 $\pm$0.4	4.0 $\pm$0.4	4.3 $\pm$0.7	5.0 $\pm$0.7

carriage reserve was not exhausted in any of the 32 patients. Despite this fact, asymmetry of regional CMRO$_2$ (with a lower value in the territory of the occluded ICA) was observed in 19 of the 24 unilateral cases. This suggests that even in some patients without any history of a prolonged neurologic deficit and with normal CT scans, there is frequently some degree of irreversible neuronal damage in the territory of an occluded ICA. This finding of metabolic depression is consistent with the concept of partial cerebral infarction proposed by Lassen (1982).

There is normally a coupled relationship between regional CBF and CBV, since at constant cerebral perfusion pressure alterations of CBF are achieved by appropriate vasodilatation or constriction. However, despite the reduction of regional CBF in these patients with ICA occlusion, mean CBV was found to be increased in the MCA territory distal to occluded vessels (see Table 1). An example of this phenomenon is shown in Fig. 2, the PET images from a patient with continuing TIAs in the territory of an occluded ICA on the right side. The increased regional CBV is likely to represent the compensatory vasodilatation which is necessary to maintain appropriate CBF in the face of reduced cerebral perfusion pressure (CPP). This interpretation is supported by the fact that highest CBV values are observed in patients with bilateral ICA occlusion (see Table 1). A recently published report of direct measurements of local CPP at

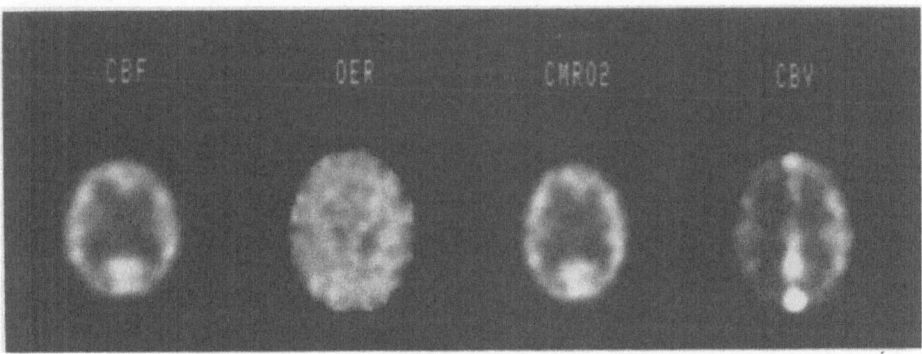

Fig. 2. PET images showing regional CBF, CMRO$_2$, OER, and CBV in a patient with right internal carotid artery occlusion. Note the focal increase of CBV in the right hemisphere

the time of bypass surgery has confirmed that the reduction of MCA pressure distal to an occluded ICA is substantially greater in patients with bilateral carotid occlusion (Spetzler et al. 1983).

In view of the normal tendency for blood flow and volume to vary in a coupled fashion, CBV being influenced by prevailing CBF as well as by CPP, CBF should ideally be taken into account when interpreting values of CBV. The ratio CBF/CBV should therefore provide a more sensitive index of regional CPP than CBV alone. This ratio has the additional advantage that it is sensitive to changes of CPP both within and below the range of autoregulation. After the point of maximal compensatory vasodilatation is reached, further reduction of CPP will be characterized not by continued increase of CBV but by a fall of CBF. Since CBF/CBV is the reciprocal of the expression for mean vascular transit time, a falling value of this ratio is literally a measure of the diminishing velocity of flow which occurs as CPP falls, even when flow itself (volume/min) remains normal.

In patients with unilateral ICA occlusion, values of CBF/CBV are reduced to a highly significant degree in the territories of occluded vessels ($P < 0.001$). The further reduction in patients with bilateral occlusion is also highly significant ($P < 0.001$). These results are shown in Table 2. Furthermore, analysis of the relationship between CBF, CBV, and OER has indicated that those few MCA regions in which OER is raised are also characterized by reduction of CBF/CBV below a critical value around 5.5 (Gibbs et al. 1983). This cut-off value of CBF/CBV therefore appears to represent that level of regional CPP below which CBF can no longer be appropriately matched to $CMRO_2$ by regional autoregulation.

Since blood flow is maintained at an appropriate level in the majority of patients with ICA occlusion (CBF matched to $CMRO_2$, OER normal), measurement of CBF alone can provide little information about the degree of reduction of CPP, and hence residual perfusion reserve. This is confirmed by the frequent observation that EC-IC bypass surgery is not invariably followed by a rise of CBF in the appropriate MCA territory (Halsey et al. 1982; De Weerd et al. 1982; Vorstrup et al. 1983). In fact if autoregulation is preserved, a postoperative rise in blood flow would be expected only in those few patients with inappropriately low CBF and raised OER. Restoration of CPP in the majority of

Table 2. Mean values ($\pm$ SD) of CBF/CBV in the four categories of MCA territory. (Statistical comparison by 2-tailed t-test)

			CBF/CBV	
Unilateral occlusion	Normal	$(n = 18)$	10.2 ± 0.8	NS
	Patent	$(n = 24)$	9.7 ± 1.01	$P < 0.001$
	Occluded	$(n = 24)$	7.8 ± 1.1	$P < 0.001$
	Bilateral occlusion	$(n = 16)$	6.4 ± 1.3	

cases should be evident only as a fall in the previously elevated CBV. This hypothesis has been borne out by our preliminary studies of patients before and after bypass surgery. In the small group accumulated so far, the only statistically significant finding has been a postoperative reduction of CBV, which in some cases is evident in the contralateral hemisphere as well as in the MCA territory supplied by the graft. This corresponds to the observations of other authors that although regional CBF values may be unremarkable, CO_2 reactivity is often reduced in the territory of occluded carotid arteries (Norrving et al. 1982) and a more normal response may be restored after EC-IC bypass surgery (Halsey et al. 1982). As an alternative to CO_2 inhalation, Vorstrup et al. (1983) have measured regional CBF before and after intravenous acetazolamide as a means of testing the capacity for further dilatation in the cerebral circulation.

Conclusions

In patients considered for EC-IC bypass surgery, the finding of focally increased OER − evidence that CBF is inappropriately low in relation to cerebral oxygen demands − indicates that regional CPP has fallen below the range of effective autoregulation. Although the natural history of this critical pathophysiologic state remains unknown, it appears a sound indication for revascularization surgery on hemodynamic grounds. However, raised OER is an infrequent finding in patients with ICA occlusion, a group which is likely to include the majority of those stroke and TIA patients whose symptoms are due to critical reduction of perfusion pressure rather than embolism.

Our findings suggest that reduction of circulatory reserve is identifiable at an earlier stage as a focal increase in CBV, reflecting compensatory vasodilatation in response to diminished CPP. Because of the interrelationship of blood flow and volume, the ratio CBF/CBV provides a more sensitive index of reserve than CBV alone. CBF itself provides little information about CPP unless it is measured before and during some form of physiologic stress test, such as CO_2 inhalation or acetazolamide infusion.

A significant rise of OER is found only when the regional CBF/CBV ratio has fallen below a critical value. This implies that even without a measurement of OER, combined analysis of regional CBF and CBV can provide useful information about cerebral perfusion reserve at all stages of decompensation in individual patients with occlusive carotid disease. Since both regional CBF and CBV (or mean vascular transit time) can be measured by isotope techniques other than PET, this approach could be applied more widely both for patient selection and for evaluation of the effects of EC-IC bypass surgery.

References

Barnett HJM (1978) Delayed cerebral ischaemic episodes distal to occlusion of major cerebral arteries. Neurology (Minneap) 28:769–774

Baron JC, Bousser MG, Rey A, Guillard A, Comar D, Castaigne P (1981) Reversal of "misery-perfusion syndrome" by extra-intracranial arterial bypass in haemodynamic cerebral ischaemia. Stroke 12:454–459

Countee RW, Vijayanathan T (1980) Intracranial embolisation via the external carotid artery: report of a case with angiographic documentation. Stroke 11:465–468

De Weerd AW, Veering MM, Mosmans PCM, Van Huffelen AC, Tulleken CAF, Jonkman EJ (1982) Effect of the extra-intracranial (STA-MCA) arterial anastomosis on EEG and cerebral blood flow. Stroke 13:674–679

Finklestein S, Kleinman GM, Cuneo R, Baringer JR (1980) Delayed stroke following carotid occlusion. Neurology (NY) 30:84–88

Frackowiak RSJ, Lenzi GL, Jones T, Heather JD (1980) Quantitative measurement of regional cerebral blood flow and oxygen metabolism in man using ^{15}O and positron emission tomography: Theory, procedure, and normal values. J Comput Assist Tomogr 4:727–736

Gibbs JM, Wise RJS, Leenders KL, Jones T (1983) The relationship of regional cerebral blood flow, blood volume, and oxygen metabolism in patients with carotid occlusion: evaluation of perfusion reserve. J Cereb Blood Flow Metabol 3:S 590

Grubb RL, Phelps ME, Raichle ME (1973) The effects of arterial blood pressure on the regional blood volume by X-ray fluorescence. Stroke 4:390–399

Halsey JH, Movawetz RB, Blanenstein UW (1982) The haemodynamic effect of STA-MCA bypass. Stroke 13:163–167

Lassen NA (1982) Incomplete cerebral infarction – focal incomplete ischemic tissue necrosis not leading to emollision. Stroke 4:522–523

Martin WRW, Baker RP, Grubb RL, Raichle ME (in press) Cerebral blood volume, blood flow and oxygen metabolism in cerebral ischaemia and subarachnoid haemorrhage: an in vivo study using positron emission tomography. Acta Neurochir

Norrving B, Nilsson B, Risberg J (1982) rCBF in patients with carotid occlusion: resting and hypercapnic flow related to collateral pattern. Stroke 13:155–162

Phelps ME, Huang S-C, Hoffman EJ, Kuhl DE (1978) Validation of tomographic measurement of cerebral blood volume with C-11 labeled carboxyhemoglobin. J Nucl Med 20:328–334

Shanbrom E, Levy L (1957) The role of systemic blood pressure in cerebral circulation in carotid and basilar artery thrombosis. Am J Med 23:197–204

Spetzler RF, Roski RA, Zabramski J (1983) Middle cerebral artery perfusion pressure in cerebrovascular occlusive disease. Stroke 14:552–555

Stark RJ, Wodak J (1983) Primary orthostatic cerebral ischaemia. J Neurol Neurosurg Psychiatry 46:883–891

Vorstrup S, Hemmingsen R, Henriksen L, Lindewald H, Boysen G, Paulson OB, Lassen NA (1983) Cerebral blood flow in patients with transient ischaemic attacks (TIA) studied with xenon 133 inhalation and emission tomography before and after reconstructive vascular surgery. J Cereb Blood Flow Metabol 3:S 592–S 593

Wise RJS, Bernardi S, Frackowiak RSJ, Legg NJ, Jones T (1983) Serial observations on the pathophysiology of acute stroke: the transition from ischaemia to infarction as reflected in regional oxygen extraction. Brain 106:197–222

Positron Emission Tomography in the Study of Parkinson's Disease

K. L. Leenders, L. Wolfson, and T. Jones

Positron emission tomography (PET) can be applied to study many conditions of the brain (Phelps et al. 1982). We have used the ECAT-II positron emission tomograph in combination with the steady state inhalation technique to measure regional cerebral blood flow (rCBF), oxygen extraction (rOER), and oxygen utilization (rCMRO$_2$) in patients with Parkinson's disease. The theoretical and practical details of our techniques are described elsewhere in this book (Gibbs et al., this volume).

Results

Cortical Regions

Although rCBF and rCMRO$_2$ are important expressions of brain tissue function, it is not anticipated that these parameters would be changed very much in Parkinson's disease. A generalized decrease might be expected as a result of diffuse or regional neuronal degeneration. Indeed, the patients we have studied so far did have lower mean values for rCBF and rCMRO$_2$ than normal volunteers. This seemed to be related to the degree of atrophy visible on their CT scan. Patients with normal CT scans ($n = 4$) had mean CMRO$_2$ values in the cortical regions of 4.1 ml/100 ml/min (not corrected for CBV). Patients with mild ($n = 8$) and moderate to marked ($n = 6$) atrophy had values of 3.5 and 3.3 ml/100 ml/min respectively. These lower values could be due to partial volume effect of the volume of measurement. However, the same gradient was seen in the region of the basal ganglia, well away from the ventricle and extracerebral spaces: the regional CMRO$_2$ values here were 3.8, 3.6, and 3.3 ml/100 ml/min respectively for the above three groups of patients.

Basal Ganglia

We have looked more specifically at the region of the basal ganglia in our patients with Parkinson's disease. In a group of six patients with unilateral symp-

MRC Cyclotron Unit, Hammersmith Hospital, Ducane Road, London W12 OHS, Great Britain

Cerebral Blood Flow and
Metabolism Measurement.
Eds. Hartmann/Hoyer
© Springer-Verlag Berlin Heidelberg 1985

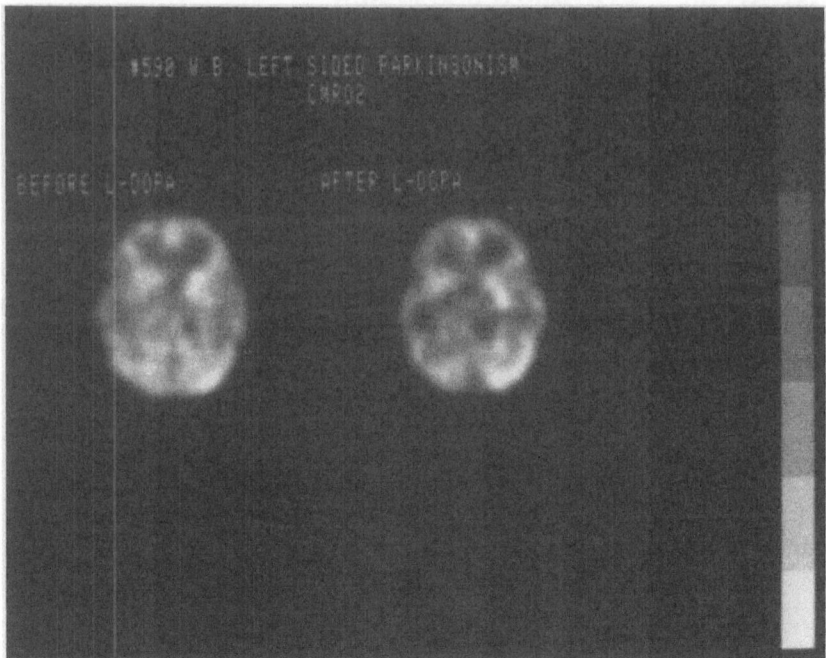

Fig. 1. A section (OM + 4.5 cm) through the brain of a hemiparkinsonian patient (left body side affected) depicting the rCMRO₂ before and 1 h after administration of L-dopa. The left side of the brain is on the left-hand side on the photograph

toms and signs, rCBF and rCMRO₂ were found to be higher in the basal ganglia territory of the predominantly affected hemisphere (contralateral to the symptomatic limbs). The side-to-side differences in the individual patients ranged from 12% to 56% (mean 19.4%, $P < 0.025$). Regional values in the more affected basal ganglia were higher than those of the normal volunteers. This is illustrated for rCMRO₂ and rCBF in Figs. 1 and 2. The corresponding cortical values did not show side-to-side differences.

Effects of L-dopa

We have also measured rCMRO₂ and rCBF before and after treatment with L-dopa. Baseline scans were carried out after withdrawal of antiparkinsonian medication for a period of 5 days. Immediately after the first study, a single oral dose of L-dopa was given and 1 h later the scans were repeated. The dose of L-dopa ranged from 300 to 1250 mg, depending on the patient's previous daily dose, to ensure that a clear clinical response would occur. At several stages the clinical disability was assessed according to the NYU disability score for Parkinson's disease (Lieberman 1980). In total 18 patients had a technically valid baseline scan and 14 a valid pre and post L-dopa scan. Of these 11 were successfully studied again several weeks later when on regular optimal L-dopa

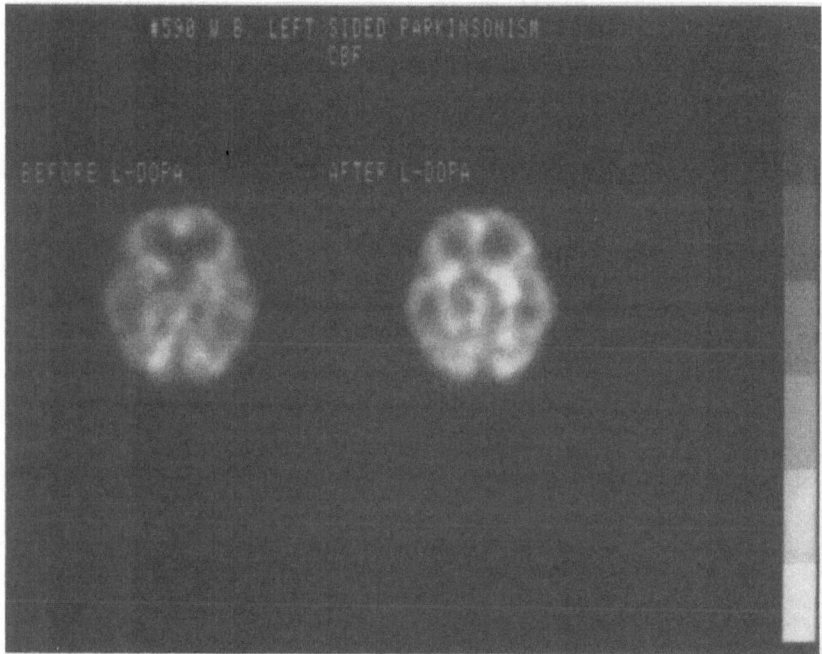

Fig. 2. A section (OM + 4.5 cm) through the brain of the same hemiparkinsonian patient as in Fig. 1 depicting the rCBF before and 1 h after administration of L-dopa. The scale is identical for both images

therapy. Six patients had predominantly unilateral disease. Six age-matched volunteers were also studied before and after administration of L-dopa.

A separate group of five patients studied before and after L-dopa were also given 60 mg domperidone, a peripheral dopamine receptor antagonist.

The clinical disability score showed a good correlation with the duration of the disease ($r = 0.65$; $P < 0.005$). The clinical disability improved considerably following L-dopa, the mean NYU score falling from 661 ± 363 to 303 ± 268 ($P < 0.001$; paired t-test).

After administering L-dopa, there was a marked increase of rCBF ($P < 0.005$) in all regions of the brain except the cerebellum. This increase was slightly greater in the basal ganglia territory than in the cortical regions. The rise of rCBF was not a response to increased metabolic demand, since cerebral oxygen consumption remained unchanged after L-dopa. As a result of this dissociated rise of rCBF, the oxygen extraction ratio decreased throughout the brain. Although the oxygen utilization did not change in the basal ganglia after L-dopa, the clinical disability had markedly improved after 1 h. After several weeks of optimal continuous oral L-dopa treatment of our patients, the rCBF had returned almost to baseline levels. At this time the clinical disability scores were as good as after the acute L-dopa study. The effect of L-dopa is illustrated in Figs. 1−5. The group of normal volunteers showed the same pattern of response as the patients. In those patients given domperidone in addition to L-dopa, the increase in rCBF was completely abolished.

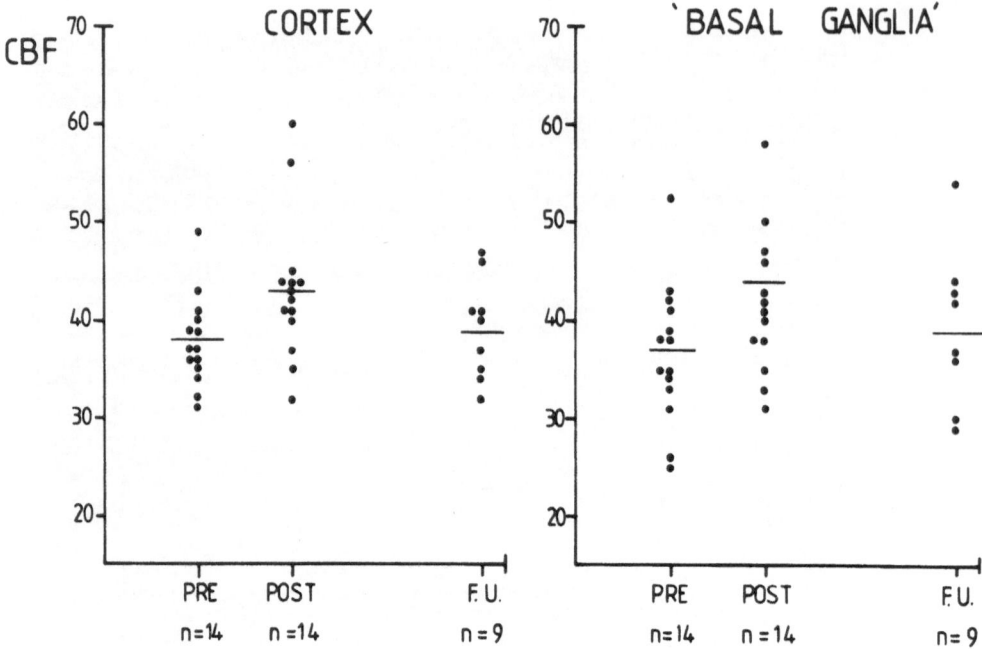

Fig. 3. The effects of L-dopa in the group of Parkinsonian patients on rCBF

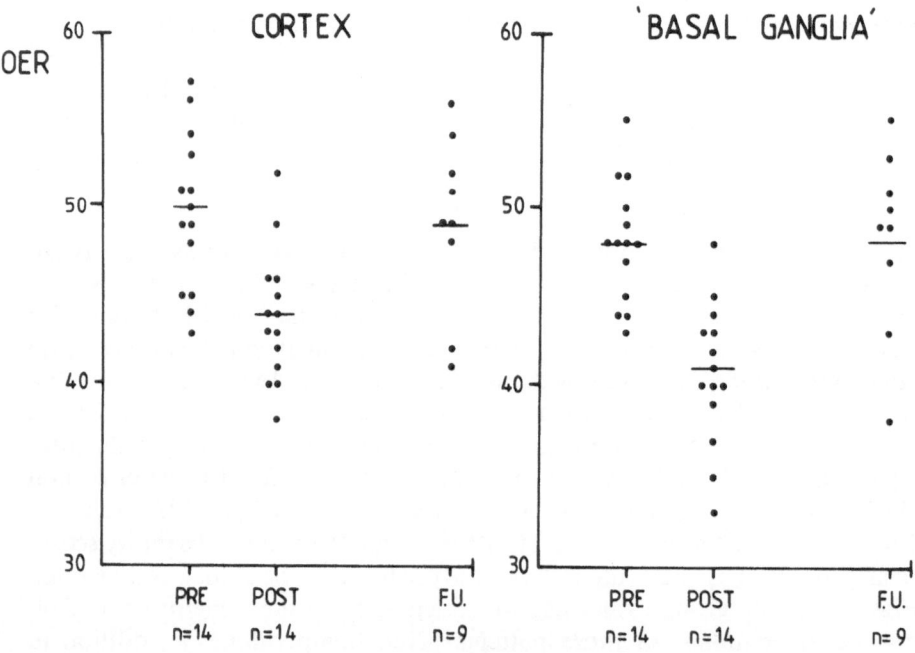

Fig. 4. The effects of L-dopa in the group of Parkinsonian patients on rOER

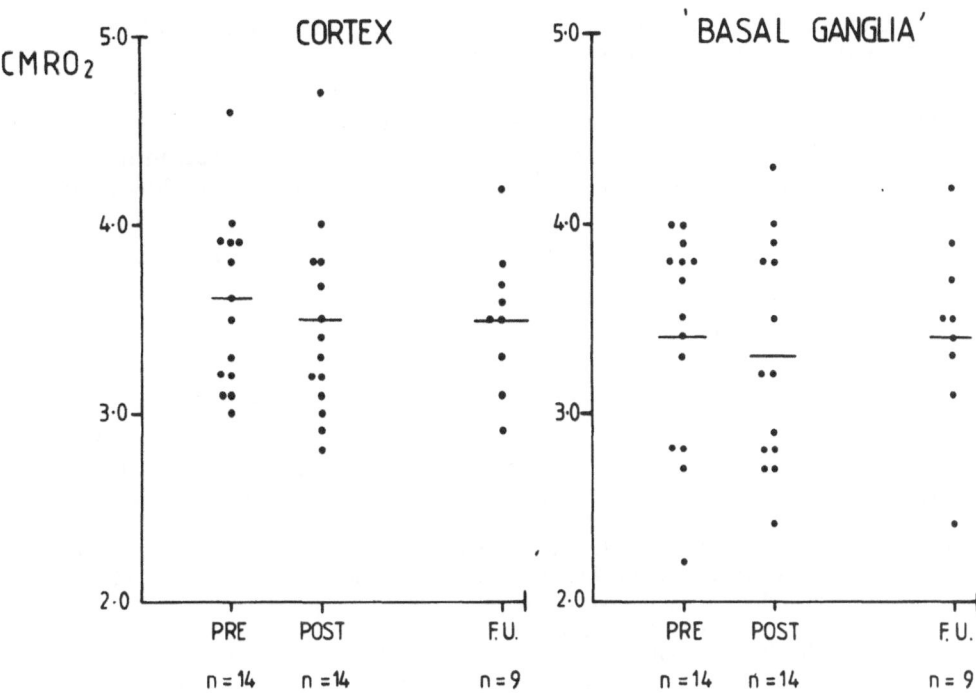

Fig. 5. The effects of L-dopa in the group of Parkinsonian patients on rCMRO₂

Discussion and Conclusions

We have found, rather surprisingly, a raised rCBF and rCMRO₂ in the basal ganglia contralateral to the symptomatic limbs in the group of patients with unilateral Parkinson's disease. These data suggest that the specific degeneration of the nigrostriatal pathway leads to a disinhibition of oxygen metabolism and, accordingly, of the rCBF in the basal ganglia territory. The limited spatial resolution of our PET scanner makes it impossible to indicate which part of the basal ganglia generates this increased metabolic activity. Animal studies (Wooten 1981) have shown that experimental lesions of the substantia nigra lead to increased glucose utilization in the globus pallidus.

Locally and systemically administered L-dopa or dopamine agonists alter the glucose metabolism in a complex manner in the basal ganglia nuclei in animals (Brown 1978; Brown 1983; McCulloch 1982). It has been suggested that this metabolic effect of dopamine agonists causes an increase of regional cerebral blood flow (McCulloch 1982). A direct influence of dopaminergic agonists on the cerebral blood vessel wall has also been shown (Edvinsson 1978). We have shown in man that orally administered L-dopa produces a marked increase of regional cerebral blood flow in all parts of the brain except the cerebellum, without a concomitant rise in oxygen metabolism. Accordingly, the oxygen extraction fraction decreases significantly. The rise in rCBF is blocked by domperidone and is independent of the clinical effect of L-dopa, since the clinical improvement was maintained after several weeks of treatment whereas the rCBF had fallen almost to its baseline value.

We suggest that L-dopa has at least two separate effects at the same time. One is the influence on the neurotransmitter system of certain nerve cells, which gives rise to the favourable clinical response. The other is a direct influence on the cerebral blood vessel wall, resulting in a rise of regional blood flow which is independent of cerebral oxygen demands. We therefore conclude that, when administered in sufficiently large doses, L-dopa is a potent cerebral vasodilator.

References

Brown LL, Wolfson LI (1978) Apomorphine increases glucose utilisation in the substantia nigra, subthalamic nucleus and corpus striatum of rat. Brain Research 140:188−193

Brown LL, Wolfson LI (1983) A dopamine-sensitive striatal efferent system mapped with [14] deoxyglucose in the rat. Brain Research 261:213−229

Edvinsson L, Hardebo JE, McCulloch J, Owman C (1978) Vasomotor response of cerebral blood vessels to dopamine and dopaminergic agonists. In: Cervos-Navarro et al (eds) Advances in neurology, vol. 20. Raven Press, New York, p 85−96

Lieberman A, Dziatolowski M, Gopinathan G, Kupersmith M, Neophytides A, Korein J (1980) Evaluation of Parkinson's disease. In: Goldstein M et al (eds) Ergot compounds and brain function: neuroendocrine and neuropsychiatric aspects. Raven Press, New York, p 277−286

McCulloch J, Kelly PAT (1982) Effects of apomorphine on the relationship between local cerebral glucose utilisation and local cerebral blood flow (with an appendix on its statistical analysis). J Cereb Blood Flow and Metabol 2:487−499

Phelps ME, Mazziotta JC, Huang S-C (1982) Study of cerebral function with positron computed tomography. J Cereb Blood Flow and Metabol 2:113−162

Wooten GF, Collins RC (1981) Metabolic effects of unilateral lesion of the substantia nigra. J of Neuroscience 1 (3):285−291

The Effects of Dexamethasone in Brain Tumor Patients Measured with Positron Emission Tomography

K. L. LEENDERS, R. P. BEANEY, and D. J. BROOKS

Dexamethasone is widely used in patients with brain tumors and perifocal edema (Weinstein 1972). It has generally been argued that the beneficial clinical influence of dexamethasone is via a direct effect on the edema itself. Possible mechanisms include diminution of edema formation and a reduction of the water content of the edematous region (Yamada 1979; Reulen 1972; Reid 1982). Some investigators have suggested that dexamethasone decreases intracranial pressure (ICP) with a concomitant immediate improvement of clinical symptoms (Brock et al. 1976; Alberti 1978). Others have found an increase in rCBF after dexamethasone therapy and suggested this is secondary to an improvement in cerebral function (Buttinger 1982; Reulen 1972). A direct action on the blood vessel wall by corticosteroids has been recognized for some time (Axelrod 1983; Altura 1966).

We have used the PET technique to measure regional blood flow (rCBF), oxygen extraction (rOER), oxygen utilization (rCMRO$_2$), and blood volume (rCBV) in patients with cerebral tumors before and after dexamethasone. The first findings will be discussed here.

Ten patients with brain tumors (seven secondaries, two astrocytomas, and one lymphoma) and accompanying perifocal brain edema were scanned both before and 1–5 days after dexamethasone treatment. A loading dose of 20 mg dexamethasone was given intravenously, and thereafter 4 mg orally was given four times daily.

Technically the scans were difficult to perform, as the duration of one scan is about 2 h and most patients, especially before dexamethasone treatment, were restless, making a steady state (see the paper of J. Gibbs in this volume for technique) difficult to obtain. In addition movement artifacts occurred. Of the above-mentioned ten patients, rCBF measurements of the contralateral cortex are available in all ten cases but for technical reasons rCMRO$_2$ and rCBV data has had to be rejected in certain cases.

Clinically all the patients improved subjectively: headaches disappeared and the general feeling of well-being improved substantially. As the patients selected for study were not very ill, for the technical reasons outlined above, it is not surprising that no big clinical changes were noted. The few patients who had focal neurologic deficits did not show improvement of these within the few days in between the two scans.

MRC Cyclotron Unit, Hammersmith Hospital, Ducane Road, London W12 OHS, Great Britain

Cerebral Blood Flow and
Metabolism Measurement.
Eds. Hartmann/Hoyer
© Springer-Verlag Berlin Heidelberg 1985

r CBF

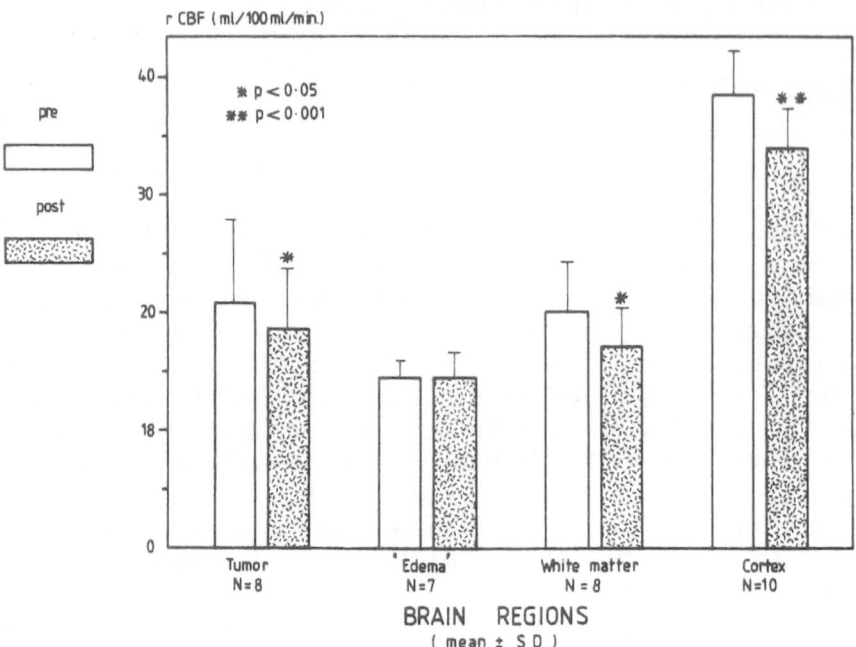

Fig. 1. The effects of dexamethasone on rCBF

rOER

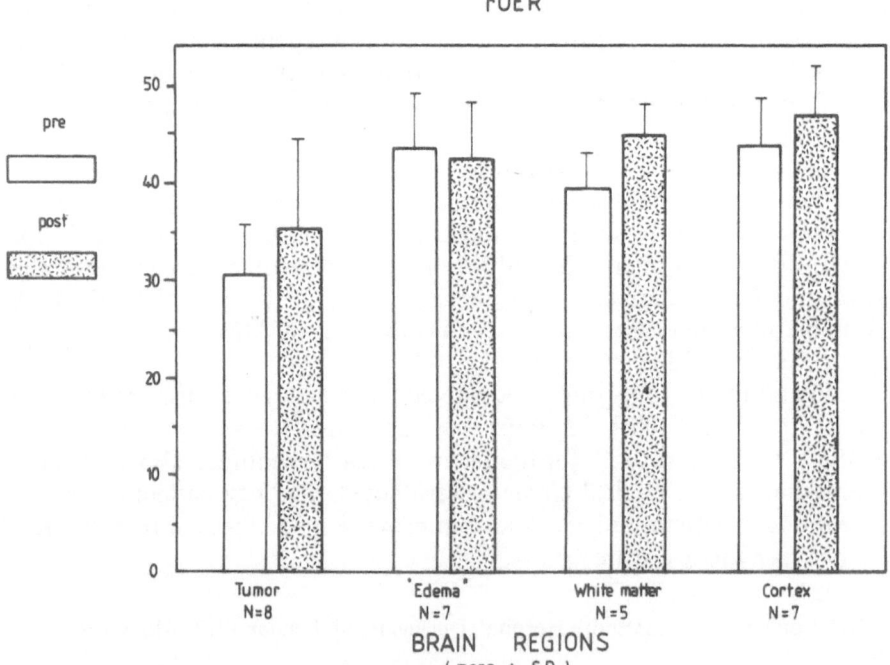

Fig. 2. The effects of dexamethasone on rOER

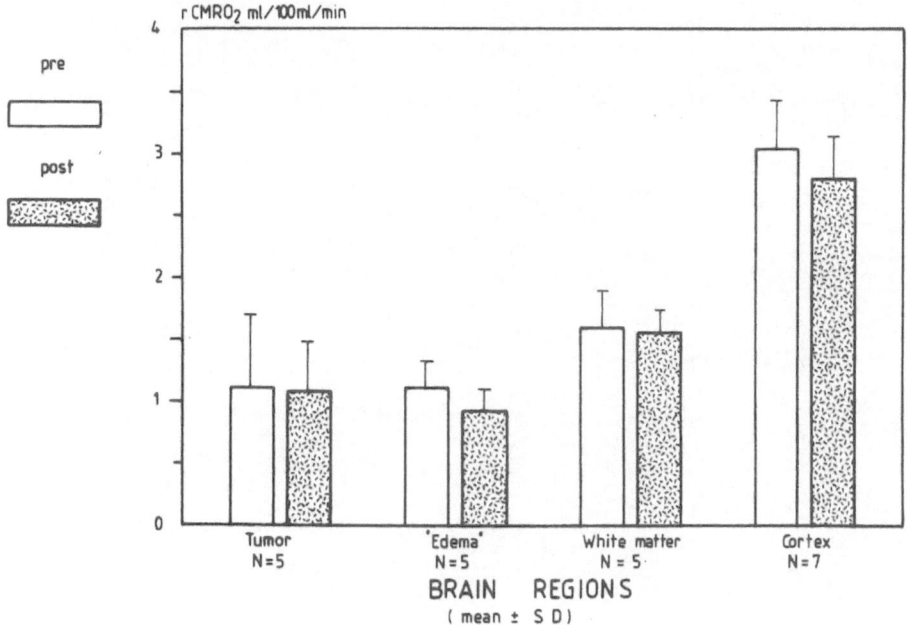

Fig. 3. The effects of dexamethasone on rCMRO$_2$

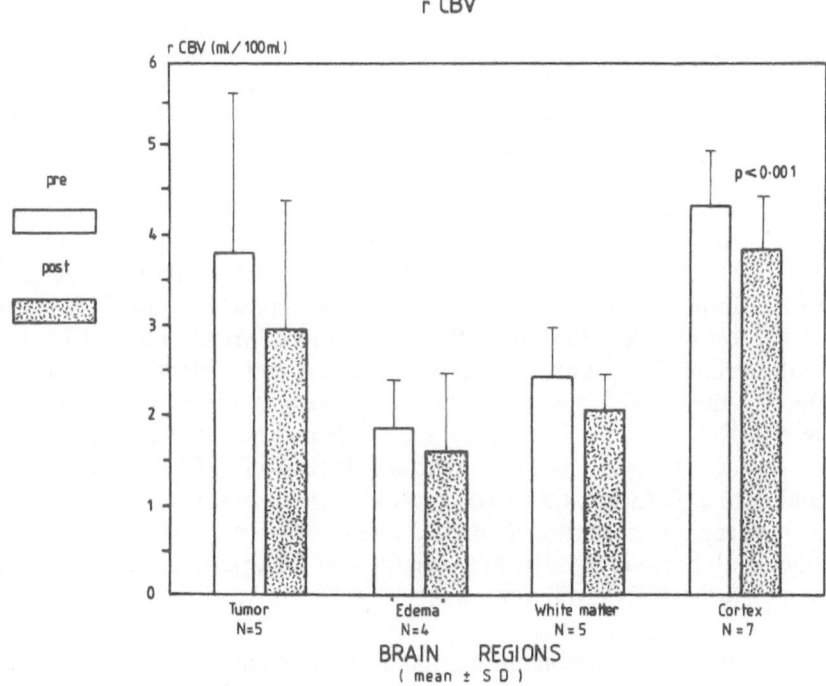

Fig. 4. The effects of dexamethasone on rCBV

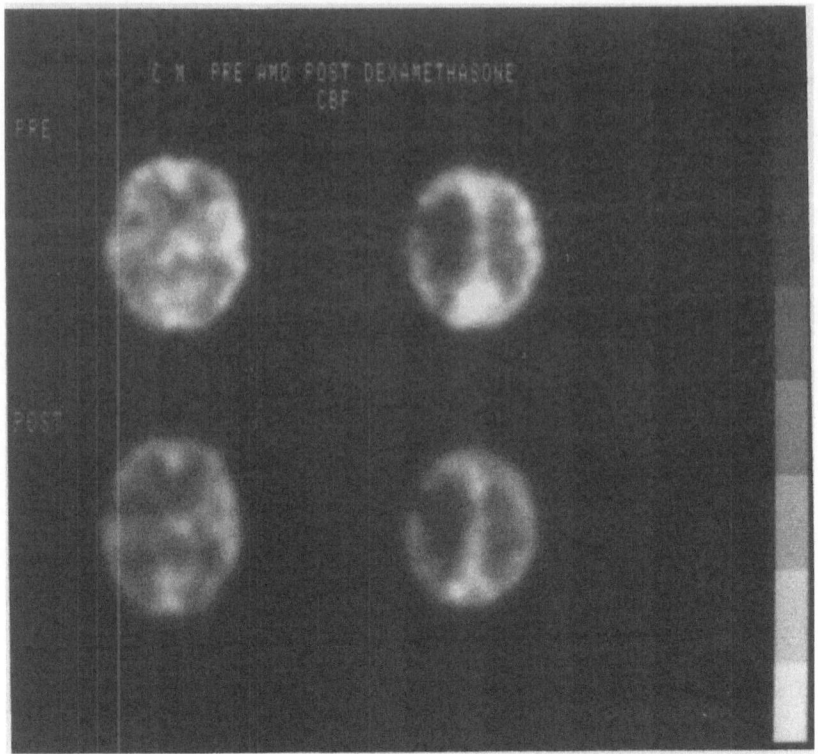

Fig. 5. Sections depicting the rCBF in the brain of a patient with a left frontoparietal second-ary. The images on the *right-hand side* are cutting through the tumor and surrounding edema. The *left-hand side* image is 2 cm lower. The *top* of each image is the front of the brain and the *left side* of each image is the left side of the brain. After dexamethasone (= lower two images) there is a diffuse decrease of the rCBF

Our scan results so far are summarized in Figures 1–4 and illustrated in Figs. 5 and 6. Prior to commencement of dexamethasone therapy, the rCBF and rCMRO$_2$ values of contralateral cortex and white matter were ± 10% lower for the tumor group compared with a normal control group. Edematous regions of interest had very low values for rCBF and rCMRO$_2$, but rOERs comparable with normal (i.e., contralateral) white matter and cortical gray matter, suggest-ing that the increased water content in such regions leads to a coupled reduction of rCMRO$_2$ and rCBF. An increased water and Na$^+$ content has been described as characteristic of vasogenic edema (Hossmann 1979 and 1981). After dexa-methasone treatment the rCBF decreased in all brain regions by 10% – 14% ex-cept in regions of perifocal edema. The rCBV decreased in parallel with blood flow in all brain regions by between 6% and 22%, particularly in the tumor terri-tory. The rCMRO$_2$ did not change significantly, although there was a trend towards a decrease.

Previous investigations in man and animals have shown an increase in CBF after dexamethasone (Buttinger 1982; Reulen 1972) and methylprednisolone treatment (Yamada 1983). Such studies, however, used ^{133}Xe washout tech-

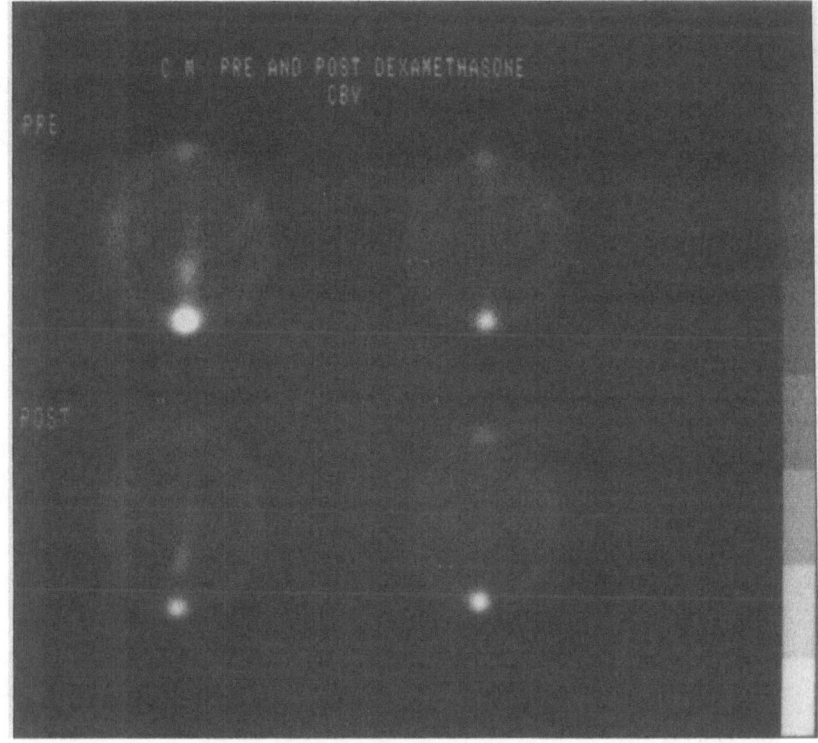

Fig. 6. The same sections of the same patient as shown in Fig. 6, but now depicting the rCBV. The blood volume decreases diffusely after dexamethasone

niques to measure rCBF and far larger doses of steroids. Objections may be raised concerning the blood-brain barrier (BBB) permeability towards water which is known to be reduced by dexamethasone (Reid 1983). This would give rise to artifactually lower measured rCBF values after drug treatment as our technique of rCBF measurement is dependent on passage of $H_2^{15}O$ across the BBB. However, we have calculated that for a 10% decrease in rCBF with our method, an 80% decrease of PS product would be necessary. Animal studies have shown a decrease of only 30% of the PS product for water after dexamethasone treatment in high dosage (Reid 1983).

Another source of error could be a change in the partition coefficient for water after dexamethasone treatment. This is unlikely as in normal brain tissue no changes in water content have been found following such dexamethasone therapy (Yamada 1979; Pak Hoo Chan 1983).

Conclusions

We conclude that dexamethasone has a rapid generalized effect on cerebral vasculature increasing vascular resistance and hence lowering both cerebral blood flow and blood volume. The resulting decrease in cerebral blood volume may in turn lead to a reduction in intracranial pressure and thus relief of clinical symptoms.

There may also be a separate effect on perifocal edema, which arises after several days of treatment. We did not see local signs of ischemia in edematous regions as assessed by their rOER in spite of low rCBF values. Further work needs to be done to validate our preliminary findings.

References

Alberti E, Hartmann A, Schirtz H-D, Schreckenberger F (1978) The effect of large doses of dexamethasone on the cerebrospinal fluid pressure in patients with supratentorial tumors. J neurol 217:173−181

Altura BM (1966) Role of glucocorticoids on local regulation of blood flow. Am J Physiol 211:1393−1397

Axelrod L (1983) Inhibition of prostacyclin production mediates permisive effect of glucocorticoids on vascular tone. The Lancet i:904−906

Brock M, Wiegand H, Zillig C, Zywietz C, Mock P, Dietz H (1976) The effect of dexamethasone on intracranial pressure in patients with supratentorial tumours. In: Pappius HM, Feindel W (eds) Dynamics of brain edema. Berlin, Heidelberg, New York, Springer Verlag pp 330−336

Buttinger C, Hartmann A, von Kummer R, Menzel J (1982) The effect of high doses of dexamethasone on cerebral blood flow in patients with cerebral tumours. In: Hartmann A, Brock M (eds) Treatment of cerebral edema, Springer Verlag, Berlin Heidelberg, pp 132−138

Hossmann K-A, Blöink M (1982) Blood flow and regulation of blood flow in experimental peritumoral edema. Stroke 12 (2):211−217

Hossmann K-A, Wechseler W, Wilmes F (1979) Experimental peritumorous edema. Acta neuropathologica (Berl) 45:195−203

Pak Hoo Chan, Fishmann RA, Caronna J, Schmidley JW, Prioleau G, Lee J (1983) Induction of brain edema following intracerebral injection of arachidonic acid. Ann Neurol 13:625−632

Reid AC, Teasdale GM, McCulloch J (1983) The effects of dexamethasone administration and withdrawal on water permeability across the blood-brain barrier. Ann Neurol 13:28−31

Reulen HJ, Hadjidimos A, Schürmann K (1972) The effect of dexamethasone on water and electrolyte content and on rCBF in perifocal brain edema in man. In: Reulen HJ, Schürmann K (eds) Steroids and brain edema. Berlin Heidelberg New York, Springer pp 239−252

Weinstein JD, Toy FJ, Jaffe ME, Goldberg HI (1972) The effect of dexamethasone on brain edema in patients with metastatic brain tumours. Neurology 23:121−129

Yamada K, Bremer AM, West CR (1979) Effects of dexamethasone on tumor-induced brain edema and its distribution in the brain of monkeys. J Neurosurg 50:361−367

Yamada K, Ushio Y, Hayakawa T, Arita N, Hamada N, Mogami H (1983) Effects of methyl prednisolone on peritumoral brain edema. J Neurosurg 59:612−619

Pharmacologic Studies in Man with PET: An Investigation Using ^{11}C-Labeled Ketanserin, a 5 HT$_2$ Receptor Antagonist

J. C. Baron[1,2], Y. Samson[1], C. Crouzel[1], M. Berridge[1], L. Chretien[1], P. Deniker[3], D. Comar[1], and Y. Agid[4]

By allowing in vivo quantitative autoradiography, positron emission tomography (PET) offers unique new opportunities in pharmacologic research. Two different approaches have been used. One of them is to measure the effects of a given pharmacologic manipulation on various physiologic parameters (e.g., cerebral blood flow, oxygen consumption, glucose utilization) regionally in humans (Rougemont et al. 1983; Leenders et al. 1983). The other, which uses labeled drugs to study their regional distribution and pharmacokinetics in brain in vivo, has several différent applications: (1) to measure the fate of the labeled drug in normal and diseased brain, i.e., penetration, distribution, and equilibrium (Comar et al. 1979a; Baron et al. 1983; Ramsay 1983; Yamamoto et al. 1983); (2) to use the labeled drug as a radioligand to estimate the density and function of specific receptor sites (see below); (3) to use a pharmacologically active labeled neurotransmitter precursor to map a specific neurotransmitter system (Garnett et al. 1983); and (4) to use the physicochemical properties of some classes of pharmacologic agents to derive a physiologic parameter (e.g., CBF, pH) from knowledge of the labeled drug's uptake and kinetics in brain.

One application has been the study of specific receptor sites in vivo by means of radioligands injected intravenously. Using PET, the feasibility of the approach was first demonstrated by Comar et al. (1979 b) in a study of the benzodiazepine (BZD) receptors in the baboon's brain using ^{11}C-flunitrazepam, but the displaceable (specifically bound) tracer was somewhat low compared with the nondisplaceable (nonspecifically bound) fraction. This problem was not encountered in the baboon studies of brain BZD receptors using ^{11}C-RO 151788 (Mazière et al. 1983) and of cardiac muscarinic receptors using ^{11}C-MQNB (Mazière et al. 1981). Although encouraging results were reported using ^{11}C-pimozide in humans (Baron et al. 1983), the study of central dopaminergic (DA) receptors has been hampered by radiochemical difficulties; the availability of more suitable DA ligands, such as ^{11}C-spiroperidol (Fowler et al. 1982), ^{18}F-haloperidol (Zanzonico et al. 1983), ^{11}C-methyl-spiperone (Wagner et al. 1983), and ^{76}Br-bromospiperone (Mazière et al. 1984) should prove more successful.

We wish to present here the results of a pioneer study of the serotonin receptors in the human brain in vivo using ^{11}C-ketanserin, a novel antagonist of

1 Service Hospitalier Frédéric Joliot, C.E.A., Département de Biologie, F-91406 Orsay
2 Clinique des Maladies du Système Nerveux, La Salpétrière, F-75013 Paris
3 Clinique de Santé Mentale et de Thérapeutique, Hôpital Sainte-Anne, F-75014 Paris
4 Clinique de Neurologie et de Neuropsychologie, La Salpétrière, F-75013 Paris

Cerebral Blood Flow and
Metabolism Measurement.
Eds. Hartmann/Hoyer
© Springer-Verlag Berlin Heidelberg 1985

the 5 HT$_2$ receptors (Leysen et al. 1981). (Present evidence suggests that the 5 HT$_2$ receptors have a specific physiologic and pharmacologic role, whereas no function has been attributed to the 5 HT$_1$ receptors) (Ilien et al. 1982). Ketanserin has high affinity for the 5 HT$_2$ receptors, moderate affinity for the histamine (H$_1$) and α_1 adrenergic sites, weak affinity for the DA sites, and practically no affinity for other receptors, including the 5 HT$_1$ receptors (Leysen et al. 1981). Specific binding of ketanserin to rat brain homogenates in vitro shows regional variations, with highest binding in frontal cortex, undetectable binding in cerebellum, and intermediate binding in temporal cortex and striatum (Leysen et al. 1982). Similar trends of regional variations have also been found in human brain postmortem (Schotte et al. 1983) Consonant with these in vitro data, significant retention of radioactivity was found in the frontal cortex for up to 1 h after i.v. injection of tracer doses of ^{3}H-ketanserin in rats, while marked washout was seen in cerebellum, and intermediate kinetics observed in striatum (Laduron et al. 1982). Part of this in vivo binding in the frontal cortex was saturable, displaceable, and preventable by unlabeled competitors, and was hence taken to represent specific binding to 5 HT$_2$ receptors (Laduron et al. 1982).

Characterization of serotonin receptors in vivo in man would be highly desirable in view of the suggested impairment of the central 5 HT systems in pathophysiologic conditions like depression (Coppen and Wood 1982), autism (Geller et al. 1983), schizophrenia (Rodnight et al. 1983), ethanol tolerance (Klanna 1981), Alzheimer's dementia (Bower et al. 1983), Parkinson's disease (Scatton et al. 1984), myoclonus (Van Woert 1981), migraine (Sicuteri et al. 1974), and pain syndromes (Jessel 1982). More specifically, increased density of 5 HT$_2$ receptors in the cerebral cortex of suicide victims has been reported (Stanley et al. 1983).

Patients and Methods

1. *Patients:* Eleven patients were studied. Control studies were done on seven patients without any neuropsychiatric condition or medication known to interfere with the 5 HT receptors. Four patients (treated schizophrenics) were studied 2 h after intramuscular injection of unlabeled chlorpromazine in therapeutic doses (75 mg, 235 µmol). Chlorpromazine (CPZ) is an efficient 5 HT$_2$ antagonist with an inhibition constant (Ki) of 3.3 nM for specific ^{3}H-ketanserin binding to rat frontal cortex in vitro, and a 50% antagonism (ED$_{50}$) of ~ 0.8 µmol/kg for mescaline-induced head twitches in rats (Leysen et al. 1982). It was hoped, therefore, that the large dose of CPZ used here would completely occupy the 5 HT$_2$ receptors and prevent any in vivo specific binding of ^{11}C-ketanserin. Among the potential candidates for accomplishing this (Leysen et al. 1982), CPZ was selected because it is the only parenterally injectable drug that was both of common therapeutic use and ethically sound.

2. *Labeling* of ketanserin with ^{11}C (T 1/2 = 20.3 min) followed the scheme published previously (Berridge et al. 1983); 3.2 − 25.6 mCi were injected intravenously as a slow bolus, and specific radioactivity ranged from 75 to

175 mCi/μmol (mean injected mass: 129 nmol). Venous blood was sampled at intervals after injection and counted for radioactivity subsequently.

3. Using the ECAT II single-slice positron tomography, 14 sequential scans of the same head level were collected starting at end-injection and up to t = 73 min. To maintain roughly similar statistics despite ^{11}C decay, the acquisition time was progressively expanded from 2 to 10 min. Total events per scan ranged from 0.15 to 0.9 million counts, but were lower in late scans of some studies. The head level was selected to study simultaneously the frontal cortex (expected to have the highest density of $5 HT_2$ receptors) and the cerebellum (lowest density expected) to optimize the characterization of specific in vivo binding of ^{11}C-ketanserin. Two control studies, however, were performed at a head level suited to provide data from cerebral gray and white matter, hence missing the cerebellum.

4. *Data analysis:* Two large ROIs, one each covering bilaterally the frontal cortex and the cerebellum, were placed on the 7th scan of each study; the time course of the relative radioactivity (corrected for ^{11}C decay, and expressed in percent of injected dose/liter brain, % ID) within the 2 ROIs across the 14 scans was computer calculated. Then, the mean % ID at each time was calculated for both the control group ($n = 5$) and the CPZ-pretreated group ($n = 4$). The same calculation was performed for the frontal/cerebellum radioactivity ratio, and for both the frontal/blood and the cerebellum/blood radioactivity ratios.

Standard tests for linearity showed that the early part (between t = 120 s and t = 1500 s) of the mean % ID time-activity curves could be accurately described by a straight line. The frontal and cerebellar time-activity curves of each *individual* study were therefore subjected to a linear least-square fit, which yielded kinetic slopes in % ID · s^{-1}. Subsequently, mean frontal and cerebellar kinetic slopes, and mean frontal minus cerebellum slope difference were calculated for each study groups, and compared by standard *t*-test procedures.

Results

1. *Blood curve:* Following a highly variable peak of activity at 2 min, the blood kinetics quickly reached a stable plateau at time $t \simeq 20$ min. The blood pharmacokinetics were not affected significantly by CPZ pretreatment.

2. *PET images* (Fig. 1): In both controls and CPZ-pretreated patients, the early PET images after ^{11}C-ketanserin injection showed a ^{11}C distribution similar to that of any perfusion tracer, with higher activity in gray than in white matter structures. In controls, this early pattern of roughly equal activity in frontal cortex and cerebellum progressively changed to a late appearance of higher activity in frontal cortex relative to cerebellum. Such a redistribution of ^{11}C activity over time was not so apparent in CPZ-pretreated subjects, in whom both structures remained more comparable with regard to their relative ^{11}C activity.

3. *Mean kinetics* (Table 1 and Fig. 2) confirmed these differences between controls and CPZ-pretreated subjects. In the former, the cerebellar kinetics showed a constant washout, but there was accumulation (up to t = 260 s) and

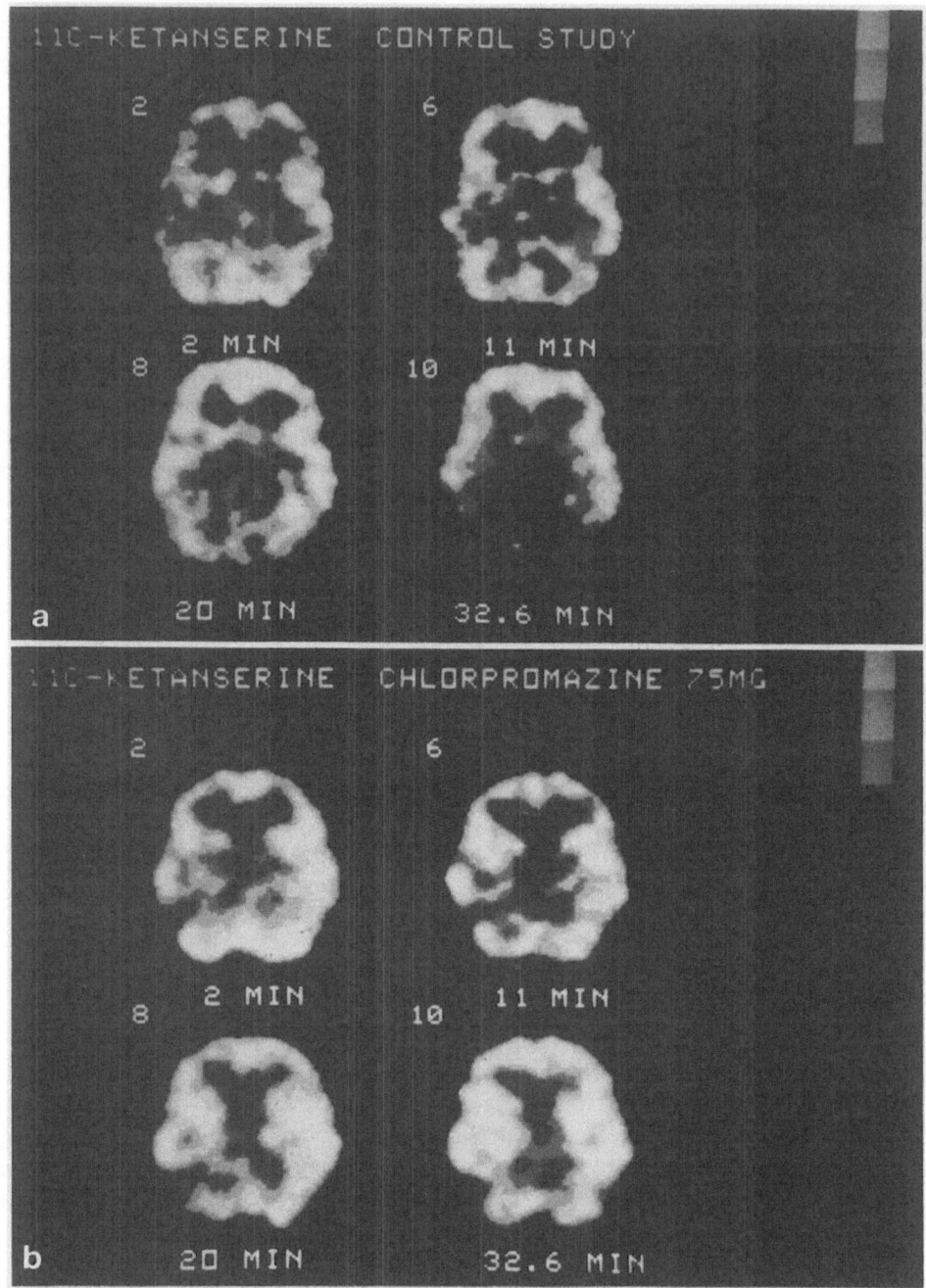

Fig. 1. a, b Sequential PET images of ^{11}C-ketanserin distribution in brain obtained in a control study (**a**) and in a chlorpromazine (CPZ) pretreatment study (**b**). Out of the 14 scans taken, four were selected (scans # 2, 6, 8, and 10 taken at the times indicated). The whiter the shade of gray, the higher the relative radioactive concentration. In both studies, earlier images (t = 2 min) clearly display cerebral cortex and white matter (upper and lateral part of images) and cerebellum (lower part of images). In the control study, radioactivity is retained in the cerebral cortex but progressively declines in cerebellum. In the CPZ pretreatment study, however, cerebral cortex and cerebellum remain roughly similar with respect to their relative radioactive concentration

^{11}C-KETANSERIN

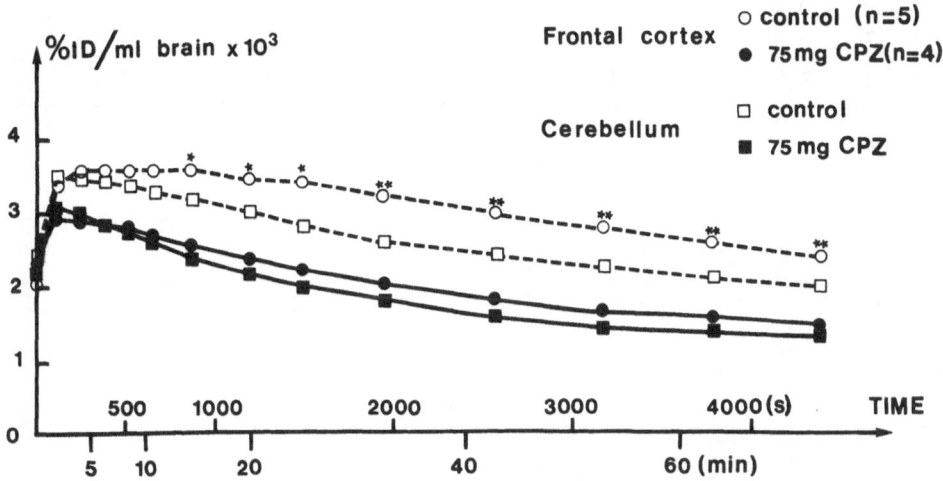

Fig. 2. Time-course of mean ^{11}C concentration (expressed in % ID per liter brain) in frontal cortex and cerebellum in the control studies ($n = 5$) and the CPZ pretreatment studies ($n = 4$). In controls, frontal cortex radioactivity increased up to t $\simeq$ 5 min, remained in plateau for the next 10 min, and washed out subsequently, while continuous washout occurred in cerebellum. In CPZ-pretreated subjects, radioactivity washed out from both structures from 2 min on, at a slightly faster rate in cerebellum than in frontal cortex. Differences in brain uptake between controls and pretreated subjects were significant only in frontal cortex (* = $P < 0.05$; ** = $P < 0.01$)

retention (up to t = 880 s) of ^{11}C activity in the frontal cortex, followed by a progressively steeper washout. In CPZ-pretreated patients, however, marked washout of ^{11}C activity from both structures was apparent early, and was grossly of a similar rate in both structures, but still slightly steeper in cerebellum than in frontal cortex. At t = 63 min, the frontal/cerebellum mean activity ratios were 1.26 and 1.12 in controls and pretreated subjects, respectively.

The tissue/blood activity ratios were highly variable among subjects; they showed highest values at times < 25 min that ranged from 3.6 to 7.4 in controls and from 2.7 to 12.7 in CPZ-pretreated subjects (frontal cortex). At later times, the tissue/blood ratios fell progressively, indicating that true equilibrium was never reached.

4. *Individual kinetics* (Table 2): The mean frontal slope was significantly different from the mean cerebellar slope in controls ($P < 0.001$) but not in CPZ-pretreated subjects; and the frontal slopes, but not the cerebellar slopes, differed significantly between control and pretreated subjects. Moreover, the mean slope difference was highly significant in controls ($P < 0.01$), but only marginally so in pretreated patients ($P < 0.05$).

Table 1. Mean results[a]

Time (s)	0	130	260	390	520	660	880	1190	1500	1960	2560	3180	3780	4400
Frontal (% ID) (Controls) n=5	1.99 ±0.96	3.34 ±0.75	3.49 ±0.78	3.53 ±0.78	3.52 ±0.73	3.54 ±0.71	3.53 ±0.72	3.44 ±0.68	3.38 ±0.70	3.38 ±0.64	2.95 ±0.64	2.74 ±0.58	2.53 ±0.55	2.37 ±0.48
Cerebellum (% ID) (Controls) n=5	2.64 ±0.75	3.43 ±0.78	3.44 ±0.76	3.40 ±0.73	3.33 ±0.79	3.23 ±0.75	3.13 ±0.74	2.98 ±0.75	2.78 ±0.76	2.57 ±0.76	2.39 ±0.76	2.20 ±0.73	2.07 ±0.69	1.95 ±0.70
Frontal (% ID) (CPZ) n=4	2.24 ±0.49	2.91 ±0.48	2.86 ±0.46	2.82 ±0.44	2.75 ±0.42	2.66 ±0.40	2.52 ±0.33	2.35 ±0.26	2.20 ±0.25	2.01 ±0.21	1.79 ±0.19	1.62 ±0.19	1.52 ±0.18	1.43 ±0.15
Cerebellum (% ID) (CPZ) n=4	2.19 ±0.89	3.09 ±0.50	2.96 ±0.45	2.83 ±0.43	2.71 ±0.41	2.58 ±0.38	2.36 ±0.33	2.14 ±0.31	1.97 ±0.28	1.77 ±0.27	1.56 ±0.26	1.43 ±0.24	1.36 ±0.24	1.29 ±0.21

[a] Results are means of five control and four CPZ pretreatment studies, and are expressed in percent of injected dose per 10^3 ml brain (% ID)

Table 2. Individual results

Study	CPZ 75 mg i.m.	^{11}C-ketanserin			Frontal slope[a]	Cerebellum slope[a]	Slope difference[a]
		mCi	Spec. act (mCi/μmol)	Mass (nmol)			
1	−	25.0	−	−	+0.19	−3.07	+3.26
2	−	11.1	−	−	−0.81	−5.07	+4.26
3	−	15.4	75	205	−0.74	−6.55	+5.81
4	−	3.2	−	−	−1.18	−6.93	+5.75
5	−	10.1	89	113	+1.12	−3.40	+4.52
Mean (±SD)		13.0 (±8.0)			−0.28 (±0.93)[b]	−5.00 (±1.76)[c]	+4.72 (±1.08)[d]
6	+	8.5	103	81	−6.23	−9.70	+3.47
7	+	9.5	175	54	−1.87	−3.92	+2.05
9	+	23.7	115	206	−7.70	−9.55	+1.85
10	+	19.7	152	130	−5.95	−10.21	+4.26
Mean (±SD)		15.4 (±7.5)	136 (±33)	118 (±67)	−5.44 (±2.50)[e]	−8.34 (±2.96)[f]	+2.91 (±1.15)[g]

[a] Expressed in % ID×10^4/10^3 s; [b] $P < 0.001$ with respect to c and $P < 0.01$ with respect to e; [c] Not significantly different from f; [d] $P < 0.01$ with respect to zero and $P < 0.05$ with respect to g; [e] Not significantly different from f; [f] $P < 0.05$ with respect to zero; [g] Not significantly different from f

Discussion

The immediate, marked brain uptake of ^{11}C-ketanserin found in humans after i.v. injection indicates absence of complete trapping in lungs and easy penetration across the blood-brain barrier, as suggested in previous animal studies (Laduron et al. 1982; Berridge et al. 1983). Another reflection of the high lipophilicity of the drug is its marked penetration in brain (peak brain/blood ratios of $\sim 3-13$) and, presumably, its perfusion-like initial distribution (Fig. 1). Equilibrium between brain and blood radioactivities was never quite reached: while blood ^{11}C radioactivity plateaued from $t \simeq 20$ min on, a marked decline in radioactivity was observed in every brain structure analyzed, starting at ~ 2, 10, 15, and 15 min in cerebellum, white matter, frontal cortex, and visual cortex, respectively (data not shown). This behavior suggests that drug metabolites appear in brain tissue and/or in blood (Laduron et al. 1982) earlier and to a higher extent in man than in rats and rabbits (Janssen Pharmaceuticals 1983).

The present study was designed to evaluate the use of ^{11}C-ketanserin in the characterization of the central 5 HT$_2$ receptors in humans in vivo. In vitro binding data on both rat and human frontal cortex, such as high affinity (nM range), reasonable specificity, and fairly high specific over nonspecific binding ratio (Leysen et al. 1981, 1982; Schotte et al. 1983), should make ketanserin a suitable candidate for in vivo 5 HT$_2$ binding studies. In addition, 5 HT$_2$ receptors are relatively abundant in frontal cortex and almost absent in the cerebellum, allowing the latter to serve as an internal standard to detect specific binding in the former. Laduron et al. (1982) studied the pharmacokinetics of tracer doses of ^{3}H-ketanserin injected intravenously in rats. They found retention of radioactivity in the frontal cortex up to $t = 60$ min, and continuous washout from cerebellum, a kinetic difference resulting in a frontal/cerebellar ratio of 2.7 at $t = 60$ min; at later times, however, radioactivity declined markedly in the frontal cortex and reached the cerebellum level at $t = 8$ h. That the retention seen in frontal cortex represented specific ^{3}H-ketanserin binding was documented by saturation and displacement experiments, and by its inhibition by the competitors pipamperone, mianserin, and methysergide in a dose-dependent way (Laduron et al. 1982). We decided, therefore, to apply this approach to humans using ^{11}C-ketanserin and PET.

In controls, we found statistically significant differences between frontal cortex and cerebellar ^{11}C-ketanserin kinetics consistent with, although less conspicuous than, the above-mentioned rat studies (Laduron et al. 1982), that is a 20-min plateau (followed by marked washout) in frontal cortex and a continuous washout in cerebellum (Figs. 1 and 2, Table 1). Pretreatment with CPZ in doses large enough to achieve almost full occupation of the 5 HT$_2$, α_1, and H$_1$ receptors (Leysen et al. 1984) resulted in an immediate washout from frontal cortex ($P < 0.01$ with respect to controls) but did not affect significantly the cerebellar kinetics (Fig. 2, Table 2); likewise, the frontal, but not the cerebellar, ^{11}C uptake decreased significantly (Fig. 2), and the 60 min frontal/cerebellum ratio decreased from 1.26 in controls down to 1.12 in CPZ studies. Taken together, these results strongly support the working hypothesis that specific binding of ^{11}C ketanserin to human frontal cortex could be demonstrated in vivo.

Despite these encouraging results, it is clear from our data that the difference between frontal cortex and cerebellar kinetics is much less striking in humans with ^{11}C-ketanserin than it is in rats with ^{3}H-ketanserin (Laduron et al. 1982), in terms of both duration of the frontal cortex plateau and frontal/cerebellum ratio. This may be due to at least four factors acting alone or in combination. Drug metabolism is one such factor: in addition to possible species differences in drug disposition within brain tissue, for which no data are available, it has been found that rats form a blood metabolite with high affinity for the $5\,HT_2$ receptors (thus allowing further specific binding to frontal cortex), while humans form metabolites inactive at the $5\,HT_2$ site (Janssen Pharmaceutical 1983). Species differences in binding characteristics could be a second factor: in vitro studies disclosed a lower density of $5\,HT_2$ receptors and a higher nonspecific binding in frontal cortex of humans compared with that of rats (Schotte et al. 1983). Thirdly, Schotte et al. (1983) reported some specific binding of ^{3}H-ketanserin to human cerebellum that was displaceable by methysergide; since the latter is ineffective at histamine receptor sites (Leysen et al. 1981), the occurrence of additional binding of ^{11}C-ketanserin to cerebellar H_1 sites in our studies must also be considered, although they are in low density in the human cerebellum (Chang et al. 1979). A final factor that could have contributed to the less satisfactory results in our human controls comes from the fact that, due to limited specific radioactivity, the amount of ketanserin injected resulted in a drug concentration in frontal cortex (calculated peak value $\sim 4\,nM$) sufficiently large to considerably enlarge the nonspecifically bound fraction.

The persistence, in CPZ-pretreated subjects, of a marginally significant retention of ^{11}C-ketanserin in frontal cortex relative to cerebellum (Fig. 2, Table 2) needs consideration. One possibility is that the dose of CPZ used did not fully occupy the receptors to ketanserin; this is unlikely, however, judging from the inhibitory potency of CPZ on $5\,HT_2$, α_1, H_1, and DA receptor sites in vitro (Leysen 1984). Another explanation would incriminate higher blood flow, and hence faster clearance of ^{11}C-ketanserin, in cerebellum than in frontal cortex (Lebrun-Grandié et al. 1983).

In addition, pretreatment with CPZ increased the rate of washout of ^{11}C-ketanserin from cerebellum (Fig. 2, Table 2). Although the difference from control studies was not quite significant, this finding, if true, may have resulted from any of the following unwanted effects of CPZ: (1) prevention of specific binding of ^{11}C-ketanserin to cerebellum (see earlier discussion); (2) alteration in peentration, metabolism, and disposition of the drug in cerebellum; and (3) changes in the nonspecific binding characteristics of ketanserin. Phenomena similar to effects 2 and 3 were considered previously to explain several unexpected properties of neuroleptics on the binding of labeled neuroleptics to brain membranes in vitro (Leysen 1984) and in vivo (Laduron et al. 1982; Zanzonico et al. 1983), including one human PET study (Baron et al. 1983). In the case of ketanserin, pretreatment with pipamperone or haloperidol significantly elevated the radioactive concentration in cerebellum of rats 1 h after i.v. injection of ^{3}H-ketanserin (Laduron et al. 1983), suggesting that such "nonspecific" effects may be of frequent occurrence in in vivo situations.

In summary, the present investigation has demonstrated the feasibility of studying the central 5 HT$_2$ receptor sites in humans in vivo. It has also highlighted several important methodological limitations. To become a practical tool in pathophysiologic research, however, the use of radioligands with better selectivity to the 5 HT$_2$ receptors and lower nonspecific binding than ketanserin may prove more satisfactory.

References

Baron JC, Roeda D, Munari C, Crouzel C, Chodkiewicz JP, Comar D (1983) Brain regional pharmacokinetics of ^{11}C-labeled diphenyl hydantoïn: positron emission tomography in humans. Neurology, 33:580−588

Baron JC, Comar D, Zarifian E, Crouzel C, Mestelan G, Loo H, Agid Y (1983) An in vivo study of the dopaminergic receptors in the brain of man using ^{11}C-pimozide and positron emission tomography, in: "Functional radionuclide imaging of the brain", P. L. Magistretti ed., Raven Press, NY, pp 337−345

Berridge M, Comar D, Crouzel C, Baron JC (1983) ^{11}C-labelled ketanserin, a selective serotonin S$_2$ antagonist. J Label Comp Radiopharm 20:73−78

Bowen DM, Allen SJ, Benton JS et al (1983) Biochemical assessment of serotoninergic and cholinergic dysfunction and cerebral atrophy in Alzheimer's disease: J Neurochem 11:266−272

Chang RSL, Tran VT, Snyder SH (1979) Heterogeneity of Histamine H$_1$ − receptors: species variations in ^{3}H-mepyramine binding of brain membranes. J Neurochem 32:1653−1663

Comar D, Zarifian E, Verhas M et al. (1979a) Brain distribution and kinetics of ^{11}C-chlorpromazine in schizophrenics: positron emission tomography studies. Psychiat Res 1:23−29

Comar D, Mazière M, Godot JM, Berger G, Soussaline F, Menini C, Arfel G, Naquet R (1979 b) Visualisation of ^{11}C-flunitrazepam displacement in the brain of the live baboon. Nature (London), 280:329−331

Coppen A, Wood K (1982) 5-hydroxytryptamine in the pathogenesis of affective disorders. In: Serotonin and biological psychiatry, B.T. Ho et al., eds, Raven Press, NY, pp 249−258

Fowler JS, Arnett CD, Wolf AP et al. (1982) ^{11}C-spiroperidol: synthesis, specific activity determination, and biodistribution in mice. J Nucl Med 23:437−445

Garnett ES, Firnau G, Nahmias C (1983) Dopamine visualized in the basal ganglia of living man. Nature (London), 305:137

Geller E, Ritvo ER, Freeman BJ, Yuwiler A (1982) Preliminary observations on the effect of fenfluramine on blood serotonin and symptoms in three austistic boys. New Engl J Med 307:165−169

Ilien B, Schotte A, Laduron PM (1982) Solubilized serotonin receptors from rat and dog brain. FEBS Lett., 138:311−315

Janssen Pharmaceutical (1983) Preclinical research report

Jessel TM (1982) Pain Lancet 2:1084−1087

Khanna JM, Kalant H, Lê AD, Leblanc AE (1981) Role of serotergic and adrenergic systems in alcohol tolerance. Prog Neuro-Psychopharmacol 5:459−465

Laduron PM, Janssen PFM, Leysen JE (1982) In vivo binding of ^{3}H-ketanserin on serotonin S$_2$-receptors in rat brain. Eur J Pharmacol 81:43−48

Lebrun-Grandié P, Baron JC, Soussaline F, Loc'h C, Sastre J, Bousser MG (1983) Coupling between regional blood flow and oxygen utilization in the normal human brain. Arch Neurol 40:230−236

Leenders K, Wolfson L, Gibbs J, Wise R, Jones T, Legg N (1983) Regional cerebral blood flow and oxygen metabolism in Parkinson's disease and their response to Ldopa. J Cereb Blood Flow Metabol 3, suppl. 1:S 488−S 489

Leysen JE (1984) Receptors for neuroleptic drugs, in: "Advances in human psychopharmacology", vol. 3, G. D. Burrows and J. S. Wektey eds., JAI Press, Greenwich Conn. 315−316

Leysen JE, Awouters F, Kennis L, Laduron PM, Vandenberk J, Janssen PAJ (1981) Receptor profile of R 41 468 (ketanserin) a novel antagonist at 5 HT$_2$ receptors. Life Sci, 28:1015

Leysen JE, Niemegeers CJE, Van Nueten JM, Laduron PM (1982) ^{3}H-ketanserin (R 41 468), a selective ligand for serotonin$_2$ receptor binding sites. Binding properties, brain distribution and functional role. Molec Pharmacol 21:301 – 304

Mazière M, Comar D, Godot JM, Collard P, Cepeda C, Naquet R (1981) In vivo characterization of myocardium muscarinic receptors by positron emission tomography. Life Sci 29:2391 – 2397

Mazière M, Prenant C, Sastre J, Crouzel M, Comar D, Hantraye P, Kaïsima M, Grubert P, Naquet R (1983) ^{11}C-RO 15 1788 et ^{11}C-flunitrazépam, deux coordinats pour l'étude par tomographie par positons des sites de liaison des benzodiazépines. C.R. Acad Sci Paris, 286:871 – 876

Mazière B, Loc'h C, Hantraye P et al. (1984) ^{76}Br-bromospiroperidol: a new tool for quantitative in vivo imaging of neuroleptic receptors. Life Sci 35:1349 – 1356

Ramsay RE (1983) Valproate brain tissue kinetics determined by PET. Neurology 33, Suppl 2:147

Rodnight R (1983) Schizophrenia: some current neurochemical approaches. J Neurochem 41:12 – 21

Rougemont D, Baron JC, Collard P, Bustany P, Comar D, Agid Y (1983) Local cerebral metabolic rate of glucose (lCRMGlc) in treated and untreated patients with Parkinson's disease. J Cereb Blood flow metabol, vol. 3, suppl. 1:S 504 – S 505

Scatton B, Javoy-Agid F, Rouquier L. Dubois B, Agid Y (1985) Monoamines in the cerebral cortex of control and Parkinsonian subjects (submitted)

Schotte A, Maloteaux JM, Laduron PM (1983) Characterization and regional distribution of serotonin S$_2$-receptors in human brain. Brain Res, 276:231 – 235

Sicuteri F, Anselmi B, Fanciullacci M (1974) The serotonin theory of migraine. Adv Neurol 4:383 – 394

Stanley M, Mann JJ (1983) Increased serotonin-2 binding in frontal cortex of suicide victims. Lancet 1:214 – 216

Van Woert MH, Hwang EC (1981) Role of brain serotonin in myoclonus in: "Neurotransmitters, seizures and epilepsy". P. L. Morselli et al., eds., Raven Press, NY, p 239 – 249

Wagner HN, Burns HD, Dannals RF et al (1983) Imaging dopamine receptors in the human brain by positron tomography. Science 221:1264 – 1266

Yamamoto YL, Diksic M, Sako K, Arita N, Feindel W, Thompson CJ (1983) Pharmacokinetic and metabolic studies in human malignant glioma in: "Functional radionuclide imaging of the brain", P. L. Magistretti, ed., Raven Press, NY: 327 – 335

Zanzonico PB, Bigler RE, Schmall B (1983) Neuroleptic binding sites: specific labeling in mice with ^{18}F-haloperidol, a potential tracer for positron emission tomography. J Nucl Med 24:408 – 416

Local Rates of Cerebral Protein Synthesis in the Gerbil and Monkey Brain

W. Bodsch, K. Takahashi, B. Grosse Ophoff, and K. A. Hossmann

In autoradiographic studies of cerebral metabolism a radioactively labeled precursor is commonly used to trace the events of neurochemical reactions following brain uptake of the isotopic compound. Therefore, knowledge of the biochemical mechanisms operative within the observation period, i.e., the labeling period, is essential for any attempt to quantify reaction rates. These may be predicted to a certain extent from the integration of the specific radioactivity input function to the system and from the amount of radioactivity in brain per unit mass of tissue. The validity of this approach is largely dependent on the complexity of the mechanism under investigation and may, therefore, require more or less extensive additional measurements or assumptions.

Experimental animal systems offer the opportunity for further biochemical measurements in determining full descriptions of the inherent cerebral metabolic events. In determining the cerebral protein synthesis rates (CPSR) in mammalian brains it is a prerequisite to realize several metabolic steps (Bodsch 1983b) between brain uptake of a radioactive amino acid and its stable incorporation into cerebral proteins.

The CPSR Method

The biochemical, mathematical, and autoradiographic validation of the total procedure will be described in detail elsewhere (Bodsch 1985) and can be summarized as follows: radioactive large neutral amino acids (an equimolar but nonequispecific activity mixture of carboxyl-^{14}C-labeled phenylalanine, tyrosine, isoleucine, leucine, and methionine) is introduced as an i.v. bolus into the circulation at zero time and allowed to incorporate until 45 min, i.e., the time of termination of the experiment. Two minutes before termination the same mixture of amino acids in the ^{3}H-labeled form (the specific activity of the ^{3}H-amino acids being about 100-fold that of the ^{14}C-amino acids in the initial bolus) is continuously infused intravenously. Upon termination the brain is quickly frozen and alternate 20-μm and 200-μm sections are cut for autoradiography and tissue sampling techniques. When 20-μm coronal sections are first subjected to LKB ultrofilm autoradiography, the exposure of the film is a func-

Max-Planck-Institut für neurologische Forschung, Abteilung für experimentelle Neurologie Ostmerheimer Straße 200, D-5000 Köln 91

Cerebral Blood Flow and
Metabolism Measurement.
Eds. Hartmann/Hoyer
© Springer-Verlag Berlin Heidelberg 1985

tion of the amount of free ^{3}H-labeled amino acids per unit mass of brain tissue. Following ^{3}H-autoradiography, sections are immersed so as to elute free amino acids completely from the tissue, and a second exposure of the treated section to Kodak NMB film is a function of ^{14}C-labeled amino acids incorporated into proteins exclusively. Finally, two separate autoradiographs are obtained from the same brain section illustrating ^{3}H-regional amino acid influx (rAAI) and ^{14}C-amino acid incorporation into proteins. The rAAI autoradiogram is frequently used to measure the unidirectional transport constant K1 (Blasberg et al. 1983) in terms of flow-independent blood-brain barrier permeability characteristics of large neutral amino acids (Hawkins et al. 1983). This is extremely helpful for CPSR quantifications under pathophysiologic conditions. The reaction constant for metabolism of amino acids, k3, is determined in the tissue from the amount of radioactive amino acids metabolized and the enzymatic activity of the enzyme responsible for the derived amino acid metabolism. k4 and k5, the reaction constants for aminoacylation, are determined in the tissue by measurements of the aminoacyl-transfer RNA pool and the charging and releasing parameters of these transfer RNAs. The blood- brain barrier efflux constant, k2, is then calculated from known K1, k3, k4, k5, and the half-life of the free amino acid pool, q. These parameters and the measurement of radioactive amino acids present in proteins (Bodsch and Hossmann 1983a) allow the determination of k6, the incorporation constant. Finally, the integral of the plasma-specific radioactive curve is added to the relation of reaction constants and together with the amount of ^{14}C-radioactivity per unit mass of tissue the rate of cerebral protein synthesis (CPSR) is obtained. The biochemically determined CPSR value is correlated with the optical density of the desired region in the ^{14}C-autoradiograph, and digitalized optical densities (Goochee et al. 1980) are converted to quantitative image reconstructions.

Five-Minute Ischemia in Mongolian Gerbils

Mongolian gerbils were subjected to bilateral carotid occlusion according to the model described by Kirino (1982) for 5 min in order to evoke postischemic maturation of hippocampal lesions. The morphological changes produced by bilateral clamping of the arteries were confined to the hippocampus and revealed a delayed destruction of the CA1 neurons (Suzuki et al. 1983), occurring after 3 days. The first histopathologic changes in CA neurons were observed 3 h after 5-min ischemia. Because dissociations between regional cerebral blood flow and glucose utilization were reported as most pronounced at 1 h after recirculation (Suzuki et al. 1983), we investigated local rates of protein synthesis at 1, 2, and 4 h of recirculation by the CPSR method. Figure 1 illustrates quantitative image reconstructions of rAAI and CPSR from the same brain section of a control animal. Protein synthesis in the dentate gyrus and the CA neurons (CA1 − CA4) of the hippocampus is about $120 - 145$ nmol g^{-1}min^{-1}. While no changes of CPSR in the hippocampus were observed 1 h after recirculation following 5-min ischemia, a marked reduction in the capacity of CA1 neurons to

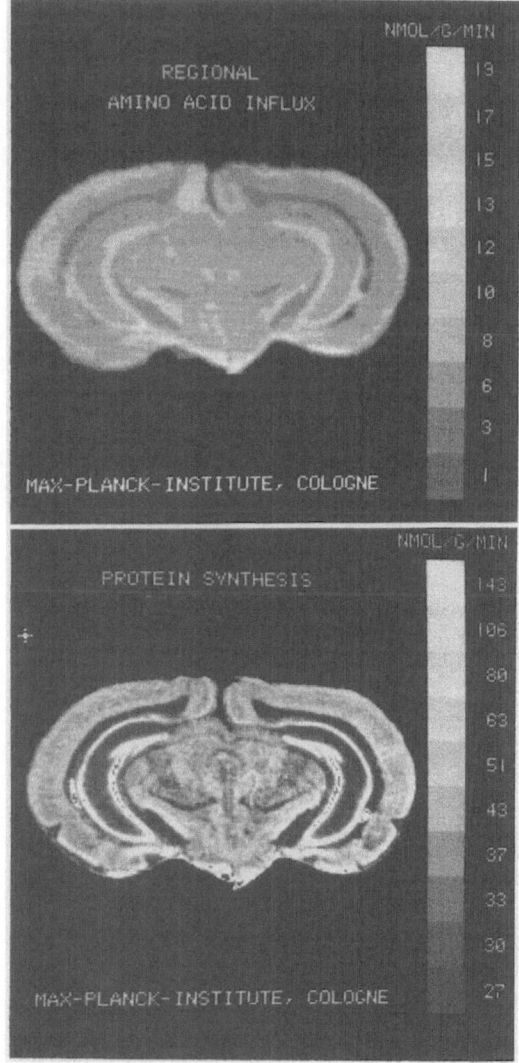

Fig. 1. Quantitative image reconstructions of ³H-regional amino acid influx (rAAI) (*top*) and ¹⁴C-cerebral protein synthesis rate (CPSR) (bottom) from the same brain section of a mongolian gerbil (control animal, halothane anesthesia)

synthesize proteins was evident at 2 h (Fig. 2); CPSR values in CA1 were 47 ± 6 nmol $g^{-1}min^{-1}$. In the fourth hour after recirculation CPSR values in CA1 decreased to 17 ± 4 nmol $g^{-1}min^{-1}$ (Fig. 2) and the protein synthesizing activity of the neurons did not recover up to the longest recirculation time of 4 days. It is of interest that protein synthesis is affected relatively late after recirculation, i.e., at between 1 and 2 h, in this model. For up to 1 h CA1 neurons preserve their protein synthesis capacity, indicating that 5-min ischemia in the gerbil is itself not the onset of reduction in CPSR. However, because CA1 neurons remain functionally active – with respect to protein synthesis – the marked decrease in CPSR between 1 and 2 h may be the result of recirculation effects.

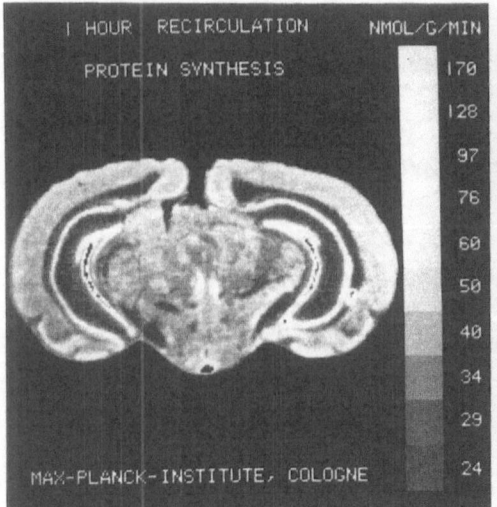

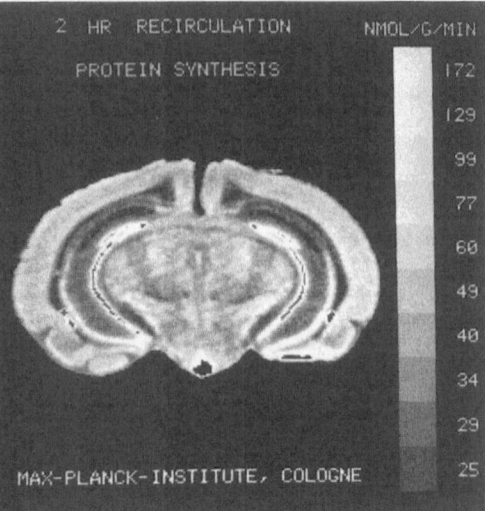

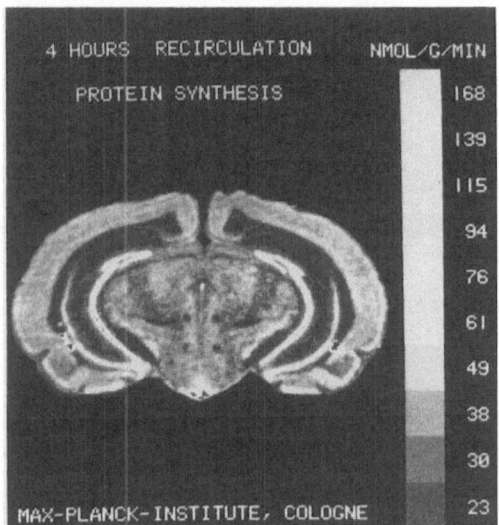

Fig. 2. Quantitative image reconstructions of CPSR from ¹⁴C-auto-
radiographs of mongolian gerbil brains.
After 5-min bilateral occlusion of
common carotid arteries: recirculation periods were 1 h (*top*), 2 h
(*middle*), and 4 h (*bottom*)

One-Hour Ischemia in Monkeys

Some of the differences between shortlasting and prolonged ischemia (Hossmann 1983) may also be a consequence of the viability of cellular key proteins.
The so-called selective vulnerability of certain neuronal regions would probably
be related to a minimal set of proteins present in neurons in order to maintain
or recover functional cellular activity. The well-described model of 1-h complete ischemia in the monkey brain (Hossmann and Zimmermann 1974) was in-

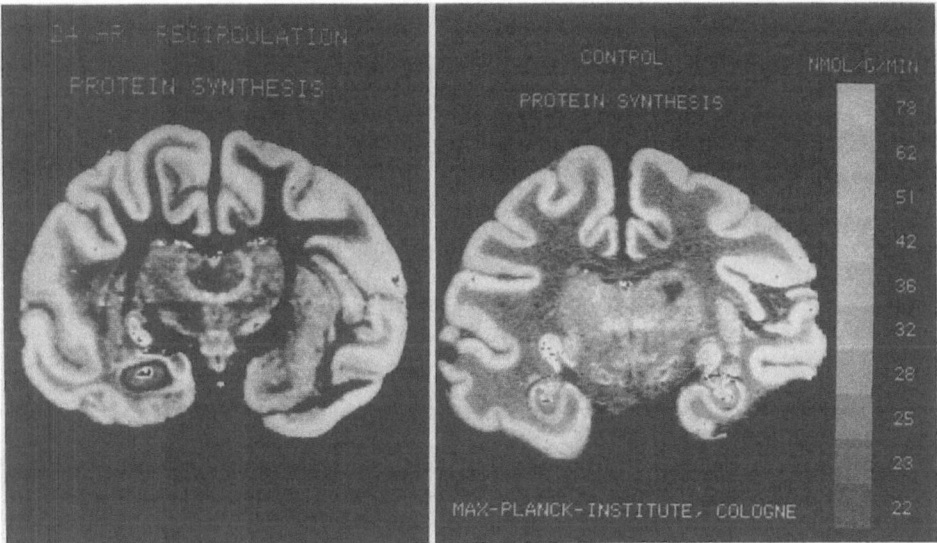

Fig. 3. Quantitative image reconstructions of CPSR from monkey brains of a control animal (*right*) and 24 h after start of recirculation, following 1-h total ischemia

vestigated by the CPSR method. An excellent correlation between the recovery of the cortical EEG and regional protein synthesis was observed. Twenty-four hours after recirculation, following 1-h total ischemia, ECoG recordings revealed nearly no differences from the control ECoG, and the recovery of CPSR in the cortex, geniculate body, and hippocampus returned to values about 90% that of controls (Fig. 3). Several thalamic regions recovered more slowly. The striking parallel appearance of electrical activity and functional reactivation of the protein synthesis machinery with prolonged recirculation periods is different from the shortlasting ischemia in gerbils. This difference may be masked by the time required to obtain functional protein synthesis in these two models and/or species. However, longer recirculation periods of more than 4 days in the gerbil may also result in protein synthesis recovery. With 1-hour total ischemia in cats it has been reported (Kleihues and Hossmann 1971) that shortly after the onset of recirculation a dissociation of ribosomal subunits occurs, which − after longer recirculation periods − continuously reassociate to functional ribosomes and thereby recover protein synthesis activity.

The degree of cellular damage (Siesjö 1981) occurring from shortlasting or prolonged ischemia is either the result of the ischemic or the recirculation period, respectively. With regard to the viability or the recovery of the protein-synthesizing system, the results reported here force us to speculate that damage of cellular protein synthesis determines neuronal death or reversible activation of cell function. Since the CPSR method has been designed to proceed by biochemical analysis following local autoradiographic results (Bodsch and Hossmann 1983b), the mechanisms regulating cerebral protein synthesis may be investigated (Bodsch et al. 1983).

References

Blasberg RG, Patlak CS, Fenstermacher JD (1983) Selection of experimental conditions for the accurate determination of blood-brain transfer constants from single-time experiments: a theoretical analysis. J Cereb Blood Flow Metabol 3:215−225

Bodsch W (1985) The double-label method for the measurement of local cerebral protein synthesis in rat: Biochemical, mathematical and autoradiographic validation using methionine as a model amino acid. In preparation

Bodsch W (1983) Prerequisites for measurement of cerebral protein synthesis. In: Heiss W-D, Phelps ME (eds) Positron emission tomography of the brain. Springer, New York, p 229−231

Bodsch W, Hossmann K-A (1983a) A quantitative regional analysis of amino acids involved in rat brain protein synthesis by high performance liquid chromatography. J Neurochem 40:371−382

Bodsch W, Hossmann K-A (1983b) The double-label method of local protein synthesis: Autoradiography and protein biochemistry of in vivo labeled proteins. J Cereb Blood Flow Metabol 3, Suppl 1:S468−S469

Bodsch W, Mies G, Hossmann K-A (1983) Local rates of protein synthesis and analysis of specific hippocampal proteins in the gerbil brain following ischemia. J Neurochem 41, Suppl 1:S137

Goochee C, Rasband W, Sokoloff L (1980) Computerized densitometry and color coding of ^{14}C-deoxyglucose autoradiographs. Ann Neurol 7:359−370

Hawkins RA, Mans AM, Biebyck JF (1982) Amino acid supply to individual cerebral structures in awake and anesthetized rats. Am J Physiol 241:E1−E11

Hossmann K-A (1982) Treatment of experimental cerebral ischemia. Review. J Cereb Blood Flow Metabol 2:275−297

Hossmann K-A, Zimmermann V (1974) Resuscitation of the monkey brain after 1 h complete ischemia. I.Physiological and morphological observations. Brain Res 81:59−74

Kirino T (1982) Delayed neuronal cell death in the gerbil hippocampus following ischemia. Brain Res 239:57−69

Kleihues P, Hossmann K-A (1971) Protein synthesis in the cat brain after prolonged cerebral ischemia. Brain Res 35:409−418

Siesjö BK (1981) Cell damage in the brain: a speculative synthesis. J Cereb Blood Flow Metabol 1:155−185

Suzuki R, Yamaguchi T, Kirino T, Orzi F, Klatzo I (1983) The effects of 5-minute ischemia in Mongolian gerbils: I. Blood-brain barrier, cerebral blood flow, and local cerebral glucose utilization changes. Acta Neuropathol (Berl) 60:207−216

Measurement of Tissue PO_2 on the Brain Surface: Clinical Application of the Polarographic Method

R. Schultheiss[1], F. Assad[1], E. Leniger-Follert[2], G. Pfeiffer[3], H. Wassmann[1], and R. Wüllenweber[1]

A multiwire surface electrode was used during neurosurgical operations for the measurement of local tissue PO_2 in brain cortex and tumors. This clinical application of the polarographic method is still experimental, whereas in many animal experiments it has been used with good results (Leniger-Follert et al. 1975; Grote et al. 1980).

The principle of polarographic PO_2 measurement is the reduction of molecular oxygen on a negatively polarized platinum electrode. Reduction of oxygen results in an electric current which is proportional to the amount of oxygen diffusing to the cathode. We used a multiwire surface electrode as described by Kessler and Luebbers (1966). It contains eight platinum wires of 15 μm diameter, infused in glass, allowing eight separate simultaneous PO_2 measurements in the tissue. A ring of Ag/AgCl is used as a reference electrode. An electrolyte solution of 0.2 M KCl impregnated in a cellophane layer connects the cathode with the anode. Both electrodes are covered by a common membrane which is oxygen permeable according to the Clark principle. Thus, the measuring circle is isolated from the measuring medium. To avoid a compression of capillaries and tissue a special support is used, allowing a careful balance and free movement of the electrode.

Luebbers et al. (1969) have shown that it is important to judge tissue oxygen supply be a statistical analysis of repeat measurements and not by single values. This can be done most easily by PO_2 histograms, which allow an immediate comparison of tissue PO_2 under different conditions or in different areas. By means of electronic data processing a fast analysis of the achieved values was possible; thus the examination time could be lowered somewhat, so that the operative procedure was not prolonged unduly. Details of the equipment used are described by Hauss (1982).

The performance of an operative procedure on the brain is a precondition for these measurements, because the electrode has to be placed directly on to the cortex after opening the dura. Thus, the physiologic conditions of the patient are defined by those of a general anesthesia. Arterial PCO_2 values were kept deliberately around 30 mmHg and definitely above 25 mmHg (in order to exclude marked circulatory responses to alterations of PCO_2). Arterial values of PO_2 had to be kept stable, as tissue PO_2 is correlated to arterial PO_2. A level of

1 Neurochirurgische Universitätsklinik, D-5300 Bonn
2 Max-Planck-Institut für Systemphysiologie, D-4600 Dortmund
3 Institut für Anaesthesiologie der Universität, D-5300 Bonn

Cerebral Blood Flow and
Metabolism Measurement.
Eds. Hartmann/Hoyer
© Springer-Verlag Berlin Heidelberg 1985

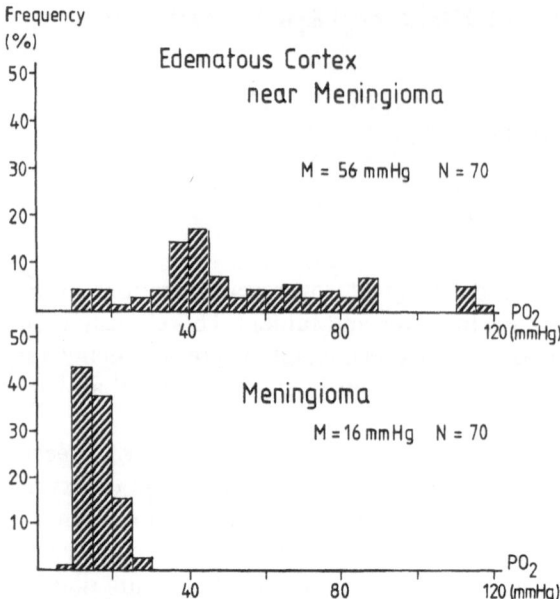

Fig. 1. Tissue oxygen distribution curve (PO$_2$ histogram) of meningioma and surrounding edematous cortex. Narrow grouping of values around a low mean value in the meningioma; slight left of an otherwise rather normal curve in the edematous cortex

around 170 mmHg proved to be most easily reproducible in most patients. If necessary, blood loss was replaced.

The possibilities of the method as well as our preliminary results may be summarized by the following examples:

Meningiomas show a very constant feature: a narrow grouping of PO$_2$ values around a low mean value of 16 mmHg (Fig. 1). As meningiomas are known to be rather vascularized tumors, this finding is unexpected. The edematous brain cortex surrounding the meningioma gives quite different patterns in the histograms, which constantly show a left shift to more hypoxic PO$_2$ values as compared with normal brain tissue.

Gliomas offer a very varying picture: narrow grouping, broad scattering, as well as a left shift of an otherwise normal looking histogram. Definite conclusions cannot yet be drawn, but we had the impression that the appearance of the histogram might be correlated to the histology of the tumor.

A very constant phenomenon could be demonstrated in patients with arteriovenous malformations: Histograms taken in cortical areas adjacent to the AVM before and after extirpation of the lesion constantly showed an increase of the mean value and a right shift. This proves better oxygenation of the cortex neighboring the AVM after the surgical excision (Fig. 2). The phenomenon of "oxygen steal," caused by the AVM, corresponds to the well-known results of angiographic, physiologic, and recent NMR studies, proving a CBF steal phenomenon.

After surgical performance of an extra-intracranial arterial bypass, PO$_2$ was measured in the cortex during the opening of the temporal vessel at the end of

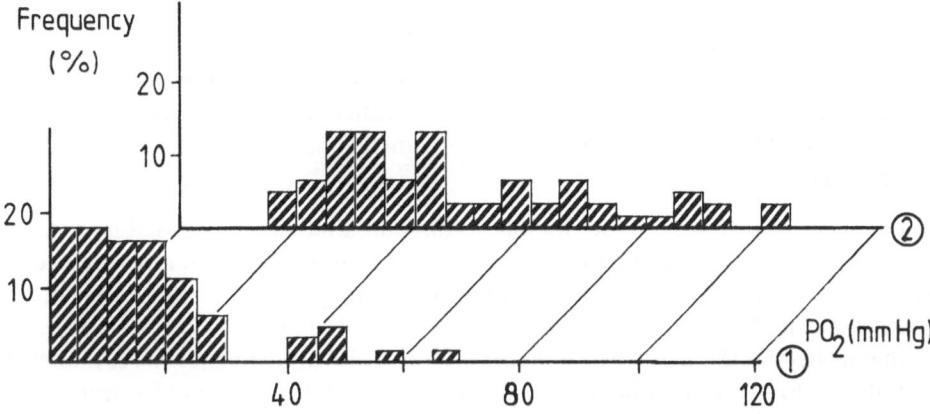

Fig. 2. PO₂ histograms taken before (*1*) and after (*2*) extirpation of an arteriovenous mal-formation in adjacent cortex. Change of mean value from 17 mmHg to 48 mmHg after ex-tirpation; normalization of hypoxic and left-shifted curve before the operation

the operation (Fig. 3). Immediately after removing the clamp from the tempo-ral vessel, PO₂ values go up markedly in the neighboring cortex supplied by the cortical vessel taken for the anastomosis. It has to be underlined that these are only casual observations; for a correct analysis it would be necessary to perform histograms after the opening of the cortical branch and after the opening of the temporal vessel. Up to now, this seemed a bit too dangerous to us because of the time needed for the first histogram, when a thrombosis of the still occluded temporal branch of the anastomosis might appear.

Local tissue PO₂ is influenced by several parameters:
1. Arterial PO₂, which is the result of many components itself.
2. Metabolism, i.e., oxygen consumption in a tissue. This is a factor which is still difficult to determine with physiologic methods in vivo, although recent developments show that with certain techniques it is possible even with the

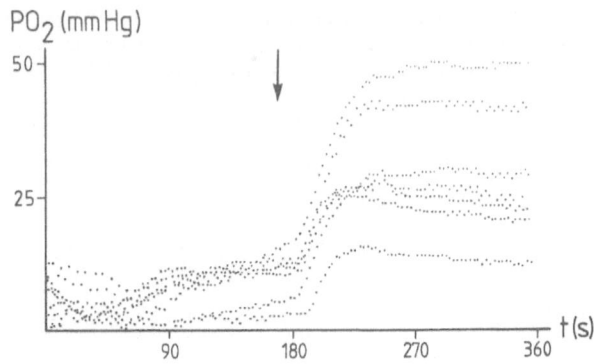

Fig. 3. Recording of the eight channels of a multiwire surface PO₂ electrode during the open-ing of the temporal vessel after an extra-intracranial arterial bypass operation. Note the in-crease in PO₂ values after opening of the vessel (indicated by the *arrow*)

electrode we used. In the tumor examinations it would be necessary to determine this factor in order to be able to evaluate the results.

3. Microcirculation. It is important that it is possible to judge the state of microcirculation from PO_2 measurements if both other factors are kept constant. In our examples, this situation was given during the measurements at AVM operations.

Clinical applications other than experimental studies seem possible, e.g., to prove a cortical artery to be the feeding vessel of an AVM. From our results one would expect a rise of PO_2 after temporary clipping of a feeding vessel, but a fall if a nutrient artery of the brain cortex were occluded.

In conclusion, the described method has the following advantages: It makes it possible to follow rapid changes of oxygen tension in brain tissue. Examinations as in the extra-intracranial arterial bypass case could not be performed with any other method except other electrode methods, as concerning time resolution. Physiologic observations accompanying manipulative procedures during operations can be performed, e.g., during controlled hypotension or hemodilution. The "disadvantage" of having to expose the cortex in fact appears a real advantage, considering the method can be used during operative procedures, in contrast to other methods (like PET or NMR) used for similar studies.

References

Grote J, Krüger-Biegner M, Schubert R, Zimmer K (1980) Significance of brain surface PO_2 measurements in determining oxygen supply conditions in the brain cortex. In: Kimmich HP (ed) Monitoring of vital parameters during extracorporeal circulation. Proc Int Conf, Nijmegen. Karger, Basel, pp 101–105

Hauss J, Schönleben K, Spiegel HU (1982) Therapiekontrolle durch Überwachung des Gewebe-PO_2. Eine tierexperimentelle und klinische Studie. (Aktuelle Probleme in der Angiologie Band 41). Huber, Bern Stuttgart Wien

Kessler M, Lübbers DW (1966) Aufbau und Anwendungsmöglichkeiten verschiedener PO_2-Elektroden. Pflügers Arch ges Physiol 291:82

Leniger-Follert E, Lübbers DW, Wrabetz W (1975) Regulation of local tissue PO_2 of the brain cortex at different arterial O_2 pressures. Pflügers Arch 359:81–95

Lübbers DW (1969) The meaning of the tissue oxygen distribution curve and its measurement by means of Pt-electrodes. Progr Resp Res 3:112

Quantitative Measurement of CBF in Single Pial Vessels: A Comparison of Techniques

L. M. AUER[1] and F. HAYDN[2]

The oldest technique for the investigation of the cerebral circulation is the cranial window technique. The cranium and the dura are opened and one is free to observe the surface vessels and the bloodstream within them.

The value of this method has been appreciated until today, and it is used in order to directly observe vessel reaction and regulatory functions. There has, however, always been justified criticism of conclusions and conjectures about cerebral blood flow drawn from such observations. A hypothesis has been put forward that, under special circumstances, intraparenchymal vessels can behave contrarily to the pial vessels in a compensatory way, thus jeopardizing a net effect on tissue perfusion. This hypothesis has in fact been confirmed under conditions of small variations of caliber.

The problem can be solved by measuring quantitative flow in the vessel under observation by means of calculation from two parameters, diameter and velocity. This "volume-flow" indicates flow quantity within a single vessel. The procedure excludes the possibility of erroneous results due to variations in blood viscosity, which one would risk by calculating flow with the aid of Poisseiulle's equation.

Internal diameters can be easily measured through the microscope by means of a micrometer scale; new technologies allow rapidly repeated or even continuous analysis (Auer and Haydn 1979; Baez 1966; Gotoh et al. 1982).

The technical problem is greater for the measurement of blood velocity within a pial vessel. Using slit noise analysis, Wiederhelm (1966) was able to measure flow velocity in very small vessels; Wayland and Johnson (1967) and Intaglietta et al. (1971) used photodetectors, a dual slit system, and a cross-correlation procedure. Röckemann and Plesse (1972) evaluated selected wavelength and superposed oscillations with the aid of a grating technique. Koyama et al. (1982) developed a laser Doppler system, also used for continuous monitoring of blood velocity within single pial vessels; this technique does not, however, allow calibration in vivo, thus providing only relative values.

Several groups have recently tried to solve the problem of simultaneous registration of vessel caliber changes and flow velocity: Busija et al. (1982) used automatic though intermittent measurement of absolute diameters. Relative changes in flow velocity were registered by means of a pulsed Doppler technique: a microcrystal was placed underneath large pial arteries of around

1 Universitätsklinik für Neurochirurgie, A-8036 Graz
2 Institut für Elektronik, Technische Universität, A-8010 Graz

Cerebral Blood Flow and
Metabolism Measurement.
Eds. Hartmann/Hoyer
© Springer-Verlag Berlin Heidelberg 1985

500 µm diameter. Since a calibration of the Doppler signal is not possible, this method provides intermittent data at short intervals on relative changes of flow within the vessel under observation. These flow values correlated highly ($r = 0.94$) with comparative measurements of flow in the underlying cortex using the microsphere technique. The limitations of this method are that only very large vessels can be used for observations and the additional invasiveness caused by placing the Doppler crystal under the vessel.

Gotoh et al. (1982) used the analysis of a video signal similar to our own technique to obtain absolute continuous values of vessel diameters and their variation. Flow velocity is intermittently obtained using TV densitometry and bolus injection of saline into the internal carotid artery retrogradely via the lingual artery. The intermittently obtained absolute values of flow velocity were compared with the values from an in vitro system, and a high correlation was found ($r = 0.98$).

Description of Present Techniques

In our own laboratory two different approaches have recently been tried to obtain continuous and absolute information on flow velocity changes. Diameter variations are continuously registered using the multichannel videoangiometer technique (Auer and Haydn 1979); with this method gray value changes on a single TV line as they occur between the bright cortical surface and a dark pial vessel are used for a calibrated oscillometer, the number of oscillations per unit of time of which are a measure of diameter. The image of the pial surface seen through a closed cranial window (Auer 1978) can be stored by a videotape recorder. Multiple replay thus allows analysis of many vessel portions. This technique has routinely been used in several hundred experiments (Fig. 1). The scanning by the TV pick up tube is done horizontally line by line; the blood vessel has to be positioned at an angle of 90 ° to the TV scan lines to get the true vessel diameter. This can be obtained by rotating the TV camera, mounted on the microscope. Simultaneous measurement of three diameters is possible during one videotape-run. All diameters with positions different from 90 ° have to be calculated. Measurement of the angle between the vessel and the TV scan lines can be done automatically from the TV signal. The numerical value is put into the computer and used for calculation of the true vessel diameter. The diameter signal can be used for analogue display on a penwriter (Fig. 2) as well as for calculation of volume-flow together with velocity.

Slow flow velocity in small vessels up to 40 µm diameter was measured by means of a cross-correlation technique of video signals; the latter were obtained from a video-window of a defined length and width placed over the vessel portion under observation. The change in red blood cells and plasma gaps in a blood-vessel results in an optical density wave form at the intersection of the TV scan lines with the longitudinal axis of the vessel. This optical density wave form is sampled at two different successive moments, stored, and calculated off line. The calculation of the optical wave form shift is done by spatial cross-

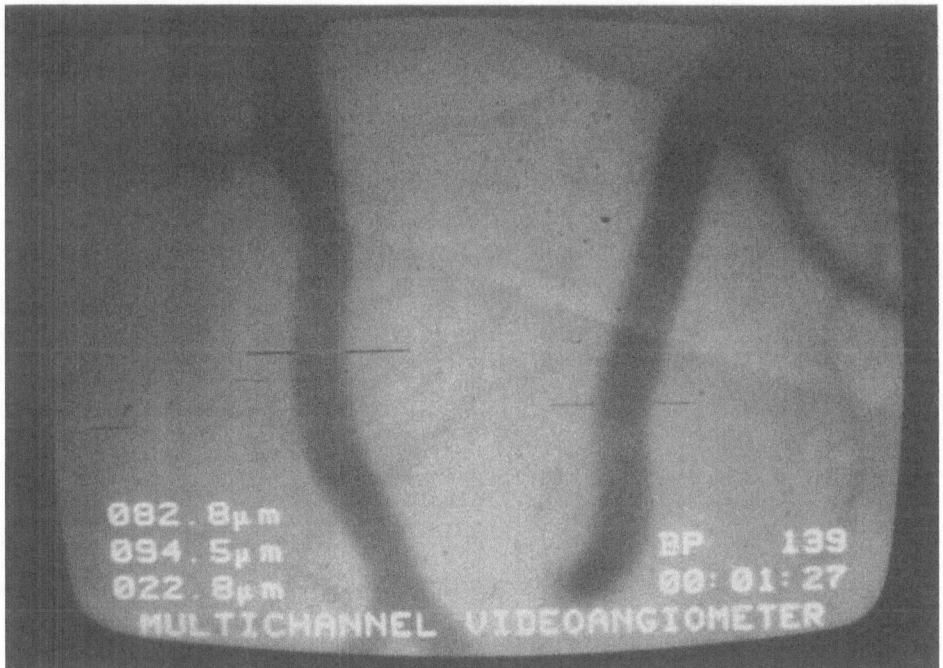

Fig. 1. TV-microscope picture of pial vessels in vivo. Diameters are continuously recorded on vessel portions marked by a *black bar* (display in μm at *lower left*)

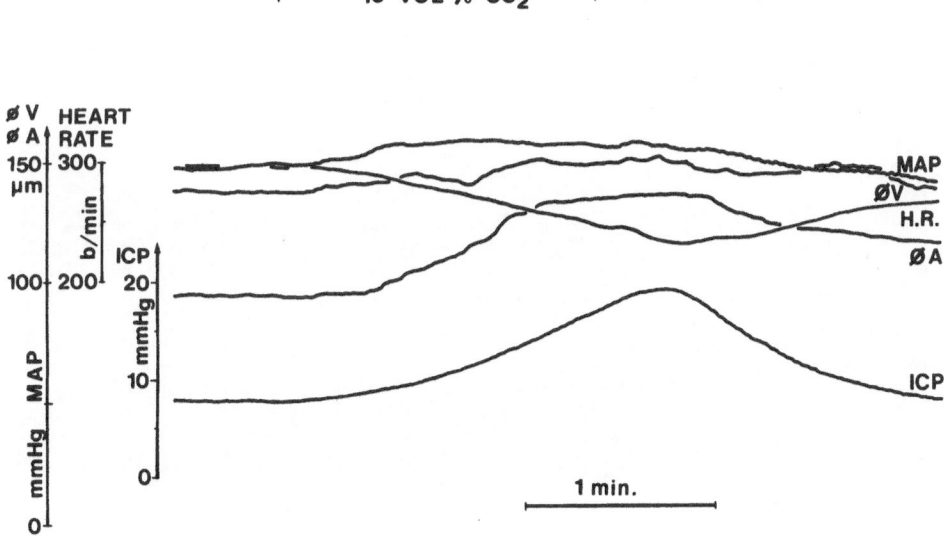

Fig. 2. Recording of diameter changes of a 95-μm pial artery ($\varnothing A$) and a 140-μm pial vein ($\varnothing V$) during an episode of hypercapnia

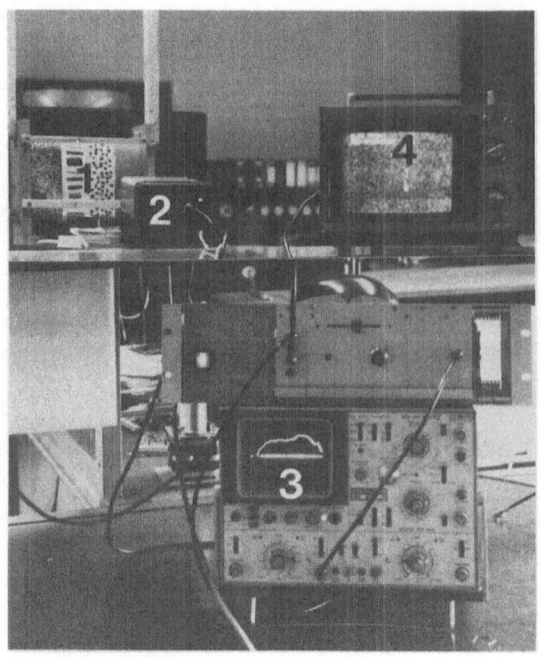

a

Fig. 3. a Experimental setup for
continuous estimation of flow velocity by means of TV techniques.
1, rotating samples for simulation of
granular flow; *2*, TV camera; *3*,
oscilloscopic picture; *4*, TV picture
of granular flow with video-window.
b Magnified TV picture with video-
window and display of flow velocity
in mm/s

b

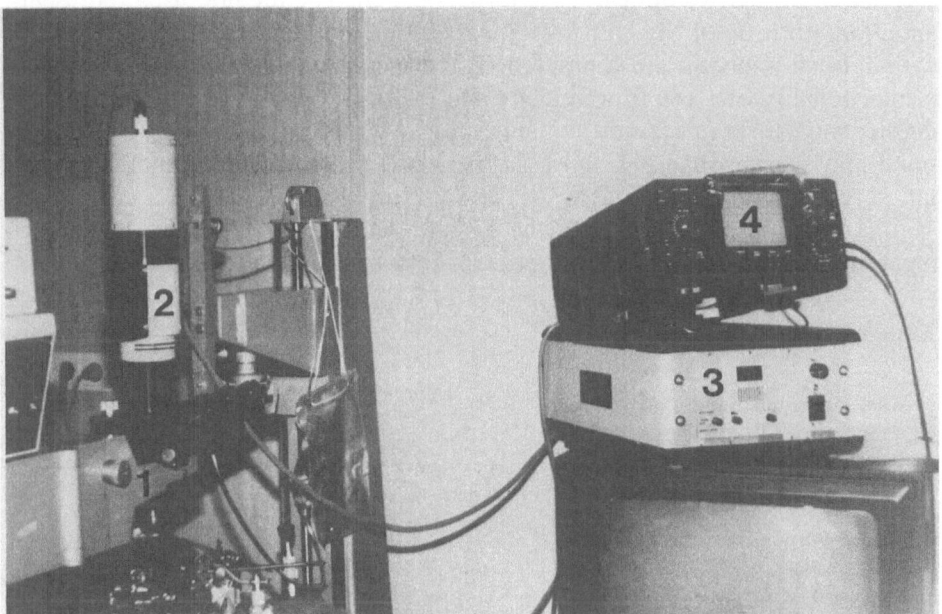

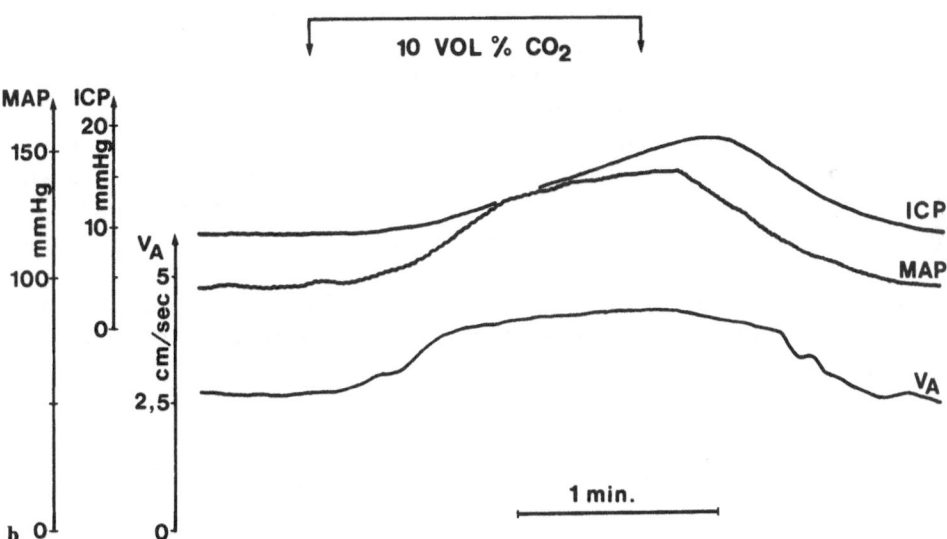

Fig. 4. a Experimental setup of optical correlation technique for measurement of flow velocity. *1*, microscope; *2*, photometer; *3*, correlator with digital display of velocity; *4*, oscillometer. **b** Penwriter recording of blood velocity in a pial artery (V_A) (drawing) in cm/s

correlation of the two optical wave forms (Fig. 3). The method used represents an automatization of the known frame-to-frame method where pictures, taken with defined intervals, are compared. The principle of this method is the direct proportionality between flow velocity and the displacement of a certain optical density wave form in a fixed time interval. A TV picture can be seen as the temporal and spatial quantitation of the recorded process. The temporal quantitation results from the frequency of the TV frames (i.e., 25/50 per second). The spatial quantitation results from the distance of the TV scan lines. The movement of an object down the TV screen can be described as follows:

$$V[mm/s] = \frac{x \cdot d}{n \cdot \Delta t} [mm/s]$$

v = velocity of the object
x = number of TV scan lines
d = distance between TV scan lines
n = number of frames
Δt = time between frames

Errors of this method are *smaller* than errors of the well-known two-window method (Wayland and Johnson 1967 b), because here a distance is measured. The error in measuring the distance depends on inexact calculation of X_{max} by spatial correlation. X_{max} corresponds to τ_{max} when measuring time. The smaller the similarity of the signal of the optical density wave form between two samples, the less exact is X_{max}. As the similarity grows with Δt, Δt has to be chosen as small as possible in accordance with the time required for calculation. The quality of velocity measurements also depends on the contrast between red blood cells and plasma gaps. Disturbing reflections can be eliminated by special polarization techniques (Figs. 2, 3).

For flow velocity measurement in larger vessels, a cross-correlation technique using grating and photometric analysis was used (Leitz) (Fig. 4). Again, a window of a defined length and width can be placed into the vessel under observation in the microscope. The signal for evaluation is obtained by a photometer.

By calculating flow from volume and velocity, absolute and continuous values of flow through the vessel portion under observation will be obtainable with both techniques.

References

Auer LM (1978) The pathogenesis of hypertensive encephalopathy. Acta Neurochir Suppl 27:1−111
Auer LM, Haydn F (1979) Multichannel videoangiometry for continuous measurement of pial microvessels. Acta Neurol Scand 60 [Suppl 72], 208−209
Baez S (1966) Recording of microvascular dimensions with an image splitter television microscope. J Appl Physiol 21:299−301
Busija DW, Marcus ML, Heistad DD (1982) Pial artery diameter and blood flow velocity during sympathetic stimulation in cats. J Cereb Blood Flow Metab 2:363−367

Gotoh F, Muramatsu F, Fukuuchi Y, Okayasu H, Tanaka K, Suzuki N, Kobari M (1982) Video camera method for simultaneous measurement of blood flow velocity and pial vessel diameter. J Cereb Blood Flow Metab 2:421–428

Intaglietta M (1971) Pulsatile velocity components in the omental microvasculature. Proc. 6th Europ Conf Microcirculation, Aalborg 1970, pp 74–76 Karger Basel–New York

Koyama T, Horimoto M, Mishina H, Asakura T (1982) Measurement of blood flow velocity by means of a laser Doppler microscope. Optik 61:411–426

Rückemann W, Plesse GJ (1973) Registration of erythrocyte velocity by selecting wavelength and superposition of oscillations. Proc. 7th Europ Conf Microcirculation, Aalborg 1972, pp 50–54, Karger Basel-New York

Wayland H, Johnson PC (1967) Erythrocyte velocity measurement in microvessels by correlation method. Proc. 4th Europ Conf Microcirculation, Cambridge 1966, Bibl. anat., vol. 9, pp 160–163, Karger Basel-New York

Wayland H, Johnson PC (1967b) Erythrocyte velocity measurements in microvessels by a two slit photometric method. J Appl Physiol 22:333–337

Wiederhielm CA (1966) Photospectrum analysis technique. Methods in Medical Research, vol. 11, pp 212–216, Yearbook Med Publ Chicago

Regional Cerebral Blood Flow in Experimental Chronic Hypoxic Hypoxia

H.-M. Mayer, E. Fritschka, and J. Cervós-Navarro

Introduction

Acute hypoxia increases cerebral blood flow (CBF), overriding the vaso-constrictive effect of the associated hyperventilation and hypocapnia (Kety and Schmidt 1948). However, during longer periods of hypoxia, the homeostasis of additional CBF regulators involving metabolic needs, hemodynamic and hemorrheologic factors as well as structure and function of cerebral vessels may be considerably changed. We therefore studied regional CBF in cat brain after a period of stepwise adaptation to decreasing inspiratory oxygen contents as a model of chronic hypoxic hypoxia.

Methods

Adult mongrel cats (body weight $3.0 - 3.5$ kg) were exposed to stepwise de-creasing oxygen concentrations (from 21 vol% through 15, 12, and 10 down to 8 vol%) over a period of 4 months. During this time, cats were held in Plexiglas cabins which were supplied by various mixtures of humidified room-air and ni-trogen, to keep oxygen contents at the desired levels (Fig. 1).

Hourly gas-exchanges were adjusted to keep CO_2-content below 0.5 vol%. The control group was held under similar conditions but supplied with room-air only. Hematocrit values were determined regularly at the start of the exper-iment as well as before each oxygen-decreasing step. At the end of the last ex-perimental period (i.e., 40 days at 8 vol% oxygen) regional CBF and cardiac output were measured in nitrous-oxide anesthesia. Chronic hypoxic cats and control cats were measured under normoxic (paO_2, 100 mmHg) as well as under hypoxic (paO_2, 30 mmHg) conditions. During the measurements, $paCO_2$, sys-temic arterial pressure (SAP), and body temperature were kept within the nor-mal ranges. Measurements of regional blood flow distribution were carried out using radioactively labeled microspheres according to the method described by Fritschka et al. (1981) and Heiss and Traupe (1981). Microspheres 15 µm in di-ameter and labeled with scandium 46, ruthenium 103, cerium 141, and chro-mium 51 were used in each experiment. Brain tissue samples were taken from

Neurochirurgische Klinik, Freie Universität Berlin, Klinikum Steglitz, Hindenburgdamm 30
D-1000 Berlin 45

Cerebral Blood Flow and
Metabolism Measurement.
Eds. Hartmann/Hoyer
© Springer-Verlag Berlin Heidelberg 1985

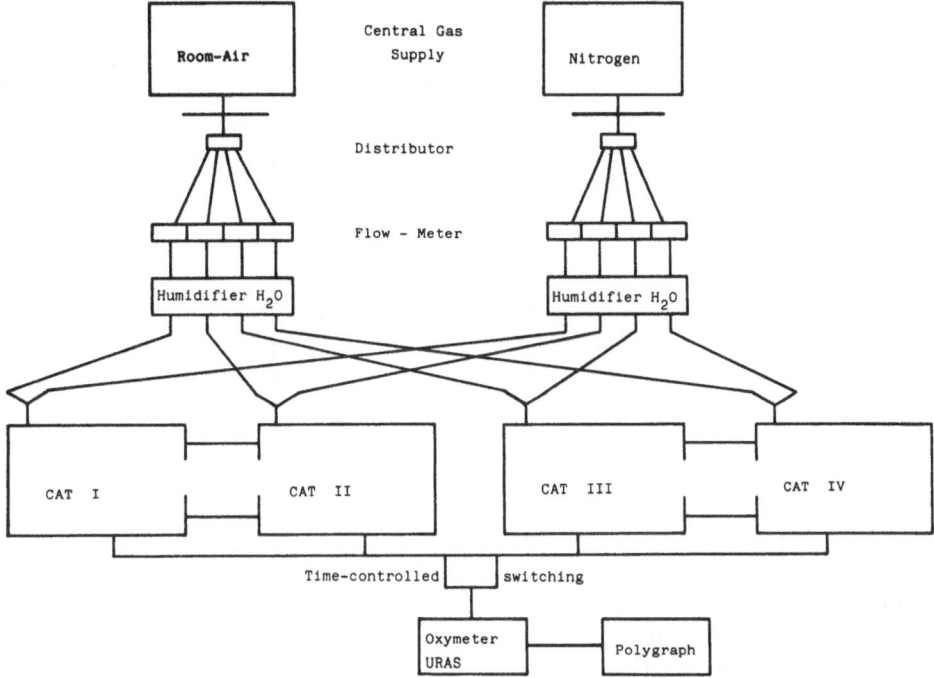

Fig. 1. Experimental arrangement for chronic exposure of cats to decreasing inspiratory oxygen contents

gray and white matter, weighed, and then counted in a gamma-counter. Regional CBF values were calculated in terms of ml/100 g wet weight/min as well as in terms of % of cardiac output/100 g. Using the microsphere method for CBF measurements offers the following advantages over other techniques:

1. Higher resolution of regional CBF as compared with the xenon clearance method
2. More areas atraumatically measurable as compared with the hydrogen clearance method
3. Possibility of simultaneous measurement of cardiac output and distribution of flow as % of cardiac output in different organs as well as in different areas of one organ
4. Possibility of sequential measurements in anesthetized as well as in nonanesthetized animals.

Results and Discussion

1. *Morphology*. Macroscopic and microscopic examination of the brain of adapted cats revealed striking hyperemia with an obvious increase of microvessels (arterioles, capillaries, and venules) on cross-sections. However, since there were only slight signs of vasoproliferation, such as glomerulus-like vessel formations, this result should be confirmed morphometrically.

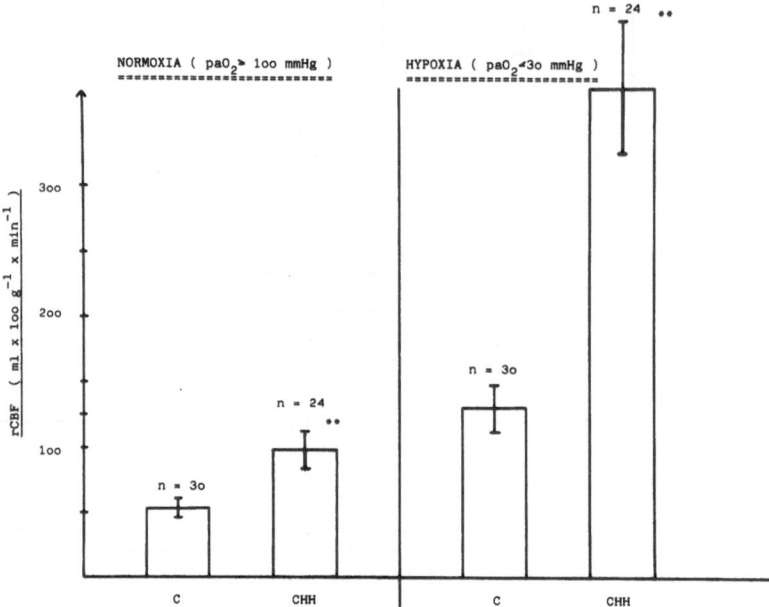

Fig. 2. Regional cerebral blood flow in normoxia and hypoxia in control cats (C) and hypoxia-adapted cats (CHH). $n =$ number of tissue samples. Mean $\pm$ SEM are given. * $P < 0.05$, ** $P < 0.01$; unpaired t-test. Gray matter

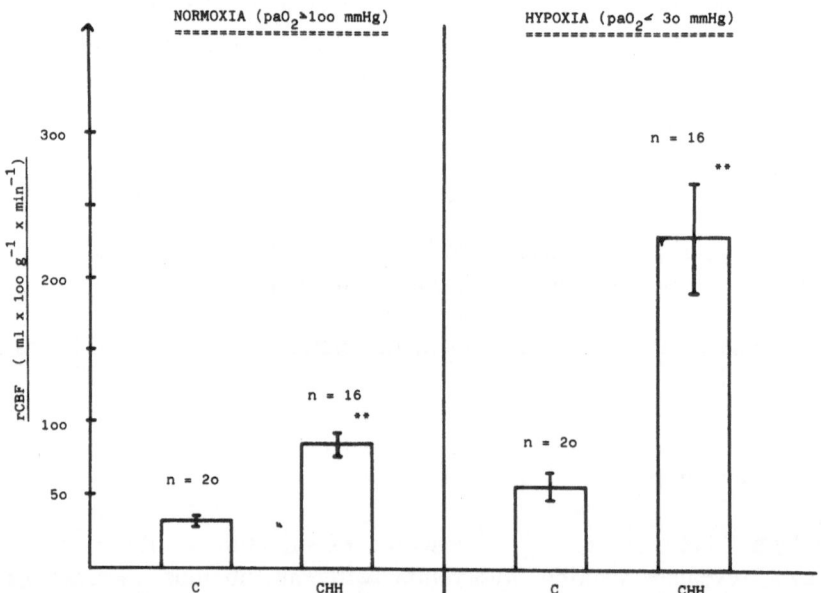

Fig. 3. Regional cerebral blood flow in normoxia and hypoxia in control cats (C) and hypoxia-adapted cats (CHH). $n =$ number of tissue samples. Mean $\pm$ SEM are given. * $P < 0.05$. ** $P < 0.01$; unpaired t-Test. White matter

Table 1. Regional cerebral blood flow in hypoxia and normoxia in cats adapted to chronic-hypoxic hypoxia (CHH) and control cats (C). Means ± SEM are given

| | Hypoxia $paO_2 < 30$ mmHg | | | | Normoxia $paO_2 > 100$ mmHg | | | |
| | CBF gray | | CBF white | | CBF gray | | CBF white | |
	ml/100 g/min	% CO/100 g	ml/100 g/min	% CO/100 g	ml/100 g/min	% CO/100 g	ml/100 g/min	% CO/100 g
CHH $n=5$	$373.6\pm51.6**$ $n=24$	34.2 ± 4.4 $n=24$	$228.6\pm42.3**$ $n=16$	18.9 ± 4.5 $n=16$	$99.2\pm13.1**$ $n=24$	16.2 ± 2.8 $n=24$	$85.5\pm8.2**$ $n=16$	12.4 ± 1.4 $n=16$
C $n=5$	129.2 ± 6.5 $n=30$	23.9 ± 5.2 $n=30$	56.2 ± 7.7 $n=20$	28.4 ± 8.4 $n=20$	51.5 ± 5.8 $n=30$	10.6 ± 1.7 $n=30$	34.1 ± 2.4 $n=20$	6.1 ± 0.6 $n=20$

$* P<0.05, \quad ** P<0.01$; unpaired t-test

2. *Blood Viscosity.* Besides fibrinogen as a blood-serum constituent, he-matocrit, representing the corpuscular part of blood, is the main factor de-termining blood viscosity (Grotta et al. 1982). While hematocrit of chronic hy-poxic cats increased continuously to 56% at the end of the last experimental period (i.e., 40 days at 8 vol% oxygen), hematocrit of controls remained at a mean level of 31.9%. Hematocrit values of more than 53% have previously been shown to lead to a global decrease of rCBF in the rat (Iadecola et al. 1983). Since it is known that vasoconstriction due to initial hyperventilation is over-ridden by hypoxic vasodilatation (Betz 1965), increasing hematocrit might be the main factor counteracting a chronic increase of rCBF.

3. *Cerebral Blood Flow* (Figs. 2, 3). Values of rCBF in gray and white matter of the experimental and control groups are shown in Table 1. During normoxia as well as in hypoxia, CBF of hypoxia-adapted animals is significantly higher as compared with controls. This can be shown in gray and white matter. Whereas the possibility to increase CBF in hypoxia is between 84% and 150% in controls, chronic hypoxic cats can increase their flows by between 167% and 280%, though starting from an initially higher CBF value (Table 1; Figs. 2, 3) with an altered blood viscosity. Since cardiac output of hypoxia-adapted ani-mals is significantly higher during hypoxia, rCBF was also analyzed as % of cardiac output per 100 g wet weight (Table 1). Under normoxic conditions, brains of hypoxia-adapted cats receive a higher percentage of cardiac output in gray as well as in white matter. Though there is an increase of rCBF/% cardiac output/100 g in hypoxia in both experimental groups, values do not differ sig-nificantly from each other.

Conclusions

1. Experimental animals exposed to chronic hypoxic hypoxia show only slight morphological signs of vasoproliferation.

2. Blood viscosity is severely increased by high hematocrit values of more than 56%.

3. Using the microsphere method, rCBF is absolutely higher in animals adapted to chronic hypoxia in normoxia as well as in hypoxia. This finding can be correlated to an obviously increased number and diameter of brain microvessels.

4. There is a decrease of CBF in adapted cats while returning from hypoxic to normoxia, but absolute values are still higher than in controls. This leads to the assumption that restoration of normal vascular tone might be only partially preserved.

5. There is a marked increased of cardiac output in chronic hypoxic hypoxia in the presence of partially disturbed autoregulation which may possibly be in-volved in the high CBF values measured. We are therefore currently investigat-ing the autoregulatory capacity of these animals in hemorrhagic hypotension also.

References

Betz E (1965) Adaptation of regional cerebral blood flow in animals exposed to chronic alterations of PO_2 and PCO_2. Acta Neurol Scand 41 Suppl 14:121–128

Fritschka E, Artigas J, Shigeno T, Minguillon C, Cervos-Navarro J (1981) Regional cerebral blood flow after occlusion of the middle cerebral artery in cats: Correlation of blood flow and specific gravity at 24 to 48 hours post occlusion. In: J Cervos-Navarro, E Fritschka (eds) Cerbral microcirculation and metabolism. Raven press New York, p 433–441

Grotta J, Ackerman R, Correia J, Fallick G, Chang J (1982) Whole blood viscosity parameters and cerbral blood flow. Stroke 13:296–301

Heiss WD, Traupe H (1981) Comparison between hydrogen clearance and microsphere technique for rCBF measurement. Stroke 12:161–167

Iadecola C, Reis DJ, Underwood M, Fieschi C (1983) Regional cerebral blood flow in relationship to variations in hematocrit in rat. J Cerebr Blood Flow and Metabolism 3, Suppl 1:S636–637

Kety SS, Schmidt CF (1948) Effects of altered arterial tensions of carbon dioxide and oxygen in cerebral blood flow and cerebral oxygen consumption in young men. J Clin Invest 27:484

Nuclear Magnetic Resonance for Vessel Anatomy, Blood Flow, and Metabolism Studies

A. GANSSEN

The first work directed toward the application of nuclear magnetic resonance (NMR) to medical problems was probably done in 1956 at the Laboratory of Technical Development, National Heart Lung and Blood Institute (Bethesda, Ma.) by R.L. Bowman (1), who was already interested in the noninvasive measurement of blood flow in vivo. Several schemes for the measurement of blood flow with NMR were published by J.R. Singer in 1959 [2] and 1960 [3].

A clinical application of NMR to the measurement of human blood flow has not been published as yet in spite of several proposals for improvement [4–6]. Since the development of NMR imaging methods [6–9], and especially since the installation of an increasing number of clinical NMR imaging units, interest in NMR blood flow measurement and flow imaging has been growing. Prior to the discussion of flow effects some basic facts concerning NMR should be recapitulated.

All atomic nuclei containing odd numbers of protons or neutrons or of both possess a mechanical spin and a magnetic moment.

Figure 1 is a table of the NMR parameters of some atomic nuclei which are important in the human organism. The most important is the hydrogen nucleus because of its abundance and easy detection. When exposed to an ambient magnetic field, these nuclei will adjust by assuming magnetic energy states separated by an energy gap which is proportional to the magnetic field applied (Fig. 2).

In condensed matter like a liquid or a solid there will be a dynamic interaction between neighboring particles given by the thermal motion at temperatures above 0 °K. An equilibrium condition between the magnetic and the thermal energy of the atomic system will be established according to the Boltzmann factor.

The originally randomly oriented spins will come to a new equilibrium distribution after exposure to a magnetic field, with a speed which is proportional to the strength of the mutual intraction – or, in other words – to the coupling with the atomic and molecular lattice. The time behavior of the magnetization is described by the exponential function:

$$M_z = M_0 \left(1 - e^{-\frac{t}{T_1}}\right)$$

Siemens AG, UB Med, Henkestraße 127, D-8520 Erlangen

Cerebral Blood Flow and
Metabolism Measurement.
Eds. Hartmann/Hoyer
© Springer-Verlag Berlin Heidelberg 1985

Isotope	Spin Unit $\frac{h}{2\pi}$	Energy Levels	NMR – Frequency in MHz at 1T	Relative sensitivity at constant B_0	Natural abundance in %
^{1}H	$-1/2$, $+1/2$, $J=1/2$	$+\mu B_0$ / $-\mu B_0$	42.577	1.00	99.985
^{2}H	-1, $+1$, $J=1$	$+\mu B_0$ / 0 / $-\mu B_0$	6.536	0.0096	0.015
^{7}Li	$-3/2$, $-1/2$, $+1/2$, $+3/2$, $J=3/2$	$+\mu B_0$ / $+1/3\,\mu B_0$ / $-1/3\,\mu B_0$ / $-\mu B_0$	16.547	0.294	92.57
^{13}C	$-1/2$, $+1/2$, $J=1/2$	$+\mu B_0$ / $-\mu B_0$	10.705	0.0159	1.108
^{19}F	$-1/2$, $+1/2$, $J=1/2$	$+\mu B_0$ / $-\mu B_0$	40.055	0.834	100
^{23}Na	$-3/2$, $-1/2$, $+1/2$, $+3/2$, $J=3/2$	$+\mu B_0$ / $+1/3\,\mu B_0$ / $-1/3\,\mu B_0$ / $-\mu B_0$	11.262	0.0927	100
^{31}P	$-1/2$, $+1/2$, $J=1/2$	$+\mu B_0$ / $-\mu B_0$	17.235	0.066	100

Nuclear properties important for NMR of nuclei of possible interest in medical applications

Fig. 1. Important NMR data of some stable isotopes

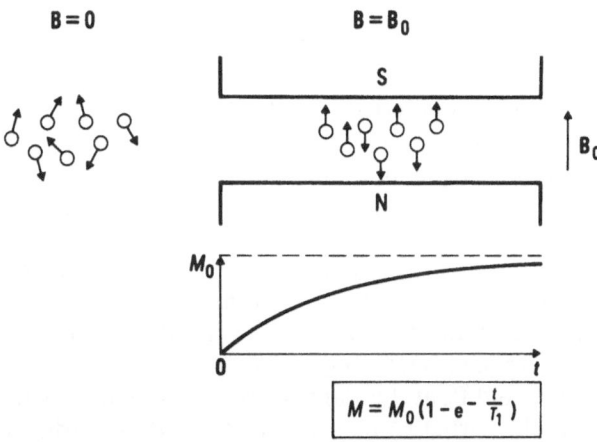

Time dependence of the nuclear magnetization M in a static magnetic field B_0.

The time constant T_1 represents the longitudinal nuclear magnetic relaxation time or spin lattice relaxation time.

Nuclear magnetization in static magnetic field B_0

Fig. 2. The time dependence of the nuclear magnetization. The characteristic time constant for the nuclear magnetization process is the spin-lattice relaxation time T_1

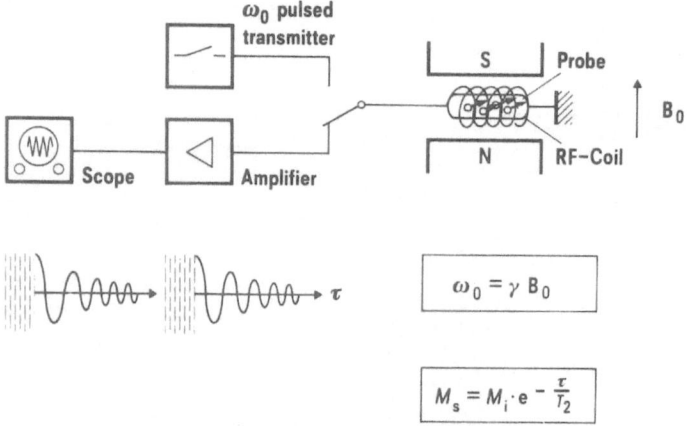

Time dependence of the transverse nuclear magnetization rotating with the Larmor frequency after R.F. – pulse excitation.

The time constant T_2 is the transverse nuclear magnetic relaxation time or spin – spin – relaxation time, which characterizes the R.F. – signal decay or "Free induction decay".

Fig. 3. The NMR experiment. The nuclei, i.e. protons in water, are contained within an RF coil which is localized within a static magnetic field B_0. After RF excitation at the characteristic precession frequency $\omega = \gamma\, B_0$ the protons continue to precess around B_0, thereby exciting an RF voltage within the RF coil which can be amplified and displayed as the NMR signal. Attenuation of the NMR signal with time takes place with the characteristic time constant T_2, the spin-spin relaxation time

with the characteristic time constant T_1, which is the longitudinal or spin lattice relaxation time of the nuclear magnetization.

For hydrogen nuclei or protons within the human body T_1 can be in the order of 1 s in the case of the cerebrospinal fluid or blood plasma and less then 1 ms for protons bound to macromolecules.

The nuclear magnetization of a material in a static magnetic field can be meesured by applying a precisely controlled amount of RF energy to the nuclei of interest and then measuring the RF signal radiated as these nuclei relax to their original state (Fig. 3). The RF field, B_1, is transmitted perpendicular to the static field B_0 at the Larmor frequency, which is the natural precession frequency of the particular nuclei in the B_0 field. According to physical principles that also apply to the gyroscope, the precession frequency ω_0 is proportional to the force tending to change the direction of the spins. This force is given by the magnetic field B_0, and the gyromagnetic ratio γ, which is different for each nuclear species. Accordingly, $\theta_0 = \gamma\, B_0$.

The total of all the nuclear magnetic moments can be viewed as a single net magnetization vector for the material reaching maximum value after sufficiently long exposure (several times T_1) to the static field B_0. An efficient NMR measurement is performed by applying just enough RF energy with the field B_1 at the RF frequency ω_0 to rotate the net magnetization vector into a

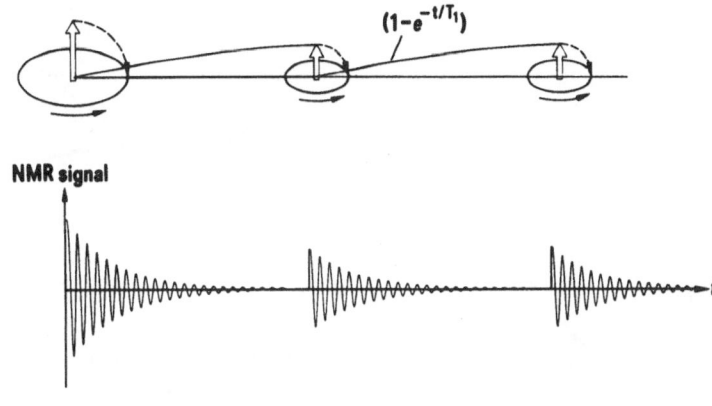

Sequence of 90° RF pulses

Fig. 4. Pulse series produced by periodic excitation of the nuclear spin system. When the repetition time between pulses $T_{rep} < T_1$ the initial NMR signal amplitude decreases

plane perpendicular to the static field B_0, then turning off the RF power and measuring the RF signal radiated by the material at the frequency ω_0. The observed signal amplitude is proportional to the number of spins within the sample. The RF signal decays with the transverse relaxation time constant, T_2, which is also called the "spin-spin" relaxation time. In pure low viscosity liquids, T_2 can be close to T_1. Usually, however, T_2 is less than T_1.

Compared to other substances within the living body the relaxation times of blood and CSF are long (at 10 MHz: T_1 for blood ~ 0.8 s, and brain $0.3-0.4$ s), thus facilitating the discrimination of the liquid components against the surrounding tissue without artifical contrast enhancement.

If a series of RF pulses is being applied to the spin system with a time interval which is short compared with the relaxation time T_1, according to the usual NMR imaging procedure, then the spin system cannot be remagnetized totally in the time interval between the pulses. The result is a reduced NMR signal, as we see in Fig. 4. If we perform the hypothetical experiment of supplying new polarized protons to the NMR apparatus, the signal amplitude will grow with flow velocity approaching asymptotically the initial amplitude (i.e., observed for long times between pulses) and then fall as the velocity becomes so high that unmagnetized protons enter the measurement volume. We performed this type of flow experiment 16 years ago and observed the predicted variation of signal amplitude with flow velocity: first an increase of signal and then a decrease with increasing flow velocity (Fig. 5).

A two-dimensional NMR imaging experiment takes place within a single slice of the body. Slice selection is accomplished by applying simultaneously a linear field gradient and a 90 ° RF pulse of proper shape and frequency in such away that all the static nuclear spins within the slice will contribute to the

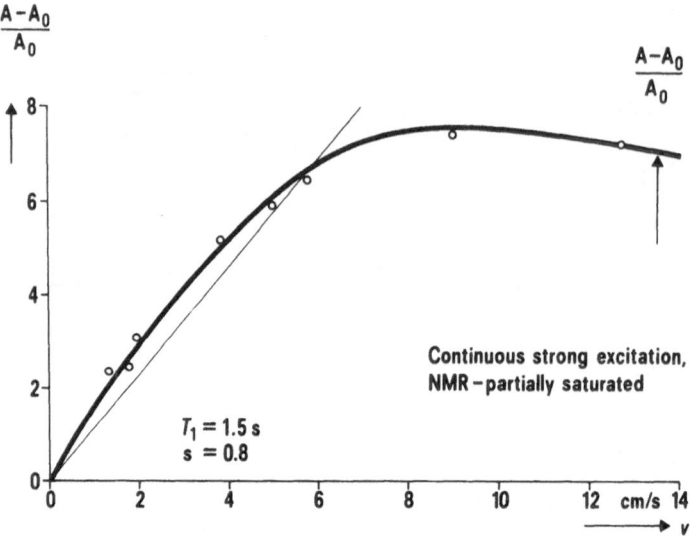

Fig. 5. Change in the relative NMR signal amplitude as a function of flow velocity for $T_{rep} < T_1$. There is an increase in signal amplitude until all nuclei within the coil volume are replaced by new fully magnetized protons

signal (Fig. 6). If the repetition time of the RF pulses is short compared with T_1 then the signal amplitude is reduced by partial saturation during previous RF pulses. For this situation, flow through the slice will increase the number of fully magnetized spins (i.e., those not affected by previous pulses) and increase the signal until the whole flow cross-section within the slice is filled with magnetized protons. As flow velocity increases still more, the NMR signal decreases because the flowing protons no longer remain in the plane of slice long enough to complete a full imaging cycle.

This can be tested with a phantom, a container filled with water penetrated by parallel pipes through which water flows at different velocities (Fig. 7). In the transaxial NMR image, which was taken with a repetition time short compared with the relaxation time T_1 of water, an increase in brightness corresponding to an increase in NMR signal can be seen at low flow velocities – corresponding to the inflow of spins with the magnetization unspoiled by previous RF pulses.

Figure 8 shows the NMR signal strength as a function of flow velocity obtained with transaxial imaging of this phantom from the investigation of Dr. Deimling of our NMR group. The curve agrees well with the function obtained by computer simulation.

In Figs. 9 and 10 examples of transaxially displayed veins with a comparatively slow blood flow are depicted. The sagittal sinus (Fig. 9) can easily be seen due to what is sometimes called "paradox enhancement" of the NMR signal. Figure 10 shows the lateral lacunae of the sagittal sinus.

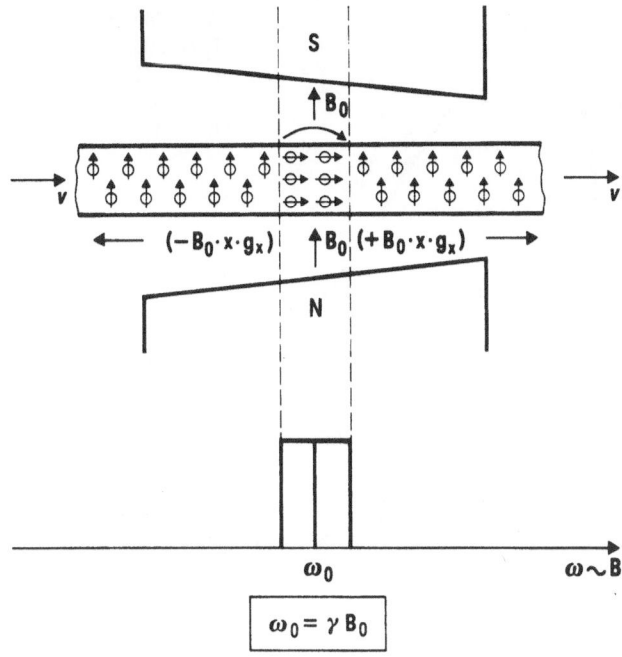

Fig. 6. Slice selection perpendicular to flow. For $T_{rep} < T_1$ there can be an increase in NMR signal amplitude until all "saturated" protons within the selected slice are substituted by the newly entering "unsaturated" protons

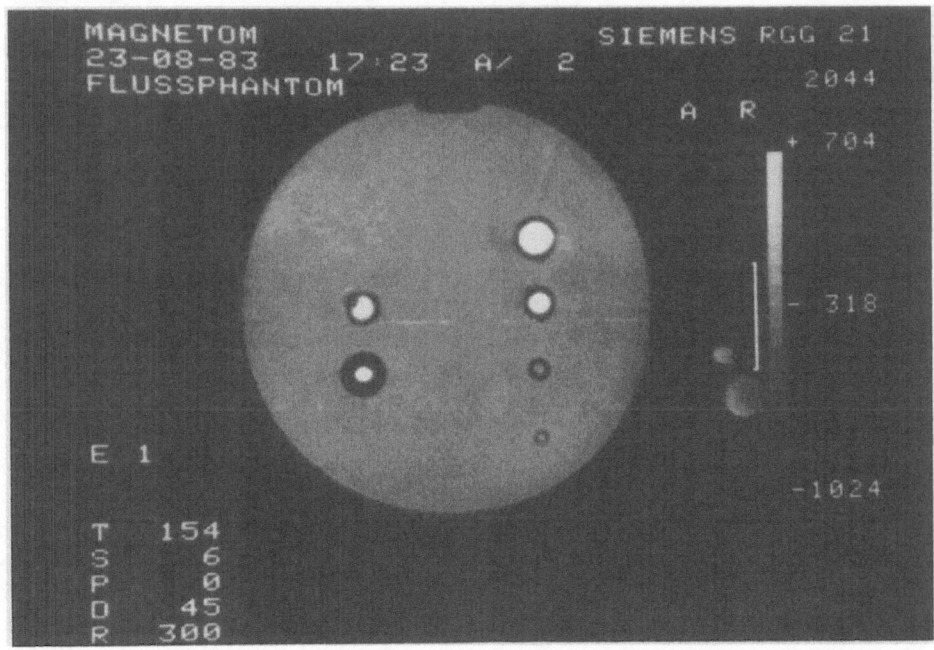

Fig. 7. NMR image of flow phantom with different flow velocities within the four tubes. For slow flow velocities an increase in brightness can be observed

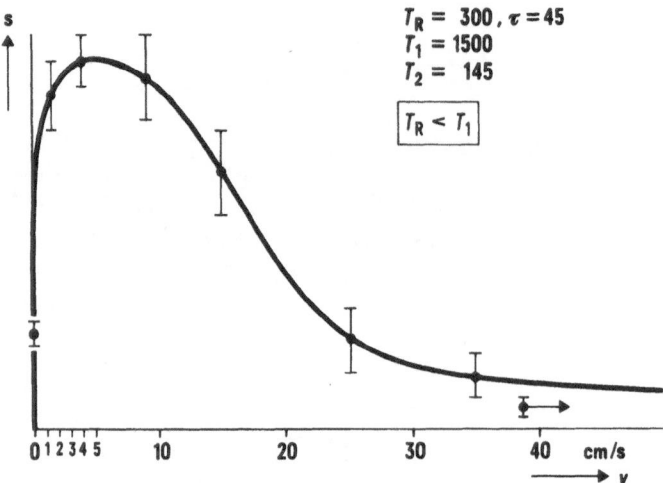

Fig. 8. NMR signal amplitude as function of flow velocity in NMR imaging experiment ($T_{rep} < T_1$). With faster flow the NMR signal decreases since the protons are no longer exposed to a full imaging cycle

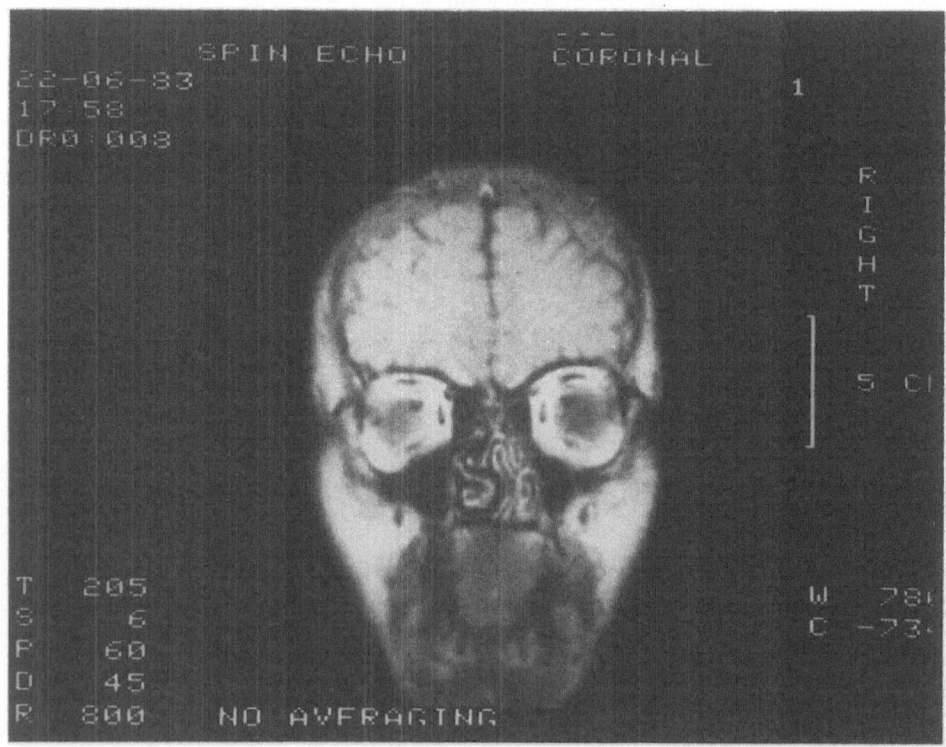

Fig. 9. Coronal head scan with the slowly flowing blood within the sagittal sinus appearing bright. (By courtesy of Dr. Gado, Mallinckrodt Institute, St. Louis, USA)

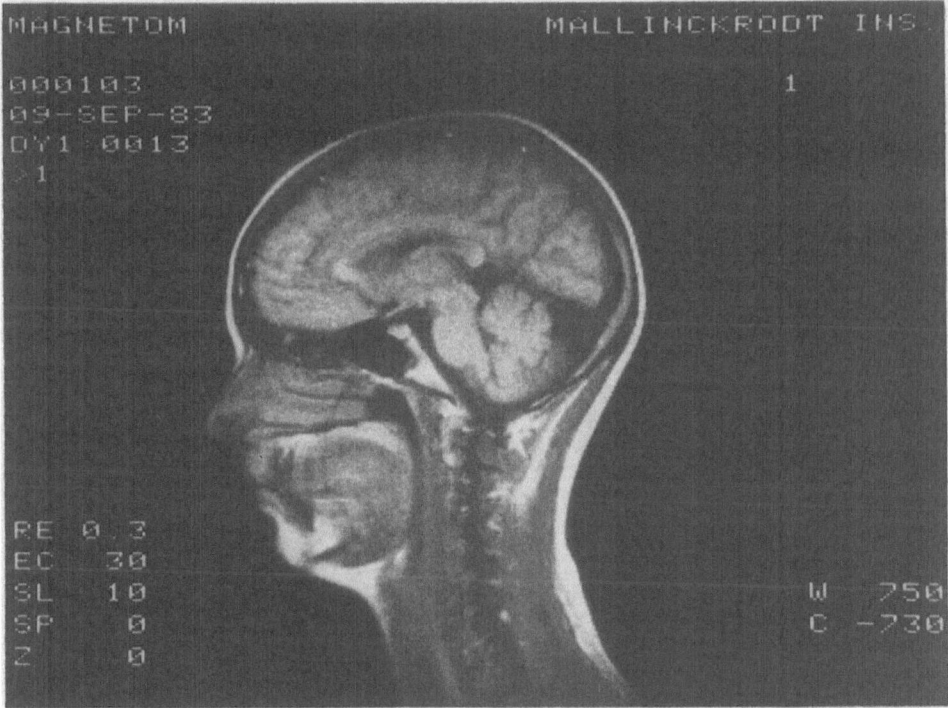

Fig. 10. Sagittal head scan with the slow venous blood flow through the lateral lacunae showing up bright (By courtesy of Dr. Gado, Mallinckrodt Institute, St. Louis, USA)

Figures 11 and 12 represent transaxial scans through the upper neck. The blood flow in all major blood vessels is displayed with sufficient contrast relative to the surrounding tissue. All arteries appear black. Only the comparatively slow flow in the jugular veins shows up light.

For a quantitative blood flow determination the NMR scan could be performed ECG-triggered with a successive digitalized signal read out, which can be converted into flow velocity via the signal/flow-velocity function.

For the longitudinal imaging of blood flow the contrast mechanism is different. Figure 13 shows the pulse sequence for the "spin echo mode" which is most often used for NMR imaging. The formula for the signal strength of the first and the second echo signal is also given.

After a selective 90 ° pulse there is a series of 180 ° pulses causing a periodic refocussing of the phasing of the spin precession showing up as "spin echos" in the case of stationary protons (9). Whenever the spins move during this train of pulses within a field distribution which may be arbitrary or dominated by intentionally provided field gradients, a phase refocussing of the spin precession will not occur. In this case the spin echo train will decay much faster, thus pro-

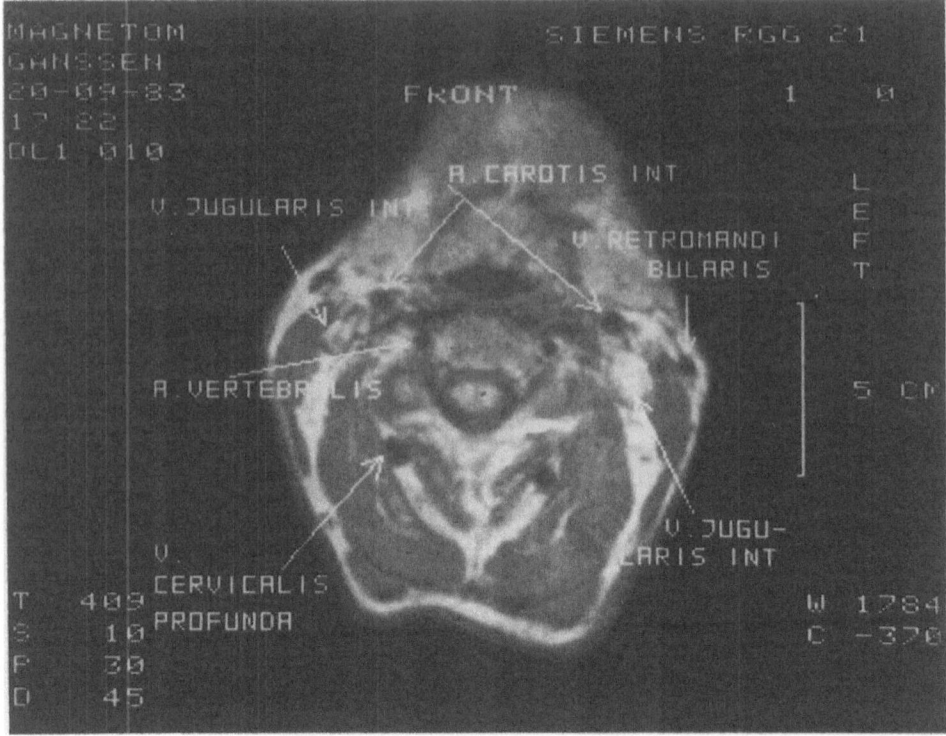

Fig. 11. Transaxial scan through the human neck. All arteries appear black. The slow flow within the jugular vein appears light

viding a sensitive means of measuring spin motion with respect to the field distribution. In the formula for the signal amplitude the described effect gives rise to an effective shortening of the relaxation time T_2. For a given magnetic field distribution the signal will decay faster as flow velocity increases, as can be seen in longitudinal water phantom scans made by Dr. Deimling (Fig. 14).

A cylindrical water tube is positioned in a container filled with water. The flow velocities in the tube are 0, 3, 7, and 35 cm/s.

Figures 15 and 16 give examples of the imaging of the carotis artery. Figure 16 is a scan which was done by Dr. Huk of the Neurological Hospital of Erlangen-Nürnberg University.

Figure 17 shows the a. vertebralis especially well. The jugular vein in Fig. 18 appears with less contrast, probably because of the considerably slower blood flow.

A complete follow-up of the vessels will be possible when true three-dimensional NMR imaging becomes a reality. At present a simultaneous multislice technique with up to 15 parallel slices has been accomplished.

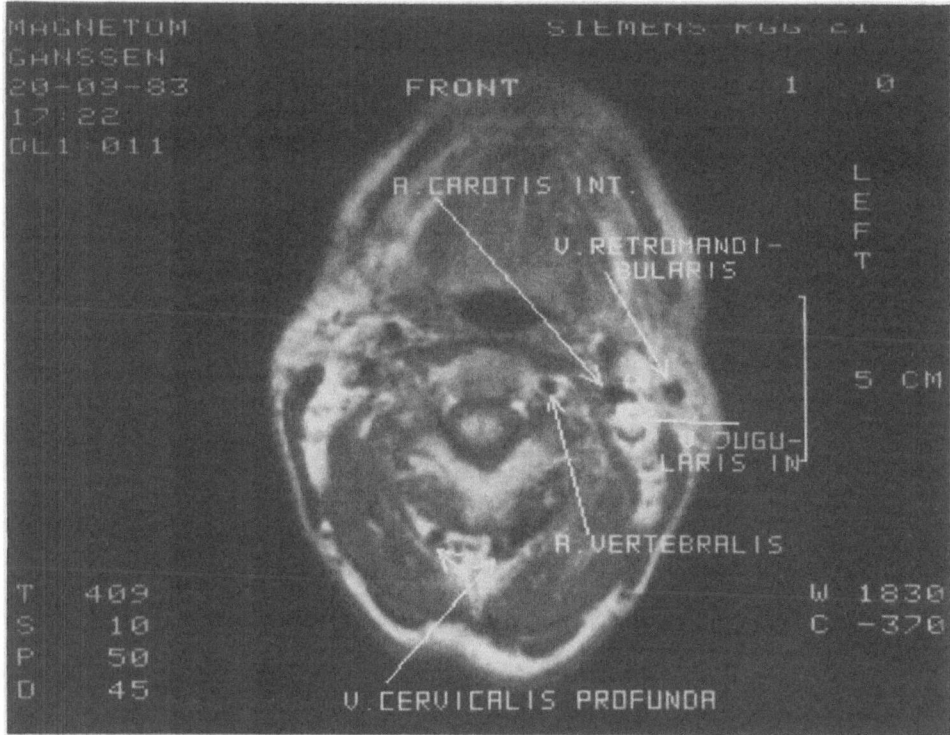

Fig. 12. Transaxial scan through the human neck. All major blood vessels are visible

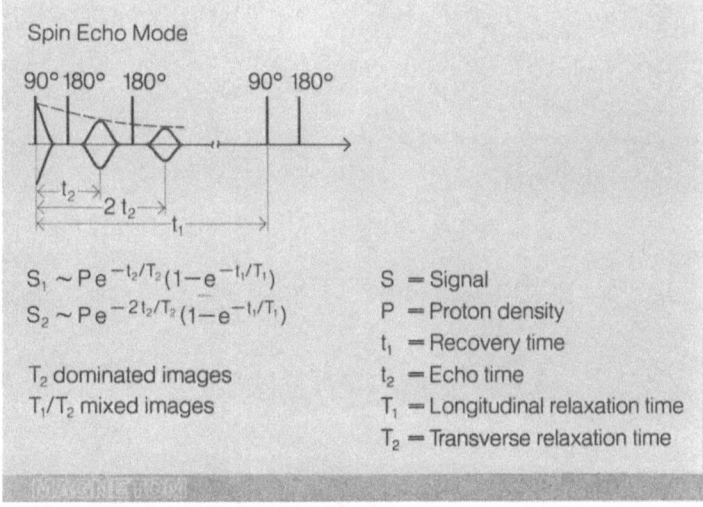

Fig. 13. Spin echo series usually applied in NMR imaging

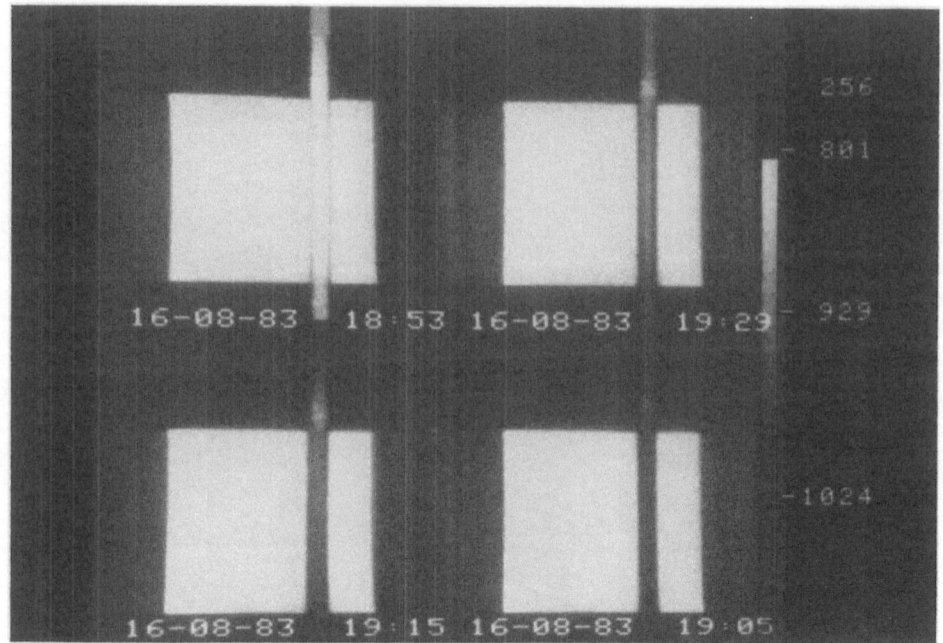

Fig. 14. NMR images of flow phantom with the flow within the displayed slice for the flow velocities 0, 3, 7, and 35 cm/s

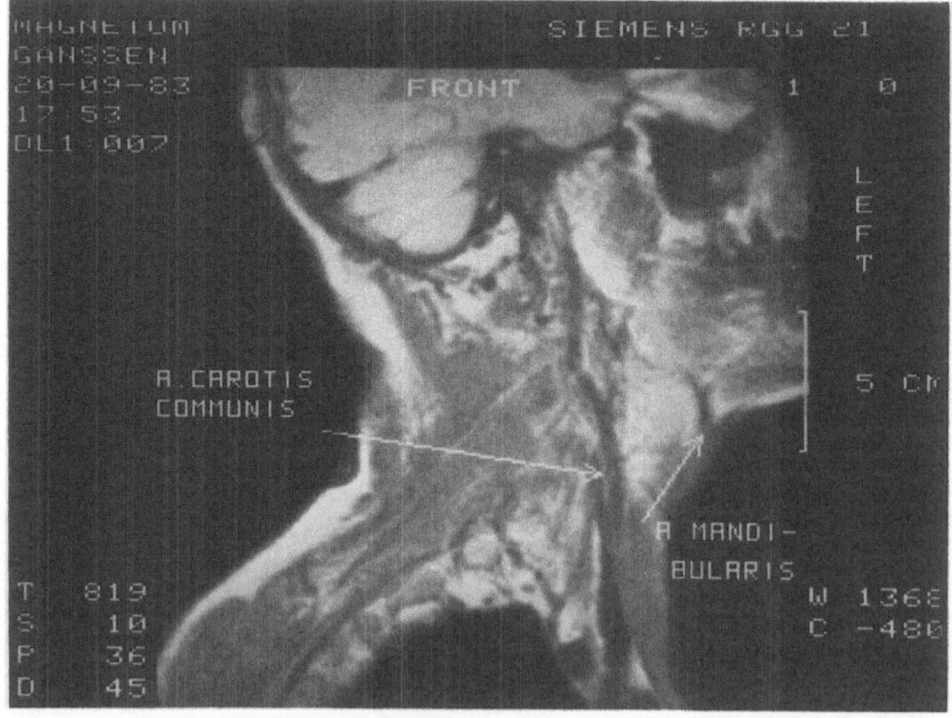

Fig. 15. Sagittal neck scan with 10 mm slice thickness with the a. carotis communis clearly visible

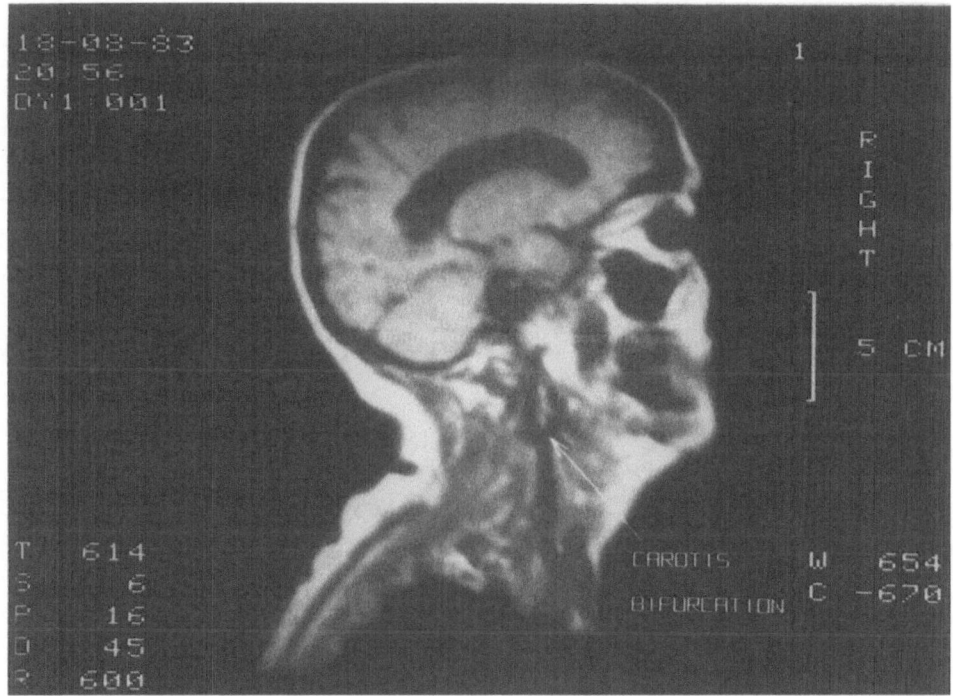

Fig. 16. Sagittal head and neck scan with 6 mm slice thickness with the carotid bifurcation clearly visible. The imaging was done in cooperation with Dr. Huk of the University Neurological Hospital, Erlangen

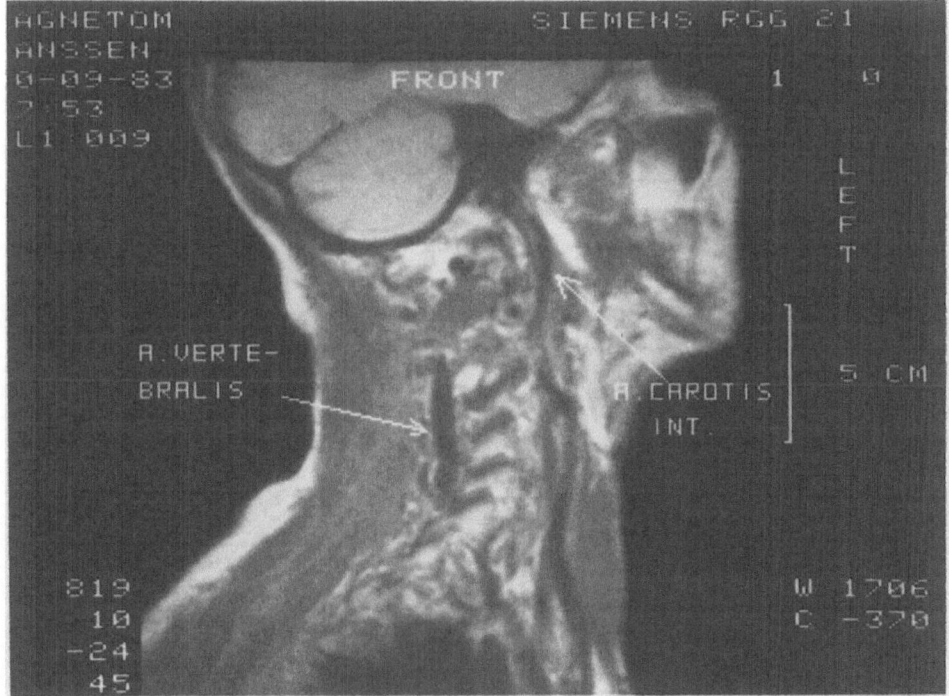

Fig. 17. Sagittal neck scan with the a. vertebralis and the a. carotis showing up

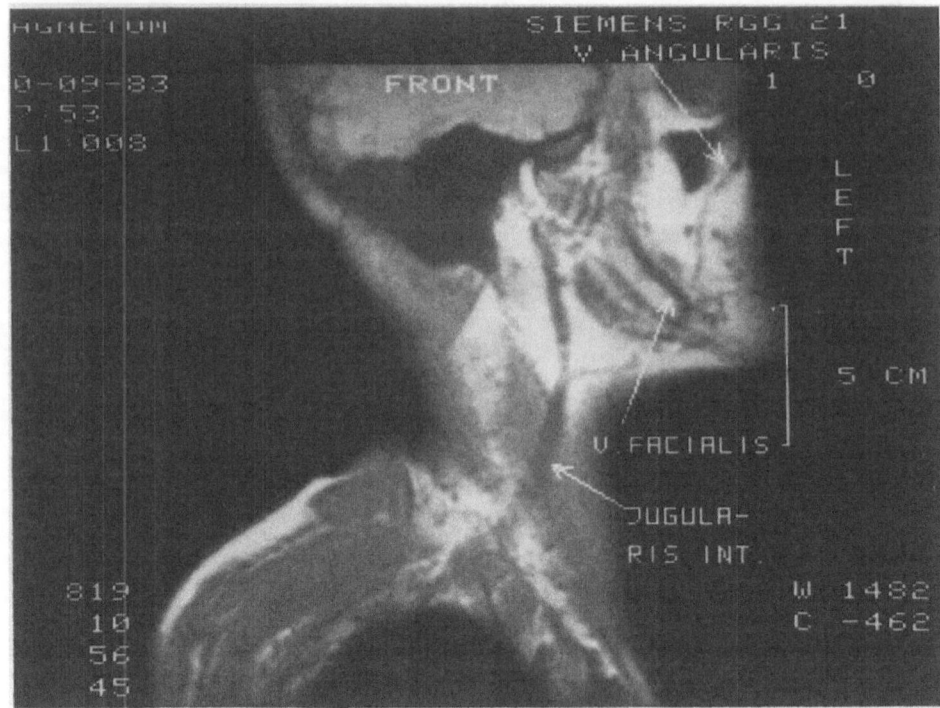

Fig. 18. Sagittal neck scan with the jugular and facial veins visible

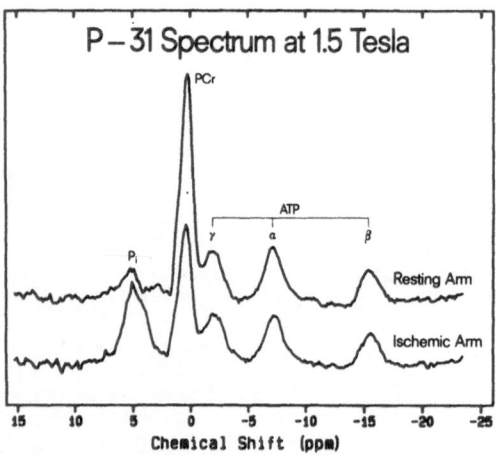

Fig. 19. P 31-NMR spectra of a human arm under resting and ischemic conditions taken within a 1-m bore, 1.5 T-superconducting magnet

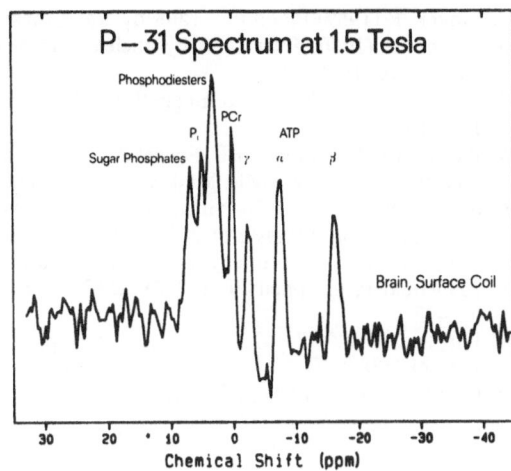

Fig. 20. P 31-NMR spectrum of human brain within a 1-m bore, 1.5 T-superconducting magnet

A survey of the diagnostic possibilities of NMR with respect to brain physiology would be incomplete without mentioning the application of high resolution NMR spectroscopy for in vivo study of brain metabolism.

Figures 19 and 20 show the high resolution phosphorous spectra obtained in man within a 1-m bore superconducting magnet at 1.5 T, and 24 MHz on the arm muscle and the brain. There is a remarkable difference in the composition of metabolic components of skeletal muscle and the brain which has still to be investigated.

In this short survey the subject of NMR application to the investigation of cerebral blood flow and cerebral metabolism has only been touched upon: at present it is still a very early state.

Acknowledgment. The author wishes to acknowledge the help received in the preparation of this paper from Dr. Michael Deimling, Dipl. Phys. Helmut Neumann, and cand. med. Gerhard Reuther.

References

1. Bowman RL, Kudravcev V (1959) Blood flowmeter utilizing nuclear magnetic resonance. IRE Trans Med Electr 6:267–269
2. Singer JR (1959) Blood flow rates by NMR measurements. Science 130:1652–1653
3. Singer JR (1960) Flow rates by nuclear and electron paramagnetic resonance methods. J Appl Phys 31:125–238
4. Ganssen A (1967) Elektromagnetische Hochfrequenzspule für Diagnostikeinrichtungen. BDP 1566148

2. Singer JR (1959) Blood flow rates by NMR measurements. Science 130:1652–1653
3. Singer JR (1960) Flow rates by nuclear and electron paramagnetic resonance methods. J Appl Phys 31:125–238
4. Ganssen A (1967) Elektromagnetische Hochfrequenzspule für Diagnostikeinrichtungen. BDP 1566148
5. Battocletti JH, Halbach RE, Sances A, et al (1979) Flat crossed coil detector for blood flow measurement using NMR. Med and Biol eng and Comp 17.183
6. Singer JR (1978) NMR diffusion and flow measurements and an introduction to spin phase graphing. J Phys E 11:281–291
7. Lauterbur PC (1973) Image formation by induced local interactions. Nature 242 190
8. Mansfield P, Grannell PK (1973) NMR diffraction in solids. J Phys C: Solid State Phys 6:L422
9. Hinshaw WS (1974) Spin mapping: the application of moving gradients to NMR. Phys Lett. 48A:78
10. Carr HY, Purcell EM (1954) Phys Rev 94:630

The Investigation of Structure and Metabolism by In Vivo NMR

G. K. Radda[2], P. Styles[2], P. B. Bore[2], G. Galloway[2], R. J. Ordidge[1],
R. E. Gordon[1], W. E. Timms[1], and S. B. Prime[1]

The development of wide-bore superconducting magnets has enabled nuclear magnetic resonance (NMR) to be performed on living tissue within both animals and human beings. The application of high resolution NMR spectroscopy to biology and medicine has been established by several research groups. NMR techniques can be used by biochemists and clinicians to noninvasively study metabolism in a variety of tissues. The phosphorus nucleus in particular can provide a detailed description of changes occurring in intracerebral, intracellular constituents before, during, and after severe ischemic events, as shown by Delpy et al. (1982). The physiologic potential of NMR spectroscopy has now been combined with the diagnostic capability of proton NMR imaging in a single instrument. The authors would like to report on the use of one such machine which has been developed by Oxfrod Research Systems in order to study the human head and torso.

The system incorporates a high field superconducting magnet with a clear bore of 60 cm, and which has been energized to a field strength of 1.6 Tesla. Spatial localization, in order to obtain NMR spectra from a particular tissue volume within a whole living animal, can be provided by the combined use of a surface radiofrequency coil, with a magnetic field profiling technique known as topical magnetic resonance (TMR). The surface coil is placed over the region of interest which must previously have been positioned in the "sensitive region" of the magnet. Positioning of the subject is facilitated by the use of proton NMR imaging.

Figure 1 shows the distribution of phosphorus resonances as a function of frequency obtained from human brain tissue. The axis is calibrated in parts per million (ppm) relative to the resonant frequency of phosphocreatine. The spectrum was obtained in 8½ min from the brain of one of the authors (GG) using a 6-cm surface coil placed over the frontoparietal suture. The peaks can be assigned from left to right as (a) sugar phosphate, (b) inorganic phosphate, (c) a peak which is as yet unassigned but may be from certain phosphodiesters, (d) phosphocreatine, and (e), (f), and (g), the γ, α, and β phosphates of adenosine triphosphate (ATP) respectively. Figure 1 also shows a proton NMR image of the same volume of the head. The NMR image demonstrates the volume of material excited by the surface coil, and therefore the volume from which the

---

1 Oxford Research Systems, 28 Nuffield Way, Abingdon Oxon OX14 1RY, Great Britain
2 Dept. of Biochemistry, Oxford University, South Parks Road, Oxford OX1 3QU, Great Britain

Cerebral Blood Flow and
Metabolism Measurement.
Eds. Hartmann/Hoyer
© Springer-Verlag Berlin Heidelberg 1985

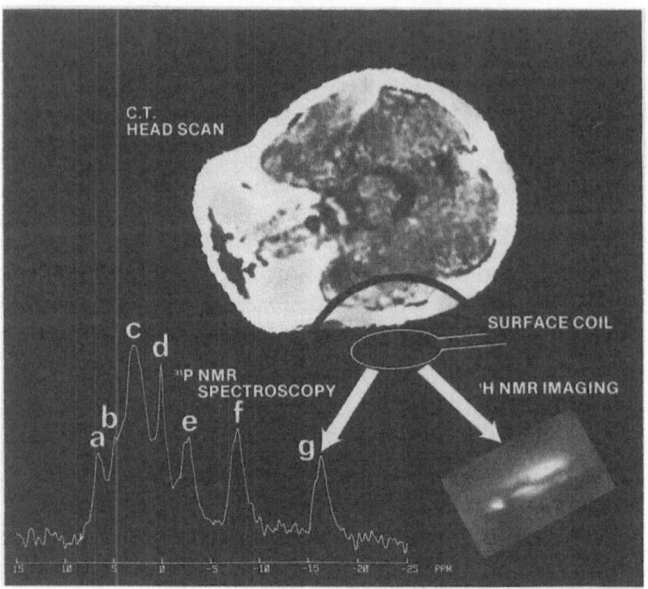

Fig. 1. Phosphorus spectrum and proton NMR image of a region of the head localized by the use of a 6-cm surface coil. The CT head scan demonstrates the position of the surface coil. The phosphorus peaks can be assigned as (*a*) sugar phosphates, (*b*) inorganic phosphate, (*c*) a peak which may indicate the level of certain phosphodiesters, (*d*) phosphocreatine, and (*e*), (*f*), and (*g*) the γ, α, and β phosphates of adenosine triphosphate respectively

phosphorus signal was derived. The CT scan is shown in order to demonstrate the position of the coil. The image is the central plane of an eight-slice three-dimensional data set and shows a cross-sectional slice cutting perpendicular to the plane of the coil. The subdural space between the scalp and the brain can be seen clearly. The pH value of the brain tissue can also be calculated by measurement of the relative positions of the phosphocreatine and inorganic phosphate peaks.

The NMR imaging capability can also be used to produce full two-, or even three-dimensional head images in order to aid placement of the patient as well as to provide clinical information. Figure 2 shows NMR images taken from a 32-plane three– dimensional data set. The lowest two scans show full cross-sectional images which cut the head at the level of the eyes. The data can also be reformatted to provide coronal sections, some of which are shown in the top half of Fig. 2. The sections cut through the eyes in coronal planes which proceed from the back to a region in the center of the eyes. The slice thickness for the cross-sectional images is 5 mm, and the resolution is approximately 1.75 mm. The images show proton density and were obtained using a three-dimensional Fourier Transform imaging technique. The limited extent of the coronal sections along the vertical axis demonstrates that the imaging experiment is only applied to a thick slab of material, which limits the required imaging time to approximately 15 min.

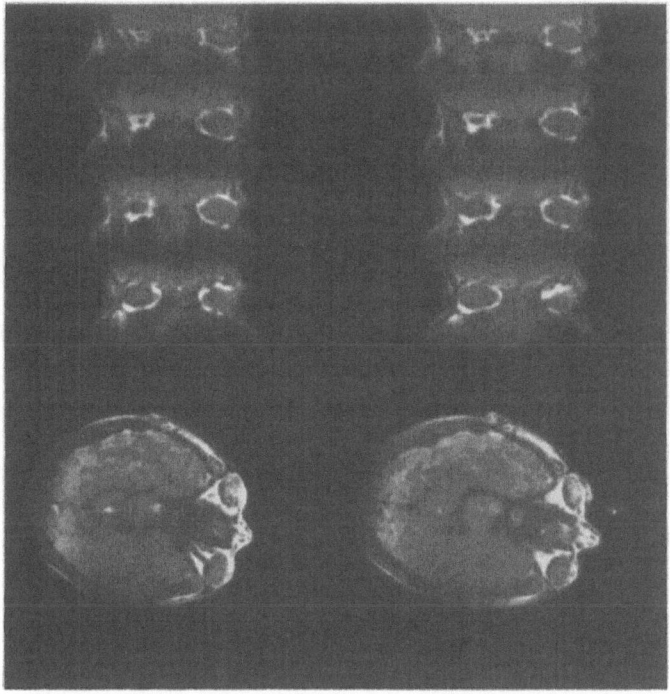

Fig. 2. Proton NMR images of the human head taken from a three-dimensional data set. The upper eight slices show coronal sections through the eyes progressing from the back to a plane in the center of the eyes. The lower two scans show full cross-sectional head slices taken from the same data set

The information these basic energy substrates provide, and also the associated pH measurement, can be used to monitor various abnormal conditions and events such as ischemia, ischemic exercise, the condition of organs, and McArdle's syndrome, as shown by Radda et al. (1981). Phosphorus NMR has already been used by Cady et al. (1983) to study cerebral metabolism in newborn infants. In addition, both fluorine and carbon NMR can provide invaluable biochemical information. For example, fluorinated inhalation anesthetics can be observed in the brains of live mammals using ^{19}F NMR, as shown by Wyrwicz et al. (1983). Clearance of these anesthetics from the brain and other tissues can be detected and followed for as long as 98 h after termination of anesthesia. The noninvasive observation of biochemistry and physiology without appreciable alteration of the processes under investigation opens new possibilities for improving our understanding of biochemistry and in particular cerebral metabolism. The combined use of spectroscopy and imaging offers new insight into many areas of biology and medicine.

References

Cady EB, Costello AM de L, Dawson MJ, Delpy DT, Hope PL, Reynolds EOR, Tofts PS, Wilkie DR (1983) Non-invasive investigation of cerebral metabolism in newborn infants by phosphorus nuclear magnetic resonance spectroscopy. Lancet (i); 1059

Delpy DT, Gordon RE, Hope DL, Parker D, Reynolds EOR, Shaw D, Whitehead M (1982) Non-invasive investigation of cerebral ischaemia by phosphorus nuclear magnetic resonance. Paediatrics. 70:310

Radda GK, Chan L, Bore PB, Gadian DG, Ross BD, Styles P, Taylor D (1981) Clinical applications of ^{31}P NMR. Proceedings of the International Symposium on Nuclear Magnetic Resonance at Bowman Gray School of Medicine, Winston-Salem, North Carolina, USA 159

Wyrwicz AM, Pszenny MH, Schofield JC, Tillman PC, Gordon RE, Martin PA (1983) Non-invasive observation of fluorinated anaesthetics by ^{19}F NMR in rabbit brain. Science, Volume 22, 428:430

Correlative Studies of the Brain with Positron Emission Tomography, Nuclear Magnetic Resonance, and X-Ray Computed Tomography

A. Alavi[1], J. C. Leonard[1], J. Chawluk[1], R. A. Zimmerman[1], R. W. Dann[1], J. Alavi[1], W. Edelstein[2], P. Bottomley[2], R. Redington[2], and M. Reivich[1]

Introduction

During the past decade the central nervous system (CNS) has witnessed an unanticipated explosion in tomographic techniques. Based upon principles developed by Cormack, Oldendorf, Kuhl, and Hounsfield (1–4), tomographic scanning systems have, for the first time, allowed the detailed study of the human brain's structure and function in vivo.

X-ray computed tomography (XCT) is firmly established as an essential diagnostic step in the evaluation of any CNS disorder. Nuclear magnetic resonance (NMR) imaging, which has only been in serious preliminary clinical use since 1980, has already demonstrated the potential for better resolution than XCT in certain neurologic conditions, e.g., demyelinating diseases, posterior fossa lesions, and low grade tumors. Positron emission tomography (PET), limited to major research centers in the United States, Europe, and Japan, nevertheless has begun providing valuable insight into normal and abnormal brain function.

Rather than following a linear evolution, with each tomographic method superseding its predecessor, XCT, PET, and NMR are developing concurrently, providing complementary information in many instances. This paper will review the applications of each of these imaging techniques to representative neurologic disorders (stroke, brain, tumor, epilepsy, dementia, and psychiatric disease), concentrating on the newer investigative tools, PET and NMR.

Methods and Materials

During the past 10 months we have studied 30 patients with the following diagnoses: 13 with primary brain tumors, 6 with dementia, 2 with acute and 5 with old infarcts, 3 with focal seizure disorders, and 3 with psychiatric disorders (2 patients with multi-infarct dementia are included in the old infarct category). All patients underwent XCT, NMR, and PET examinations within a short period of time.

---

1 University of Pennsylvania School of Medicine, 3400 Spruce Street, Philadelphia, PA 19104, USA
2 General Electric Corporate Research and Development Center, Schenectady, NY, USA
Supported by NIH, NS 14867-01, NIH, AG 03934-01

Cerebral Blood Flow and
Metabolism Measurement.
Eds. Hartmann/Hoyer
© Springer-Verlag Berlin Heidelberg 1985

XCT

The majority of patients were examined with GE 8800 or GE 9800 CT scanners. The brain was imaged with and without contrast enhancement.

PET

18F-2-fluoro-D-glucose ([18F]FDG) was used to determine local cerebral metabolic rates for glucose (LCMRgl) in all subjects studied. A radial artery catheter was inserted under local anesthesia. $100-115\,\mu\text{Ci/kg}$ of [18F]FDG were administered intravenously as a bolus. Blood samples were drawn from the radial artery to monitor the time course of the [18F]FDG and glucose. Blood was drawn every 15 s for the first minute, every minute until 10 min, every 5 min until 30 min, and every 15 min until the end of the study. Forty minutes after the injection of the radiopharmaceutical the head was positioned securely in the scanner. Transaxial tomographic images were obtained with canthomeatal line 20 ° hyperextended in relation to the perpendicular line to the horizontal plane. The tissue activity of [18F]FDG was measured using a PET scanner.

A modified version of the PETT-V system [5] was employed to obtain brain images of local cerebral metabolic rates for glucose. This machine generates simultaneously seven cross-sectional images of the brain along the rostrocaudal axis. Two scans of $12-20$ min duration are obtained, with a minimum of $1.5-2$ million counts per slice collected. Between scans the patient is displaced along the rostrocaudal axis by one-half slice thickness (8.8 mm), so that the total study consists of 14 images at 8.8-mm intervals.

Every day the system is calibrated by scanning a phantom consisting of six wedges containing known amounts of ^{68}Ga. The regression line relating PET number from the images to measured concentration allows the determination of tissue concentration in the patient images. Metabolic rates for glucose are calculated using a modified version [6] of the 18FDG method described by Reivich et al. [7].

For each patient the global, or "wholebrain," metabolic rate is determined by averaging data from eight slices, four from each scan, extending from the lowest slice containing occipital cortex to the cingulate gyrus. For each slice a large region of interest is placed to determine the total counts in the brain, and another elliptic region of interst is placed at the edge of the brain to determine its area. The effective PET number is then calculated by dividing the total counts by the brain area. The metabolic rate is determined using average values for gray and white rate constants and assuming a $50-50$ gray-white mixture. The average of eight sections is the "wholebrain" value, and is used as a reference point which is relatively insensitive to slight differences in positioning.

Regional metabolic data are generated using an atlas overlay system [8], which defines approximately 50 structures covering the entire cerebral cortex, white matter, basal ganglia, and cerebellum. For each structure the region/ wholebrain ratio is calculated as a measure of relative activity, and left-right asymmetries are assessed using the laterality index, according to the following formula:

$$[(R - L)/(R + L)] \times 200$$

In calculating metabolic rates, the rate constants and lumped constant values used are those actually measured by our group [9]. All data are automatically entered into a database system which allows a wide variety of subsequent statistical tests to be applied.

NMR

A 0.12-T (5.1 MHz) developmental resistive-magnet NMR proton imaging system designed and built by GE was used for this study. The bore size at the aperture of the imager is 20 in. (50.8 cm); this narrows to 18 in. (45.7 cm) centrally to accommodate the radiofrequency (RF) receiver coils for body imaging. For brain imaging a tightly fit head coil was used. RF shielding is not provided internally, within the magnet; rather, the unit is housed in a 10×15 ft $(3 \times 4.6$ m), 90 dB, RF-shielded room. This form of RF shielding allows a technician to be in direct attendance during the course of the examination, both to reduce patient anxiety and to administer care. NMR imaging on our unit is performed with a spin-warp technique, which has been described in detail [10]. In this form of NMR imaging, while spatial encoding is performed along the x-axis, spatial information also is phase-encoded on the y-axis, and a two-dimensional (2D) Fourier transform is performed. This method of imaging is advantageous in that it is less sensitive to local magnetic field inhomogeneities. Both planar (2D) and multiplanar (3D) data collection are available. Data are acquired and processed on a 128×128 matrix which is then interpolated to 256×256 for purposes of display. Slice thickness is 1.3 cm.

Imaging time per slice depends on the repetition time (TR), the number of data averages per pixel value, and the matrix size. In general, scan times vary from 2 to 4 min for a planar (2D) image. In the multiplanar mode, data for eight images typically are obtained in 10 min. We commonly use multiplanar imaging for examination of the head.

As initially configured, both saturation recovery (SR) and inversion recovery (IR) data collection schemes are available. Although saturation recovery (SR) with long repetition times has been reported as a method of obtaining proton-density-weighted scans, saturation recovery techniques with short repetition times may provide T_1-weighted information as well. Inversion recovery scanning is another method commonly used to provide T_1 weighted information.

A comparison of the effectiveness of these two pulse sequences in NMR imaging has been described recently: Edelstein et al. [11] found that when compared for similar imaging times, saturation recovery is superior to inversion recovery in distinguishing between tissues of the same proton density but of different T_1s. Our initial experience confirmed their findings, and as a result, T_1-weighted images on our system typically were performed with saturation recovery pulse sequences and a 0.15-s repetition rate (TR). The short TR permitted shorter scan time and increased patient acceptance of the examination procedure.

Results I

Thirteen brain tumor patients were studied. All had primary tumors, mostly astrocytomas. In each case, the XCT scan outlined the tumor size and location accurately, as far as could be ascertained from surgical or autopsy results. The NMR image, although it detected an abnormality in every case, did not appear to be superior to XCT scan in most cases. The tumor margins were more accurately determined by the enhanced CT scan than by the NMR. However, the extent of peritumoral edema was more clearly seen with NMR. Another advantage of NMR was its ability to delineate clearly a tumor in the brainstem which was not as easily identifiable on XCT. This patient had symptoms and signs of a slowly growing brainstem lesion. The NMR image demonstrated a huge brainstem mass on the sagittal sections. In another patient with a mixed astrocytoma-oligodendroglioma, the NMR was very helpful in delineating the true anatomic location of the tumor on sagittal section studies. Two low grade tumors showed no enhancement on XCT scan. Grade III and IV tumors showed significant enhancement after contrast infusion. One patient with a slowly growing hemangiopericytoma showed significant enhancement with contrast. NMR was unable to distinguish low grade tumors from high grade malignancies.

The PET studies were done at various times pre- and postsurgery or radiotherapy. Nevertheless, certain conclusions could be drawn from the data. The low-grade tumors, with a prolonged clinical course, were hypometabolic. These included the brainstem tumor (unbiopsied) mentioned above, as well as a mixed astrocytoma-oligodendroglioma. The grade III or IV gliomas had increased metabolic rates compared with surrounding brain. Occasionally a necrotic area of tumor (as seen on CT scan) was hypometabolic. Some patients were studied during or after radiation therapy and chemotherapy. These tumors were sometimes less active than preoperatively. Several unusual tumors were studied: a pituitary tumor (unbiopsied) was metabolically very active; a primitive neuroectodermal tumor was also active; but a recurrent hemangiopericytoma, which was densely contrast enhancing on CT, was metabolically inactive. In general, the metabolic activity correlated with the degree of contrast enhancement on CT. Exceptions included the hemangiopericytoma and the tumors studied during or after radiation therapy.

In two patients with acute stroke the NMR appeared within normal limits although XCT obtained earlier was abnormal in one. In the other patient the early XCT scan was normal and became abnormal when repeated later. PET scan was clearly abnormal in both cases which were obtained close in time to both the XCT and NMR scans. The abnormality was more widespread on the PET than on the XCT scan. Old infarcts were clearly seen on both XCT and NMR studies although some of the defects were better seen on coronal sections obtained with NMR imaging. PET was unable to delineate small infarcts, specifically those located in the white matter.

NMR and XCT did not demonstrate any abnormalities in patients with psychiatric disorders although PET showed certain abnormalities which will be analyzed in greater detail in the near future.

One patient with a seizure disorder had positive results on all three studies. In one patient all three examinations appeared within normal limits. The third patient was found to have an abnormality only on the PET scan.

In four patients with senile dementia, NMR and XCT scans showed ventricular dilatation and cortical atrophy although the latter appeared more pronounced on the NMR study (see below). In two patients with MID, both XCT and NMR showed small infarcts. These were more clearly seen on the coronal section obtained with NMR imaging.

Results II. Calculation of Metabolic Rates in Cerebral Atrophy; Correction for Cerebrospinal Fluid Spaces

We are currently investigating the effect of CSF spaces on regional and global metabolic rate calculations using PET [12]. Our initial studies concentrated upon the influence of cerebral atrophy on measurements of gluocse metabolism in a group of demented patients and normal aged and young controls [13]. To date we have determined cerebral metabolic rates for glucose (CMRglu) in 12 patients with the clinical diagnosis of dementia, 4 elderly controls, and 12 young controls. The normal young and aged controls were carefully screened by a neurologist and a psychiatrist, and found to be free of mental or physical disorders. All PET studies were done under similar resting, unstimulated conditions (eyes open, ears unplugged, dimmed lights, usual ambient noise). Data analysis has been completed on the cohort described in Table 1.

For each patien the global, or "wholebrain," metabolic rate was determined as described above.

In addition, all elderly subjects (demented patients and controls) underwent XCT and/or proton-NMR transaxial scanning at the time of their PET study.

Table 1. Demographic data on demented and normal aged subjects studied by PET and [18F] FDG

Group	No. of subjects	Age range	Mean age ± 1 SD
1. Dementia (total of # 2, 3, 4 below)	9	54–80	64 ± 9*
2. Alzheimer's	7	54–80	62 ± 10*
3. Multi-infarct	1	69	
4. Parkinsonian	1	74	
5. Elderly controls	2	48–54	51 ± 4*
6. Young controls	9	19–32	24 ± 5

* $P < 0.005$ compared with young normals, two-tailed Student's t-test

Table 2. Brain volume (%) by subject group

1. Dementia (9)	84.8±3.8*, **
2. Alzheimer's (7)	85.4±4.2*
3. Multi-infarct (1)	83.9
4. Parkinsonian (1)	82.0
5. Elderly controls (2)	92.4±2.6
6. Young controls	92.5±2.5
	(from Yamaura et al. [46])

* $P < 0.0005$ compared to young controls
** $P < 0.05$ compared to elderly controls

XCT scans were done on a GE 8800 or GE 9800 unit, with a spatial resolution of 0.64 mm × 0.64 mm and 5 mm slice thickness. NMR scans were performed with a GE 0.12 Tesla unit (1 mm × 1 mm × 13 mm). In order to quantitate the degree of atrophy, all anatomic (XCT or NMR) images were digitized using a Vidicon camera (Dage) interfaced to a 6-bit digitizer. The images were displayed on a high resolution color monitor (Conrac) utilizing an image analysis system (Grinnell Systems). With these elements the resolution of digitization of the projected image is at least three times greater than that of the intrinsic XCT resolution. Areas of interest were then highlighted subjectively via a data tablet interface, using nonoverlapping contiguous slices extending from the lowest level containing occipital cortex rostrally to the cingulate gyrus. These levels correspond to the PET levels used to calculate "wholebrain" (global) metabolic rates with [18F]FDG. The highlighted areas of interest on each slice coresponded to [1] the ventricular cavity, [2] the sulcal and cisternal spaces, and [3] the entire intracranial cavity. The number of pixels designated as ventricle or sulci on each slice is expressed as a percentage of the total number of pixels in that slice's intracranial cavity of interest, and these values are summated over all levels to obtain relative volumetric estimates of ventricular size, sulcal size, and residual brain tissue [intracranial cavity minus (ventricle and sulci)]. The brain volume estimates by this method are presented in Table 2.

These data show that significantly greater cerebral atrophy is present in demented patients compared with young and older controls.

Table 3. Change in global metabolic rates (% increase) after correction for cerebral atrophy

1. Dementia	18.2±5.3*, **
2. Alzheimer's	17.6±5.9*
3. Multi-infarct dementia	19.1
4. Parkinsonian dementia	22.0
5. Elderly controls	8.3±3.0
6. Young controls	8.1±0.1

* $P < 0.0005$ compared to young controls
** $P < 0.05$ compared to elderly controls

Table 4. Comparison of volumetric estimates from NMR and XCT scans (transaxial plane) on three patients (V, ventricle; S, sulei)

(a) Patient X, right subcortical infarct

	NMR	XCT
% V	4.5	4.2
% S	13.2	5.2
% V+S	17.7	9.4
% Brain tissue	82.3	90.6

(b) Patient Y, Parkinsonian dementia

	NMR	XCT
% V	8.6	8.1
% S	10.6	8.6
% V+S	19.2	16.7
% Brain tissue	80.8	83.3

(c) Patient Z, severe Alzheimer's disease

	NMR	XCT
% V	12.5	7.8
% S	27.4	16.8
% V+S	39.9	24.6
% Brain tissue	60.1	75.4

Table 5. Summary of Table 4 (mean $\pm$ SD, $n = 37$)

	NMR	XCT	Difference
% Ventricle	8.5 ± 4.0	6.7 ± 2.2	1.8%
% Sulci	17.1 ± 9.0	10.2 ± 6.0	6.9%

Since the in-slice resolution of current PET scanners (including the PETT V unit employed in our studies) does not discriminate sulci from cortex, nor ventricle from white matter, routine regional and global metabolic rate calculations are contaminated by contributions from metabolically inactive CSF spaces. Metabolism in brain tissue is therefore underestimated. Our next step was to recalculate global metabolic rates after correction for CSF ("atrophy"). These data are presented in Table 3, and demonstrate a significantly greater change in metabolic rates for demented individuals compared with controls.

Thus caution should be used in interpreting PET data showing decreased metabolic activity in dementia, since a significant portion of such a decrement may be due to cerebral atrophy, which is demonstrated by XCT scan.

The above atrophy correction was performed using XCT scans, since more of our PET subjects had been studied by that modality. We have preliminary data, however, demonstrating larger estimates of sulcal volume and atrophy in general by proton-NMR imaging (Tables 4, 5).

No definite conclusions can be drawn from this small sample, but theoretical considerations such as the possibility for greater CSF-parenchyma contrast, and the absence of bony artifacts over the convexities lead us to suspect that NMR may prove superior to XCT in estimating the degree of sulcal enlargement (peripheral or cortical atrophy) in aged and demented individuals. We are currently investigating the feasibility of phantom studies to help determine which imaging modality comes closest to providing "true" volumetric estimates of cerebral atrophy.

Discussion

With the current state of the art, XCT remains the tomographic procedure of choice in critical care neurology. XCT now allows the easy recognition of acute cerebral hemorrhage with the distinctive hyperdense appearance of fresh intraparenchymal blood. Noninvasive detection of intracerebral bleeding permits prudent use of anticoagulants in cases of stroke-in-evolution [14]. Although spectral-shift artifact resulting from close approximation of parenchyma to bone has limited XCT detection of small posterior fossa lesions, the presence of significant mass effect or hemorrhage in the brainstem-cerebellar region can be rapidly determined and, where indicated, treated with surgical expediency.

In ischemic stroke XCT may remain normal for up to 72 h after the ictus with small infarctions [15]. It appears that NMR proton imaging will allow the earlier detection of small infarctions. Animal studies indicate a 39% detection rate for NMR imaging alterations 3 h after symptomatic carotid ligation [16]. In humans, infarctions less than 24 h old are readily demonstrated as low intensity areas (prolonged T_1 values) on inversion recovery (IR) scans [17]. Direct comparison of XCT and NMR imaging in clinical studies of cerebral infarction suggests that NMR is more sensitive in defining the extent of pathologic changes [18], and in detecting these changes at an earlier time postictus. Buonanno et al., in their study of 25 stroke patients, found no XCT lesions which were missed on NMR scanning. NMR demonstrated abnormalities, however, in two patients with normal XCT studies [17]. One of their patients had a normal XCT scan at 12 h, positive NMR imaging (IR) at 18 h, and finally and XCT study demonstrating a large hemispheric infarction at 48 h [19]. The second patient's inversion recovery NMR image was positive at 1 week, while XCT was normal 2 days and 1 week after the acute event.

As noted above, XCT of posterior fossa abnormalities, especially in attempting to demonstrate small bland infarcts, is limited by bone-parenchyma juxtapositions and resultant beam-hardening (spectral shifts) artifacts. Since no NMR signal is detected in bone due to the fixed nature of the hydrogen protons, the brainstem and cerebellum are imaged without resultant artifact. In a series of eight patients with posterior circulation infarcts, abnormalities were demonstrated in each case on IR NMR imaging [20]. One of us (J.B.C.) has seen a diabetic patient with an acute third nerve palsy, whose NMR scan clearly demonstrated dorsal midbrain infarction before the evolution of additional brainstem signs.

Several groups have reported NMR studies demonstrating acute intracerebral hemorrhage [17, 18, 20]. The higher signal intensity from the hemorrhage on spin-echo (SE) and IR images appears to reflect shortened T_1 values, whereas surrounding edema is charcterized by a prolonged T_2 relaxation time. Once the acute clot has lysed, XCT attenuation in the hemorrhagic region may return to near normal, while iron-containing heme breakdown products continue to produce abnormal NMR images by virtue of their magnetic properties.

The increased sensitivity of NMR imaging to early pathologic changes in acute stroke underlies its potential for directing therapeutic interventions, as well as leading to a better understanding of the sequence and timing of pathophysiologic events.

Early reports of PET studies in stroke using $^{15}O_2$ and [18 F] FDG indicated that, like NMR, PET could demonstrate abnormalities earlier than XCT, and with a wider topographic distribution [21, 22]. Emphasis initially was placed on showing the extent of lesions, or seeking rather simplistic correlations between the degree of hypometabolism and the ultimate viability of ischemic tissue. Many stroke patients exhibited the phenomenon of "luxury perfusion," with PET studies disclosing regional cerebral blood flow (rCBF) in excess of local metabolic needs [23, 24]. During their first 2 weeks postinfarction, some patients demonstrated what Baron et al. termed "misery perfusion," i.e., a relative increase in tissue oxygen metabolism compared with rCBF [24]. Preliminary attempts to predict recovery from ischemia suggest that oxygen metabolic rates ($CMRO_2$) less than 1.25 ml O_2/min/100 g brain tissue indicate a poor clinical outcome, presumably secondary to completed infarction [25].

These findings have been elaborated upon, primarily by the MRC Cyclotron Unit at Hammersmith, and a more detailed model of perfusion-metabolism dynamics in cerebral ischemia has been hypothesized [26]. Five of six patients studied with $^{15}O_2$, $C^{15}O_2$, and PET within 12 h after onset of stroke symptoms showed an increase in oxygen extraction (OEF) in the affected region compared with surrounding brain. Twenty-four to 72 h after clinical ictus, 12 of 14 patients had a low OEF, indicating probable tissue death (infarction). Wise et al. concluded that the early period of high oxygen extraction represented transient ischemia, yielding in 12 − 24 h to completed infarction. Furthermore, in instances of persistent ischemia ("misery perfusion"), increases in the OEF have been observed, implying a compensatory mechanism, with the $rCMRO_2$ (absolute regional metabolic rate) maintained at a normal level to prevent infarction. Concurrent PET measurements of rCBF are decreased, and cerebral blood volume (also directly determined by emission tomography) is increased, indicating maximal regional compensatory vasodilatation. One may speculate that such situations where homeostasis has been pushed to its limits warrant therapeutic intervention, and several PET centers have reported restitution of these parameters ($CMRO_2$, CBF, and CBV) to normal levels after extracranial-intracranial bypass surgery [27 − 29].

PET has also provided interesting data on distal effects of focal infarction. These effects, termed "diaschisis" by Von Monakow [30], may explain stroke symptoms and signs for which no previous pathologic or morphologic correlates had been found. Kuhl et al. reported hypofunctional zones in cerebral cortex, striatum, and thalamus ipsilateral to, but remote from, the infarcted region; all of these structures appeared normal on XCT scans [22]. Cerebral defects of metabolism and flow contralateral to areas of infarction, so-called "mirror foci," were observed by Lenzi et al. [31] and felt to represent interruption of transhemispheric connections. "Crossed cerebellar diaschisis" was first described on PET images by Baron et al., who felt it to be an acute transient phenomenon [32]. Our investigations also suggest that cerebellar hypometabolism contralateral to cerebral infarction is most prominent early in the clinical course [33]. Other PET centers, while confirming its occurrence, contend that crossed cerebellar diaschisis can be demonstrated months or even years after acute infarction [31, 34]. Cerebellar hypometabolism in these instances may arise from interruption of cerebrocerebellar pathways and could serve to explain changes in limb tone after acute infarction or the well-recognized clinical entity of ataxic hemiparesis [35].

The role of PET in expanding our knowledge of pathophysiologic changes during cerebral ischemia is still in its infancy. Further creative applications of PET technology based on sound biochemical principles will continue to supply additional insights into cerebral ischemic conditions. This may in turn result in a more scientific and appropriate management of this disorder.

The first clinical reports of XCT scanning indicated that atrophy demonstrable on pneumoencephalography was also seen on XCT images [37]. A number of groups subsequently looked at linear, planimetric (single slice areas), or subjective (rank order) estimates of ventricular dilatation and cortical atrophy by XCT [38–42]. Diverse methodologies, including manner of XCT analysis, type of cognitive tests, and selection of control groups, led to conflicting conclusions. These studies generally showed increasing ventricular volume with age [37–41]. Sulcal atrophy was an inferior discriminator, presumably due to convexity artifacts, arbitrary measurement techniques, and relatively inferior scanner resolution [39, 42, 43].

Attempts to associating atrophic changes with cognitive impairment were even less successful. Several centers found no correlation between the degree of dementia and the extent of cerebral atrophy [37, 44, 45].

These initially disappointing data have been improved upon with the advent of better resolution XCT scanners, along with sophisticated computer programs for analysis of CSF volumes and brain parenchyma characteristics. Yamaura et al. were the first to report relative volume estimates from XCT scans of CSF spaces, examining 228 normal subjects between 2 and 89 years old [46]. They found a significant increase in atrophy beginning at age 50, in agreement with previous autopsy findings [47]. The same center refined its technique for volumetric analysis of CSF spaces (excluding cortical sulci, however), and applied a dementia rating scale to the study population [48]. Their finding that atrophy correlated inversely with mental function is weakened by an associated direct correlation between atrophy and age. Gado et al., using a similar method for estimating relative ventricular and sulcal volumes, noted a highly significant separation of controls from mildly demented patients on the basis of XCT scan atrophic changes [49]. Measurements of ventricular volumes in the two clinical groups did not overlap, in contrast to their previous study which employed a linear quantification of ventricular size [40]. A less clear-cut but still significant separation of demented patients from age-matched controls was obtained by George and co-workers, who measured only ventricular volumes while ignoring sulcal changes [50]. Since their correlations were still relatively weak, they further speculated that ventricular enlargement only in part reflects the underlying pathology of Alzheimer's disease.

Given the prominence of cortical changes seen histopathologically, one might expect to find an increase in sulcal volumes before ventricular dilatation becomes apparent in Alzheimer's disease. Arai et al. recently reported a highly significant increase in cortical atrophy measured volumetrically when young patients (54 years old at onset) with mild Alzheimer's disease were compared with age-matched controls [51]. The increase in ventricular volume was less dramatic until cognitive and functional disturbances became severe. Their results suggested that cortical atrophy was detectable in early (mild) disease, and that ventricular dilatation was a later (? secondary) occurrence [51]. If their hypothesis is correct, then an imaging modality which more accurately demonstrates sulcal enlargement would in theory prove superior to XCT in diagnosing early dementia.

Unlike XCT, proton-NMR scanning can image brain surface without interference from beam hardening (bone) artifacts. In addition, pulse sequences can be modified to maximize the contrast between adjacent structures with different proton environment characteristics, such as CSF and gray matter. Preliminary studies by our group suggest a greater sensitivity in detecting cortical atrophy with NMR compared to XCT [12, 13]. The validity of these measurements awaits appropriate phantom studies, the feasibility of which is being explored at our institution [52, 53].

Frackowiak et al. have studied aging [54] and dementia [55] using the ^{15}O equilibrium imaging technique. These authors report a progressive fall during aging of both CBF and $CMRO_2$. This fall could be associated with atrophy associated with age in that the PET technique is sensitive to and measures the activity in a volume of tissue. If that volume of tissue includes more CSF and less perfused brain as a function of age, this would cause a decline in both CBF and $CMRO_2$ as determined by PET. A much quicker decline of $CMRO_2$ and CBF was demonstrated in ten demented patients over a period of 6 months [56]. The declines were approximately 30% for both variables. The couple between flow and metabolism remained intact, however, during this 6-month period. No differences were noted between degenerative (Alzheimer) dementia and multi-infarct dementia groupings of patients whose diagnoses were made on the basis of ischemia score and XCT findings. A negative correlation was found between the severity of dementia and $CMRO_2$, which was significant at the 0.001 level. Decreases in both CBF and $CMRO_2$ of 20%−40% were noted in comparison with age-matched controls.

Foster et al. studied 13 right-handed patients with Alzheimer's disease using the [^{18}F]FDG technique [56]. For this project three subjects were selected with predominant language deficits, and four with disproprotionate failure of visuoconstructive function; the other six suffered primarily from memory loss. The group with language deficits showed marked reduction in LCMRgl in the left frontal, temporal, and parietal regions. Those with predominant visuoconstructive dysfunction demonstrated a focus of hypometabolism in the right parietal cortex. Patients with memory loss had no significant asymmetry of metabolic activity in cortical regions. In all subjects a good correlation was noted between verbal competency and left frontal and temporal area metabolic activity. Visuoconstruction test performance was linked to glucose utilization in the right parietal lobe.

Our preliminary data in patients with dementia show that both XCT and NMR are excellent techniques in detecting structural lesions that can cause dementia. The degree of cortical atrophy can be better estimated by NMR imaging than by XCT scan. Both techniques are useful in delineating ventricular dilatation. Small infarcts in patients with multi-infarct dementia are seen well with both XCT and NMR although the latter images in coronal sections are able to demonstrate these lesions more distinctly. PET techniques offer the unique opportunity of visualizing regional functional abnormalities. Also this technique is capable of quantifying regional function and metabolism.

Kuhl et al. studied 17 patients with partial epilepsy using the [^{18}F]FDG and ^{13}NH$_3$ techniques [57]. All patients underwent scalp EEG recording at the time of imaging. The partial seizure pattern was diagnosed on clinical or EEG grounds. These patients experienced seizure episodes that did not respond well to adequate drug therapy.

[^{18}F]FDG scans obtained interictally were able to detect dysfunctional brain zones considered most likely to be responsible for seizures in patients with partial epilepsy. These areas usually appeared normal on the XCT scan. In 12 of 15 patients who had focal or unilateral EEG abnormalities, broad regions of cortical hypometabolism and hypoperfusion were demonstrated on the PET scans. These areas corresponded well to the EEG lateralization and localizations. In five of six patients who underwent temporal lobectomy, the [^{18}F]FDG scan correctly detected the pathologically confirmed lesion as a hypometabolic region, and the surgical removal of the lesion resulted in marked clinical improvement. In this study group only five of 17 epileptic patients had abnormalities seen on the XCT scans. All were atrophic abnormalities. In only three patients did the atrophic lesions seen on the XCT scan coincide with both EEG and PET localization. In the other two patients, while the abnormalities on the EEG and PET were shown in the temporal lobe, the XCT atrophy was seen in the frontal region. In both patients the presence of a lesion in the anterior temporal lobe was confirmed after temporal lobectomy. In patients with clearly defined focal atrophic lesions on the XCT of at least 2 cm in extent, the metabolic depressions were maximal, usually more than 58%.

When studied during an active seizure, the metabolism and perfusion in the cortical epileptic focus were found to increase to about twice normal. These same areas were hypometabolic and hypoperfused in the interictal state. It is already known that in humans during seizure activity, increased neuronal metabolism is accompanied by a rise in regional CBF [58–61]. Increases in CBF of two to ten times normal in the epileptic focus during spontaneous seizures in humans have been recorded [62–64]. In one patient decreases in both metabolism and perfusion in the area surrounding the seizure focus were noted. These depressed values were attributed to "surround inhibition," which has been defined electrophysiologically for the penicillin-induced focus by Prince and Wilder [65].

Our studies are inconclusive at this time, but the preliminary data indicate that NMR, like XCT, may not be capable of demonstrating the seizure focus unless significant structural abnormalities have occurred. PET imaging is probably superior to both XCT and NMR in this regard.

Widespread availability of XCT since the mid-1970s has dramatically increased the clinician's ability to arrive at more precise diagnoses when confronted with an intracranial mass lesion. Even before the arrival of current high resolution scanners the ability of XCT to delineate tumors as small as 5 mm in diameter had been demonstrated [66]. At present XCT is routinely used to detect and characterize intracranial tumors [67]. Unfortunately XCT may not be able to detect early lesions or low grade tumors [68]. There may be discrepancies between histology and predictive XCT diagnosis; assessment of tumor necrosis or change of histologic type may be difficult [68].

Preliminary data from several centers indicate that NMR imaging may be superior to XCT scanning in the detection of low grade, infiltrating lesions, especially those in the posterior fossa [69 – 72]. In general NMR is as sensitive as XCT in the detection of brain tumor [73]. Using the presently available NMR imaging techniques the tumor boundaries are not as well defined as with contrast-enhanced XCT scan [73]. With the introduction of newer techniques and the use of NMR contrast enhancing agents, this difficulty may be overcome. The contrast between the surrounding edema and the adjacent brain is usually higher with NMR than with XCT. Pituitary neoplasms are clearly visualized by NMR and this modality will probably play a major role in the evaluation of patients with disorders of this gland.

DiChiro et al. [74] from the National Institutes of Health have reported the use of [18F]FDG imaging in the evaluation of cerebral gliomas in 23 patients. These authors have shown a positive correlation between the [18F]FDG activity as measured by PET and the histologic grade of the gliomas. The metabolic activities for high grade tumors generally were significantly higher than for low grade tumors. However, because of the incomplete knowledge (such as a measured value for the lumped constant) of the glucose metabolism in these tumors absolute quantification is not possible; therefore, they have used the terms "hot" and "cold" to indicate relative metabolic activity of these lesions. The high grade glioma appears to be "hot," and conversely low grade tumors appear to be "cold" on these images. An interesting finding is the suppression of cortical glucose metabolism in tissue adjacent to the tumor by as much as 50% relative to contralateral cortex in preoperative scans of patients with brain tumors. This does not seem to be explained simply by peritumor edema. In another study these investigators were able to differentiate tumor recurrence from tumor necrosis following radiation therapy [75]. The former showed increased metabolic activity at the site of abnormality on the XCT scan while the latter revealed decreased uptake of [18F]FDG. This differential diagnosis is not possible with XCT scans [76].

Our preliminary data confirm what has been reported in the literature. Both NMR and XCT techniques are quite sensitive in detecting brain neoplasms.

The tumor borders are more clearly defined on the XCT images than on the NMR scans. XCT is superior to NMR in determining the histologic grade of the tumor. Our data with PET imaging corroborate the findings described by DiChiro [74]. We believe PET should play a role in the evaluation of the majority of patients with brain tumor. PET is of considerable value in assessing the effects of both chemo- and radiation therapy.

Conclusions

Our data indicate that NMR imaging provides excellent delineation of the structural abnormalities of the brain. This is achieved without the administration of contrast enhancing agents and there have been no adverse side effects. Because of these advantages NMR techniques will continue to be refined and its role in the diagnosis of CNS disorders explored extensively using different imaging sequences to obtain T_1, T_2, and proton density weighted images.

PET imaging, on the other hand, demonstrates regional function and metabolism but with less detailed resolution than is obtainable by either XCT or NMR. The metabolic information may be crucial in the management of patients without structural abnormalities of the brain. For example, in patients with seizure disorders, anatomic imaging generally reveals no abnormalities while functional scans with PET may show seizure foci. Also in patients with brain tumors, PET can predict the histologic grade (high vs. low) and identify tissue necrosis (spontaneous or radiation induced). In addition, the extent of functional abnormalities is best demonstrated by PET imaging. Therefore, PET studies play a complementary role to XCT and NMR in the management of CNS disorders.

References

1. Cormack AM (1963) Representation of a function by its line integrals, with some radiological applications. J Appl Physics 34:2722
2. Oldendorf WH (1961) Isolated flying-spot detection of radiodensity discontinuities; displaying the internal structural pattern of a complex object. IRE-Trans Bio-Med Elect BME 8:68−72
3. Kuhl DE, Edwards RQ (1968) Reorganizing data from the transverse section scans of the brain using digital processing. Radiology 91:975−983
4. Hounsfield GN (1973) Computerized transverse axial scanning (tomography). I. Description of system. Br J Radiol 46:1016−1022
5. Ter-Pogossian M, Mullani N, Hood J (1978) Design considerations for a positron emission tomograph (PETT V) for imaging of the brain. J Comput Assist Tomogr 2:539−544
6. Phelps M, Huang S, Hoffman E, Selin C, Sokoloff L, Kuhl D (1979) Tomographic measurements of local cerebral glucose metabolic rate in humans with (F-18) 2-fluoro-2-deoxy-D-glucose: Validation of method. Ann Neurol 6:371−388
7. Reivich M, Kuhl D, Wolf A, Greenberg J, Phelps M, Ido T, Cassella V, Fowler J, Hoffman E, Alavi A, Sokoloff L (1979) The ^{18}F fluorodeoxyglucose method for measurement of local cerebral glucose utilization in man. Circulation Research 44:127−137

8. Dann R, Muehllehner G, Rosenquist (1982) Computer-aided data analysis of ECT data. J Nuc Med 24:P82
9. Reivich M, Greenberg J, Alavi A, Wolf A, Fowler J, Arnett C, Ferrieri R, Atkins H (1984) The evaluation of the lumped constant for deoxyglucose and fluorodeoxyglucose in man. J Nucl Med (abstract submitted)
10. Edelstein WA, Hutchison JMS, Johnson G, Redpath T (1980) Spin warp NMR imaging and applications to whole-body imaging. Phys Med Biol 25:751−756
11. Edelstein WA, Bottomley PA, Hart HR, Smith LS (1983) Signal, noise and contrast in nuclear magnetic resonance (NMR) imaging. J Comput Assist Tomogr 7:391−401
12. Chawluk J, Alavi A, Dann R, Hurtig H, Kushner M, Reivich M, Zimmerman RA (1984) Volumetric correction of cerebral metabolic rates in dementia and normal aging. Neurology 34,[4 Suppl] (abstract)
13. Chawluk J, Alavi A, Dann R, Kushner MJ, Hurtig H, Zimmerman RA, Reivich M, PET measurements of cerebral metabolism corrected for CSF contributions. J Nucl Med: in press (abstract)
14. Ruff RL, Dougherty JH Jr (1981) Evaluation of acute cerebral ischemia for anticoagulant therapy: Computed tomography of lumbar puncture. Neurology 31:736−740
15. Zimmerman RA: Personal communication
16. Buonanno FS, Pykett IL, Brady TJ et al (1983) Proton NMR imaging in experimental ischemic infarction. Stroke 14:178−184
17. Buananno FS, Kistler JP, DeWitt KR et al (1983) Nuclear Magnetic resonance imaging in central nervous system disease. Sem Nucl Med 13:329−338
18. Brant-Zawadzki M, Davis PL, Crooks LE, et al (1983) NMR demonstration of cerebral adnormalities: Comparison with CT. AJR 140:847−854
19. Pykett IL, Buonanno FS, Brady TJ, Kistler JP (1983) True three-dimensional nuclear magnetic resonance neuro-imaging in ischemic stroke: Correlation of NMR, X-ray CT, and pathology. Stroke 14:173−177
20. Doyle FH, Gore JC, Pennock JM, et al (1981) Imaging of the brain by nuclear magnetic resonance. Lancet 2 (8237):53−57
21. Ackerman RH, Alpert NM, Correia JA, et al (1979) Correlations of positron emission scans with TCT scans and clinical course. Acta Neurol Scand (Suppl 72) 60:230−231
22. Kuhl DE, Phelps ME, Kowell AP, et al (1980) Effects of stroke on local cerebral metabolism and perfusion: mapping by emission computed tomography of 18 FDG and 13 NH3. Ann Neurol 8:47−60
23. Ackerman RH (1981) Positron imaging in stroke disease. In: Cerebral vascular diseases: Research conferences on cerebral vascular disease, 12th ed. Moossy J and Reinmuth DM (eds.) New York: Raven Press pp 67−72
24. Baron JC, Bousser MG, Comar D, et al (1981) Noninvasive tomographic study of cerebral blood flow and oxygen metabolism in vivo. Eur Neurol 20:273−284
25. Lenzi GL, Frackowiak RSJ, Jones T (1981) Regional cerebral blood flow (CBF), oxygen utilization (CMRO$_2$), and oxygen extraction ratio (OER) in acute hemispheric stroke. J Cerebr Blood Flow Metabol 1 (Suppl 1):S504−505
26. Wise RJS, Bernardi S, Frackowiak RSJ, et al (1983) Serial observations on the pathophysiology of acute stroke: the transition from ischaemia to infarction as reflected in regional oxygen extraction. Brain 106:197−222
27. Baron JC, Bousser MG, Rey A, et al. (1981) Reversal of focal "misery-perfusion syndrome" by extracranial bypass in hemodynamic cerebral ischemia. Stroke 12:454−459
28. Grubb RL Jr, Ratcheson RA, Raichle ME, et al (1979) Regional cerebral blood flow and oxygen utilization in superficial temporal-middle cerebral artery anastomosis patients. J Neurosurg 50:733−741
29. Yamamoto YL, Little J, Thompson C et al (1979) Positron tomography with krypton-77 for evaluation of topographical rCBF changes following EC-IC bypass surgery. Acta Neurol Scand (Suppl 72) 60:522−523
30. von Monakow C (1914) Die Lokalisation in Großhirn und der Abbau der Funktion durch kortikale Herde. Wiesbaden, J. F. Bergmann

31. Lenzi GL, Frackowiak RSJ, Jones T, et al (1981) CMRO$_2$ and CBF by the oxygen-15 inhalation technique: Results in normal volunteers and cerebrovascular patients. Eur Neurol 20:285−290

32. Baron JC, Bousser MG, Comar D, et al (1981) Crossed cerebellar disaschisis: A remote functional depression secondary to supratentorial infarction of man. J Cereb Blood Flow Metabol 1 (Suppl 1):S500−S501

33. Kushner MJ, Alavi A, Reivich M, et al (1984) Contralateral cerebellar hypometabolism following cerebral insult. A PET study. Neurology (in press)

34. Martin WRW, Raichle ME (1983) Cerebellar blood flow and metabolism in cerebral hemisphere infarction. Ann Neurol 14:168−176

35. Fisher CM (1978) Ataxic hemiparesis: A pathologic study. Arch Neurol 35:126−128

36. Paxton R, Ambrose J (1974) The EMI scanner. A brief review of the first 650 patients. Br J Radiol 47:530−565

37. Earnest MP, Heaton RK, Wilkinson WE, Manke WF (1979) Cortical atrophy, ventricular enlargement and intellectual impairment in the aged. Neurology 29:1138−1143

38. Kasniak AW, Garron DC, Fox JH, Bergen I, Huckman M (1979) Cerebral atrophy, EEG slowing, age, education and cognitive functioning in suspected dementia. Neurology 29:1273−1279

39. Jacoby RJ, Levy R, Dawson JM (1980) Computed tomography in the elderly: I. The normal population. Br J Psychiatr 136:249−255

40. Hughes CP, Gado M (1981) Computed tomography and aging of the brain. Radiology 139:391−396

41. Barron SA, Jacobs L, Kinkel WR (1976) Changes in size of normal lateral ventricles during aging determined by computerized tomography. Neurology 26:1011−1013

42. Jacoby RJ, Levy R (1980) Computed tomography in the elderly 2. Senile dementia: Diagnosis and functional impairment. Br J Psychiatr 136:256−269

43. Roberts MA, Caird FI, Grossart KW, Steven JL (1976) Computerized tomography in the diagnosis of cerebral atrophy. J Neurol Neurosurg Psychiatr 39:909−915

44. Claveria LE, Moseley IF, Stevenson JR (1977) The clinical significance of "cerebral atrophy" as shown by C.A.T. In: The First European Seminar on Computerized Axial Tomography in Clinical Practice, duBoulay GH, Moseley IF (eds), Springer-Verlag, Berlin pp 213−217

45. Hughes CP, Gado M (1981) Computed tomography and aging of the brain. Radiology 139:391−396

46. Yamaura H, Ito M, Kubota K, Matsuzawa T (1980) Brain atrophy during aging: A quantitative study with computed tomography. J Gerontol 35:492−498

47. Miller AKH, Alston RL, Corsellis JAN (1980) Variation with age in the volumes of grey and white matter in the cerebral hemispheres of man: Measurements with an image analyser. Neuropath Appl Neurol 6:119−132

48. Ito M, Hatazawa J, Yamaura H, Matsuzawa T (1981) Age-related brain atrophy and mental deterioration − a study with computed tomography. Br J Radiol 54:384−390

49. Gado M, Hughes CP, Danziger W, Chi D, Jost G, Berg L (1982) Volumetric measurements of the cerebrospinal fluid spaces in demented subjects and controls. Radiology 144:535−538

50. George AE, deLeon MJ, Rosenbloom S, Ferris SH, Gentes C, Emmerich M, Kricheff II (1983) Ventricular volume and cognitive deficit: A computed tomographic study. Radiology 149:494−498

51. Arai H, Kogayashi K, Ikeda K, Nagao Y, Ogihara R, Kosaka K (1983) A computed tomography study of Alzheimer's disease. J Neurol 229:69−77

52. Herman GT, personal communication

53. Kundel HL, personal communication

54. Frackowiak RSJ, Lenzi GL, Jones T, Heather JD (1980) Quantitative measurement of regional cerebral positron emission tomography. Theory, procedure and normal values. J Comput Assist Tomogr 6:727−736

55. Frackowiak RSJ, Pozzilli C, Legg NJ, DuBoulay GH, Marshal J, Lenzi GL, Jones T (1981) A prospective study of regional cerebral blood flow and oxygen utilization in dementia

using positron emission tomography and oxygen-15. J Cerebr Blood Flow Metabol (Suppl 1) 1:S453

56. Foster NL, Chase TN, Fedio P, Patronas NJ, Brooks RA, DiChiro C (1983) Alzheimer's disease: Focal cortical changes shown by positron emission tomography. Neurology 33:961–965

57. Kuhl DE, Engel J, Phelps ME, Selin C (1980) Epileptic patterns of local metabolism and perfusion in human determined by emission computed tomography of 18-FDG and 13-NH3. Ann Neurol 8:348–360

58. Broderson P, Paulson JB, Bolwig TG (1973) Cerebral hyperemia in electrically induced seizures in man. Arch Neurol 20:334–338

59. Meyer JS, Gotoh F, Favale E (1966) Cerebral metabolism during epileptic seizures in man. Electroencephalogr Clin Neurophysiol 21:10–22

60. Penfield W, Von Santha K, Cirriani A (1939) Cerebral blood flow during induced epileptiform seizures in animals and man. J Neurophysiol 2:257–267

61. Posner JB, Plum F, Van Posnak A (1969) Cerebral metabolism during electrically induced seizures in man. Arch Neurol 20:388–395

62. Ingvar DH (1973) Regional cerebral blood flow in focal cortical epilepsy. Stroke 4:359–360

63. Hougaard K, Oikawa T, Sveinsdottir E, Skinhof E, Ingvar DE, Lassen NA (1976) Regional cerebral blood flow in focal cortical epilepsy. Arch Neurol 33:527–535

64. Ingvar DH (1975) rCBF in focal cortical epilepsy. In: Cerebral circulation and metabolism, edited by T. W. Langfitt, L. C. McHenry, Jr., M. Reivich, and H. Wollman, pp 361–363, Springer-Verlag, New York

65. Prince DA, Wilder BJ (1967) Control mechanisms in cortical epileptogenic foci. Arch Neurol 16:194–202

66. Abrams HL, McNeil BJ (1978) Medical implications of computed tomography (part one) NEJM 298:255–261

67. Weisberg L, Nice C, Katz M (1984) Cerebral computed tomography. A test – atlas, 2nd edition, W. B. Saunders Co., Philadelphia

68. Lilja A, Bergstrom K, Spannare B, Olsson Y (1981) Reliability of computed tomography in assessing histopathological features of malignant supratentorial gliomas. J Comput Assist Tomogr 5:625–636

69. Bydder GM, Steiner RE, Young IR, et al (1982) NMR imaging of the brain: 140 cases. AJR 139:215–236

70. Bydder GM, Steiner RE (1982) NMR imaging of the brain. Neuroradiology 23:231–240

71. McGinnis B, Brady TJ, New PFJ et al. Nuclear magnetic resonance (NMR) imaging of tumors in the posterior fossa. J Comput Assist Tomogr (in press)

72. Brady TJ, Buonanno FS, New PFJ et al (1982) Comparison of NMR imaging to X-ray CT in detecting brain tumors, RSNA, 68th Scientific Assembly and Annual Meeting, Chicago

73. Zimmerman RA: Personal communication

74. DiChiro G, DeLapaz RL, Brooks RA, Sokoloff L, Kornblith PL, Smith BH, Patronas NJ, Kufta CV, Kessler RM, Johnston GS, Manning RG, Wolf AP (1982) Glucose utilization of cerebral gliomas measured by [^{18}F] fluorodeoxyglucose and positron tomography. Neurology 32:1323–1329

75. Patronas NJ, DiChiro G, Brooks RA, DeLapaz RL, Kornblith PL, Smith BH, Rizzoli HV, Kessler RM, Manning RG, Channing M, Wolf AP, O'Connor CM (1982) [^{18}F]fluorodeoxyglucose and positron emission tomography in the evaluation of radiation necrosis of the brain. Radiology 144:885–889

76. Van Dellen JR, Danziger A (1978) Failure of computerized tomography to differentiate between radiation necrosis and cerebral tumour. S Afr Med J 53:171–172

Nuclear Magnetic Resonance (NMR) Imaging and Spectroscopy in Experimental Brain Edema in the Rat

H. M. Bartkowski, J. Bederson, M. Nishimura, M. Brant-Zawadzki, K. Moon, S. Longar, and L. Pitts

Introduction

High resolution in vivo NMR images are now obtainable and noninvasively provide both anatomic and biochemical information [1]. The quantitation of brain edema, the effects of cerebral blood volume upon NMR relaxation values, and the reliability of imager longitudinal and transverse relaxation times are several aspects of NMR imaging and spectroscopy which have not been adequately explored. These factors can be investigated by in vitro NMR spectroscopy, in vivo NMR imaging, microgravimetry and wet/dry weights, cerebral blood volume determination, and phospholipid assay in experimental brain edema.

Methods

Sixty young adult Sprague-Dawley rats of either sex weighing approximately 350 g each were used in this study. The rats were divided into six groups: control, fluid percussion brain injury [2], cold lesion [3], infarct [4], abscess [5], and tumor [6]. Those animals in the percussion brain injury, cold lesion, and infarct groups were studied under chloral hydrate anesthesia 24 h after insult; the abscess and tumor animals were studied on the 7th and 11th day, respectively, following the implantations. Cerebral blood volume measurements were done in several animals of the control, cold lesion, and percussion injury groups utilizing a chromium 51 technique [7]. NMR imaging was performed with a small animal imager utilizing an external magnetic field of 0.35 Tesla and obtaining five contiguous tomographic sections, each 4.8 mm thick. Immediately after NMR imaging was completed, the animals were euthanized with a barbiturate overdose, and the brains removed. Coronal slices were made at the level of the lesion and gray and white matter samples were dissected for the following determinations: T1 using a pulse spectrometer at 0.4 Tesla and 17.43 MHz with an inversion recovery sequence, water contents by microgravimetry [8] and wet/dry weights, phospholipid phosphorus [9], and histology. Statistical analyses

---

Department of Neurosurgery and Neuroradiology, University of California, San Francisco, USA

Cerebral Blood Flow and
Metabolism Measurement.
Eds. Hartmann/Hoyer
© Springer-Verlag Berlin Heidelberg 1985

were carried out to ascertain the existence, if any, of correlations among the various parameters studied.

Results

Based on *t*-test of means, all groups studied had prolonged T1 and T2 values of the ipsilateral hemisphere compared with controls ($P < 0.001 - 0.05$) (Table 1). Spectroscopic T1 values differed from imager T1 values by 25% – 70% in terms of absolute numbers, with control, cold lesion, and tumor values demonstrating the greatest differences. However, correlative trends were similar. The contralateral hemispheres of the cold lesion and percussion injury groups were found to have significantly prolonged T2 values compared with controls ($P < 0.001$). Significantly increased water contents were found in the brain tissue adjacent to the particular lesions in all groups ($P < 0.001 - 0.05$). Water contents were not increased in the contralateral hemispheres. A linear relationship existed between increasing water content and increasing T1 and T2 relaxation values within individual brain edema groups. Comparison of percent change in T1 and T2 values with water content among the different brain lesion

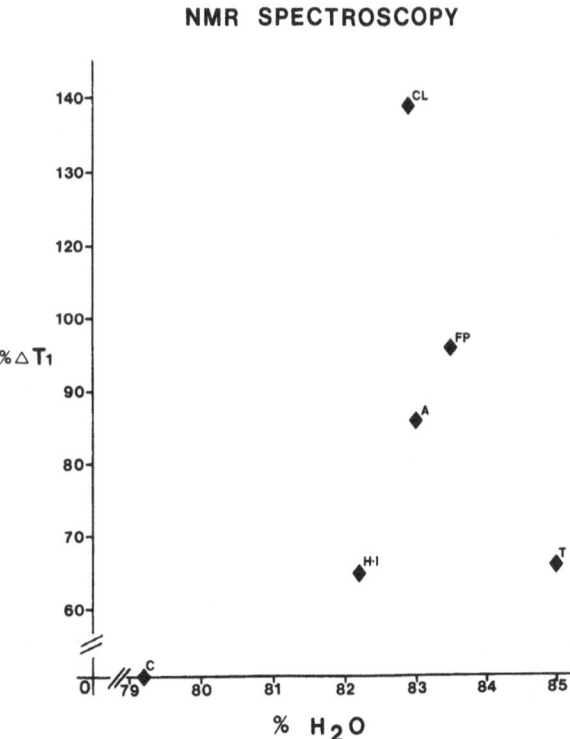

Fig. 1. In vitro percent change in T1 relaxation time of experimental groups compared with controls

Table 1. Mean values of T1 and T2 relaxation times

	Imager		Spectrometer	
	T1	T2	T1 (white)	T1 (gray)
Control ($n=5$)	1007 ± 122	48 ± 3	332 ± 147	341 ± 76
Impact ($n=8$)	1138 ± 188	60 ± 4	787 ± 420	897 ± 308
Cold ($n=6$)	1455 ± 143	60 ± 2	623 ± 117	825 ± 337
Infarct ($n=8$)	1253 ± 347	55 ± 3	1128 ± 155	735 ± 456
Abscess ($n=5$)	1069 ± 148	52 ± 2	687 ± 333	747 ± 72
Tumor ($n=9$)	1515 ± 591	58 ± 7	295 ± 86	662 ± 83

T1 = longitudinal relaxation time; T2 = transverse relaxation time

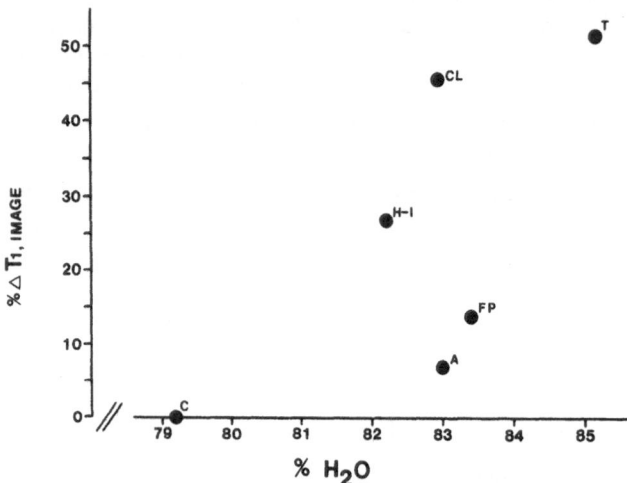

Fig. 2. In vivo percent change in T1 relaxation time of experimental groups compared with controls

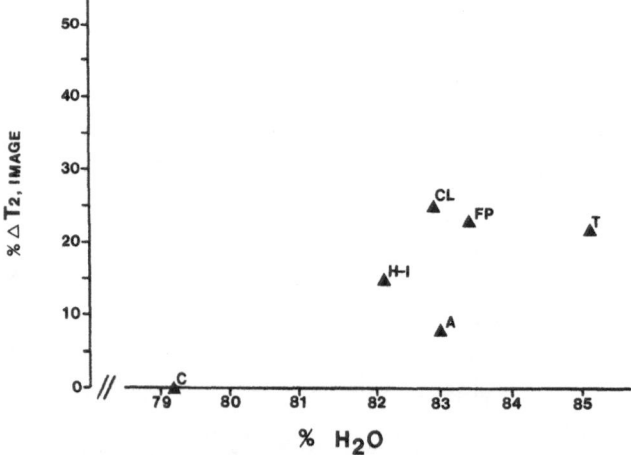

Fig. 3. In vivo percent change in T2 relaxation time of experimental groups compared with controls

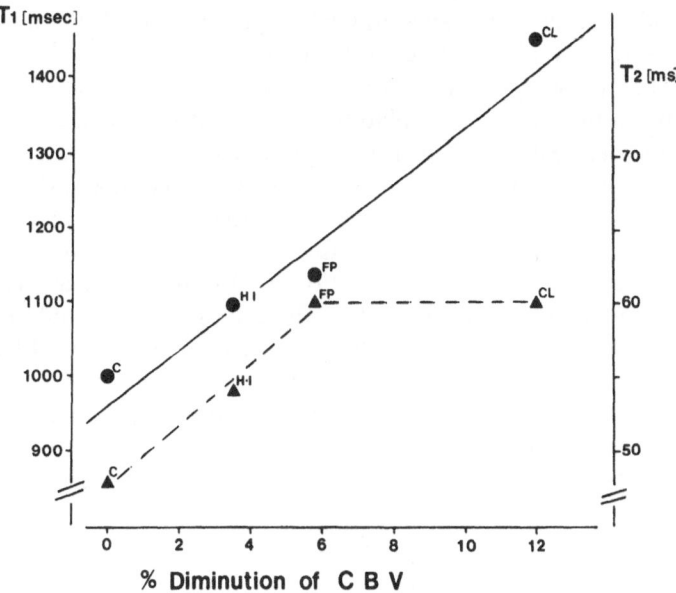

Fig. 4. Correlation of T1 and T2 relaxation times with decreasing cerebral blood volume

groups revealed no direct relationship between water content and relaxation values (in vitro T1, in vivo T1, in vivo T2; see Figs. 1, 2, and 3, respectively). Phospholipid phosphorus determinations in the cold lesion and fluid percussion groups were significantly altered ($P < 0.05$ and $P < 0.01$, respectively) from control values. Cerebral blood volume studies revealed a 4%–14% decrease compared with controls. Decreasing cerebral blood volume correlated linearly with increasing T1 relaxation values and nonlinearly with increasing T2 relaxation values (Fig. 4). Histological sections confirmed the presence of individual lesions and edema.

Discussion

The study demonstrated that although a linear relationship existed between increasing water content and increasing T1 and T2 prolongations within individual experimental groups, this relationship did not hold true when one group was compared with another, indicating that very different T1 and T2 values may be gotten for relatively similar water contents. Thus one should be cautious in trying to quantitate the amount of brain edema based upon T1 and T2 values, especially in the clinical setting where various combinations of edema mechanisms may be operating. A decreasing cerebral blood volume correlated linearly with an increasing T1 prolongation and nonlinearly with an increasing T2 prolongation, suggesting that changes in CBV may significantly affect T1 and T2 relaxation values. Prolongation of T2 in the contralateral hemispheres of the cold and percussion injury groups without an increase in water content and without any change in CBV suggests that NMR may be detecting changes not readily identifiable by standard laboratory methods, i.e., changes in the "state of water" in the tissue, specifically, bound versus free water. It is known that multiple sclerosis plaques are beautifully seen on NMR images although the basic pathology of multiple sclerosis plaques is that of myelin degradation without any significant increases in water content. Altered phospholipid contents in the two groups assayed adds to the speculation that lipids may certainly affect T1 and T2 relaxation values as well as NMR images.

We conclude that prolongations of T1 and T2 are associated with increases in brain water content. However, the relationship is nonlinear and quantification of brain edema is not possible solely on the basis of T1 and T2 prolongation. Other factors are involved, for instance, changes in CBV, as we have shown. In addition, NMR parameters may be altered by changes in the molecular environment of water protons, i.e., bound versus free water, and in brain phospholipids.

Summary

Experimental data from our laboratory on cerebral brain edema and NMR imaging and spectroscopy suggest that although a linear relationship exists between increasing water content and increasing longitudinal (T1) and transverse (T2) relaxation time prolongations within individual brain lesion groups, this relationship does not hold true when various types of brain lesion groups are compared with each other. This indicates that very different T1 and T2 values may be gotten for relatively similar water contents. Thus, one should be cautious in trying to quantitate brain edema based solely upon T1 and T2 values, especially in the clinical setting where various combinations of edema formation mechanisms may be operating. Factors other than water content may affect T1 and T2 relaxation values; for example, our data indicate that a decreasing cerebral blood volume (CBV) is correlated directly with increasing T1 and T2 relaxation times.

References

1. Bydder GM, Steiner RE, Young IR, et al (1982) Clinical NMR imaging of the brain: 140 cases. Amer J Radiol 139:215–236
2. Sullivan HG, Martinez J, Becker DP, et al (1976) Fluid percussion model of mechanical brain injury in the cat. J Neurosurg 45:520–534
3. Klatzo I, Piraux A, Laskowski EJ (1958) The relationship between edema, blood-brain barrier and tissue elements in local brain injury. J Neuropath Exp Neurol 17:548–564
4. Plum F, Posner JB, Alvord EC (1963) Edema and necrosis in experimental cerebral infarction. Arch Neurol 9:563–570
5. Britt RH, Enzmann DR, Yeager AS (1981) Neuropathological and computerized tomographic findings in experimental brain abscess. J Neurosurg 55:590–603

6. Barker M, Hoshino T, Gurcay O, et al (1973) Development of an animal brain tumor model and its response to therapy with BCNU. Cancer Res 3030:976–986
7. Albert SN, Hirsch EF, Economopoulos B, et al (1968) Triple-tracer technique for measuring red-blood-cell, plasma and extracellular-fluid volume. J Nuc Med 9:19–23
8. Shigeno T, Brock M, Shigeno S, et al (1982) The determination of brain water content: microgravimetry versus drying-weighing method. J Neurosurg 57:99–107
9. Chan PH, Yurko M, Fishman RA (1982) Phsopholipid degradation and cellular edema induced by free radicals in brain cortical slices. J Neurochem 38:525–531

The Capabilities of Ultrasonic Diagnostic Procedures in Cerebrovascular Disease

G.-M. von Reutern and H. Kapp

Ultrasonic methods are used to examine the hemodynamics and morphology of the blood vessels that supply the brain. Direct information on the circulation in the brain itself is not obtained. In clinical use, these ultrasonic methods should provide an answer to the question as to why strokes or ischemic attacks occur or whether they are threatened. Recognition of those vascular disorders which reduce the volume flow is, in this respect, not sufficient. The number of strokes in cases with normal volume flow, due to embolic complications of arteriosclerosis or to sudden thrombosis in an existing noncritical narrowing of the vessel, are by far in the majority. Ultrasonic methods provide the clinician with information both on changes of varying severity of the vessel wall and on the hemodynamic effects of this. They are complex and reliable clinical methods, and not simply screening tests. Where necessary, they are supplemented with angiography.

Technical Principles

Ultrasound which is sent into the body is reflected at interfaces. The artery walls are particularly strong reflecting interfaces. The reflections from the blood cells in the vessel are about 40−60 dB weaker than those from the vessel walls. This difference makes it possible to depict the vessel walls as an image. Two basically different methods are available to obtain information from the reflected ultrasound: the pulse echo method for imaging of vessels, and the Doppler method for evaluating the flow.

With the *pulse echo method* (Fig. 1) short trains of pulses are sent in rapid sequence, and in the intervals the intensity and transit time of the reflected signals are measured. The return time is a measure of the depth from which the signal is reflected. A-scan refers to time plotted on the x-axis against intensity on the y-axis. B-scan refers to a two-dimensional image made up of several sound beams. A reflection is shown as a spot with varying brigthness according to the intensity of the reflection. Such a sectional image can be achieved by the movement of the transmitter or the parallel use of several transmitters.

Abteilung klinische Neurologie und Neurophysiologie der Universität Freiburg, Hansastraße 9, D-7800 Freiburg

Cerebral Blood Flow and
Metabolism Measurement.
Eds. Hartmann/Hoyer
© Springer-Verlag Berlin Heidelberg 1985

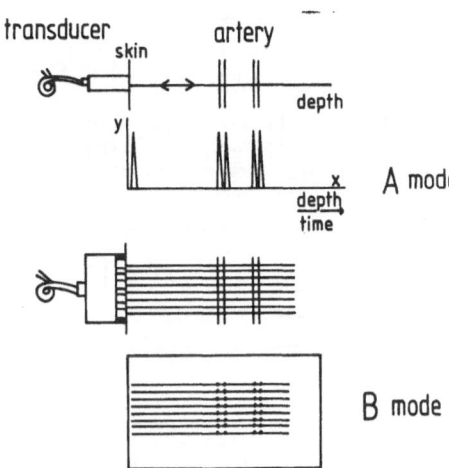

Fig. 1. Echo impulse technique. For explanation see text

With the *Doppler method,* it is not the transit time of the reflected ultrasound (i.e., the site of a reflector), but the speed of the reflector in relation to the transmitter that is measured by using the Doppler effect. If the reflector is approaching the transmitter/receiver, then the received frequency is higher than the transmitted frequency. The difference between the two is the Doppler frequency (Doppler shift), which in the insonation of blood vessels is proportional to the velocity of the blood cells and is dependent on the angle of insonation. We distinguish between continuous wave (CW) Doppler and pulsed Doppler. With CW Doppler, the tip of the probe contains separate transmitting and receiving crystals, which continuously transmit and receive. All particles which move within the beam of the ultrasound contribute to the resulting measurement. Thus this measurement gives no information as to the depth at which it is made. With pulsed Doppler a transmitter is used which acts as a receiver in the intervals between transmitting. With this technique the position and size of the sample volume can be selected by means of electronic gating. Only the signal of a particular time interval between the transmitted pulse is processed. Through the known speed of ultrasound in tissue, the depth and size of the sample volume is therefore defined.

Figure 2 is a schematic diagram of a Doppler instrument with directional measurement. After amplification the received signals are compared with the transmitted signals in two detectors, one of which receives a signal with a phase shift of 90 °. This permits the determination of the flow direction. The Doppler signals contain a spectrum of frequencies which correspond to the varied flow velocities of the blood cells. This spectrum contains the maximum information. For the cheapest and most common documentation, frequencies are detected by a zero-crossing system. The continuous registration of an electronically average mean value produces the well-known velocity pulse curves. Independent of this registration, the spectrum is evaluated by the ear of the operator via a loudspeaker. The operator can hear more than the curve can document. In searching for an artery or in following the course of an artery, the operator is guided by the acoustic signal. Less common, because of its expense, is registration of the

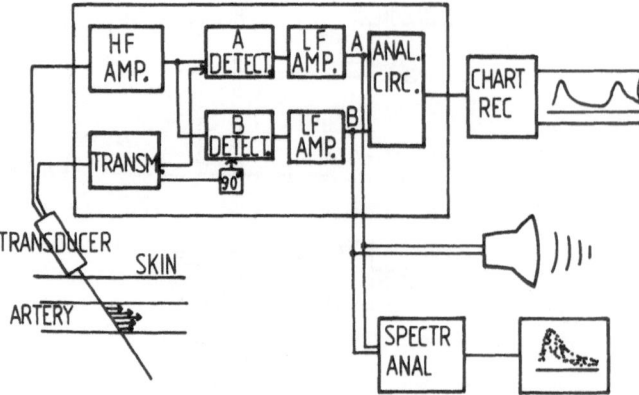

Fig. 2. Schematic diagram of Doppler instruments with detection of flow direction (modified after Reid). For explanation see text

spectrum in real time. To permit this type of documentation, the frequency is plotted against time. The intensity of a frequency band represents the number of insonated particles flowing at a specific speed. Intensity is reproduced as dot density, gray scale or color coded (Fig. 5).

Doppler flow-imaging permits the imaging and documentation of the course of a vessel. For this, the Doppler transducer is held in a scanning arm at a fixed angle. The positions of received Doppler signals are registered on a storage oscilloscope with the help of position translating circuits. By scanning manually, a diagram of the flowing column of blood is built up. This method does not permit an anatomically exact image of the arteries; rather it should be considered an additional sophisticated method of documentation which does not basically increase the diagnostic possibilities.

The most complex of all systems is the *duplex scanner*. It produces cross-section views (B-scan) of the pulsating artery in real time and carries out a simultaneous Doppler measurement. Figure 3 for example, shows a schematic dia-

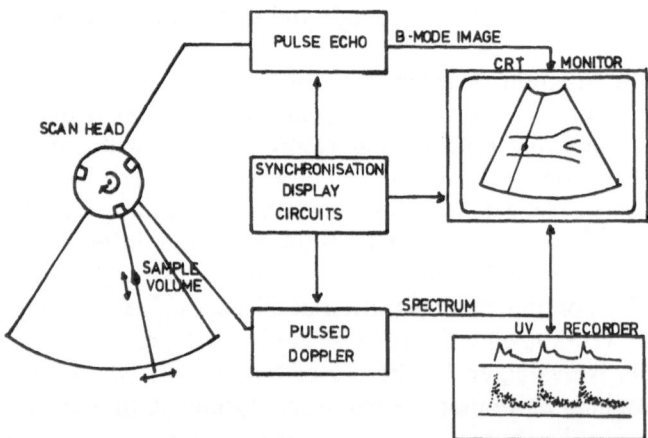

Fig. 3. Schematic diagram of a duplex scanner. For explanation see text

gram of a mechanical sector scanner. Three rotating transducers produce a real time B-scan. The rotor can be stopped, the picture frozen on the monitor, and a transducer then used for a Doppler measurement. These instruments thus produce images of the vessel wall and their movement can be observed. Additionally, the velocity of blood is measured in the imaged vessels.

The measurement of volume flow will not be dealt with here, as this is the subject of the contribution by H. R. Müller et al. in this symposium.

In the following the application of the methods and their results in the various vessel areas will be discussed.

Carotids

The first ultrasonic examinations of the brain-supplying arteries were made with continuous wave Doppler, which can still be considered the standard method today. The examination of the carotid system began historically with the insonation of branches of the ophthalmic artery, which are end-arteries of the carotid artery. Today, this "doppler ophthalmic test" is just one criterion among many. Used on its own, it is very unreliable. Nevertheless, this test (and

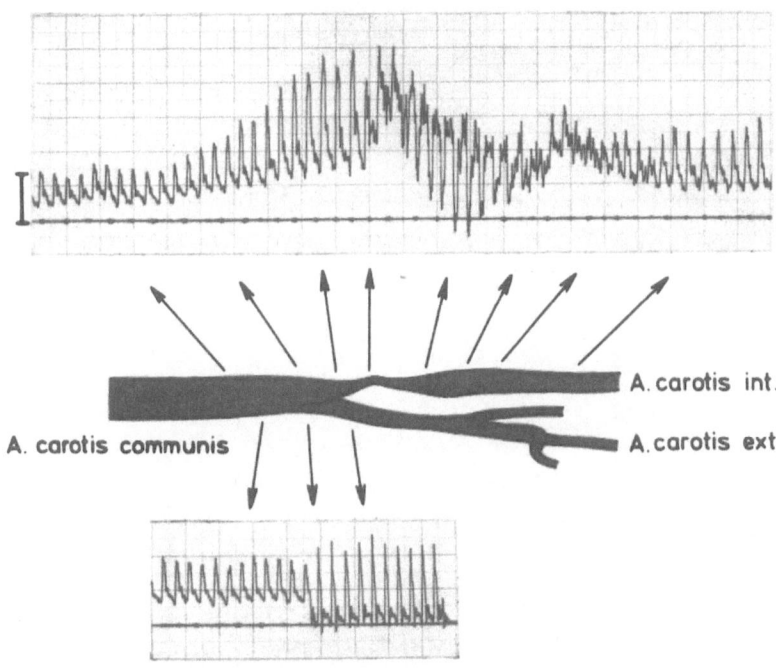

Fig. 4. CW Doppler sonography image of a carotid bifurcation with stenosis of the internal carotid artery. The narrowing is characterized by increasing flow velocity, followed by turbulences which are documented in the irregularities of the curve. The probe is shifted slowly along the vessel during recording. Semischematic vessel image after angiogram (Büdingen et al. 1982)

not the complete Doppler examination) is still encountered, together with oph-
thalmoplethysmography and phonoangiography in so-called test batteries. This
appears to have little purpose. The low specificity and sensitivity of this test is
perhaps the reason why Doppler sonography is today still regarded by many as
an inexact screening method. Significantly more exact results are obtained
through combination of the ophthalmic examination with direct insonation of
the arteries in the neck. This can be carried out either with a hand-held probe
or with flow imaging. With equal experience in the methods, results are about
the same. Although only flow velocity is measured, information is also obtained
on the morphology of the vessel. The blood flow is slower at the wider parts
and, conversely, faster at the narrow parts. This is shown in Fig. 4 on a carotid
bifurcation with a stenosis of the internal carotid artery. The difficulty in Doppler sonography is the correct identification of the vessels. This is achieved
mainly by a range of compression tests. Additionally, the external and internal

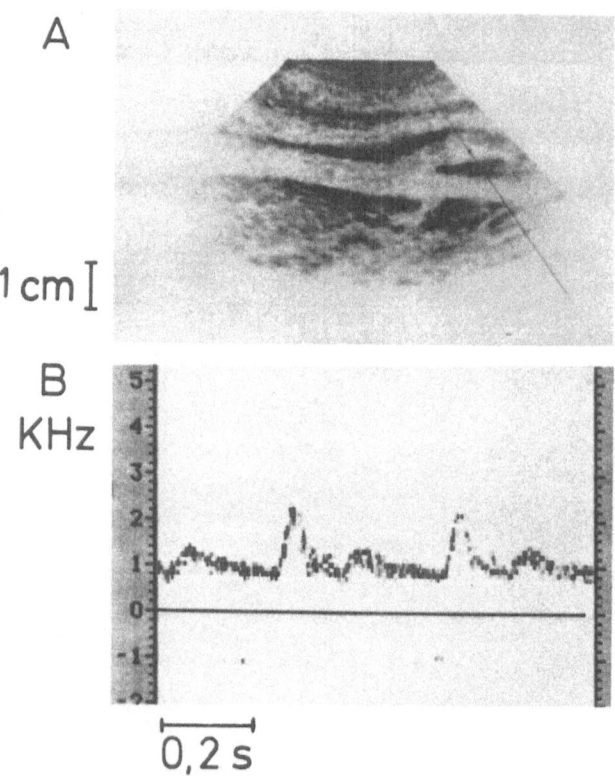

Fig. 5a, b. Duplex scan examination of a normal carotid bifurcation. **a** B-mode image of the
common carotid artery (left), internal carotid (above) and external carotid artery (below). The
direction of the Doppler-beam is marked by the inserted line, the location of the sample volume is indicated by a bright dot. **b** Spectrum of flow velocities in the internal carotid artery. The
position of the sample volume is shown in **a**

carotids can be distinguished by their different pulse forms. A detailed description of these criteria, which always permit a correct identification is given elswhere (Büdingen et al. 1977; Büdingen et al. 1982; Pourcelot 1975; von Reutern and Thron 1980). An occlusion of the internal carotid artery produces a reduced speed of flow in the common carotid artery and a Doppler signal cannot be obtained in the internal carotid artery. A stenosis is recognizable by a defined area of high frequencies, followed by turbulences and a poststenotic signal of reduced amplitude. Both the turbulences and the acceleration of flow are best documented with the spectrum analyzer. The determination of the degree of stenosis is based on hemodynamic criteria. One parameter, for example, which correlates well with the degree of stenosis is the systolic peak frequency in the region of the stenosis (Spencer and Reid 1979). Also, nearly occluding stenoses can be demonstrated and correctly differentiated from occlusions.

The accuracy of the Doppler diagnosis of an occlusion of the internal carotid artery is about 90% in the hands of an operator who applies the method correctly but does not have very much experience. Good laboratories achieve an accuracy of over 95% (Büdingen et al. 1982; von Reutern and Thron 1982). CW Doppler sonography is still the most reliable method for middle – and high-grade stenoses (only the latter reduce volume flow) and occlusions, even though it requires the minimum of technical investment. However, low-grade vessel wall disease may have little significant effect even on local flow patterns, and its diagnosis is therefore unsatisfactory. This is an indication for the additional use of the duplex scanner (Fell et al. 1981; von Reutern and Büdingen 1981). In contrast to angiography, the vessel wall itself is imaged (Fig. 5), including in cases of occlusion. The paper of M. Hennerici will deal with the further possibilities of this method.

Vertebral Arteries

Standard CW Doppler sonography is still the most informative ultrasonic examination of the vertebral arteries. A sector scan of the proximal part of the vertebral arteries is sometimes succesful, but anatomic variations and bony structures do not allow a regular and complete imaging.

The following diagnoses can be made reliably with CW Doppler sonography: subclavian steal, basilar artery occlusion, and occlusion or ostial stenosis of the vertebral artery of greater than 50%. The diagnosis of subclavian steal is possible upon examination of the atlas loop of the vertebral artery. For the diagnosis of occlusions and stenoses, additional examination of the proximal segment of the vertebral arteries is necessary. This requires clear differentiation of the vertebral arteries from the numerous other vessels of this region. The manipulation of the probe and compression tests needs a great deal of manual dexterity and practice. This is probably the reason why examination of the vertebral arteries is not generally used. It is, at least as a complement to the carotid findings, of significance. Thus a synoptic evaluation of four brain-supplying arteries is possible. This is helpful in differential diagnosis, since it can be difficult to differentiate between carotid and vertebrobasilar insufficiency on the basis of the symptoms alone. Occlusions in both the vertebral and carotid area represent a particularly serious finding (Hennerici et al. 1982).

Intracranial Arteries

The examination of the intracranial arteries has until now been prevented by the lack of an instrument that could enable insonation through the bone, which severely weakens the ultrasonic beam. Only total occlusion in the area of the carotid or the main trunk of the middle cerebral arteries could be recognized by observing the differences in the velocity of flow between the two carotids (von Reutern et al. 1977). Aaslid, Markwalder, and Nornes were the first to report direct transcranial Doppler examination of the cerebral arteries (Aaslid et al. 1982). They used a pulsed Doppler technique with a low frequency (2 MHz). In recent months we have been able to gather experience with a prototype of this instrument. In the Neurosurgical Department of Freiburg University Clinic, J. M. Gilsbach and his colleagues are testing applications to demonstrate cerebrovascular spasms, while we in the Neurological Clinic are more interested in the demonstration of intracranial occlusions and evaluation of the effect of extracranial occlusions on the intracranial flow.

Also in this vascular area, the difficulty lies in the identification of the vessels. It is not sufficient to know the depth at which one is measuring (pulsed Doppler). The vessels of the circle of Willis are too close together to be differentiated by depth measurement alone. Compression tests must be additionally applied. These permit identification of the vessel and simultaneously provide information on the collateral function of the circle of Willis. Figure 6 shows the examination of the anterior cerebral artery. If the common carotid artery on the same side is compressed, then the direction of flow in this artery reverses. An incomplete compression can produce an oscillating flow. In patients with carotid occlusion it is possible to demonstrate a cross-flow over the anterior communicating artery in the same way. Also the posterior cerebral arteries and the basilar bifurcation can be found from the temporal approach. If those arteries are examined under carotid compression then an increase in flow or no change is observed. In this way they can be distinguished from the middle cerebral arteries. At the same time their collateral function is demonstrated. Doppler examination of the neck arteries alone usually (but not always) permits the demonstration of occlusions of the main trunk of the middle cerebral artery. This is, however, impossible if an additional extracranial stenosis of the carotid artery is present. Direct transcranial insonation can demonstrate an additional stenosis or an occlusion of the main trunk of the middle cerebral artery in such cases. Another question to be considered is the hemodynamic effect of an extracranial occlusion.

It is not certain what role transcranial Doppler examination will play in routine clinical application. It has, however, undoubtedly expanded the diagnostic possibilities of Doppler sonography significantly. A further possibility of

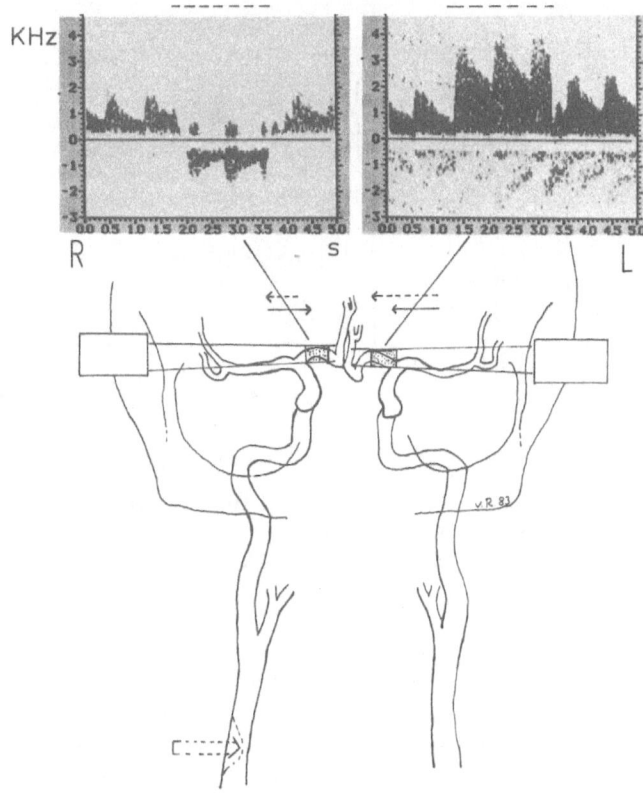

Fig. 6. Transcranial Doppler sonography (2 MHz pulsed). Examination of anterior cerebral arteries of both sides. Compression of the right common carotid artery (→) produces a reversal of flow in the anterior cerebral artery of the same side and an increase in flow (with unchanged direction) in the left anterior cerebral artery. Flow away from the probe is shown by an upward deflection of the pulse curve in relation to the zero-line

examining the intracranial vessels will be covered by Dr. J. M. Gilsbach in his contribution on intraoperative Doppler sonography with miniaturized probes.

Summary

Different ultrasonic methods are available for individual vascular areas. The continuous wave Doppler is the most universally used and produces reliable results in the examination of the carotid and vertebral systems. Examination of the carotid can be enhanced by the combination of B-scan and Doppler sonography (duplex scan). Low-grade stenoses and plaques are particularly well reproduced by this method. The possibility of missing a pathologic, and possibly operable lesion after continuous wave Doppler and duplex examination is small. The basal cerebral arteries could until recently be only unsatisfactorily evaluated with ultrasound. A breakthrough has been achieved with the new transcranial pulsed Doppler sonography.

The ultrasonic methods therefore offer an excellent addition to those methods which examine the circulation and metabolism of the brain, in that

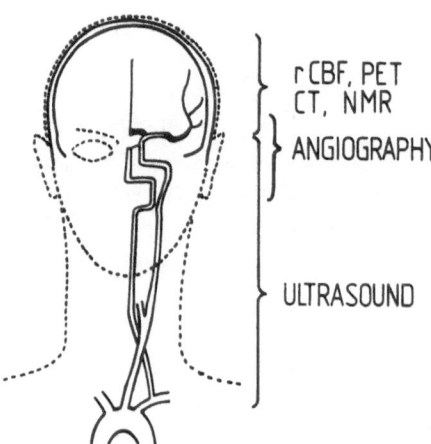

rCBF, PET
CT, NMR

ANGIOGRAPHY

ULTRASOUND

Fig. 7. Main areas of indication of the various examination methods in cerebrovascular disturbances

they represent the anatomy and function of the arteries from the aortic arch to the circle of Willis (Fig. 7).

With regard to patient management, ultrasonic methods can be applied with the following advantages:

− The examination of all patients with cerebrovascular disease and those who, due to nontypical symptoms, are merely suspected of such a disease
− The first provisional categorization of patients into those in whom vascular surgery appears necessary and those who, prior to further measures, should undergo additional neurologic, cardiologic, or hemostatic examination
− The possibility of frequent follow-up examinations

The indication for angiography is present, especially preoperatively, when the section of the internal carotid artery at the base of the skull requires to be imaged.

Angiography of the intracranial arteries is often indicated for differential diagnostic purposes. Venous digital substraction angiography does not meet these requirements.

In common with other methods, application of ultrasonic methods in inexperienced hands does more harm than good. Unfortunately, this is not seldom the case, and it is associated with both a lack of the time necessary to practice the method and the willingness to provide sufficient equipment and personnel. From the point of view of the patient, however, such effort and equipment is a useful investment.

Acknowledgment. We thank Dr. Alec Eden for translating the manuscript.

References

Aaslid R, Markwalder TM, Nornes H (1982) Noninvasive transcranial Doppler ultrasound recording of flow velocity in basal cerebral arteries. J Neurosurg 57:769−774

Büdingen HJ, Reutern von GM, Freund HJ (1977) Diagnosis of cerebro-vascular lesions by ultrasonic methods. International Journal of Neurology 11:206–218

Büdingen HJ, Reutern von GM, Freund HJ (1982) Doppler-Sonographie der extrakraniellen Hirnarterien. Thieme, Stuttgart

Fell G, Philips DJ, Chikos PM, Harley JD, Thiele BL, Strandness DE (1981) Ultrasonic duplex scanning for disease of the carotid artery. Circulation 64:1191–1195

Hennerici M, Rautenberg W, Mohr ST (1982) Stroke from symptomless extracranial arterial disease. The Lancet 27:1180–1183

Pourcelot L (1975) Indications de l'ultrasonographie Doppler dans l'étude des vaisseaux périphéries. L'Année du Praticien 25:4671–4680

Reutern von GM, Thron A (1980) Dopplersonographie des carotides. Résultats et causes d'erreurs. Actualité d'Angeilogie 5/1:9–15

Reutern von GM, Büdingen HJ (1981) Möglichkeiten und Grenzen der Dopplersonographie an den extrakraniellen Hirnarterien. Ultraschall 2:35–42

Reutern von GM, Voigt K, Ortega-Suhrkamp E, Büdingen HJ (1977) Dopplersonographische Befunde bei intracraniellen vaskulären Störungen. Differentialdiagnose zu Obliterationen der extracraniellen Hirnarterien. Arch Psychiat Nervenkr 223:181–186

Spencer MP, Reid JM (1979) Quantification of carotid stenosis with continuous-wave (C-W) Doppler ultrasound. Stroke 10:326–330

Intracranial Doppler Sonography During Surgery

J. GILSBACH

Introduction

Reports on intracranial Doppler sonographic measurements of blood flow velocity in dissected brain arteries have been rare (Friedrich et al. 1980; Hitchon et al. 1979; Moritake et al. 1980; Nornes et al. 1979; Nornes and Grip 1980). The main reason for this is that commercial instruments and probes were not suited for use on tiny vessels. A high frequency Doppler velocity meter, specially developed for microsurgery with miniaturized and sterilizable probes, now makes it possible to record vessels as small as one-tenth of a millimeter, even in the narrow space at the base of the brain. The method provides a direct, atraumatic, repeatable control of the surgical measures during aneurysm, bypass, and angioma operations on the manipulated vessel itself. Patency, flow direction, and flow disturbances can be assessed on the basis of velocity distribution and changes.

Furthermore, with the Doppler method a number of hemodynamically interesting observations can be made, beyond those associated with the operative measures themselves.

Method

Using a pulsed Doppler ultrasound velocity meter (MF 20, Eden Medical Electronic, Überlingen) with a transmission frequency of 20 MHz and probes having an external diameter of 2 or 3 mm, the studies were carried out on dissected vessels under the surgical microscope. In addition to the acoustic evaluation, signals were processed via a built-in zero-crosser with spatial and time mean frequency registration. Polaroid documentation of the Doppler spectrum was provided by an additional real time spectrum analyzer.

Normal Findings

In cases of asymptomatic aneurysms and small basal tumors, normal brain vessels could be recorded under open operative conditions. The flow patterns resembled those of the internal cervical carotid artery (Fig. 1). The time

Neurochirurgische Universitätsklinik, D-7800 Freiburg

Cerebral Blood Flow and
Metabolism Measurement.
Eds. Hartmann/Hoyer

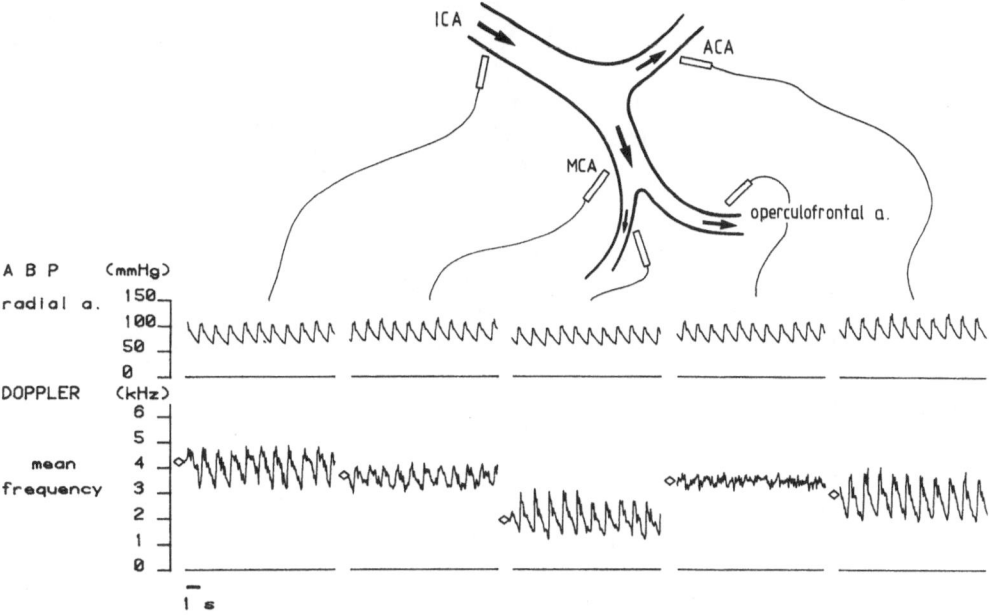

Fig. 1. Flow pattern of a normal intracranial vessel tree in a case of a small extracerebral basal tumor. Note the high, unmodulated flow in the operculofrontal branch due to hyperemia after discontinuation of retraction of the frontal lobe. (Gilsbach 1983)

averaged mean Doppler frequencies in the intracranial internal carotid artery and in the middle cerebral artery were about the same (1.5–6, average 3.3 kHz) and, at an incident angle of 55 °, corresponded to mean velocities of 22 cm/s. The anterior system, however, with a velocity of 19 cm/s (0.8–4.4, average 2.8 kHz), was clearly slower.

Superficial Temporal Artery – Middle Cerebral Artery Anastomoses

Intraoperative Doppler sonographic examinations were carried out on 40 anastomoses between the superficial temporal artery (STA) and the middle cerebral artery (MCA). Ninety percent of the anastomoses were primarily open (Fig. 2). In 10%, stenoses or occlusions had to be remedied by mechanical manipulation or by reopening the suture. Hemodynamically ideal anastomoses showed no irregularities or acceleration, either at the site of the temporary clip or in the anastomotic region itself, and had a bidirectional flow in the recipient artery (i.e., in addition to an orthograde distal flow there was a retrograde proximal flow towards the Sylvian fissure).

The proximal recipient artery can be used as a standard for the efficiency and necessity of an anastomosis. A high, retrograde flow only occurs if there is a marked pressure difference between the recipient and donor branches. Accord-

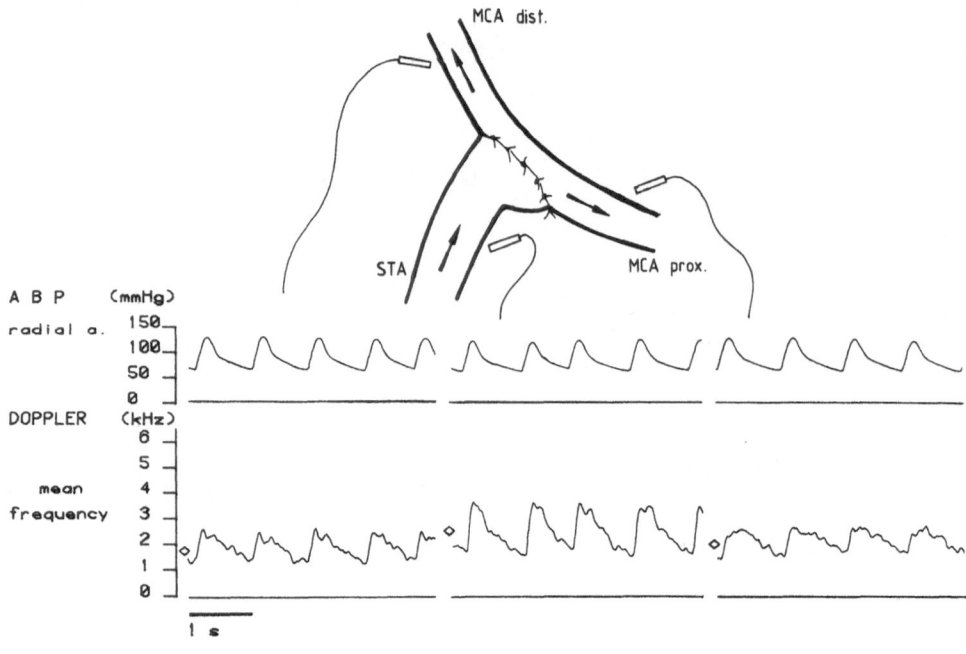

Fig. 2. Flow pattern of a well functioning extracranial-intracranial anastomosis between the parietal branch of the superficial temporal artery (*STA*) and a branch of the middle cerebral artery (*MCA*). Note the reversed proximal flow, directed to the Sylvian fissure (Gilsbach 1983)

ingly, these findings are not established in open cervical arteries (see below) or in unsatisfactory sutures.

The recipient temporal branches of the MCA had preanastomotic flow velocities that corresponded to Doppler frequencies of 0.2 to 3.6 kHz, which could be divided roughly into two separate groups: (a) frequencies above 1 kHz, which are typical for vessels of comparable size and location among healthy persons and which are regarded as normal, and (b) frequencies under 0.8 kHz with sometimes vein-like flow patterns which were not detectable in normal cases. An interesting finding was that half the patients with asymptomatic occlusions had pathologically slow velocities and consequently a good bypass flow. This means that perhaps these patients profit from the bypass in hemodynamic stress situations. On the other hand, almost half the patients with transient ischemic attacks showed normal flow velocities and only a fair bypass function (Table 1). In these cases the question must be asked as to whether there may have been causes other than the occlusions and if perhaps these operations are not as necessary as we supposed.

In cases of bilateral carotid artery occlusions, the velocities in the recipient artery were always pathologic. In unilateral occlusions only 64% of the velocities were pathologic and in cases of unilateral stenoses, only 50%. The patients with occlusive and stenotic processes and having normal Doppler findings were predominantly asymptomatic or had only transient ischemic attacks (Table 2).

Table 1. Clinical findings in the patients with occlusive disease and time averaged mean Doppler frequencies of the recipient MCA branch before anastomosis

Clinical diagnosis	Mean Doppler frequencies in the recipient MCA branch before anastomosis ($n = 40$)	
	Slow (0.2–0.8 kHz)	Normal (1–3.4 kHz)
CS severe	6 (86%)	1 (14%)
CS mild	13 (100%)	0 (0%)
Prind	4 (80%)	1 (20%)
TIA	5 (56%)	4 (44%)
Asymptomatic	3 (50%)	3 (50%)

Table 2. Angiographic diagnosis of the occlusive disease and time averaged mean Doppler frequencies of the recipient MCA branch before anastomosis

Angiographic diagnosis	Mean Doppler frequencies in the recipient MCA branch before anastomosis ($n = 40$)	
	Slow (0.2–0.8 kHz)	Normal (1–3.4 kHz)
BICO	6 (100%)	0 (0%)
ICO+ICS	14 (68%)	2 (32%)
ICO	9 (64%)	5 (36%)
CSS	2 (50%)	2 (50%)

Staged Carotid Ligation and Bypass in Cases of Inoperable Giant Aneurysms

The Doppler sonographic recording of the cranial recipient vessel during staged ligation of the cervical vessel shows the effect of carotid ligation on the local hemodynamics. In our five patients the flow velocity never decreased to less than two-thirds of the initial value before ligature of the cervical carotid artery. In these five patients the effect of a bypass with open cervical vessels could be observed. In four of the five patients the donor artery took part only in the distal flow without retrograde flow in the proximal anastomotic branch. It started only after an artificial stenosis of the cervical carotid artery that we performed using a Selverstone clamp.

Aneurysm, Delayed Operation

Thirty-one patients were operated on 1–12 weeks after subarachnoid hemorrhage. One-third of them had normal flow patterns, at the latest at the end of

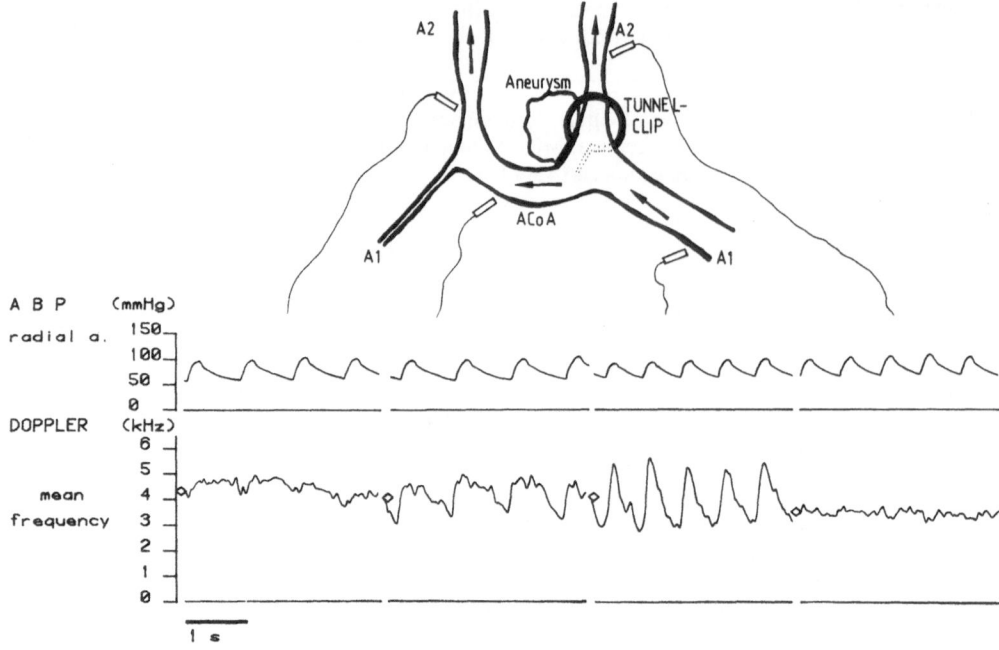

Fig. 3. Flow pattern after clipping of an aneurysm of the anterior communicating artery (*ACoA*). Note the accelerated and irregular flow in the spastic parts of the ascending anterior cerebral artery (*A2*). (Gilsbach 1983)

the operation. Some of them developed temporary local spastic reactions to the manipulation. These vasospasms subsided after manipulation had been discontinued and topical vasodilatators were applied. Artificial hypotension and temporary clipping also caused transient flow pattern alterations by reducing the resistance.

Two-thirds of the patients had permanent local accelerations and irregularities in one or more parts of the circle of Willis. These cases were interpreted as vasospasms, which correlated well with angiograms taken directly pre- or postoperatively (Fig. 3). The outer appearance of these vasospasms was often normal and the detection of the narrowing was only possible by means of Doppler findings, including a profile through the vessel lumen. In this type of vasospasm, the topical application of papaverin or nimodipine had no marked effect. The quantification of the stenosis by our Doppler sonographic equipment was restricted because the filter arrangement and the zero crosser underestimated the highest velocities. Qualitatively, however, a distinction could be made between moderate and mild reductions of the lumen area.

Aneurysms, Early Operation

Twenty-two patients were operated on within 72 h after the last subarachnoid hemorrhage. Only 36% of them had normal flow patterns in the investigated vessels of the circle of Willis. Fourteen percent showed local acceleration and irregularities due to lumen reductions. All the patients suffered more than one recent previous bleeding. Fifty percent of the patients had vessels with a normal outer aspect and normal inner diameters but with an increased flow velocity with a relatively high diastolic component in all the investigated vessels.

These findings were interpreted as signs of reduced peripheral resistance with resulting hyperemia. Two-thirds of these findings were accompanied by severe subarachnoid bleeding. In contrast, two-thirds of the cases with moderate and minor bleeding had normal Doppler findings.

The pathomechanism of the hyperemia is unclear. To our knowledge, a corresponding result in cerebral blood flow measurement is not common. Angiographically, however, enlarged vessels in the acute stage after the bleeding are known and can be proved by our findings. Transcranial Doppler sonography can help answer the question as to whether the hyperemia is present preoperatively or whether it begins with the trepanation which reduces the intracranial pressure.

Final Remarks

Our experiences show that the use of intraoperative Doppler sonography in neurovascular procedures makes it possible for the first time to control intraoperatively microvascular measures on the vessel itself. Furthermore, one must keep in mind that Doppler sonography can be used to obtain normal and pathologic hemodynamic data on intracranial vessels – something which up to now has been virtually impossible.

References

Friedrich H, Hänsel-Friedrich G, Seeger W (1980) Interaoperative Dopplersonographie an Hirngefäßen. Neurochirurgia 23:89–98

Gilsbach J (1983) Intraoperative Doppler sonography in neurosurgery. Springer Wien

Hitchon PW, Kassell NF, McDonnell DE (1979) The Doppler ultrasonic flowmeter as an adjunct to operative management of cerebral arteriovenous malformations. Surg Neurol 11:345–347

Moritake K, Handa H, Yonekawa Y, et al (1980) Ultrasonic Doppler assessment of hemodynamics in superficial temporal artery – middle cerebral artery anastomosis. Surg Neurol 13:249–257

Nornes H, Grip A (1980) Hemodynamics aspects of cerebral arteriovenous malformations. J Neurosurg 53:456–464

Nornes H, Grip A, Wickeby P (1979) Intraoperative evaluation of cerebral hemodynamics using directional Doppler technique. Part 2, saccular aneurysms. J Neurosurg 50:570–577

Arteriovenous Shunts and Results
of Doppler Ultrasound in Supraaortic Vessels

W. von Kalckreuth[1], W. Hillesheimer[2], and G.-M. von Reutern[1]

Continuous wave (CW) Doppler ultrasound is now a widely used non-invasive method for recognizing flow-reducing lesions in the arteries feeding the brain (Spencer and Reid 1981). CW Doppler ultrasound is also capable of locating arteriovenous (AV) shunts in conjunction with increased flow velocity (fistulas, AV malformations, tumors).

The diagnosis of an AV shunt can be based on four criteria:
1. Direct recording of the shunt itself
2. Unilaterally increased flow velocity
3. Pathologic collateral pathway
4. Pathologic venous drainage

Patients and Methods

Doppler records from January 1976 until March 1983 have been reviewed retrospectively. Seventy-four patients were found to have AV shunts. Twelve had to be excluded from this study for various reasons (postoperative results only, incomplete angiographic studies, patient refused further investigations). The diagnoses in the remaining 62 patients are listed in Table 1.

Table 1. Patients ($n = 62$) and diagnoses

Cerebral angioma	16
Fistula of the lateral sinus	14
Carotid-cavernous fistula	8
Large meningioma	6
Venous/cavernous angioma	5
Iatrogenic/traumatic AV fistula	4
Glomus tumor	3
Other tumor	3
Fistula of sup. sagittal sinus	1
Malformation of vein of Galen	1
Angioma of the scalp	1

1 Universität Freiburg, Abteilung Neurologie, Hansastraße 9, D-7800 Freiburg
2 Universität Freiburg, Sektion Neuroradiologie, Hansastraße 9, D-7800 Freiburg

Cerebral Blood Flow and
Metabolism Measurement.
Eds. Hartmann/Hoyer
© Springer-Verlag Berlin Heidelberg 1985

A 4 MHz CW Doppler device was used in combination with a strip chart recorder. Side-to-side differences of flow velocity were evaluated by analogue tracing. Only six patients with fistulas of the lateral sinus had no angiogram (for reason, see below).

Results

The results are summarized in Table 2.

Four *iatrogenic or traumatic AV fistulas* in the neck were recorded directly. Figures 1 and 2 represent a typical finding.

Fourteen *fistulas of the lateral sinus* (Djindjian and Merland 1978) were found. This lesion is characterized by very high flow velocity in the ipsilateral occipital artery and multiple other pathologic vessels in the retroauricular region. In eight angiographically controlled patients 100% sensitivity was achieved. Diagnosis in the other six patients was made by Doppler alone. Due to the high reliability angiography was dropped, because no aggressive treat-

Table 2. Doppler results in AV lesions of supraaortic vessels

Lesions	Doppler
AV-fistula, iatrogenic/traumatic	Recordings of fistula itself (100%)
Fistula of the lateral sinus	Occipital a. with increased FV[a] (100%)
Carotid-cavernous fistula	Abnormal orbital venous drainage (100%)
Cerebral angioma	Unilateral increased FV (37.5%)
Tumors, highly perfused	Rarely increased FV in feeding arteries
Venous/cavernous angioma	No positive findings

[a] Flow velocity

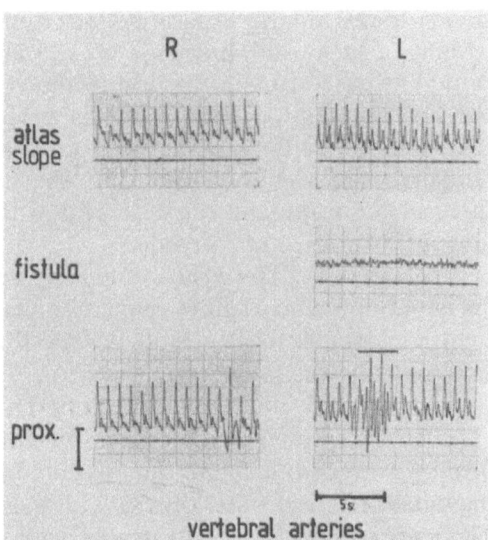

Fig. 1. Iatrogenic AV fistula of the left vertebral artery due to unintended puncture. Fistula clearly recorded by CW Doppler ultrasound

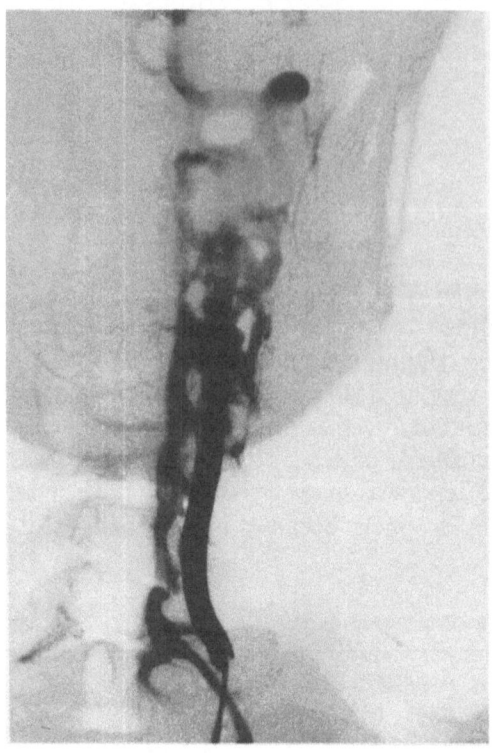

Fig. 2. The corresponding angiogram.
Site of fistula difficult to see

ment was planned. A *glomus tumor* had to be ruled out by E.N.T. examination as two of the three patients had a similar flow pattern in the supplying occipital artery.

Eight *carotid-cavernous fistulas* occurred. Six were traumatic in origin, and two were spontaneous fistulas fed by branches of the external carotid artery (ECA). CW Doppler ultrasound was highly sensitive but not specific. All had pathologic venous drainage via the orbita. 100% specifity was not achieved because one case of malformation of the vein of Galen and another of fistula of the superior sagittal sinus also had orbital venous drainage. After successful treatment the pathologic venous flow disappeared (Büdingen et al. 1978). Unilaterally increased flow velocity in the supplying carotid artery was found in three of the traumatic cases. Increased flow velocity in ECA feeding a spontaneous fistula was not recorded.

Findings in the 16 *cerebral angiomas* were not sensitive but specific. Diagnosis is only possible if flow velocity is very high in one carotid artery (six patients) and pathologic orbital venous flow is excluded. Bilaterally increased flow velocity was found in five patients. Four patients had a steal phenomenon in the ipsilateral supratrochlear artery (reduced flow velocity).

As an exceptional finding some ECA-fed intracranial tumors such as *large meningiomas* or other rare tumors may have increased flow velocity. Six meningiomas and three other tumors with these Doppler characteristics were found. There are no positive results in *cavernous or venous angiomas*.

Discussion

Literature about Doppler diagnosis in AV shunts of supraaortic vessels is rare. Aminoff 1983 does not mention CW Doppler ultrasound as a noninvasive diagnostic possibility. Büdingen et al. 1982 describe Doppler features of cavernous fistulas, fistulas of the lateral sinus, and cerebral angiomas in detail. It was shown that the Doppler diagnosis of fistulas of the lateral sinus and of traumatic AV fistulas in the neck is reliable. Angiography can be dropped if embolization or operation is not intended. It is surprising that a side-to-side difference in cerebral angioma is not a constant finding. Such differences may be reduced by collateral flow through the circle of Willis and by adaptation of the width of the vessels to volume flow. Therefore flow velocities may be bilaterally within the range of normal subjects. In consequence only 37.5% of cerebral angiomas were correctly predicted.

Any pathologic Doppler result is an indication for follow-up studies. The course is followed noninvasively. After successful treatment the disturbed hemodynamics have to return to normal. Thus a control angiogram was seldom necessary.

Any physician using CW Doppler ultrasound should know the features of AV lesions, otherwise unusual results may not be interpreted correctly.

Summary

In a retrospective study Doppler records of 74 patients with arteriovenous shunts of the supraaortic vessels have been reviewed. In 62 of them the role of continuous wave Doppler ultrasound in such lesions could be evaluated. The diagnosis of a fistula of the lateral sinus ($n = 14$) by Doppler ultrasound is possible if a glomus tumor ($n = 3$) can be excluded by other means. In traumatic fistulas of the neck ($n = 4$) the fistula itself is recorded. In cavernous fistulas ($n = 8$) the results are suited to control successful treatment. Only about one-third of cerebral angiomas ($n = 16$) have detectable side-to-side differences of flow velocity in the feeding carotid arteries.

References

Aminoff MJ (1983) Angiomas and fistulae involving the nervous system. In: Ross Russell RW (ed) Vascular disease of the central nervous system. 2nd edition, Churchill Livingstone, Edinburgh p 296–323

Büdingen HJ, Gilsbach J, von Reutern GM (1978) Doppler-sonographische Therapie- und Verlaufskontrolle einer katheter-okkludierten Cavernosus-Fistel. Arch Psychiat Nervenkr 266:19

Büdingen HJ, von Reutern GM, Freund HJ (1982) Doppler-Sonographie der extrakraniellen Hirnarterien. Thieme, Stuttgart New York

Djindjian R, Merland JJ (1978) Super-selective arteriography of the external carotid artery. Springer, Berlin Heidelberg

Spencer MP, Reid JM (eds) (1981) Cerebrovascular evaluation with Doppler ultrasound. M. Nijhoff, The Hague Boston London

Noninvasive Detection of Intracranial Vertebrobasilar Lesions in Patients with Severe Brainstem Disturbances and Coma

E. B. RINGELSTEIN and H. ZEUMER

Introduction

Recently, continuous wave Doppler sonography has been accepted as a reliable, noninvasive diagnostic tool for the detection of vascular lesions within the *extracranial* brain supplying arteries (Büdingen et al. 1982). Besides our own publications (Ringelstein and Zeumer 1982; Ringelstein 1984; Ringelstein et al. 1983; Ringelstein et al. 1985), no reports exist concerning its application for the evaluation of *intracranially* located vertebrobasilar occlusions and high-grade stenoses. The following study was performed in order to evaluate the diagnostic reliability of Doppler ultrasound for the detection of such lesions within the "intradural vertebrobasilar artery" (Thompson et al. 1978).

Material and Methods

The hindbrain circulation should be examined in the following way (see details in Ringelstein 1984 and Ringelstein et al. 1985): Insonation of the mastoidal slope of the vertebral artery is performed in both directions. The probe is then drawn through the supraclavicular fossa in a lateromedial direction. Simultaneously, the submastoidal region has to be compressed quickly in order to modulate the vertebral signal. With the help of this rhythmic modulation of the velocity profile, the vertebral artery can be identified in the lateral neck region before entering the bony canal of the transverse processes (cf. Fig. 1).

Intracranial lesions, if hemodynamically relevant, regularly induce characteristic alterations of the blood flow profile within the proximal, *extracranial* part of the vessel. This is due to severely increased peripheral flow resistance and leads to an abrupt deceleration of the initially accelerated blood column. The flow is reduced to zero during diastole. In cases with sufficient systolic heart ejection and complete distal arterial occlusion, the flow profile is characterized by a reflux phenomenon during early diastole (Yoneda et al. 1974; Ringelstein et al. 1985). These alterations of the flow profile appear during extracranial registration of vertebral blood flow if an occlusion of the intradural segment of the vertebral artery or of the basilar artery is present (Fig. 1).

Rheinisch-Westfälische Technische Hochschule, Abteilung Neurologie, Neuklinikum, D-5100 Aachen

Cerebral Blood Flow and
Metabolism Measurement.
Eds. Hartmann/Hoyer
© Springer-Verlag Berlin Heidelberg 1985

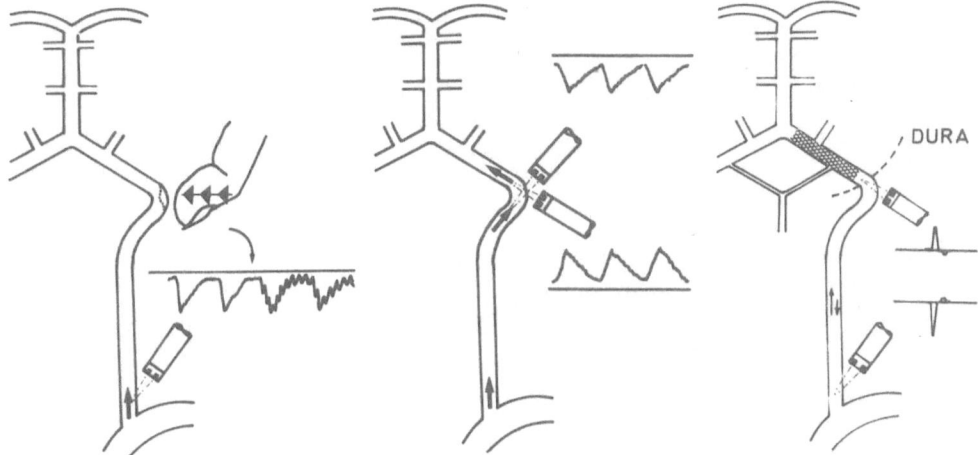

Fig. 1. CW Doppler examination technique of the vertebrobasilar system with flow profiles in normal and pathologic conditions.
Middle diagram. During insonation of the mastoidal slope of the vertebral artery in both directions, normal flow profiles can be registered with a gradual decrease of flow velocity during late systole and a basic blood flow during diastole. The direction of the curves depends on the position of the probes
Left diagram. Quick, strong, rhythmic tapping of the submastoidal region induces oscillations which are superimposed on the normal flow profile. This allows identification of the vertebral artery during insonation at its origin. The velocity profile shows the same characteristics as in the middle diagram
Right diagram. Intracranial blockage of the vertebral blood flow due to vertebrobasilar occlusion significantly alters the flow profile. After blood acceleration during early systole, the blood column is immediately decelerated during late systole and arrests completely during diastole. Regularly, reflux of the blood column occurs due to the elasticity of the thrombus and the vascular system. This alteration of the curve is defined as a "high resistance flow profile"

On admission, 21 patients with alarming brainstem disturbances and/or coma underwent both ultrasound examination of the hind circulation as well as subsequent vertebrobasilar angiography. Additionally, clinical and CT examinations were performed. The ultrasound findings led to a preliminary angiologic diagnosis with angiography serving as a control.

According to the ultrasound findings, the patients were allocated to the following subgroups:

Subgroup I: Bilateral occlusion of the intracranial vertebral arteries and/or basilar thrombosis ($n = 7$). All of these patients revealed the same striking alterations of the vertebral flow pattern as evidenced in Fig. 2. In this illustrative case, basilar thrombosis is visible and led to a large necrosis of the pons. Doppler flow signals were absent on the right and revealed the typical "high resistance flow profile" at both the mastoidal slope and the origin of the left vertebral artery.

Subgroup II: Unilateral intracranial vertebral artery occlusion ($n = 7$). In all of these cases, the vertebral artery on one side was open and Doppler sonography revealed a normal flow signal. On the opposite side, a typical "high re-

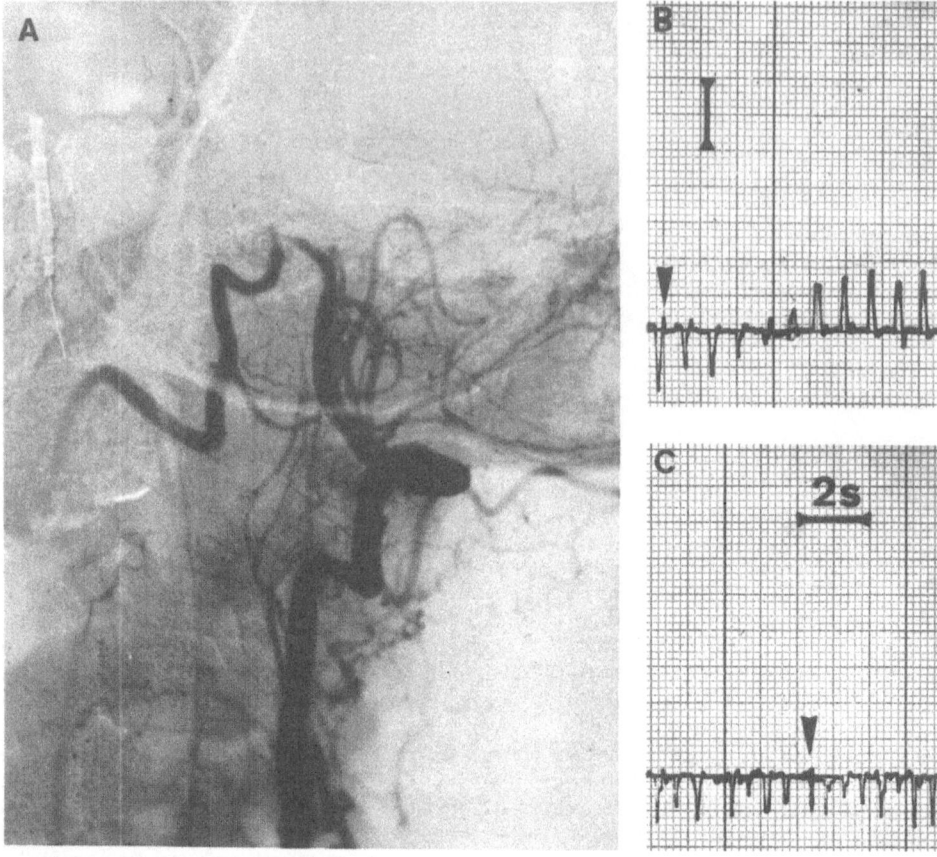

Fig. 2a – c. CT and angiologic findings in basilar thrombosis. Complete thrombosis of the basilar trunk is obvious (**a**) and led to perpendicular flow phenomena in the left vertebral artery (**b, c**) (*arrow heads* indicate reflux of blood column; *vertical bar* = 20 cm/s). No signal was back-scattered from the right vertebral artery. The alterations of the left-sided flow profile indicate high peripheral flow resistance

sistance flow profile" could be found within the vertebral artery with slight reflux of the blood column during diastole. The diagnoses were all confirmed by angiography. In the cases of unilateral occlusion of the distal vertebral artery, occurrence of a dorsolateral medullary syndrome (Wallenberg's Syndrome) was the rule (Fisher et al. 1961; Ringelstein et al. 1983).

Subgroup III: Those in whom angiography did not reveal circumscribed stenosing lesions of the intradural vertebrobasilar artery ($n = 6$). The diagnosis of basilar thrombosis of intracranial vertebral artery occlusion had already been rejected by normal Doppler sonographic findings on admission. These subjects suffered from one of the following illnesses: drug intoxication, extreme elongation and dilatation of the basilar artery due to severe hypertension, spontaneous brainstem hemorrhage, and lacunar stroke of the brainstem as part of a generalized cerebral microangiopathy. Basilar artery ectasia and tortuosity did not induce alterations of the blood flow profiles.

Subgroup IV: Normal findings of Doppler sonography ($n = 1$). In this single patient, however, angiography of the posterior circulation revealed an approximately 70% smooth hourglass-shaped stenosis of the midbasilar artery. Laboratory tests indicated syphilitic angiitis of the central nervous system. However, stenosis of the midbasilar trunk was not "critical" and Doppler findings remained normal.

In four of the patients with bilateral thrombosis of the intradural vertebrobasilar artery, thrombolysis could be achieved with the use of intravertebrally applied streptokinase, giving favorable results in three cases (Zeumer et al. 1983). The close relationship between sonographic and angiographic findings during the course of local intraarterial fibrinolysis could be evaluated in all of the patients who underwent this kind of therapy (Ringelstein et al. 1983, 1985).

Discussion and Conclusions

As has already been pointed out by others (von Reuthern 1983), the diagnosis during Doppler examinations of the brain supplying arteries must primarily be based on the acoustic perception of a well-trained ear and cannot be grounded on the curves on the strip charts.

Some limitations of noninvasive screening of the vertebrobasilar system became apparent during this study and should be mentioned here. Upstream from the lesions, secondary flow changes can only be recorded if the following two conditions are fulfilled: (1) the lesions have reached a high degree of stenosis, and (2) there are no greater collateral branches which drain the affected vessel upstream from the blockage.

The advent of therapeutic approaches in patients suffering from thrombosis of the intradural vertebrobasilar artery has become a challenge for the establishment of noninvasive, handy, and reliable diagnostic procedures (Ringelstein and Zeumer 1982; Zeumer et al. 1983). In the study presented here, the evaluation of Doppler sonography for noninvasive detection of hemodynamically relevant intracranial lesions of the vertebrobasilar system could be substantiated for both the primary diagnosis of such lesions and their intraindividual follow-up during local fibrinolysis.

References

Büdingen HJ, von Reutern M, Freund HJ (1982) Dopplersonographie der extracraniellen Hirnarterien. Grundlagen, Methodik, Fehlermöglichkeiten, Ergebnisse. Thieme Stuttgart, New York

Fisher CM, Karnes WE, Kubik CS (1961) Lateral medullary infarction. The patterns of vascular occlusion. J Neuropathol exp Neurol 20: 323–377

Reuthern GM von (1983) Capabilities of ultrasonic-diagnostic procedures in cerebrovascular disease. International Symposium on Measurement of Cerebral Blood Flow and Cerebral Metabolism in Man. September 28–October 1, 1983, Heidelberg (FRG)

Ringelstein EB, Zeumer H (1982) The role of CW Doppler sonography in the diagnosis and management of basilar and vertebral artery occlusion, with special reference to its application during local thrombolysis. J Neurol 228:161–170

Ringelstein EB, Zeumer H, Hündgen R, Meya U (1983) Angiologische und prognostische Beurteilung von Hirnstamminsulten mit Hilfe klinischer, dopplersonographischer und neuroradiologischer Befunde. Dtsch Med Wschr 108:1625–1631

Ringelstein EB (1984) Ultraschalldiagnostik am vertebrobasilären Kreislauf. I. Diagnose intrakranieller vertebrobasilärer Thrombosen mit Hilfe der konventionellen Dopplersonographie. Ultraschall 5:215–223

Ringelstein EB, Zeumer H, Poeck K (1985) Non-invasive diagnosis of intracranial lesions in the vertebrobasilar system. A comparison of Doppler sonographic and angiographic findings. Stroke 1985 (in press)

Thompson JR, Simmons CR, Hasso AN, Hinshaw jr DB (1978) Occlusion of the intradural vertebrobasilar artery. Neurorad 14:219–229

Yoneda S et al (1974) To and fro movement and external escape of carotid arterial blood in brain death cases. A Doppler ultrasonographic study. Stroke 5:707–713

Zeumer H, Ringelstein EB, Hacke W (1983) Lokale Fibrinolysetherapie bei Thrombose der A. basilaris und distaler subtotaler Stenose der A. vertebralis. In: Trübestein G, Etzel F (eds) Fibrinolytische Therapie. Schattauer, Stuttgart – New York 1983, pp 317–322

Carotid Blood Flow Measurement by Means of Ultrasonic Techniques: Limitations and Clinical Use

H. R. MÜLLER[1], E. W. RADUE, A. SAIA, C. PALLOTTI, and M. BUSER

Compared with regional cerebral blood flow, volume flow through the pre-cerebral arteries has thus far received little clinical attention. The main reason is that until very recently no appropriate method was available for its noninvasive measurement. But even after the development of ultrasonic flowmeters designed for transcutaneous use (Baker 1970; Peronneau et al. 1970; Doriot et al. 1975; Keller et al. 1976; Furuhata et al. 1978; Gill 1979; Fish 1981) the interest in cranial blood flow measurement remained rather limited.

The fact that with the present state of the art Doppler flow measurement in the neurovascular field is largely confined to the common carotid artery appears to have prevented most centers from getting involved in this type of investigation. Moreover, no normal values based on sufficiently large series have been published to date, and the data obtained on pilot studies varied considerably according to the particular flowmeter and method used (Table 1).

Table 1. Data available from the literature on normal common carotid volume flow rates as measured with ultrasonic Doppler techniques

Author	Technique, equipment	n	Age (yrs)	Flow (ml/min)
Olson and Cooke 1975	CW + echo, prototype	7		325
Keller et al. 1976	Pulsed, prototype	22		300 480
Uematsu and Yang 1981	CW + echo			
Uematsu et al. 1983	QFM Hayashi Denki	46	5 – 20	510 ± 78[a]
			21 – 40	479 ± 90[a]
			41 – 60	433 ± 69[a]
			> 60	402 ± 29[a]
Fujishiro and Yoshimura 1982	CW + echo QFM Hayashi Denki	140	11 – 20	576 ± 102[a]
			21 – 70	570 486[a] Decr. w. age
			> 70	450 ± 66[a]
Payen et al. 1982	Pulsed, Echovar, Alvar	11[b]		387 ± 183
Simon et al. 1982	Pulsed, Echovar, Alvar	11		429 ± 32

[a] Original data indicated in ml/s
[b] Measurements in 11 patients anesthetized for cerebral angiography and suffering from subarachnoid hemorrhage or cerebral trauma

1 Neurologische Universitätsklinik, Kantonsspital, Petersgraben 4, CH-4031 Basel

Cerebral Blood Flow and
Metabolism Measurement.
Eds. Hartmann/Hoyer
© Springer-Verlag Berlin Heidelberg 1985

After ultrasonic Doppler equipment for transcutaneous carotid blood flow measurement became commercially available, we undertook a comparative in vitro and in vivo evaluation of two different systems, and studied the question of usefulness of common carotid blood flow measurement for clinical purposes.

Equipment

The two instruments evaluated are the QFM system developed by Furuhata et al. (1978) and the MAVIS scanner of Fish (1981). For detailed information on these systems, the reader is referred to the literature (Furuhata et al. 1978; Uematsu 1981; Uematsu and Yang 1981; Fish 1981). Only a brief description can be given in this article.

QFM System

This is a flowmeter using a dual-beam continuous wave technique for determining mean instantaneous flow velocity, and the impulse echo technique for measuring the inner diameter of the vessel. The transducers of both the echo and the Doppler system are contained in one and the same probe. Their arrangement is demonstrated in Fig. 1 a. The formula in Table 2 shows how true velocity is calculated from the two Doppler shift signals received in spite of the angle of incidence remaining unknown to the investigator.

The transmitted frequency of the echo system is 6 MHz, the pulse repetition rate 10 kHz. The transmitted frequency of the Doppler system is 5 MHz. Doppler shifts arising from reverse flow are eliminated by means of a crystal filter, to the effect that only forward flow is measured.

For transcutaneous measurement the probe is placed over the vessel to be investigated. By carefully changing the site and angulation of the probe with regard to the skin surface, a position is searched for with which the vessel wall echoes monitored on the screen have the highest amplitudes, indicating a beam

Table 2. Transducer arrangement in QFm probe and formula for calculating true velocity

Transducer arrangement in QFM probe:

E = Echo transmitter/receiver, at 90°
R_a = Doppler receiver a, at 55°
T = Doppler transmitter, at 65°   ← 10°
R_b = Doppler receiver b, at 75°   ← 10°

Formula for calculating velocity:

$$v = \frac{a}{h}\sqrt{1 + \cot 10° - \frac{b/a}{\sin 10°}} \qquad\qquad h = \frac{2 \cdot f}{c}$$

v = blood flow velocity
a = Doppler frequency obtained from R_a
b = Doppler frequency obtained from R_b

f = transmitted frequency
c = ultrasound velocity

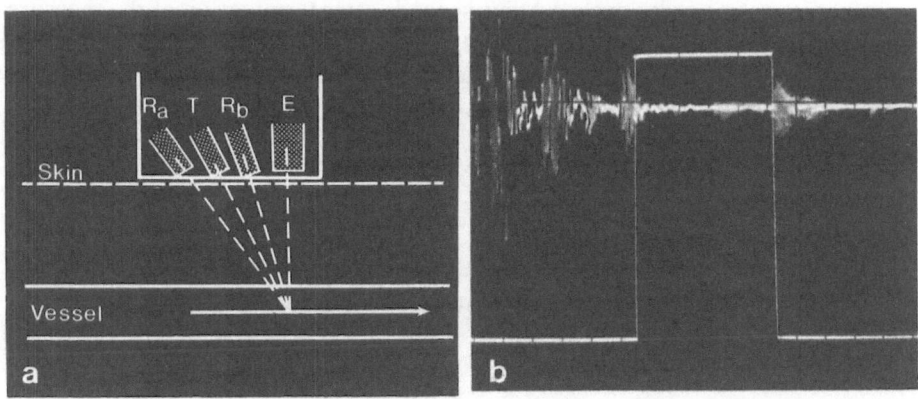

Fig. 1. a Transducer arrangement in QFM probe (see Table 2). **b** Echo display of near and far vessel wall echo with tracking gate appropriately set

incidence of 90 °C. The diameter of the blood column is then measured by the operator by moving a tracking gate to the innermost component of both the near and the far wall echo (Fig. 1 b). The tracking gate locks onto these two echoes and tracks their transversal motion throughout the pulse cycle.

During the pulsatile movement of the vessel wall, the diameter of the lumen is measured every 2 ms, and so is the mean velocity as determined from the two Doppler shift signals. The R-wave of the EKG is used for triggering. Data are

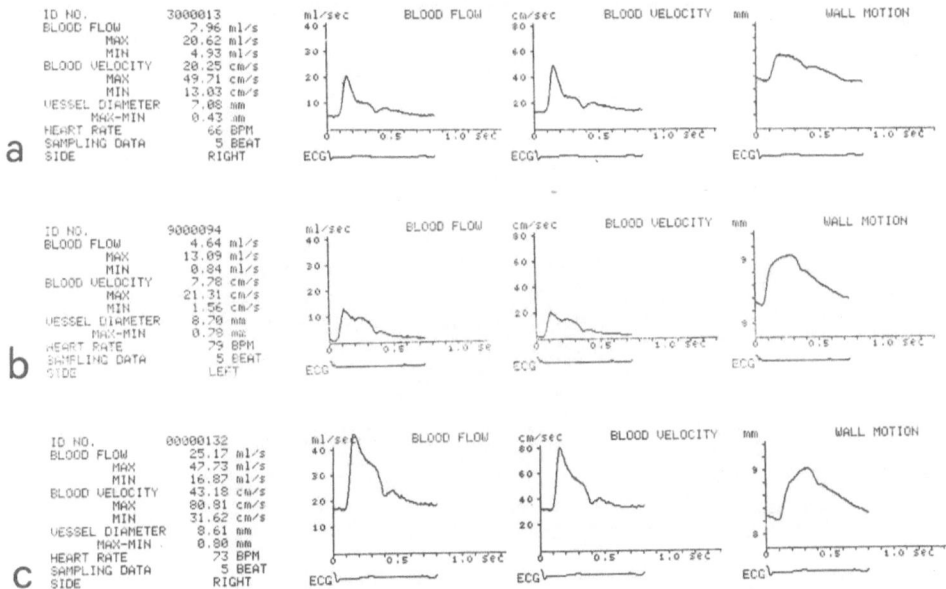

Fig. 2. a – c QFM printout displaying flow data as measured on a common carotid (**a**) in a healthy subject, (**b**) in a patient having right internal carotid occlusion, and (**c**) in a patient having a dural AV malformation fed by both external carotids

collected over five pulse cycles. A few seconds after termination of the measurements, the calculation process is accomplished, and averaged digital and analog data, including flow, velocity, and vessel diameter are displayed on a hard copy (Fig. 2).

MAVIS Scanner

In contrast to the QFM system this is a multichannel pulse Doppler instrument. The Doppler shift frequencies measured from its 30 range gated channels cor-

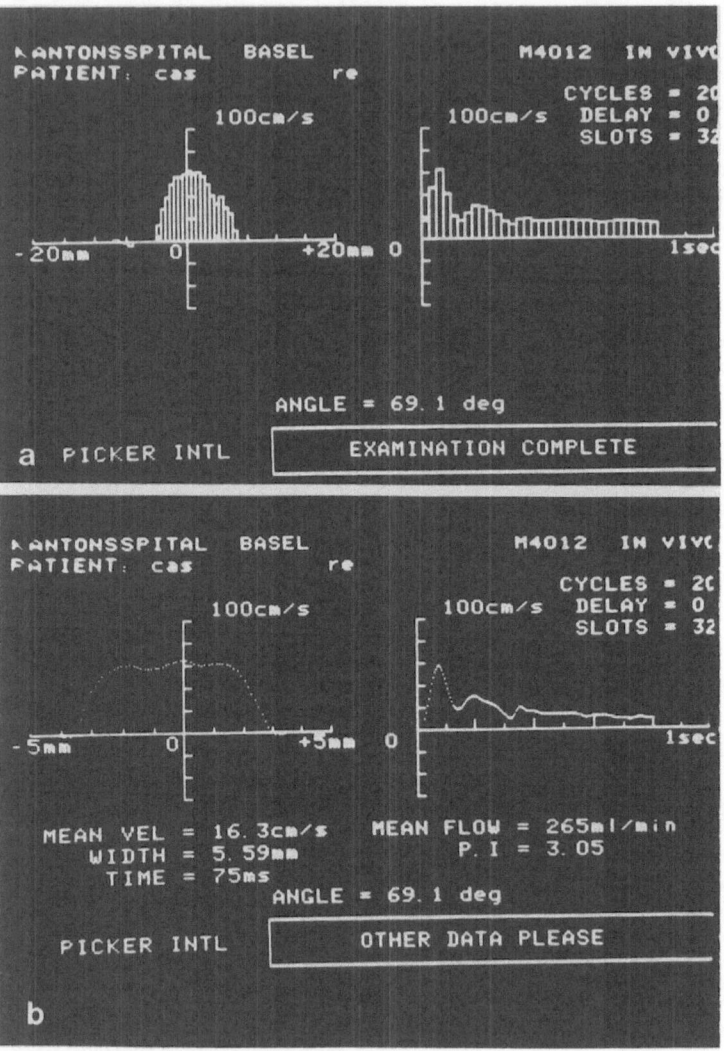

Fig. 3. a, b Flow data display of MAVIS scanner showing peak systolic velocity profile, spatial mean velocity versus time curve and digital angle and flow information

respond to the velocities of blood passing through sample volumes contiguously spaced along the beam. This technique not only allows one to obtain the velocity information necessary for flow calculation, but also to measure the diameter of the blood column without using the vessel wall echoes. The transmitted frequency is 5 MHz.

The angle of beam incidence, which must be known both for converting the Doppler shift into true velocity and for calculating the orthogonal diameter of the blood column from the oblique diameter measurement, is determined using the imaging facility of the instrument: two transverse images of the blood column are first made by scanning the vessel 5 mm upstream and 5 mm downstream of the flow measuring site. From the information furnished by the potentiometers of the scanning arm and from the Doppler information obtained with the beam being directed through the center of the two cross-sections, the vessel axis is determined by a microprocessor. This allows the computer to also calculate the angle of beam incidence at the flow measuring site, and hence to determine true flow velocity.

Volume flow is calculated assuming a uniform velocity within a semiannulus extending to 90 ° on either side of the volume sampled by each channel, and by summing up the flows through the individual semiannuli. Measurements are made over a number of pulse cycles. During a subsequent measuring period noise only is registered for subtraction. Averaged results including analog display of the velocity profiles throughout the pulse cycle are available within a few minutes after completion of measurement (Fig. 3).

This method has the advantage over the continuous wave technique used with the QFM system that it is less affected by blood streaming through nearby vessels and that the diameter measurement cannot be biassed by the operator. Its accuracy seems, however, to be limited by the requirement of a small beam width with regard to the vessel diameter and by the assumption made as to the symmetry of the velocity profile. Owing to the relatively long duration of the investigation, the MAVIS measurements are also considerably prone to movement artifacts.

In Vitro Evaluation of the QFM System and the MAVIS Scanner

The two systems were tested on a flow rig (Fig. 4) designed for both continuous and pulsed flow. A Sephadex solution (particle size $10-40$ µm, $\sim 10^7$ particles/ 1000 ml) was used as circulating medium. The flow pattern was monitored by means of a Kranzbühler Doppler 762 and a Carolina Medical Electronics Sonacolor 6200 spectrum analyzer (Fig. 5). Hydrostatic pressure was kept constant be means of a relay controlling the pump according to the level of a swimmer placed on the surface of the fluid in the reservoir. Changes in real volume flow were produced by graduated clamping of the tube proximally and distally to the measuring site.

Measurements were done on a distensible Silicone caoutchouc tube having a wall thickness of 0.9 mm and a diastolic internal diameter of 6 mm. With the

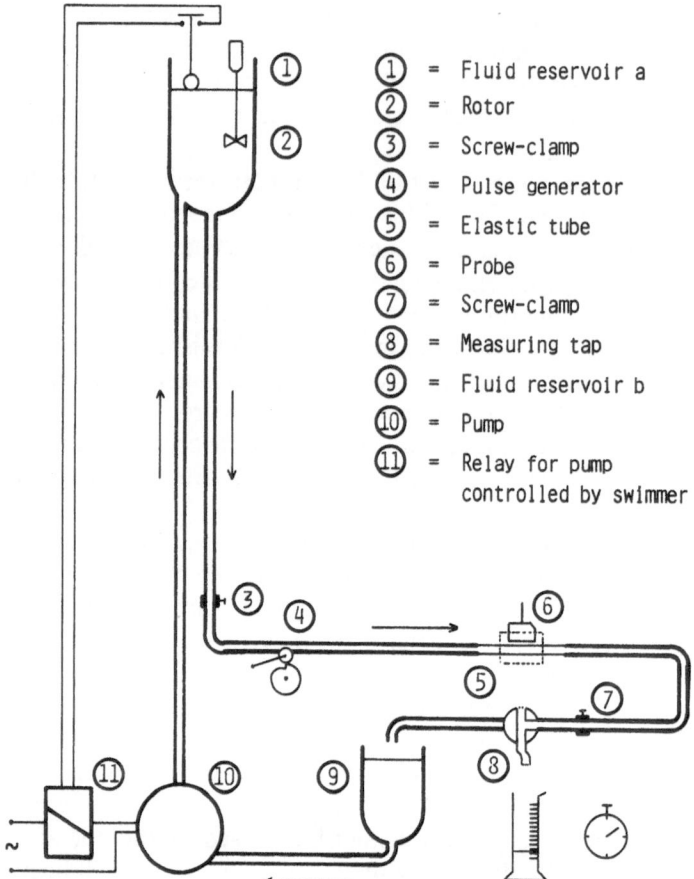

① = Fluid reservoir a
② = Rotor
③ = Screw-clamp
④ = Pulse generator
⑤ = Elastic tube
⑥ = Probe
⑦ = Screw-clamp
⑧ = Measuring tap
⑨ = Fluid reservoir b
⑩ = Pump
⑪ = Relay for pump
      controlled by swimmer

Fig. 4. Diagram of flow rig used for in vitro tests

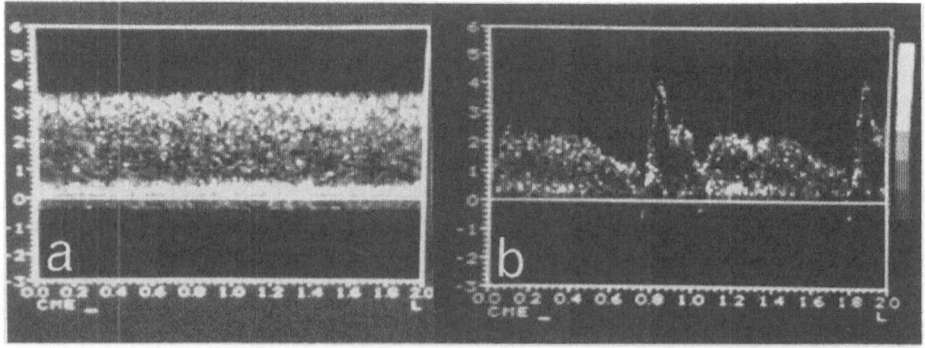

Fig. 5. a, b Sonagrams (**a**) of continuous flow (**b**) of pulsatile flow produced on the flow rig

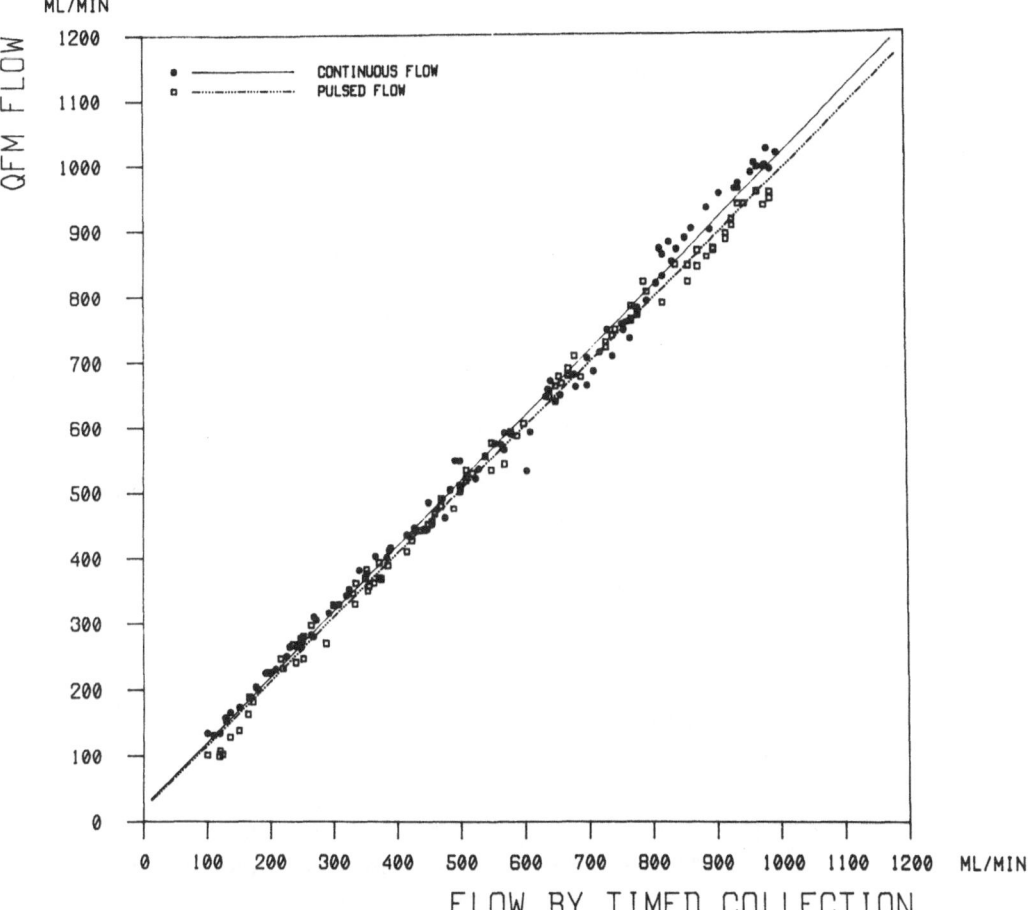

Fig. 6. Plot of Doppler flow versus time collected flow. QFM measurements, Sephadex

MAVIS scanner measurements were carried out through a waterpath of 15 mm. For the QFM measurements the water tank at the measuring site was emptied, and a 15-mm layer of Aquasonic was used as coupling medium. With both systems, a probeholder controlled by micrometer screws was used for adjusting the probe.

With both types of equipment 100 measurements were carried out with each continuous and pulsatile flow, true flow being adjusted to a range of values between 100 and 1000 ml/min. All measurements were done during timed collection of the outflow in a graduated cylinder.

The results are shown in Figs. 6 and 7. Figure 8 demonstrates an additional plot of MAVIS measurements made with human blood having a hematocrit of 45%.

There was an excellent linear correlation of the QFM measurements with the true flow on both the continuous and the pulsatile flow experiment with a $P < 0.001$ for both. With pulsatile flow there was a slight overestimation of

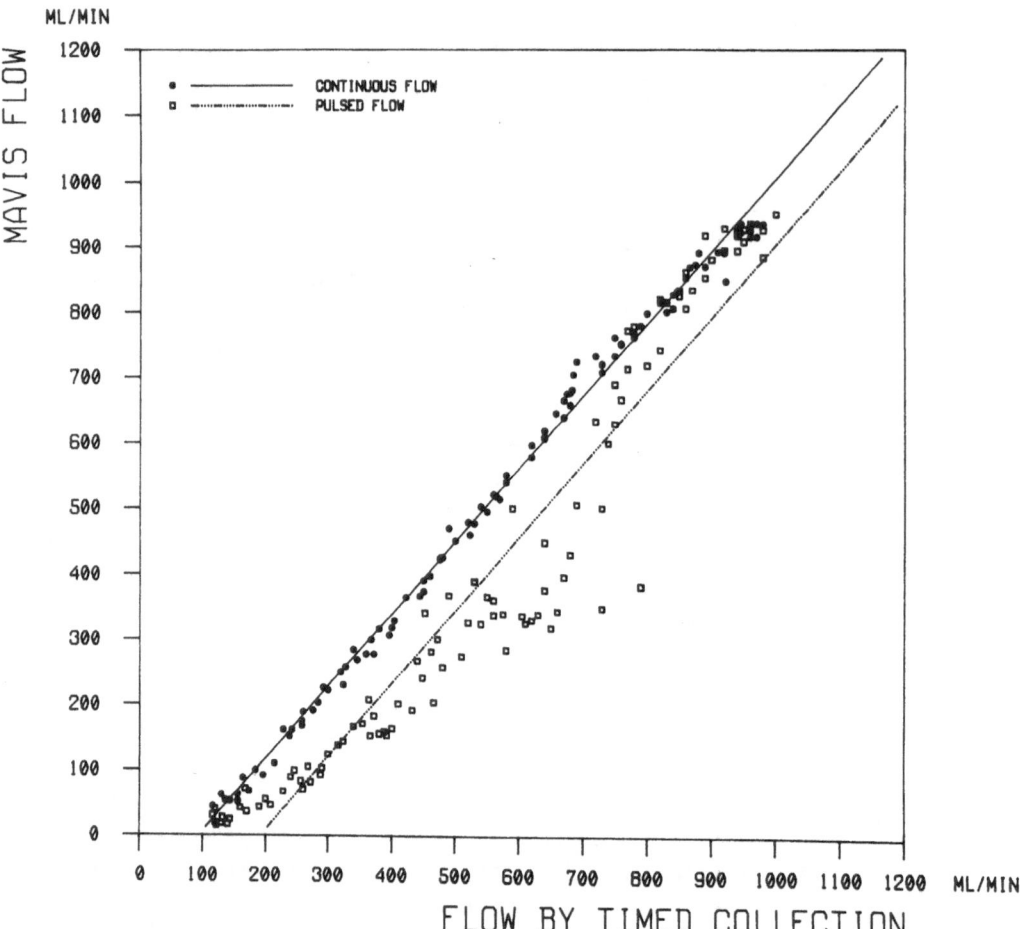

Fig. 7. Plot of Doppler flow versus time collected flow. MAVIS measurements, Sephadex

flow up to 800 ml/min (mean 2%, but 6.5% for true flows between 200 and 300 ml/min). True flows over 800 ml/min were slightly underestimated.

With the MAVIS scanner the correlation was linear only for continuous flow. A linear approximation for pulsatile flow within the clinically relevant range of 100−500 ml/min true flow was: MAVIS flow = 0.73 times true flow minus 85 ml, signifying a considerable underestimation. It was thought that with the circulating medium used, owing to the relatively low number of reflectors, the velocity information fed to the MAVIS computer may not be sufficient and that the computer program of this instrument may be too rigid to cope with the flow patterns arising with pulsatile streaming of Sephadex through our flow rig. For this reason, an additional set of MAVIS measurements, limited to clinically relevant flow volumes, was carried out using human blood as circulating medium. On this experiment the correlation of ultrasonically measured to true flow proved to be linear ($P < 0.001$), but there remained a considerable underestimation (mean 15%).

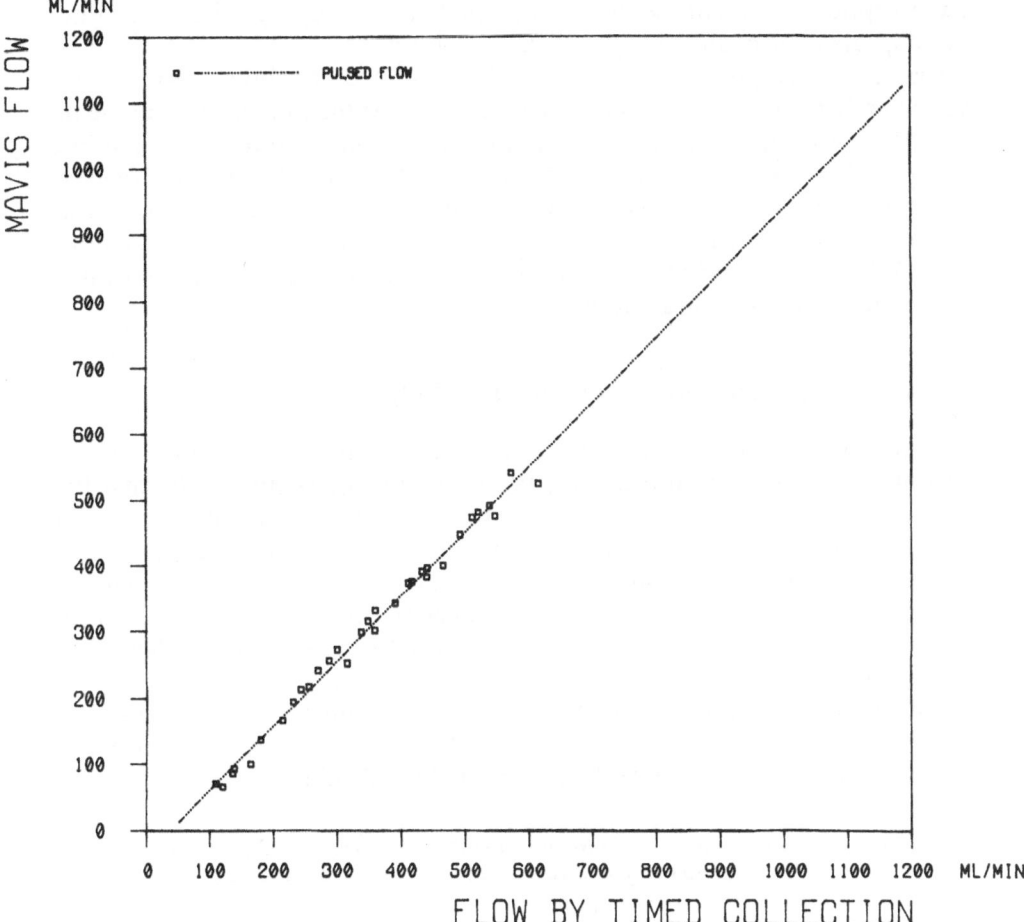

Fig. 8. Plot of Doppler flow versus time collected flow. MAVIS measurements, blood

In Vivo Evaluation of the QFM System and the MAVIS Scanner

Arteries Accessible to Examination

The Doppler system of the *QFM flowmeter* is not range gated. Its use must, therefore, obviously be limited to such arteries and such measuring sites where no major contamination of the signal through Doppler shifts arising from overlying vessels is to be expected. It is this consideration which appeared to us to exclude the internal carotid and the vertebral arteries from reliable measurements, since at the sites where these arteries could be investigated there is too high a risk of picking up, in an uncontrollable way, signals from external carotid branches.

But even when we neglected this source of error and tried systematically to carry out measurements on the internal carotid and vertebral arteries in some 50 volunteers, these were, due to the size of the probe, unsuccessful with a very few exceptions. The clinical use of the method in neurovascular diagnostics thus appears to be confined to the common carotid.

Similarly, attempts to measure internal carotid and vertebral artery flow using the *MAVIS scanner* were mostly unsuccessful. At the internal carotid the investigation was only possible with very low bifurcations. However, even in case of a technically successful measurement, the reliability of the result remains doubtful. The reason is that in the vicinity of the bifurcation, the flow profile cannot be expected to be symmetric as is assumed for the flow calculation using the MAVIS program. MAVIS flow measurements therefore also seem to be limited to the common carotid artery.

Comparative Measurement of Common Carotid Flow

In order to determine the normal range of common carotid volume flow as measured with the two instruments, 100 healthy subjects (five males and five females in age groups of 5 years between 16 and 65 years) were investigated. On two different sessions each of the two flowmeters was used to take three measurements on each common carotid. During the investigation the volunteer was in the relaxed supine position, having been so for 5 – 10 min before the examination. Examinations were done proximally to the thyroid cartilage, with the head of the patient slightly turned to the contralateral side.

The mean values as determined with the two instruments are shown in Table 3. For convenient comparison of the results, the QFM measurements, displayed by the equipment in ml/s, have been converted to ml/min.

Table 3. Common carotid flow by QFM and MAVIS in 100 healthy subjects (five males and five females in each age group of 5 years. Three measurements on each carorid)

Age group (years)	Method	Mean flow (ml/time unit)			
		Total R + L		Side difference	
		Males	Females	Males	Females
16–25	MAVIS ml/min	631 ±159	527 ± 91	82 ±74	60 ±34
	QFM { ml/min	973 ±142	870 ± 95	68 ±50	80 ±53
	ml/s	16.2± 2.1	14.5± 1.6	1.1± 0.8	1.3± 0.9
26–35	MAVIS ml/min	584 ± 93	547 ±146	62 ±48	54 ±42
	QFM { ml/min	980 ±125	936 ± 63	79 ±49	65 ±56
	ml/s	16.3± 2.1	15.7± 1.0	1.3± 0.8	1.1± 0.9
36–45	MAVIS ml/min	595 ±119	495 ± 96	67 ±50	66 ±40
	QFM { ml/min	911 ±112	906 ±154	57 ±44	46 ±45
	ml/s	15.2± 1.9	15.1± 2.5	0.9± 0.7	0.8± 0.7
46–55	MAVIS ml/min	683 ± 99	516 ± 94	83 ±67	48 ±37
	QFM { ml/min	902 ±106	894 ± 78	77 ±65	55 ±47
	ml/s	15.0± 1.8	14.9± 1.3	1.3± 1.1	0.9± 0.8
56–65	MAVIS ml/min	575 ±125	560 ±114	77 ±78	66 ±51
	QFM { ml/min	1001 ±110	916 ±128	104 ±79	60 ±43
	ml/s	16.7± 1.8	15.3± 2.1	1.7± 1.3	1.0± 0.7

The first, and rather intriguing result was the very considerable difference between the measurements obtained with the two methods, those accomplished with the MAVIS-scanner amounting to an average of only 62% of the QFM values and also lying well below those published from experiments using other pulsed wave Doppler equipment (Table 1).

With the age groups chosen there was a marked tendency for the flows to be lower in subjects from 36 to 55 years than in the younger groups. However, flow again assumed higher values in the subsequent decennium. A similar evolution was found with the MAVIS scanner, but was limited in this experiment to the females. Males in the age group from 46 to 55 years showed the highest mean carotid flow as measured with the MAVIS equipment.

A feature common to the QFM and MAVIS measurements was that significantly higher flow rates were found in men than in women ($P < 0.02$). However, while this sex difference was as much as 14% in the MAVIS study, it was only 5% in the flow values obtained with the QFM system.

A further observation which was, however, only made in the QFM series, and not confirmed with the MAVIS scanner measurement, was a linear correlation of common carotid flow measurement with stature ($P < 0.01$).

The intraindividual variation of the QFM measurements, including physiologic changes of flow in time, was within a standard deviation of 5.2%. For the MAVIS scanner this value was 6.7%.

As expected from theoretical considerations, there was considerable interindividual variation of flow and side difference which most likely is due to variations in the anatomy of the circle of Willis and in the external carotid supply. The lowest values measured in the QFM series were around 6 ml/s, and only in six carotids were values slightly below this limit. The highest value was 11.7 ml/s. Only five carotids had a flow over 10 ml/s. Side difference was over 3 ml/s in one and over 2.5 ml/s in four cases. Only in one case was the side difference over 20% of the total right plus left flow, while in eight cases it was over 15%. With a mean total right plus left flow of 15.5 ml/s the 95% interval of confidence (mean $\pm$ 2SD) ranged from 11.7 to 19.3

As a preliminary conclusion these findings indicate that only with QFM flow values beyond a range of 6–11 ml/s and side differences of more than 20% of the total right plus left flow can the presence of a hemodynamically significant vascular pathology be expected with reasonable probability. The possible value of total right plus left common carotid flow as a diagnostic parameter remains to be discussed.

Clinical Evaluation of the QFM System

The reason for limiting the clinical evaluation to the QFM system was not only the better in vitro results obtained with this equipment compared with the MAVIS measurements, but even more so the impracticability of comparative studies using each method in patients. In fact, the duration of a bilateral common carotid flow measurement using the MAVIS scanner has often proved to be more than 1 h. Furthermore, if reproducible results are to be obtained with this technique, a considerable amount of patient collaboration is required in order to avoid swallowing (regularly producing an increase of flow over a few seconds) and other movements.

Flow measurements with the QFM system on the other hand are much more straightforward, allowing most investigations to be carried out within 10–15 min. It therefore seemed logical for practical reasons to give priority to the QFM technique in clinical evaluation.

Over 600 patients, including a wide range of cerebrovascular pathology, have so far been examined by us with the QFM system. While a detailed evaluation is still in progress, the clinical part of this paper will essentially deal with the findings obtained in unilateral obstructive carotid disease, and with the effects on common carotid flow observed after carotid endarterectomy and EC/IC bypass operations. For discussion of possible clinical applications of the method, first experiences with pre- and postoperative measurements in AV malformations will also be reported.

Common Carotid Flow Measurement in Obstructive Carotid Disease

Tables 4 and 5 show the results obtained in a series of patients having either total occlusion or high grade stenosis of one of the internal carotid arteries, while the contralateral carotid as well as the vertebrobasilar system were without significant obstructive pathology.

With total occlusion of the internal carotid in only 3 of the 23 cases evaluated, homolateral common carotid flow was within the normal range, which in two of them was obviously due to a very strong ophthalmic collateral. But even in these three cases there was a marked side difference of flow, exceeding the

Table 4. Common carotid flow in 23 patients having unilateral internal carotid occlusion, Ophthalmic collateral: (+)= to orbit only, + =angiographically demonstrated to siphon, + + =to middle cerebral artery

File no.	Ophth. coll.	Flow (ml/s) Homolat.	Flow (ml/s) Contralat.	File no.	Ophth. coll.	Flow (ml/s) Homolat.	Flow (ml/s) Contralat.
8103	–	1.6	10.9	5866	(+)	4.3	8.7
8142	–	1.8	7.5	6050	+	4.3	7.0
8009	–	2.7	8.2	5050	+ +	4.5	9.2
4909	(+)	2.8	8.3	7957	+	4.6	13.3
4219	–	2.9	7.5	7140	–	4.7	8.3
8741	–	3.0	8.5	7097	–	4.9	8.4
7370	(+)	3.6	9.5	7262	+	5.2	8.7
8411	–	3.7	7.7	7236	(+)	5.8	11.6
4856	–	3.7	6.4	4814	–	6.1	11.0
8016	+	3.7	9.8	5961	+ +	6.3	10.3
7150	(+)	3.9	9.7	6049	+ +	7.6	10.3
8112	(+)	4.3	9.9				

Table 5. Common carotid flow in 21 patients having unilateral high grade internal carotid stenosis. Sten. % refers to % reduction of diameter on lateral view angiogram with the 100% measurement being taken 2.0 cm downstream. Only the two patients marked with an asterisk had an ophthalmic collateral

File no.	Sten. %	Flow (ml/s)		File no.	Sten. %	Flow (ml/s)	
		Homolat.	Contralat.			Homolat.	Contralat.
7511	80	2.7	6.1	5781	67	6.0	7.5
8193	59	3.3	4.3	7139*	78	6.1	8.0
7708	62	3.7	7.5	8771	73	6.1	12.2
7277	90	3.8	10.9	7391	69	6.3	9.0
4800	60	4.0	8.9	7669	56	6.3	9.3
7649	84	4.9	7.5	7394	63	6.4	10.0
7922	75	4.9	6.6	7699*	50	6.4	6.8
7342	65	5.0	10.0	7407	75	6.6	8.1
8576	50	5.2	8.0	7890	64	6.9	6.9
7125	76	5.6	9.8	8745	84	7.4	10.5
7660	75	5.9	10.5				

upper limit of 20% of the total right plus left flow accepted for normals in two cases and approaching it in the third (side difference 15%). While the impact of the ophthalmic collateral on common carotid flow can clearly be recognized from Table 4, there were still four cases where in spite of common carotid flow being as high as 4.7−6.1 ml/s, no ophthalmic contribution to the cerebral circulation could be demonstrated on angiography. The relatively high flow in these four cases must, therefore, be explained by an unusually high external carotid supply, including collateral circulation to the forehead, a region which is normally supplied by the internal carotid.

The findings are even more intriguing with internal carotid stenosis. Though Table 5 includes only unilateral stenoses showing on angiography a diameter reduction of the vessel of at least 50%, in almost half of the cases homolateral common carotid flow was within normal limits. Among these ten cases side difference exceeded 20% only in three, another three showing values between 17% and 19%. Accepting the lesions as actually being hemodynamically significant, those findings falling within the normal range both as to homolateral flow and side difference are possibly to be explained not only by collateral external carotid supply, but also by variations of the circle of Willis.

Pre- and Postoperative Measurements in Neurovascular Surgery

Table 6 shows the pre- and postoperative flow data of seven patients undergoing *carotid endarterectomy* or balloon dilation for internal carotid stenosis. This table demonstrates that not only can an increase in homolateral common carotid flow be seen after successful operation of a hemodynamically significant stenosis (case 1, 2, 5, 6) but the method also allows recognition of reversal of a carotid steal as present in cases 1 and 2. In cases 3 and 7 no significant change in

Table 6. Pre- and postoperative common carotid flow in seven patients with unilateral internal carotid stenosis. Notice the effect of redistribution on flow value of the contralateral side in cases no. 1 and 2, and evidence of postoperative carotid occlusion in case no. 4

Case no.	File no.	Age (years)	Side	Treatment	Comm. car. flow (ml/s)			
					Preop.		Postop.	
					R	L	R	L
1	7076	55	R	Endarterectomy	3.8	10.9	6.7	8.7
2	7196	68	R	Endarterectomy	4.9	12.3	8.0	8.1
3	7407	59	R	Endarterectomy	6.6	8.1	6.7	8.9
4	7649	69	R	Endarterectomy	4.9	7.5	2.5	9.4
5	8123	62	L	Endarterectomy	11.5	5.7	10.6	7.7
6	8385	73	L	Endarterectomy	6.3	5.7	8.1	8.6
7	7251	55	L	Dilation	8.4	7.2	9.7	8.6

flow was produced by the intervention. In case 4, showing no clinical evidence of a complication, the postoperative Doppler examination clearly demonstrated internal carotid occlusion.

The effect of 32 *EC/IC bypass operations* on common carotid flow is demonstrated on Fig. 9. Four different patterns of evolution can be observed: in a first group of patients (nos. 1–11) all showing marked carotideal cross flow preoperatively, the operation resulted in a redistribution of flow, the increase

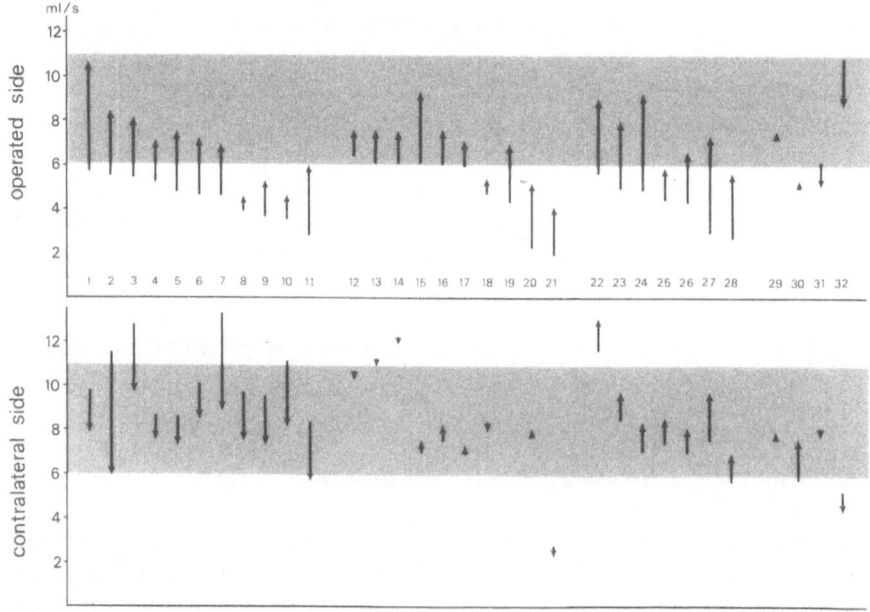

Fig. 9. Impact of STA/MCA bypass on common carotid flow as measured with the QFM system. *Arrows* indicate evolution from pre- to postoperative flow. Gray area indicates normal range

occurring on the operated side along with a decrease on the contralateral side, which obviously indicates a reversal of the carotid steal.

In a second group (nos. 12–21) also showing an increase of common carotid flow on the operated side, there was no significant change on the contralateral side. This appears to indicate an increase of total hemispheric blood flow, unless there was a redistribution with regard to a vertebrobasilar collateral. Not having performed postoperative angiography, we cannot exclude this mechanism in those cases (nos. 13, 20, 21) where a vertebrobasilar collateral was demonstrated on the preoperative angiograms.

While a clear increase in common carotid flow on the operated side could be seen in the third group of patients (nos. 22–28), there was a mostly minor increase on the contralateral side as well. A logical explanation of this phenomenon is that a decrease in external carotid pressure on the operated side, produced by the bypass, had resulted in collateral external carotid cross flow from the nonoperated side.

The fourth group of our series consists of patients with poorly functioning (nos. 29, 30) and nonfunctioning bypasses (nos. 31, 32). No uniform response on the nonoperated side could be seen in these cases.

Though our experience with the use of common carotid flow measurement for assessing the effect of surgical and neuroradiologic *interventions in AV malformations* and fistulas is still rather limited, Table 7 clearly demonstrates the value of the method when used for this purpose. While a dramatic bilateral effect on common carotid flow can be seen in cases no. 1, 2, and 6, having AV malformations mainly fed by the carotid artery, in case no. 3, where this artery contributed little to the perfusion of an AV malformation located in the posterior cerebral artery territory, the flow value measured after selective emboli-

Table 7. Pre-and postinterventional common carotid flow in patients treated for AV malformations and fistulas

Case no.	File no.	Age (yrs)	Diagnosis	Treatment	Comm. car. flow (ml/s)			
					Preop.		Postop.	
					R	L	R	L
1	4783	10	AV malform. parietal R	Embolization + operation	21.0	14.5	9.3	8.9
2	5956	46	Dural AV malform. bioccip.	Embolization + operation	25.2	14.5	9.8	8.9
3	7687	26	Facial AV malformation R	Embolization	20.6	18.6	14.3	20.3
4	8228	26	AV malform. occipital R	Operation	9.3	6.3	6.6	7.2
5	8267	34	Carot. cav. sin. fistula	Balloon occlusion	10.5	15.9	10.8	10.6
6	8981	42	AV malform. parietal R	Embolization	11.1	15.6	6.9	10.6

Table 8. Effects of embolization of external carotid branches, and of operation in a case of dural AV malformation. Case no. 2, Table 7

Date		Comm. carot. R	Flow (ml/s) L
9-12-81		25.2	14.5
21-01-82	Embolization R		
22-01-82		7.9	17.3
22-01-82	Embolization L		
22-03-82		9.5	15.4
20-04-82		12.1	18.6
1-07-82	Embolization RL		
1-07-82		10.4	11.4
10-11-82		15.6	17.1
17-11-82	Embolization RL		
18-11-82		14.7	7.3
25-11-82	Embolization R		
25-11.82		10.0	10.8
6-11-82		12.7	11.0
8-12-82	Operation		
10-12-82		10.4	7.4
15-12-82		9.8	8.9

zation was only 30% below the preinterventional measurement. Similarly, the decrease was of only minor degree in case no. 4, where two feeding external carotid artery branches were occluded in a facial AV malformation. Case no. 5 demonstrates the impact on common carotid flow of the balloon occlusion of a carotid to cavernous sinus fistula.

The evolution of case no. 2 in Table 7, undergoing a series of treatments with embolization of external carotid branches before extirpation of the AV malformation, is shown in Table 8. After each session the impact on common carotid flow was investigated, and redistribution, as well as the effects of repeated opening of ever new feeding channels, can be recognized.

Discussion

In the *in vitro experiments* reported in this paper it could be demonstrated that the QFM system is highly accurate on measuring not only steady but also pulsatile flow of $100-1000$ ml/min. With blood rather than a Sephadex solution as circulating medium the correlation of ultrasonically determined to true pulsatile flow was also linear when measuring with the MAVIS scanner. Within the range of $100-500$ ml/min examined in this experiment, flow was, however, underestimated by 15%, compared with a mean overestimation of 2% in the QFM experiments.

The use of either instrument for transcutaneous *carotid volume flow measurement* appears to be limited, for anatomic reasons, to the common carotid, where astonishingly the MAVIS measurements amounted only to a mean of 62% of the flow values obtained with the QFM system. While part of this difference may, from the evidence of the in vitro results, be due to a systematic underestimation by the MAVIS technique, other explanations need to be considered.

Systematic studies demonstrating that MAVIS flow measurement does not depend on the tissue depth of the vessel examined have so far not been carried out. It remains, therefore, to be proven that with the given beam parameters of the MAVIS scanner the accuracy of flow measurement remains comparable to the in vitro experiment, regardless of the considerable variation of the tissue depth in which the common carotid is located. Other factors which must be further discussed and experimentally studied include asymmetries of the velocity profiles not affecting the QFM measurements but obviously doing so with the MAVIS scanner and possibly producing a systematic error of flow calculation.

On the other hand, an overestimation of flow by the QFM system, when used for in vivo measurements, cannot be excluded. While with the silicone caoutchouc tube used to simulate the blood vessel in our in vitro tests, the inner wall echo was quite unambiguous, this is not always the case with human carotids. In fact, in a number of instances we observed a very weak innermost echo, possibly attributable to the true endothelium/blood interface, to which the tracking gate would not lock but rather jump to the neighboring more peripherally situated high energy echo. Setting the gate to an intramural rather than to the vessel wall/blood column interface echo obviously leads to an overestimation of the inner diameter and thus of the calculated volume flow. Parasite Doppler signals picked up from superimposed vessels are unlikely to play a substantial role when measuring common carotid flow, but their interference must still be considered when discussing the possible reasons for an overestimation of flow.

There is, however, a good argument for accepting the QFM measurements as being much closer to true flow than those obtained with the MAVIS scanner. Yoshimura et al. (1982) recently reassessed, in a group of 10 healthy volunteers, common carotid flow distribution to the internal and external carotid arteries using a modified QFM system equipped with a miniaturized probe. They found the ratio of common:internal:external carotid flow to be 1.54:1.00:0.56, signifying that the internal carotid receives 63.7% and the external carotid 36.3% of the common carotid blood. Based on these findings, the internal carotid volume flow calculated from our own data would be 63.7% of 7.74 ml/s = 4.93 ml/s or 295.8 ml/min. If the some 20% of carotids with the posterior cerebral artery originating from this vessel are disregarded and the posterior cerebral artery territory is estimated to be about 20% of the hemisphere, this value represents 80% of the hemispheric blood flow. Dividing it by 80% of an estimated mean hemisphere weight of 600 g, i.e. 480 g, it equals a mean regional blood flow of 61.6 ml/100 g/min. This figure is only about 14% higher than those determined for normal regional cerebral blood flow using the ^{133}Xe inhalation technique [54.3 ± 9.2 according to Meyer et al. (1978) and 53.7 ± 2.2 in the group of 44 volunteers studied by Melamed et al. (1980) and having a similar age distribution to our group's]. Contrariwise, the 64% of the QFM value computed as an equivalent to the mean of our MAVIS flows amounts only to

39.4 ml/100 g/min, which is some 34% less than the ^{133}Xe reference value and would have to be considered as pathologic when taking this latter as a standard.

Notwithstanding the considerable difference of absolute flow values found using the QFM and MAVIS techniques, it seems remarkable that in our material with both methods significantly higher flow rates were found in men than in women (5% with the QFM and 14% with the MAVIS system). This appears to be well explained by the sex difference in brain weight (Müller et al. 1983), which according to Ho (1980) amounts to 10%.

Limiting the further discussion to the results obtained with the QFM system, to which our clinical experience is confined, the mean normal common carotid flow values measured by us are in fair agreement with those reported by Uematsu et al. (1981, 1983) and Fujishiro and Yoshimura (1981) (see Table 1). Our data did not, however, reveal a continuous decrease with age as observed by these authors and also shown by studies of regional cerebral blood flow (Naritomi et al. 1979; Melamed et al. 1980) but rather, after a drop of flow around the mid-forties, an increase occurred in the age group of 56 to 65 years. How far, given the limited number of measurements, these findings reflect a true evolution of common carotid flow with age remains to be proved by studying a larger normal series extending to age groups beyond 65 years.

Discussing the possible *clinical value* of common carotid flow measurement, some comments must be made on the wide range of normal values found (6 – 11 ml/s for an individual vessel and up to 20% of total right plus left flow for the side difference). One of the reasons for this is obviously the variations of the circle of Willis. Variations in external carotid flow may be a further important factor in bringing about the wide range of absolute flow values.

As to the side difference of flow, this criterion for pathology may well considerably add to the specificity of the method for diagnosis of obstructive carotid disease, but it almost certainly reduces its sensitivity for recognizing bilateral processes of this type. It would therefore appear that the total right plus left common carotid flow should be considered as another criterion, allowing the diagnosis not only of such states of low perfusion which are due to bilateral obstructive pathology but also of others related to low cardiac output or increased blood viscosity.

But even when basing the interpretation of findings on the three criteria mentioned, and possibly others to be worked out upon further analysis of our material, the method cannot be expected to be sensitive and specific enough to be used on its own for the purpose of screening. This becomes sufficiently clear if one considers the cases observed where, with a very active ophthalmic collateral, common carotid flow remained within normal limits in spite of total internal carotid occlusion. Common carotid volume flow measurement should therefore always be used in conjunction with conventional Doppler sonography, including registration of common carotid, supratrochlear, and vertebral artery velocity pulse curves and a careful search for stenosis signals and signs of local turbulence in the region of the carotid bifurcation (Müller 1981). To this basic noninvasive investigation as well as to cerebral angiography, carotid volume flow measurement adds what is certainly clinically valuable information.

While the considerable interindividual variability of common carotid flow is undoubtedly a serious limiting factor in the diagnostic use of the method, intraindividual physiologically and methodologically determined changes are by far not of the same degree, though by no means negligible. In our normal series, the 95% interval of confidence calculated from the data obtained on three consecutive measurements carried out on each of the 200 carotids investigated was 0.79 ml/s, signifying that in 95% of all arteries the three measurements were within ± 11% of their mean. Though this indicates a fair reproducibility of the measurements we have adopted the practice of always taking three measurements in clinical investigations as well and to average the results in order to minimize diagnostic errors caused through physiologic fluctuations and minor technical mistakes.

Uematsu et al. (1983) calculated a standard deviation of 10.6% for the results of six measurements done on each of five volunteers, the examinations being spread over 3 weeks. Accepting from these data that readings on repeat examinations can fluctuate over a range of ± 21.2% (2 SD), one must certainly be careful when interpreting the results of any single control examination, be it for monitoring the spontaneous course of the disease or to assess the effect of therapy. Nevertheless, tentative use of common carotid flow measurement for quantifying the effect of vascular surgery has convinced us of the utility of the method for this purpose. In fact, from the data collected to date, the assessment of operative results may be the most valuable use of the method while it is still confined, for technical reasons, to the common carotid.

Measuring common carotid flow prior to and after the operation in patients undergoing endarterectomy for carotid stenosis allows noninvasive assessment of the hemodynamic effect of the surgical treatment, and in particular recognition of possible redistribution of contralateral carotid flow. Similarly, the results of EC/IC bypass surgery can be monitored, including the measurement, not reported in this paper, of common carotid differential flow (Müller and Gratzl 1980a).

With this technique common carotid flow is subsequently measured with and without compression of the donor branch of the external carotid. In patients having no ophthalmic collateral, the difference in flow produced through the compression may be considered as representing the bypass flow. If an ophthalmic collateral is present, however, external carotid blood may escape during compression through this collateral channel (Müller and Gratzl 1980b), so that differential common carotid flow is less than true bypass flow. Nevertheless, differential common carotid flow has proved to be a valuable parameter for semiquantitatively assessing the bypass function even in these cases.

Important hemodynamic information is also obtained with surgery, balloon occlusion, and embolization of pathologies having arterioveinous shunts. This is most clearly shown in our material with a case of AV malformation, where a stepwise procedure of treatment was chosen. After each occlusion of one or more feeding arteries, the effect on common carotid flow could be assessed, and so could the evolution of collateral circulation between the individual interventions. Follow-up in patients in whom the extirpation of an AV malformation is not possible and in whom other forms of treatment, such as embolization or radiotherapy, are used therefore seems a particularly suitable indication for the application of this method, allowing the early diagnosis of shunt persistency and opening of collateral feeding channels. With this noninvasive assessment, further angiography can be limited and timed according to the hemodynamic evolution.

Summary

A continuous wave (QFM) and a pulsed Doppler ultrasonic flowmeter were tested in vitro, using a Sephadex solution as circulating medium. The measurements done with the QFM system showed an excellent linear correlation to true flows from $100-1000$ ml/min with both continuous and pulsatile streaming. With the MAVIS scanner this correlation was linear only with continuous flow. However, when using blood as circulating medium, a linear correlation was found for pulsed flow within the range of $100-500$ ml/min, to which this experiment was limited. Within this range the QFM system overestimated flow by a mean of 2%, while with the MAVIS scanner there was an underestimation of 15%.

On comparative measurement of common carotid flow in 100 healthy volunteers the mean right plus left flow as determined with the QFM system was 15.5 ± 1.9 ml/s $= 930 \pm 114$ ml/min but only 565 ± 118 ml/min when measured with the MAVIS scanner. Considering the distribution of common carotid flow to the external and internal carotid and an estimate of the mean weight of the brain tissue irrigated through this latter artery, the QFM values appear to be in better agreement with the accepted normal values of regional cerebral blood flow.

The wide range of QFM normal values ($6-11$ ml/s for an individual carotid), at least partly attributable to variations of the circle of Willis, considerably limits the diagnostic value of the method. Based on the findings in the normal group and in a series of patients suffering from obstructive carotid disease or AV malformations, three criteria of pathology are proposed: (1) unilateral flow below 6 or over 11 ml/s; (2) a side difference of more than 20% of total right plus left flow; and (3) total right plus left flow below 11.7 or over 19.3 ml/s (95% confidence limit).

Experiences with pre- and postoperative QFM common carotid flow measurements in patients having carotid endarterectomy or balloon dilation ($n = 7$), EC/IC bypass surgery ($n = 32$), and surgery or neuroradiologic interventions for AV malformations and fistulas ($n = 6$) demonstrate that this method is of particular value for noninvasively assessing the effect of surgery and treatment with catheter techniques.

Acknowledgments. Prof. H. J. Zweifel, Buchs SG, and Dipl. Ing. W. Müller, Aarau, are gratefully acknowledged for their collaboration on designing the flow rig. We also would like to thank the mechanical department of our hospital (Head: Mr. H. Hoppler) for its realization.

References

Baker DW (1970) Pulsed ultrasonic Doppler flowmeter. Biological and engineering applications. IEEE Trans Sonics Ultrason SU-17:65

Borodzinski K, Filipczynski L, Novicki A, Powaowski T (1976) Quantitative transcutaneous measurements of flow in the carotid artery by means of pulse and continuous wave Doppler methods. Ultrasound Med Biol 2:189−193

Doriot PA, Casty M, Milakara B, Anliker M, Bollinger A, Siegenthaler W (1975) Quantitative analysis of flow conditions in simulated vessels and large human arteries and veins by means of ultrasound. In: Kazner E, de Vlieger M, Müller HR, McCready VR (eds) Ultrasonics in medicine. Proc 2nd congr on ultrasonics in med. Excerpta Medica. Internat Congr Ser 363, Elsevier Amsterdam, Oxford, p 160−168

Fish PJ (1981) A method for transcutaneous blood flow measurement − accuracy considerations. In: Kurjak A, Kratochwil A (eds) Recent advances in ultrasound diagnosis. Proc 4th Europ Congr on Ultrasonis in Med. Excerpta Medica. Internat Congr Ser 553, Elsevier North Holland, Amsterdam, Oxford, Princeton, p 110−115

Fujishiro K, Yoshimura S (1981) Haemodynamic changes in carotid blood flow with age. Jikeikai Med J 29:125−138

Furuhata H, Kanno R, Kodairo K, Aoyagi T, Hayashi J, Matsumoto H, Yoshimura S (1978) An ultrasonic blood flow measuring system to detect the absolute volume flow. Jpn J Med Electron Biol Eng 16 (Suppl):334

Gill RW (1979) Pulsed Doppler with B-mode imaging for quantitative blood flow measurement. Ultrasound Med Biol 5:223−235

Ho K, Roessmann U, Straumfjord V, Monroe G (1980) Analysis of brain weight. I. Adult brain weight in relation to sex, race and age. Arch Pathol Lab Med 104:635−639

Keller HM, Meier WE, Anliker M, Kumpe DA (1976) Noninvasive measurement of velocity profiles and blood flow in the common carotid artery by pulsed ultrasound. Stroke 7:370−377

Melamed E, Lavy S, Bentin S, Cooper G, Rinot Y (1980) Reduction of regional cerebral blood flow during normal aging in man. Stroke 11:31−35

Meyer JS, Ishihara N, Deshmukh VD, Naritomi H, Sakai T, Hsu MC, Pollack P (1978) Improved method for noninvasive measurement of regional cerebral blood flow by [133]xenon inhalation. Part I: Description of method and normal values obtained in healthy volunteers. Stroke 9:195−205

Müller HR (1981) Evaluation of cranial blood flow by means of ultrasonic Doppler techniques. In: Barnett H, Paoletti P, Flamm E, Brambilla G (eds) Cerebrovascular diseases: New trends in surgical and medical aspects. Elsevier North Holland, Amsterdam, Oxford, Princeton, p 77−90

Müller HR, Gratzl O (1980a) Quantitative Funktionsprüfung des ATS/ACM-Bypasses mittels eines neuartigen Ultraschall-Flowmeters. Ultraschall 1:217−222

Müller HR, Gratzl O (1980b) Interaction of superficial temporal/middle cerebral artery bypass with the ophthalmic collateral. Stroke 11:11−12

Müller HR, Radue EW, Saia A, Pallotti C (1983) Doppler ultrasonic measurement of carotid flow. (Letter to the Editor) Ultrasound Med Biol 9:L91−L95

Naritomi H, Meyer JS, Sakai F, Yamaguchi F, Shaw T (1979) Effects of advancing age on regional cerebral blood flow. Studies in normal subjects and subjects with risk factors for atherothrombotic stroke. Arch Neurol 36:410−416

Olson RM, Cooke JP (1975) Human carotid artery diameter and flow by a noninvasive technique. Med Instrum 9:99−102

Payen DM, Levy BI, Menegalli DJ, Layat YI, Levenson JA, Nicolas FM (1982) Evaluation of human hemispheric blood flow based on non-invasive carotid blood flow measurement using the range-gated Doppler technique. Stroke 13:392−398

Peronneau P, Hinglais J, Pellet M, Léger F (1970) Vélocimètre sanguin par effect Doppler à émission ultra-sonore pulsée. A. Description de l'appareil. Résultats. L'Onde Electrique 50:3−18

Simon A, Levenson J, Safar M, Diebold B, Peronneau P (1982) Non-invasive pulsed Doppler
    measurement of blood flow: Investigation of internal carotid stenosis. Abstr 3rd meeting of
    WFUMB. Ultrasound Med Biol 8 (Suppl 1):180
Uematsu S (1981) Determination of volume of arterial blood flow by an ultrasonic device. J
    Clin Ultrasound 9:209−216
Uematsu S, Yang A (1981) Transcutaneous measurement of carotid artery volume flow by an
    ultrasonic device. In: Cohen BA (ed) IEEE 1981. Frontiers of Engineering in Health Care.
    Proc 3rd Annual Conf Engg in Med and Biol Soc of IEEE, p 154−158
Uematsu S, Yang A, Preziosi TJ, Kouba R, Toung TIK (1983) Measurement of carotid blood
    flow in man and its clinical application. Stroke 14:256−266
Yoshimura S, Furuhata H, Suzuki N, Kodaira K, Hirota H (1982) A method for the quantita-
    tive and noninvasive measurement of blood flow volume in internal and external carotid
    arteries and vertebral artery. Jikeikai Med J 29:197−208

Ultrasonic Diagnosis of Carotid Artery Disease: Imaging and Combined Transcutaneous Flow Measurements

M. Hennerici and U. Trockel

Continuous-wave (CW) Doppler examination is known to represent a reliable noninvasive method for estimation of the amount, extent, and location of extracranial carotid lesions producing obstructions of the carotid lumen greater than 50% (Hennerici et al. 1981; Büdingen et al. 1982). However, in the absence of widespread flow alterations, it fails to detect hemodynamically non-significant lesions characterizing the early stage of the disease. Although the sensitivity of this method may be extended by a detailed interpretation of disturbances in the flow pattern as revealed by audiospectrum analysis of the Doppler signals (Reneman and Spencer 1979), a number of limitations restrict its clinical applicability: firstly, spectral broadening is used to indicate disturbances in the flow patterns; however, it occurs not only in abnormal turbulent flow but also under several physiologic conditions (e.g., in healthy children and young adults). Secondly, audiospectrum analysis does not reveal any information about the arterial velocity profile, that is, the distribution over the cross-sectional area of the blood vessel at discrete time intervals, changing considerably during each cardiac cycle. Thirdly, no differentiation can be achieved between flat and ulcerated plaques, the latter representing possible sources of cerebral emboli. Finally, changes of the arterial flow volume cannot be examined. Among a variety of new techniques developed for a refined diagnosis of cerebrovascular disease, we used a new instrument combining a high resolution (10 MHz) imaging system with a specially designed multichannel pulsed Doppler blood flow measuring device (Green et al. 1977; Hennerici and Freund 1984; Taenzer et al. 1981). The advantages and limitations for a clinical diagnosis of this system versus CW Doppler examination were compared in patients with different stages of carotid disease.

Methods

CW Doppler examination of the carotid and vertebral arteries in the neck was performed using a directional CW Doppler device (Debimètre ultrasonique directionnel Delalande) with an unfocussed (4 MHz) beam according to previous reports (Büdingen et al. 1982; Hennerici et al. 1981). A specially designed Duplex system was used for ultrasonic imaging and measurement of

Neurologische Klinik der Universität Düsseldorf, Moorenstraße 5, D-4000 Düsseldorf 1

Cerebral Blood Flow and
Metabolism Measurement.
Eds. Hartmann/Hoyer
© Springer-Verlag Berlin Heidelberg 1985

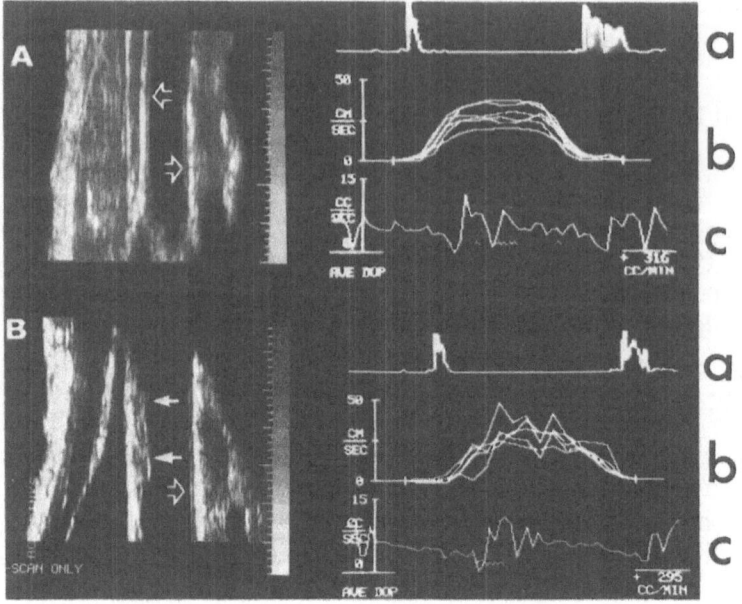

Fig. 1a, b. High-resolution B-mode longitudinal sections (*left side*) and corresponding flow (cm/s) and volume (cc/s) pattern displays (*right side*) of a patient with a history of repetitive TIAs of the left hemisphere but normal angiograms. **A** Normal right carotid artery with typical continuous intima reflections (⇒) and laminar flow velocity profiles (b) during both systole and diastole. **B** Left carotid artery with intimal flat extended fibrosis of the anterior wall (⇄) and corresponding undulating flow velocity profiles (b), in particular during systole, indicating turbulent instead of laminar flow. Volume flow was 316 and 295 cc/min respectively

flow velocity profiles and flow volume of the extracranial carotid arteries. The imaging component consists of a linear mechanical, real-time B-scan (lateral resolution 0.5 mm, axial resolution 0.35 mm). The probe of the 16 range-gated 5 MHz pulsed Doppler system is fixed laterally to the image transducer. Each of the 16 measurement volumes occupies 1 mm³. Figure 1 shows the B-mode images and the display of the Doppler data consisting of the flow velocity patterns (b) across the arterial diameter (a) and the flow volume measurements (c) for each cardiac cycle. The averaged distribution of the velocity profiles at selected intervals within each single cardiac cycle can successively be displayed. In addition, instantaneous quantitative volume flow measurements may be computed using a special algorhithm. In vitro tests revealed volume flow estimate variances of only ± 10% (Hennerici and Freund 1984).

Results

Table 1 summarizes the results obtained in 134 patients (193 carotid arteries) in comparison with arteriography. Since the major disadvantage of the Duplex

Table 1. Comparison of arteriographic results with CW Doppler and Duplex system examinations in 134 patients (193 carotid arteries)

Arteriogram (n)	CW Doppler				Duplex system				
	Normal	Stenosis <50%	Stenosis >50%	Occlusion	Normal	Stenosis <50%	Stenosis >50%	Occlusion	Equivocal
Equivocal (6)	6				6				
Normal (59)	**57**	2			**37**	12			10
<50% Stenosis (61)	51	**10**			6	**45**			10
>50% Stenosis (57)			**55**	2			**43**		14
Occlusion (10)				**10**			2	**5**	3

system used is a high rejection rate (19%) of technically unsatisfactory scans, CW Doppler examination yields the most comprehensive information for the detection of obstructive carotid lesions. However, provided sufficient carotid display in several horizontal and vertical sections can be achieved, Duplex system analysis considerably improves the detection as well as the morphological and hemodynamic evaluation of nonstenotic atherosclerosis, being superior even to arteriography. Figure 1 demonstrates an example of a flat intimal fibrosis of the arterial wall of the left carotid artery (B), corresponding with abnormally undulating flow velocity profiles particularly during systole, indicating turbulent instead of laminar flow patterns, in a patient with repetitive TIAs of the left hemisphere but normal angiograms. For comparison, the B-scan and laminar flow velocity profiles of the normal right carotid artery are shown (A). Flow volumes are similar on either side.

Flow abnormalities are not necessarily attached to morphologically detectable lesions and may well be observed even in their absence, e.g., in the presence of "sonolucent plaques" characterized by the same acoustic impedance as flowing blood. On the other hand, plaques may not always cause significant flow pattern changes but leave flow conditions unchanged. Thus Duplex system analysis offers the unique possibility to study further the relation between morphological and hemodynamic alterations in carotid atherosclerosis. This is particularly helpful for prospective studies considering the natural history of atherosclerosis during follow-up.

Another advantage of the system used consists in the possibility of quantifying volume flow with sufficient accuracy for repetitive examination of the carotid arteries in the neck. Figure 2 shows an example of a 28-year-old patient studied repetitively during and after a right hemispheric migraine accompagnée. During and even a fortnight after the original migraine attack, volume flow of the right common carotid (R) was reduced to less than 50%, while volume flow of the contralateral common carotid was increased (L), indicating a sufficient collateral capacity. Four weeks after the onset of symptoms both carotids revealed a similar flow within normal limits (N). Flow velocity patterns

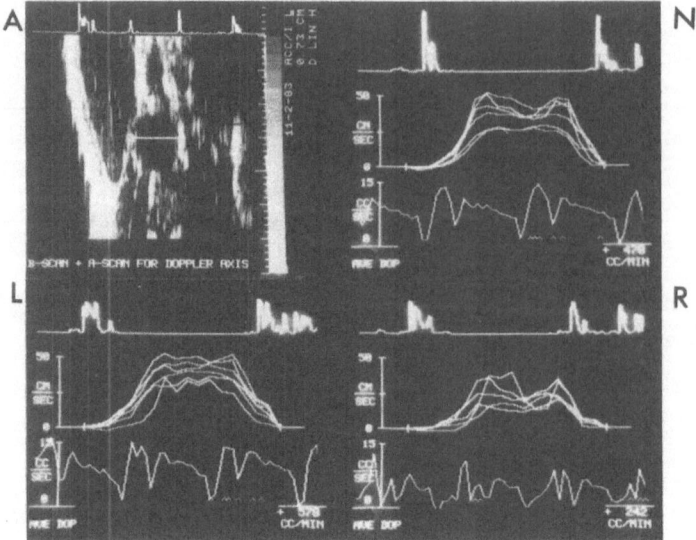

Fig. 2. High-resolution B-mode cross section of a normal common carotid artery in a 28-year-old patient with an ipsilateral hemispheric migraine accompagneé (A). Volume flow measurements during the original attack revealed a considerable reduction within this vessel (R) accompanied by an increased volume flow within the contralateral common carotid artery (L). Three weeks later flow volumes were similar on either side (N). Note that the velocity profiles remained relatively preserved for all volume flow conditions

remained relatively preserved, showing a dichotomic distribution which represents a physiologic variation of the normal flow pattern near the bifurcation. No morphological abnormalities could be detected from either B-scan or extra- and intracranial angiography of the right brain supplying arteries. Among a total of 35 patients with typical attacks of migraine accompagneé, all five patients studied during their attack revealed a considerable volume flow reduction within the common carotid artery ipsilateral to the symptomatic hemisphere, as did about one-third of those studied just after an attack. This may be interpreted as a result of intracranial vasospasms inducing a significant increase of peripheral vascular resistances. Therefore this method may similarly be used in studies of patients with subarachnoid hemorrhages for preoperative follow-up of major cerebral vasospasms.

In conclusion, the combination of a high resolution imaging system with a multigated pulsed Doppler device may provide important clinical information in patients with cerebrovascular disease. Handling difficulties and the rejection rate of technically unsatisfactory examinations (about 20%) restrict its clinical usefulness to a group of selected patients in whom information additional to that obtained from CW Doppler and angiography is required.

Acknowledgment. This study was supported by the Deutsche Forschungsgemeinschaft SFB 200/D2

References

Büdingen HJ, von Reutern G-M, Freund H-J (1982) Dopplersonographie der extrakraniellen Hirnarterien. Thieme Stuttgart – New York

Green PS, Taenzer JC, Ramsey SD, Holzemer JF, Suarez JR, Marich KW, Evans TC, Sandok BA, Greenleaf JF (1977) A real-time ultrasonic imaging system for carotid arteriography. Ultrasound Med Biol 3:129–142

Hennerici M, Freund H-J (1984) Efficacy of CW-Doppler and Duplex-system examinations for the evaluation of extracranial carotid disease. J Clin Ultrasound 12:155–161

Hennerici M, Aulich A, Sandmann W, Freund H-J (1981) Incidence of asymptomatic extracranial arterial disease. Stroke 12:750–758

Reneman RS, Spencer MP (1979) Local Doppler audiospectra in normal and stenosed carotid arteries in men. Ultrasound Med Biol 5:1–11

Taenzer JC, Burch DJ (1981) Real-time Doppler blood flow measurements: qualitative and quantitative. 6th International Symposion on Ultrasonic Imaging and Tissue Characterization (abstract)

Measurement of Blood Flow by Ultrasound in Healthy Subjects and in Patients with Cerebrovascular Disease

H. Weiss[1], A. Haass[1], K. Schimrigk[1], and E. Wenzel[2]

Introduction

The most common diagnostic method employed to determine the degree of arteriosclerotic change in the internal carotid artery is angiography. The often encountered discrepancy between clinical and angiographic findings demonstrates that a mere morphological characterization of a stenosis is not sufficient. The relevant functional parameter of a hemodynamically significant arteriosclerotic lesion is the blood flow through it. In our investigations we determined the effective flow in the stenosed vessel and the collateral flow in the unaffected vessel. The aim of our investigations was twofold: to quantify the functional degree of a stenosis and to establish a relationship between the measured flow rates and neurologic deficits as well as the clinical course.

Methods

Instrumentation: For the transcutaneous blood flow measurements we used a 5-MHz, 30-channel, pulsed Doppler system with imaging facility, which feeds frequency and range information to a flow and a position computer (MAVIS, described by Fish 1981).

Subjects:
- Fourteen test persons without any signs of cerebrovascular disease, with normal CW Doppler findings, normal heart function, normal blood pressure, and normal pulse rate (age 36 – 78 years; mean 51.1 ± 12.8)
- Six patients with unilateral stenosis of the internal carotid artery of less than 50% (age 33 – 68 years; mean 52.7 ± 13.3)
- Fourteen patients with stenosis of more than 50% (age 38 – 68 years; mean 56.2 ± 9.7)
- Ten patients with occlusion of the internal carotid artery (age 16 – 63 years; mean 42.7 ± 15.3).

Determination of the degree of stenosis: We defined the degree of stenosis as the maximum percentage reduction in vessel diameter at a point within the stenotic

---

1 Abteilung Neurologie, Universität des Saarlandes, D-6650 Homburg
2 Abteilung für Klinische Haemostaseologie und Transfusions-Medizin, D-6650 Homburg

Cerebral Blood Flow and
Metabolism Measurement.
Eds. Hartmann/Hoyer
© Springer-Verlag Berlin Heidelberg 1985

segment compared with the diameter 3 cm distal of that point as measured by micrometer from X-ray contrast angiograms.

Measuring procedure: With patients supine, flow measurements were performed on the common and internal carotid arteries of both sides four times on average. For the internal carotid artery, the most distal extracranial part of the vessel close to the mandibula was chosen.

Significance levels were calculated by *t*-test.

Results

Control Group

The mean volume flow in the internal carotid artery was about 190 ml/min on the right as well as on the left side, i.e., 65% of the common carotid artery flow of about 300 ml/min on each side (Table 1). Flow in the internal carotid artery showed an age dependence. A significant difference ($P = 0.07$ and 0.08) of about 30 ml/min existed in patients older ($n = 7$) and younger ($n = 7$) than 50 years. In the over-50 group the average flow was 176 ml/min (SD 24) on the right and 179 (SD 27) on the left side; in the under-50 group it was 206 (SD 32), and 208 (SD 30) ml/min respectively.

Group with Stenosis of Less than 50%

In six patients with angiographically defined stenosis of the internal carotid artery of less than 50% no flow reduction in the stenosed vessel was observed. The flow patterns of the internal and common carotid artery did not differ much from those in healthy subjects. Individual variation increased as shown by higher standard deviations (Table 1).

Table 1. Volume flow (ml/min) in the internal and common carotid artery

Groups	ICA		CCA		CCA:ICA	
	Right/ stenosed	Left/non- stenosed	Right/ st. side	Left n. st. side	Right/ stenosed	Left/non- stenosed
Healthy subjects ($n = 14$)	191.0 (±31.3)	193.4 (±31.1)	296.2 (±33.6)	298.9 (± 40.0)	1:0.65	1:0.65
Stenosis < 50% ($n = 6$)	185.5 (±58.1)	176.9 (±59.0)	298.8 (±76.2)	293.0 (± 74.3)	1:0.61	1:0.59
Stenosis > 50% ($n = 14$)	82.0 (±49.5)	202.7 (±55.8)	214.6 (±72.2)	312.7 (± 74.4)	1:0.39	1:0.65
Occlusion ($n = 10$)	–	256.0 (±93.9)	141.0 (±45.5)	374.2 (±107.3)	–	1:0.70

Table 2. Volume flow (ml/min) and flow distribution in the internal and common carotid artery in stenosis of 50% and more

Degree of stenosis in ICA	Internal carotid artery		Common carotid artery		CCA : ICA	
	Stenosed	Non-stenosed	St. side	N. st. side	Stenosed	Non-stenosed
50%–69% (n=6)	135.5* (±31.1)	233.2 (±54.6)	249.7 (±52.4)	333.1 (±63.6)	1:0.55	1:0.70
70%–90% (n=6)	66.9* (±17.2)	180.9 (±63.8)	208.0 (±85.8)	301.3 (±101.3)	1:0.37	1:0.61
>90% (n=2)	20 (±0)	196.00 (±24.5)	151.5 (±87.0)	304.67 (±34.41)	1:0.16	1:0.64

* $P = 0.005$

Group with Stenosis of More than 50%

In fourteen patients, a reduction of the vessel diameter of 50% or more caused a significant decrease in mean flow with 82.0 ml/min in the stenosed internal carotid artery, as compared with 202.7 ml/min in the nonstenosed vessel ($P = 0.001$). The flow rate in this unaffected internal carotid artery was not significantly higher than in the control group. Similar conditions, but to a lesser degree, were found in the common carotid artery: the mean flow was 214.6 ml/min on the affected and 318.2 ml/min on the unaffected side ($P = 0.001$).

Splitting this group into three subdivisions – stenosis from 50% to 69%, from 70% to 90%, and more than 90% – revealed further significant differences in internal carotid artery flow (Table 2). Flow variation decreased to the same degree as the stenosis increased.

There exists a negative correlation of -0.84 between flow and degree of stenosis. Nevertheless the residual flow and thus the relevant dimension of a stenosis is not solely a function of a reduction in diameter in the individual case. Patients with the same degree of stenosis show considerable differences in flow, especially in lower grade stenoses. It was particularly noticeable that patients with complete recovery from cerebral infarction had a mean residual flow (107 ml/min) in the affected vessel which was nearly twice as high as that in patients with incomplete recovery (59 ml/min), even though, at 6%, the mean degree of stenosis did not differ much between these two groups. Similar results were found in patients with slight and severe strokes where the flow was reduced by more than half.

Group with Occlusion

Ten patients with an occlusion demonstrated a significant ($P = 0.02$) flow increase of about 25% in the contralateral internal carotid artery (256.0 ml/min), and somewhat less in the common carotid artery (374.2 ml/min). The 141.0 ml/min mean flow in the common carotid artery of the occluded side was less than half the flow in the opposite vessel (Table 1).

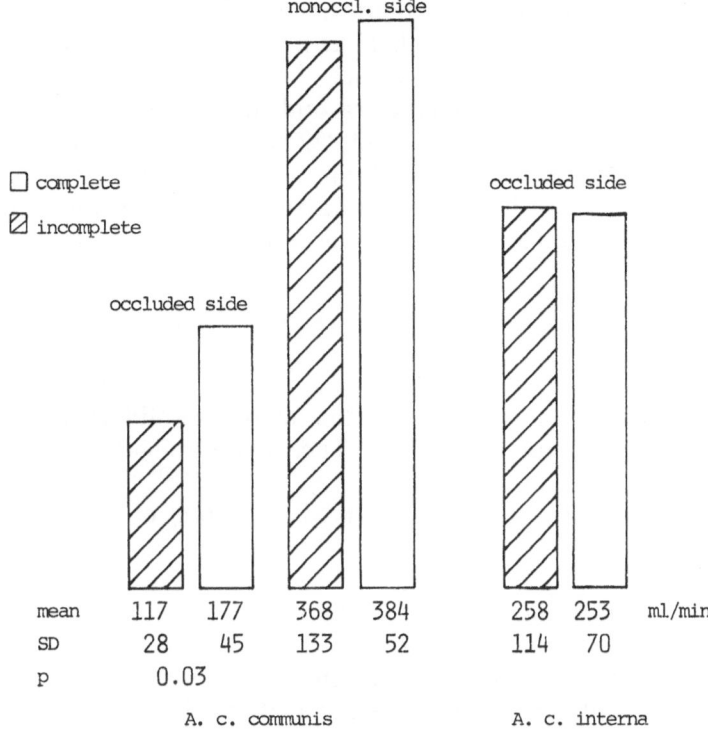

Fig. 1. Occlusion of the internal carotid artery: volume flow (ml/min) in patients with complete and incomplete recovery from neurologic deficits

A significant increase in residual common carotid artery flow on the occluded side was observed in patients with complete clinical recovery compared with the flow in patients with incomplete recovery (Fig. 1). This is probably due to collateralization via the ophthalmic artery in the first group, in which periorbital Doppler showed a strong flow reversal, whereas in the group with enduring deficits a slight flow reversal with orthograde flow after compression of the temporal artery was detected in only one patient. It should be noted that both groups differed in the same way with respect to initial deficits, size of infarction in cranial computer tomography, and cerebral scintigrams.

Discussion

The accuracy of the method presented in this paper had been assessed in vitro (Fish 1981). Considering our measurements in the vertebral arteries showing flow rates of about 145 ml/min each, a mean cerebral blood flow of 670 ml/min can be assumed. Comparing this result with global cerebral blood flow measurements by the Kety-Schmidt method a slight underestimation of our in vivo measurements could be supposed: in his survey Lübbers (1972) cites five authors who had determined a mean cerebral blood flow of about

750 ml/min (assuming a brain weight of 1400 – 1500 g). Similar results were presented by Lassen et al. (1966), comparing the intraarterial xenon clearance technique with this method.

Investigations on internal carotid artery flow are rare. Yoshimura et al. (1982) measured a flow rate of 322 ml/min ± 79.8 in ten healthy subjects with the QFM Doppler system. A mean cerebral blood flow of 1025 ml/min, i.e., 20% of HMV (5 liters/min), could be calculated by adding their measured values of vertebral artery flow, which seems to be an overestimation.

Summary

1. The ca. 190 ml/min flow in the internal carotid artery represents 65% of the common carotid artery flow. There exists a slight decrease in flow with age.

2. No flow reduction was observed in stenoses of the internal carotid artery of less than 50%.

3. Flow through a more than 50% stenosed internal carotid artery decreased with an increasing degree of stenosis ($r = -0.84$). Mean flow was reduced to 82.0 ml/min. Considerable differences at the same degree of stenosis demonstrated that flow through a stenosed vessel not only is determined by diameter reduction but is a function of complex influences whose effects can be sufficiently described by flow measurment distal of the stenosis. The close relationship of residual flow to size of infarction and to remission of symptoms underlines the importance of this method of evaluating a stenosis. Collateral flow over the nonaffected internal carotid artery seems to have far less influence on the clinical course.

4. Patients with occlusion of the internal carotid artery showed a significant flow increase in the common and therefore in the external carotid artery on the affected side in all cases of complete recovery. This is probably due to collateralization via the ophthalmic artery.

References

Fish PJ (1981) A method of transcutaneous blood flow measurement – accuracy considerations. In: Kurjak A, Kratochwil A (eds) Recent advances in ultrasound diagnosis 2. Excerpta medica, Amsterdam – Oxford – Princeton, p 110 – 115

Lassen NA et al (1966) Human cerebral blood flow measured by two inert gas techniques. Comparison of the Kety-Schmidt method and the intra-arterial injection method. Circulat Res 19:681

Lübbers DW (1972) Physiologie der Gehirndurchblutung. In: Gänshirt H (ed) Der Hirnkreislauf. Thieme, Stuttgart, p 214 – 260

Yoshimura S et al. (1982) A method for the quantitative and noninvasive measurement of blood flow volume in internal and external carotid arteries and vertebral artery. Jikeikai Med J 29:197 – 208

Ultrasonic Doppler Assessment of Hemodynamics and Flow Volume in Cerebrovascular Neurosurgery

H. Wassmann and G. Fischdick

Introduction

During cerebrovascular surgical procedures such as carotid endarterectomy and extra-intracranial arterial bypass (EIAB) operations it is important to determine the blood flow, the change of blood flow volume, and the patency of an indwelling shunt. Since 1978 we have used Doppler vessel clamps and have tested their accuracy and practicability of use during surgery.

Methods

We use a directional ultrasonic Doppler flowmeter system, Delalande, with a transmission frequency of 4 MHz. Intraoperative measurements were performed by Doppler vessel clamps with a diameter of 2, 5, and 10 mm (Fig. 1). The accuracy of these Doppler clamps was tested by comparison with electromagnetic flow measurements. In three dogs ten consecutive measurements were made with calibrated electromagnetic flowmeters on dissected arteries with blood pressure maintained at a constant level. Furthermore the Doppler probes with a diameter of 2, 5, and 10 mm were tested on dissected isolated arteries perfused by a pump with heparinized blood and varing the flow velocity. In 48 patients (13 women, 35 men; mean age 52 years) suffering from a segmental stenosis of the internal carotid artery near the bifurcation of about 90% in 16 cases, of 70% in 22 cases, and of about 50% in 10 cases, the flow was determined pre-, intra-, and postoperatively. In addition the function and flow volume of an internal shunt used in 12 patients was checked.

In 86 patients (mean age 53 years, 64 patients with unilateral and 4 with bilateral carotid artery occlusion, 8 with occlusion of the middle cerebral artery, 7 with supraclinoidal stenosis of the carotid artery, and 3 with Moya Moya disease) undergoing EIAB surgery, the 2-mm Doppler clamp was used to check bypass patency and to determine blood flow volume passing through the anastomosis immediately after completion of the procedure.

Neurochirurgische Universitätsklinik D-5300 Bonn-Venusberg

Cerebral Blood Flow and
Metabolism Measurement.
Eds. Hartmann/Hoyer
© Springer-Verlag Berlin Heidelberg 1985

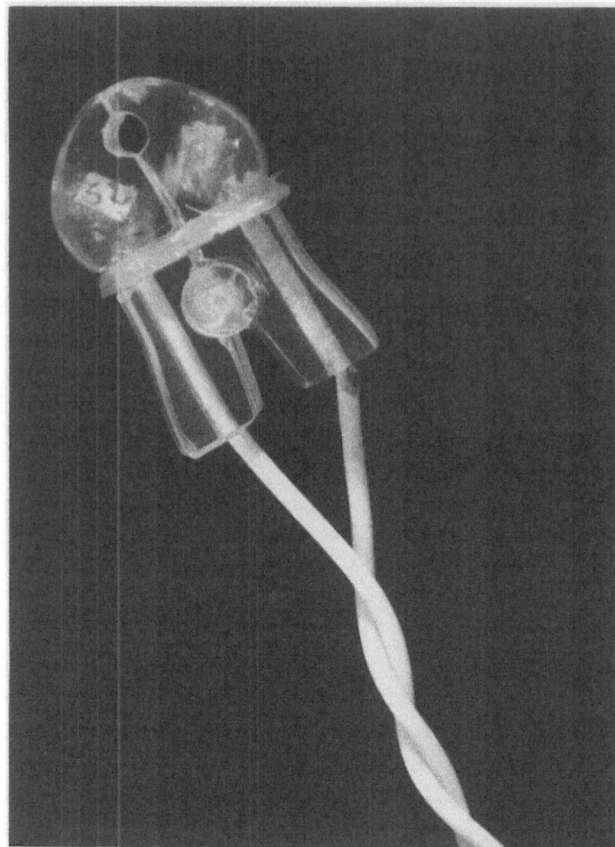

Fig. 1. Doppler clamp with a diameter of 2 mm

Results

Determination of Accuracy of Doppler Vessel Clamps

Two-, 5-, and 10-mm Doppler clamps were placed on a dissected dog artery. Each probe was repositioned ten times and the flow measurements were recorded. These values were compared with electromagnetic flow measurements during a constant systolic blood pressure of 120 mmHg. The deviation from these values was ± 12.8% to ± 15.2% (Table 1). During in vitro examinations the perfusion of the arteries was changed and the deviation of the Doppler values from the real flow volume was determined, showing values of ± 11% to ± 17%.

Intraoperative Use of Doppler Vessel Clamps

In a 45-year-old patient with right-sided transient ischemic attacks, reduced flow over the left supratrochlear artery and the left common and internal ca-

Table 1. Variation of Doppler values from electromagnetic and in vitro measurements

Doppler clamp	Blood flow volume	Variation of Doppler values
Electromagnetic flow measurements:		
2 mm	52 ml	±12.8%
5 mm	188 ml	±14.3%
10 mm	753 ml	±15.2%
In vitro measurements:		
2 mm	9.4– 97.8 ml	±17.0%
5 mm	59 –330 ml	±17.2%
10 mm	188 –753 ml	±11.0%

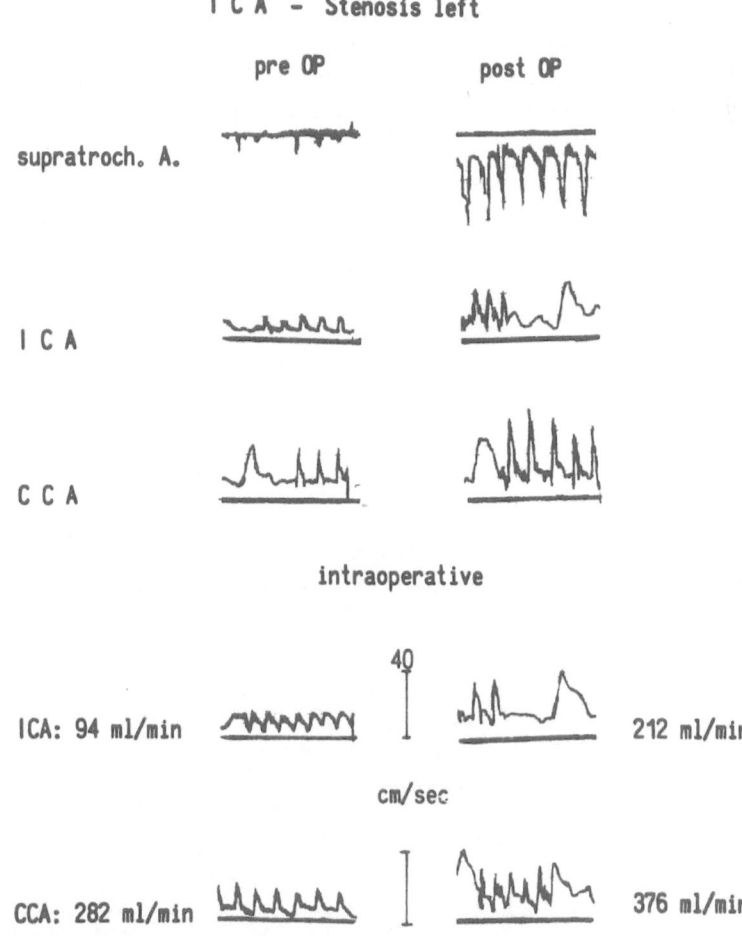

Fig. 2. Pre-, intra-, and postoperative Doppler examinations in a patient with stenosis of the left internal carotid artery (*ICA*). *CCA*, common carotid artery

rotid artery was found, as well as turbulence over the carotid bifurcation (Fig. 2).

Angiography showed a severe segmental stenosis of the internal carotid artery. Intraoperatively we recorded a flow of 282 ml/min in the left common carotid artery and one of 94 ml/min in the left internal carotid artery.

After carotid endarterectomy Doppler studies revealed a 125% increase of blood flow volume in the internal carotid artery. Postoperative angiography showed a patent carotid artery. Carotid perfusion after endarterectomy in 48 patients with segmental stenosis of the internal carotid artery showed an average increase of 56 ml/min.

The highest mean increase occurred in the 90% stenosis group (82 ml/min) and the lowest in the 50% stenosis group (37 ml/min). In 12 cases the use of an internal shunt during endarterectomy was deemed necessary. We determined patency and blood flow through the shunt, which ranged from 52 to 97 ml/min.

A Doppler vessel clamp was also used in EIAB surgery. In a 46-year-old patient with right internal carotid artery occlusion a mean velocity over the parietal branch of the right superficial temporal artery of 16 cm/s was measured. After performing the EIAB, flow through the anastomosis of 52 ml/min was determined intraoperatively by Doppler probe. The increase in diastolic Doppler values was observed as a sign of cerebral perfusion resistance. Ten days after

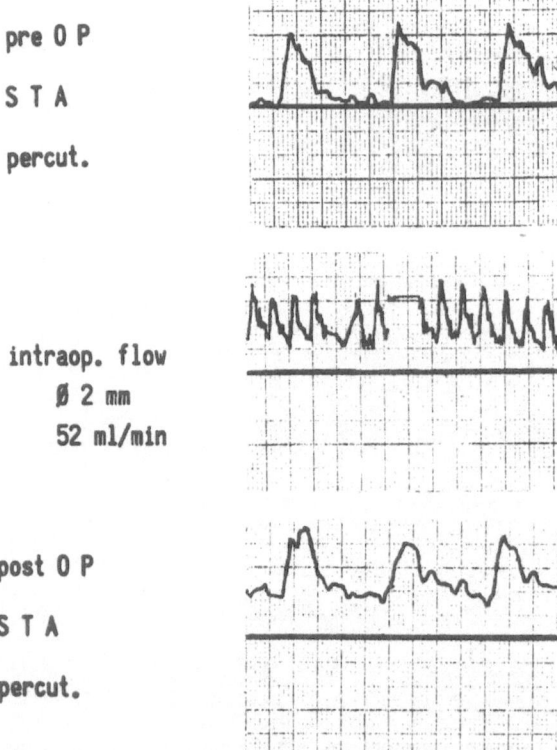

pre 0 P

S T A

percut.

intraop. flow
Ø 2 mm
52 ml/min

post 0 P

S T A

percut.

Fig. 3. Pre-, intra-, and postoperative Doppler examinations in a patient with occlusion of the right internal carotid artery undergoing bypass surgery

surgery an increase in velocity of 85% was registered over the bypass artery. Furthermore Doppler sonography of the temporal artery showed an elevated diastolic pressure like the internal carotid artery (Fig. 3) [1, 2]. We examined a total of 86 patients intraoperatively during EIAB surgery and found blood flow values through the bypass ranging from 3.7 to 65 ml/min with a mean value of 36 ml/min. The highest mean value of 44.6 ml/min was found in patients with bilateral carotid artery occlusion. In patients with unilateral occlusion a mean value of 43.4 ml/min was determined, and the lowest mean value of 23.8 ml/min was found in patients with Moya Moya disease.

Summarizing Remarks

Investigations by other workers and our own experiments concerning the accuracy of blood flow volume determinations by Doppler vessel clamps have shown that even minor fluctuations in vessel diameter can lead to deviations of 50% from the real flow volume due to the square relationship between the velocity and volume [3, 4]. We found a discrepancy between Doppler probe values in vitro and in vivo of up to ± 17%. Nevertheless, the intraoperative use of this simple method seems useful since the relative change in flow after carotid endarterectomy and the patency of the internal shunt are well demonstrated.

Additionally, our mean values of increase of blood volume after endarterectomy are very similar to those determined by electromagnetic flowmeter by Wiberg and Nornes [5], who found a mean value rising from 96.4 to 178.3 ml/min after removal of the arteriosclerotic plaque in 160 operations.

In EIAB surgery intraoperative flow measurements not only enable us to check bypass patency but also provide a baseline from which the flow through the anastomosis can be established. This bypass flow is highest in patients with bilateral carotid artery occlusion, probably as a sign of difference between extra- and intracranial arterial blood pressure.

Intraoperative Doppler flow volume measurements seem to us to be a suitable method for evaluating cerebrovascular changes in patients undergoing endarterectomy or EIAB surgery.

References

1. Gottwald K (1981) Doppler-Sonographie der extrakraniellen hirnversorgenden Gefäße. EEG-Labor 3:121−137
2. Hopman H et al. (1976) Doppler-Sonographie bei mikrovaskulärem Bypass. Neurochirurgia 19:190−196
3. Messmer K et al. (1968) Flow tracings of three types of flowmeters in acute dog experiments. In: Cappelen I (ed) New findings in blood flowmetry. Oslo, p 136−145
4. Thal HU, Leem W (1979) Animal experiments with Doppler flow transducers. Acta Neurochirurgica 28:275−277
5. Wiberg J, Nornes H (1983) Effects of carotid endarterectomy on blood flow in the internal carotid artery. Acta Neurochirurgica 68:217−226

CW Doppler Sonography Versus Angiography in the Determination of Patency After Extra-Intracranial Bypass Operations

H.-U. THAL and T. MICUS

Introduction

The evaluation of patency during extra-intracranial bypass procedures depends on the personal experience of the surgeon. Postoperatively the rate of patent anastomoses is demonstrated angiographically, although, as pointed out by Fox and Allock (1978) and Reisner et al. (1980), this investigation has a high risk for patients with cerebrovascular occlusive disease, especially after extra-intracranial bypass operations. Therefore the need for noninvasive methods is obvious.

Materials and Methods

Between 1 January 1979 and 31 December 1981, 97 patients underwent a superficial temporal − middle cerebral artery bypass procedure. 104 operations were performed. The diagnoses, depending on the neurologic signs and symptoms, were as follows: 25 TIA, 16 PRIND, and 47 completed strokes, of which 26 involved minor neurologic deficits and 21 severe completed stroke. Nine patients had other diagnoses (Table 1). All patients had an arteriogram preoperatively. The angiographic investigation showed no pathologic evidence in three cases,

Table 1. Diagnosis depending on neurologic signs and symptoms

	No. of cases
TIA	25
PRIND	16
SE	0
CS mild	26
CS severe	21
Other	9
Total	97

Neurochirurgische Klinik der Universität Düsseldorf, Moorenstraße 5, D-4000 Düsseldorf

Cerebral Blood Flow and
Metabolism Measurement.
Eds. Hartmann/Hoyer
© Springer-Verlag Berlin Heidelberg 1985

Table 2. Diagnosis depending on the angiographic investigation

	No. of cases
Stenosis of CCA + ICA	10
MCA	11
Occlusion of CCA + ICA	38
MCA	10
Multiple lesions	13
Vertebrobasilar insufficiency	2
Miscellaneous	10
No pathologic evidence	3
Total	97

stenosis of the common carotid artery and internal carotid artery in ten, and stenosis of the middle cerebral artery in 11. Occlusions were found in 48 patients; in 38 the lesion was in the common carotid and internal carotid artery while in 10 it was in the middle cerebral artery. The vertebrobasilar system was affected in two cases. Thirteen patients had multiple lesions and ten other operative indications (Table 2). All underwent Doppler sonography preoperatively and postoperatively before discharge and during the control period (Fig. 1).

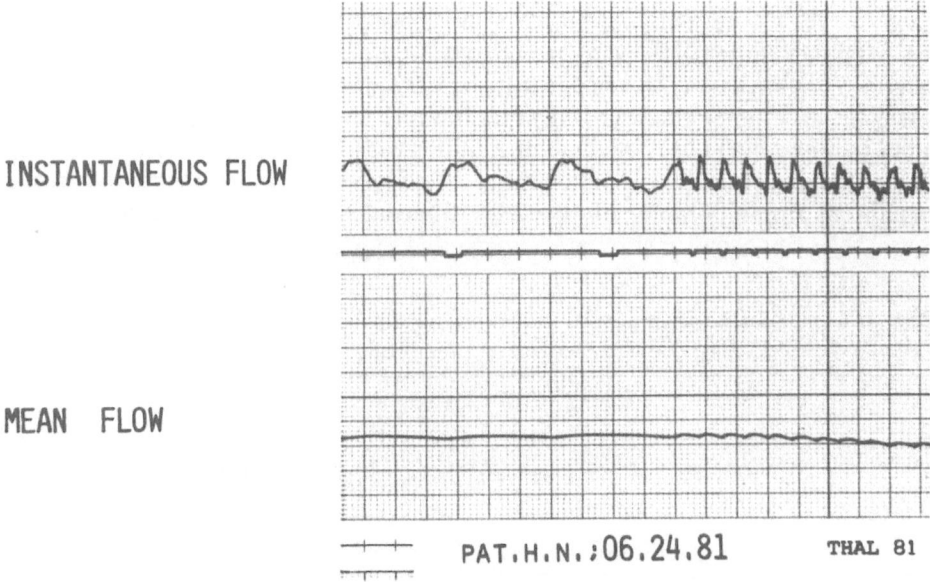

INSTANTANEOUS FLOW

MEAN FLOW

PAT.H.N.:06.24.81   THAL 81

Fig. 1. Postoperative percutaneous CW Doppler sonogram postoperatively

Results

In two patients the preoperative Doppler was very valuable because of inadequate angiograms of the external carotid artery. A comparison with intraoperative Doppler provided no evidence because of the impossibility of measuring the inner diameter of the vessel. The intraoperative flow measurements made it possible to determine the more suitable branch of the STA before cutting the artery. After the preparation, flow was reduced according to arterial spasm. Nimodipin given locally was able to eradicate spasm. After suturing the anastomosis a typical reduction in flow was seen, but the number of cases does not allow statistical evaluation. When performing Doppler sonography intraoperatively after the anastomosis had been performed we ensured that it was open. In one case it seemed to be open − the vessel showed the signs of patency listed by Acland (1972) − but no flow was to be seen or heard with the Doppler. After reopening the anastomosis we saw that one suture had taken the opposite intima. After correcting this failure we observed a normal flow. With a new 20-MHz Doppler and miniprobes this can be done easily Gilsbach and Harders (1983).

Postoperative Doppler controls were carried out 1 week after the operation using a CW Doppler pencil probe and showed a reduction in velocity compared with the opposite STA. Four to eight weeks later the reduction of velocity diminished and after 6 − 12 months velocity and flow increased. Forty-six patients agreed to undergo postoperative angiography, but many refused this investigation (Table 3). In a subgroup of 21 patients 16 had a patent anastomosis in the Doppler sonogram and four showed a weak and pendulous flow. One occluded bypass was seen in this group. In the arteriogram 14 anastomoses were patent, six were uncertain, and one looked occluded in a common carotid angiogram. As we know from Ausman and Diaz (1980), Moritake et al (1980), and Hopman et al (1977) even the arteriogram does not really prove whether an arterial bypass is patent or not, for it depends on the cardiovascular state, the place of the catheter, and dynamic alterations during the bolus injection. On the other hand we also know of some difficulties with the Doppler: With the pencil probe one can occlude a functioning anastomosis or measure a flow towards a remaining frontal branch of the STA. During angiography vasospasm and thrombosis can occur or simulate an occluded bypass.

Table 3. Correlation between doppler and arteriography postoperatively

	CW Doppler	Angiography	
Patent	16	14	Patent
Patent with weak or pendulous flow	4	6	Uncertain
Occluded	1	1	Occluded
Total		21	

Overall 91% of the bypasses were patent angiographically – 75% showed a good filling of cortical branches of the middle cerebral artery, while 16% were patent but showed only rare cortical branches. Doppler sonography in this subgroup showed an identical occlusion rate of 9%. The rate of absolute positive patency was slightly higher, at 78%. In 12% the ultrasound showed a weak and pendulous flow, which indicated a patent but insufficient anastomosis.

Conclusion

In conclusion we can state that a postoperative arteriogram is no longer indicated after a normal extra-intracranial bypass procedure just to determine the patency of the arterial shunt. Patency can easily and reliably be proven by Doppler sonography when one has some experience of the technique. In doubtful Doppler sonograms the arteriogram often fails as well. Patients must be informed of the high risk involved in doing an arteriogram in the presence of occlusive cerebrovascular disease, and about 50% then refuse this invasive investigation postoperatively. In the future other noninvasive methods like the sector scan, the serial CT, nrCBF measurements, or PET may prove the answer to this problem.

References

Acland R (1972) Signs of patency in small vessel anastomosis. Surgery Vol 72 No 5:744–748

Ausman JI, Diaz FG (1980) Correlation of noninvasive Doppler and angiographic evaluation of extra-intracranial anastomoses: In: Microsurgery for cerebral ischemia. Peerless SJ, McCormick CW (eds) Springer Berlin Heidelberg New York, p 125–127

Fox AJ, Allock JM (1978) Angiography of the external carotid to internal carotid anastomosis. Neuroradiology 16:104–107

Gilsbach JM, Harders A (1983) Intraoperative Doppler sonography. In: Jensen HP, Brock M, Klinger M (eds) Advances in neurosurgery, vol. 11. Springer, Berlin Heidelberg New York

Hopman H, Gratzl O, Schmiedek P, Schneider I (1977) Doppler sonographic control of microvascular bypass function. In: Schmiedek P (ed) Microsurgery for stroke. Springer, Berlin Heidelberg New York, p 230–232

Mortake K, Handa H, Yonekawa Y, Nagata I (1980) Ultrasonic Doppler assessment of hemodynamics in superficial temporal artery-middle cerebral artery anastomosis. Surg Neurol Vol 13:249–257

Reisner H, Samec P, Zeiler K (1980) On the complication rate of cerebral angiography. Neurosurg Rev 3:23–29

Subject Index